The Lizards, Crocodiles, and Turtles of Honduras

Systematics, Distribution, and Conservation

JAMES R. MCCRANIE

Bulletin of the Museum of Comparative Zoology

Special Publications Series, No. 2

All photos and images are by the author unless otherwise credited.

The Lizards, Crocodiles, and Turtles of Honduras

Systematics, Distribution, and Conservation

JAMES R. MCCRANIE

Harvard University Museum of Comparative Zoology

2018

Distributed by Harvard University Press

Cambridge, Massachusetts, and London, England

Printed in the United States of America

ISBN 978-0-674-98416-5

PREFACE

When I started working seriously on the taxonomy and geographic distribution of the amphibians and reptiles of Honduras in 1976, pristine forest was easily reachable either by car, a four-wheel-drive vehicle, walking, or by riding horses or mules. Those days are gone forever, as very little forest in good condition remains in the country today. Deforestation began to become serious during the early 1980s. That deforestation has accelerated to the point where at least 90% of pristine forest that was in the country in 1976 is now lost. The forest devastation began to increase significantly in the last few years (around 2005) under the last years of control of Administración Forestal del Estado, Corporación Hondureña de Desarrollo Forestal, Tegucigalpa, the Honduran group responsible for protection of its forest and fauna until 2008. That deforestation has drastically increased under the watch of the Instituto Nacional de Conservación y Desarrollo Forestal, Áreas Protegidas y Vida Silvestre (ICF), Tegucigalpa, that took over in March 2008, even in the World Heritage Biosphere Reserve Río Plátano (Reserva del Hombre y la Biósfera del Río Plátano). Flying from La Ceiba to Puerto Lempira will take you above the heart of that Biosphere Reserve and will reveal the increasing devastation having taken place, especially since about 2010, in that reserve, with that devastation currently close to its center.

In 2010, I began working with colleagues gathering molecular samples of reptiles, which has already helped discover six previously undescribed cryptic reptile species. I also had ideas of where to target several other taxa that certainly represent complexes of undescribed cryptic species. I made a list of localities where I was sure undescribed species occurred among several species complexes. I also had field trips planned to collect tissues from specimens for molecular study of various populations I suspected represented new cryptic species. Those suspicions came from carefully examining and taking morphological data for some specimens in my personal collection and in U.S. museums. Those plans of discovering new species were stopped by the current governmental agency in 2014 when they began to deny new collecting permits.

I need to stress that my taxonomic and geographical distribution work requires fieldwork and the need to collect select species at select localities. The days of "general collecting" are long gone, and the few select specimens I have collected over the last 10 years (up to 2013) can in no way damage the populations of the few species involved, especially when compared with the accelerated habitat destruction that is going on in Honduras. My collecting only requires select species to continue to gain knowledge of species taxonomy and distribution in the country. During the most 10 recent years, most individuals of most species have been released on spot. My only desire is to collect specimens from species complexes that are likely undescribed, before they become extinct when their few remaining forested habitats are gone. Looking back at my earlier work on amphibians, at least five of the new species I discovered have apparently gone extinct. If coworkers and I had not been allowed to do that fieldwork, those species would have disappeared from the planet without any human knowledge that they formerly existed. I fear there are other undiscovered species awaiting discovery that will now suffer that fate. That extinction rate can only be expected to increase with the current uncontested and illegal forest destruction taking place in Honduras.

James R. McCranie
Miami, Florida, USA
8 October 2015, updated 20 December 2016

THE LIZARDS, CROCODILES, AND TURTLES OF HONDURAS. SYSTEMATICS, DISTRIBUTION, AND CONSERVATION

JAMES R. McCRANIE[1,2]

CONTENTS

[1] Smithsonian Research Associate, 10770 SW 164th Street, Miami, Florida 33157 (jmccrani@bellsouth.net).

[2] Dedication: To the memory of my *suegro* Emiliano Green (Warunta to his many friends), who was not only my father-in-law, but also a valued friend. I miss you, man, especially remembering those long boat rides and walks we made together.

ABSTRACT. I discuss the taxonomy, distribution, and conservation of the lizards, crocodiles, and turtles occurring in the Central American country of Honduras. This is the fourth and final work in this series documenting the amphibians and reptiles of that country. I propose three new species: one of *Laemanctus* and two of the *Sceloporus malachiticus* species complex of the *S. formosus* group. I also suggest several other taxonomic changes, including elevating *Laemanctus waltersi*, *Leiocephalus varius*, *Sceloporus schmidti*, and *Kinosternon albogulare* to valid species. I also place *Ctenosaura praeocularis* in the synonymy of *C. quinquecarinata* and *Laemanctus serratus mccoyi* in the synonymy of *L. serratus*. All of those decisions are documented with specimens examined (with the exception of *L. s. mccoyi*) and data and were reviewed by experts on those groups. Several geographically wide-ranging species appear to represent more than one species (i.e., *Phyllodactylus tuberculosus*, *Sphaerodactylus continentalis*, *Basiliscus vittatus*). Also discussed is the uncontested, illegal forest destruction occurring throughout the country, including in areas designated as national parks and other types of reserves. The reader is urged to refer to the materials and methods section for an explanation of the procedures used in this work. In total, 107 species of lizards, 2 species of crocodiles, and 17 species of turtles are discussed. Full species accounts of 86 nominal forms are included herein, since a recent review of the anoles had included full species accounts for 39 of those species (one anole species recently added not included in a species account).

INTRODUCTION

Meyer and Wilson (1973) provided an annotated checklist of the lizards, crocodiles, and nonmarine turtles of Honduras. Those authors included a total of 59 known species for the country. That total represented 48 lizard species, 2 crocodile species, and 9 turtle species. The Meyer and Wilson (1973) effort represented "a revised version of a doctoral thesis" completed by the first author in 1969. Thus, the majority of the work for that thesis was finished some 50 years ago. Needless to say, much progress has occurred in those 50 years regarding our understanding of the systematics and taxonomy of those reptile groups. Those 50 years have also seen significant fieldwork taking place in Honduras, especially at moderate and intermediate elevations, including in isolated cloud forest areas. In that

work in broadleaf humid forests away from roads, whether at moderate or intermediate elevations, the author quickly began to discover species of amphibians and reptiles new to science. Also, that fieldwork at those elevations was expanded to regions in Honduras close to the international border with other countries, as well as into the vast (at that time) lowland rainforests of the Mosquitia in northeastern Honduras. That extended fieldwork also resulted in the documentation of various amphibian and reptilian species previously unknown from Honduras.

An idea of the scope of that additional fieldwork can be gained by the following discussion. Meyer and Wilson (1973) listed 2,351 specimens of lizards and close to 100 more of crocodiles and nonmarine turtles in museum collections around the world from Honduras. However, those authors made no effort to examine those specimens they listed. A serious and misleading example of not examining those specimens is that Meyer and Wilson (1973) listed the holotype of *Sceloporus schmidti*, but that specimen was already lost about 1 year after its description in 1927. Currently, close to 14,000 Honduran specimens of those three groups reside in museum collections, almost all of which I have examined. Those additional specimens, including marine turtles and specimens from the Swan Islands and Cayos Miskitos, not treated by Meyer and Wilson (1973) have resulted in the knowledge that 126 species of those three groups (107 lizard, 2 crocodile, and 17 turtle species) currently are documented to occur in Honduras (with several additional lizard species descriptions in press or in preparation), as opposed to the 59 species recognized by Meyer and Wilson (1973). Thus, those 50 years of additional study of these three reptile groups have been productive for increasing our knowledge about them, but the rate of forest devastation over those same 50 years have crucially destroyed any hope that those types of increase in knowledge will ever be duplicated in the future. Instead, the future for all of the Honduran native flora and fauna is in dire straits.

MATERIALS AND METHODS

First and foremost, it needs to be made clear that I use the Linnaean taxonomic system throughout this paper and try to adhere to the International Code of Zoological Nomenclature (ICZN, 1999; hereafter, the Code) and its principles. I do not use the PhyloCode but realize that several recent authors have more and more become inclined to confuse the two systems with each other. This is the fourth and final installment in this series documenting the amphibian and reptilian faunas of Honduras.

Vidal and Hedges (2009) provided a revised classification of the squamates, including infraorders and superfamilies. Six unranked taxa I include as equivalents of infraorders were proposed for lizards that occur in Honduras: Cordylomorpha Vidal and Hedges (2009: 134, for Xantusioidea); Gekkomorpha Fürbringer (1900: 607, as Geckonomorpha; Vidal and Hedges, 2009: 134, included the Honduran superfamilies Eublepharoidea and Gekkonoidea); Neoanguimorpha Vidal and Hedges (2009: 133, for Honduran Anguidae and Diploglossidae); Neoiguania Vidal and Hedges (2009: 134, for New World Iguanoidea); Scincomorpha Camp (1923: 298, 313, used for all Scincoidea by Vidal and Hedges, 2009: 134); and Teiformata Vidal and Hedges (2005: 1005, for Gymnophthalmidae + Teiidae (but see remarks in that section below). Vidal and Hedges (2009) also recognized seven lizard superfamilies that occur in Honduras: Xantusioidea Baird (1859a: 254; as Xantusidae); Eublepharoidea Boulenger (1883: 308; as Eublepharidae); Gekkonoidea Gray (1825: 198; as Gecktoidae); Anguioidea Gray (1825: 201; as Angudidae); Iguanoidea Bell (1825: 206; as Iguanidae); Scincoidea Gray (1825: 201; as Sincidae); and Gymnophthalmoidea. For the last mentioned, Goicoechea et al. (2016) demonstrated that Gymnophthalmidae is an older name than Teiidae; (also see Remarks in that section below), thus proposed using the superfamily name Gymnophthalmoidea in place of Teiioidea. Hedges and Conn (2012: 29) added the superfamily Lygosomoidea Mittleman (1952: 3; as Lygosominae) for part of the former Scincoidea (both superfamilies are represented in Honduras). The Vidal and Hedges (2005, 2009) and Hedges and Conn (2012) classifications, as slightly modified, are followed in this work at the time of its last major revision (1 March 2016, with some known subsequent literature published during April–December 2016 added, plus a few important publications up to close to mid-2017 are briefly discussed), with the caveat that those superfamily classifications are not fully accepted (i.e., Pyron et al., 2013; Reeder et al., 2015). Also, those superfamily classifications and relationships are far from settled. Using the Vidal and Hedges (2005, 2009) classification has the benefit of grouping in this work seemingly morphologically related taxa, instead of having some seemingly morphologically similar taxa widely separated because of molecular only suggestions.

The general statements of geographical distribution of each family, genus, and species as they occur outside and within Honduras, along with content of each group from family to species, are presented for living species only. Extinct species known only as fossils are not included in the Geographical Distribution accounts and total number of species for each group. The format for groups follows those used by McCranie (2011a) for snakes and McCranie and Köhler (2015) for anoles.

I have not provided hemipenial descriptions in the species accounts of lizards (Squamata) as was done for snakes (McCranie, 2011a) and anoles (McCranie and Köhler, 2015). Few of the lizard groups, with the exception of the anoles, have Honduran specimens represented in collections with fully everted hemipenes, and little research has been done on hemipenial characters of most of those lizard groups. However, I have included references to published illustrations of male reproductive organs of various lizard species in the Illustrations sections. Penis descriptions or illustrations of crocodiles and turtles are rarely given in the literature. Some exceptions regarding those groups occurring in Honduras are Medem (1981, 1983) and Zug (1966), for crocodiles and turtles, respectively. The lack of penis descriptions for turtles was unexpected given the plethora of turtle books published, most of which were authored by professional scientists.

Abbreviations used for measurements and morphometrics in species descriptions and some keys are as follows: AGL, axilla-groin length; BL, greatest length of turtle bridge; CH, maximum height of turtle carapace; CL, maximum length of turtle carapace; CW, maximum width of turtle carapace; EEL, eye-ear distance; FLL, forelimb length; HD, greatest depth of head; HL, head length; HLL, hind limb length; HW, head width; L, length; LCA, Spanish for head length; LHC, Spanish for snout-vent length; MSH, same as CH; PL, plastron length; PW, plastron transverse width; SEL, snout-eye distance; SHL, shank length; SL, snout length; S-OL, snout-occipital distance; SVL, snout-vent length; SW, snout width; TAH, greatest height of tail; TAL, tail length; TAW, greatest width of tail; TOL, total length; TYMH, greatest height of tympanum; TYML, greatest length of tympanum; and W, width. Most of those abbreviations are explained in more detail in the Supplemental Information.[1] Additionally, other terms are explained in a glossary, and collecting localities are included in a gazetteer, both of which are in the Supplemental Information.

[1] 10.3099/0027-4100-15.1.1.s1

In this work, I propose three new species, one of *Laemanctus* and two of the *Sceloporus malachiticus* complex of the *S. formosus* species group. I also suggest several taxonomic changes, including elevating *Laemanctus waltersi*, *Leiocephalus varius*, *Sceloporus schmidti*, and *Kinosternon albogulare* to species. I also place *Ctenosaura praeocularis* in the synonymy of *C. quinquecarinata* and *Laemanctus serratus mccoyi* in the synonymy of *L. serratus*.

Museum acronyms used follow those of Sabaj (2016). An exception is FN (Field Number), which is used for specimens still in my personal collection. Color codes and names (capitalized) used in color in life sections for many species follow Smithe (1975–1981), whereas a few follow Köhler (2012). When the latter was used, Köhler is acknowledged when a color and code are identified at first use. Noncapitalized color names in life (and without a color code) are those color notes not represented in either of those two guides for color. I also note that the Smithe (1975–1981) guide is long out of print; thus, the user could consult Köhler (2012), which is a reprint of the Smithe colors, with some additional colors included. The number in brackets in each Specimens Examined section represents the number of specimens listed in Meyer and Wilson (1973).

The ICZN (1999) regulates family-group names down to "subspecies" level. Dubois and Bour (2012: 154) argued that interpretation of the Code (ICZN, 1999) is that "the valid nomen of a family-series taxon is the senior one, among all those potentially available for the taxon" (Principle of Priority; Article 23). Those authors also wrote "The family nomenclatures of rather few zoological groups have been surveyed extensively for the valid nomina of taxa and especially for their valid authorships and dates" (p. 155). Also, original spellings used by those describers are almost never used in the taxonomic literature.

A feature not included in previous volumes of this series on the amphibians and reptiles of Honduras (McCranie, *In* McCranie and Wilson, 2002; McCranie, 2011a; McCranie and Köhler, 2015), is the inclusion of authors and dates of publication for each family-group name up to ordinal level. I give the original spelling the author(s) used if different from the current spelling. My literature searches revealed many errors and inconsistencies regarding author(s) and date(s) of publication in family-group taxonomy for amphibian and reptilian groups occurring in Honduras. I tried to adhere to two principles of the Code (Priority [Article 23] and Coordination [Articles 36, 43, 46]) for family-group and nonregulated higher level names. However, confusion and contradictions in the literature with those family and higher level names, and the fact that some of those original names were Latinized, non-Latinized, or violated some articles of the Code, have made some of those decisions difficult to reach and justify. Vernacular and/or names without scientific basis or intention should not be considered valid in family-group taxonomy (but see Articles 11.7 and 40.2.1 of the Code), as is the case with specific and generic names regulated by the Code.

McCranie (2015a) published a checklist of the amphibians and reptiles of Honduras that included authors and dates of publications for each family name included therein, but some usages were erroneous or not in the interest of stability (authors of some higher level names are changed herein). Errors certainly remain herein regarding authors and dates used for some family-group names. Also, the author(s) of some of those names remain controversial. I would appreciate being notified of errors, or of different opinions, found in this volume.

To make this work more "user friendly," I added page numbers to most literature citations in the Remarks sections where the information given was originally published (e.g., Kluge, 1969: 32). That information is given because I know how frustrating it can be for an interested user needing or wanting convenient access to the original source of data.

The systematics section of this work treats the orders Squamata, except that the Dactyloidae is only briefly treated because that family was covered by McCranie and Köhler (2015), and the snakes are not included because they were treated by McCranie (2011a). An exception is *Norops wermuthi*, which was not included in McCranie and Köhler (2015) and is treated fully in this work. The Crocodylia and then the Testudinata follow the Squamata. Within each order, each infraorder is treated (if used by Vidal and Hedges, 2009, or other recent taxonomists), followed then by superfamilies, families, genera, and species. Generic names are treated as plural, species names are considered singular. All information provided is based on Honduran specimens, unless otherwise noted. The systematics sections for each species follow the format of each snake and anole species as treated in McCranie (2011a) and McCranie and Köhler (2015), respectively. The cut-off date for adding literature of substance was 1 March 2016, with a few exceptions in the final review (minor changes up through December 2016).

One reviewer stressed the importance in a historical sense to comment on the correct authorship of the first volume in this series on the amphibians of Honduras (McCranie and Wilson, 2002). All specimens of amphibians were examined solely by me, all data were also taken only by me, and all (with one exception) of the writing for the entire book was done only by me, with help with the Spanish translation of the identification keys as acknowledged. The one exception was "The Environment" which was written by Wilson. Therefore, the correct way to cite authorship of "The Amphibians of Honduras" is "Wilson, *In* McCranie and Wilson, 2002" for reference to the environment section and "McCranie, *In* McCranie and Wilson, 2002" for all other sections of the book. Therefore any errors in that book outside of the environment section are completely mine, including any errors with the original data.

McCranie (2011a) recently included a revised discussion of the environment of Honduras, including a general description of the country, the physiography, the climate, and the forest formations. The interested reader is referred to that book for information on those subjects.

Color photographs in life of all 86 species, including a single anole species with a full species account herein not included in McCranie and Köhler (2015), are included (but see Dactyloidae below). Dichotomous identification keys, in both English and Spanish, are included to help the interested user identify any specimen in hand. Those keys also include 197 black and white photographs.

An asterisk (*) following a species or genus name in a key signifies that that particular form has not been reported from Honduras, but probably occurs somewhere in the country. Additionally, an asterisk (*) before an author's name means I was unable to find a copy of that work (three works not seen).

Descriptions of bilateral parts of each species are treated as one side only (data usually taken from left side), because bilateral structures are usually similar to each other. When possible confusion or uncertainties could occur, the phrase "(per side)," or something similar, is inserted in the species description.

AN UPDATE ON THE HISTORY OF REPTILIAN STUDY OF SPECIES OCCURRING IN HONDURAS

McCranie (2011a) included a summary on the history of the study of reptiles occurring in Honduras. That summation was current through about the middle of 2010. Subsequently, much has been accomplished regarding the species, genera, family-level, and infraorder-level taxonomy of those Honduran reptiles. Some of that recent work has combined molecular data with morphological data sets into more meaningful phylogenetic studies that resulted in the discovery of cryptic species and previously unknown relationships among the reptiles.

McCranie (2011a) recognized three snake subfamilies among the former Colubridae occurring in Honduras: Colubrinae (Oppel, 1811a: 376; as Colubrini); Dipsadinae Bonaparte (1838a: 124; as Dipsadina); and Natricinae Bonaparte (1838a: 124; as Natricina). The last two names were repeated verbatim by Bonaparte (1838b: 392). Also, as frequently occurring in the literature when citing family-level names in older literature, Vidal et al. (2007: 185) cited the incorrect Bonaparte publication (Bonaparte, 1840: 285) for the introduction of the names Dipsadinae and Natricinae. McCranie (2011a) had used Colubridae in the traditional sense, following the proposed taxonomy of Lawson et al. (2005) over those of others who would elevate those subfamilies to families (i.e., Vidal and Hedges, 2009). Vidal and Hedges (2009) had proposed that those three snake subfamilies of the Colubridae covered by McCranie (2011a) were best treated as families, and that suggestion is now followed here (also see Vidal et al., 2010). Additionally, Pyron et al. (2011: 341) discovered the genus *Scaphiodontophis* Taylor and Smith (1943: 302) formed a clade within the traditional Colubridae; thus, Pyron et al. named that clade the Scaphiodontophiinae. Zaher et al. (2012) performed a new phylogenetic study that included *Scaphiodontophis* and some Old World genera previously postulated to be close relatives of *Scaphiodontophis*, and their results suggested transferring the Scaphiodontophiinae to the newly elevated family Sibynophiidae Dunn (1928: 20). Following Zaher et al. (2012; also see Zaher et al., 2009; Grazziotin et al., 2012), I recognize the following four families of the traditional "Colubridae" as occurring in Honduras: Colubridae; Dipsadidae (including the Xenodontinae); Natricidae; and Sibynophiidae (also see Chen et al., 2013: 259). In another recent family change Pyron et al. (2014: 254) returned the Ungaliophiidae to its former subfamily classification (Ungaliophiinae McDowell, 1987: 25). Pyron et al. (2014: 254) also placed that subfamily in the boid family Charinidae Gray (1849: 84; as Charinina).

The following pertinent changes in snake generic names have been proposed subsequent to McCranie (2011a): *Ramphotyphlops braminus* (Daudin, 1803c: 279) placed in *Indotyphlops* by Hedges et al. (2014: 37; also see Pyron and Wallach, 2014: 56); *Typhlops costaricensis* Jiménez and Savage (1962: 199), *T. stadelmani* Schmidt (1936: 48), and *T. tycherus* Townsend, Wilson, Ketzler, and Luque-Montes (2008: 20) all placed in *Amerotyphlops* by Hedges et al. (2014: 43; also see Pyron and Wallach, 2014: 45); all Honduran species of *Rhadinaea* Cope (1863: 100), except *R. decorata* (Günther, 1858: 35), placed in *Rhadinella* Smith (1941a: 7) by Myers (2011: 28); *Pseustes poecilonotus* (Günther, 1858: 100) questionably placed in *Phrynonax* Cope (1862d: 348) by Jadin et al. (2013: 261); and *Pelamis* Daudin (1803c: 357) placed in *Hydrophis* Latreille (1801: 193, *In* Sonnini and Latreille, 1801) by Sanders et al. (2013: 584).

Proposed snake name changes at the species level include: *Boa constrictor* Linnaeus (1758: 215) = *B. imperator* Daudin (1803a: 150; see Hynková et al., 2009: 630; Reynolds et al., 2014: 208); *Dendrophidion clarki* (Dunn, 1933b: 78) = *D. rufiterminorum* Cadle and Savage (2012: 23); *Dendrophidion vinitor* Smith (1941b: 74) = *D. apharcybe* Cadle (2012: 222); *Lampropeltis triangulum* (Lacepède, 1789: 86) = *L. abnorma* (Bocourt, 1886, plate 39, *In* A. H. A. Duméril et al., 1870–1909b; see

Ruane et al., 2014: 247); *Tantilla olympia* described as a new species by Townsend, Wilson, Medina-Flores, and Herrera-B. (2013: 194); *Tantilla psittaca* described as a new species by McCranie (2011b: 38; for *T. taeniata* Bocourt, 1883: 587, *In* A. H. A. Duméril et al., 1870–1909a [in part]); *Oxyrhopus petola* (Linnaeus, 1758: 225) = *O. petolarius* (Linnaeus, 1758: 225; see Savage, 2011: 224); *Xenodon rabdocephalus* (Wied, 1824: col. 668) tentatively = *X. angustirostris* W. Peters (1864: 390; see Myers and McDowell, 2014: 89); *Bothriechis guifarroi* described as a new species by Townsend, Medina-Flores, Wilson, Jadin, and Austin (2013: 85, reported as *B. marchi* [Barbour and Loveridge, 1929b: 1, in part] by McCranie, 2011a: 487); *Agkistrodon bilineatus* (Günther, 1863: 364) = *A. howardgloydi* Conant (1984: 135; see Porras et al., 2013: 61, who also suggested that the Copán, Honduras, photograph only record might represent *A. bilineatus*; that opinion was based on published photographs, thus is not supported by an actual specimen); *Cerrophidion wilsoni* Jadin et al. (2012: 461, reported as *C. godmani* [Günther, 1863: 364] in McCranie, 2011a). All those suggested name changes are accepted, but it needs to be noted that some of those systematic proposals are better supported than others (i.e., *Cerrophidion wilsoni* cannot be distinguished from *C. godmani* [*sensu stricto*] based on any single morphological character). Additionally, Myers and McDowell (2014: 68) placed *Enuliophis* McCranie and Villa (1993: 262) in the synonymy of *Enulius* Cope (1871: 558), but I disagree with that decision. The four named species of *Enulius* as recognized by Wallach et al. (2014: 270) are, as far as known, unique among the Dipsadidae in having uniform, minute spines covering the hemipenis. Adding *Enuliophis*, with its large hemipenial spines, to *Enulius* would eliminate that *Enulius* synapomorphy. The long, fragile tail, considered a synapomorphy without a phylogenetic analysis by a few coworkers, contained by those two genera is present in two other genera occurring in Honduras; thus, that character cannot be a synapomorphy uniting any or all of those four genera with long, fragile tails. McCranie and Hedges (2016) revised the taxonomy of the Central American species of threadsnakes *Epictia*, recognizing *E. ater* (Taylor, 1940b: 536) and *E. phenops* (Cope, 1875: 128) as occurring on mainland Honduras, in addition to the Caribbean insular *E. magnamaculata* (Taylor, 1940b: 532). Wallach (2016) followed that taxonomy, with the exception of proposing the new species *E. martinezi* for one population from southern Honduras. McCranie, Valdés Orellana, and Gutsche (2013: 289) reported the first record of *Tantilla vermiformis* (Hallowell, 1861: 484) from Honduras. With these updates, the current known snake fauna consists of 12 families, 68 genera, and 143 species (see McCranie, 2015a; Wallach, 2016).

McCranie and Köhler (2015) treated the 39 species of the family Dactyloidae Fitzinger (1843: 63) that they considered part of the Honduran fauna. Sunyer, García-Roa, and Townsend (2013: 103) added *Norops wermuthi* Köhler and Obermeier (1998: 129) to the Honduran anole fauna (despite not depositing the purported voucher specimen in the indicated museum) after it was too late to add information of substance in McCranie and Köhler (2015). McCranie and Köhler (2015) recognized one species of *Anolis* Daudin (1802: 50) and 38 of *Norops* Wagler (1830a: 149) as occurring in the country. Subsequently, after adding *N. wermuthi* as a species account in this work, Köhler, Townsend, and Petersen (2016) revised the *N. tropidonotus* group and described two new species as occurring in Honduras, *N. mccraniei* and *N. wilsoni*, and considered *N. tropidonotus* to occur outside of Honduras. These changes are reflected in the anole section of the text and in the tables in this work, but no

species descriptions of these two new species are included herein (the *N. tropidonotus* account in McCranie and Köhler, 2015, will need to be consulted). McCranie and Hedges (2012, 2013a,b,c) combined morphological and molecular data sets in efforts to discover cryptic species among lizard complexes occurring in isolated populations on the Honduran Bay Islands. Those studies resulted in the description of five new species (*Phyllodactylus paralepis, Sphaerodactylus alphus, S. guanajae, S. leonardovaldesi*, and *S. poindexteri*) and the resurrection of two species (*Sphaerodactylus millepunctatus* Hallowell, 1861: 480, and *Cnemidophorus ruatanus* Barbour, 1928: 60) from synonymy. McCranie and Hedges plan further studies on Honduran lizards of the genus *Marisora* Hedges and Conn (2012: 119) and studies of the *Sceloporus malachiticus* complex are planned by Eric N. Smith and McCranie. Harvey et al. (2012) provided a detailed morphological study of the Teiidae (also see McCranie and Gotte, 2014); Hedges and Conn (2012) and Hedges (2014) provided new arrangements of the skink fauna; Pyron et al. (2013) provided a new classification of the lizard and snake fauna; and Pinto-Sanchez et al. (2015) considered *Marisora* a synonym of *Mabuya*, and *M. brachypoda* and *M. roatanae* synonyms of *M. unimarginata* (Pinto-Sanchez et al., 2015, suggestions not followed; also see *Marisora* section). Goicoechea et al. (2016) published a new taxonomic arrangement of the Gymnophthalmidae and Teiidae, which is followed herein.

Solís et al. (2014) published an updated list of the amphibians and reptiles, but that work also contained numerous errors. McCranie (2015a) produced a new checklist that updated the Solís et al. (2014) list, made corrections to that list, and added authors, dates, and pagination of family-group and higher level names. A companion bibliography to the updated list was also published (McCranie, 2015b).

McMahan et al. (2015: 525) discussed the "ever-rising popular view of the superiority of molecular data over morphological data." Those authors went on to discuss some incongruence among those data sets in squamate phylogeny. They concluded that no superiority of molecular data over morphological data existed in their phylogenetic reconstructions of squamate evolution. Harrington et al. (2016) provided a critical response to the McMahan et al. (2015) conclusions; thus, the superiority of either dataset over the other remains a contentious issue.

CLASS REPTILIA LAURENTI, 1768

Three orders of (nonavian) reptiles, the Squamata Oppel (1811a: 376; lizards and snakes), the Crocodylia Wagler (1830a: 130; crocodiles), and the Testudinata Behn (1760: Tabula Generalis; turtles) occur in Honduras. McCranie (2011a) treated the snakes known at that time from Honduras, and McCranie and Köhler (2015) treated 39 of the 41 Honduran species of the lizard family Dactyloidae currently known from the country. As currently understood, the lizards occurring in Honduras consist of 107 named species in six infraorders, eight superfamilies, 18 families, and 30 genera. The Crocodylia consists of two superfamilies, two families, two genera, and two named species, and the Testudinata contains one infraorder, four superfamilies, six families, 10 genera, and 17 named species (Table 1). Thus, 126 species of these three orders are known to occur in Honduras. Six species are human-aided introductions to Honduran territory. Those species are: all three species of *Hemidactylus*, *Norops sagrei*, *Leiocephalus varius*, and *Trachemys scripta*. All lizard species (including the anoles not containing a full species account in this work) are discussed in the distribution and conservation sections of this book following the species accounts.

KEY TO HONDURAN REPTILE ORDERS

1A. Body encased in a shell, or epidermal scutes, which encloses girdles (a leathery covering replaces shell in one species); teeth absent, jaws enclosed in a horny sheath....... .... Testudinata (turtles; p. 437)

1B. Body not encased in a shell or leathery covering; teeth present on jaws; jaws enclosed in skin covered with scales.......................... 2

2A. Cloacal opening longitudinal; secondary palate present; teeth socketed .. Crocodylia (crocodiles; p. 422)

2B. Cloacal opening transverse; secondary palate absent; teeth not socketed...... Squamata (lizards and snakes; p. 9)

CLAVE PARA LOS ÓRDENES DE REPTILIA HONDUREÑAS

1A. Cuerpo encapsulado en una concha (carapacho) cubierto de escudos duros o con una cubierta coriácea; la concha cubre las cinturas (pélvica y pectoral); dientes ausentes; mandíbulas cubiertas de un pico queratinizado........... .... Testudines (tortugas; p. 437)

1B. Cuerpo no encapsulado en una concha; dientes presentes sobre las mandíbulas; mandíbulas cubiertas de piel con escamas............. 2

2A. Abertura cloacal longitudinal; paladar secundario presente; dientes implantados en un alvéolo....... ...... Crocodylia (crocodrilos; p. 422)

2B. Abertura cloacal transversal; paladar secundario ausente; dientes no implantados en un alvéolo....... Squamata (lagartos y serpientes; p. 9)

SPECIES ACCOUNTS
ORDER SQUAMATA OPPEL, 1811A

The Order Squamata includes the lizards, snakes, and amphisbaenians (Vidal

TABLE 1.   LISTING OF THE ORDERS (O, 3), INFRAORDERS (IO, 7), SUPERFAMILIES (SF, 14), FAMILIES (F, 26), GENERA (G, 42), AND SPECIES (SP., 126) OF THE LIZARDS, CROCODILES, AND TURTLES OF HONDURAS, WITH THE NUMBERS FOR EACH CATEGORY SHOWN IN PARENTHESES. SIX SPECIES ARE HUMAN-ASSISTED INTRODUCTIONS AND ARE INDICATED AS SUCH IN THIS TABLE.

Order Squamata (6 IO, 8 SF, 18 F, 30 G, 107 sp.)
Infraorder Cordylomorpha (1 SF, 1F, 1G, 2 sp.)
  Superfamily Xantusioidea (1 F, 1 G, 2 sp.)
    Family Xantusiidae (1 G, 2 sp.)
      Genus *Lepidophyma* (2 sp.)
        Lepidophyma flavimaculatum
        Lepidophyma mayae
Infraorder Gekkomorpha (2 SF, 4 F, 7 G, 21 sp.)
  Superfamily Eublepharoidea (1 F, 1 G, 1 sp.)
    Family Eublepharidae (1 G, 1 sp.)
      Genus *Coleonyx* (1 sp.)
        Coleonyx mitratus
  Superfamily Gekkonoidea (3 F, 6 G, 20 sp., 3 introduced)
    Family Gekkonidae (1 G, 3 sp., all introduced)
      Genus *Hemidactylus* (3 sp.; all introduced)
        Hemidactylus frenatus
        Hemidactylus haitianus
        Hemidactylus mabouia
    Family Phyllodactylidae (2 G, 4 sp.)
      Genus *Phyllodactylus* (3 sp.)
        Phyllodactylus palmeus
        Phyllodactylus paralepis
        Phyllodactylus tuberculosus
      Genus *Thecadactylus* (1 sp.)
        Thecadactylus rapicauda
    Family Sphaerodactylidae (3 G, 13 sp.)
      Genus *Aristelliger* (2 sp.)
        Aristelliger sp. A
        Aristelliger nelsoni
      Genus *Gonatodes* (1 sp.)
        Gonatodes albogularis
      Genus *Sphaerodactylus* (10 sp.)
        Sphaerodactylus alphus
        Sphaerodactylus continentalis
        Sphaerodactylus dunni
        Sphaerodactylus exsul
        Sphaerodactylus glaucus
        Sphaerodactylus guanajae
        Sphaerodactylus leonardovaldesi
        Sphaerodactylus millepunctatus
        Sphaerodactylus poindexteri
        Sphaerodactylus rosaurae
Infraorder Neoanguimorpha (1 SF, 2 F, 3 G, 6 sp.)
  Superfamily Anguioidea (2 F, 3 G, 6 sp.)
    Family Anguidae (2 G, 3 sp.)
      Genus *Abronia* (2 sp.)
        Abronia montecristoi
        Abronia salvadorensis
      Genus *Mesaspis* (1 sp.)
        Mesaspis moreletii
    Family Diploglossidae (1 G, 3 sp.)
      Genus *Diploglossus* (3 sp.)
        Diploglossus bivittatus

TABLE 1.   CONTINUED.

        Diploglossus montanus
        Diploglossus scansorius
Infraorder Neoiguania (1 SF, 6 F, 10 G, 64 sp.)
  Superfamily Iguanoidea (6 F, 10 G, 64 sp.)
    Family Corytophanidae (3 G, 9 sp.)
      Genus *Basiliscus* (2 sp.)
        Basiliscus plumifrons
        Basiliscus vittatus
      Genus *Corytophanes* (3 sp.)
        Corytophanes cristatus
        Corytophanes hernandesii
        Corytophanes percarinatus
      Genus *Laemanctus* (4 sp.)
        Laemanctus julioi sp. nov.
        Laemanctus longipes
        Laemanctus serratus
        Laemanctus waltersi
    Family Dactyloidae (2 G, 41 sp., 1 sp. introduced)
      Genus *Anolis* (1 sp.)
        Anolis allisoni
      Genus *Norops* (40 sp., 1 sp. introduced)
        Norops amplisquamosus
        Norops beckeri
        Norops bicaorum
        Norops biporcatus
        Norops capito
        Norops carpenteri
        Norops crassulus
        Norops cupreus
        Norops cusuco
        Norops heteropholidotus
        Norops johnmeyeri
        Norops kreutzi
        Norops laeviventris
        Norops lemurinus
        Norops limifrons
        Norops loveridgei
        Norops mccraniei
        Norops morazani
        Norops muralla
        Norops nelsoni
        Norops ocelloscapularis
        Norops oxylophus
        Norops petersii
        Norops pijolense
        Norops purpurgularis
        Norops quaggulus
        Norops roatanensis
        Norops rodriguezii
        Norops rubribarbaris
        Norops sagrei (mainland and Bay Islands populations introduced)
        Norops sminthus
        Norops uniformis
        Norops unilobatus
        Norops utilensis
        Norops wampuensis
        Norops wellbornae
        Norops wermuthi

TABLE 1. CONTINUED.

Norops wilsoni
Norops yoroensis
Norops zeus
Family Iguanidae (2 G, 7 sp.)
 Genus *Ctenosaura* (6 sp.)
  Ctenosaura bakeri
  Ctenosaura flavidorsalis
  Ctenosaura melanosterna
  Ctenosaura oedirhina
  Ctenosaura quinquecarinata
  Ctenosaura similis
 Genus *Iguana* (1 sp.)
  Iguana iguana
Family Leiocephalidae (1 G, 1 sp., introduced)
 Genus *Leiocephalus* (1 sp., introduced)
  Leiocephalus varius (introduced)
Family Phrynosomatidae (1 G, 5 sp.)
 Genus *Sceloporus* (5 sp.)
  Sceloporus esperanzae sp. nov.
  Sceloporus hondurensis sp. nov.
  Sceloporus schmidti
  Sceloporus squamosus
  Sceloporus variabilis
Family Polychrotidae (1 G, 1 sp.)
 Genus *Polychrus* (1 sp.)
  Polychrus gutturosus
Infraorder Scincomorpha (2 SF, 3 F, 4 G, 7 sp.)
 Superfamily Lygosomoidea (2 F, 2 G, 5 sp.)
  Family Mabuyidae (1 G, 2 sp.)
   Genus *Marisora* (2 sp.)
    Marisora brachypoda
    Marisora roatanae
  Family Sphenomorphidae (1 G, 3 sp.)
   Genus *Scincella* (3 sp.)
    Scincella assata
    Scincella cherriei
    Scincella incerta
 Superfamily Scincoidea (1 F, 2 G, 2 sp.)
  Family Scincidae (2 G, 2 sp.)
   Genus *Mesoscincus* (1 sp.)
    Mesoscincus managuae
   Genus *Plestiodon* (1 sp.)
    Plestiodon sumichrasti
Infraorder Teiiformata (= Gymnoformata; see text; 1 SF, 2 F, 5 G, 7 sp.)
 Superfamily Gymnophthalmoidea (2 F, 5 G, 7 sp.)
  Family Gymnophthalmidae (1 G, 1 sp.)
   Genus *Gymnophthalmus* (1 sp.)
    Gymnophthalmus speciosus
  Family Teiidae (4 G, 6 sp.)
   Genus *Ameiva* (1 sp.)
    Ameiva fuliginosa
   Genus *Aspidoscelis* (2 sp.)
    Aspidoscelis deppii
    Aspidoscelis motaguae
   Genus *Cnemidophorus* (1 sp.)
    Cnemidophorus ruatanus

TABLE 1. CONTINUED.

 Genus *Holcosus* (2 sp.)
  Holcosus festivus
  Holcosus undulatus
Order Crocodylia (2 SF, 2 F, 2 G, 2 sp.)
 Superfamily Alligatoroidea (1 F, 1 G, 1 sp.)
  Family Alligatoridae (1 G, 1 sp.)
   Genus *Caiman* (1 sp.)
    Caiman crocodilus
 Superfamily Crocodyloidea (1 F, 1 G, 1 sp.)
  Family Crocodylidae (1 G, 1 sp.)
   Genus *Crocodylus* (1 sp.)
    Crocodylus acutus
Order Testudinata (1 IO, 4 SF, 6 F, 10 G, 17 sp., 1 sp. introduced)
 Infraorder Cryptodira (4 SF, 6 F, 10 G, 17 sp., 1 sp. introduced)
  Superfamily Chelonioidea (2 F, 5 G, 5 sp.)
   Family Cheloniidae (4 G, 4 sp.)
    Genus *Caretta* (1 sp.)
     Caretta caretta
    Genus *Chelonia* (1 sp.)
     Chelonia mydas
    Genus *Eretmochelys* (1 sp.)
     Eretmochelys imbricata
    Genus *Lepidochelys* (1 sp.)
     Lepidochelys olivacea
   Family Dermochelyidae (1 G, 1 sp.)
    Genus *Dermochelys* (1 sp.)
     Dermochelys coriacea
  Superfamily Chelydroidea (1 F, 1 G, 2 sp.)
   Family Chelydridae (1 G, 2 sp.)
    Genus *Chelydra* (2 sp.)
     Chelydra acutirostris
     Chelydra rossignonii
  Superfamily Kinosternoidea (1 F, 2 G, 3 sp.)
   Family Kinosternidae (2 G, 3 sp.)
    Genus *Kinosternon* (2 sp.)
     Kinosternon albogulare
     Kinosternon leucostomum
    Genus *Staurotypus* (1 sp.)
     Staurotypus triporcatus
  Superfamily Testudinoidea (2 F, 2 G, 7 sp., 1 sp. introduced)
   Family Emydidae (1G, 3 sp., 1 sp. introduced)
    Genus *Trachemys* (3 sp., 1 introduced)
     Trachemys emolli
     Trachemys scripta (introduced)
     Trachemys venusta
   Family Geoemydidae (1 G, 4 sp.)
    Genus *Rhinoclemmys* (4 sp.)
     Rhinoclemmys annulata
     Rhinoclemmys areolata
     Rhinoclemmys funerea
     Rhinoclemmys pulcherrima

Figure 1.   Moveable eyelids absent. *Hemidactylus frenatus.* USNM 581894 from Tegucigalpa, Francisco Morazán (introduced population).

and Hedges, 2009), with the last mentioned not represented in Honduras. Lizards and snakes are easily separated from each other in Honduras by the presence (lizards) or absence (snakes) of functioning limbs. All, except one Honduran lizard, *Gymnophthalmus speciosus* (Hallowell, 1861: 484), have 5 digits on the forelimbs and all have 5 digits on the hind limbs (versus only 4 digits on forelimbs in *G. speciosus*). All squamates have paired copulary organs called hemipenes (Vitt and Caldwell, 2014). In recent molecular phylogenies based on nuclear genes, snakes and iguanian and anguimorph lizards form a clade within the Squamata (i.e., Vidal and Hedges, 2009: 134), thus challenging the preexisting theory that either amphisbaenians or dibamid lizards are closest relatives of snakes (also see Pincheira-Donoso et al., 2013, who recovered lizards as being paraphyletic with regard to snakes). Thus, separation of snakes and lizards into two suborders is not substantiated on evolutionary grounds.

KEY TO HONDURAN LIZARD INFRAORDERS

1A. No moveable eyelids (Fig. 1[2]) 2
1B. Moveable eyelids present (Fig. 2). . . 4
2A. Dorsal surface of head covered with small to granular scales (Fig. 3). . . . . Gekkomorpha (in part) (p. 29)
2B. Dorsal surface of head with enlarged, platelike scales (Fig. 4). 3
3A. Scales of venter large, squarish, juxtaposed, and platelike (Fig. 5); 5 digits on forelimbs present (Fig. 2) . . . . Cordylomorpha (p. 18)
3B. Scales of venter large, smooth, imbricate, and cycloid (Fig. 6); 4 digits on forelimbs present (Fig. 6) . . . . Teiformata (in part) (p. 370)
4A. Dorsal surface of head covered with small to granular scales or mixture of granules and enlarged tubercles (Fig. 2) . . . . Gekkomorpha (in part) (p. 29)

---

[2] All photographs by James R. McCranie, unless otherwise noted by giving the name of that photographer. SVL and sex of animal given when known.

Figure 2.   Moveable eyelids present and small to granular dorsal head scales present, with mixture of tubercles and granules, and five digits on forelimbs. *Coleonyx mitratus*. USNM 580245 from Garroba Island, Valle.

4B. Dorsal surface of head with enlarged scales or plates, never covered with granular scales 5

5A. Dorsal surface of anterior part of head with 2 pairs of scales (anterior and posterior internasals) between rostral scale and first unpaired plate (Fig. 7) ....

.... Neoanguimorpha (p. 134)

5B. Dorsal surface of anterior part of head with fewer than 2 pairs of scales between rostral scale and first unpaired plate (Fig. 8), or dorsal surface of anterior part of head (snout region) with small, irregular scales..................... 6

6A. Dorsal surface of body in all but one species covered by uniform, rather large cycloid scales (Fig. 9)

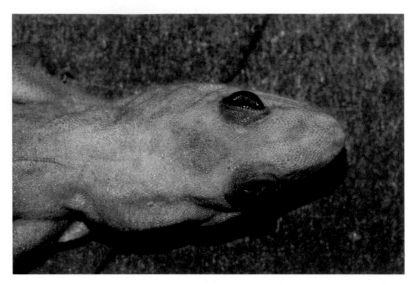

Figure 3.   Small to granular dorsal head scales. *Hemidactylus frenatus*. USNM 581894 from Tegucigalpa, Francisco Morazán (introduced population).

similar in size and shape to those of venter; one exception has medial middorsal row enlarged, but remainder of dorsals similar in size and shape of those on venter; lower eyelid present or absent (Figs. 10, 11) .... Scincomorpha (p. 334)

6B.   Dorsal surface of body covered by granular (Fig. 12) or small scales that are not similar in size and shape to those of venter (Fig. 13); lower eyelid window absent (Fig. 11) 7

Figure 4.   Enlarged, platelike dorsal head scales. *Lepidophyma flavimaculatum*. USNM 580386 from San José de Texíguat, Atlántida.

Figure 5. Large, squarish, juxtaposed, and platelike ventral scales, ground color cream, but can have dark brown flecking or mottling. *Lepidophyma flavimaculatum.* USNM 580386 from San José de Texíguat, Atlántida.

7A. Venter of body covered by large, rectangular, smooth platelike scales (Fig. 13) ....
.... Teiformata (in part) (p. 370)

7B. Venter of body not covered by large rectangular, platelike scales, ventral scales can be keeled or smooth, but never rectangular-shaped.......... Neoiguania (p. 162)

CLAVE PARA LOS INFRAORDENES DE LAGARTIJAS HONDUREÑAS

1A. Sin párpados móviles (Fig. 1) 2

1B. Con párpados móviles (Fig. 2) 4

2A. Escamas pequeñas o granulares presentes en la parte superior de la cabeza (Fig. 3) ....
..... Gekkomorpha (en parte) (p. 29)

Figure 6. Large, smooth, imbricate, and cycloid (rounded distal ends) ventral scales, and four digits on forelimbs. *Gymnophthalmus speciosus.* USNM 579592 from Puerto Lempira, Gracias a Dios.

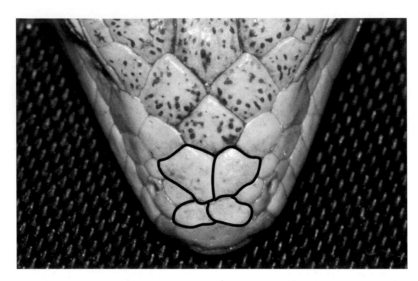

Figure 7.   Dorsal head scales enlarged with two pairs of scales (outlined) between rostral and first unpaired platelike scale. *Abronia salvadorensis*. USNM 520002 from Zacate Blanco, Intibucá.

2B. Escamas grandes (como placas) presentes en la parte superior de la cabeza (Fig. 4) 3

3A. Escamas ventrales grandes, cuadrangulares, yuxtapuestas, y en forma de placa (Fig. 5); 5 dígitos presentes en las extremidades anteriores (Fig. 2).................. .... Cordylomorpha (p. 18)

3B. Escamas ventrales grandes, lisas, imbricadas y de forma cicloidea (Fig. 6); 4 dígitos presentes en las extremidades anteriores (Fig. 6) .. .. Teiformata (en parte) (p. 370)

4A. Escamas en la parte superior de la cabeza de pequeñas a granulares, o mezcla con escamas granulares o tuberculares (Fig. 2)............. ..... Gekkomorpha (en parte) (p. 29)

4B. Escamas en la parte superior de la cabeza grandes o en forma de placas, nunca cubierta con escamas granulares 5

5A. Superficie dorsal anterior de la cabeza con 2 pares de escamas (internasales anteriores y posteriores) entre la escama rostral y la primera placa impar (Fig. 7)..... .... Neoanguimorpha (p. 134)

5B. Superficie dorsal anterior de la cabeza con menos que 2 pares de escamas entre la escama rostral y la primera placa impar (Fig. 8), o cubierta con escamas pequeñas y irregulares en el región del hocico.. 6

6A. Superficie dorsal del cuerpo (excepto en una especie) cubierta por escamas poco grandes, de forma cicloidea y de tamaño uniforme (Fig. 9), parecidas a las escamas ventrales; excepto una especie que posee la hilera dorsal media con escamas más grandes que el resto de las escamas dorsales; también esos escamas grandes y similar en tamaño y forma a las escamas ventrales; el párpado inferior con un gran disco translucido no dividido presente o ausente (Figs. 10, 11) .... Scincomorpha (p. 334)

6B. Superficie dorsal del cuerpo no cubierta por escamas poco grandes de forma cicloidea (Fig. 12) y de tamaño uniforme parecidas a las escamas ventrales (Fig. 13); el

Figure 8.   Dorsal head scales enlarged with fewer than two pairs of scales (outlined) between rostral and first unpaired platelike scale. *Mesoscincus managuae*. USNM 580382 from Punta Novillo, Valle.

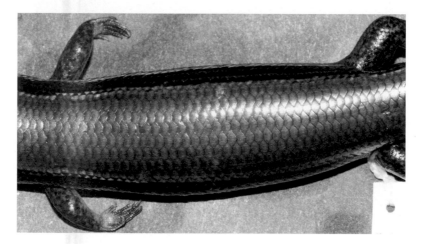

Figure 9.   Uniform, large cycloid dorsal scales. *Marisora roatanae*. USNM 589205 from near Turquoise Bay, Roatán Island, Islas de la Bahía.

Figure 10.   Lower eyelid window present (outlined), with single median scale. *Marisora roatanae*. USNM 589205 from near Turquoise Bay, Roatán Island, Islas de la Bahía.

párpado inferior siempre ausente
(Fig. 11) 7
7A. Superficie ventral del cuerpo cu-
     bierto con escamas grandes, rec-
     tangulares y lisas (Fig. 13) . . . .
     . . . . Teiformata (en parte) (p. 370)
7B. Superficie ventral del cuerpo no
     cubierto con escamas grandes,
     rectangulares; escamas ventrales
     quilladas o lisas, pero nunca de

forma rectangular lisas . . . .
. . . . Neoiguania (p. 162)

Infraorder Cordylomorpha Vidal and Hedges, 2009

Following the suggested classification of Vidal and Hedges (2009: 134), this infra-order includes the families Cordylidae Mertens (1937: 8), Gerrhosauridae Boulenger (1884: 122), and Xantusiidae Baird

Figure 11.   Lower eyelid window absent, divided into 5–6 scales (outlined). *Mesoscincus managuae*. USNM 579864 from near Orealí, El Paraíso.

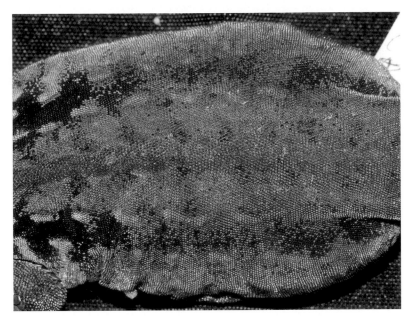

Figure 12.   Granular dorsal scales present. *Holcosus festivus.* USNM 578849 from Los Pinos, Cortés.

(1859a: 254; as Xantusidae), with only the last family occurring in Honduras. Vidal and Hedges (2009) placed the first two families in the Cordyloidea and Xantusiidae in the superfamily Xantusioidea.

Superfamily Xantusioidea Baird, 1859a

The superfamily Xantusioidea includes only the family Xantusiidae (see Vidal and Hedges, 2009: 134) in Honduras. Externally, the Honduran species of Xantusioidea

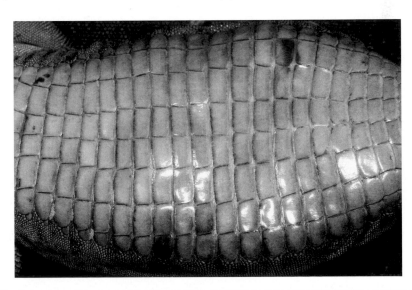

Figure 13.   Enlarged, rectangular ventral scales present. *Holcosus festivus.* USNM 578849 from Los Pinos, Cortés.

can be distinguished from all remaining Honduran lizards by the combination of lacking moveable eyelids and having large, platelike, square-shaped, juxtaposed ventral body scales. One of the two named species occurring in Honduras is known to be viviparous and the second is almost certainly viviparous.

Family Xantusiidae Baird, 1859a

This lizard family is found only in the Western Hemisphere, where it occurs from southern Utah, southern Nevada, central California, and central and western Arizona, USA, southward through much of Baja California and northwestern Sonora, Mexico. Populations also occur on the Channel Islands of California, in the central Chihuahuan Desert region and the Sierra Madre Oriental, Mexico, and from Michoacán and Veracruz, Mexico, southward to the Canal Zone region of Panama. An isolated monotypic genus also occurs in eastern Cuba. Three genera comprising 35 named species are included in this family, with two named species placed in a single genus occurring in Honduras.

Remarks.—Males of Honduran members of the Xantusiidae can be difficult to distinguish from females without making dissections, thus my xantusid descriptions are of unsexed adults.

Genus *Lepidophyma* A. H. A. Duméril, 1851

Lepidophyma A. H. A. Duméril, 1851: 137, *In* A. M. C. Duméril and Duméril, 1851 (type species: *Lepidophyma flavimaculatus* A. H. A. Duméril, 1851: 138, *In* A. M. C. Duméril and Duméril, 1851, by monotypy).

Geographic Distribution and Content.—This genus ranges from central Nuevo León and western Michoacán, Mexico, southward to central Panama. There are 19 recognized species, two of which are known to occur in

Honduras (but see Species of Probable Occurrence).

Remarks.—Terminology for head scales used herein follow that of Gauthier et al. (2008), and other scale terminology follow that of Bezy and Camarillo R. (2002). Species of *Lepidophyma* have rounded pupils (see Bezy and Camarillo R., 2002: 14), not "vertically elliptical pupils" as stated by Savage (2002: 498). The morphological information for *L. smithii** Bocourt (1876: 402) in the *Lepidophyma* key below is from Bezy (1989a) and Bezy and Camarillo R. (2002).

Etymology.—The generic name *Lepidophyma* is derived from the Greek *lepidos* (scale) and *phyma* (tumor, growth), in reference to the numerous large tubercles on the dorsal surfaces of the body and tail.

KEY TO HONDURAN SPECIES OF THE GENUS *LEPIDOPHYMA*

1A. Two pretympanic scales (both sides summed; Fig. 14); about 36–39 lateral tubercles; venter of adults with much dark pigment (Fig. 15) .... *mayae* (p. 26)

1B. Four to 12 pretympanic scales (both sides summed; Fig. 16); about 19–33 lateral tubercles; venter of adults cream with dark brown flecking or mottling (Fig. 5).. 2

2A. Pale parietal spot usually present on posterior third of interparietal scale in adults; usually an externally visible parietal foramen; 23–34 total femoral pores (occurs on Atlantic versant in northern Honduras) .... *flavimaculatum* (p. 22)

2B. Pale parietal spot absent on interparietal scale in adults; consistently lacks externally visible parietal foramen; 16–24 total femoral pores (expected to occur on Pacific versant in extreme southwestern Honduras)................. *smithii**

Figure 14.   One pretympanic scale present per side (outlined). *Lepidophyma mayae.* USNM 573973 from near Río Liston, Santa Bárbara.

CLAVE PARA LAS ESPECIES HONDUREÑAS DEL GÉNERO *LEPIDOPHYMA*

1A. Dos escamas pretimpánicas (ambos lados incluidos; Fig. 14); aproximadamente 36–39 tubérculos laterales presentes; patrón de co-

loración ventral de los adultos oscuro (Fig. 15)........ *mayae* (p. 26)

1B. De 4 a 12 escamas pretimpánicas (ambos lados incluidos; Fig. 16); aproximadamente 19–33 tubérculos laterales presentes; patrón de

Figure 15.   Darkly pigmented venter in adults. *Lepidophyma mayae.* USNM 573973 from near Río Liston, Santa Bárbara.

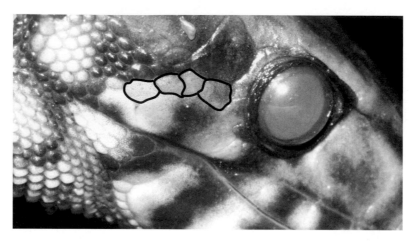

Figure 16.   More than one pretympanic scale present per side (outlined). *Lepidophyma flavimaculatum*. USNM 578856 from Lancetilla, Atlántida.

coloración ventral de los adultos crema, salpicado o moteado de pardo oscuro (Fig. 5) 2

2A. Mancha pálida usualmente presente en la tercera parte de la escama interparietal en adultos; usualmente con una foramen parietal visible externo; 23–34 poros femorales en totale (se encuentran en el lado del Atlántida en el norte de la regíon continental)........

.... *flavimaculatum* (p. 22)

2B. Mancha pálida usualmente ausente en la escama interparietal en adultos; consistemente sin foramen parietal visible externo; 16–24 poros femorales totales (potencialmente en la vertiente del Pacifico en el extremo sudoeste de Honduras)........................ *smithii**

Lepidophyma flavimaculatum A. H. A. Duméril, 1851

Lepidophyma flavimaculatus A. H. A. Duméril, 1851: 138, *In* A. M. C. Duméril and Duméril, 1851 (holotype, MNHN 782 [see Guibé, 1954 :58]; type locality: "Province du Peten [Amér. centrale]").

Lepidophyma flavimaculatum: A. H. A. Duméril, 1852b, plate 17; Meyer, 1969: 283; Wilson and Meyer, 1969: 146; Meyer and Wilson, 1973: 34; Hahn and Wilson, 1976: 179; Wilson et al., 1979a: 25; Wilson and McCranie, 1998: 16; Espinal et al., 2001: 106; Wilson et al., 2001: 135; McCranie, 2005: 20; McCranie and Castañeda, 2005: 15; McCranie et al., 2006: 135; Townsend, 2006a: 35; Wilson and Townsend, 2006: 106; Canseco-Márquez et al., 2008: 67; Sinclair et al., 2009: 1348; García-Vázquez et al., 2010: 54; Townsend and Wilson, 2010b: 692; Townsend et al., 2012: 102; McCranie and Solís, 2013: 243; Noonan et al., 2013: 110; Solís et al., 2014: 133 (in part); McCranie, 2015a: 365.

Lepidophyma flavomaculatum [sic]: Werner, 1896: 346; O'Shea, 1986: 45; Mahler and Kearney, 2006: 30.

Lepidophyma flavimaculatum complex: Bezy, 1989a: 76; Bezy and Camarillo R., 2002: 16.

Lepydophyma [sic] *flavimaculatum*: Espinal, 1993, table 3.

Lepidophyma mayae: Castañeda and Marineros, 2006: 3.8.

Geographic Distribution.—*Lepidophyma flavimaculatum* occurs at low and moderate elevations on the Atlantic versant from southern Veracruz, Mexico, to central Panama and on the Pacific versant in central Panama. In Honduras, this species occurs across the northern portion of the mainland.

Description.—The following is based on 20 unsexed adults (LSUMZ 88071, 88092; SMF 77762; UF 142464–67; USNM 94417–18, 94422, 563278–81, 563288–90, 565511, 573969, 573971). *Lepidophyma flavimaculatum* is a moderate-sized lizard (maximum SVL 127 mm [Bezy and Camarillo R., 2002]; 125 mm SVL in largest Honduran specimen [UF 142467]); dorsal head scales enlarged, platelike, consisting of a single rostral (visible from above), paired internasals, a single frontonasal, paired prefrontals, a single median frontal, paired frontoparietals, paired parietals, a single interparietal, paired postparietals, and 3 supratemporals on each side; pale interparietal spot usually present (absent in USNM 573969); 6 supralabials, 2 located posterior to eye; second supralabial posterior to eye 0.3–0.5 times height of first supralabial posterior to eye; usually 1 scale row separating first supralabial posterior to eye from supratemporal one, rarely first supratemporal enlarged and contacting first supralabial posterior to eye; 1–2 preoculars; moveable eyelid absent; pupil circular; 2–4 postoculars; 6–12 (8.7 ± 1.7) total pretympanic scales; 1 large mental; 4 pairs of chinshields, each bordering lip line, first 2 pairs large, in contact medially, third on each side about half size of second, fourth on each side about half size of third; 44–54 (48.9 ± 3.1) gulars medially in line between gular fold and midventral contact of second chinshields; 154–194 (175.0 ± 9.6) middorsal scales between level above cloaca to postparietals; 17–30 (22.7 ± 3.4) enlarged tubercles in right paravertebral row from levels above axilla and groin; 19–31 (26.0 ± 2.8) lateral tubercles on right side of body between levels of axilla and groin; 31–39 (34.8 ± 2.6) ventral scales between cloacal scale and gular fold; 23–34 (28.7 ± 4.6) total femoral pores in 19; 21–30 (25.1 ± 2.5)

subdigital scales on Toe IV of hind limb on 38 sides; 3–4 (usually 4) interwhorls on dorsal surface and 1 on ventral surface of tail near base; SVL 61–125 (86.3 ± 15.2) mm; TAL/SVL 1.11–1.60 in 18; SHL/SVL 0.13–0.17; HL/SVL 0.20–0.26.

Color in life of an adult (USNM 563281): dorsal surface of head olive brown; lateral surface of head olive brown with Sulphur Yellow (57) markings; dorsal surfaces of body, limbs, and tail dark chocolate brown with Sulphur Yellow spots and pale brown tips on some tubercles; ventral surfaces of head, body, limbs, and tail Straw Yellow (56); iris brown.

Color in alcohol: dorsal surface of body dark brown to black, with distinct cream to pale brown spots, some tubercles pale brown tipped; dorsal surface of tail brown to black, with some scales pale brown; dorsal surface of head brown to dark brown, but paler brown than body; supralabials dark brown with pale brown vertical bars; infralabials dark brown with white to cream vertical bars; gulars pale brown; belly cream with dark brown flecking or mottling; subcaudal surface cream to pale brown with dark brown mottling.

Diagnosis/Similar Species.—*Lepidophyma flavimaculatum* is distinguished from all remaining Honduran lizards, except *L. mayae*, by the combination of lacking moveable eyelids, having enlarged platelike scales on top of the head, having 5 digits on the forelimbs, and having large, squarish juxtaposed platelike scales on the venter. *Lepidophyma mayae* has two total pretympanic scales when both sides combined, has 36–39 lateral tubercles in Honduran specimens, has smaller pale brown dorsal spots, and has more darkly marked ventral surfaces (versus 6–12 total pretympanic scales, 19–31 lateral tubercles in Honduran specimens, distinct pale brown dorsal spots, and ventral surfaces cream with dark brown flecking or mottling in adult *L. flavimaculatum*).

Plate 1.   *Lepidophyma flavimaculatum.* USNM 342396. Olancho: Quebrada Las Cantinas.

Illustrations (Figs. 4, 5, 16; Plate 1[3]).— Álvarez del Toro, 1983 (adult); Bezy, 1973 (adult), 1989a (adult), 1989b (adult); Bezy and Camarillo R., 2002 (adult); Calderón-Mandujano et al., 2008 (adult); Campbell, 1998 (adult); A. H. A. Duméril, 1852b (adult, head scales); A. H. A. Duméril et al., 1870–1909b (adult, head scales, midventral scales, anal and thigh scales); Guyer and Donnelly, 2005 (adult, juvenile venter); Köhler, 1999b (adult), 2000 (adult, head scales), 2001b (adult), 2003a (adult, head scales), 2008 (adult, head scales); Köhler and Seipp, 1998 (adult); Lee, 1996 (adult, head scales), 2000 (adult, head scales); McCranie et al., 2006 (adult); Savage, 1963 (head scales), 2002 (adult, throat coloration); Stafford and Meyer, 1999 (adult); Taylor, 1955 (adult; as *L. ophiophthalmum* Taylor, 1955: 558); Townsend et al., 2012 (adult); Vitt and Caldwell, 2014 (adult).

Remarks.—Bezy (1989a) and Bezy and Camarillo R. (2002) provided systematic reviews of *Lepidophyma flavimaculatum* based on morphology. Bezy and Camarillo R. (2002: 19) wrote "the term complex is used for the aggregation of *Lepidophyma flavimaculatum* populations to denote that it may contain more than 1 species. The populations in Panama and most of Costa Rica lack males but do not appear to differ significantly in allozymes, karyotype, scalation, or color pattern from the populations to the northwest that contain males." Sinclair et al. (2009) recovered a phylogeny based on molecular data that shows those populations of *L. flavimaculatum* to represent a single evolutionary species that is sister to *L. reticulatum* Taylor (1955: 551) of Costa Rica and western Panama. Noonan et al. (2013: 114), also based on molecular data, placed *L. flavimaculatum* in "a southern clade composed of six species," including one that was undescribed at that time.

Natural History Comments.—*Lepidophyma flavimaculatum* is known from near sea level to about 1,400 m elevation in the Lowland Moist Forest and Premontane Wet Forest formations and peripherally in the Lowland Dry Forest formation. This species

[3] All photographs by James R. McCranie, unless otherwise noted by giving the name of that photographer. SVL and sex of animal given when known.

is primarily nocturnal and is usually found in forested areas that are well shaded during the day. It can be especially common in old cacao fields and has been found active at night in leaf litter, low on tree trunks, and in karsted limestone outcrops. Several were active during midday on the floor of a concrete building that had all windows and doors shuttered, thus providing artificial nighttime conditions. One was basking on the ground next to a hole under a rock on an overcast morning (about 11:00 a.m.). Some daytime retreats include under and inside rotten logs, inside rotten tree stumps, in leaf litter, in piles of discarded cacao husks, and under slabs of concrete. It seems to be active throughout the year. Extralimital females are reported to give birth to 1–9 young in April, June, or July (Telford and Campbell, 1970; Álvarez del Toro, 1983; Lee, 1996; Miralles, 2004; Goldberg, 2009c). This species feeds on various insects, including termites (Lee, 1996) and centipedes, spiders, and ants (literature summarized in Savage, 2002). Populations in Honduras are apparently sexually reproducing, whereas those from Panama and portions of Costa Rica are all females that reproduce asexually (Telford and Campbell, 1970; Bezy, 1989a; Bezy and Camarillo R., 2002; Sinclair et al., 2009). A histological examination of gonadal material from *L. flavimaculatum* from Costa Rica by Goldberg (2009c: 59) found a sex ratio of 6 adult males and 26 adult females. Goldberg (2009c: 61) concluded that "*L. flavimaculatum* in Costa Rica may consist of a mixture of sexually reproducing and parthogenetic populations (based on female-biased sex ratios), mating occurs in late summer, females produce live young in late spring at the beginning of the wet season, and only a portion of females produce offspring in a given year." There are no known external sexually dimorphic characters that distinguish between the sexes in *L. flavimaculatum*. Since I did not examine any specimens internally, I cannot offer any conclusions about Honduran populations being bisexual or parthogenetic.

Etymology.—The name *flavimaculatum* is derived from the Latin *flavis* (yellow) and *maculatus* (spot, stain, speckle) and refers to the yellow spots on the dorsum of this species.

Specimens Examined (153, 4 skeletons, 1 cleared and stained [C&S] [38]; Map 1[4]).—**ATLÁNTIDA**: Corozal, LSUMZ 21521; mountains S of Corozal, LACM 47825–30, 47835, LSUMZ 21514–20; Estación Forestal CURLA, USNM 563278, 578854; La Liberación, USNM 578858; Lancetilla, MCZ R29394–95, USNM 578855–57; near Pico Bonito Lodge, USNM 579614–15; Piedra Pintada, LACM 47834; Puerto Arturo, MCZ R21741; Quebrada de Oro, USNM 508444; San José de Texíguat, USNM 580386; about 80 km ESE of Tela, FMNH 13002 (C&S), 13003; near Tela, USNM 101278; Tela, MCZ R21814, USNM 94417–22. **COLÓN**: Cerro Calentura, CM 64539; near Las Champas, BMNH 1985.1283; Quebrada Machín, USNM 563279–80. **COMAYAGUA**: 1.6 km N of Pito Solo, LSUMZ 52503; near Pito Solo, LSUMZ 33661. **COPÁN**: La Castellona, USNM 573970; Los Achiotes, USNM 573969; below Quebrada Grande, SMF 79115; Quebrada Grande, UF 142464–66, 166350–54; Río Amarillo, USNM 570403, 579610; San Isidro, UF 142467. **CORTÉS**: 3.2 km N of Agua Azul, LSUMZ 28510; Amapa, AMNH 70459; Buenos Aires, SMF 77762; near Chamelecón, MCZ R29886–87; 1.6 km SE of El Jaral, LACM 47831–33; 1.6 km W of El Jaral, LACM 45260, LSUMZ 24217, 88071; El Paraíso, UF 144708–09; Finca Naranjito, KU 194255; Los Pinos, UF 166349, USNM 573971, 579619–20; near Peña Blanca, LSUMZ 88092, USNM 243387, 281198; W of San Pedro Sula, MCZ R29396; about 1 km SSE of Tegucigalpita, USNM 563281–85; about 3 km SSE of Tegucigalpita, SMF 79124. **GRACIAS A**

---

[4] Map legends are as of 12 January 2017.

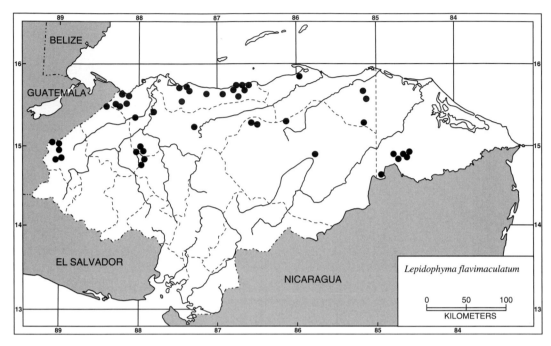

Map 1.   Localities for *Lepidophyma flavimaculatum*. Solid circles denote specimens examined.

DIOS: Bodega de Río Tapalwás, USNM 563288; Cerro Wahatingni, USNM 563290; Caño Awalwás, USNM 563286, 563289; Hiltara Kiamp, USNM 563287; San San Hil, USNM 570404; Warunta Tingni Kiamp, USNM 565511. **OLANCHO**: 4.5 km SE of Catacamas, LACM 47823–24; Cuaca, UTA R-53226, 53675–80, USNM 579616–18; Montaña del Ecuador, USNM 344796; Quebrada El Pinol, USNM 344795; Quebrada Las Cantinas, USNM 342396; near Río Cuaca, USNM 579611–13. **SAN-TA BÁRBARA**: Buena Vista, USNM 580791; tributary of Río Listón, USNM 573972. **YORO**: Los Indios, FMNH 21788 (7), MCZ R38903–19, UMMZ 77840 (10); San José de Texíguat, USNM 578859–60; Subirana Valley, MCZ R38901–02. "HON-DURAS": FMNH 22299 (skeleton), UF 42894 (skeleton), 43113 (skeleton), 76232 (skeleton), 80335, 124768.

Other Records.—"HONDURAS": (Werner, 1896).

Lepidophyma mayae Bezy, 1973

Lepidophyma mayae Bezy, 1973: 1 (holo-type, KU 59554; type locality: "near Chinajá, elev. 140 m, Depto. Alta Verapaz, Guatemala"); McCranie, 2005: 20; McCranie et al., 2006: 218; McCranie and Valdés Orellana, 2011b: 241; Solís et al., 2014: 133 (in part); McCranie, 2015a: 365.

Geographic Distribution.—*Lepidophyma mayae* occurs at low and moderate eleva-tions on the Atlantic versant from the Sierra de Los Cuchumatanes in west-central Gua-temala eastward to the Maya Mountains of southern Belize and southeastward to northwestern Copán and northwestern San-ta Bárbara in western Honduras. In Hon-duras, this species is known from two localities about 5–7 km from the Guatema-lan border.

Description.—The following is based on two unsexed adults (UF 142392; USNM

Plate 2.   *Lepidophyma mayae.* USNM 573973, SVL = 80 mm. Santa Bárbara: Río Listón.

573973). *Lepidophyma mayae* is a moderate-sized lizard (maximum SVL about 90 mm [Campbell, 1998, but without supporting data; however, Bezy and Camarillo R., 2002, in their summation of the systematics of the species gave the maximum SVL as 75 mm]; 80 mm in largest Honduran specimen [USNM 573973]); dorsal head scales enlarged, platelike, consisting of a single rostral (visible from above), paired supranasals, a single frontonasal, paired prefrontals, a single median frontal, paired frontoparietals, paired parietals, a single interparietal, paired occipitals, and three supratemporals on each side; pale interparietal spot present; 6–7 supralabials, 2 located posterior to eye; second supralabial posterior to eye, 0.9–1.3 times height of first supralabial posterior to eye; first supralabial posterior to eye separated from first supratemporal by 1 scale row; preocular single; moveable eyelid absent; pupil circular; 2 postoculars, upper narrow, elongate; 2 total pretympanic scales; 1 large mental; 4 pairs of chinshields, each bordering lip line, first 2 pairs large, in contact medially, third on each side about half size of second, fourth on each side about half to two-thirds size of

third; 42–45 (43.5) gulars medially in line between gular fold and midventral contact of second chinshields; 161–162 (161.5) middorsal scales between level above cloacal scale and postparietals; 44–45 (44.5) enlarged tubercles in right paravertebral row between levels of axilla and groin; 36–39 (37.5) lateral tubercles on right side of body between levels of axilla and groin; 38–39 (38.5) ventral scales between cloacal scale and gular fold; 26–29 (27.5) total femoral pores; 23–25 (23.8 ± 1.0) subdigital scales on Toe IV of hind limb; 4 interwhorls on dorsal surface and 1 on ventral surface of tail near base; SVL 70–80 (75.0) mm; TAL/SVL 1.29–1.64; SHL/SVL 0.15 in both; HL/SVL 0.20 in both.

Color in life of an adult (USNM 573973; Plate 2): dorsal surface of body Dark Grayish Brown (20) with indistinct Buff (24) spots and some tubercle tips pale brown; top of head Dark Grayish Brown; dorsal surface of tail Cinnamon-Brown (33) with Dark Grayish Brown mottling; supralabials and infralabials Burnt Umber (22) with yellowish brown streaks; enlarged scales of venter of head yellowish white with Burnt Umber mottling and blotches;

small scales of remainder of ventral surface of head Buff with Sepia (219) spotting; chest and belly Clay Color (123B) with yellowish tinge and Sepia cross-stripes, Clay Color of posterior belly with orange tinge; subcaudal surface Cinnamon Rufous (40) with Sepia and Dark Grayish Brown blotches; iris Tawny (38).

Color in alcohol: dorsal surface of body dark brown to almost black, with small cream to pale brown spots, some tubercles also pale brown tipped; dorsal surface of tail brown to black, with pale brown spotting; dorsal surface of head dark brown, but paler brown than body; supralabials dark brown with narrow pale brown vertical bars; infralabials dark brown with white vertical bars; gulars cream with dark brown mottling; belly cream with considerable amount of dark brown mottling, especially along anterior edge of each ventral scale; subcaudal surface pale brown with dark brown mottling.

Diagnosis/Similar Species.—*Lepidophyma mayae* is distinguished from all remaining Honduran lizards, except *L. flavimaculatum*, by the combination of lacking moveable eyelids, having enlarged platelike scales on top of the head, having 5 digits on the forelimbs, and having large, squarish, juxtaposed platelike scales on the venter. *Lepidophyma flavimaculatum* has 6–12 pretympanic scales when both sides combined, has 19–31 lateral tubercles in Honduran specimens, has larger pale brown dorsal spots, and adults have cream ventral surfaces with dark brown flecking or weak mottling (versus two pretympanic scales with both sides combined, 36–39 lateral tubercles, smaller pale dorsal spots, and ventral surfaces cream with extensive dark brown mottling in *L. mayae*).

Illustrations (Figs. 14, 15; Plate 2).— Bezy, 1973 (adult); Bezy and Camarillo R., 2002 (adult); Campbell, 1998 (adult, head pattern and scales); Köhler, 2003a (head scales), 2008 (head scales); Lee, 1996 (head scales), 2000 (adult, head scales); Stafford and Meyer, 1999 (adult).

Remarks.—Bezy and Camarillo R. (2002) provided a systematic review of *Lepidophyma mayae* based on morphology. Bezy and Camarillo R. (2002) considered *L. mayae* to be closely related to *L. flavimaculatum*. However, the phylogenetic analyses based on molecular data performed by Sinclair et al. (2009) found *L. mayae* to be sister to a group of three species from Oaxaca, Mexico. Noonan et al. (2013: 114), also based on molecular data, placed *L. mayae* in "a southern clade composed of six species," including one that was undescribed at that time. McCranie (2005) reported *L. mayae* from Honduras, but neither Köhler (2008: 174), Sinclair et al. (2009, fig. 2), nor Noonan et al. (2013: 114) included *L. mayae* as occurring in Honduras.

Natural History Comments.—*Lepidophyma mayae* is known from 435 m to 1,040 m elevation in the Lower Montane Moist Forest and Premontane Wet Forest formations. One was basking next to a pile of logs on an overcast morning (10:00 a.m.) in July in a clearing about 10 m from secondary forest. Another was inside a rotten log overhanging a stream in primary broadleaf forest in November. *Lepidophyma flavimaculatum* was also collected at both *L. mayae* localities. Nothing has been published on reproduction of *L. mayae*, but like all remaining xantusids, it almost certainly is viviparous. Campbell, 1998: 130, gave the general statement that this species "gives birth to live young" but did not provide supporting data. Also, nothing has been published on diet, but it is probably similar to that known for *L. flavimaculatum*.

Etymology.—The specific name *mayae* refers to the Maya, the indigenous group in the region of the type locality.

Specimens Examined (2 [0]; Map 2).— **COPÁN**: San Isidro, UF 142392. **SANTA**

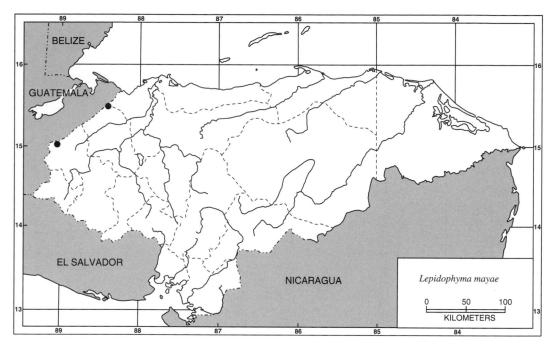

Map 2.   Localities for *Lepidophyma mayae*. Solid circles denote specimens examined.

BÁRBARA: tributary of Río Listón, USNM 573973.

Infraorder Gekkomorpha Fürbringer, 1900

The infraorder Gekkomorpha, as recognized by Vidal and Hedges (2009: 134), contains the superfamilies Eublepharoidea Boulenger (1883: 308) and Gekkonoidea Gray (1825: 198), both of which have representatives in Honduras. Vidal and Hedges (2009: 134) also placed the newly named Pygopodomorpha in an unranked taxon sister to the Gekkomorpha. There are no representatives of the Pygopodomorpha in the Western Hemisphere.

KEY TO HONDURAN SUPERFAMILIES OF THE
INFRAORDER GEKKOMORPHA

1A. No moveable eyelids (Fig. 1) ....
.... Gekkonoidea (p. 34)
1B. Moveable eyelids present (Fig. 2)
.... Eublepharoidea (p. 29)

CLAVE PARA LAS SUPERFAMILIAS HONDUREÑAS
DEL INFRAORDEN GEKKOMORPHA

1A. Sin párpados móviles (Fig. 1) ....
.... Gekkonoidea (p. 34)
1B. Con párpados móviles presentes
(Fig. 2) ...... Eublepharoidea (p. 29)

Superfamily Eublepharoidea Boulenger, 1883

This superfamily contains only the Eublepharidae according to the classification of Vidal and Hedges (2009). Externally, the single Honduran member of this superfamily can be distinguished from all remaining Honduran lizards by the combination of having moveable eyelids and the dorsal surface of the head covered with granular scales. Members of this superfamily are oviparous.

Family Eublepharidae Boulenger, 1883

This family of lizards has a disjunct distribution circumglobally in tropical and some temperate areas. In the Western Hemisphere, it is distributed from the southwestern United States (southern California, southern Nevada, southwestern Utah, western and southern Arizona, southern New Mexico, and southwestern to south-central Texas) southward into Baja California and northwestern and north-central to northeastern mainland Mexico, and from southern Mexico and Central America south to south-central Costa Rica and possibly Panama. In the Eastern Hemisphere, the family is distributed in eastern and western tropical Africa, from eastern Iraq and easternmost Turkey to eastern India, southern Thailand through Borneo to Sanana Island, the Ryu Kyu Islands of Japan, and islands in the Gulf of Tonkin, China. Six genera containing 30 named species are included in this family, with one species occurring in Honduras.

Remarks.—Females of the Honduran member of this family average slightly larger than the males, but the largest specimen examined is a male.

Genus *Coleonyx* Gray, 1845a

Coleonyx Gray, 1845a: 162 (type species: *Coleonyx elegans* Gray, 1845a: 163, by monotypy).

Geographic Distribution and Content.— This genus occurs from southwestern Utah, southern Nevada, and adjacent California, USA, southward to southern Sinaloa and southwestern Durango, Mexico, and to the southern tip of Baja California del Sur, Mexico. It also occurs from southern New Mexico to south-central Texas, USA, southward to northeastern Durango and central Nuevo León, Mexico, and from central Veracruz and southern Nayarit, Mexico, to south-central Costa Rica and possibly Panama. Eight named species are recognized, one of which occurs in Honduras.

Remarks.—Dial and Grismer (1992: 180) claimed that *Coleonyx elegans* Gray (1845a: 163) occurred in northern Honduras, citing

Klauber (1945) as the source. However, Klauber (1945: 191) included British Honduras (= Belize), not Honduras, in the range of *C. elegans*. No specimens of that taxon are known from Honduras. Scale terminology used herein for *Coleonyx* generally follow that of Klauber (1945).

Etymology.—The name *Coleonyx* is derived from the Greek words *koleos* (sheath) and *onyx* (fingernail, talon, or claw) and refers to the sheathed claws of the type species of the genus.

Coleonyx mitratus (W. Peters, 1863a)

Brachydactylus mitratus Peters, 1863a: 42 (holotype, ZMB 4598 [see Bauer et al., 1995: 54]; type locality "Costa Rica").
Eublepharis dovii: Werner, 1896: 345.
Coleonyx mitratus: Schmidt, 1928a: 194; Klauber, 1945: 199; Neill and Allen, 1962: 83; Meyer, 1969: 201; Meyer and Wilson, 1973: 10; Gundy and Wurst, 1976: 115; Kluge and Nussbaum, 1995: 20; Köhler, 1997b: 13; Köhler, 1997e: 20; Köhler, 1998b: 140; Köhler, 1998d: 375; Monzel, 1998: 155; Wilson and McCranie, 1998: 15; Köhler, 2000: 45; Lundberg, 2000: 8; Espinal et al., 2001: 106; Wilson et al., 2001: 135; Köhler, 2003a: 67; Powell, 2003: 36; Köhler et al., 2005: 101; McCranie et al., 2005: 74; Castañeda, 2006: 28; Mahler and Kearney, 2006: 29; McCranie et al., 2006: 216; Wilson and Townsend, 2006: 105; Conrad, 2008: 164; Köhler, 2008: 71; Townsend and Wilson, 2010b: 692; McCranie, 2014: 292; McCranie and Valdés Orellana, 2014: 45; McCranie et al., 2014: 100; Solís et al., 2014: 130; McCranie, 2015a: 365; McCranie and Gutsche, 2016: 864.
Coleonyx elegans: Dunn and Emlen, 1932: 26.

Geographic Distribution.—*Coleonyx mitratus* occurs at low and moderate elevations from extreme eastern Guatemala to central Costa Rica on the Atlantic versant and from southeastern Guatemala to south-central Costa Rica on the Pacific versant. In Honduras, this species is known from scattered localities in about the western two-thirds of the country, including several islands in the Golfo de Fonseca in the Pacific Ocean. It is also known from Utila Island, in the Bay Islands (Islas de la Bahía) in the Caribbean.

Description.—The following is based on ten males (FMNH 5044, 5046, 5048–49; SMF 77980; USNM 34794, 565396, 565399, 570109, 570112) and ten females (FMNH 5037, 5039, 5041, 5043, 5045, 5047, 5051; USNM 565397, 570110, 579576). *Coleonyx mitratus* is a moderate-sized gecko (maximum recorded SVL 97 mm [Savage, 2002, but without providing supporting data]; 80 mm in largest Honduran specimen [SMF 77980, a male]); dorsal surface of head covered by small scales, except rostral enlarged and pair of slightly enlarged internasal scales present, with enlarged tubercles present posteriorly; moveable eyelid present; pupil vertically elliptical; 6–7 supralabials and infralabials to level below center of eye; first infralabial squarish; no field of small scales posterior to internasal, those scales granular and similar to remainder of dorsal head scales; dorsal body scales granular, with 18–24 (20.6 ± 1.9) longitudinal, very irregular rows of enlarged tubercles; ventral body scales flat, smooth, imbricate, 41–59 (51.1 ± 4.6) along midline between levels of axilla and groin; 11–15 (12.4 ± 0.9) subdigital lamellae on Digit IV of forelimb; 14–19 (16.4 ± 1.7) subdigital lamellae on Digit IV of hind limb; subdigital lamellae single, uniform; claws partially hidden by two laterally compressed terminal digital scales; webbing absent between digits; femoral pores absent in both sexes; 4–8 (6.6 ± 1.3) precloacal pores in males in continuous, upside down V-shaped row; 6–8 (6.9 ± 0.8, n = 9) analogous (to precloacal pores) scales with indentations in females; 1 cloacal spur with blunt distal end present per side in males; SVL 47.4–80.0 (66.6 ± 9.8) mm in males, 58.0–78.5 (70.4 ± 7.1)

mm in females; HW/SVL 0.12–0.14 in males, 0.12–0.13 in females; HL/SVL 0.22–0.24 in males, 0.21–0.23 in females; TAL/SVL 0.78–0.96 in six males, 0.63–0.96 in seven females; SEL/EEL 1.02–1.35 in males, 1.04–1.18 in females.

Color in life of a subadult male (USNM 570109): dorsal surface of body Tawny (38) with dark brown outlined pale brown crossbars on body, arm of anterior crossbar extending anteriorly above tympanum to posterior edge of eye, that arm becoming white near eye; dorsal surface of head Tawny; dorsal surface of tail Tawny with Fuscous (21) outlined white crossbars; ventral surfaces of head, body, and limbs pale pink; subcaudal surface white anteriorly, becoming pale brown distally; iris pale brown.

Color in alcohol: dorsal surface of body brown with 3 cream, chevron-shaped, crossbars in juveniles and some adults, body cream in other adults, with dark brown linear blotches on each side of vertebral area, these linear blotches sometimes joined across vertebral area; lateral surface of body of adults cream with dark brown spots or mottling; dorsal surface of head pale brown with brown linear spots, those markings on snout and below eye forming variously shaped lines; brown line originating posterior to eye and looping on posterior part of nape to form continuous stripe; dorsal surface of tail cream with 8–17 dark brown crossbands; ventral surfaces of head and body essentially uniformly cream; subcaudal surface cream with lateral extension of dark brown dorsal crossbands.

Diagnosis/Similar Species.—The combination of having moveable eyelids, having the dorsal surface of the head covered with granular scales, and having granular dorsal body scales with 18–24 longitudinal, irregular rows of enlarged tubercles distinguishes *Coleonyx mitratus* from all remaining Honduran lizards.

Illustrations (Fig. 2; Plate 3).—Grismer, 1988 (adult); Kliment, 2011 (adult, juvenile,

Plate 3. *Coleonyx mitratus.* USNM 580245. Valle: Garroba Island, near San Carlos.

hatchling, ventral surface of foot); Kluge, 1975 (adult); Köhler, 1998b (adult), 2000 (adult, toe scales), 2001b (adult), 2003a (adult, toe scales), 2008 (adult, toe scales); Köhler et al., 2005 (adult, toe scales); McCranie et al., 2005 (adult, subdigital lamellae); Pianka and Vitt, 2003 (adult); Powell, 2003 (adult); Sasa and Solórzano, 1995 (adult); Savage, 2002 (adult); Taylor, 1956b (juvenile); Villa et al., 1988 (adult); Werner, 1896 (adult; as *Eublepharis dovii*).

Remarks.—*Coleonyx mitratus* is apparently most closely related to *C. elegans* (see Dial and Grismer, 1992: 187). The single record of *C. mitratus* from Panama is based on the holotype of the nominal form *Eublepharis dovii* Boulenger (1885a: 233, a junior synonym of *C. mitratus*), described from a single specimen without precise locality data. Savage (2002: 482) said that no further specimens of *C. mitratus* have been collected in Panama, but "it is possible that the species occurs there."

Seeing *Coleonyx mitratus* from both versants in Honduras leaves me with the impression that the Atlantic versant populations differ somewhat in dorsal color pattern from those of the Pacific versant.

Unfortunately, I did not take enough color notes or photographs to qualify those observations. Klauber (1945: 202) also discussed color differences in Honduran specimens from San Pedro Sula, Cortés, compared with Pacific versant specimens from Nicaragua and Costa Rica.

Natural History Comments.—*Coleonyx mitratus* is known from near sea level to about 1,400 m elevation in the Lowland Moist Forest, Lowland Dry Forest, Lowland Arid Forest, Premontane Wet Forest, and Premontane Dry Forest formations. These lizards were found from March to August and in October under logs, tree bark, and rock piles by day and crossing dirt roads and trails or crawling in rocky and sandy areas at night. The FMNH series collected by Schmidt in March 1923 was taken at night, with those from Hacienda Santa Ana and west of San Pedro Sula found on the ground "mostly along a path following a hydroelectric penstock down a hill from 1000 ft. to 500 ft. elevation. The vegetation was scrubby, low forest interspersed with cohune palms" (Klauber, 1945: 203). The Laguna Ticamaya FMNH specimens collected in May 1923 by

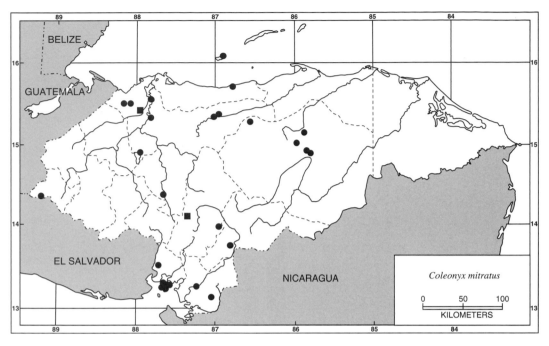

Map 3.   Localities for *Coleonyx mitratus*. Solid circles denote specimens examined and solid squares those for accepted records.

Schmidt "were found in dry leaves on the forest floor" (Klauber, 1945: 203). Adult females deposit one or two eggs per clutch and are capable of producing multiple clutches per year (Köhler, 1997f, and references cited therein). Apparently nothing has been published on diet in this species, but like that of the remaining species of *Coleonyx*, it probably feeds on a variety of small arthropods.

Etymology.—The name *mitratus* is Latin (wearing a headband) alluding to the pale crossband on the posterior portion of the head.

Specimens Examined (57 + 7 skeletons [20]; Map 3).—**ATLÁNTIDA**: 10.8 km W of La Ceiba, TCWC 21965. **CHOLUTE-CA**: 36 km SE of Choluteca, TCWC 30126–27; Choluteca, MSUM 4641. **COMAYA-GUA**: Lo de Reina, UTA R-41268. **CORTÉS**: Hacienda Santa Ana, FMNH 5037–39, MCZ R31625 (skeleton); Lago de Yojoa, UF 11626 (skeleton); Laguna Tica-

maya, CM 8120 (was FMNH 5050), FMNH 5051, 5053 (skeleton); San Pedro Sula, MCZ R38793; W of San Pedro Sula, FMNH 5041, 5043–49. **EL PARAÍSO**: El Rodeo, UNAH 5656. **FRANCISCO MORAZÁN**: El Zamorano, MCZ R49752–53, UMMZ 148902–03 (both skeletons). **ISLAS DE LA BAHÍA**: Isla de Utila, Utila, SMF 77980, USNM 570109. **OCOTEPEQUE**: near Antigua, FMNH 283716. **OLANCHO**: 1 km WNW of Catacamas, LACM 47855; 4.5 km SE of Catacamas, LSUMZ 21419–20; El Ocotal, USNM 579575; Montaña del Ecuador, USNM 344794; 10.6 km S of San Esteban, KU 200561. **VALLE**: near Amapala, FN 256922–23 (still in Honduras because of permit problems), CM 158365–66; Playona Exposición, USNM 579581; Punta Copalillo, UNAH (1); Punta Novillo, UNAH (1), USNM 580246; near San Carlos, UNAH (1), USNM 580245; 5.6 km ESE of El Salvador border, UMMZ 131854. **YORO**: El Progreso, UMMZ 58371; 4.7 km

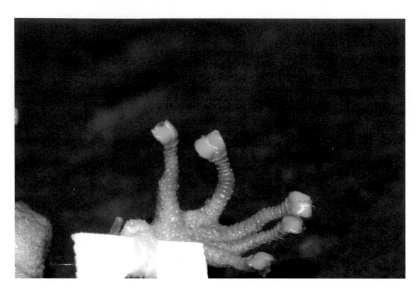

Figure 17.   Distal subdigital lamella expanded into paired leaflike pad and webbing absent between digits. *Phyllodactylus palmeus.* UF 149596 from Roatán Island (no other data), Islas de la Bahía.

ESE of San Lorenzo Arriba, USNM 565396–99, 570110–11; about 3 km S of San Lorenzo Arriba, USNM 570112, 579576–80. "HONDURAS": UF 42755 (skeleton), 71788 (skeleton), 124692.

Other Records (Map 3).—**CORTÉS**: La Lima (Meyer, 1969). **FRANCISCO MORAZÁN**: El Zamorano, MCZ R48673 (Meyer and Wilson, 1973; specimen now lost). "HONDURAS": (Werner, 1896).

Superfamily Gekkonoidea Gray, 1825

The superfamily Gekkonoidea includes the families Gekkonidae, Phyllodactylidae, and Sphaerodactylidae (Vidal and Hedges, 2009), all of which have members in Honduras, although the Honduran members of the Gekkonidae are recent anthropogenic introductions. Externally, the Honduran members of this superfamily can be distinguished from all remaining Honduran lizards by the combination of lacking moveable eyelids and having the dorsal surface of the head covered with small to granular scales. The Honduran members of this superfamily have fragile, easily damaged skin, and easily loose their

tail. Members of this superfamily are oviparous and have rigid shelled eggs.

Remarks.—Kluge (2001) and de Lisle et al. (2013) provided recent lists of the genera and species of all Gekkonoidea worldwide.

KEY TO HONDURAN FAMILIES OF THE
SUPERFAMILY GEKKONOIDEA

1A. Distal subdigital lamella expanded into paired leaflike pad (Fig. 17), or distalmost lamellae divided and separated by skin (Fig. 18), also with webbing present between toes (Fig. 18) . . . . . Phyllodactylidae (p. 59)

1B. Distal pair of subdigital lamella not expanded into leaflike pad, remaining lamellae divided or not, not separated by skin (Fig. 19); distinct toe webbing absent 2

2A. Subdigital lamellae of Digit IV of hind limb expanded with 1 or more lamellae divided into 2 rows (Fig. 19); claw protruding from medial portion of single distalmost lamella . . . . Gekkonidae (p. 35)

Figure 18. Distalmost subdigital lamellae divided and separated by skin, and webbing present between digits. *Thecadactylus rapicauda*. USNM 573100 from Rus Rus, Gracias a Dios.

2B. Subdigital lamellae single, expanded, forming pad (Fig. 20), or subdigital lamellae only slightly expanded with claws displaced laterally by slightly more expanded terminal lamella (Fig. 21), or minimally expanded subdigital lamellae with claws not displaced laterally from terminal lamella on each digit (Fig. 22) . . . . Sphaerodactylidae (p. 78)

CLAVE PARA LAS FAMILIAS HONDUREÑAS DE LA SUPERFAMILIA GEKKONOIDEA

1A. Par distal de lamelas subdigitales expandidas en una almohadilla en forma de hoja (Fig. 17) o lamelas más distales divididas y separadas por piel (Fig. 18); membrana interdigital presente entre los dedos (Fig. 18). Phyllodactylidae (p. 59)

1B. Par distal de lamelas subdigitales no está expandidas como una almohadilla en forma de hoja; las otras lamelas divididas o no divididas, y nunca separado por piel

(Fig. 19); sin membrana interdigital presente 2

2A. Lamelas subdigitales del Dígito IV de las extremidades posteriores agrandadas y con 1 o más lamelas divididas en dos hileras (Fig. 19); garra sobresaliente en la parte median de la lamela más distal . . . . Gekkonidae (p. 35)

2B. Lamela subdigitales simple, agrandada, como 1 cojinete (Fig. 20), o lamelas subdigitales ligeramente agrandadas con garras desplazadas lateralmente por 1 lamela distal (Fig. 21), o lamela subdigital agrandada mínimamente, con garras no desplazadas lateralmente por lamelas subdigitales terminales en cada dígito (Fig. 22) . . . . Sphaerodactylidae (p. 78)

Family Gekkonidae Gray, 1825

This is a predominately Eastern Hemisphere gecko family that occurs from south of about 42°N latitude from Europe to Middle Asia to the Pacific Ocean, plus further north in parts of Middle Asia, to

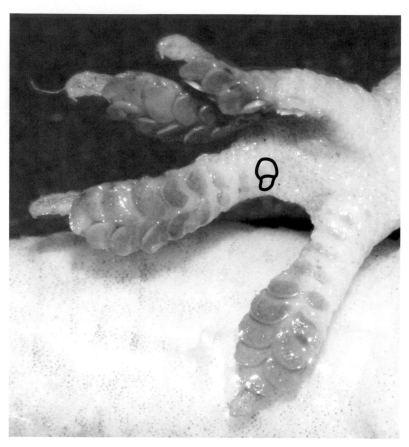

Figure 19.   Expanded subdigital lamellae with one or more lamellae divided into two rows and divided lamellae on Digit IV of hind limbs reach origin of digits (basal pair outlined). *Hemidactylus frenatus*. USNM 580772 from Finca Nakunta, Gracias a Dios (introduced population).

the southern tips of Africa and Australia. In the Western Hemisphere it occurs in eastern South America from Uruguay through Brazil into Argentina (some of which populations are definitely introduced), plus French Guiana, Suriname to eastern Venezuela, and along most of the length of the Amazon River. It also occurs in northern Colombia and western Venezuela, Trinidad and Tobago and other islands off the coast of northern South America, the Lesser Antilles, Puerto Rico and the Virgin Islands, Hispaniola, Jamaica, and Cuba. Many of those populations are recent introductions associated with humans. It is not certain whether some of the South American populations are natural or associated with early human traffic (pre-Columbus) between the Eastern and Western hemispheres; most likely both situations occur there, with those doubts pointing to a need for a thorough molecular study of many South American populations. The family has also been recently introduced across the southern USA and on the Hawaiian Islands and in Mexico and Central America. About 51 genera containing about 945 named species are recognized. One genus containing three named species has recently been introduced in Honduras, but one species (*H. frenatus*) appears to be

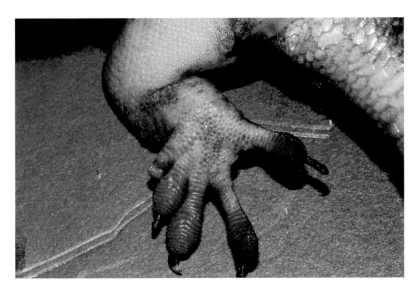

Figure 20.   Claws not displaced laterally, subdigital lamellae expanded, single, forming expanded pads on digits II–V. *Aristelliger* sp. A. USNM 581324 from Cayo Vivorillo Grande, Gracias a Dios.

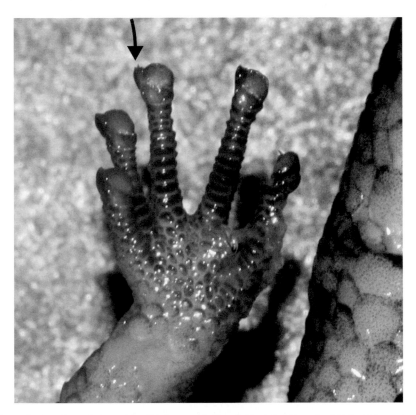

Figure 21.   Claws displaced laterally (arrow) by slightly expanded terminal subdigital lamella of digits. *Sphaerodactylus rosaurae.* UF 91320 from near Oak Ridge, Roatán Island, Islas de la Bahía.

Figure 22.   Digits with only slightly expanded subdigital lamellae, no lamellae divided, except terminal lamella containing claws diagonally divided, and claws not displaced laterally. *Gonatodes albogularis.* USNM 581899 from El Faro, Choluteca.

quickly replacing the others (also see Species of Probable Occurrence).

Remarks.—The Principle of Coordination (Articles 36, 43, 46; ICZN, 1999) dictates that Gray (1825: 198) is the author of the superfamily name Gekkonoidea.

KEY TO HONDURAN GENERA OF THE FAMILY GEKKONIDAE

1A.   Enlarged tubercles present among dorsal and lateral granules of body and tail (Fig. 23); interiormost digit

Figure 23.   Dorsal and lateral surfaces of body and tail with enlarged tubercles present among granules. *Hemidactylus frenatus.* USNM 581896 from Savannah Bight, Guanaja Island, Islas de la Bahía (introduced population).

Figure 24.   Digit I of hind limbs with free, clawed terminal phalanx, disc pads elongated, and subdigital lamellae on Digit IV of hind limbs fail to reach origin of digits. *Hemidactylus mabouia*. USNM 570548 from Savannah Bight, Guanaja Island, Islas de la Bahía (introduced population).

of hind limb with free, clawed terminal phalanx (Fig. 24) and disc pads elongated (Fig. 24) .... *Hemidactylus* (p. 40)

1B. Enlarged tubercles absent among dorsal and lateral granules of body and tail (Fig. 25); interiormost digit of hind limb without a claw, or

Figure 25.   Dorsal and lateral surfaces of body without tubercles scattered among granules. *Gehyra mutilata*.* UF 79916 from Philippines.

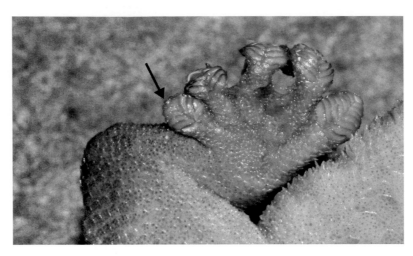

Figure 26.   Digit I of hind limbs without claws (arrow), or with only minute claw on each terminal lamella, and toe pads oblong. *Gehyra mutilata*.* UF 79916 from Philippines.

with only a minute claw (Fig. 26) and disc pads oblong-shaped, not elongated (Fig. 26) .... *Gehyra**

CLAVE PARA LOS GÉNEROS HONDUREÑOS DE LA FAMILIA GEKKONIDAE

1A. Tubérculos agrandados presentes entre las escamas granulares que cubren el cuerpo y la cola dorsal y lateralmente (Fig. 23); extremidades posteriores con 1 garra libre en la lamela terminal en el dígito más interior (Fig. 24); almohadillas de los dígitos de extremidades posteriores alargadas (Fig. 24) ...
.... *Hemidactylus* (p. 40)

1B. Tubérculos agrandados ausentes entre las escamas granulares que cubren el cuerpo y la cola dorsal y lateralmente (Fig. 25); extremidades posteriores sin una garra libre en la lamela terminal en el dígito más interior, o con una garra pequeña en la falange terminal del dígito interior (Fig. 26); almohadillas de los dígitos oblongas, no alargadas (Fig. 26) .... *Gehyra**

Genus *Hemidactylus* Oken, 1817

Hemidact[ylus] Oken, 1817: col. 1182 [see Remarks] (type species: *Gecko tuberculosus* Daudin, 1802: 158 [a senior synonym of *Gecko mabouia* Moreau de Jonnès, 1818: 138; but see Powell, Crombie, and Boos, 1998: 674.5; Kluge, 2001: 13], by subsequent designation of Stejneger, 1907: 172; but see Bauer, 1994: 113, who said "by monotypy").

Geographic Distribution and Content.— The members of this speciose genus are predominantly an Eastern Hemisphere group that is widespread through Africa (except Sahara and temperate and arid regions of southern Africa), the Mediterranean region, Asia (exclusive of northern Palearctic zone), Madagascar, Japan, Philippines, Indo-Australia Archipelago, and Oceania (Eastern Hemisphere portion from Bauer, 1994: 114). In the Western Hemisphere, this genus is found in eastern South America from Montevideo, Uruguay, Buenos Aires area of Argentina (definitely introduced) through Brazil, French Guiana, Suriname, and Guyana to eastern Venezuela and along most of the length of the Amazon

River in Brazil, Ecuador, Colombia, Bolivia, and Peru. It is also found in northern Colombia and western Venezuela plus Trinidad and Tobago and several satellite islands and is widely distributed through the Lesser Antilles, Puerto Rico and the Virgin Islands, Hispaniola, and Cuba (at least most of those populations are introduced, whereas others are possibly, but not likely, native). Members of this genus are also widely introduced in the southern USA from southern Florida (including the Florida Keys) across 27 states to southern California, in Mexico, Central America (including numerous islands on both versants), the Bahama Islands, Grand Cayman, Jamaica, and the Turks and Caicos. The genus contains about 130 named species, three of which have been introduced into Honduras (but see Remarks and Natural History Comments for each species of *Hemidactylus*, especially those of *H. frenatus*).

Remarks.—The species of *Hemidactylus* in the Western Hemisphere arrived by anthropogenic dispersal or by transatlantic dispersal, depending on the species (Gamble et al., 2011: 235), although it is not clear if those of much of South America are native or more likely represent early human-aided introductions (pre-Christopher Columbus). Several species in South America are considered to be endemic to that region. They need to be studied by molecular and morphological methods to understand whether they are valid species or are only conspecific with their Eastern Hemisphere counterparts. Apparently, males grow to a larger size than do females in the three species occurring in Honduras.

The terminology and methods of counting and measuring the characters use herein for the species of *Hemidactylus* descriptions follow those of Kluge (1969).

Etymology.—The name *Hemidactylus* is formed from the Greek prefix *hemi-* (half) and Greek word *daktylos* (fingers or toes).

The name refers to the majority of the enlarged scales on the ventral surfaces of the digits being paired.

KEY TO HONDURAN SPECIES OF THE GENUS *HEMIDACTYLUS*

1A. Subdigital lamellae on Digit IV of hind limb fail to reach base of digit between digits III and IV (Fig. 24) .... *mabouia* (p. 54)
1B. Subdigital lamellae on Digit IV of hind limb usually reach base of digit between digits III and IV (Fig. 19) 2
2A. Distinct tubercles present on upper surfaces of hind limbs (Fig. 27) and above ear openings (Fig. 28) .... *haitianus* (p. 50)
2B. Distinct tubercles absent on upper surfaces of hind limbs (Fig. 29) and above ear opening (Fig. 30) .... *frenatus* (p. 42)

CLAVE PARA LAS ESPECIES HONDUREÑAS DEL GÉNERO *HEMIDACTYLUS*

1A. Lamelas subdigitales de Dígito IV de las extremidades posteriores no se extendiénden hasta la base del dígito entre los dígitos III–IV (Fig. 24) .... *mabouia* (p. 54)
1B. Lamelas subdigitales de Dígito IV de las extremidades posteriores usualmente extendiénden hasta la base del dígito entre los dígitos III–IV (Fig. 19) 2
2A. Tubérculos distintivos presentes sobre las superficies dorsales de las extremidades posteriores (Fig. 27) y arriba de la abertura ótica (Fig. 28) .... *haitianus* (p. 50)
2B. Tubérculos distintivos ausentes en las superficies dorsales de las extremidades posteriores (Fig. 29) y por arriba de la abertura ótica (Fig. 30) .... *frenatus* (p. 42)

Figure 27. Dorsal surfaces of hind limbs with distinct tubercles present. *Hemidactylus haitianus*. USNM 572074 from Comayagüela, Francisco Morazán (introduced population).

Hemidactylus frenatus Schlegel, 1836

Hemidactylus frenatus Schlegel, 1836: 366, *In* A. M. C. Duméril and Bibron, 1836 [see Remarks] (two syntypes under the number MNHN 5135 [see Guibé, 1954: 6; also see Remarks]; type locality restricted to: "Java" [see Pope, 1935: 460; also see Remarks] and by syntypes designation [see Guibé, 1954: 6; also see Loveridge, 1947: 127]); Franklin, 2000: 53; Lundberg, 2000: 3; Köhler, 2001a: 57; Köhler, 2003a: 71; Powell, 2003: 36; McCranie et al., 2005:

Figure 28. Distinct tubercles present above each ear opening. *Hemidactylus haitianus*. USNM 572074 from Comayagüela, Francisco Morazán (introduced population).

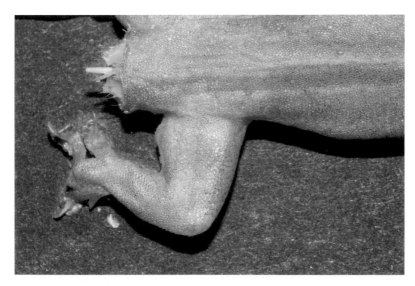

Figure 29.   Dorsal surfaces of hind limbs without distinct tubercles. *Hemidactylus frenatus*. USNM 581894 from Tegucigalpa, Francisco Morazán (introduced population).

76; Lovich et al., 2006: 14; McCranie et al., 2006: 109; Townsend, 2006a: 35; Wilson and Townsend, 2006: 105; Townsend et al., 2007: 10; Wilson and Townsend, 2007: 145; Köhler, 2008: 75; Muelleman et al., 2009: 452; Lovich et al., 2010: 113; Townsend and Wilson, 2010b: 692; McCranie and Rovito, 2011: 241; Pasachnik, 2011b: 391; Valdés Orellana et al., 2011a: 240; McCranie and Solís, 2013: 240; McCranie, 2014: 292; McCranie and Valdés

Figure 30.   Distinct tubercles absent above each ear opening. *Hemidactylus frenatus*. USNM 581894 from Tegucigalpa, Francisco Morazán (introduced population).

Orellana, 2014: 44; Solís et al., 2014: 130; McCranie, 2015a: 365; McCranie and Gutsche, 2016: 865; Heyborne and Mahan, 2017: 437; McCranie et al., 2017: 267.

Hemidactylus spp.: Ferrari, 2008: A14, *In* Anonymous, 2008.

Hemidactylus brookii: Köhler and Ferrari, 2011b: 240 (in part).

Geographic Distribution.—*Hemidactylus frenatus* is native to the Indoaustralian Archipelago, southern and southeastern Asia, and various islands in the Pacific (see Bauer, 1994: 120–121). It is also widely introduced and established in Mexico, Central and South America (including numerous islands on both versants), Florida, Texas, and the Hawaiian Islands, USA, the Galapagos Islands, Cuba, Hispaniola, the Mariana Islands (Guam), East Africa, islands in the Indian and South Pacific oceans, and parts of Australia. In Honduras, this species (but see Remarks) has become widely introduced into cities, towns, and villages on the mainland. It is also known from Islas de Guanaja, Roatán, and Utila, and on the Cayo Cochinos in the Bay Islands, Isla Grande in the Swan Islands, and numerous islands in the Golfo de Fonseca.

Description.—The following is based on 12 males (USNM 565400, 570123–25, 570127, 570135, 573083, 573089, 573091, 579591, 581894–95) and 11 females (USNM 570126, 570134, 570142–43, 570145, 573082, 573084, 573087, 573090, 573092, 581896). *Hemidactylus frenatus* is a moderately small gecko (maximum SVL 65 mm [Bauer and Sadlier, 2000]; 60 mm SVL in largest Honduran specimen [USNM 570127, a male]); dorsal surfaces of head, body, and limbs covered with minute granules with 5–8 transverse rows of small, trihedral tubercles on body; 4–12 longitudinal tubercles in paravertebral row; tubercles absent on upper surface of hind limb and above ear opening; moveable eyelid absent;

pupil vertically elliptical; 15–22 (17.6 ± 1.6) loreal scales on 45 sides; 0 or 1 (usually 0) cheek tubercle; auricular scales absent; 9–13 supralabials, 8–10 to level below mideye; 8–10 infralabials, 6–8 to level below mideye; 1 pair of postmentals, each contacting infralabial 1; 1 pair of enlarged chinshields, separated medially by granular scales, each chinshield contacting infralabial 2, sometimes also contacting posterior edge of infralabial 1; 8–10 (8.9 ± 0.7) enlarged subdigital lamellae on Digit IV of hind limb, 2–7 divided medially (except distalmost lamella single), enlarged lamellae reaching origin of digit; nonretractile, visible claw located in terminal phalanx; webbing absent between digits; usually pair of enlarged postcloacal scales present in males; 18–32 (27.4 ± 4.0) femoral-precloacal pores present in males; 0 interprecloacal poreless scales in males; 1–6 tubercles in fourth caudal whorl; 5–9 scales between caudal whorls; SVL 42.8–59.7 (53.0 ± 5.3) mm in males, 42.0–49.9 (47.0 ± 2.5) mm in females; TAL/SVL 0.64–1.21 in eight males, 0.72–1.13 in nine females; HW/SVL 0.10–0.17 in males, 0.14–0.17 in females; HL/SVL 0.20–0.26 in males, 0.24–0.27 in females; SEL/SVL 0.07–0.12 in males, 0.10–0.12 in females; SEL/EEL 1.14–1.50 in males, 1.11–1.74 in females.

Color in life of an adult male (USNM 581894): all dorsal surfaces Tawny Olive (223D) without distinct markings, except tail with pale yellowish brown mottling, especially distally; ventral surface of head cream, those of body and tail Buff-Yellow (53); ventral surfaces of expanded digits grayish white; iris grayish brown; pupil black with cream rim. Color in life of another adult male (USNM 581895): dorsal ground color of body and head Pale Horn Color (92) with some tiny brown flecking; dorsal surface of tail Straw Yellow (56); chin and throat pinkish brown; belly Straw Yellow; anterior third of subcaudal surface Buff-Yellow (53), distal two-thirds Pale Horn Color; ventral surfaces of dilated digits brown with silver pads; iris golden brown with brown punctuations, gold rim around pupil. Color in life of an adult female (USNM 573084): dorsal surfaces of head, body, and limbs Light Drab (119C) with dark brown flecking; dorsal surface of tail pale brown with dark brown flecking; all ventral surfaces (except dark brown lamellae) pale brown, belly also heavily flecked with dark brown, subcaudal surface also with dark brown tiny spots, especially distally; iris golden brown with brown flecking. Color in life of another adult female (USNM 570134): dorsal ground color Drab-Gray (119D); venter Pearl Gray (81); toe pads dark gray; iris pale brown. Color in life of a third adult female (USNM 579592): dorsum Smoke Gray (45) with Brownish Olive (29) dorsolateral stripes beginning posterior to eye and extending length of body onto anterior third of tail; similar middorsal stripe present on body; belly Pale Horn Color (92); subcaudal surface Pale Horn Color with indistinct Brownish Olive mottling distally; toe pads grayish brown; iris Mikado Brown (121C).

Color in alcohol: dorsal surfaces usually pale brown without distinct markings other than brown flecking, but an adult collected in its daytime retreat (and preserved within minutes) is uniformly dark brown; ventral surfaces of head and body cream, that of limbs cream, except subdigital lamellae pale gray; subcaudal surface cream with sparse, brown flecking, except specimen with dark dorsal surfaces with much dark brown flecking on subcaudal surface.

Diagnosis/Similar Species.—The combination of lacking moveable eyelids and having the dorsal surface of the head covered with granular scales distinguishes *Hemidactylus frenatus* from all other Honduran lizards, except the remaining Gekkonoidea. *Thecadactylus rapicauda* has 13 or more divided subdigital lamellae on Digit IV of the fore- and hind limb, has the distalmost subdigital lamellae separated by skin, and has distinct webbing between the digits (versus distalmost lamella single with fewer than 6 lamellae divided and digital webbing absent in *H. frenatus*). *Aristelliger* have a digital pad with all subdigital lamellae distinctively expanded and single (versus one or more lamellae divided in *H. frenatus*). *Gonatodes albogularis* lacks distinctively expanded subdigital lamellae and has a circular pupil (versus lamellae expanded and pupil vertically elliptical in *H. frenatus*). *Phyllodactylus* have the distalmost subdigital lamella paired, expanded and forming a leaflike pad (versus distalmost lamella single and not forming leaflike pads in *H. frenatus*). *Sphaerodactylus* have only the terminal subdigital lamella distinctly enlarged, the terminal lamella displacing the claws laterally, and all Honduran *Sphaerodactylus* have a circular pupil (versus claw in terminal phalanx free from expanded lamella and pupil vertically elliptical in *H. frenatus*). *Hemidactylus haitianus* has distinct tubercles on the upper surface of the hind limb and above the ear opening (versus distinct tubercles absent on upper surface of hind limb and above ear opening in *H. frenatus*). Subdigital lamellae of Digit IV on the hind limb in *H. mabouia* does not reach the origin of the digit (versus subdigital lamellae of Digit IV reaching origin of digit in *H. frenatus*).

Illustrations (Figs. 1, 3, 19, 23, 29, 30; Plate 4).—Alemán and Sunyer, 2015 (adult); Bauer and Sadlier, 2000 (adult, tail scales, chin region scales, subdigital lamellae); Calderón-Mandujano et al., 2008 (adult); Henkel and Schmidt, 1991 (adult); Köhler, 2000 (adult, subdigital lamellae), 2001b (adult), 2003a (adult, chin scales, subdigital lamellae), 2008 (adult, chin scales, subdigital lamellae); Köhler et al., 2005 (adult, subdigital lamellae); Lee, 1996 (adult, subdigital lamellae), 2000 (adult, subdigital lamellae); Lemos-Espinal, 2015 (adult); Lemos-Espinal and Dixon, 2013 (adult); Lundberg, 2000 (adult); McCranie and Solís, 2013 (adult); McCranie et al., 2005 (adult, subdigital lamellae), 2006 (adult,

Plate 4.   *Hemidactylus frenatus.* MCZ R191070, adult female, SVL = 51 mm. Gracias a Dios: Isla Grande, Swan Islands.

subdigital lamellae), 2017 (adult); Meshaka et al., 2004 (adult); Savage, 2002 (chin scales, basal subcaudal scales); Stafford and Meyer, 1999 (adult, subdigital lamellae); Sunyer, Nicholson et al., 2013 (adult); Zug, 1991 (hind foot), 2013 (adult, hind foot).

Remarks.—De Lisle et al. (2013: 115) listed 11 syntypes of *Hemidactylus frenatus* in the MNHN collection. Loveridge (1947: 127) stated that the type locality for *H. frenatus* was "Java (restricted)" which has led some authors (i.e., Wermuth, 1965: 74; Hardy and McDiarmid, 1969: 111) to say that Loveridge (1947) had made that type locality restriction. Actually, Pope (1935: 460) made that restriction when he said the type locality of *H. frenatus* "is hereby restricted to Java." Smith, M. A. (1935: 94) also listed only "Java" as the type locality of this species (Schlegel, *In* A. M. C. Duméril and Bibron, 1836: 365, had listed several localities for his *H. frenatus*). Taylor (1953: 1549) further restricted the *H. frenatus* type locality to "Batavia, Java," but without supportive evidence. Bauer (1994: 115) included some information on the syntypes and the entire original type series.

Kraus (2009: 242) used Taylor (1940a: 444) as the source for *Hemidactylus frenatus* collected in Mexico on the New World mainland for specimens collected during the 1920s. Taylor (1922: 55) had previously said that *H. frenatus* had been collected in "Mexico." Also, Stejneger (1907: 174) reported *H. frenatus* had "recently been found in western Mexico" (probably based on USNM specimens collected by E. Palmer in 1895 at Acapulco, Guerrero; also see Farr, 2011: 267, who erroneously credited Pope, 1935, as the first to report *H. frenatus* from the New World). Burt and Myers (1942: 285) committed another error concerning the first collection of this species from Mexico when they reported on a series of *H. frenatus* from the Pacific coast town of Acapulco, Guerrero, collected in December 1939, which they claimed represented "the first record of this species from the New World." Lever (2003: 70) also erroneously said that Grant (1957) claimed to have said that *H. frenatus* "was first introduced to Mexico at Acapulco de Juaréz on the Pacific coast in 1955." Grant did not say that, only saying that he had found the species at Acapulco in December 1955. Lever (2003:

70) also erroneously said Pope (1935) wrote that *H. frenatus* was established in Acapulco at least by the 1920s. Lever (2003) and Kraus (2009) also do not agree on when *H. frenatus* was first introduced onto the Hawaiian Islands. Lever (2003: 73) gives 1951 as the timing of that event, whereas Kraus (2009: 245) says "1940s," with both authors giving the same literature source for those data (Hunsaker, 1967). Hunsaker (1967: 121–122) said, "*H. frenatus* was first observed in Kailua 20 miles N of Honolulu, Oaku, in June 1951," but also said that one of the reasons *H. frenatus* occurred on the Hawaiian Islands "was the establishment of the species on the islands due to the shipping of cargo and materials during World War II."

Molecular studies in Carranza and Arnold (2006: 537) recovered significant divergence between *Hemidactylus frenatus* populations from Colombia and Hawaii and those from Asia. Preliminary molecular data suggests multiple origins of the various Honduran populations (S. Blair Hedges, unpublished data). Bauer et al. (2010: 348–349), also based on molecular data, found that Eastern Hemisphere *H. frenatus* consists of a complex of several species. The limited Honduran molecular data also suggests more than one species of *H. "frenatus"* is involved in those Honduran populations.

This human commensal is quickly expanding its range in various areas of the Eastern and Western hemispheres (see Farr, 2011, for a through review of Mexican introductions) where it is known to be displacing resident geckos at many localities (also see Carranza and Arnold, 2006: 532; Kraus, 2009: 71; Niewiarowski et al., 2012: 195–196; and below).

Natural History Comments.—*Hemidactylus frenatus* is known from near sea level to 1,340 m elevation in the Lowland Moist Forest, Lowland Dry Forest, Lowland Arid Forest, Premontane Moist Forest, Premontane Dry Forest, and Lowland Dry Forest

(West Indian Subregion) formations. This introduced and nocturnal species was collected on walls of buildings, trees, rocks, and rock walls from February to August and from October to December and has become extremely abundant and widespread since 1998, the year it was first collected in Honduras (Franklin, 2000). Unlike Mexico (see above), the first confirmed Honduran localities for this species are from the Caribbean versant. It can now be found on both versants in virtually every mainland village and city below about 1,000 m elevation, and occasionally above that point. It has also recently spread to Utila, Roatán, and Guanaja in the Honduran Bay Islands and on Pacific islands in the Golfo de Fonseca. Also, the species was introduced to Isla Grande in the Swan Islands by 2007, when it was first collected on that island (Ferrari, *In* Anonymous, 2008). A disturbing fact is that numerous specimens have also been seen or collected in natural vegetation well away from human-made structures, including several small islands in the Golfo de Fonseca where no humans live. I suspect that the massive rebuilding beginning after Hurricane Mitch in 1998 helped speed dispersal of this species complex throughout the mainland and Bay Islands. Although *H. frenatus* is primarily active at night or during the day in dark buildings with daytime light conditions blocked, it frequently leaves its daytime retreats during afternoons when the sun heats its daytime retreats to above tolerable temperatures. In those instances, the lizards sit inactive in fully exposed conditions such as walls of houses (both inside and outside) and curtains covering windows. Marcellini (1971) documented activity patterns of an introduced population from San Luís Potosí, Mexico, and summarized his findings as "Geckos emerge from diurnal retreats at dusk and numbers increase to a high at midnight, remaining high until they return to diurnal retreats at dawn" (p. 631). The species has a voice and both sexes, including adults and subadults, can make distress calls, and apparently males make a distinctive clicking sound during courtship and territorial activities (personal observation; but see Marcellini, 1974, who attributed three distinct call types to this species in a population from San Luís Potosí, Mexico). Females of this species deposit two eggs per clutch with an extensive breeding season in the tropics (Fitch, 1970; Köhler, 1997f, and references listed in those works, although that literature likely contains errors of lizard species identifications), and females can produce more than one clutch per year. Krysko et al. (2003) reported finding multiple egg clutches of this species in a communal nest in March in the Florida Keys, USA. Eggs of two other species of Gekkonoidea were also present. *Hemidactylus frenatus* frequently forage around night lights, where they can be seen feeding on a wide variety of insects and other arthropods attracted to those lights. Díaz Pérez et al. (2012) reported a population in Colombia fed mainly on arthropods, with Hemiptera being the most frequent food item. Meshaka et al. (2004: 84) reported "invertebrates, especially flies (Diptera) and roaches (Dictyoptera)" from the stomachs of an introduced population in Florida, USA. Poulin et al. (1995) reported *H. frenatus* in Panama feeding on a solution of refined sugar from a hummingbird feeder. This species is also cannibalistic and will eat young of other gecko species (see Frenkel, 2006; Alemán and Sunyer, 2015).

This invasive species appears to be a serious threat to the native and nocturnal *Phyllodactylus palmeus* populations on Roatán and Utila Islands in the Bay Islands. Walls of several buildings on those two islands where the native *P. palmeus* was common during the 1990s and early 2000s are now populated exclusively by *Hemidactylus frenatus* (also see the *P. palmeus* account). Also, *H. frenatus* was found in September 2012 commonly to occur in natural habitat away from buildings on Utila Island. Both *H. frenatus* and the native *P. palmeus* were found on the same coconut palms and hardwood trees well away from human structures, with the *Hemidactylus* being much more frequently encountered than the native *Phyllodactylus* at that time. *Hemidactylus frenatus* might have also displaced the native *P. tuberculosus* in edificarian situations in the departments of Colón, El Paraíso, and Valle in southern Honduras, because all of the numerous buildings searched from 2004 to 2013 were occupied solely by *H. frenatus*. *Hemidactylus frenatus* was also common away from edification situations on all islands visited in the Golfo de Fonseca from 2010 to 2013 and on the mainland in southwestern Ocotepeque in 2012, thus possibly posing additional threats to the native *P. tuberculosus*. *Phyllodactylus tuberculosus* remained common on the Valle islands during that time, but only the *Hemidactylus* was seen in southern Ocotepeque, including well away from edificarian situations. *Hemidactylus frenatus* also seems to have critically and negatively affected the endemic gecko *Aristelliger nelsoni* on Isla Grande in the Swan Islands (see McCranie et al., 2017). The *Hemidactylus* was first collected on Isla Grande in 2007, and by December 2012 it was extremely common throughout the island in both edificarian and natural situations. On the other hand, the *Aristelliger* proved difficult to find in 2012. Cole, N. C. et al. (2005) also documented the negative impacts of *H. frenatus* on native island populations of geckos on the Mascarene Islands.

Both *Hemidactylus frenatus* and *H. mabouia* were present on walls of a house in Trujillo, Colón, in 1998. However, shortly thereafter (about 2000), *H. mabouia* had apparently been displaced from those walls by *H. frenatus*. The same occurrence appears to have taken place on Isla Grande on the Swan Islands (Islas del Cisne); also see Natural History Comments for *H.*

haitianus). Two *Hemidactylus* specimens were collected on Isla Grande in 2007, one of which was *H. frenatus* and the other *H. mabouia*. However, by December 2012, literally hundreds of *H. frenatus* were seen on Isla Grande, but only one specimen of *H. mabouia* was seen at that time and was only seen away from edificarian situations (also see Natural History Comments for *H. mabouia*). The same thing appears to have happened in Tegucigalpa, Francisco Morazán, where *H. haitianus* was the gecko species on walls of a house in 1999 and 2006, whereas by 2012 the gecko on those walls was *H. frenatus* (also see Natural History Comments for *H. haitianus*).

Species of *Hemidactylus* are also known to have replaced each other at other localities because of competition between those ecologically analogous house geckos (see Rödder et al., 2008, and references cited therein). Rödder et al. (2008) also predicted that future climatic conditions might also favor *H. frenatus* over *H. mabouia*. Niewiarowski et al. (2012) also demonstrated that *H. frenatus* is displacing a resident species of gecko (*Lepidodactylus lugubris* [A. M. C. Duméril and Bibron, 1836: 304]) on the island of Moorea in French Polynesia. That gecko is itself an invasive species in parts of the world. Niewiarowski et al. (2012: 194) "demonstrate[d] that comparatively high maximal locomotor speeds [in *H. frenatus* versus *L. lugubris*] may contribute to the observed success of *H. frenatus* over resident geckos in French Polynesia, and possibly in other areas where they [*H. frenatus*] have been introduced." Kraus (2009), in a literature review, also discussed the displacement ability of *H. frenatus* to native geckos.

Etymology.—The name *frenatus* is formed from the Latin word *frenum* (bridled) and Latin suffix *-atus* (provided with or having the nature of). The name pertains to the dusky streaks on the sides of the head between the eyes and the nostrils.

Specimens Examined (139 [0]; Maps 4, 5).—**ATLÁNTIDA**: El Naranjal, USNM 579591; La Ceiba, USNM 570123–25, UTA R-42657; Pico Bonito Lodge, SMF 92849; Río Viejo, USNM 570126–27; San Marcos, USNM 570128; Tela, UF 144713. **CHOLUTECA**: Choluteca, USNM 570129; La Fortuna, SDSNH 72726. **COLÓN**: Trujillo, USNM 536063. **COPÁN**: 0.5 km W of Copán, USNM 570130; 1.0 km W of Copán, USNM 570131–32; Copán, FMNH 282557–59, UF 166346; Río Lindo, USNM 573883. **CORTÉS**: near Agua Azul, UF 166341–42; Agua Azul, USNM 570133; Cofradía, UF 144710–11; El Paraíso, UF 144712, 144714; Los Naranjos, UF 166345; Los Pinos, UF 166340, 166343, USNM 573092–93; Oropéndolas, UTA R-41280; San Isidro, UF 166348; San Pedro Sula, USNM 573084; Tegucigalpita, FMNH 283644–45. **EL PARAÍSO**: Danlí, USNM 570134–38; El Rodeo, USNM 580766; 1 km S of Orealí, USNM 580771. **FRANCISCO MORAZÁN**: Amarateca, USNM 570139; Tegucigalpa, USNM 570140, 579592–96, 581894. **GRACIAS A DIOS**: Cauquira, USNM 573082; Finca Nakunta, USNM 580772; Isla Grande, Swan Islands, MCZ R191070, 191124, 191127–30, 191136, 191140–41, 191147–54, SMF 90456; Puerto Lempira, USNM 570141–44; Tánsin, USNM 573085–86; Warunta, USNM 573882; Yahurabila, USNM 573087–90. **ISLAS DE LA BAHÍA**: Isla de Guanaja, Posada del Sol ruins, FMNH 283582; Isla de Guanaja, Savannah Bight, FMNH 283575, USNM 581895–96; Isla de Roatán, Port Royal, USNM 578821; Isla de Roatán, Sandy Bay, USNM 580264–66; Isla de Roatán, 6.6 km E of West End, MVZ 263826; Isla de Utila, east coast near Trade Winds, FMNH 283647; Isla de Utila, Utila, SMF 79856. **LA PAZ**: Potrerillos, UNAH (1). **LEMPIRA**: Gracias, FMNH 283709–11. **OCOTEPEQUE**: Nuevo Ocotepeque, FMNH 283707–08; Río Lempa near Antigua, FMNH 283646. **OLANCHO**: El

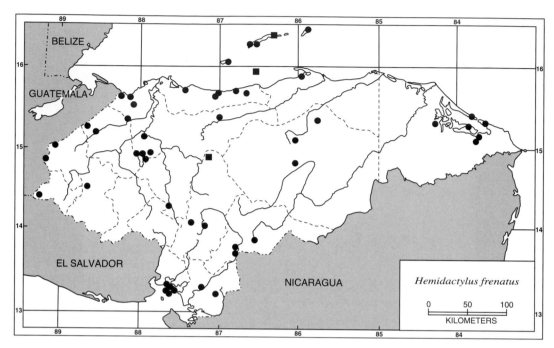

Map 4.   Localities for *Hemidactylus frenatus*. Solid circles denote specimens examined and solid squares those for other records. *Hemidactylus frenatus* is also known to occur on Isla Grande, Swan Islands, in the Atlantic Ocean (see Map 5).

Carbón, USNM 579590; Tulín, USNM 579597. **SANTA BÁRBARA**: Cerro Negro, USNM 573091; Sula, USNM 573083. **VALLE**: Amapala, USNM 565827; Coyolitos, FN 256888 (still in Honduras because of permit problems); Isla Comandante, USNM 580259–61; Isla Conejo, UNAH (1); Isla de Las Almejas, USNM 580256–58; Isla El Pacar, USNM 580767–68; Isla Inglesera, USNM 580247, 580250; Isla Tigrito, USNM 580769–70; Playona Exposición, USNM 565828–29, 580248–49; Punta Novillo, USNM 580262–63; near San Carlos, USNM 580251–55, FN 256907–08 (still in Honduras because of permit problems). **YORO**: Pino Alto, UF 166347; Río San Lorenzo, USNM 579598; near San Lorenzo Abajo, USNM 565400, 570145; Yoro, UF 166344.

Other Records (Map 4).—**FRANCISCO MORAZÁN**: Marale (Townsend et al., 2007). **ISLAS DE LA BAHÍA**: Cayo Cochino Mayor (Muelleman et al., 2009);

Cayo Cochino Menor, MPM Herpetological Photographs P741–743.

Hemidactylus haitianus Meerwarth, 1901

Hemidactylus brookii haitianus Meerwarth, 1901: 17 (syntypes, formerly ZMH 1500, 2250 [2], now destroyed [see Powell and Maxey, 1990: 493.2; Powell et al., 1996: 67]; type locality: "Haiti" and "Port au Prince").

Hemidactylus haitianus: Powell and Parmerlee, 1993: 54; McCranie, 2015a: 365.

Hemidactylus brookii: Köhler, 2003a: 71; Köhler, 2008: 75; Townsend and Wilson, 2010b: 692.

Hemidactylus brooki: Köhler et al., 2009: 451.

Hemidactylus angulatus: Solís et al., 2014: 130.

Geographic Distribution.—In the Western Hemisphere, *Hemidactylus haitianus* is found on Hispaniola, Cuba, and Puerto Rico

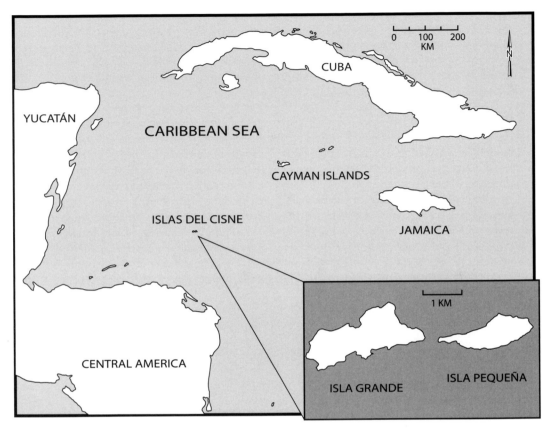

Map 5. Map showing the locality of the Swan Islands (Islas del Cisne), the only known locality for *Aristelliger nelsoni* and *Sphaerodactylus exsul*, plus the only known Honduran locality for *Leiocephalus varius* and *Ameiva fuliginosa*. Additionally, *Hemidactylus frenatus*, *H. mabouia*, and *Iguana iguana* are also known to occur on these islands. Three sea turtles (*Caretta caretta, Chelonia mydas,* and *Eretmochelys imbricata*) have also been reported from the waters surrounding the islands. Map modified from Morgan (1985).

(see Remarks) and presumably in northern coastal Colombia southward in the Río Magdalena system. Isolated recent introductions are also known from Mexico, central Panama, and Honduras. In the Eastern Hemisphere, the species occurs in western Africa, including Cameroon, Bioko Island, and Equatorial Guinea. In Honduras, *H. haitianus* was introduced, and at one time established, in Comayagüela, Francisco Morazán, in the south-central portion of the country (but see Remarks for *H. frenatus*).

Description.—The following is based on one male (USNM 572074) and one female (SMF 81770). *Hemidactylus haitianus* is a moderately small gecko (maximum SVL 68 mm [Kluge, 1969]; 60 mm SVL in largest Honduran specimen [USNM 572074, a male]); dorsal surfaces of head, body, and limbs covered with minute granules with 13–19 transverse rows of small, trihedral tubercles on body; 18 longitudinal tubercles in paravertebral row; distinct tubercles present on upper surface of hind limb and above ear opening; moveable eyelid absent; pupil vertically elliptical; 16–18 (16.8 ± 1.0) loreal scales (per side); 4–7 cheek tubercles; 1–3 auricular scales; 9–10 supralabials, 8 to level below mideye; 8 infralabials, 7 to below mideye; 1 postmental, contacting infralabial 1 on each side; 1 pair of enlarged chinshields, each separated medially by granular scales, each chinshield contacting infralabial 2, sometimes also contacting posterior edge of infralabial 1; 7–9 (8.0 ± 0.8) enlarged subdigital lamellae on Digit IV of hind limb, 2–6 divided medially (but not distalmost lamella), enlarged lamellae reaching origin of digit; nonretractile claw located in terminal phalanx, claw free from expanded terminal lamella; webbing absent between digits; 1 enlarged postcloacal scale present in each cloacal corner of male; 28 total femoral-precloacal pores present in male; femoral-precloacal pores absent in female; 3 interprecloacal poreless scales in male; 5–6 tubercles in fourth caudal whorl; 4–6 scales between caudal whorls; SVL 60.0 mm in male, 58.5 mm in female; TAL/SVL 1.07 in male (tail incomplete in female); HW/SVL 0.17 in male, 0.15 in female; HL/SVL 0.26 in male, 0.27 in female; SEL/SVL 0.12 in male, 0.11 in female; SEL/EEL 1.57 in male, 1.35 in female.

Color in alcohol: dorsal surfaces pale brown with small dark brown spots on body and anterior half of tail, posterior half of tail with 10 brown crossbands, crossbands becoming darker brown distally; ventral surfaces of head and body cream, that of limbs cream, except lamellae pale gray; subcaudal surface cream with brown flecking and mottling, markings becoming stronger distally.

Diagnosis/Similar Species.—The combination of lacking moveable eyelids and having the dorsal surface of the head covered with granular scales distinguishes *Hemidactylus haitianus* from all other Honduran lizards, except the remaining Gekkonoidea. *Aristelliger* have a digital pad with distinctively expanded, single subdigital lamellae (versus one or more lamellae divided in *H. haitianus*). *Gonatodes albogularis* lacks distinctively expanded subdigital lamellae and has a circular pupil (versus lamellae distinctly expanded and pupil vertically elliptical in *H. haitianus*). *Phyllodactylus* have the distalmost subdigital lamella expanded to form paired leaflike pads (versus distalmost lamella single, not forming leaflike pad in *H. haitianus*). *Sphaerodactylus* have only the terminal subdigital lamella distinctly enlarged, with that terminal lamella displacing the claws laterally, and all Honduran species also have circular pupils (versus claws in terminal phalanx free from expanded lamella, claw not displaced laterally, and pupil vertically elliptical in *H. haitianus*). *Thecadactylus rapicauda* has 13 or more divided subdigital lamellae on Digit IV of the fore- and hind limb, with the distalmost ones separated by skin, and distinct webbing present between the digits (versus distalmost lamella single, with fewer than 6 lamellae divided, and webbing absent between digits in *H. haitianus*). *Hemidactylus frenatus* lacks distinct tubercles on the upper surface of the hind limb and above the ear opening (versus distinct tubercles present on upper surface of hind limb and above ear opening in *H. haitianus*). Subdigital lamella of Digit IV on the hind limb in *H. mabouia* do not reach the origin of the digit (versus subdigital lamella of Digit IV on hind limb reaching origin of digit in *H. haitianus*).

Illustrations (Figs. 27, 28; Plate 5).—Cochran, 1941 (head, dorsal, ventral, and subdigital scales; as *H. brookii*); Grant, 1932 (adult; as *H. brookii*); Kluge, 1969 (adult, lateral scales of region between anterior margin of ear opening and posterior extreme of angle of mouth, subdigital lamellae; as *H. b. haitianus*); Köhler, 2003a (adult, chin scales; as *H. brookii*), 2008 (adult, chin scales; as *H. brookii*); Powell and Maxey, 1990 (adult; as *H. b. haitianus*); Powell and Parmerlee, 1993 (adult); Powell et al., 1996 (adult); Rivero, 1998 (adult, juvenile, subdigital lamellae; as *H. b. haitianus*).

Remarks.—Carranza and Arnold (2006) studied mitochondrial DNA sequences of

Plate 5.   *Hemidactylus haitianus*. Dominican Republic: Isla Alto Velo. Photograph by S. B. Hedges.

various populations of *Hemidactylus*, including two specimens of *H. haitianus* (as *H. brookii haitianus*) from western Cuba. The two Cuban populations nested among populations of *H. angulatus* Hallowell (1854a: 63) from Africa. Furthermore, they clustered closely with a sample of *H. angulatus* from Bioko Island, Equatorial West Africa. Therefore, Carranza and Arnold (2006: 539, 542) suggested that *H. b. haitianus* might be conspecific with *H. angulatus*. Powell (1993) had suggested elevating *Hemidactylus b. haitianus* to species level (also see Powell and Parmerlee, 1993). Weiss and Hedges (2007) reported sequencing DNA samples of *H. haitianus* from Hispaniola (the type locality of *H. haitianus* is on Haiti), Puerto Rico, and Cuba. Because their data showed virtually no molecular variation in the Greater Antilles populations studied and they clustered very closely with *H. angulatus* from Bioko Island, Equatorial Guinea, Africa, Weiss and Hedges (2007) placed *H. haitianus* in the synonymy of *H. angulatus*. Weiss and Hedges also concluded those

geckos likely reached the Greater Antilles by the slave trade that occurred for several hundred years between West Africa and the Greater Antilles. Subsequently, Bauer et al. (2010) suggested that the name *H. haitianus* be retained for the time being for the Dominican Republican and Cuban populations. Thus, I am using *H. haitianus* for the Honduran population, although realizing that the type locality contains a nonnative population that is a result of human traffic, instead of the polytypic *H. brookii* Gray (1845b: 153) of Asia (Bauer et al., 2010: 348–349, also recovered *H. brookii* as a member of a species group distinct from *H. angulatus* or *H. haitianus*). DNA data of *H. haitianus* from the Caribbean studied thus far are identical to that of a population of *H. angulatus* from western Africa (Weiss and Hedges, 2007: 411), but I continue to use *H. haitianus* for the present because it is expected that African *H. angulatus* is a complex of several cryptic species (see Bauer et al., 2010). Honduran *H. haitianus* populations need to be studied with molecular methods. Finally, Moravec et al. (2011)

demonstrated the presence of *H. haitianus* in coastal Cameroon, Africa, based on mitochondrial DNA.

Kluge (1969) presented a systematic review of Western Hemisphere populations of *Hemidactylus haitianus* (as *H. brookii haitianus* and presumably *H. b. leightoni* Boulenger, 1911: 19, as well). Powell and Maxey (1990; as *H. brookii*) provided a redescription and included an extensive literature review. Also see Remarks for *H. mabouia*.

Köhler (2003a, 2008) mapped a second Honduran locality for this species (La Ceiba, Atlántida) based solely on a photograph he had seen. There are no specimens available from La Ceiba; thus, that record is rejected. Köhler and Ferrari (2011b) reported *H. haitianus* from Isla Grande in the Swan Islands based on SMF 90456–57. Reexamination of those two specimens reveals one (SMF 90456) is *H. frenatus* and the other (SMF 90457) represents *H. mabouia*.

Natural History Comments.—*Hemidactylus haitianus* is known only from 950 m elevation in the Premontane Dry Forest formation. The Honduran specimens of this introduced species were collected on walls of a single building at night in June and July 1999 and 2006. However, all geckos collected on those same walls in 2012 proved to be *H. frenatus*. I have not found *H. haitianus* and do not know if an established population remains in Honduras due to the *H. frenatus* ability to displace *H. haitianus*. Thus, I have not heard its voice, but it is likely that both sexes emit distress calls and males likely produce the clicking sounds during courtship and territorial displays like other *Hemidactylus* introduced to Honduras are known to do (see Natural History Comments for *H. frenatus*). Powell et al. (1990) reported adult females of this species from Hispaniola usually deposit two eggs per clutch but rarely deposit one egg per clutch. *Hemidactylus haitianus* eggs are sometimes communally deposited and are

tolerant of exposure to salt water (Henderson and Powell, 2009, and references cited therein), thus no doubt assisting in transoceanic introductions. This species feeds on a variety of insects and other invertebrates and frequently forages around night lights. See Henderson and Powell (2009) for a list of insect prey recorded on Hispaniola. I suspect that in time the Caribbean Island *H. haitianus* populations will be negatively affected by *H. frenatus* populations on those islands.

Etymology.—The name *haitianus* is derived from Haiti and the Latin *-anus* (pertaining to, belonging to) in reference to the type series having been taken on Haiti.

Specimens Examined (3 [0]; Map 6).— **FRANCISCO MORAZÁN**: Comayagüela, SMF 81770–71, USNM 572074.

Hemidactylus mabouia (Moreau de Jonnès, 1818)

Gecko mabouia Moreau de Jonnès, 1818: 138 (holotype, MNHN 6573 [see Guibé, 1954: 6; Kluge, 1969: 32; also see Remarks]; type locality: "Amerique, dans les contrées continentales qui avoisinent au midi l'archipel des Antilles, et qu'il est également répandu dans les îles même de l'archipel, depuis la Trinité jusqu'à la Jamaïque." Type locality restricted to "Antilles" by A. M. C. Duméril and Duméril, 1851: 39; further to "St. Vincent" by Stejneger, 1904: 599 [but see Kluge, 1969: 31–32]).

Hemidactylus mabouia: A. M. C. Duméril and Bibron, 1836: 362 (in part); McCranie and Wilson, 2000: 113; Köhler, 2003a: 71; McCranie et al., 2006: 216; Wilson and Townsend, 2006: 105; Köhler, 2008: 75; Gutsche and McCranie, 2009: 112; McCranie and Valdés Orellana, 2014: 45; Solís et al., 2014: 130; McCranie, 2015a: 365; McCranie et al., 2017: 268 (in part).

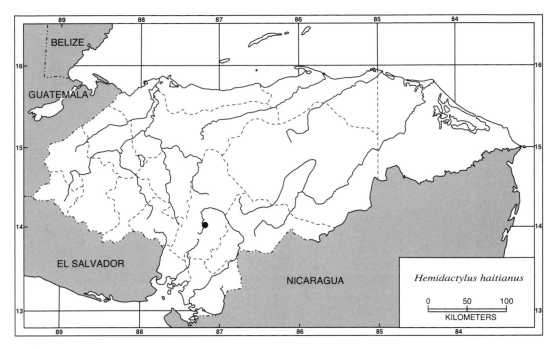

Map 6.   Locality for *Hemidactylus haitianus*. Solid circle denotes specimens examined from the same locality.

Hemidactylus brookii: Ferrari, 2008: A14, *In* Anonymous, 2008; Köhler and Ferrari, 2011b: 240 (in part).

Geographic Distribution.—In the Western Hemisphere, *Hemidactylus mabouia* is found in eastern South America from Montevideo, Uruguay, through Brazil, French Guiana, and Suriname to Venezuela and along most of the length of the Amazon River in Brazil, Ecuador, Colombia, Bolivia, and Peru. It also occurs on many islands off the north coast of South America and throughout the Lesser Antilles. Those southern populations might be natural, but the populations from the more northerly islands in the Lesser Antilles are probably introductions. The species is known to be introduced and established in central and southern Florida (including the Florida Keys), USA, Mexico, Honduras, Costa Rica, Panama, the Bahamas, Cuba, Grand Cayman, Hispaniola, Jamaica, Puerto Rico and the Virgin Islands, the Turks and Caicos, and islands off the north coast of Venezuela. In the Eastern Hemisphere, this species occurs "in Africa south of 10°N latitude [although absent from arid and temperate regions of southwestern Africa], Ascension Island, and on Madagascar and islands in the Mozambique Channel" (Powell, Crombie, and Boos, 1998: 674.2, but also see p. 674.6). In Honduras, this species has been introduced on the north coast at Trujillo, Colón, on Isla de Guanaja in the Bay Islands, and on Isla Grande (or Big Swan Island) of the Swan Islands (but see Remarks).

Description.—The following is based on four males (MCZ 192116; SMF 90457; USNM 570548; ZMB 73625) and four females (USNM 536064, 570549, 581897; ZMB 76324). *Hemidactylus mabouia* is a moderately small gecko (maximum SVL in Western Hemisphere 68 mm [Kluge, 1969]; 64 mm SVL in largest Honduran specimen [USNM 570549, a female]); dorsal surfaces of head, body, and limbs covered with minute granules with 9–14 transverse rows of small, trihedral tubercles on body; 11–15 longitudinal tubercles in paravertebral row; a few tubercles present on upper surface of hind limb; no tubercles present above ear opening; moveable eyelid absent; pupil vertically elliptical; 15–22 (17.1 ± 1.8) loreal scales (per side); 0–5 cheek tubercles; auricular scales absent; 10–13 supralabials, 8–11 to level below mideye; 8–11 infralabials, 6–8 to level below mideye; paired postmentals, each contacting infralabial 1; paired enlarged chinshields, separated medially by granular scales, each chinshield contacting infralabial 2, sometimes also contacting posterior edge of infralabial 1; 3–8 (6.6 ± 1.3) enlarged subdigital lamellae on Digit IV of hind limb, 4–7 divided medially (but not distalmost lamella), enlarged lamellae of Digit IV not reaching origin of digit; nonretractile claw located in terminal phalanx, not laterally displaced, claw free; webbing absent between digits; enlarged postcloacal scales absent; 21–33 (28.0 ± 5.3) total precloacal pores in males, precloacal pores absent in females; zero interprecloacal poreless scales in males; 3–6 tubercles in fourth caudal whorl; 6–14 scales between caudal whorls; SVL 53.1–62.0 (57.6 ± 3.7) mm in males, 49.5–63.6 (56.9 ± 5.8) mm in females; TAL/SVL 0.92–1.18 in two males, 0.90–1.07 in three females; HW/SVL 0.14–0.16 in males, 0.13–0.16 in females; HL/SVL 0.25–0.30 in males, 0.23–0.26 in females; SEL/SVL 0.11 in males, 0.08–0.11 in females; SEL/EEL 1.24–1.41 in males, 1.24–1.70 in females.

Color in alcohol: all dorsal surfaces pale brown; dorsal surface of body uniformly brown, or with a darker brown V-shaped mark present posteriorly; dorsal surface of tail with or without darker brown crossbands; ventral surfaces of head and body cream, those of limbs cream, except subdigital lamellae pale gray to medium gray; subcaudal surface cream with brown flecking distally.

Diagnosis/Similar Species.—The combination of lacking moveable eyelids and having the dorsal surface of the head covered with granular scales distinguishes *Hemidactylus mabouia* from all other Honduran lizards, except the remaining Gekkonoidea. *Aristelliger* have a distinctively expanded digital pad with single subdigital lamellae (versus one or more lamellae divided in *H. mabouia*). *Gonatodes albogularis* lacks distinctively expanded subdigital lamellae and has a circular pupil (versus lamellae expanded and pupil vertically elliptical in *H. mabouia*). *Phyllodactylus* have the distal pair of subdigital lamella expanded to form a paired leaflike pad (versus distalmost lamella single in *H. mabouia*). *Sphaerodactylus* have only the terminal subdigital lamella more than minimally enlarged, that terminal subdigital lamella displacing the claw laterally, and all Honduran species have a circular pupil (versus claws in terminal lamella not displaced laterally, claws extending from expanded lamella, and pupil vertically elliptical in *H. mabouia*). *Thecadactylus rapicauda* has 13 or more divided subdigital lamellae on Digit IV of the fore- and hind limb, with the distalmost ones separated by skin, and has distinct webbing between the digits (versus distalmost lamella single, with fewer than 6 lamellae divided, and digital webbing absent in *H. mabouia*). Subdigital lamellae of Digit IV on the hind limb of *H. frenatus* and *H. haitianus* reach the origin of the digit (versus subdigital lamellae of Digit IV on hind limb not reaching origin of digit in *H. mabouia*).

Illustrations (Fig. 24; Plate 6).—Avila-Pires, 1995 (adult, head scales, dorsal tubercles, subdigital lamellae); Bauer, 2013 (adult); Cochran, 1941 (head, dorsal, ventral, subdigital, and tail scales); Grant, 1932 (adult, juvenile); Hoogmoed, 1973 (adult, head scales); Kluge, 1969 (adult, lateral scales of region between anterior margin of ear opening and posterior extreme of angle of mouth, subdigital lamellae); Köhler, 2000

Plate 6.   *Hemidactylus mabouia*. MCZ R192116, adult male, SVL = 57 mm. Gracias a Dios: Isla Grande, Swan Islands.

(subdigital lamellae), 2003a (adult, chin scales, subdigital lamellae), 2008 (adult, chin scales, subdigital lamellae); McCranie et al., 2017 (adult); Meshaka et al., 2004 (adult); Murphy, 1997 (adult, subdigital lamellae); Powell, Crombie, and Boos, 1998 (adult); Powell, Collins, and Hooper, 1998 (subdigital lamellae), 2012 (tail scales, subdigital lamellae); Rivas Fuenmayor et al., 2005 (head, subdigital lamellae); Rivero, 1998 (adult, subdigital lamellae); Schmidt, 1928c (head and subdigital scales); Stejneger, 1904 (head, tail, and toe scales); Vanzolini, 1978 (subdigital lamellae); Vitt, 1996 (adult); Vitt and Caldwell, 2014 (adult).

Remarks.—Kluge (1969: 31–32), based on his examination of the holotype (MNHN 6573) on which Moreau de Jonnès' name *Gecko mabouia* was founded, concluded that the specimen actually represented *Hemidactylus brookii.* Kluge (1969: 32) suggested that in the best interest of nomenclatural stability "the true identity and probable geographic origin of Moreau de Jonnès' holotype (probably Cartagena, Colombia)" be ignored. Kluge (1969: 32) also recommended "that the name *mabouia*

(Moreau de Jonnès) be retained for the Lesser Antilles and South American *Hemidactylus* species populations" that he (Kluge, 1969: 29) diagnosed as *H. mabouia.* Powell, Crombie, and Boos (1998: 674.5) stated that Kluge's conclusions "should be reviewed and ruled upon by the ICZN." Much more molecular work is needed on both New World and Old World *Hemidactylus* species before the correct names for the various New World populations can be settled. Kluge (1969) presented a systematic review of the Western Hemisphere populations of *H. mabouia.* Powell, Crombie, and Boos (1998) provided a redescription and included an extensive review of the literature. De Lisle et al. (2013: 122) stated that A. M. C. Duméril and Duméril (1851) designated MNHN 6573 as the lectotype of *H. mabouia*, but those authors only restricted the type locality without giving a museum number. If the specimen Moreau de Jonnès described as *Gecko mabouia* actually represented *H. brookii* as suggested by Kluge (1969: 31–32), the correct nomenclature of the species involved would be further complicated.

Molecular studies by Carranza and Arnold (2006: 540–541) recovered minimum divergence between samples studied over much of the range of this species, including introduced populations from Argentina, Brazil, Trinidad and Tobago, Puerto Rico, and Florida. Thus, Western Hemisphere populations of *Hemidactylus mabouia* appear to represent recent introductions by man. However, Bauer et al. (2010: 348) recovered a phylogeny based on molecular data suggesting South African *H. mabouia* was probably a species distinct from the U.S. specimen they included in their study. Molecular study of MCZ R192116 from Isla Grande in the Swan Islands indicates the specimen is the result of a recent human introduction (S. B. Hedges, in litteris, 29 April 2013).

Natural History Comments.—*Hemidactylus mabouia* is known from near sea level to 12 m elevation in the Lowland Moist Forest and Lowland Dry Forest (West Indian Subregion) formations. The first Honduran specimen of this introduced species was collected at night on walls of a house in Trujillo, Colón, in June 1998. However, the species apparently no longer occurs at that locality, with *H. frenatus* now being common on those walls. Powell, Crombie, and Boos (1998) suggested that *H. frenatus* might displace *H. mabouia*, which seems to be the case at Trujillo (also see Rödder et al., 2008, and references cited therein). On the other hand, Meshaka et al. (2004) stated *H. mabouia* displaces other hemidactylines in Florida, including *H. frenatus*, which might indicate actually more than one species of *H. "mabouia"* exists in the Western Hemisphere. Four specimens of *H. mabouia* were collected in May 2007 on the walls of a hotel in Savannah Bight on Isla de Guanaja, Bay Islands, and several others were seen at that time but were not collected. In November 2011, three *Hemidactylus* were collected on the same walls, two of which are *H. frenatus* and the other *H. mabouia*. Numerous other *Hemidactylus* were seen at that time but not collected. However, the only specimen of *Hemidactylus* preserved in September 2012 from the walls of that hotel proved to be *H. frenatus*. This species was seen foraging around night lights, where they were feeding on insects attracted to those lights. Both *H. mabouia* and *H. frenatus* were collected on walls of buildings on Isla Grande in the Swan Islands in April and May 2007, but by December 2012, only one *H. mabouia* was collected there, whereas literally hundreds of *H. frenatus* were seen (also see McCranie et al., 2017). Both males and females of *H. mabouia* produce distress calls, and males make a clicking sound during courtship and territorial activity (personal observation). Females of this species deposit multiple clutches of two eggs with an extensive breeding season in the tropics and subtropics (Fitch, 1970; Köhler, 1997f; Powell, Crombie, and Boos, 1998; Meshaka et al., 2004, and references cited in those works). Communal nesting has also been documented (Krysko et al., 2003; Meshaka et al., 2004), including with other genera of Gekkonoidea. Henderson and Powell (2009) reviewed the literature for West Indian populations and provided data on diet in this insectivorous species. Inturriaga and Marrero (2013) studied feeding ecology of this introduced species in Cuba and found those populations feed mainly on nonflying arthropods. Bonfiglio et al. (2006) reported that a Brazilian population fed mainly on arthropods but also documented cannibalism (also see Costa-Campos and Furtado, 2013). Dornburg et al. (2011) also found remains of a native, nocturnal *Gonatodes* in the stomach of one *H. mabouia* on Curaçao; that *H. mabouia* population is recently introduced. A treefrog of the genus *Scinax* Wagler was reported preying on a juvenile *H. mabouia* in Brazil (Zanchi-Silva and Borges-Nojosa, 2017). Oliveira Nogueira et al. (2013) provided a list of the published records of predation on this species. The various

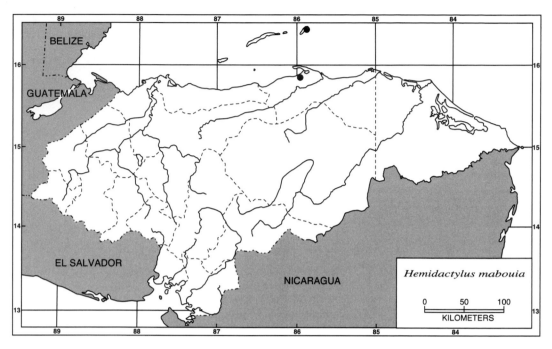

Map 7.   Localities for *Hemidactylus mabouia*. Solid circles denote specimens examined. This species also occurs on Isla Grande, Swan Islands, in the Atlantic Ocean (see Map 5).

populations studied for reproduction and diet mentioned above probably represent more than one species masquerading under the name *H. mabouia*.

Although *Hemidactylus frenatus* apparently has replaced *H. mabouia* at many localities (see above and *H. frenatus* account), *H. mabouia* is known to have replaced the native species of *Phyllodactylus* on Curaçao and Bonaire in edificarian situations, but not in nonedificarian situations (Buurt, 2006). It is not known if the Honduran Savannah Bight population of *H. mabouia* displaced those of the native *Phyllodactylus paralepis* on the walls of the hotel (before the *Hemidactylus* invasions, *P. paralepis* and *P. palmeus* were common on human structures on the Bay Islands) where the *Hemidactylus* is now known, but *P. paralepis* was common in nonedificarian situations near Savannah Bight in 2011–2012.

Etymology.—The name *mabouia* is derived "from the language of savage tribes of Septentrional America, and refers to everything that induces disgust or fear" (translated from Lacepède, 1788: 379, by A. Miralles, personal communication, 28 November 2006).

Specimens Examined (8 [0]; Maps 5, 7).— **COLÓN**: Trujillo, USNM 536064. **GRACIAS A DIOS**: Isla Grande, Swan Islands, MCZ R192116, SMF 90457. **ISLAS DE LA BAHÍA**: Isla de Guanaja, Savannah Bight, USNM 570548–49, 581897, ZMB 73624–25.

Family Phyllodactylidae Gamble, Bauer, Greenbaum, and Jackman, 2008.

This family of geckonid lizards is found in the Western Hemisphere from southern California, USA, to the southern tip of Baja California del Sur, Mexico, and from Sonora and the Yucatán Peninsula, Mexico, to northwestern Chile and southeastern Argentina all the way to the Strait of Magellan. It also occurs on the Galapagos Islands, Ecuador, on the islands off northern South America, throughout the Lesser Antilles to Puerto Rico and the Virgin Islands, and on several smaller islands off the Caribbean coast of southeastern Mexico and Central America. In the Eastern Hemisphere, this family occurs across northern Africa, the Arabian Peninsula, and central Asia. Nine genera containing about 135 named species are included in this family, with two genera containing four named species occurring in Honduras.

Remarks.—Males and females of the Honduran members of *Phyllodactylus* are similar in size to each other, whereas males of *Thecadactylus* are larger than are females.

KEY TO HONDURAN GENERA OF THE FAMILY PHYLLODACTYLIDAE

1A. Distalmost pair of subdigital lamella expanded to form leaflike pad (Fig. 17); webbing absent between digits on hind limbs (Fig. 17) . . . .
. . . . *Phyllodactylus* (p. 60)
1B. All subdigital lamellae divided, distal ones separated by skin (Fig. 18); webbing present between digits on hind limbs (Fig. 18) . . . .
. . . . *Thecadactylus* (p. 73)

CLAVE PARA LOS GÉNEROS HONDUREÑOS DE LA FAMILIA PHYLLODACTYLIDAE

1A. Par distal de lamelas subdigitales expandida en una almohadilla en forma de hoja (Fig. 17); membrana interdigital ausente entre los dedos en las extremidades (Fig. 17) . . . .
. . . . *Phyllodactylus* (p. 60)
1B. Todas del lamelas subdigitales divididas y las más distales separadas por piel (Fig. 18); membrana interdigital presente entre los de-

dos en las extremidades posteriores (Fig. 18) . . . . *Thecadactylus* (p. 73)

Genus *Phyllodactylus* Gray, 1828

Phyllodactylus Gray, 1828:3 (type species: *Phyllodactylus pulcher* Gray, 1828: 3, by monotypy).

Geographic Distribution and Content.—This genus occurs from southern California, USA, to the southern tip of Baja California del Sur, Mexico, from Sonora, Mexico, to Costa Rica, and from northwestern Ecuador to northwestern Chile on the Pacific versant, including the Galapagos Islands and numerous islands in the Golfo de Fonseca of Honduras. It occurs on the Atlantic versant from Quintana Roo, Mexico, and Belize (and some offshore islands) to eastern Guatemala, the Bay Islands, Honduras, and from northeastern Colombia to eastern (and many offshore islands) and southern Venezuela (via the Río Orinoco), Aruba, Bonaire, Curaçao, Barbados, and Puerto Rico. Bauer et al. (1997) placed the Eastern Hemisphere species of *Phyllodactylus* into several genera, thus restricting *Phyllodactylus* to the Western Hemisphere. About 55 named species are included in this genus, three of which occur in Honduras.

Remarks.—De Lisle et al. (2013: 204) recorded *Phyllodactylus insularis* Dixon (1960: 9) from the Honduran Bay Islands. Those Bay Island specimens represent *P. palmeus* or *P. paralepis*, depending on the island.

Scale terminology and methods of counting scales used herein follow those of Dixon (1960, 1964, 1968) as slightly modified by McCranie and Hedges (2013b).

Etymology.—The name *Phyllodactylus* is derived from the Greek words *phyllon* (leaf) and *daktylos* (finger, toe) and refers to the leaflike terminal subdigital lamella.

Figure 31.   Outer postmental on each side contacting both first and second infralabials (outlined). *Phyllodactylus tuberculosus.* USNM 580936 from Isla El Pacar, Valle.

KEY TO HONDURAN SPECIES OF THE GENUS
PHYLLODACTYLUS

1A. Outer postmental on each side usually contacting both first and second infralabials (Fig. 31); 20 or fewer scales in midorbital region; 24 or fewer paravertebral tubercles from head to tail (occurs on mainland in southern Honduras).
.... *tuberculosus* (p. 69)

1B. Outer postmental on each side contacting only first infralabial (Fig. 32); 20 or more scales in midorbital region; 35 or more paravertebral tubercles from head to tail (occurs on Bay Islands) 2

2A. Forty-one to 53 paravertebral tubercles from head to tail; 16–17 dorsal tubercle rows across midbody............... *paralepis* (p. 66)

2B. Thirty-five to 43 paravertebral tubercles from head to tail; 11–15 dorsal tubercle rows across midbody................ *palmeus* (p. 62)

CLAVE PARA LAS ESPECIES HONDUREÑAS DEL
GÉNERO *PHYLLODACTYLUS*

1A. Postmental externa en cada lado usualmente tocando la primera y el segunda infralabial (Fig. 31); 20 o menos escamas en la región medio-orbital; 24 o menos tubérculos paravertebrales desde la cabeza hasta la cola (se encuentran en la parte continental del sur de Honduras) .... *tuberculosus* (p. 69)

1B. Postmental externa en cada lado tocando solamente la primera infralabial (Fig. 32); 20 o más escamas en la región medio-orbital; 35

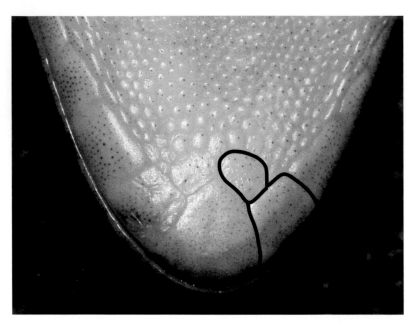

Figure 32.   Outer postmental on each side contacting only first infralabial (outlined). *Phyllodactylus palmeus*. USNM 570146 from Roatán Island, Flowers Bay, Islas de la Bahía.

o más tubérculos paravertebrales desde la cabeza hasta la cola (se encuentran solamente en las Islas de la Bahía) 2

2A. De 41–53 tubérculos paraverte-brales desde la cabeza hasta la cola; 16–17 hileras de tubérculos dorsales en la parte media del cuerpo . . . . *paralepis* (p. 66)

2B. 35–43 tubérculos paravertebrales desde la cabeza hasta la cola; 11–15 hileras de tubérculos dorsales en la parte media del cuerpo . . . .
. . . . *palmeus* (p. 62)

Phyllodactylus palmeus Dixon, 1968

Phyllodactylus tuberculosus: Günther, 1890: 80, *In* Günther, 1885–1902.
Phyllodactylus insularis: Dixon, 1964: 78; Echternacht, 1968: 151 (in part).
Phyllodactylus palmeus Dixon, 1968: 419 (in part) (holotype, LSUMZ 16986 [see Remarks]; type locality: "0.5 km N. Roatan [sic], Isla de Roatan [sic], ca.

25 mts."); Meyer, 1969: 204 (in part); Peters and Donoso-Barros, 1970: 224; Meyer and Wilson, 1973: 11 (in part); Wilson and Hahn, 1973: 104 (in part); MacLean et al., 1977: 4; Hudson, 1981: 377; Rossman and Good, 1993: 8; Wilson and Cruz Díaz, 1993: 17; Köhler, 1994a: 4; Köhler, 1995b: 102; Köhler, 1996c: 25; Cruz Díaz, 1998: 29, *In* Bermingham et al., 1998; Köhler, 1998d: 375; Monzel, 1998: 155 (in part); Lundberg, 2000: 3; Grismer et al., 2001: 135; Lundberg, 2001: 27; Wilson et al., 2001: 136 (in part); Lundberg, 2002a: 7; Gutsche and Ohl, 2003: 48; Köhler, 2003a: 74 (in part); Powell, 2003: 36; Wilson and McCranie, 2003: 59 (in part); McCranie et al., 2005: 78 (in part); Wilson and Townsend, 2006: 105 (in part); Köhler, 2008: 79 (in part); McCranie and Hedges, 2013b: 58; de Lisle et al., 2013: 207; McCranie and Valdés Orellana, 2014: 45; Solís et al., 2014: 131; McCranie, 2015a: 365.

Sphaerodactylus millepunctatus: Monzel, 1998: 156 (in part).

Geographic Distribution.—*Phyllodactylus palmeus* occurs on the Cayos Cochinos and the islands of Barbareta, Morat, Roatán, and Utila, all in the Bay Islands, Honduras.

Description.—The following is based on ten males (FMNH 283556, 283559; KU 203121, 203123–25, 220098; USNM 570146, 570151, 570154) and ten females (FMNH 283557–58; KU 203122, 220099; USNM 570147–48, 570150, 570152–53, 570156). *Phyllodactylus palmeus* is a moderate-sized gecko (maximum recorded SVL 79 mm [KU 220098, a male]); dorsal surface of head covered with granular scales, 24–30 (28.0 ± 1.7) granules across snout between third supralabials; 11–14 (12.5 ± 1.0) loreals (per side); 20–28 (24.8 ± 2.3) midorbital scales; 5–6 (usually 6) supralabials and infralabials to levels below mideye; 6–8 (6.9 ± 0.8) scales bordering internasals posteriorly; moveable eyelid absent; pupil vertically elliptical, reticulated; 2–3 (rarely 3) enlarged postmentals, outer postmental on each side contacting infralabial 1; 6–9 (7.3 ± 0.9) scales bordering postmentals posteriorly; 11–15 (13.4 ± 1.0) transverse rows of keeled tubercles among small dorsal body scales at midbody; 35–43 (39.3 ± 2.1) tubercles in paravertebral row from rear of head to base of tail; 23–33 (27.6 ± 2.5) tubercles in paravertebral row between levels of axilla and groin; 4–7 (5.2 ± 1.1) tubercles across base of tail; enlarged dorsal tubercles separated by 1–3 granules; ventral scales smooth, imbricate, 52–63 (59.0 ± 3.3) along midline from posterior edge of throat to cloacal opening, 20–30 (26.5 ± 2.0) rows across midventer; subdigital lamellae narrow, single, except terminal pair divided, as is row just proximal to terminal pair, terminal pair greatly expanded to form leaflike pad; nonretractile claw present at distal end of leaflike pad; webbing absent between digits; 11–16 (12.8 ± 1.2) subdigital lamellae on Digit IV of hind limb, 9–11 (10.1 ± 0.8) on Digit IV of forelimb; femoral and precloacal pores absent in both sexes; SVL 56.3–78.9 (65.9 ± 8.2) mm in males, 46.2–72.1 (60.8 ± 7.3) mm in females; HW/SVL 0.13–0.17 in males, 0.14–0.16 in females; HL/SVL 0.22–0.26 in males, 0.24–0.25 in females; HW/HD 1.21–1.53 in males, 1.18–1.53 in females; TAL/SVL 1.00–1.15 in four males, 0.92–1.10 in seven females; SEL/EEL 1.20–1.44 in males, 1.09–1.54 in females.

Color in life of an adult female (USNM 570152): dorsal surface of body Drab (27) with Olive-Brown (28) linear mottling suggestive of stripes; dorsal surface of head Drab with Olive-Brown mottling; dorsal surfaces of fore- and hind limb Drab with Olive-Brown crossbands; dorsal surface of tail mottled with cream and Olive-Brown; lateral surface of head with Olive-Brown pre- and postocular stripes on Drab ground color; ventral surfaces of head, body, and limbs pale yellowish brown; subcaudal surface mottled cream and Olive-Brown; digit pads pale gray; iris Clay Color (26) with black reticulations.

Color in alcohol: dorsal ground color pale brown to brown, with darker brown linear mottling usually present on body; dorsal tubercles pale brown; pre- and postocular stripes brown, with postocular stripe usually continuing onto lateral portion of body posterior to forelimb insertion; pale brown stripe (paler than dorsal ground color) present dorsal to pre- and postocular brown stripes; supralabials ventral to eye vary from brown, to brown mottled with white, to white; ventral surfaces of head and body cream to dirty white, with most scales lightly flecked with dark brown; subcaudal surface more heavily flecked with dark brown than ventral surface of body.

Diagnosis/Similar Species.—The combination of lacking moveable eyelids, having the dorsal surface of the head covered with granular scales, and having the distal pair of subdigital lamella expanded to form a paired leaflike pad will distinguish *Phyllodactylus*

Plate 7.   *Phyllodactylus palmeus.* FMNH 283558, adult female, SVL = 58 mm. Islas de la Bahía: Roatán Island, Palmetto Bay.

palmeus from all other Honduran lizards, except the remaining *Phyllodactylus*. *Phyllodactylus paralepis* has 0 or 1 granule separating the enlarged dorsal tubercles and has 16–17 dorsal tubercles across the midbody (versus 1–3 granules separating enlarged dorsal tubercles and 11–15 rows of dorsal tubercles across midbody in *P. palmeus*). *Phyllodactylus tuberculosus* has 13–20 scales in the midorbital row, has 16–18 transverse dorsal tubercles at midbody, and has the outer postmental on each side, usually contacting both the first and second infralabials (versus 20–28 scales in midorbital row, 11–15 transverse dorsal tubercles at midbody, and outer postmental not contacting second infralabial in *P. palmeus*).

Illustrations (Figs. 17, 32; Plate 7).— Dixon, 1968 (adult); Gutsche and Ohl, 2003 (adult); Lundberg, 2002a (adult); McCranie and Hedges, 2013b (dorsal tubercles); McCranie et al., 2005 (adult); Powell, 2003 (adult).

Remarks.—Dixon (1968) considered *Phyllodactylus palmeus* to be most closely related to *P. insularis* of Half Moon Cay and Long Cay off the coast of Belize. Molecular data from *P. insularis* is needed to test that scenario. *Phyllodactylus palmeus* and *P. paralepis* were placed in the newly formed *P. palmeus* species group by McCranie and Hedges (2013b: 56). It seems likely that the insular *P. insularis* also belongs to that group.

De Lisle et al. (2013: 207) gave an erroneous holotype number for *Phyllodactylus palmeus*.

Natural History Comments.—*Phyllodactylus palmeus* is known from near sea level to 30 m elevation in the Lowland Moist Forest formation. This nocturnal species is active throughout the year. It was formerly frequently active on walls of buildings (see below) but remains active on coconut palm trunks, on thorn palm trunks, and on mangrove tree trunks. It has been found inactive by day in termite nests, in holes in mangrove trees, beneath tree bark, and in rotten logs and other debris on the ground, especially coconut palm debris. Grismer et al. (2001) reported finding *P. palmeus* under rocks, logs, palm fronds, and other ground debris in both disturbed and relatively undisturbed situations. This species was formerly common on walls of at least one hotel on Isla de Utila in 2001 but had been

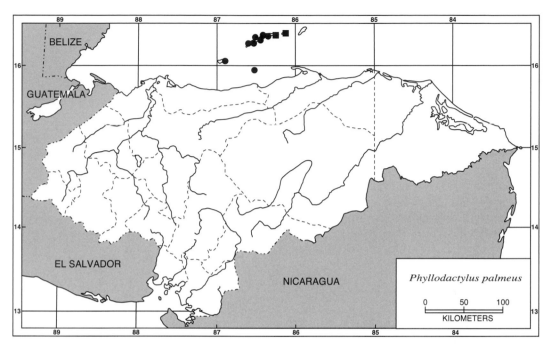

Map 8.   Localities for *Phyllodactylus palmeus*. Solid circles denote specimens examined and solid squares represent those for accepted records.

replaced by the subsequently introduced *Hemidactylus frenatus* by 2004. The same had also occurred on walls of the Iguana Station on Utila by 2004. *Phyllodactylus palmeus* was also common on the walls of a hotel in Sandy Bay, Isla de Roatán, in the 1990s to at least 2004. However, by 2010, the native *Phyllodactylus* had been completely replaced on those walls, and by 2012 an enormous number of the invasive *H. frenatus* were on those walls. The name of that hotel in 2012 was Pirates Inn, but that hotel ownership and name have changed several times over the years. Also, Powell (2003), based on one short trip, said *H. frenatus* was replacing native *P. palmeus* on buildings on Utila Island. Additionally, *H. frenatus* appeared to be replacing *P. palmeus* away from edificarian situations on Utila in September 2011. Females of this species deposit two eggs per clutch in communal nests (Dixon, 1968; that data probably also includes the recently de-

scribed *P. paralepis*) and probably multiple clutches per year throughout an extended reproductive season. I found communal nests of *P. palmeus* inside mangrove tree hollows on Isla de Utila. Nothing has been published on diet of *P. palmeus*. Gutsche and Ohl (2003) reported an arthropod preying on an adult *P. palmeus* on Isla de Utila.

Etymology.—The name *palmeus* is formed from the English word palm and the Latin suffix *-eus* (made of, having the quality of), and alludes to the purported, but not necessarily true, habitat preference of this species for palms.

Specimens Examined (93 [23]; Map 8).— **ISLAS DE LA BAHÍA**: Cayo Cochino Mayor, near La Ensenada, KU 220098–99; Cayo Cochino Mayor, UTA R-53953; Isla de Roatán, Fiddlers Bight, UMMZ 142650; Isla de Roatán, Flowers Bay, USNM 570146–47; Isla de Roatán, Fort Key, UMMZ 142649 (2); Isla de Roatán, near French Harbor,

UF 28560–61; Isla de Roatán, Gibson Bight, LSUMZ 33782–83; Isla de Roatán, W of Oak Ridge, UTA R-10707 (formerly MCZ 89385), 10722 (formerly MCZ 89384); Isla de Roatán, near Oak Ridge, TCWC 52410, UTA R-10708, 55240–43; Isla de Roatán, Palmetto Bay, FMNH 283556, 283558, MVZ 267203; Isla de Roatán, 1 km E of Pollytilly Bight, FMNH 282617–20; Isla de Roatán, near Port Royal Harbor, LSUMZ 33787–91, TCWC 52411–13; Isla de Roatán, near Port Royal, USNM 578824–27; Isla de Roatán, 0.5–1.0 km N of Roatán, TCWC 24016; Isla de Roatán, about 3 km N of Roatán, LSUMZ 16986–94; Isla de Roatán, about 3.2 km W of Roatán, CM 64514; Isla de Roatán, about 1.6 km W of Roatán, LSUMZ 33792; Isla de Roatán, about 4.8 km W of Roatán, LSUMZ 22350–51; Isla de Roatán, near Roatán, CM 57185, 64515, UF 28458, 28541–42; Isla de Roatán, Roatán, LSUMZ 22335–37; Isla de Roatán, Rocky Point, USNM 570148–50; Isla de Roatán, 1.2 km E, 0.4 km S of Sandy Bay, KU 203121–22; Isla de Roatán, near Sandy Bay, USNM 570151–53; Isla de Roatán, Sandy Bay, KU 203123–26, LSUMZ 33784–86; Isla de Roatán, 6.6 km E of West End, MVZ 263856–58; Isla de Roatán, West End, USNM 578822–23; Isla de Roatán, West End Point, USNM 570154–55; Isla de Roatán, West End Town, USNM 570156–57; "Isla de Roatán," UF 149596; Isla de Utila, 2 km S of Rock Harbor, SMF 77108; Isla de Utila, east coast near Trade Winds, FMNH 283557, 283559; Isla de Utila, Utila, UF 28398; "Isla de Utila," SMF 77109–10. "no other data": MVZ 52402.

Other Records (Map 8).—**ISLAS DE LA BAHÍA**: Cayo Cochino Menor (Lundberg, 2002a); Isla Barbareta (personal sight records); Isla de Roatán, 0.5–1.0 km N of Roatán, LSUMZ 16986–92 (Meyer and Wilson, 1973; "outstanding loan to Jaime Villa" as of February 2009); Isla de Roatán, about 3 km N of Roatán, LSUMZ 16993–94 (Meyer and Wilson, 1973; "outstanding loan to Jaime Villa" as of February 2009); Isla

Morat, LSUHC 3706–07, LSUPC 4769–81 (Grismer et al., 2001).

Phyllodactylus paralepis McCranie and Hedges, 2013b

Phyllodactylus insularis: Echternacht, 1968: 151 (in part).
Phyllodactylus palmeus Dixon, 1968: 419 (in part); Meyer, 1969: 204 (in part); Meyer and Wilson, 1973: 11 (in part); Wilson and Hahn, 1973: 104 (in part); Köhler, 1998d: 382; Monzel, 1998: 155 (in part); Köhler, 2000: 44; Wilson et al., 2001: 136 (in part); Lundberg, 2002b: 7; Köhler, 2003a: 74 (in part); Wilson and McCranie, 2003: 59 (in part); McCranie et al., 2005: 78 (in part); Wilson and Townsend, 2006: 105 (in part); Köhler, 2008: 79 (in part).
Sphaerodactylus millepunctatus: Monzel, 1998: 156 (in part).
Phyllodactylus paralepis McCranie and Hedges, 2013b: 53 (holotype, FMNH 283552; type locality: "Savannah Bight, 16.29078°, −85.50300°, Isla de Guanaja, Islas de la Bahía, Honduras, 15 m elevation"); McCranie and Valdés Orellana, 2014: 45; Solís et al., 2014: 131; McCranie, 2015a: 365.

Geographic Distribution.—Phyllodactylus paralepis occurs on Isla de Guanaja on the Bay Islands, Honduras.

Description.—The following is based on four males (FMNH 283552; KU 101377; USNM 580288–89) and four females (FMNH 283553–54; USNM 565401, 580290). *Phyllodactylus paralepis* is a moderate-sized gecko (maximum recorded SVL 70 mm [USNM 580288, a male]); dorsal surface of head covered with granular scales, 20–29 (25.1 ± 3.1) across snout between third supralabials; 12–16 (14.3 ± 1.6) loreals (per side); 20–25 (22.9 ± 1.6) midorbital scales; 6–8 (usually 6) supralabials to level below mideye; 5–6 (usually 6) infralabials to level below mideye; 5–8 (6.5 ± 0.9) scales bordering internasals posteriorly; moveable eyelid absent; pupil vertically

Plate 8.   *Phyllodactylus paralepis*. FMNH 283552, adult male, SVL = 60 mm. Islas de la Bahía: Guanaja Island: Savannah Bight.

elliptical, reticulated; 2 enlarged postmentals, outer postmental on each side contacting infralabial 1 (no contact with infralabial 2); 4–8 (6.3 ± 1.4) scales bordering postmentals posteriorly; 16–17 (usually 16) transverse rows at midbody of keeled tubercles among small dorsal body scales; 41–53 (47.3 ± 3.7) tubercles in paravertebral row from posterior edge of head to base of tail; 29–36 (31.4 ± 2.3) tubercles in paravertebral row between levels of axilla and groin; 6–8 (7.0 ± 1.1) tubercles across base of tail; enlarged dorsal tubercles separated by 0–1 granules; ventral scales smooth, imbricate, 53–61 (57.6 ± 3.4) along midline from posterior edge of throat to cloacal scale, 25–28 (26.9 ± 1.0) rows across midventer; subdigital lamellae narrow, single, except terminal pair divided, as is row just proximal to terminal pair, terminal pair greatly expanded to form paired leaflike pad; nonretractile claw present at distal end of leaflike pad; webbing absent between digits; 12–14 (12.6 ± 0.7) subdigital lamellae on Digit IV of hind limb, 9–12 (10.6 ± 1.0) on Digit IV of forelimb; femoral and precloacal pores absent in both sexes; SVL 55.6–70.4 (60.8 ± 6.7) mm in males, 60.1–

63.3 (61.8 ± 1.5) mm in females; HW/SVL 0.15–0.16 in both sexes; HL/SVL 0.25–0.26 in males, 0.21–0.27 in females; HW/HD 1.32–1.66 in males, 1.33–1.78 in females; TAL/SVL 1.04–1.15 in two males, 0.95–1.21 in three females; SEL/EEL 1.23–1.24 in males, 1.11–1.44 in females.

Color in life of adult male holotype (FMNH 283552; Plate 8): dorsal surface of body pale greenish brown with Sepia (219) mottling and Mikado Brown (121C) tubercles; top and lateral surface of head similar in color to body, except Sepia pigment forming longitudinal lines on snout and supralabials; dorsal surface of tail brown with Sepia crossbands; dorsal surfaces of fore- and hind limb pale brown with Sepia and Raw Umber (123) mottling and crossbands; chin, throat, and belly pale brown with Raw Umber mottling on anterior half of belly and Sepia mottling and blotches on posterior half of belly; ventral surfaces of fore- and hind limb pale brown, except sole and palm pinkish brown; digital pads dirty white to white; iris golden brown.

McCranie and Hedges (2013b: 54–55) described color in alcohol of the adult male holotype (FMNH 283552): "dorsal ground color pale tan with narrow dark brown, incomplete medially, reticulated crossbands; dorsolateral paired dark brown lines; dorsal surfaces of limbs pale tan with dark brown reticulated crossbands; top of head pale brown with reticulated dark brown lines; side of head with dark brown postnasal and postorbital lines; tail tannish brown with narrow reticulated dark brown crosslines; supralabials cream with small dark brown spots on those anterior to eye; ventral surfaces of head and body nearly immaculate cream; venter of limbs nearly immaculate cream, except that posterior ventrolateral edges mottled with dark brown; palms, soles, and digits tan; subcaudal surface cream with dark brown mottling on anterior third, becoming crossbanded with dark brown on tan ground color on distal half." The paratypes have a somewhat more muted dorsal pattern in alcohol than does the holotype, consisting of scattered dark brown spotting and mottling; otherwise, they are similar in color to that described above for the holotype. A juvenile (FMNH 283555; SVL 27.1 mm) is very similar in color in alcohol to the holotype, except the dark dorsal crossbands on the anterior third of the body are solid, thus more distinct than in the holotype.

Diagnosis/Similar Species.—The combination of lacking moveable eyelids, having the dorsal surface of the head covered with granular scales, and having the distal pair of subdigital lamellae expanded to form a paired leaflike pad distinguishes *Phyllodactylus paralepis* from all other Honduran lizards, except the remaining *Phyllodactylus*. *Phyllodactylus palmeus* has 1–3 granules separating the enlarged dorsal tubercles, has 35–43 paravertebral tubercles from the rear of head to the base of tail, and has 11–15 transverse dorsal tubercles at midbody (versus 0 or 1 granule separating enlarged dorsal tubercles, 41–53 paravertebral tubercles from rear of head to base of tail, and 16–17 transverse dorsal tubercles at midbody in *P. paralepis*). *Phyllodactylus*

tuberculosus has 13–20 scales in the midorbital row, has 26–33 tubercles in the paravertebral row from rear of head to base of tail, and has each outer postmental usually contacting both the first and second infralabials (versus 20–25 scales in midorbital row, 41–53 paravertebral tubercles, and outer postmental not contacting second infralabial in *P. paralepis*).

Illustrations (Plate 8).—Köhler, 2000 (adult; as *P. palmeus*), 2003a (adult; as *P. palmeus*), 2008 (adult; as *P. palmeus*); McCranie and Hedges, 2013b (adult, dorsal tubercles).

Remarks.—*Phyllodactylus paralepis* is apparently most closely related to *P. palmeus* (McCranie and Hedges, 2013b). Tissues of *P. insularis* of Belize (also an island form) are needed to test that species' relationship with *P. paralepis*, *P. palmeus*, and *P. tuberculosus*. McCranie and Hedges (2013b: 56) placed *P. paralepis* in the newly formed *P. palmeus* species group.

Natural History Comments.—*Phyllodactylus paralepis* is known from sea level to 30 m elevation in the Lowland Moist Forest formation. This nocturnal gecko was formerly found on the walls of buildings in Savannah Bight before the invasion of the species of *Hemidactylus*. Those *Hemidactylus* appear to have replaced *P. paralepis* in edificarian situations in recent years. However, *P. paralepis* remains common in nonedificarian situations on Guanaja Island, but in my experience *H. frenatus* will quickly become adapted to nonedificarian habitats on Guanaja Island as has happened on Roatán, Utila, and islands in the Golfo de Fonseca. Other places of nocturnal activity include on walls in a cave and on coconut and thorn palms. Its diurnal hiding places include termite nests, beneath tree bark, and especially within bases of palm fronds and associated coverings. It can also be seen under and within coconut palm debris on the ground and occasionally in Sea Grape (*Cocoloba uvifera*) leaf litter. Echternacht (1968: 151; as *P. insularis*) reported finding

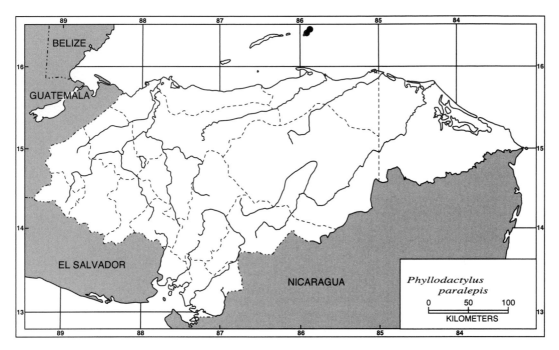

Map 9.   Localities for *Phyllodactylus paralepis*. Solid circles denote specimens examined.

one under "loose palm bark about 1.5 m above ground." Some of the reproductive data reported for *P. palmeus* by Dixon (1968) might apply to *P. paralepis*. Nothing has been reported on its diet.

Etymology.—The name *paralepis* is formed from the Greek *para* (meaning near) and *lepis* (scale) referring to closely spaced dorsal tubercles in this Guanaja Island endemic.

Specimens Examined (21 [10]; Map 9).— **ISLAS DE LA BAHÍA**: Isla de Guanaja, El Bight, USNM 565401; Isla de Guanaja, East End, FMNH 283553, USNM 580288–90; Isla de Guanaja, SE shore opposite Guanaja, LACM 38514–15, LSUMZ 22402–03; Isla de Guanaja, La Playa Hotel, LACM 38516–20, MVZ 52402; Isla de Guanaja, Posada del Sol Hotel ruins, FMNH 283554–55, 283590; Isla de Guanaja, Savannah Bight, FMNH 283552; Isla de Guanaja, near Savannah Bight, CM 64513; "Isla de Guanaja," KU 101377.

Phyllodactylus tuberculosus Wiegmann, 1834a

Phyllodactylus tuberculosus Wiegmann, 1834a: 241 (lectotype, ZMB 412A [designated by A. Bauer and Günther, 1991: 297; also see Remarks]; type locality: "Californien" [restricted to "the village of California, Nicaragua" by Dixon, 1960: 4]); Dunn and Emlen, 1932: 26; Dixon, 1960: 4; Meyer, 1969: 204; Meyer and Wilson, 1973: 11; Wilson and McCranie, 1998: 16; Wilson et al., 2001: 136; Lovich et al., 2006: 15; Wilson and Townsend, 2007: 145; Townsend and Wilson, 2010b: 692 (in part); McCranie and Hedges, 2013b: 58; Solís et al., 2014: 131; McCranie, 2015a: 365; McCranie and Gutsche, 2016: 866.

Phyllodactylus tuberculosus tuberculosus: Dixon, 1964: 22; Meyer, 1966: 174.

Thecadactylus rapicauda: Townsend and Wilson, 2010b: 692 (in part).

Geographic Distribution.—*Phyllodactylus tuberculosus* occurs at low and moderate elevations from southern Sonora and Chihuahua, Mexico, to central Costa Rica on the Pacific versant and from coastal Quintana Roo, Mexico, and Belize (and a few offshore islands) to Guatemala via the Río Motagua Valley on the Atlantic versant (but see Remarks). The species is known in Honduras on both versants, but only from the central and southern portions of the country. *Phyllodactylus tuberculosus* also occurs on numerous islands in the Golfo de Fonseca.

Description.—The following is based on 11 males (LSUMZ 33668–69, 38813; SDSNH 72761; USNM 570160, 580269–70, 580279, 580281–82, 580933) and 13 females (CAS 152992; KU 194256, 209313; LSUMZ 33667, 33671–72, 38814; UF 124699; USNM 570158–59, 580934, 580936, 580938). *Phyllodactylus tuberculosus* is a relatively large gecko (maximum recorded SVL 100 mm [Dixon, 1964]; 66 mm SVL in largest Honduran specimen [USNM 580269, a male]); dorsal surface of head covered with granular scales, 18–26 (21.4 ± 2.5) scales across snout between third supralabials; 5–8 (6.9 ± 0.8) scales bordering paired internasals, internasals usually in contact medially; 10–15 (12.3 ± 1.6) loreals (per side); 13–20 (17.1 ± 2.4) midorbital scales; 5–8 (most often 6 or 7) supralabials to level below mideye; 5–6 (usually 5) infralabials to level below mideye; moveable eyelid absent; pupil vertically elliptical; 2–3 (usually 2) enlarged postmentals, outer postmental usually contacting first and second infralabials on each side, occasionally contacting first infralabial only; 6–9 (6.9 ± 1.0) scales bordering postmentals posteriorly; 5–8 (6.9 ± 0.8) scales bordering internasals posteriorly; 16–18 (16.8 ± 1.0) transverse rows of keeled tubercles (midbody) among granular dorsal body scales; enlarged tubercles separated by 0–3 granules; 26–33 (29.4 ± 2.2) tubercles

in paravertebral row from posterior edge of head to base of tail in 14; 18–24 (20.9 ± 1.6) tubercles in paravertebral row between levels of axilla and groin in 14; 5–8 (7.4 ± 1.1) dorsal tubercles across base of tail in 14; ventral scales smooth, imbricate, 55–70 (61.4 ± 3.9) along midline from posterior to throat to cloacal scale, 28–31 (28.6 ± 1.1) rows across midventer; subdigital lamellae narrow, single, except penultimate one divided just proximal to terminal pair (pad); terminal pair greatly expanded to form leaflike pad; nonretractile claw present distally on leaflike pad; webbing absent between digits; 10–13 (11.6 ± 0.7) subdigital lamellae on Digit IV of hind limb; 9–12 (10.4 ± 0.9) subdigital lamellae on Digit IV of forelimb on 26 sides; femoral and precloacal pores absent in both sexes; SVL 56.2–65.9 (60.1 ± 2.5) mm in males, 51.3–62.7 (58.0 ± 3.7) mm in females; HW/SVL 0.11–0.18 in males, 0.15–0.17 in females; HL/SVL 0.19–0.27 in males, 0.23–0.28 in females; TAL/SVL 0.77–0.88 in five males, 0.87–1.05 in five females; SEL/EEL 1.20–1.61 in males, 1.02–1.71 in females.

Color in life of an adult female (USNM 570159): all dorsal surfaces pale brown with some Sepia (219) and Cinnamon (123B) tubercles, remaining tubercles similar in color to that of adjacent ground color; lateral surface of head similar in color to that of dorsal surface, except Sepia pre- and postocular stripes present; ventral surface of head pale yellowish brown; ventral surface of body pale yellowish brown with grayish tone; ventral surfaces of all digits grayish brown, except all terminal pads white; iris reticulated with golden brown and dark brown.

Color in alcohol: dorsal ground color gray to grayish brown, with black blotches or mottling present on body; most dorsal tubercles pale gray, except those involved in black markings also black; pre- and postocular black stripes present, with postocular stripe usually continuing onto lateral portion of body to about level of forelimb;

Plate 9.   *Phyllodactylus tuberculosus.* USNM 579600, adult female. Choluteca: Finca Monterrey.

posterior third of tail occasionally strongly banded with white and dark brown, but tail usually tannish brown without significant contrasting markings; supralabials mostly pale gray to white, dark brown to black mottling also present on some supralabials; ventral surfaces of head and body white to pale gray; subcaudal surface white to pale gray, usually heavily marked with dark brown.

Diagnosis/Similar Species.—The combination of lacking moveable eyelids, having the dorsal surface of the head covered with granular scales, and having the distal pair of subdigital lamella expanded to form a paired leaflike pad will distinguish *Phyllodactylus tuberculosus* from all other Honduran lizards, except the other *Phyllodactylus*. *Phyllodactylus palmeus* and *P. paralepis* have 20 or more scales in the midorbital row and have the outer postmental contacting only the first infralabial (versus 13–20 scales in midorbital row and outer postmental usually contacting both first and second infralabials in *P. tuberculosus*).

Illustrations (Fig. 31; Plate 9).—Álvarez del Toro, 1983 (adult); Calderón-Manduja-no et al., 2008 (adult); Dixon, 1964 (mid-body scales, head scales); Köhler, 2000 (adult), 2001b (adult), 2003a (adult, mid-body scales), 2008 (adult, midbody scales); Köhler et al., 2005 (adult, subdigital lamellae); Lee, 1996 (adult), 2000 (adult); McCranie et al., 2005 (subdigital lamellae); Savage, 2002 (adult, chin scales, subdigital lamellae); Stafford and Meyer, 1999 (adult); Taylor, 1942 (head scales, caudal scales, hind limb scales; as *P. ventralis* O'Shaugh-nessy, 1875: 263), 1956b (adult; as *P. eduardofischeri* Mertens, 1952a: 88); Villa et al., 1988 (adult).

Remarks.—Dixon (1964), in a systematic review of *Phyllodactylus tuberculosus*, rec-ognized four poorly defined subspecies, with the nominate one occurring in Hon-duras. These four geographic races were interpreted by Dixon (1964) to differ slightly in scale count averages, ventral coloration, and adult size. Honduran popu-lations of *P. tuberculosus* are apparently much smaller in SVL (maximum known SVL 65.9 mm) than are some populations studied by Dixon (1964; maximum SVL 100 mm), such a great distance so as to possibly suggest that more than one cryptic species is contained in this geographically widespread

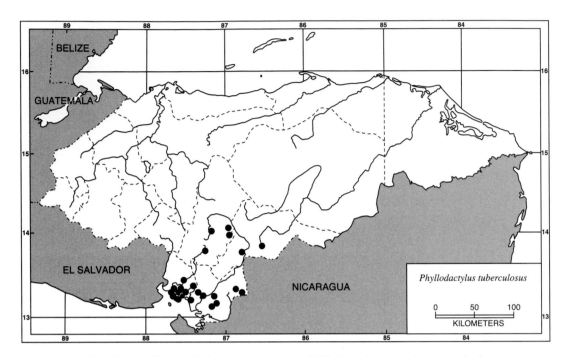

Map 10.   Localities for *Phyllodactylus tuberculosus*. Solid circles denote specimens examined.

species, some populations of which seem to be isolated (i.e., Quintana Roo, Mexico, and Belize populations). The name *P. magnus* Taylor (1942: 99) would be available for those large specimens in the populations from southwestern coastal Mexico (type locality in Guerrero). Dixon (1964) did not directly designate ZMB 412A as the lectotype of *P. tuberculosus*, only designating a type locality.

This species is in the *Phyllodactylus tuberculosus* species group (Dixon, 1964).

Natural History Comments.—*Phyllodactylus tuberculosus* is known from near sea level to 1,200 m elevation in the Lowland Dry Forest, Lowland Arid Forest, Premontane Moist Forest, and Premontane Dry Forest formations. This nocturnal species was active on rock walls used as fencerows, in trees, in rock outcrops, under concrete bridges, and formerly on walls of buildings and is active throughout the year under favorable conditions. *Hemidactylus frenatus*

appears to have replaced all Honduran *Phyllodactylus* populations on buildings. During the day this species remains hidden in the vicinities of its nighttime active places. Females deposit multiple clutches of two eggs per clutch, with the reproductive season lasting throughout the year in warm climates (see references in Dixon, 1964; Fitch, 1970). Dibble et al. (2007) reported Orthoptera, larval Coleoptera and Lepidoptera, and an Aranea in stomachs of *P. tuberculosus* from Mexico.

Etymology.—The name *tuberculosus* is derived from the Latin word *tuberculum* (small lump) and the Latin suffix *-osus* (fullness or abundance) and refers to the numerous tubercles on the skin of this species.

Specimens Examined (107 + 2 skeletons [14]; Map 10).—**CHOLUTECA**: 1.0 km N of Cedeño, LSUMZ 33667, 33687; 1.6 km N of Cedeño, KU 209312–13; 31.0 km NE of Cedeño, LSUMZ 33669–72; 1.9 km E of

Choluteca, UMMZ 117500 (7); Choluteca, LACM 64254, LSUMZ 33668, MSUM 4640; El Madreal, USNM 570158–61; El Ojochal (Botija), SDSNH 72758–59; Finca Monterrey, USNM 579599–600; La Fortuna, SDSNH 72742; Finca La Libertad, SDSNH 72761; Quebrada del Horno, SDSNH 72760. **EL PARAÍSO**: Danlí, BYU 18836; El Rodeo, USNM 580932–33; Orealí, USNM 589164. **FRANCISCO MORAZÁN**: near El Zamorano, AMNH 70365–66; El Zamorano, AMNH 70367, MCZ R49756; 3.2 km SE of Sabanagrande, TCWC 19185; San Antonio de Oriente, AMNH 70364; Tegucigalpa, BYU 16992–94, 18199–201, 18233, CAS 152992, LSUMZ 24130, 38813–14, UF 124698–99, USNM 60498, 133029. **VALLE**: near Amapala, USNM 580270–72; Amapala, MCZ R49748–50; Isla Comandante, UNAH (1); Isla Conejo, USNM 580267–69; Isla de La Vaca, USNM 580285; Isla de Las Almejas, USNM 580286; Isla de Pájaros, UNAH (2); Isla El Coyote, USNM 580279–82; Isla El Pacar, USNM 580935–36; Isla Inglesera, USNM 580273–77; Isla Tigrito, USNM 580938; Isla Violín, USNM 580934; Isla Zacate Grande, KU 194256–57, LSUMZ 36580–81; La Laguna, SDSNH 72762–64, USNM 580278; Nacaome, LACM 47301; Playona Exposición, USNM 579601–02; Puerto Salamar, LSUMZ 16049; Punta El Molino, USNM 580937, FN 256912–13 (still in Honduras because of permit problems); Punta Novillo, UNAH (4), FN 256918 (still in Honduras because of permit problems); San Carlos, USNM 580283–84; 11.9 km SSW of San Lorenzo, TCWC 61763. "HONDURAS": UF 42749 (skeleton), 43120 (skeleton), 124700–01, USNM 580287.

Other Records.—**VALLE**: Amapala, ZMH, unnumbered (Dixon, 1964).

Genus *Thecadactylus* Oken, 1817

Thecadactylus Oken, 1817: col. 1182 (see Remarks) (type species: *Gecko laevis*

Daudin, 1802: 112 [= *Gekko rapicauda* Houttuyn, 1782: 323], by monotypy).

Geographic Distribution and Content.—This genus occurs from the Yucatán Peninsula, Mexico, to Bolivia and Amazonian Brazil on the Atlantic versant and from northwestern Costa Rica to northwestern Peru on the Pacific versant (also see Geographic Distribution for *T. rapicauda*). Three named species are recognized, one of which occurs in Honduras.

Remarks.—Scale terminology for dorsal and head scales used herein for this genus follow those as defined in Schwartz and Crombie (1975).

Etymology.—The name *Thecadactylus* is derived from the Latin *theca* (case, container, envelope, sheath) and the Greek *daktylos* (finger, toe). The name refers to the diagnostic sheathed claws on the fingers and toes, the sheaths of which separate the distal subdigital lamella on all digits.

Thecadactylus rapicauda (Houttuyn, 1782)

Gekko rapicauda Houttuyn, 1782: 323 (neotype, RMNH 16267 [designated by Bergmann and Russell, 2007: 354]; type locality by neotype designation: "Republeik Surinam"; see Remarks).
Thecadactylus rapicaudus: Gray, 1845b: 146; Werner, 1896: 345; P. W. Smith, 1950: 55; Meyer, 1966: 174; Köhler, 1996c: 26; Köhler, 1996g: 66; Köhler, 1998d: 375; Lundberg, 2000: 8; Mahler and Kearney, 2006: 30; Köhler and Vesely, 2011: 107.
Thecadactylus rapicauda: Günther, 1859: 211; Meyer, 1969: 210; Meyer and Wilson, 1973: 13; O'Shea, 1986: 33; O'Shea, 1989: 16; Monzel, 1998: 156; Wilson and McCranie, 1998: 17; Köhler, 2000: 53; Köhler, McCranie, and Nicholson, 2000: 425; Nicholson et al., 2000: 30; Wilson et al., 2001: 137; Russell and Bauer, 2002: 753.2; Powell, 2003: 36; McCranie and Castañeda, 2005: 14; McCranie et al., 2005: 84; McCranie et al., 2006: 111; Wilson and

Townsend, 2006: 105; Bergmann and Russell, 2007: 364; Townsend and Wilson, 2010b: 692 (in part); Daza and Bauer, 2012: 23; McCranie and Solís, 2013: 242; McCranie and Valdés Orellana, 2014: 45; Solís et al., 2014: 131 (in part); McCranie, 2015a: 365.

Geographic Distribution.—*Thecadactylus rapicauda* occurs at low and moderate elevations from the outer end of the Yucatán Peninsula and northern Chiapas, Mexico, to much of Amazonian South America (except the southwestern portion) on the Atlantic versant and from northwestern Costa Rica to western Ecuador on the Pacific versant. An apparently isolated locality also occurs on the Pacific versant of northwestern Peru. The species also occurs on many islands in the Lesser Antilles, on St. Croix and Necker Islands in the Virgin Islands, on Trinidad and Tobago, on several small islands off the coast of Venezuela, Islas del Maíz, Nicaragua, Isla de Utila, Honduras, and the Islas de las Perlas, Panama (see Remarks). In Honduras, this species is known from across the northern portion of the mainland as well as from Utila Island in the Bay Islands.

Description.—The following is based on ten males (FMNH 13005; UF 150290; USNM 570215–16, 570222, 570227–29, 570231, 570233) and ten females (FMNH 5036; SMF 77098–99; USNM 570220, 570224–25, 570234, 573099–100, 579609). *Thecadactylus rapicauda* is a large gecko (maximum recorded SVL 126 mm [Vitt and Zani, 1997]; 98 mm in largest Honduran specimen measured [SMF 77099, a female]); dorsal surfaces of head and body covered with small, mostly homogeneous scales; 18–35 (27.0 ± 3.9) head scales in a straight line along midline of snout from median internasal to level of anterior edge of preorbital fold; 19–31 (25.2 ± 2.7) loreal scales (per side); 10–14 supralabials, 7–10 to level below mideye; 8–12 infralabials, 6–8 to level below mideye; moveable eyelid absent; pupil vertically elliptical; 2 enlarged postmentals; 44–66 (52.2 ± 5.9) dorsal granules on head from rostral to level at posterior edge of orbit; 149–169 (156.8 ± 5.8) scales around midbody; ventral scales slightly larger than dorsal body scales, 98–124 (112.9 ± 9.0) along midline between level of axilla and precloacal granules; transition between ventral and lateral scales gradual; 15–19 (16.9 ± 1.0) divided subdigital lamellae on Digit IV of forelimb, distal half separated by skin; 17–21 (18.4 ± 0.9) divided subdigital lamellae on 39 sides of Digit IV of hind limb, distal half separated by skin; retractile claws in sheaths located in skin between distalmost lamellae; distinct webbing present between all digits; femoral and precloacal pores absent in both sexes; pair of enlarged postcloacal scales present in males, located laterally; SVL 75–95 (87.4 ± 6.2) mm in males, 71–98 (85.5 ± 9.0) mm in females; HW/SVL 0.15–0.17 in both sexes; HL/SVL 0.23–0.25 in males, 0.23–0.26 in females; TAL/SVL 0.57–0.84 in nine males, 0.51–0.89 in eight females; SEL/EEL 1.06–1.52 in males, 1.34–1.53 in females.

Color in life of an adult male (USNM 570233): dorsal surfaces of head, body, and tail Pale Horn Color (92) with series of irregular butterfly-shaped Sepia (119) middorsal blotches continuing to base of tail where markings become irregularly circular, top of head also with Sepia blotching; dorsal surfaces of fore- and hind limb Pale Horn Color with olive gray reticulations; ventral surfaces of head, body, and tail Straw Yellow (56) with scattered olive green flecking on belly and more concentrated dark olive green blotching on underside of tail; toe lamellae white; iris pale olive green on periphery, bronze with black reticulations elsewhere. Color in life of a subadult (USNM 570232): dorsal surface of body Smoke Gray (45) with Dark Drab (119B), Sepia (119) outlined irregular blotches on either side of midline; dorsal surface of tail Drab-Gray (119D) with irregular Sepia crossbands that become more coalesced

Plate 10.   *Thecadactylus rapicauda*. UF 150290, adult male, SVL = 92 mm. Yoro: Quebrada San Lorenzo.

distally; dorsal surface of head Drab-Gray with brown mottling; side of head with cream temporal stripe; dorsal surface of forelimb mottled with cream and brown; dorsal surface of hind limb mottled with cream and dark brown; chin Chamois (123D) grading to Tawny Olive (223D) on venter of body, Chrome Orange (16) dots present posteriorly; undersides of fore- and hind limb Tawny Olive with scattered Chrome Orange dots; subcaudal surface Drab-Gray with obscure, diffuse pale cross-bands anteriorly and alternating cream and dark brown bands posteriorly; subdigital lamellae brilliant white; iris bronze with black reticulations.

Color in alcohol: all dorsal surfaces pale brown to brown, with dark brown mottling or suggestion of lines on head, body, limbs, and nonregenerated portion of tail; all ventral surfaces cream to pale brown, with brown flecking on most scales of head, body, and limbs; dark brown mottling also present on nonregenerated subcaudal surface.

Diagnosis/Similar Species.—The combination of lacking moveable eyelids and having the dorsal surface of the head covered with granular scales distinguishes *Thecadactylus rapicauda* from all other Honduran lizards, except the remaining Gekkonoidea. *Thecadactylus rapicauda* is the only Honduran Gekkonoidea species with more than 12 divided subdigital lamellae on Digit IV of the fore- and hind limb, with several distal lamellae separated by skin, and having distinct webbing between the digits.

Illustrations (Fig. 18; Plate 10).—Acevedo, 2006 (adult); Acosta-Chaves et al., 2015 (adult); Álvarez del Toro, 1983 (adult); Avila-Pires, 1995 (adult, head scales, subdigital lamellae); Bergmann and Russell, 2003 (subdigital lamellae); Calderón-Mandujano et al., 2008 (adult); Campbell, 1998 (adult); Dowling and Duellman, 1978 (hemipenis); Dowling et al., 1971 (hemipenis); Guyer and Donnelly, 2005 (adult); Henkel and Schmidt, 1991 (adult); Hoogmoed, 1973 (adult, head scales); Köhler, 2000 (adult, subdigital lamellae), 2001b (adult, subdigital lamellae), 2003a (adult, subdigital lamellae), 2008 (adult, subdigital lamellae); Lee, 1996 (adult, subdigital lamellae), 2000 (adult, subdigital lamellae); de

Lisle et al., 2013 (adult); McCranie et al., 2005 (adult, subdigital lamellae), 2006 (adult, subdigital lamellae); Murphy, 1997 (adult, subdigital lamellae); Pianka and Vitt, 2003 (adult); Powell, 2003 (adult); Russell and Bauer, 2002 (adult, juvenile, subdigital lamellae); Savage, 2002 (adult, chin scales, subdigital lamellae); Stafford and Meyer, 1999 (adult, subdigital lamellae); Taylor, 1956b (adult); Vitt, 1996 (adult).

Remarks.—Russell and Bauer (2002) reviewed the systematics of *Thecadactylus rapicauda* (*sensu lato*) and also provided an extensive literature review on the species complex. Mijares-Urrutia and Arends R. (2000) said karyological differences across the extensive geographical distribution of *T. rapicauda* suggested more than one species might be involved, and Kronauer et al. (2005) recovered a phylogeny based on mitochondrial DNA data that also suggested more than one species might be involved. Subsequently, Bergmann and Russell (2007: 351) described the southwesternmost populations as *T. solimoensis*, and Köhler and Vesely (2011: 99) described the population from Saint Maarten in the Lesser Antilles as *T. oskrobapreinorum*. However, the study by Gamble et al. (2011) also suggested that more cryptic species were represented in this widespread species complex (also see supplemental material in Gamble et al., 2012).

De Lisle et al. (2013: 264) gave "Paramaribo," Suriname, as the type locality of *Thecadactylus rapicauda*, but no previous author had designated that locality as the type locality.

Henderson and Powell (2009) thought that some of the West Indian populations of *Thecadactylus rapicauda* might represent human introductions.

Lundberg (2001) inaccurately listed this species from Roatán in the Bay Islands.

Natural History Comments.—*Thecadactylus rapicauda* is known from near sea level to 750 m elevation in the Lowland Moist Forest, Lowland Dry Forest, and Lowland Arid Forest formations and peripherally in the Premontane Wet Forest formation. This primarily nocturnal, primarily arboreal species is probably active throughout the year. It is active on wooden walls and thatched roofs of buildings and *champas*, on tree trunks and logs, inside tree hollows, on palm trunks, on karsted limestone boulders, and on the ground near buildings. Although primarily nocturnal, I have also found it active in low light situations on large tree trunks during the day in canopy-covered broadleaf rainforest. Meyer (1966) reported one Honduran female contained a single egg. Elsewhere, clutch size has been reported as either one or two eggs (see Russell and Bauer, 2002, and references cited therein), but one egg per clutch was reported for Amazonian populations of this species (Vitt and Zani, 1997). This species (or complex) apparently can reproduce throughout the year (Henderson and Powell, 2009, and references cited therein). I have found communal nests of four or five eggs of this species, including both fresh eggs and eggshells of hatched individuals, in shallow tree holes in June and July. This species (or complex) is apparently primarily insectivorous, but also will consume other invertebrates and small vertebrates (Henderson and Powell, 2009, and references cited therein; also see Acosta-Chaves et al., 2015: 197).

Etymology.—The name *rapicauda* is derived from the Latin *rapum* (turnip) and *cauda* (tail) and alludes to the noticeably swollen shape of the regenerated tail in this species. Neill and Allen (1962) and Russell and Bauer (2002) discussed why *T. rapicauda* is the correct name for this species, and not the frequently used *T. rapicaudus*.

Specimens Examined (75 + 8 skeletons + 1 egg [24]; Map 11).—**ATLÁNTIDA**: Coleman Plantation, UMMZ 62523; Estación Forestal CURLA, USNM 570216; 7.4 km SE of La Ceiba, USNM 570215; 11.3 km W of La Ceiba, TCWC 21967; La Ceiba, INHS 4477; Lancetilla, AMNH 70447; about 80

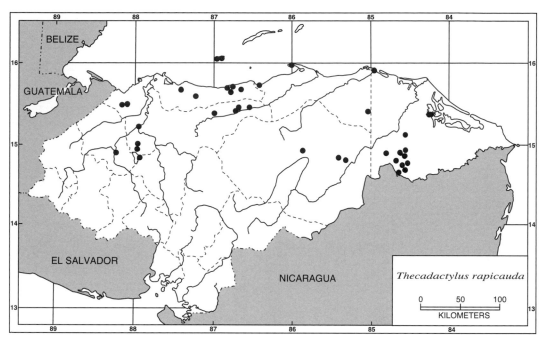

Map 11.   Localities for *Thecadactylus rapicauda*. Solid circles denote specimens examined.

km ESE of Tela, FMNH 13005, 13006 (skeleton). **COLÓN**: 4.8 km SE of Balfate, AMNH 58666–67; Balfate, AMNH 58616–20, 58622–23; Puerto Castillo, AMNH 37905; vicinity of Río Cuyamel, SMF 85940–41. **CORTÉS**: 4.8 km N of Agua Azul, LSUMZ 28509; Amapa, LACM 45376; Los Pinos, UF 166402; Río Santa Ana, FMNH 5036; 3.2–4.8 km W of San Pedro Sula, LACM 47854, TCWC 19186–87; San Pedro Sula, MCZ R38794; 7.2 km ENE of Villanueva, LACM 47853. **GRA-CIAS A DIOS**: Bachi Kiamp, FMNH 282599, USNM 573099, FN 257026, 257032 (still in Honduras because of permit problems); Bodega de Río Tapalwás, UF 150290, USNM 570217; Cerro Wahatingni, USNM 570218; Crique Ibantara, USNM 570219; Hiltara Kiamp, USNM 570220; Karasangkan, USNM 570221; Palacios, BMNH 1985.1282; Rus Rus, USNM 570222, 573100; Samil, USNM 579608; San San Hil, USNM 570223–25; Sisinbila,

USNM 579609; Warunta Tingni Kiamp, USNM 570226. **ISLAS DE LA BAHÍA**: Isla de Utila, near Iguana Station, SMF 79871; Isla de Utila, 1 km S of Rock Harbour, SMF 77098; Isla de Utila, west end of island, SMF 77099. **OLANCHO**: 3 km NW of Catacamas, LACM 47850; Las Delicias, USNM 570227; Matamoros, SMF 80824–25, USNM 570228–31. **SANTA BÁRBARA**: 7 km N of Santa Bárbara, TCWC 23624–25. **YORO**: 0.5 km N of Coyoles, LSUMZ 21417–18; 41 km WSW of Olanchito, UF 90047; Rancho San Lorenzo, LACM 47851; about 5 km ESE of San Lorenzo Arriba, USNM 570232; about 3 km S of San Lorenzo Arriba, USNM 570233–34; 4.7 km ESE of San Lorenzo Arriba, USNM 565407; Sopametepe, LACM 47852. "HONDURAS": AMNH 59722–23, FMNH 209454–56 (skeletons), UF 51549 (skeleton), 55067 (skeleton), 55158 (skeleton), 57738 (skeleton), 85373, 91952, 99722, 99723 (egg), USNM 101834.

Other Records.— "HONDURAS": (Werner, 1896); USNM 220204 (skeleton).

Family Sphaerodactylidae Underwood, 1954

This gecko lizard family occurs in the Western Hemisphere in southeastern Florida, USA, the Bahama Islands, the Greater and Lesser Antilles, Trinidad and Tobago, and other islands off the coast of northern South America. The mainland occurrence in Middle and South America is from central Veracruz and south-central Oaxaca, Mexico, to Bolivia and Brazil east of the Andes and Ecuador west of the Andes. This family also occurs on many small islands off both coasts of Middle America, including the isolated Swan Islands, Honduras. In the Eastern Hemisphere, this family occurs in northwestern and eastern costal regions of Africa, the Mediterranean areas of France and Italy, the Arabian Peninsula, and central Asia. Twelve genera and about 220 named species are included in this family, with three genera containing 13 species occurring in Honduras.

Remarks.—Males of Honduran species of *Aristelliger* are significantly larger than females, males of Honduran *Gonatodes* are slightly larger than females, and females of Honduran *Sphaerodactylus* are slightly larger than males.

Key to Honduran Genera of the Family Sphaerodactylidae

1A. Digits with all subdigital lamellae distinctively expanded, single, subdigital lamellae forming pads on digits II–V (Fig. 20) . . . . *Aristelliger* (p. 78)
1B. Digits without distinctively expanded subdigital lamellae, digits slender or only slightly expanded throughout their length (Figs. 21, 22) 2
2A. Claws displaced laterally by slightly expanded terminal subdigital lamella (Fig. 21); superciliary spine present (Fig. 33) . . . . *Sphaerodactylus* (p. 91)
2B. Claws not displaced laterally (Fig. 20); superciliary spine usually absent, occasionally present; digits with very slightly expanded subdigital lamellae, digits slender throughout length (Fig. 22) . . . . *Gonatodes* (p. 86)

Clave para los Géneros Hondureños de la Familia Sphaerodactylidae

1A. Dígitos con lamelas subdigitales bien expandidas, independientes, con almohadillas en dígitos II–V (Fig. 20) . . . . *Aristelliger* (p. 78)
1B. Dígitos sin lamelas subdigitales bien expandidas, dígitos delgados o ligeramente expandidos a todo lo largo (Figs. 21, 22) 2
2A. Garras desplazadas lateralmente por las lamelas subdigitales terminales ligeramente expandidos en cada dígito (Fig. 21); espina superciliar presente (Fig. 33) . . . . *Sphaerodactylus* (p. 91)
2B. Garras no desplazadas lateralmente (Fig. 20); espina superciliar usualmente ausente, pero puede estar ocasionalmente presente; dígitos sin lamelas subdigitales expandidas, lamelas y dígitos delgados a todo lo largo (Fig. 22) . . . . *Gonatodes* (p. 86)

Genus *Aristelliger* Cope, 1862a

Aristelliger Cope, 1862a: 496 (type species: *Aristelliger lar* Cope, 1862a: 497, by subsequent designation of Dunn and Dunn, 1940: 72 [also see Remarks]).

Geographic Distribution and Content.— The genus *Aristelliger* occurs on Jamaica, Hispaniola, Inagua in the Bahamas, on the Cayman Islands, the Caicos Islands, Navassa Island, the Swan Islands, San Andrés and Providencia, and on numerous islands off the

Figure 33. Superciliary spine present (arrow) on each side. *Sphaerodactylus alphus*. FMNH 283663 from East End, Guanaja Island, Islas de la Bahía.

coasts of Quintana Roo, Mexico, Belize, Honduras, and Nicaragua. Several mainland localities also exist in Quintana Roo, Mexico, and in Belize. Nine named species and one undescribed species are recognized. One named (plus one unnamed) species occur on Honduran territorial islands (but see Species of Probable Occurrence).

Remarks.—De Lisle et al. (2013: 16) said *Aristelliger praesignis* was the type species of *Aristelliger* "by monotypy." However, that is not the case, as Cope (1862a) included two species in his description of *Aristelliger*. Tamsitt and Valdivieso (1963: 137) and de Lisle et al. (2013: 16) erroneously stated that *A. georgeensis* (Bocourt, 1873: 41, *In* A. H. A. Duméril et al., 1870–1909a) occurs on the Swan Islands, Honduras. The *Aristelliger* species on those islands is *A. nelsoni* Barbour (1914: 258; see below).

Scale counts for the species of *Aristelliger* used herein generally follow the methods of Schwartz and Crombie (1975).

Etymology.—The name *Aristelliger* is derived from the Greek prefix *ari-* (intensive, very), the Latin word *stella* (star), and the Latin suffix *-ger* (bear, carry, have). The name apparently alludes to the pale scapular ocelli typical of *A. lar*, one of the two species included in the new genus by Cope (1862a).

KEY TO HONDURAN SPECIES OF THE GENUS
ARISTELLIGER

1A. Combined lamellae number 59–75 on Digit IV of fore- and hind limb .... sp. A (p. 80)
1B. Combined lamellae number 47–57 on Digit IV of fore- and hind limb .... *nelsoni* (p. 82)

CLAVE PARA LAS ESPECIES HONDUREÑAS DEL GÉNERO *ARISTELLIGER*

1A. Número combinado de lamelas subdigitales de 59–75 en Dígito IV en las extremidades anteriores y posteriores............. sp. A (p. 80)

1B. Número combinado de lamelas subdigitales de 47–57 en Dígito IV en las extremidades anteriores y posteriores............ *nelsoni* (p. 82)

Aristelliger sp. A

Aristelliger. Bauer and Russell, 1993a: 565.1 (in part).
Aristelliger georgeensis: Bauer and Russell, 1993b: 568.1; Lee, 1996: 182; Lee, 2000: 170; Solís et al., 2014: 132; McCranie, 2015a: 366.

Geographic Distribution.—*Aristelliger* sp. A is known to occur only on two nearby islands on the Cayos Vivorillos (also called Bancos de Vivorillos or Becerros) off the extreme northeastern coast of Honduras (but see Remarks).

Description.—The following is based on three males (USNM 581324–25, 581328) and seven females (KU 228743; USNM 581320–22, 581327, 581329–30). *Aristelliger* sp. A is a moderately large gecko (maximum recorded SVL 95 mm [USNM 581324, a male]); dorsal surface of head covered with small scales, 16–27 (20.8 ± 3.7) along midline between median internasal to level of anterior edge of eye; moveable eyelid absent; pupil vertically elliptical; 1–3 slightly enlarged superciliary scales; 15–19 (16.7 ± 1.3) loreals (per side); 6–9 supralabials and 5–7 infralabials (per side); 2–4 (rarely 3 or 4) enlarged postmentals; dorsal body scales mostly granular, homogeneous (some can be conical and heterogeneous), 31–40 (36.4 ± 3.3) in paravertebral row in distance equal to nostril-anterior edge of eye distance; ventral scales small but larger than dorsals, smooth, imbricate, 58–77 (64.9 ± 5.3) along midline between level of axilla and cloacal scale; ventral body scales posterior to hind limb larger than those on belly; 107–142 (124.1 ± 11.5) scales around midbody; transition between lateral and ventral scales rather gradual; subcaudal scales single; 13–19 (15.4 ± 1.7) single subdigital lamella on expanded pad on Digit IV of forelimb on 19 sides, 14–19 (16.4 ± 1.3) single subdigital lamella on expanded pad on Digit IV of hind limb; combined lamella number of fore- and hind limbs 59–75 (63.9 ± 5.3) in nine; small, asymmetrical adhesive plates adjacent to claw present on only Digit I of both fore- and hind limb; nonretractile claw located on terminal lamella, claw free from enlarged pad; webbing absent on all digits; enlarged postcloacal scales absent; precloacal and femoral pores absent; SVL 75.6–95.4 (82.7 ± 11.0) mm in males, 63.5–74.8 (70.1 ± 4.7) mm in females; HW/SVL 0.15 in males, 0.15–0.16 in females; HL/SVL 0.24–0.25 in both sexes; TAL/SVL 1.03–1.50 in two males, 1.28 in two females; SEL/EEL 1.29–1.49 in males, 1.32–1.68 in females.

Color in life of an adult male (USNM 581324): middorsal surface of body mottled brown and dark brown with numerous pale green scales; lateral surface of body with numerous Smoke Gray (44) scales, also with Ferruginous (41) and dark brown reticulations; top and side of head mottled brown and greenish brown with Smoke Gray scales; chin and throat brown with pale green scales; belly pale brown; subcaudal surface at base greenish brown (remainder of tail regenerated); subdigital lamellae gray, remainder of ventral surface of feet brown; iris Sepia (119), pupil golden brown with gold rim. Another adult male (USNM 581325; Plate 11) is similar in color to that of USNM 581324, except the dorsal surface of the unregenerated tail was dark reddish brown with brown crossbands and pale green and reddish brown scales.

Color in alcohol: dorsal surface of body pale brown to medium brown with series of incomplete, dark brown blotches, or darker brown mottling down midline; paler brown spotting or mottling also present laterally on body in some; dorsal surface of tail pale brown with dark brown, incomplete crossbands; dorsal surface of head slightly darker brown than body, also usually with pale

Plate 11.   *Aristelliger* sp. A. USNM 581325, adult male, SVL = 77 mm. Gracias a Dios: Cayo Becerro Grande (Cayos Vivorillos).

brown spotting or flecking; ventral surface of head brown with darker brown mottling (except mottling absent in midgular region), also usually with brown spotting or flecking; ventral surface of body cream with brown flecks on each scale; subcaudal surface cream anteriorly, becoming mostly brown for distal third; subdigital lamellae gray. Juveniles have conspicuous dark brown horizontal and longitudinal lines on the dorsal and lateral surfaces of the head, a distinct dark brown scapular mark, more prominent dark brown blotches and lines on the dorsal surface of the body, and distinct dark brown crossbands on distal third of dorsal surface of tail.

Diagnosis/Similar Species.—The combination of lacking moveable eyelids and having dorsal surface of the head covered with granular scales distinguishes *Aristelliger* sp. A from all other Honduran lizards, except the remaining Gekkonoidea. *Aristelliger* is the only Honduran genus of Gekkonoidea with all lamellae single and located on a much-expanded pad. *Aristelliger nelsoni* has 47–57 combined subdigital lamellae on Digit IV of the fore- and hind limbs (versus 59–75 such lamellae in *Aristelliger* sp. A).

Illustrations (Fig. 20; Plate 11).—None previously published.

Remarks.—*Aristelliger* sp. A is a member of the *A. georgeensis* species complex and is being described as a new species by personnel of the Hedges laboratory. This *Aristelliger* species was previously identified as *A. georgeensis* (i.e., Bauer and Russell, 1993b: 568.1; McCranie, 2015a: 366). Additionally, a Honduran Coast Guard officer told me in June 2014 that he has seen lizards, similar to those on the Cayos Vivorillos, on the Honduran island of Cayo Gorda, a small island lying about 15–20 km E of the Cayos Vivorillos. Thus, the possibility exists that a species of *Aristelliger* also occurs on Cayo Gorda.

This soon to be described species was first reported from Honduras as *Aristelliger georgeensis* by Bauer and Russell (1993b: 568.1).

Natural History Comments.—*Aristelliger* sp. A is known only from near sea level in the Lowland Dry Forest (West Indian Subregion) formation. Numerous specimens of this species (adults, subadults, juveniles) were seen during the day on two small islands in the Cayos Vivorillos (Cayo Grande and Cayo Pequeño) in the month of May. A third island (not named on available maps, but visible to the north from Cayo Grande) was searched but did not appear to have suitable habitat for this species; as a result, no geckos were seen. Individuals encountered were under the top layer of leaves and other debris on the ground and inside a concrete pit, beneath coconut palm debris both on the ground and on the palms themselves. The top level of leaves on the ground and other ground debris on the big island had salt water under about the second level of that debris. Because of strong winds, no attempt was made to search for active individuals after nightfall on that trip; thus, we did not hear its voice. Another specimen (KU 228743) was collected in the month of April, but nothing else was recording on the habitat where it was collected. Members of the *A. georgeensis* complex are oviparous with females of *A. irregularis* Cope (1885: 387) collected in October on Isla Cozumel, Mexico, containing ovarian follicles (Lee, 1996). Lee (1996) found ants and beetles in stomachs of *A. irregularis* and *A. georgeensis* he examined, and he also said those populations feed on a variety of arthropods. The taxonomy of this species complex is under study and several names will change, including one or two used here (S. B. Hedges, personal communication). Anoles have also been reported as food items of *A. irregularis* (see references in Lee, 1996; as *A. georgeensis*). Because of the high costs of gasoline, the uninhabited Cayos Vivorillos are not currently frequently visited by fishermen from the mainland, usually being visited only during Easter week. However, rising sea levels and the current "king" high tides are a serious threat to the continued existence of these *Aristelliger*, as well as to the islands themselves and their rich seabird fauna.

Etymology.—This species remains undescribed, but a description is planned (see above).

Specimens Examined (12 [0]; Map 12).— **GRACIAS A DIOS**: Cayo Vivorillo Grande, KU 228743, USNM 581320–26; Cayo Vivorillo Menor, USNM 581327–30.

Aristelliger nelsoni Barbour, 1914

Aristelliger nelsoni Barbour, 1914: 258 (holotype, MCZ R7891; type locality: "Swan Islands, Caribbean Sea"); Regan, 1916: 15; Barbour and Loveridge, 1929a: 225; Duellman and Berg, 1962: 197; Smith et al., 1964: 42; Miller, 1966: 262; Powell and Henderson, 2012: 91, 92; McCranie, 2015a: 366; McCranie et al., 2017: 268.

Aristelliger praesignis nelsoni: Hecht, 1951: 24; Wermuth, 1965: 9; Schwartz and Thomas, 1975: 109; MacLean et al., 1977: 4; Morgan, 1985: 43; Schwartz and Henderson, 1988: 93; Bauer and Günther, 1991: 281; Schwartz and Henderson, 1991: 364; Bauer and Russell, 1993c: 571.2; Henderson and Powell, 2009: 296; de Lisle et al., 2013: 17.

Aristelliger praesignis: Hecht, 1952: 113; Schwartz and Crombie, 1975: 305; Solís et al., 2014: 132.

Aristelliger: Bauer and Russell, 1993a: 565.1.

Geographic Distribution.—*Aristelliger nelsoni* occurs on Big and Little Swan Islands of the Swan Islands, Honduras.

Description.—The following is based on ten males (MCZ R9607, 9612; USNM 494650, 494658–62, 494660, 494671) and ten females (USNM 494647–48, 494651, 494657, 494663, 494665–68, 494670). *Aristelliger nelsoni* is a moderately large gecko (maximum recorded SVL 89 mm [MCZ R9612, a male]); dorsal surface of head covered with small scales, 19–25 (21.2 ± 1.8) along midline between median internasal to level of anterior edge of eye; moveable eyelid absent; pupil vertically

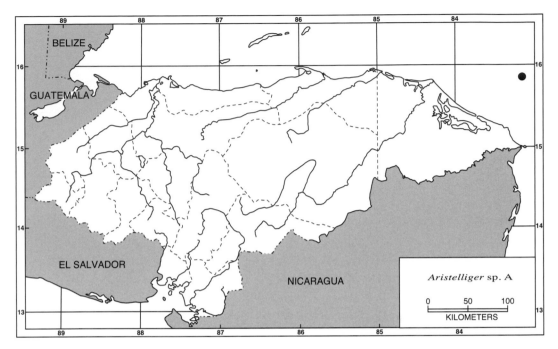

Map 12.   Localities for *Aristelliger* sp. A. Solid circle denotes specimens examined from two nearby islands.

elliptical; 1–3 slightly enlarged superciliary scales present; 15–23 (18.5 ± 1.6) loreals (per side); 6–8 supralabials; 5–6 infralabials; 2 enlarged postmentals; dorsal body scales small, mostly homogeneous, some weakly raised or with weak keels, 30–44 (36.2 ± 3.8) in paravertebral row in distance equal to nostril-anterior edge of eye distance; ventral scales small (but larger than dorsals), smooth, imbricate, 58–76 (66.6 ± 4.4) along midline between level of axilla and cloacal scale; ventral body scales posterior to hind limb larger than those on belly; 108–130 (121.3 ± 6.6) scales around midbody; transition between lateral and ventral scales rather gradual; subcaudal scales single; 10–14 (12.3 ± 0.9) single subdigital lamellae on expanded pad on Digit IV of forelimb on 39 sides; 12–16 (13.7 ± 1.1) single subdigital lamellae on expanded pad on Digit IV of hind limb; 47–57 (51.9 ± 3.3, n = 19) combined Digit IV lamellae under fore- and hind limbs; asymmetrical adhesive plates adjacent to claw only on Digit I of both fore- and hind limb; nonretractile claw located on terminal lamella, claw free from enlarged pad; webbing absent between digits; enlarged postcloacal scales absent; precloacal and femoral pores absent in both sexes; SVL 59.8–89.0 (75.3 ± 10.9) mm in males, 50.0–73.0 (57.9 ± 8.4) mm in females; HW/SVL 0.15–0.19 in males, 0.14–0.17 in females; HL/SVL 0.23–0.26 in males, 0.24–0.26 in females; TAL/SVL 1.22–1.42 in eight males, 0.93–1.33 in nine females; SEL/EEL 1.32–1.47 in eight males, 1.30–1.57 in females.

Color in life of an adult male (MCZ R191121): dorsal surface of body Beige (color 254 of Köhler, 2012) with Cinnamon-Drab (50) mottling, mottling outlined with pale brown; lateral surface of body Buff (5) with Cinnamon-Rufous (31) blotches and Dark Drab (45) mottling; top of head Drab Gray (256) with Cinnamon-Drab mottling and small white blotches and linear markings; Brussels Brown (33) pre- and

Plate 12.   *Aristelliger nelsoni.* MCZ R191146, adult female, SVL = 67 mm. Gracias a Dios: Isla Grande, Swan Islands.

postocular stripes present; dorsal surface of tail Buff with brown mottling; dorsal surfaces of fore- and hind limb Light Drab (269) with pale brown mottling, posterior surface of thigh with Salmon Color (83) blotches; belly Buff with some brown spots; subcaudal surface grayish brown with pale brown mottling; iris golden brown dorsally and outlining pupil, remainder of iris Buff (15). Color in life of an adult female (MCZ R191146; Plate 12): middorsal surface of body mottled Dark Drab (color 45 of Köhler, 2012) and Raw Umber (22); lateral surface of body Cinnamon-Drab (50) with Raw Umber (23) spotting and Kingfisher Rufous (28) blotches; top of head mottled Antique Brown (24) and Raw Umber (22), numerous white spots also present; Brussels Brown (33) postocular stripe present; side of head below eye Verona Brown (37) with pale brown spots, blotches, and lineate blotches; dorsal surface of tail mottled Dark Drab, medium brown, and pale brown; dorsal surfaces of fore- and hind limb Dark Drab with pale brown spots and mottling; belly Cinnamon-Drab with darker brown flecking; subcaudal surface mottled Dark Drab and Raw Umber (22); iris similar in

color to that of MCZ R191121. Color in life of another adult female (MCZ R191071): middorsum Pale Cinnamon (color 55 of Köhler, 2012) with Brownish Olive (276) crossbands; lateral and dorsolateral surfaces of body Chamois (84) with Raw Umber (22) lineate (vertical and longitudinal) pattern and Kingfisher Rufous (28) blotches; top of head Fawn Color (258) with white and pale brown spotting and mottling; Light Chrome Orange (76) preocular and Kingfisher Rufous postocular stripes present; lateral surface of head below ocular stripes Fawn Color with brownish white blotches; ventral surface of body Chamois with darker brown flecking; iris similar in color to that of MCZ R191121.

Color in alcohol: all dorsal surfaces of adults pale brown, except for paler brown mottling present laterally on body in some; juveniles and subadults also with pale brown ground color, but with slightly darker brown crossbands on body in some; juveniles also with distinct, black, suprascapular blotch on each shoulder; juveniles also with 2 thin, dark brown lines crossing head in front of eyes and dark brown pre- and postocular stripes; adult ventral surfaces of head, body, and tail cream to very pale brown, many ventral scales on body lightly flecked with dark brown.

Diagnosis/Similar Species.—The combination of lacking moveable eyelids and having the dorsal surface of the head covered with granular scales distinguishes *Aristelliger nelsoni* from all other Honduran lizards, except the remaining Gekkonoidea. *Aristelliger* is the only Honduran genus of Gekkonoidea with all lamellae single and located on a much-expanded pad. *Aristelliger* sp. A has 59–75 combined subdigital lamellae on Digit IV of the fore- and hind limbs (versus 47–57 such lamellae on Digit IV in *A. nelsoni*).

Illustrations (Plate 12).—McCranie et al., 2017 (adult).

Remarks.—Barbour (1914: 258) stated that there were seven paratypes of *Aristelliger nelsoni*. However, Barbour and Loveridge (1929a: 225) said that there were 15 paratypes after exchanging some with other museums, but those authors listed MCZ R9601–21 as the paratypes (= 21 numbers). Some of those inclusive numbers represent specimens that were exchanged with other museums (see Specimens Examined).

Hecht (1951: 25), without comment or supporting data relegated *A. nelsoni* to a subspecies of *A. praesignis* (Hallowell, 1857: 222). Most subsequent authors have followed that arrangement, but, as noted by Bauer and Russell (1993c) and confirmed by my study, the Swan Island populations of this complex differ from *A. praesignis* of Jamaica and the Cayman Islands in having small, flattened and generally homogeneous dorsal scales and a more pallid dorsal coloration and pattern. Also, the Swan Islands are isolated from all populations of *A. praesignis* by deep seas. Thus, I treat *A. nelsoni* as a species pending molecular studies (Powell and Henderson, 2012, listed *A. nelsoni* as a species without comment, but based on my suggestion).

Natural History Comments.—*Aristelliger nelsoni* is known from near sea level to 10 m elevation in the Lowland Dry Forest (West Indian Subregion) formation. This largely crepuscular species was reported to be common on both of the Swan Islands in March 1912 when the type series was collected (Barbour, 1914). Although it is likely the type series was collected on both Swan islands, no such data were given by Barbour (1914: 258). Specimens were also collected by a USNM expedition (probably on both islands) in February 1974 and an SMF expedition to the big island in April and May 2007. My party, on Big Swan, collected two specimens active on a fence post about 1 hour after dark, and two others were either inside a termite nest or beneath tree bark during the day, all instances in December 2012 (McCranie et al., 2017). This species has a voice, and according to the collector of the type series, "it is frequently heard croaking at dusk both among the coconut palms, in the houses, and in the woods" (Nelson, *In* Barbour, 1914: 259). My group did not hear the call of this species in December 2012. However, when we visited the Swan Islands, *A. nelsoni* did not appear common and proved difficult to find. On the other hand, the recently introduced (by 2007) *Hemidactylus frenatus* was extremely common on Isla Grande in the Swan Islands, both in edificarian and natural environments, giving the impression that the introduced *H. frenatus* was replacing the endemic *A. nelsoni*. Thus, the introduction of *H. frenatus* appears to represent a serious threat to the long-term survival of *A. nelsoni*. I did not get a chance to collect much on Isla Pequeña, only about 2 hours one morning during my visit to the Swan Islands in December 2012. Nothing has been reported on reproduction in *A. nelsoni*, but it is probably similar to the related *A. praesignis* in depositing a single egg at a time, often in communal nests (Henderson and Powell, 2009, and references cited therein). Diet of *A. nelsoni* is also unknown, but it probably feeds on cockroaches and other insects, as reported for the related *A. praesignis* (Henderson and Powell, 2009, and references cited therein).

Etymology.—The name *nelsoni* is a patronym honoring George Nelson, the collector of the type series, who also was a taxidermist at the MCZ at Harvard University.

Specimens Examined (55 + 1 skeleton [0]; Map 5).—**GRACIAS A DIOS**: Isla Grande (but some also collected on Isla Pequeña), Swan Islands, AMNH 43181 (formerly MCZ R9621), 24709 (formerly MCZ R9620), KU 47169 (formerly MCZ R9602), MCZ R7891, 9603–05, 9606 (skeleton), 9607–10, 9612, 9615, 9617–19, 161189, 191071, 191121, 191146, 191165, SMF 90458–60, UIMNH 41500 (formerly MCZ unnumbered), USNM 142360, 494647–71, ZMB 29818 (formerly MCZ R9611); "no other data," CAS 39416–18 (formerly in MCZ collection).

Genus *Gonatodes* Fitzinger, 1843

Gonatodes Fitzinger, 1843: 91 (type species: *Gymnodactylus albogularis* A. M. C. Duméril and Bibron, 1836: 415, by original designation of Fitzinger on p. 18).

Geographic Distribution and Content.—This genus ranges from southeastern Chiapas, Mexico, and eastern Guatemala southward to Bolivia and Brazil east of the Andes, and Ecuador west of the Andes. It also occurs on the Greater and Lesser Antilles and islands north of Venezuela. The genus has been introduced and established in Belize City, Belize, on the Corn Islands, Nicaragua, Grand Cayman, Aruba, Curaçao, Hispaniola, Jamaica, and Dominica, and one species at one time was introduced and established into the Miami and Key West areas, Florida, USA, but the U.S. populations are currently either extirpated or seriously declining. Thirty named species are recognized, one of which occurs in Honduras.

Remarks.—Terminology and methods of counting scales in *Gonatodes albogularis* used herein follow Rivero-Blanco (1979) and Rivero-Blanco and Schargel (2012). Kluge (1995) provided excellent drawings of the claw and adjacent scales of a representative of *Gonatodes*.

Etymology.—The name *Gonatodes* is probably derived from the Greek *gonatos* (joint, node) in reference to the sharp angle formed between phalanges II–III of the digits.

Gonatodes albogularis (A. M. C. Duméril and Bibron, 1836)

Gymnodactylus albogularis A. M. C. Duméril and Bibron, 1836: 415 (three syntypes under the number MNHN 1766 [see Guibé, 1954: 19; but Vanzolini and Williams, 1962: 489, listed three syntypes under MNHN 1776]; type locality: "l'île de la Martinique" [but see Vanzolini and Williams, 1962: 489–490; Rivero-Blanco, 1979: 41]).

Gonatodes albogularis: Boulenger, 1885a: 59; Meyer and Wilson, 1973: 10; Wilson et al., 1979a: 25; Wilson and McCranie, 1998: 16; Wilson et al., 2001: 135; Castañeda, 2002: 15; McCranie, Castañeda, and Nicholson, 2002: 25; Lovich et al., 2006: 15; McCranie et al., 2006: 108; Wilson and Townsend, 2006: 105; Townsend et al., 2007: 10; Townsend and Wilson, 2010b: 692; Valdés Orellana and McCranie, 2011b: 568; Solís et al., 2014: 132; McCranie, 2015a: 366; McCranie and Gutsche, 2016: 866.

Gonatodes fuscus: Meyer, 1969: 201; Hahn, 1971: 111.

Gonatodes albogularis fuscus: Rivero-Blanco, 1979: 225.

Geographic Distribution.—*Gonatodes albogularis* occurs at low and moderate elevations from southeastern Chiapas, Mexico, to western Colombia on the Pacific versant and from eastern Guatemala to western Venezuela on the Atlantic versant.

It also occurs on the Greater Antilles, and possibly on the Martinique Bank in the Lesser Antilles. It has been introduced and established (but see genus comments) in the Miami and Key West areas, Florida, USA, Belize City, Belize, Islas del Maíz, Nicaragua, Aruba, Curaçao, Grand Cayman (possibly established), and Hispaniola. This species occurs in both edificarian and non-edificarian situations, where in Honduras it is known from both coasts and also some interior portions of the country.

Description.—The following is based on ten males (CAS 152993; KU 209311; MCZ R163584, 163591, 163598; UF 90213; USNM 570113, 570115, 570121, 579582) and ten females (CAS 152998; CM 64432; MCZ R49945, 163589; UF 150305–06; USNM 570114, 570119–20, 579583). *Gonatodes albogularis* is a small gecko (maximum SVL 42 mm [MCZ R163591, a male]); dorsal surface of head covered with minute granules; moveable eyelid absent; pupil circular; superciliary spine usually absent, occasionally present; 11–16 (13.2 ± 1.7) snout scales in nine; 5–6 supralabials to level below mideye; 3–4 (usually 4) infralabials to level below mideye; dorsal surface of body covered with granules, body granules slightly larger than those of head, with 81–101 (89.2 ± 5.3) dorsals along midline between levels of axilla and groin; ventral scales flat, smooth, imbricate, 40–59 (50.2 ± 4.7) along midline between levels of axilla and groin; males with hypertrophied escutcheon; 12–18 (14.8 ± 1.9) single, narrow subdigital lamellae (infradigitals) on Digit III of forelimb, with 9–14 (11.5 ± 1.2) infradistals and 2–5 (3.4 ± 1.0) infraproximals; 17–24 (20.2 ± 2.2) single, narrow subdigital lamellae (infradigitals) on Digit IV of hind limb, with 13–19 (15.3 ± 1.7) infradistals and 4–8 (4.9 ± 1.3) infraproximals; sharp angle formed between phalanges II–III of digits; nonretractile claw located medially between pair of divided terminal lamella scales, but claw distinctively visible; all remaining subdigital lamellae single; webbing absent between digits; enlarged postcloacal scales absent; femoral-precloacal pores absent; subcaudal scale pattern most often B (sometimes pattern A; see Remarks); SVL 35.3–42.0 (38.1 ± 1.8) mm in males, 33.2–41.0 (36.5 ± 3.0) mm in females; TAL/SVL 1.19–1.56 in five males, 1.07–1.09 in three females; HW/SVL 0.13–0.14 in males, 0.12–0.15 in females; HL/SVL 0.22–0.24 in males, 0.21–0.25 in females; SEL/EEL 1.09–1.48 in males, 1.07–1.78 in females.

Color in life of an adult male (UNAH; Isla del Tigre): dorsal surface of body mottled with Blackish Neutral Gray (82) and Flaxflower Blue (170C); top of head Buff (24) with Flaxflower Blue punctuations; side of head same color as top of head, except large Flaxflower Blue blotch outlined by Sepia (119) present below eye; chin with large Burnt Orange (116) central blotch divided by Yellow Ocher (123C) stripe; belly Blackish Neutral Gray with pale blue punctations; iris Buff with golden rim. Color in life of an adult female (USNM 580291): dorsal surface of body Buff (24) with Sepia (119) blotches and mottling, head color same, except blotches more lineate; tops of fore- and hind limb and anterior third of tail same color as body, distal two-thirds of tail Clay Color (26) with tiny dark brown punctations; chin pale brown with dark brown punctations; belly pale brown with dark brown punctations laterally; subcaudal surface True Cinnamon (139) on anterior third, True Cinnamon with dark brown punctations on distal two-thirds; iris True Cinnamon with gold rim.

Color in alcohol sexually dichromatic: dorsal surfaces of body, tail, and limbs of males essentially uniformly dark brown; dorsal surface of head of males pale brown with dark brown flecking from between eyes posteriorly; dorsal surfaces of head, body, and tail of females brown with dark brown lines on head, dark brown spots on body, and dark brown crossbands on tail; ventral surface of head of males cream with sparse

Plate 13.   *Gonatodes albogularis.* USNM 579584, adult male. Olancho: Río Catacamas.

brown flecking, that of body and tail cream with heavy dark brown flecking, except along midline; those ventral surfaces of females with much less brown flecking than in males.

Diagnosis/Similar Species.—The combination of lacking moveable eyelids and having the dorsal surface of the head covered with granular scales distinguishes *Gonatodes albogularis* from all other Honduran lizards, except the remaining Gekkonoidea. *Aristelliger, Hemidactylus,* and *Thecadactylus* have distinctively expanded subdigital lamellae and have vertically elliptical pupils (versus all lamellae narrow and pupil circular in *G. albogularis*). *Phyllodactylus* have the terminal subdigital lamella expanded, divided, forming paired leaflike pad, and have vertically elliptical pupils (versus terminal lamella narrow, but divided, and pupils circular in *G. albogularis*). *Sphaerodactylus* have the claws displaced laterally by expanded terminal subdigital lamella and have a spinelike superciliary scale (versus claws located medially between pair of divided, nonexpanded lamella and spinelike superciliary scale usually absent in *G. albogularis*).

Illustrations (Fig. 22; Plates 13, 14).—Alemán and Sunyer, 2015 (adult); Álvarez del Toro, 1983 (adult); Conant and Collins, 1998 (adult); de Lisle et al., 2013 (adult); Guyer and Donnelly, 2005 (adult, head); Köhler, 2000 (adult), 2001b (adult), 2003a (adult, head scales, subdigital lamellae), 2008 (adult, head scales, subdigital lamellae); Köhler et al., 2005 (adult, subdigital lamellae); Lee, 2000 (adult); McCranie et al., 2006 (adult, subdigital lamellae); Meshaka et al., 2004 (adult); Murphy, 1997 (adult); Pianka and Vitt, 2003 (adult); Powell, Collins, and Hooper, 1998 (subdigital lamellae), 2012 (subdigital lamellae); Savage, 2002 (adult); Stafford and Meyer, 1999 (adult, subdigital lamellae); Sunyer, Nicholson et al., 2013 (adult).

Remarks.—Vanzolini and Williams (1962) and Rivero-Blanco (1979) studied geographic variation and proposed that the Central American populations be called *Gonatodes albogularis fuscus* (Hallowell, 1855: 33 [type locality in Nicaragua]). However, Savage (2002: 491) wrote, "There seems little basis for recognizing geographic races in this species, although differences in adult male coloration in life may have some

Plate 14.   *Gonatodes albogularis.* USNM 579585, adult female. Olancho: Río Catacamas.

geographic fidelity." Unfortunately, Savage (2002) did not provide any basis for that statement.

Rivero-Blanco and Schargel (2012) placed subcaudal patterns found in *Gonatodes* into seven types (patterns A–G). Those authors stated that *G. albogularis* had pattern A. However, the Honduran specimens examined for that character (the USNM specimens listed in the Description) most often have pattern B, a few have pattern A, and others have a variation of pattern A in which there were 2–14 single scales between the smaller and paired subcaudals (see Rivero-Blanco and Schargel, 2012, fig. 1).

Natural History Comments.—*Gonatodes albogularis* is known from near sea level to 1,000 m elevation in the Lowland Moist Forest, Lowland Dry Forest, Lowland Arid Forest, and Premontane Dry Forest formations. This diurnal species is common on walls of buildings in some villages and towns and is probably active throughout the year under favorable conditions. The species is also found under logs and other ground debris in edificarian situations and can be especially common in stacks of roofing tiles on the ground. It is also sometimes found active on logs, tree trunks, and rock walls in nonedificarian situations, but appears to be most common in edificarian situations. Active individuals are shy and jittery and can be difficult to approach close enough to capture by hand. By being largely diurnal it does not seem the normally nocturnal introduced *Hemidactylus* populations are having much of an effect on the edificarian populations of *Gonatodes*. However, Alemán and Sunyer (2015) reported an adult *H. frenatus* attempting to feed on an adult *G. albogularis* in Nicaragua at around dusk. Fitch (1973a,b) described the ecology of this species in Costa Rica; those observations showing Costa Rican populations are similar to the Honduran ones in occurring in both edificarian and nonedificarian situations. It can also be particularly common on large tree trunks. Costa Rican and Panamanian females are known to deposit multiple clutches during the year of a single egg, usually in communal nests (also see Jablonski, 2015: 195), with the reproductive period lasting much of the year on the moister Atlantic versant, whereas on the drier Pacific versant, reproduction is re-

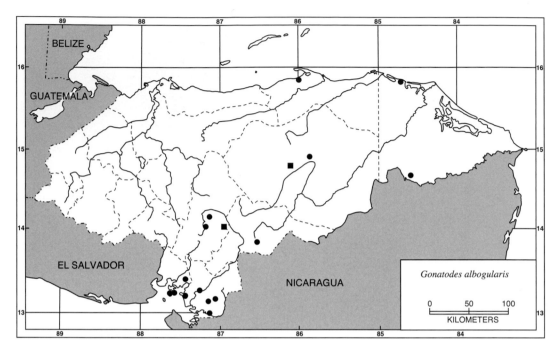

Map 13.   Localities for *Gonatodes albogularis.* Solid circles denote specimens examined and solid squares those for accepted records.

duced drastically during the strong dry season (Fitch 1970, 1973a,b; Sexton and Turner, 1971). Rösler (1998) reported on various parameters of eight eggs of this species from Costa Rica. Apparently nothing has been published on diet in this species other than it eats insects (Fitch, 1973a) and spiders (Carr, 1939).

Etymology.—The name *albogularis* is derived from the Latin words *albus* (white) and *gula* (throat) and Latin suffix *-arius* (pertaining to). The name refers to the white throats of the syntypes alluded to by A. M. C. Duméril and Bibron (1836).

Specimens Examined (109 + 1 skeleton [36]; Map 13).—**CHOLUTECA**: 1.6 km N of Cedeño, KU 209311; 36 km SE of Choluteca, TCWC 30124–25; Choluteca, LSUMZ 33665, 33679, TCWC 30123; El Faro, USNM 581898–99; Finca Monterrey, USNM 579587. **COLÓN**: Trujillo, CM

64432, KU 101376, LSUMZ 22419–26, 22431–44, 22503–12, UF 90213, USNM 570113–14. **EL PARAÍSO**: Mapachín, USNM 578912–14. **FRANCISCO MORAZÁN**: Tegucigalpa, CAS 152988–92, LACM 45246, LSUMZ 52599, 52601–04, UTA R-53999; Villa San Francisco, USNM 578915–16. **GRACIAS A DIOS**: Río Plátano, CM 59124; Rus Rus, UF 150305–06, USNM 570115–22, 579582–83. **OLANCHO**: near Río Catacamas, USNM 579584–86. **VALLE**: near Amapala, USNM 579588–89; Amapala, MCZ R49945–47, 163583–600, USNM 580291; Isla Comandante, USNM 580292–93; San Lorenzo, AMNH 70403, FMNH 5028 (2). "HONDURAS": UF 61964 (skeleton).

Other Records (Map 13).—**FRANCISCO MORAZÁN**: El Zamorano (Townsend et al., 2007). **OLANCHO**: Juticalpa (Wilson and McCranie, 1998; personal sight records).

Genus *Sphaerodactylus* Wagler, 1830a

Sphaerodactylus Wagler, 1830a: 143 (in part) (type species: *Spheriodactylus sputator* Gray, 1831: 52, *In* Gray, 1830–1831 [= *Lacerta sputator* Sparrman, 1784: 164, in part], by subsequent designation of Fitzinger, 1843: 18 [but see Remarks]).

Geographic Distribution and Content.— This genus occurs in southeastern Florida, including the Florida Keys, USA (controversy exists if some Florida populations are native or introduced by humans), the Bahama Islands, the Greater and Lesser Antilles, and Trinidad and Tobago. It also occurs from central Veracruz and south-central Oaxaca, Mexico, southward to north-central Guyana and northwestern Ecuador. The genus also occurs on many Atlantic versant islands off the coast of Middle America and several Pacific islands (Isla Gorgona, Colombia; Isla del Coco, Costa Rica; plus a few small islands near the mainland of Costa Rica and Panama). A few West Indian species have also been introduced to West Indian islands they are not native to. About 105 named species are recognized, ten of which occur in Honduran territory.

Remarks.—Fitzinger (1843: 18) designated *Sphaerodactylus sputator* Cuvier as the type species of *Sphaerodactylus* "Gray. (Cuv.)." Gray (1831: 52, *In* Gray, 1830–1831) did use the combination *Sphaerodactylus sputator*. However, Sparrman (1784: 164) is the describer of *Lacerta sputator* (= *Sphaerodactylus sputator*). King (1962: 9) concluded that Sparrman's (1784) type series of *S. sputator* was a composite of two species, the smallest of the three specimens actually representing *S. sabanus* Cochran (1938: 148).

Kraus (2009: 279) listed *Sphaerodactylus argus* Gosse (1850: 347) as introduced and established in Honduras and attributed Schwartz (1973) as the source for that listing. However, Schwartz (1973) did not list *S. argus* from Honduras. Tamsitt and Valdivieso (1963: 133) also suggested *S. argus* was known from Honduras. Werner (1896: 345) described the currently recognized *S. continentalis* (*S. millepunctatus* species group) as a subspecies of *S. argus* with a type locality of only Honduras. The Werner description is likely the original source of those inaccurate locality statements about *S. argus* occurring in Honduras.

External morphological characters used herein for the descriptions of various species of *Sphaerodactylus* follow the definitions used by Harris and Kluge (1984) and McCranie and Hedges (2012, 2013a). Kluge (1995) provided excellent drawings of the claw and adjacent scales of a representative of *Sphaerodactylus*.

Etymology.—The name *Sphaerodactylus* is formed from the Greek *sphaira* (ball) and *daktylos* (finger, toe). The name alludes to the more-or-less round scale at each digital tip.

KEY TO HONDURAN SPECIES OF THE GENUS
SPHAERODACTYLUS

1A. Middorsal zone of granular scales, sharply and distinctively differentiated from larger surrounding dorsal scales (Fig. 34) 2

1B. No distinct middorsal zone of granular scales, dorsal scales all subequal in size (Figs. 35, 36) 3

2A. Distinct white occipital spot present; maximum known SVL 41 mm .... *alphus* (p. 97)

2B. No distinct white occipital spot; maximum known SVL 38 mm ... . *rosaurae* (p. 130)

3A. Dorsal scales smooth (Fig. 35)... .... *glaucus* (p. 112)

Figure 34.   Middorsal zone of granules present. *Sphaerodactylus rosaurae*. FMNH 282674 from East End, Guanaja Island, Islas de la Bahía.

3B. Dorsal scales weakly keeled or rugose (Fig. 36)................... 4

4A. Superciliary spine at or posterior to level of mideye; third supralabial at level below anterior half of eye (Fig. 37); subcaudal scales in alternating series (Fig. 38). *dunni* (p. 106)

4B. Superciliary spine anterior to level of mideye; third or fourth supralabial at level below anterior half of eye; subcaudal scales aligned in single median row (Fig. 39)........ 5

5A. Third supralabial scale at level below anterior half of eye (Fig. 37); dorsal scales relatively large (Fig. 40), 22–30 between levels of axilla and groin .... *exsul* (p. 108)

5B. Fourth supralabial scale at level below anterior half of eye (Fig. 41); dorsal scales relatively small (Fig.

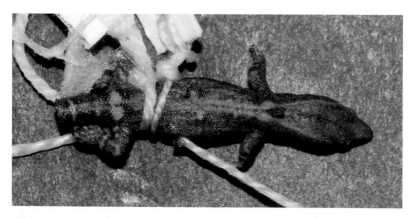

Figure 35.   Smooth dorsal scales, no medial zone of granular scales, and all middorsal scales subequal in size. *Sphaerodactylus glaucus*. USNM 570162, from Copán, near Copán.

Figure 36.   Weakly keeled or rugose and relatively small dorsal scales. *Sphaerodactylus continentalis*. UF 166399 from Los Pinos, Cortés.

36), 42–70 between levels of axilla and groin 6

6A. Short pelvic pale line absent; 42–57 (51.7) dorsal scales between levels of axilla and groin . . . .
. . . . *millepunctatus* (p. 123)

6B. Short pelvic pale line present or absent; 57 or more dorsal scales between levels of axilla and groin . . . 7

7A. Pale, short pelvic lines present in life and usually in alcohol (occurs on Roatán or Guanaja islands) 8

Figure 37.   Third supralabial at level below anterior half of each eye (outlined). *Sphaerodactylus dunni.* UF 90212 from WSW of Olanchito, Yoro.

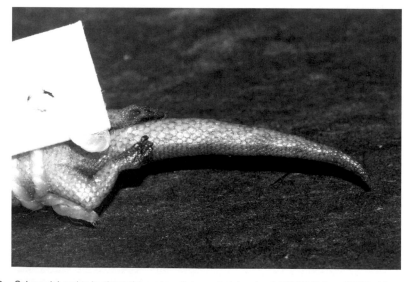

Figure 38.   Subcaudal scales in alternating series. *Sphaerodactylus dunni.* UF 90212 from WSW of Olanchito, Yoro.

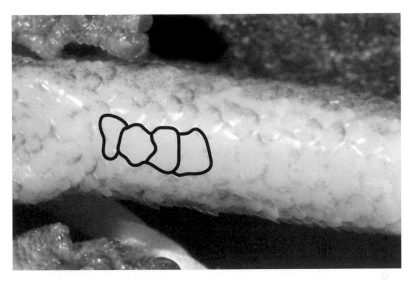

Figure 39.   Subcaudal scales aligned in single median row (outlined). *Sphaerodactylus continentalis.* UF 90210 from WSW of Olanchito, Yoro.

7B. Short pale pelvic lines absent (occurs on mainland or Utila Island) 9

8A. Pale pelvic line not dorsally crossing base of tail to connect with counterpart on other side; 7–10

subdigital lamellae on Digit IV of forelimb (occurs on Isla de Roatán) .... *leonardovaldesi* (p. 118)

8B. Pale pelvic line usually dorsally crossing base of tail to connect with counterpart on other side; 6–8

Figure 40.   Relatively large dorsal scales. *Sphaerodactylus exsul.* MCZ R191126 from Isla Grande in the Swan Islands, Gracias a Dios.

Figure 41.   Fourth supralabial at level below anterior half of each eye (outlined). *Sphaerodactylus continentalis.* UF 166399 from Los Pinos, Cortés.

subdigital lamellae on Digit IV of forelimb (occurs on Isla de Guanaja)............... *guanajae* (p. 114)

9A. HL/SVL 0.272–0.273 in females (HL not known in males); body dorsal pattern indistinct, dark spots, if present, confined to one dorsal scale (occurs only on Isla de Utila)........... *poindexteri* (p. 126)

9B. HL/SVL 0.221–0.248 in females; dark spots on body not confined to single scale (occurs on mainland from NW to south-central Honduras)............. *continentalis* (p. 100)

CLAVE PARA LAS ESPECIES HONDUREÑAS DEL GÉNERO *SPHAERODACTYLUS*

1A. La región mediodorsal de escamas granulares, distintivamente diferenciando de las escamas dorsales más grandes que la rodean (Fig. 34) 2

1B. Sin una región mediodorsal de escamas granulares, todas las escamas dorsales similares en tamaño

(Figs. 35, 36) 3

2A. Una mancha occipital blanca presente; tamaño máximo conocido 41 mm LHC............. *alphus* (p. 97)

2B. Sin una mancha occipital blanca; tamaño máximo conocido 38 mm LHC .... *rosaurae* (p. 130)

3A. Escamas dorsales lisas (Fig. 35).. .... *glaucus* (p. 112)

3B. Escamas dorsales débilmente quilladas o rugosus (Fig. 36) 4

4A. Espina superciliar sobre la parte media posterior del ojo; la tercera supralabial está por debajo de la mitad anterior del ojo (Fig. 37); escamas subcaudales en una serie alternadas (Fig. 38)... *dunni* (p. 106)

4B. Espina superciliar está en la mitad anterior de la mitad del ojo; tercera o cuarta supralabial por debajo de la mitad anterior del ojo; escamas subcaudales alineadas en una sola hilera medial (Fig. 39).............. 5

5A. Tercera supralabial está por debajo de la mitad anterior del ojo (Fig. 37); escamas dorsales relativamente grandes (Fig. 40), de 22–30 entre la axila y la ingle. *exsul* (p. 108)

5B. Cuarto supralabial está por debajo de la mitad anterior del ojo (Fig. 41); escamas dorsales relativamente pequeñas (Fig. 36), de 42–70 entre la axila y la ingle.......... 6

6A. Sin una línea pélvica pálida corta; de 42–57 (51.7) escamas dorsales entre la axila y la ingle........... .... *millepunctatus* (p. 123)

6B. Con o sin una línea pélvica pálida corta; 57 o más escamas dorsales entre la axila y la ingle............. 7

7A. Con una línea pélvica pálida corta presente en vida y usualmente en alcohol (se encuentran en las islas de Roatán o Guanaja).............. 8

7B. Sin una línea pélvica pálida corta (se encuentran en la parte continental o en la Isla de Utila)......... 9

8A. La línea pélvica pálida no cruza la base de la cola en la superficie dorsal para conectarse con la línea del otro lado; 7–10 lamelas subdigitales en el Dígito IV de las extremidades anteriores (se encuentran solamente en Isla de Roatán)...... *leonardovaldesi* (p. 118)

8B. La línea pélvica pálida usualmente cruza la base de la cola en la superficie dorsal para conectar con la línea del otro lado; 6–8 lamelas subdigitales en el Dígito IV en las extremidades anteriores (se encuentran solamente en Isla de Guanaja)........... *guanajae* (p. 114)

9A. Relación de la LCA/LHC 0.272–0.273 en hembras (LCA desconocido en machos); patrón dorsal de coloración indistinto, manchas oscuras, si están presentes, restringidas a una escama (se encuentran solamente en Isla de Utila) .... *poindexteri* (p. 126)

9B. Relación de la LCA/LHC 0.221–0.248 en hembras; patrón de coloración dorsal del cuerpo con man-chas oscuras no restringidas a una escama (se encuentran en la parte continental del noroeste al centro-sur de Honduras)................ .... *continentalis* (p. 100)

Sphaerodactylus alphus McCranie and Hedges, 2013a

Sphaerodactylus rosaurae: Wilson and Hahn, 1973: 106 (in part); Schwartz, 1975: 15 (in part); Schwartz and Garrido, 1981: 20 (in part); Köhler, 1998d: 382 (in part); Monzel, 1998: 156 (in part); Grismer et al., 2001: 135; Wilson et al., 2001: 137 (in part); Lundberg, 2002b: 12; Wilson and McCranie, 2003: 59 (in part); McCranie et al., 2005: 82 (in part); Wilson and Townsend, 2006: 105 (in part).

Sphaerodactylus alphus McCranie and Hedges, 2013a: 44 (holotype, FMNH 283672; type locality: "Savannah Bight, 16.29078°, –85.50300°, Isla de Guanaja, Islas de la Bahía, Honduras, 15 m elev."); McCranie and Valdés Orellana, 2014: 45; Solís et al., 2014: 132; McCranie, 2015a: 366.

Geographic Distribution.—*Sphaerodactylus alphus* occurs on Isla de Guanaja on the Bay Islands, Honduras.

Description.—The following is based on three males (FMNH 283663–64, 283674) and four females (FMNH 283666, 283671–73). *Sphaerodactylus alphus* is a small lizard (maximum recorded SVL 41 mm [FMNH 283666, 283671, both females]); dorsal surface of head covered by small to granular scales, scales conical in parietal region; 8–11 (9.6 ± 1.1) snout scales; 4 (rarely 3) supralabials to level below anterior portion of eye; 3 infralabials to level below mideye; superciliary spine at level of about mideye; moveable eyelid absent; pupil circular; rostral with median cleft and small posterior notch; enlarged supranasals single, separated by 1 smaller internasal; dorsal surface of body with median zone of granules (2–3

Plate 15.   *Sphaerodactylus alphus*. FMNH 283663, adult male, SVL = 40 mm, Islas de la Bahía: Guanaja Island, East End.

scales wide), that zone flanked by larger, keeled, slightly imbricate to juxtaposed scales, 26–35 (30.3 ± 3.0) larger scales along paravertebral row between levels of axilla and groin; ventral scales slightly smaller than largest dorsal scales, smooth, flat, imbricate, 31–32 (31.6 ± 0.5) between levels of axilla and groin; usually with zone of lateral scales noticeably smaller than ventral or dorsal scales; 40–50 (45.6 ± 3.7) scales around midbody; 10–13 (12.1 ± 0.9) narrow subdigital lamellae on Digit IV of hind limb, 8–11 (9.6 ± 1.0) on Digit IV of forelimb, 41–47 (43.4 ± 2.3) combined lamellae on Digit IV of fore- and hind limb; retractile claw located laterally between enlarged terminal pad and several smaller scales; webbing absent between digits; subcaudal scales aligned in single median row; femoral and precloacal pores absent in both sexes; male escutcheon slightly hypertrophied, about 12–13 scales long and about 10–15 scales wide; SVL 37.7–39.9 (38.8 ± 1.1) mm in males, 33.2–41.2 (38.5 ± 3.8) mm in females; HW/SVL 0.11–0.12 in both sexes; HL/SVL 0.22–0.25 in males, 0.23–0.25 in females; TAL/SVL 1.05 in one male,

0.88 in one female; SEL/EEL 1.05–1.16 in males, 1.06–1.13 in females.

Color in life of an adult male (FMNH 283663; Plate 15): top of head Mars Brown (223A) with small white spot on occipital region; dorsal surface of body Mars Brown with greenish gray large scales that have Mars Brown edges; top of tail pale brown with extensive Mars Brown mottling and spotting; dorsal surfaces of fore- and hind limb Mars Brown with dark brown spots; chin and throat Army Brown (219B); belly dark brown with some greenish gray scales; subcaudal surface grayish brown on original portion, dark grayish brown on regenerated portion; iris Mars Brown. Color in life of an adult female (FMNH 283672; Plate 16): dorsal surface of body with Hair Brown (119A) vertical lines and mottling separating Pale Horn Color (92) bands; nuchal crossbands alternating Hair Brown–dirty white–Hair Brown–Pale Horn Color; top of head yellowish brown with dark brown longitudinal lines to level of eyes, head posterior to eyes yellowish brown with dark brown mottling; occipital region with large Hair Brown blotch with brilliant white spot in posterior portion; lateral surface of head

Plate 16.   *Sphaerodactylus alphus*. FMNH 283672, adult female, SVL = 39 mm, Islas de la Bahía: Guanaja Island, Savannah Bight.

pale brown with dark brown vertical spots; dorsal surface of tail Pale Horn Color with Hair Brown crosslines and spots; dorsal surfaces of fore- and hind limb similar in color to top of head anterior to eyes; chin and throat brown with purple tinge; belly yellowish gray; subcaudal surface and ventral surfaces of fore- and hind limb brownish yellow with dark brown spots; iris Army Brown (219B) with golden brown ring around pupil.

Color in alcohol of the female holotype (FMNH 283672) was described by McCranie and Hedges, 2013a: 46): "dorsal surface of head pale brown, with medium brown lines; occipital blotch dark brown with dirty white spot located near posterior end; occipital dark blotch confluent medially with narrow (4–5 scale rows long) dark brown nuchal crossband, dark crossband followed by complete (3 scale rows long) dirty white crossband, that pale crossband followed by complete (4 scale rows long) dark brown nuchal crossband, that dark brown crossband followed by incomplete medially, brown crossband (2 scale rows long dorsolaterally), that brown crossband followed by complete dark brown (3 rows long) crossband, that dark crossband followed by complete dirty white (3 rows long) crossband, that pale crossband bordered posteriorly by complete dark brown (2 rows long) crossband that passes just anterior to forelimb insertion; dorsal surface of body with alternating brown, dark brown, pale brown, dark brown, brown, pale brown, dark brown, pale brown, and dark brown crossbands; dorsal surfaces of limbs pale brown with dark brown spots, dark brown blotches and narrow crossbands also present on hind limb; dorsal surface of base of tail pale brown with reticulated dark brown crossbands; ventral surfaces of head, chest, and body pale brown, without markings except for dark brown spots along lateral edges of belly; basal subcaudal surface pale brown with dark brown posterior edges in rectangular-shaped medial scales." Those authors went on to say (p. 47): "The adult female paratypes agree well in color in alcohol with that described for the holotype, except one (FMNH 283666) has the dorsal crossbands broken into dark brown spots forming incomplete crossbands. The adult male paratypes show distinct sexual dichromatic patterns as follows: dorsal surface of head brown without markings except for poorly-defined dark brown occipital blotch that surrounds pale brown occipital spot; dorsal surface of body brown with widely scattered dark brown spots; dorsal surfaces of limbs brown with indistinct darker brown spots; ventral surfaces of head, chest, and body pale to medium brown, without darker markings; subcaudal surface pale brown with darker brown outlining most scales. A juvenile (FMNH 283668) agrees well in color with the female holotype except the contrast between the dark and pale colors is more distinct, with the pale color being immaculate white."

Diagnosis/Similar Species.—The combination of lacking moveable eyelids, having the dorsal surface of the head covered with small to granular scales, having the claws displaced laterally by terminal expanded subdigital lamella, and having superciliary spines distinguishes *Sphaerodactylus alphus* from all other Honduran lizards, except the remaining *Sphaerodactylus. Sphaerodactylus alphus* and *S. rosaurae* are the only Honduran *Sphaerodactylus* having 1–3 middorsal rows of granular scales sharply and distinctively differentiated from the larger surrounding dorsal scales, but a molecular analysis shows each to be a separate evolutionary species. *Sphaerodactylus rosaurae* also lacks a distinct white occipital spot and has a known maximum SVL of 38.5 mm (versus white occipital spot distinct and maximum known SVL 41.2 mm in *S. alphus*).

Illustrations (Fig. 33; Plates 15, 16).—McCranie and Hedges, 2013a (adult); Wilson and Hahn, 1973 (adult dorsal pattern variation [Guanaja specimen only]; as *S. rosaurae*).

Remarks.—*Sphaerodactylus alphus*, along with *S. rosaurae*, are the Honduran members of the *S. scaber* species group, a group otherwise known to occur only in the West Indies (see Remarks for *S. rosaurae*).

The two LSUMZ specimens of *Sphaerodactylus alphus* listed by Wilson and Hahn (1973; as *S. rosaurae*) from Isla de Guanaja lost their tags and were destroyed when the alcohol in their jar evaporated (see list in Other Records and Remarks for *S. rosaurae*).

Natural History Comments.—*Sphaerodactylus alphus* is known from near sea level to 15 m elevation in the Lowland Moist Forest formation. This diurnal species was collected from July to September and in November but is probably active throughout the year. *Sphaerodactylus alphus* was collected in and under dead Sea Grape leaves (*Cocoloba uvifera*) and under piles of dead coconut palm fronds (*Cocos* sp.). The habitat notes provided by Wilson and Hahn (1973: 109) for *S. rosaurae* likely also include *S. alphus*. Nothing has been reported on reproduction or diet for *S. alphus*, but like other species of *Sphaerodactylus*, females probably deposit multiple clutches of a single egg, often in communal nests, and it likely feeds on small insects, including ants.

Etymology.—The name *alphus* is a Latin masculine, singular noun meaning "a white spot on the skin." The name alludes to the distinctive white occipital spot found in this species, with the spot especially distinct in females.

Specimens Examined (8 [2]; Map 14).—**ISLAS DE LA BAHÍA**: Isla de Guanaja, East End, FMNH 283663, 283668, 283671, 283673–74; Isla de Guanaja, Posada del Sol ruins, FMNH 283664; Isla de Guanaja, Savannah Bight, FMNH 283666, 283672.

Other Records (Map 14).—**ISLAS DE LA BAHÍA**: Isla de Guanaja, 2 km W of Savannah Bight, LSUMZ 21943 (see Remarks); Isla de Guanaja, SE shore, LSUMZ 21942 (see Remarks).

Sphaerodactylus continentalis Werner, 1896

Sphaerodactylus argus var. continentalis
Werner, 1896: 345 (holotype, ZIL 8880 [see Smith and Terentjev, 1963: 368; Harris and Kluge, 1984: 18, 25]; type

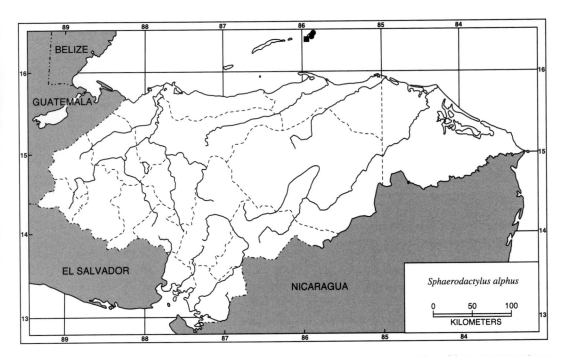

Map 14. Localities for *Sphaerodactylus alphus*. Solid circles denote specimens examined and the solid square represents an accepted record.

locality: "Honduras"); Boulenger, 1897: 20; Werner, 1899: 26.

Sphaerodactylus glaucus: Barbour, 1921b: 240 (in part).

Sphaerodactylus lineolatus: Dunn and Emlen, 1932: 26.

Sphaerodactylus argus continentalis: Dunn and Saxe, 1950: 148; Smith and Terentjev, 1963: 368.

Sphaerodactylus continentalis: Smith and Álvarez del Toro, 1962: 103; Meyer, 1969: 207 (in part); Hahn, 1971: 111; Meyer and Wilson, 1973: 12 (in part); Thomas, 1975: 195 (in part); McCranie and Hedges, 2012: 72 (in part); McCranie and Hedges, 2013a: 41; McCranie and Solís, 2013: 242; Solís et al., 2014: 132; McCranie, 2015a: 366.

Sphaerodactylus millepunctatus: Harris and Kluge, 1984: 17 (in part); Kluge, 1995: 23; Köhler, 1996g: 66; Wilson and McCranie, 1998: 17; Wilson et al., 2001: 137 (in part); McCranie and Castañeda, 2005: 14; McCranie et al., 2005: 80 (in part); Castañeda, 2006: 32; McCranie et al., 2006: 110 (in part); Townsend, 2006a: 35; Wilson and Townsend, 2006: 105 (in part); Wilson and Townsend, 2007: 145; Townsend and Wilson, 2010b: 692.

Geographic Distribution.—*Sphaerodactylus continentalis* occurs at low and moderate elevations from the Isthmus of Tehuantepec, northern Oaxaca, Mexico, to east-central Honduras on the Atlantic versant and in south-central Honduras on the Pacific versant. It also occurs on Isla Cozumel, Quintana Roo, Mexico (but see Remarks). This species is widely distributed across the northern half of Honduras to central Olancho. It is also known from a single area in south-central Honduras.

Description.—The following is based on 15 males (MCZ R191119; USNM 570165–67, 570170, 570194–95, 579972, 579974, 579979–84) and seven females (MCZ

R191113; USNM 570163, 570168, 579975–78). *Sphaerodactylus continentalis* is a tiny lizard (maximum recorded SVL 32 mm [USNM 579978, a female]); dorsal surface of head covered by small scales, at least some scales on snout and in parietal region weakly keeled; 8–13 (10.5 ± 1.3) snout scales; 4 supralabials and 3 infralabials to level below anterior half of eye; superciliary spine at level varying from anterior to mideye to about mideye; moveable eyelid absent; pupil circular; rostral with median cleft and small posterior notch; 2 supranasals, anterior largest, anterior separated medially by 0–2 (most often 2) smaller internasals; dorsal surface of body covered by slightly larger scales than on head, scales essentially homogeneous, keeled, juxtaposed to imbricate, 59–76 (64.5 ± 4.5) between levels of axilla and groin (see Remarks); ventral scales larger than dorsal scales, smooth, flat, imbricate, 31–46 (38.0 ± 3.1) between levels of axilla and groin; 64–80 (71.7 ± 4.7) scales around midbody in 21; 9–12 (10.0 ± 0.8) narrow subdigital lamellae on Digit IV of hind limb, 7–10 (8.1 ± 0.8) on Digit IV of forelimb, 33–42 (36.3 ± 2.4) combined lamellae on Digit IV of fore- and hind limb in 21; retractile claw located laterally between each enlarged terminal pad and several smaller scales; webbing absent between digits; subcaudal scales aligned in single median row; femoral and precloacal pores absent in both sexes; male escutcheon slightly hypertrophied, somewhat bell-shaped with subfemoral lateral extension, escutcheon extending slightly onto groin; 1–2 distinctively swollen granules at each corner of cloacal region in males; SVL 23.6–29.5 (26.6 ± 1.8) mm in males, 23.5–31.5 (28.9 ± 2.8) mm in females; HW/SVL 0.11–0.14 in males, 0.09–0.13 in females; HL/SVL 0.20–0.25 in males, 0.22–0.25 in females; HD/SVL 0.09–0.11 in males, 0.08–0.10 in females; TAL/SVL 0.58–1.04 in eight males, 0.81–0.87 in two females; SEL/EEL 1.04–1.35 in males, 1.03–1.23 in females.

Color in life of an adult male (USNM 570166): dorsal surface of body Ground Cinnamon (239) with dark brown spotting and mottling; dorsal and ventral surfaces of head Drab (27); dark brown suprascapular blotch present, followed by pinkish brown blotch; dorsal surface of tail Buff (24) on posterior two-thirds, anterior third with pale brown spots; lateral surface of head with distinct dark brown lines; dorsal surfaces of fore- and hind limb Ground Cinnamon with 2 pale brown bands on forelimb, pale brown spot on knee, and pale brown band on shank; belly and subcaudal surface Buff; iris medium brown with golden brown rim around pupil. Color in life of an adult female (USNM 573884): dorsal surfaces of body and tail Tawny Olive (223D) with Burnt Umber (22) flecking and spotting; pale brown dorsolateral stripe extending from axilla region onto base of tail; dorsal surface of head Buff (24) with distinct Burnt Umber stripes; distinct Burnt Umber postorbital stripe present; dorsal surfaces of fore- and hind limb Tawny Olive with 2 cream bands on forelimb, cream spot on knee, and cream band on shank; ventral surfaces of head and body Buff with medium brown flecking; subcaudal surface Buff-Yellow (53) with medium brown flecking.

Color in alcohol: dorsal surfaces of head, body, and tail brown; dark brown suprascapular and suprapelvic patches vary from absent to distinct in adults, those patches, when present, followed by white spots; dorsal surface of body with darker brown spots and/or mottling, each mark larger than 1 scale; dorsolateral pale stripe only rarely evident; posterior end of tail of adults sometimes with cream and black rings, those markings distinct in juveniles; dorsal surfaces of fore- and hind limb brown, usually with 1–2 cream bands on forelimb, cream spot on knee, and cream band on shank; dorsal and lateral surfaces of head usually with distinct darker brown lines, those lines occasionally indistinct; ventral surface of head usually lined or spotted with dark brown; venter of body cream with brown flecking; subcaudal surface similar in color to that of ventral surface of body.

Diagnosis/Similar Species.—The combination of lacking moveable eyelids, having the dorsal surface of the head covered with granular scales, having each claw displaced laterally by expanded terminal subdigital lamella, and having a superciliary spine will distinguish *Sphaerodactylus continentalis* from all other Honduran lizards, except the remaining *Sphaerodactylus*. *Sphaerodactylus continentalis* is distinguished from each other member of the genus using molecular techniques (see Remarks), as well as in slight morphological characters. *Sphaerodactylus dunni* has the superciliary spine located posterior to the level of the mideye, has the third supralabial at a level below the anterior half of the eye, and has alternating subcaudals (versus superciliary spine at level varying from mideye to anterior of mideye, fourth supralabial at level below anterior half of eye, and subcaudals in single row in *S. continentalis*). *Sphaerodactylus guanajae* has a distinct short pale brown dorsal pelvic stripe frequently crossing the base of the tail to connect with its counterpart on the other side and has 29–35 combined subdigital lamellae on Digit IV of the fore- and hind limb (versus no such short pelvic stripe and 33–42 combined subdigital lamellae in *S. continentalis*). *Sphaerodactylus leonardovaldesi* has a short pale line above each pelvis that frequently curves slightly inward on the base of tail and has smaller dorsal spots that are only 1 scale in size (versus no such pale pelvic line and dorsal spots larger than 1 scale in *S. continentalis*). *Sphaerodactylus millepunctatus* has 42–57 dorsal scales between the levels of axilla and groin (versus 59–76 in *S. continentalis*). *Sphaerodactylus poindexteri* has a slightly longer head with HL/SVL of 0.272–0.273 in females, has 58–63 scales around midbody, and has an indistinct dorsal pattern of dark brown dorsal spots confined to 1 scale (versus HL/SVL 0.221–0.248, 64–80 scales around midbody, and dark dorsal spots usually more than 1 scale in size in *S. continentalis*). *Sphaerodactylus exsul*, of the West Indian *S. notatus* species group, has the third supralabial at the level below the anterior half of eye and has 22–30 dorsal scales between the levels of axilla and groin (versus usually fourth supralabial at level below anterior half of eye and 59–76 dorsal scales in *S. continentalis*). *Sphaerodactylus glaucus* has smooth dorsal scales (versus keeled dorsals in *S. continentalis*). *Sphaerodactylus alphus* and *S. rosaurae*, of the primarily West Indian *S. scaber* species group, have 1–3 middorsal rows of granular scales sharply and distinctively differentiated from the larger surrounding dorsal scales (versus all middorsal scales similar-sized in *S. continentalis*).

Illustrations (Figs. 36, 39, 41; Plate 17).—Campbell, 1998 (adult; as *S. millepunctatus*); Harris and Kluge, 1984 (adult and juvenile pattern; as *S. millepunctatus* [in part]); Köhler 2001b (subcaudal scales; as *S. millepunctatus*), 2001c (adult; as *S. millepunctatus*), 2003a (dorsal scales; as *S. millepunctatus*); McCranie and Hedges, 2012 (adult); Savage, 2002 (adult [from Guatemala only]; as *S. millepunctatus*).

Remarks.—Harris and Kluge (1984) provided a detailed morphological and systematic revision of *Sphaerodactylus millepunctatus* that included *S. continentalis*, *S. guanajae*, and *S. leonardovaldesi* as all consumed under the name *S. millepunctatus*. McCranie and Hedges (2012) studied the *S. millepunctatus* complex in Honduras and, based on molecular and morphological data, considered *S. continentalis*, *S. guanajae*, and *S. leonardovaldesi* to represent species distinct from *S. millepunctatus*. Subsequently, McCranie and Hedges (2013a) studied molecular data from the Isla de Utila population of the *S. millepunctatus* group, which population was not included in their previous

Plate 17.   *Sphaerodactylus continentalis.* MCZ R191119, adult male, SVL = 30 mm. Atlántida: Salado Barro.

study. McCranie and Hedges (2013a) found that the Utila population represented an undescribed cryptic species, which they named S. *poindexteri*. Thus, those five species, plus S. *dunni*, make up the Honduran segment of the S. *millepunctatus* species group (McCranie and Hedges, 2012, 2013a).

Harris and Kluge (1984) reported 47–74 (58.8 ± 5.0) dorsals for a series of 57 *Sphaerodactylus continentalis* from Coyoles, Yoro. Those counts show more overlap between S. *continentalis* and S. *millepunctatus* than my counts demonstrate, including my data for a series of 12 specimens from near Coyoles (59–70, x = 63.7 ± 3.7). However, molecular data for two specimens from near Coyoles show that population to cluster with S. *continentalis*, and to be a distinct lineage compared with S. *millepunctatus* from the Mosquitia of northeastern Honduras (McCranie and Hedges, 2012, 2013a). Additionally, the molecular results presented by McCranie and Hedges (2013a) suggested that there are likely two species consumed under S. *continentalis* in Honduras. Thus, S. *continentalis* from throughout its geographical distribution is thought to remain a complex of more than one species.

Harris and Kluge (1984: 22) noted that the specimens from Isla de Cozumel, Quintana Roo, Mexico, differed in color pattern from all remaining populations of the *Sphaerodactylus millepunctatus* complex they studied. Molecular studies of that population likely will show it to represent an undescribed evolutionary species. Taylor (1956a) considered S. *lineolatus* Lichtenstein and von Martens (1856: 6) a valid species related to S. *continentalis*, but some authors consider that nominal form a synonym of S. *continentalis*.

The Cantarranas, Francisco Morazán, record listed by Dunn and Emlen (1932: 26) is either in error or the specimens are now lost. The ANSP catalogue does not list any specimens from Cantarranas.

Natural History Comments.—*Sphaerodactylus continentalis* is known from near sea level to 1,100 m elevation in the Lowland Moist Forest, Lowland Dry Forest, Lowland Arid Forest, Premontane Moist Forest, and Premontane Dry Forest formations and peripherally in the Premontane Wet Forest formation. This predomi-

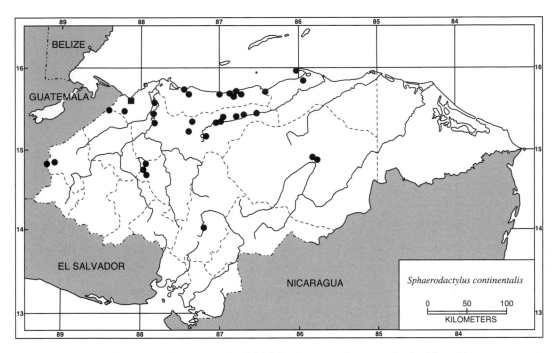

Map 15. Localities for *Sphaerodactylus continentalis*. Solid circles denote specimens examined and the solid square represents an accepted record.

nately diurnal species is probably active throughout the year, with months of collection being May to August and November and December. It is found in leaf litter and under logs well away from human habitation, as well as in edificarian situations, including inside houses and other buildings and under piles of yard debris and stacks of roofing tiles. Meyer and Wilson (1973) reported collecting a large series from banana bunches that were being washed at a packing plant in July. Hahn (1971) reported one was taken in a bromeliad in June. Nothing has been reported on reproduction in this species, but like other species of *Sphaerodactylus*, females probably deposit multiple clutches of a single egg, often in communal nests. Diet data has also not been reported for this species, but it likely feeds on small insects, including ants.

Etymology.—The name *continentalis* is derived from the Latin *continens* (continent) and *-alis* (pertaining to). The name alludes to the mainland occurrence of the holotype.

Specimens Examined (159 + 1 C&S [72]; Map 15).—**ATLÁNTIDA**: Estación Forestal CURLA, USNM 570164–68; 7.4 km SE of La Ceiba, USNM 570163; La Ceiba, USNM 55245; 11.3 km W of La Ceiba, TCWC 21966; Lancetilla, ANSP 25869, 33108–10, FMNH 21832, MCZ R32200, 39711, USNM 578829–31; Salado Barra, MCZ R19113, 19119–20; Tela, FMNH 13183, LSUMZ 24599, USNM 64173 (on bananas in Baltimore, USA), 69863, 70460–61 (all three on bananas in New Orleans, USA). **COLÓN**: Balfate, AMNH 58621; Puerto Castilla, USNM 64932 (collected on bananas); 0.5 km S of Trujillo, LACM 47809. **COMAYAGUA**: about 8 km S of La Misión, MCZ R49975; 1.6 km S of Pito Solo, LSUMZ 28528–29; Pito Solo, CM 64526. **COPÁN**: Copán, USNM 570169; Rancho El Jaral, LACM 45391. **CORTÉS**: Laguna Ticamaya, FMNH 5032–35, 5226;

Los Pinos, UF 166397–400, USNM 573096–98; Río Santa Ana, FMNH 5029. **FRANCISCO MORAZÁN**: Tegucigalpa, USNM 570170, FN 256977–78 (still in Honduras because of permit problems). **OLANCHO**: 4.5 km SE of Catacamas, LACM 47777; Catacamas, LACM 45137–39; Piedra Blanca, USNM 579972–73. **SANTA BÁRBARA**: Compañia Agricola Paradise, USNM 578828; tributary of Río Listón, USNM 573884. **YORO**: 0.5 km N of Coyoles, LACM 47779–80, LSUMZ 21441; Coyoles, LACM 47783, 47784 (C&S), 47785–808, LSUMZ 21442–74; El Progreso, UMMZ 58380; Los Indios, MCZ R38796; 21 km WSW of Olanchito, UF 90207–08; 41 km WSW of Olanchito, UF 90209–10; Rancho San Lorenzo, LACM 47781–82; San Francisco, MVZ 52401, USNM 579974; near San Lorenzo Abajo, USNM 570193, 579975–78; 5.5 km SSE of San Lorenzo Arriba, USNM 579979–82; 4.7 km ESE of San Lorenzo Arriba, USNM 565402–06, 570194–95; San Patricio, USNM 579983–84; Subirana Valley, MCZ R38795; Yoro, KU 203087. "HONDU-RAS": USNM 71733, 79951, 82573–74, 86862, 95866–67, 98909.

Other Records (Map 15).—**CORTÉS**: El Paraíso (Townsend, 2006a). "HONDU-RAS": ZIL 8880 (Werner, 1896; see Harris and Kluge, 1984).

Sphaerodactylus dunni Schmidt, 1936

Sphaerodactylus dunni Schmidt, 1936: 46 (holotype, MCZ R32199; type locality: "Naco River, near Cofradia [sic], Honduras"); M. A. Smith, 1937: 42; Barbour and Loveridge, 1946: 196; Wermuth, 1965: 166; Meyer, 1969: 207; Peters and Donoso-Barros, 1970: 252; Hahn, 1971: 111; Meyer and Wilson, 1972: 109; Meyer and Wilson, 1973: 12; Harris and Kluge, 1984: 42; Kluge, 1995: 22; Wilson and McCranie, 1998: 16; Köhler, 2000: 52; Wilson et al., 2001: 137; Köhler, 2003a: 77; Wilson and McCranie, 2003: 59; McCranie and Castañeda, 2005: 14; McCranie et al., 2006: 216; Wilson and Townsend, 2006: 105; Köhler, 2008: 81; Townsend and Wilson, 2010b: 692; Köhler and Ferrari, 2011a: 113; McCranie and Hedges, 2012: 76; Townsend et al., 2012: 109; de Lisle et al., 2013: 235; McCranie and Hedges, 2013a: 41; McCranie and Solís, 2013: 242; Solís et al., 2014: 132; McCranie, 2015a: 366.

Geographic Distribution.—*Sphaerodactylus dunni* occurs at low elevations on the Atlantic versant of northwestern to northeastern Honduras.

Description.—The following is based on four males (CM 64529; LACM 47303; UF 90211; USNM 579606) and nine females (CM 64528; LACM 47304; MCZ R32199; SMF 86283; UF 90212; USNM 579603–05, 579607). *Sphaerodactylus dunni* is a tiny lizard (maximum recorded SVL 32 mm [USNM 579603, a female]); dorsal surface of head covered by small scales, scales weakly keeled in parietal region; 7–9 (8.1 ± 0.8) snout scales in eight; 3 supralabials to level below anterior half of eye and 3 (rarely 2) infralabials to level below mideye; superciliary spine at level just posterior to mideye; moveable eyelid absent; pupil circular; rostral with median cleft and small posterior notch; 2 enlarged supranasals, with anterior supranasal sometimes fused with rostral, anterior supranasal separated from counterpart on other side by 2–3 (rarely 3) smaller internasals; dorsal surface of body covered by slightly larger scales than those of head, scales homogeneous, keeled, imbricate, 34–50 (42.4 ± 5.3) between levels of axilla and groin; ventral scales larger than dorsal scales, smooth, flat, imbricate, 27–38 (31.9 ± 3.2) between levels of axilla and groin; 53–67 (60.6 ± 4.9) scales around midbody in 12; 7–11 (9.4 ± 1.0) narrow subdigital lamellae on Digit IV of hind limb, 7–9 (7.5 ± 1.0) on Digit IV of forelimb, 31–36 (33.2 ± 1.9) combined lamellae on Digit IV of fore- and hind limb in five; retractile claw located laterally between enlarged terminal pad and several smaller scales; webbing absent between digits; subcaudal scales in alternating series; femoral and precloacal pores absent in both sexes; male escutcheon slightly hypertrophied, small, about 9 scales long and 5 scales wide; SVL 20.7–24.5 (22.9 ± 1.9) mm in males, 21.0–31.7 (25.2 ± 3.2) mm in females; HW/SVL 0.11–0.13 in males, 0.10–0.14 in females; HL/SVL 0.22–0.30 in males, 0.21–0.27 in females; HD/SVL 0.09–0.10 in two males, 0.08–0.10 in six females; TAL/SVL 0.76–0.84 in four males, 0.43–0.77 in five females; SEL/EEL 1.00–1.22 in males, 1.00–1.30 in females.

Color in life of an adult female (USNM 579605): dorsal surface of body Mikado Brown (121C) with dark brown flecking; dorsal surface of tail Kingfisher Rufous (240); nuchal band Sepia (119) with yellowish white lines; top of head True Cinnamon (139) with dark brown lines; side of head dark brown with pale yellow lines; venter of head pale brown with dirty white lines; belly pale brown; subcaudal surface slightly paler brown than top of tail; iris Mikado Brown. Meyer and Wilson (1972: 109–110) provided a composite color in life description of two adult males (LACM 47303; LSUMZ 22450; the latter now C&S): "Dorsum brown with light transverse markings, which decrease in intensity posteriorly. Nape region black, grading into orange-brown laterally and anteriorly, crossed by three white bands, which are more or less confluent with a lateral white stripe on each side. Dorsal surface of the head orange-brown to light brown, with a reddish-brown nasal blotch and a dark reddish to orange-brown parietal blotch, outlined by a Y-shaped white to pinkish-white marking, the arms of which are confluent with the light ground color of the head; lateral head color orange-brown, with pinkish-white upper labials [= supralabials]. Chin white with orange to brown reticulate or streaklike markings; throat with a mixture of white, orange, and brown scales. Venter light brown. Tail orange-brown to reddish-brown dorsally and orange ventrally. Iris red."

Color in alcohol: dorsal surfaces of head, body, and tail brown with a few darker brown spots on nape; 2 paler brown crosslines also present on nape; small darker brown spots also present above base of tail and posterior end of body; ventral surface of head pale brown with several darker brown spots and incomplete stripes, ventral surface of nape also spotted with darker brown; ventral surfaces of body and tail pale brown with dark brown flecking.

Diagnosis/Similar Species.—The combination of lacking moveable eyelids, having the dorsal surface of the head covered with granular scales, having the claws displaced laterally by the expanded terminal subdigital lamella, and having a superciliary spine distinguishes *Sphaerodactylus dunni* from all other Honduran lizards, except the remaining *Sphaerodactylus*. *Sphaerodactylus dunni* differs from those remaining Honduran *Sphaerodactylus* by the combination of having the third supralabial at the level below the anterior half of the eye, having keeled dorsal scales, having the subcaudal scales in alternating series, and usually having the superciliary spine located posterior to the level of mideye.

Illustrations (Figs. 37, 38; Plate 18).—Harris and Kluge, 1984 (snout scales, escutcheon); Köhler, 2000 (adult), 2003a (adult), 2008 (adult).

Remarks.—The only previous taxonomic study on *Sphaerodactylus dunni* is that of Harris and Kluge (1984), which included only seven specimens. Although not difficult to find where they occur, I am aware of only 16 museum specimens of *S. dunni*. This species is rather abundant at some of its known localities, but permit restrictions required that I collect only 2–3 specimens of this species, with one going into the UNAH collection. DNA sequence analysis of tissues from various *Sphaerodactylus*,

Plate 18.   *Sphaerodactylus dunni.* USNM 579607, adult female, SVL = 26 mm. Yoro: San Patricio.

including a single *S. dunni* (USNM 579605), recovered a phylogeny indicating *S. dunni* is a member of the same species group that includes *S. millepunctatus* and its allies (McCranie and Hedges, 2012: 66).

Natural History Comments.—*Sphaerodactylus dunni* is known from 60 to 280 m elevation in the Lowland Moist Forest, Lowland Dry Forest, and Lowland Arid Forest formations. This diurnal species was collected in leaf litter and under logs and rocks in April, June, and October. It is likely active during the cooler parts of the day such as early morning. Nothing has been reported on reproduction in this species, but like other *Sphaerodactylus*, it probably lays multiple clutches of one egg per clutch. No information on its diet is available, but like other species of *Sphaerodactylus*, it probably feeds on small insects, including ants.

Etymology.—The specific name *dunni* is a patronym honoring Emmett Reid Dunn, a 20th century North American herpetologist, who alerted the describer to the existence of this species (see Schmidt, 1936: 47).

Specimens Examined (14 + 2 C&S [5]; Map 16).—**ATLÁNTIDA**: 11.3 km SE of La Ceiba, LACM 47302 (C&S), 47303–04; 7.4 km SE of La Ceiba, USNM 579603; near San José de Texíguat, USNM 589163. **COLÓN**: Cerro Calentura, CM 64529, LSUMZ 22450 (C&S); 3 km WSW of La Brea, UF 90211. **CORTÉS**: near Cofradía, MCZ R32199; mountains W of San Pedro Sula, CM 64528. **GRACIAS A DIOS**: Raudal Kiplatara, SMF 86283. **YORO**: 21 km WSW of Olanchito, UF 90212; 5.5 km SSE of San Lorenzo Arriba, USNM 579606; 3.5 km ESE of San Lorenzo Arriba, USNM 579604–05; San Patricio, USNM 579607.

Sphaerodactylus exsul Barbour, 1914

Sphaerodactylus exsul Barbour, 1914: 264 (holotype, MCZ R7894; type locality: "Little Swan Island, Caribbean Sea"); Regan, 1916: 15; Barbour, 1921b: 255; Barbour and Loveridge, 1929a: 341; Cochran, 1961: 146; Wermuth, 1965: 167; Bauer and Günther, 1991: 300; Powell and Henderson, 2012: 92; McCranie, 2015a: 366; McCranie et al., 2017: 269.

Sphaerodactylus notatus exsul: Schwartz, 1966: 170; Schwartz, 1970: 90.1; Schwartz and Thomas, 1975: 158;

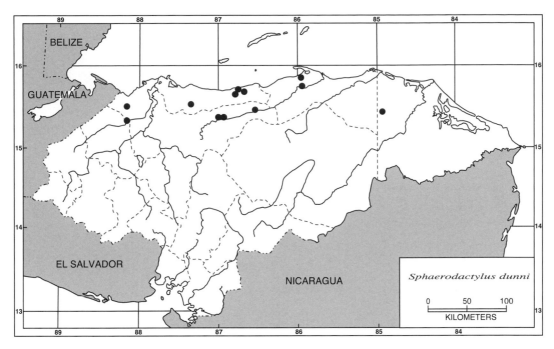

Map 16. Localities for *Sphaerodactylus dunni*. Solid circles denote specimens examined.

MacLean et al., 1977: 4; Schwartz et al., 1978: 27; Morgan, 1985: 43; Schwartz and Henderson, 1988: 187; Schwartz and Henderson, 1991: 516; Kluge, 1995: 23; Henderson and Powell, 2009: 309; de Lisle et al., 2013: 243.
Sphaerodactylus notatus: Schwartz, 1973: 142.2; Harris, 1982a: 22; Solís et al., 2014: 132.
Sphaerodactylus spp.: Ferrari, 2008: A14, *In* Anonymous, 2008.

Geographic Distribution.—*Sphaerodactylus exsul* occurs only on the Swan Islands of Honduras.

Description.—The following is based on ten males (MCZ R191090, 191142; USNM 494810, 494815–17, 494819, 494821–22, 494829) and ten females (USNM 494809, 494811–12, 494814, 494820, 494823–26, 494828). *Sphaerodactylus exsul* is a tiny lizard (maximum recorded SVL 26 mm [USNM 494823, a female]); dorsal surface of head covered by small scales, scales keeled in parietal region; 5–8 (6.3 ± 0.8) snout scales; 3 supralabials to level below anterior portion of eye and 2–3 (most often 3) infralabials to level below mideye; superciliary spine at level ranging from anterior to mideye to about mideye; moveable eyelid absent; pupil circular; rostral with median cleft, but without small posterior notch; 2 enlarged supranasals, except anterior one sometimes fused with rostral, anterior supranasal separated from counterpart on other side by 0–2 smaller internasals; dorsal surface of body covered by slightly larger scales than those of head, dorsal scales homogeneous, keeled, imbricate, 22–30 (25.1 ± 1.8) between levels of axilla and groin; ventral scales slightly smaller than dorsal scales, smooth, flat, imbricate, 25–32 (26.9 ± 2.0) between levels of axilla and groin; 38–44 (42.2 ± 1.5) scales around midbody; 8–11 (9.6 ± 0.8) narrow subdigital lamellae on Digit IV of hind limb, 6–8 (6.9 ± 0.5) on Digit IV of

Plate 19.   *Sphaerodactylus exsul.* MCZ R191125, adult male. Gracias a Dios: Isla Grande, Swan Islands.

forelimb, 30–36 (33.1 ± 1.6) combined lamellae on Digit IV of fore- and hind limb in 19; retractile claw located laterally between enlarged terminal pad and several smaller scales; webbing absent between digits; subcaudal scales aligned in single median row; femoral and precloacal pores absent in both sexes; male escutcheon slightly hypertrophied, small; SVL 22.9–25.1 (24.4 ± 0.8) mm in males, 22.2–26.1 (24.0 ± 1.0) mm in females; HW/SVL 0.10–0.12 in males, 0.11–0.12 in females; HD/SVL 0.08–0.10 in both sexes; HL/SVL 0.21–0.24 in males, 0.21–0.23 in females; TAL/SVL 0.50–0.92 in five males, 0.63–0.90 in females; SEL/EEL 1.10–1.31 in males, 1.00–1.28 in females.

Color in life of an adult male (MCZ R191125; Plate 19): dorsal surface of body Drab (color 19 in Köhler, 2012) with Dark Drab (45) spots, each spot confined to single scale; dorsal surface of head pale brownish white with numerous Brownish Olive (276) stripes and elongated blotches, nuchal area also with Sayal Brown (41) mottling; nuchal blotch absent; basal portion of top of tail similar in color to body, posterior two-thirds of top of tail mottled with Dark Drab and orange brown; dorsal surfaces of fore- and hind limb mottled with Dark Drab and Cinnamon Drab (50); ventral surfaces of head and body pale brown; ventral surfaces of toe pads white; subcaudal surface mottled pale and medium brown; iris Buff (15) with white ring around pupil. Color in life of an adult female (MCZ R191145): dorsal surface of body Clay Color (color 18 in Köhler, 2012) with Dark Brownish Olive (127) and Amber (51) spots, each spot confined to 1 scale; dorsal surface of head Cinnamon-Drab (50) with Russet (44) pre- and postocular and middorsal stripes, middorsal stripe Clay Color medially at posterior edge of head; Sepia (279) occipital blotch present, blotch outlined anteriorly by dirty white line, paired yellowish white spots present on posterolateral portion of Sepia occipital blotch; no separate nuchal blotch present; dorsal surfaces of fore- and hind limb Clay Color with Olive-Brown (278) spots and lines, Mikado Brown (42) spots also present on hind limb; basal portion of tail similar in color to that of body, posterior three-quarters of tail Mikado Brown with scattered dark brown spots; ventral surfaces of head, body, and tail pale brown, distal three-quarters of subcaudal surface Mikado Brown; iris Mikado Brown with golden brown ring around pupil.

Color in alcohol: dorsal surfaces of body and tail pale brown with darker brown spots; dark brown suprascapular patch and pale brown ocelli vary from absent to distinct in adults, both types of markings distinct in juveniles; dorsal and lateral surfaces of head of males paler brown than body, without distinct pattern; dorsal surface of head of females same ground color as body, with lateral, postocular, preocular, and middorsal darker brown stripes; ventral surfaces of head and body cream, with varying amounts of brown flecking; chin and throat usually less flecked than posterior portion of body, but throat of males vary from lightly to heavily flecked with brown; subcaudal surface more heavily flecked with brown than posterior portion of body.

Diagnosis/Similar Species.—The combination of lacking moveable eyelids, having dorsal surface of the head covered with small scales, having the claws displaced laterally by expanded terminal subdigital lamellae, and having a superciliary spine distinguishes *Sphaerodactylus exsul* from all other Honduran lizards, except the remaining *Sphaerodactylus*. *Sphaerodactylus exsul* has larger middorsal scales numbering 22–30 longitudinal rows and lacks a middorsal zone of granular scales (versus 40 or more dorsal scales, or if less than 40 dorsal scales, a distinct middorsal zone of granules present, in all other *Sphaerodactylus*).

Illustrations (Fig. 40; Plate 19).—Barbour, 1921b (adult, dorsal scales, head scales); McCranie et al., 2017 (adult).

Remarks.—Barbour (1914) described the Swan Island population as *Sphaerodactylus exsul*. Schwartz (1966) studied morphological variation in the widely distributed *S. notatus* Baird (1859a: 254) and relegated *S. exsul* to a subspecies of *S. notatus*; however, Schwartz was a strong user of subspecies, with many of those having subsequently

been elevated to species level, especially those isolated on islands. Schwartz (1973) listed *S. exsul* (as *S. notatus*) from Little Swan Island as introduced; thus, Schwartz advocated a subspecies for a perceived introduced species. However, the Swan Island *S. exsul* differs from all other populations of the *S. notatus* group examined by Schwartz (1966) in having a suprascapular spot and ocelli. Schwartz (1966) also noted that *S. exsul* has slightly higher numbers of dorsal scales between the levels of the axilla and groin and fewer numbers of scales around midbody than does *S. notatus*. *Sphaerodactylus exsul* is also smaller than other populations of *S. notatus* studied by Schwartz (1966). The Swan Island *Sphaerodactylus* appears to be native and not introduced by man. Given that there are slight differences between Swan Island populations and all others of *S. notatus* known (Schwartz, 1966, 1970) and that the Swan Islands are separated by deep seas from all other populations of the *S. notatus* species complex, *S. exsul* is treated as a species, and it is believed its ancestors reached the Swan Islands by overwater dispersal. Powell and Henderson (2012) listed *S. exsul* as a species without any documentation, but that decision was based on my suggestion without acknowledgment. Recently collected tissues of *S. exsul* are available in the MCZ collection to test that proposal of *S. exsul* representing a species.

Natural History Comments.—*Sphaerodactylus exsul* is known from near sea level to 10 m elevation in the Lowland Dry Forest (West Indian Subregion) formation. The type series was collected in March 1912. Another series was collected by an expedition from the USNM in February 1974. My group collected this species under fallen leaves, pieces of wood, and discarded tin in shaded areas in December 2012 (McCranie et al., 2017). Individuals attempt to escape when first uncovered by rapidly darting beneath the nearest cover such as leaf litter and broken limestone rock and can be difficult to capture without damaging its fragile skin. According to the collector of the type series (George Nelson), this species is "very abundant in the accumulations of humus and fallen leaves in the cavities and depressions so very common in the sharply eroded aeolian limestone" on Little Swan Island (Nelson, *In* Barbour, 1914: 264–265). Nothing has been reported on reproduction in this species, but like other species in the genus, it probably lays multiple clutches of one egg, often in communal nests. No information is available on diet of *S. exsul*, but like its close relative *S. notatus*, it probably feeds on beetles, ants, and caterpillars (see Henderson and Powell, 2009).

Etymology.—The specific name *exsul* is Latin, meaning banished person or exile. The name alludes to the isolated occurrence of this species.

Specimens Examined (136, 4 C&S, 1 egg [0]; Map 5).—**GRACIAS A DIOS**: Isla Grande, Swan Islands, MCZ R191069, 191089–92, 191125–26, 191142–45, USNM 494672–76, 494677 (egg), 494678–748, 494858–61 (all four C&S; erroneously listed as USNM 23358, 23361–63 by Kluge, 1995); Isla Pequeña, Swan Islands, AMNH 24704 (formerly MCZ R9969), CAS 39407–09 (formerly in MCZ), MCZ R7894, 9959–62, 9965–68, 9970–77, USNM 120711 (formerly MCZ R29975), 494809–33, ZMB 29811 (formerly MCZ R9963), 77700 (formerly MCZ R9964).

Sphaerodactylus glaucus Cope, 1866a

Sphaerodactylus glaucus Cope, 1866a: 192 (holotype, USNM 6572 [see Remarks]; type locality: "Near Merida [sic], Yucatan"); Harris and Kluge, 1984: 27; Wilson et al., 2001: 137; Wilson and Townsend, 2007: 145; Solís et al., 2014: 132; McCranie, 2015a: 366.

Geographic Distribution.—*Sphaerodactylus glaucus* occurs at low and moderate elevations from central Veracruz, Mexico, to extreme western Honduras on the Atlantic versant and on the Pacific versant from south-central Oaxaca to extreme southwestern Chiapas, Mexico. In Honduras, this species is known only from western Copán.

Description.—The following is based on two females (AMNH 124037; USNM 570162). *Sphaerodactylus glaucus* is a tiny lizard (maximum recorded SVL 29 mm [Harris and Kluge, 1984]; 26 mm SVL in largest Honduran specimen [AMNH 124037]); dorsal surface of head covered by small, smooth scales, but scales conical in parietal region; 6 snout scales in one; 3–4 supralabials to level below anterior half of eye; 4 infralabials to level below mideye; superciliary spine at level anterior to mideye; moveable eyelid absent; pupil circular; rostral with median cleft and tiny posterior notch; 2 enlarged supranasals, anterior supranasals separated by 1 smaller internasal; dorsal surface of body covered by smooth, juxtaposed scales, those of middorsum somewhat smaller than dorsolateral ones, 67–69 (68.0) between levels of axilla and groin; ventral scales larger than dorsal scales, smooth, flat, imbricate, 34–40 (37.0) between levels of axilla and groin; 80–85 (82.5) scales around midbody; 9–10 (9.3 ± 0.5) narrow subdigital lamellae on Digit IV of hind limb, 6 on Digit IV of forelimb in one, 31 combined lamellae on Digit IV on fore- and hind limb in one; retractile claw located laterally between enlarged terminal pad and several smaller scales; webbing absent between digits; subcaudal scales aligned in single median row; femoral and precloacal pores absent; SVL 21.8–25.9 (23.9) mm; HW/SVL 0.12–0.13; HL/SVL 0.23–0.24; HD/SVL 0.09 in one; TAL/TOL 0.88 in one; SEL/EEL 1.00–1.32.

Color in alcohol: dorsal surface of body brown with paler grayish brown middorsal swath; dark brown suprascapular V-shaped blotch present; dorsal surface of head brown with dark brown median transverse line; large, pale grayish brown blotch present dorsally at base of tail, blotch partially divided by dark brown midline;

Plate 20. *Sphaerodactylus glaucus*. SMF 99655. Mexico: Campeche, Becan. Photograph by Gunther Köhler.

ventral surface of head and nape cream with dark brown flecking, that of body cream with some dark brown flecking ventrolaterally; subcaudal surface cream with some dark brown flecking ventrolaterally in one (only small portion of tail present in remaining one, USNM 570162).

Diagnosis/Similar Species.—The combination of lacking moveable eyelids, having the dorsal surface of the head covered with granular scales, having the claws displaced laterally by the expanded distal subdigital lamellae, and having superciliary spines distinguishes *Sphaerodactylus glaucus* from all other Honduran lizards, except the remaining *Sphaerodactylus*. *Sphaerodactylus glaucus* is the only Honduran mainland species of the genus with smooth dorsal scales of equal size.

Illustrations (Fig. 35; Plate 20).—Álvarez del Toro, 1983 (adult); Barbour, 1921b (dorsal scales, head scales); Calderón-Mandujano et al., 2008 (adult); Campbell, 1998 (adult); Harris and Kluge, 1984 (dorsal color pattern variation); Köhler, 2003a (adult), 2008 (adult); Lee, 1996 (adult), 2000 (adult); Stafford and Meyer, 1999 (adult); Taylor, 1947 (adult; as *S. glaucus*; head scales and color pattern, juvenile dorsal pattern; as *S. torquatus* Strauch, 1887: 35); Villa et al., 1988 (adult).

Remarks.—Some uncertainty exists about the publication date for Cope's "Third contribution" in which the name *Sphaerodactylus glaucus* is proposed. That volume of the Proceedings is dated 1865. Most authors on gecko taxonomy (i.e., Wermuth, 1965; Harris and Kluge, 1984; Rösler, 2000) cite the work in question as published in 1865, but Kluge (1993, 2001) and de Lisle et al. (2013:236) cite the date as 1866 without comment. That the exact date of Cope's publication has not been investigated earlier and is uncertain is unfortunate, but I present evidence that it was actually published in 1866. Cope (1866a: 192) did not designate a holotype for his *Sphaerodactylus glaucus* but did provide measurements for one specimen. The description provided by Cope (1866a) also appears to represent that of a single specimen (USNM 6572 = the holotype) collected in February 1866

(Cochran, 1961: 147). So, not only was the Cope publication of *S. glaucus* published in 1866, the description also had to be written in 1866. Cope (1866b: 125) reported "several specimens" of *S. glaucus* collected by the same collector (and from the same area) as the specimen mentioned in his (1866a) description of the species. Cochran (1961: 147) stated that there were three "cotypes" of *S. glaucus* in the USNM collection (USNM 6572: collected in February 1866 = the holotype; USNM 62995–96 and one exchanged with the MCZ: collected 22 December 1866). Thus, the available data indicate the description of the holotype of *S. glaucus* was written in 1866, and Cope (1866b) was probably published in 1867. Harris and Kluge (1984: 27) accepted USNM 6572 as the holotype of *S. glaucus*. Cope (1866a) also described several other new species in his "Third contribution," but a quick check of recent literature concerning those species finds all authors list 1865 as the year Cope's description appeared. However, collection dates given by Cochran (1961) for some of Cope's material are in 1866; thus, the Cope "Third contribution" had to have some text written in 1866. Thus, I cite Cope (1866a), instead of the usual 1865 as the publication date for the type description of *S. glaucus*.

Harris and Kluge (1984) provided a thorough review of the systematics of *Sphaerodactylus glaucus*, and stated (p. 28) for *S. glaucus*, the "fourth supralabial lies below anterior half of eye." However, USNM 570162 has the third supralabial at a level below the anterior half of the eye. The redescription and illustration of *S. glaucus* in Barbour (1921b) also demonstrates the third supralabial below the level of the anterior half of the eye. The second Honduran specimen (AMNH 124037) has the fourth supralabial at a level below the anterior half of the eye; thus, that character appears variable in this species.

Sphaerodactylus glaucus has not been placed in any species group, but it is the only Middle American *Sphaerodactylus* to have smooth dorsal scales (Harris and Kluge, 1984: 28).

Natural History Comments.—*Sphaerodactylus glaucus* is known only from 600 m elevation in the Premontane Moist Forest formation. One was in leaf litter at the base of a large oak tree (*Quercus* sp.) in July. The AMNH specimen was on the grounds of a hotel in July. Lee (1996) reported that on the Yucatán Peninsula, this species is found under ground debris in edificarian situations, under palm fronds in coastal situations, and under loose tree bark and logs in forest and in and around thatched houses. Calderón-Mandujano et al. (2008) called it mainly nocturnal in Quintana Roo, Mexico; Campbell (1998) said it was diurnal or nocturnal in Guatemala; and Stafford and Meyer (1999) called it generally diurnal in Belize. Females deposit one egg per clutch and can be a communal nester (Lee, 1996, and references cited therein). Apparently nothing has been published on diet in this species other than general undocumented statements that it feeds on small arthropods like ants, termites, and spiders (Calderón-Mandujano et al., 2008) or tiny insects and spiders (Campbell, 1998).

Etymology.—The name *glaucus* is Latin (bluish green or gray) and was used by Cope (1866a: 192) for his color description of the dorsal surfaces of the holotype as "greenish stone color or glaucus."

Specimens Examined (2 [0]; Map 17).—**COPÁN**: 1 km S of Copán, USNM 570162; Copán, AMNH 124037.

Sphaerodactylus guanajae McCranie and Hedges, 2012

Sphaerodactylus continentalis: Meyer, 1969: 207 (in part); Meyer and Wilson, 1973: 12 (in part); Wilson and Hahn, 1973: 105 (in part); Schwartz and Garrido, 1981: 22 (in part).
Sphaerodactylus millepunctatus: Harris and Kluge, 1984: 17 (in part); Wilson et al., 2001: 137 (in part); Lundberg, 2002b:

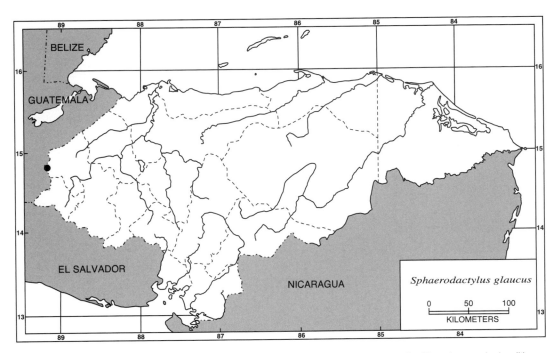

Map 17.   Localities for *Sphaerodactylus glaucus*. The solid circle denotes specimens examined from two nearby localities.

12; McCranie et al., 2005: 80 (in part); McCranie et al., 2006: 110 (in part); Wilson and Townsend, 2006: 105 (in part).

Sphaerodactylus guanajae McCranie and Hedges, 2012: 70 (holotype, USNM 580000; type locality: "East End, 16.486°N, −85.832°W, Isla de Guanaja, Islas de la Bahía, Honduras, near sea level"); McCranie and Hedges, 2013a: 41; de Lisle et al., 2013: 237; McCranie and Valdés Orellana, 2014: 45; Solís et al., 2014: 132; McCranie, 2015a: 366.

Geographic Distribution.—*Sphaerodactylus guanajae* occurs on Guanaja Island in the Bay Islands. The species is known from only three nearby localities on the northeastern end of the island.

Description.—The following is based on five males (FMNH 282334, 283676, 283679, 283681; USNM 580000) and ten females (FMNH 283675, 283678, 283680, 283686; USNM 520269, 579994–96, 579998–99).

Sphaerodactylus guanajae is a tiny lizard (maximum recorded SVL 30 mm [USNM 520269, a female]); dorsal surface of head covered by small scales, at least some scales on snout and in parietal region weakly keeled; 9–13 (10.9 ± 1.2) snout scales; 4 supralabials and 3 infralabials to level below anterior half of eye; superciliary spine at level varying from anterior to mideye to about mideye; moveable eyelid absent; pupil circular; rostral with median cleft and small posterior notch; 2 enlarged supranasals, anterior supranasal separated by 2–3 (usually 2) smaller internasals; dorsal surface of body covered by slightly larger scales than those on top of head, scales essentially homogeneous, keeled, juxtaposed to imbricate, 57–70 (62.8 ± 4.4) between levels of axilla and groin; ventral scales larger than dorsal scales, smooth, flat, imbricate, 32–44 (36.0 ± 3.1) between levels of axilla and groin; 60–77 (68.7 ± 5.4) scales around midbody in 14; 8–11 (8.8 ± 0.7) narrow subdigital lamellae on Digit IV of hind limb,

6–8 (7.4 ± 0.6) on Digit IV of forelimb, 29–35 (32.3 ± 1.8) combined lamellae on Digit IV of fore- and hind limb; retractile claw located laterally between enlarged terminal pad and several smaller scales; webbing absent between digits; subcaudal scales aligned in single median row; femoral and precloacal pores absent in both sexes; male escutcheon slightly hypertrophied, somewhat bell-shaped with subfemoral extension, escutcheon extending slightly onto groin region; pair of distinctively swollen granules at each corner of cloacal region in males; SVL 24.4–27.1 (26.0 ± 1.0) mm in males, 24.4–29.6 (26.4 ± 1.4) mm in females; HW/SVL 0.10–0.11 in males, 0.10–0.12 in females; HL/SVL 0.23–0.27 in males, 0.22–0.26 in females; HD/SVL 0.08–0.10 in both sexes; TAL/SVL 0.85–0.99 in three males, 0.67–0.82 in three females; SEL/EEL 1.11–1.30 in males, 1.10–1.57 in females.

Color in life of the male holotype (USNM 580000): dorsal surface of body Brussels Brown (121B) with pale brown mottling and scattered Sepia (119) scales; dirty white pelvic stripe extending posteriorly from level above groin to base of tail where it crosses dorsal surface of tail to connect with similar stripe on other side of tail and body; dorsal and lateral surfaces of head Sayal Brown (223C) with scattered Sepia scales forming indication of lines; dorsal surfaces of fore- and hind limb Brussels Brown with pale brown bands on lower parts of limbs and on all digits; dorsal surface of unregenerated portion of tail Sayal Brown with scattered Sepia scales and pale brown indications of crosslines; dorsal surface of regenerated portion of tail Cinnamon (39); chin pale brown; throat cream; belly pale brown with dark brown flecking; subcaudal surface of original portion of tail Flesh Ocher (132D). An adult female (USNM 579994) was similar in color in life to that of USNM 580000, except the subcaudal surface was yellowish brown.

Color in alcohol: dorsal surfaces of head, body, and tail brown; dark brown suprascapular and suprapelvic patches absent or indistinct; dorsal surface of body usually with scattered darker brown spots mostly confined to single scale per spot, those spots sometimes absent; supraocular scales dark brown; supralabials pale brown with dark brown flecking; posterior end of tail lacking distinct black rings in adults, but ring present in juveniles; dorsal and lateral surfaces of head sometimes with indistinct darker brown lines; dorsal surfaces of fore- and hind limb similar in color to body, except large pale brown spot present on knee and pale brown band present on shank; ventral surface of head cream, some having brown lines or spots, some only with indistinct brown flecking; venter of body cream with brown flecking on scale edges, especially ventrolaterally; subcaudal surface cream with brown flecking on scale edges, usually forming incomplete lineate pattern; short, pale brown pelvic line curving inward dorsally on base of tail to almost always connect with counterpart on other side.

Diagnosis/Similar Species.—The combination of lacking moveable eyelids, having the dorsal surface of the head covered with granular scales, having the claws displaced laterally by the terminal expanded subdigital lamellae, and having a superciliary spine distinguishes *Sphaerodactylus guanajae* from all other Honduran lizards, except the remaining *Sphaerodactylus*. This species is distinguished from all remaining Honduran species of *Sphaerodactylus* by results of a molecular analysis and by morphological characters. *Sphaerodactylus continentalis*, *S. millepunctatus*, and *S. poindexteri* lack a short pale line above each pelvis (versus pale line present and curving inward on base of tail to usually connect with counterpart on other side in *S. guanajae*). *Sphaerodactylus leonardovaldesi* usually has each pale pelvic line widely separated from its counterpart on the other side (versus pelvic line almost always

Plate 21.   *Sphaerodactylus guanajae*. FMNH 283675, adult female, SVL = 27 mm. Islas de la Bahía: Isla de Guanaja: Savannah Bight.

completely crossing top of base of tail to connect with counterpart in *S. guanajae*). *Sphaerodactylus millepunctatus* has 42–57 dorsal scales between levels of the axilla and groin (versus 57–70 in *S. guanajae*). *Sphaerodactylus dunni* has the superciliary spine located posterior to the level below mideye, has the third supralabial at the level below the anterior half of the eye, and has alternating subcaudals (versus superciliary spine at level of mideye to anterior of mideye, fourth supralabial at level below anterior half of eye, and subcaudals aligned in single row in *S. guanajae*). *Sphaerodactylus poindexteri* has a slightly longer head with a HL/SVL of 0.272–0.273 (versus 0.220–0.267 in *S. guanajae*). *Sphaerodactylus exsul*, of the West Indian *S. notatus* species group, has the third supralabial at the level below the anterior half of the eye and has 22–30 dorsal scales between the levels of the axilla and groin (versus usually fourth supralabial at level below anterior half of eye and 57–70 dorsal scales in *S. guanajae*). *Sphaerodactylus glaucus* has smooth dorsal scales (versus keeled dorsals in *S. guanajae*). *Sphaerodactylus alphus* and

S. rosaurae, of the mostly West Indian *S. scaber* group, have 1–3 middorsal rows of granular scales sharply and distinctively differentiated from the larger surrounding dorsal scales (versus all middorsals of similar size in *S. guanajae*).

Illustrations (Plate 21).—McCranie and Hedges, 2012 (adult).

Remarks.—*Sphaerodactylus guanajae* was known until recently from only two specimens that were considered conspecific with S. *millepunctatus*. Recently, additional specimens were collected on Guanaja, including tissue samples. Molecular and morphological studies of that material demonstrated the Guanaja population was a distinct species (McCranie and Hedges, 2012). A disturbing trend taking place on parts of Guanaja is to rake all ground debris into a pile and then burn it. That raking and burning took place at the species' type locality later the same year and the year after the holotype was collected, and no S. *guanajae* were present at the then sterile type locality.

Natural History Comments.—*Sphaerodactylus guanajae* is known from near sea

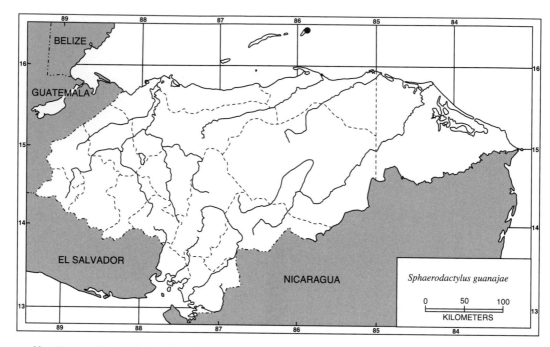

Map 18. Localities for *Sphaerodactylus guanajae*. Solid circle denotes specimens examined from nearby localities.

level to about 30 m elevation in the Lowland Moist Forest formation. This diurnal species is likely active throughout the year under favorable conditions. Active individuals were on top of leaf litter associated with Sea Grape trees (*Cocoloba uvifera*) in November, whereas others were found at the same time by raking through Sea Grape leaf litter. Also, some individuals come to the surface of leaf litter about 0.5 m in front of where raking is taking place. Therefore, it is productive to also glance at leaf litter in front of where one is raking. Specimens were also found in September under coconut palm debris on the ground and by prying off dead fronds on the sides of living coconut palms. Several others were under piles of leaves in broadleaf forest in September. Another was on a wall of an occupied house in a village in February. Wilson and Hahn (1973) reported finding one among rocks on a heavily vegetated hillside in July. Nothing has been reported on reproduction in this species, but like other species of *Sphaerodactylus*, females probably deposit multiple clutches of a single egg, often in communal nests. Feeding data has also not been reported for this species, but it likely feeds on small insects, including ants.

Etymology.—The name *guanajae* is formed from Guanaja and the Latin suffix –*ae* (derived from) in reference to the species occurring on Isla de Guanaja.

Specimens Examined (17 [1]; Map 18).— **ISLAS DE LA BAHÍA**: Isla de Guanaja, East End, FMNH 283678–79, 283681, 283686, USNM 579994–580000; Isla de Guanaja, Posada del Sol Hotel ruins, FMNH 283676, 283680; Isla de Guanaja, 2 km W of Savannah Bight, LACM 47778; Isla de Guanaja, Savannah Bight, FMNH 282334, 283675, USNM 520269.

Sphaerodactylus leonardovaldesi McCranie and Hedges, 2012

Sphaerodactylus continentalis: Meyer and Wilson, 1973: 12 (in part); Wilson and

Hahn, 1973: 105 (in part); Thomas, 1975: 195; Hudson, 1981: 377; Schwartz and Garrido, 1981: 22 (in part).

Sphaerodactylus millepunctatus: Harris and Kluge, 1984: 17 (in part); Köhler, 1994a: 4; Kluge and Nussbaum, 1995: 20; Köhler, 1995b: 102; Köhler, 1998d: 379; Grismer et al., 2001: 135; Wilson et al., 2001: 137 (in part); McCranie et al., 2005: 80 (in part); McCranie et al., 2006: 110 (in part); Wilson and Townsend, 2006: 105 (in part).

Sphaeodactylus [sic] *millepunctatus*: Lundberg, 2001: 26.

Sphaerodactylus leonardovaldesi McCranie and Hedges, 2012: 66 (holotype, FMNH 282785; type locality: "Palmetto Bay, 16.359033°, −86.486717°, Isla de Roatán, Islas de la Bahía, Honduras, near sea level"); de Lisle et al., 2013: 238; McCranie and Hedges, 2013a: 41; McCranie and Valdés Orellana, 2014: 45; Solís et al., 2014: 132; McCranie, 2015a: 366.

Geographic Distribution.—*Sphaerodactylus leonardovaldesi* occurs on Isla de Roatán and two small satellite islands in the Bay Islands.

Description.—The following is based on ten males (FMNH 282785–87, 282789; USNM 570182, 570184, 570186, 570190, 579991, 579993) and ten females (FMNH 282788, 282790–91; USNM 570183, 570185, 570189, 579988–90, 579992). *Sphaerodactylus leonardovaldesi* is a tiny lizard (maximum recorded SVL 29 mm [USNM 570185, a female]); dorsal surface of head covered by small scales, at least some scales on snout and in parietal region weakly keeled; 8–14 (11.1 ± 1.4) snout scales; 4 supralabials and 3 infralabials to level below anterior half of eye; superciliary spine at level varying from anterior to mideye to about mideye; moveable eyelid absent; pupil circular; rostral with median cleft and small posterior notch; 2 enlarged supranasals, anterior supranasal separated from counterpart on other side by 1–3 (most often 2) smaller internasals; dorsal surface of body covered by slightly larger scales than those of head, scales essentially homogeneous, keeled, juxtaposed to imbricate, 58–69 (61.7 ± 2.9) between levels of axilla and groin; ventral scales larger than dorsal scales, smooth, flat, imbricate, 29–44 (34.3 ± 4.4) between levels of axilla and groin in 19; 48–74 (62.9 ± 7.0) scales around midbody; 9–12 (9.8 ± 1.5) narrow subdigital lamellae on Digit IV of hind limb, 7–10 (8.7 ± 0.7) on Digit IV of forelimb, 33–42 (37.4 ± 2.2) combined lamellae on Digit IV of fore- and hind limb; retractile claw located laterally between enlarged terminal pad and several smaller scales; webbing absent between digits; subcaudal scales aligned in single median row; femoral and precloacal pores absent in both sexes; male escutcheon slightly hypertrophied, somewhat bell-shaped with subfemoral extension, escutcheon extending slightly onto groin region; pair of distinctively swollen granules at each corner of cloacal region in males; SVL 22.8–28.9 (25.9 ± 2.1) mm in males, 24.4–29.1 (26.1 ± 1.6) mm in females; HW/SVL 0.10–0.11 in both sexes; HL/SVL 0.21–0.27 in males, 0.21–0.26 in females; HD/SVL 0.08–0.10 in males, 0.09–0.10 in females; TAL/SVL 0.74–0.93 in four males, 0.76–1.00 in four females; SEL/EEL 1.04–1.26 in males, 1.05–1.67 in females.

Color in life of the adult male holotype (FMNH 282785): dorsal surface of body Mikado Brown (121C) with Vandyke Brown (121) small spots and some flecking; dorsal surface of head Ground Cinnamon (239) with Vandyke Brown stripes and reticulations; dorsal surfaces of fore- and hind limb Mikado Brown with golden brown spot on knee and band on shank; dorsal surface of tail brownish yellow with Vandyke Brown mottling, Vandyke Brown band present distally on tail followed by pinkish brown tail tip; ventral surfaces of head and body cream with scattered brown flecking; subcaudal surface pink; iris yellow above and below pupil, golden brown anterior and posterior to pupil. Color in life of an adult female (FMNH 282791): dorsal surface of body Flesh Color (5) with indistinct, scattered brown flecking; top of head same color as body, except with one poorly indicated, interrupted pale brown dorsolateral stripe; posterior nape region cream; short dorsal cream pelvic line, line curving inward distally, line bordered dorsally by brown; dorsal brown mark present between inward curved portion of pale lines; dorsal surfaces of fore- and hind limb Flesh Color with poorly indicated pinkish brown spot on knee and band on shank; dorsal surface of tail pinkish brown with small pink spots; subcaudal surface pink; iris yellowish pink above and below pupil, pinkish brown anterior and posterior to pupil. Color in life of a juvenile (USNM 579987): dorsal surfaces of body, head, and most of tail Sepia (119) with scattered pale brown dorsal flecks, Salmon Color (106) ring also present near tail tip; subcaudal surface otherwise reddish brown.

Color in alcohol: dorsal surfaces of head, body, and tail brown; dark brown suprascapular and suprapelvic patches absent or indistinct; dorsal surface of body with scattered darker brown spots, each spot usually confined to single scale; supraocular scales dark brown; dorsal and lateral surfaces of head sometimes with indistinct darker brown lines, except supralabials pale brown with dark brown flecking; dorsal surfaces of fore- and hind limb similar in color to body, except large pale brown spot present on knee and pale brown band present on shank; posterior end of tail lacking distinct black ring in adults, black ring present in juveniles; chin and throat cream, some have brown lines or spots, some only with indistinct brown flecking; ventral surface of body cream with brown flecking on scale edges, especially ventrolaterally; subcaudal surface cream with brown flecking on scale edges, usually forming incomplete lineate pattern; short,

pale brown pelvic line almost always present, line curving slightly inward dorsally on base of tail, but each component widely separated medially by dark brown blotch or brown pigment.

Diagnosis/Similar Species.—The combination of lacking moveable eyelids, having the dorsal surface of the head covered with granular scales, having the claws displaced laterally by expanded terminal subdigital lamella, and having a superciliary spine distinguishes *Sphaerodactylus leonardovaldesi* from all other Honduran lizards, except the remaining *Sphaerodactylus*. *Sphaerodactylus leonardovaldesi* differs from all those other *Sphaerodactylus* by phylogenetic molecular analyses, as well as in slight morphological characters. *Sphaerodactylus continentalis* has 64–80 (71.7) scales around midbody (versus 48–74 [62.9] in *S. leonardovaldesi*). *Sphaerodactylus dunni* has the superciliary spine located posterior to the level of the mideye, has the third supralabial at a level below the anterior half of the eye, and has alternating subcaudals (versus superciliary spine varying from level of mideye to anterior of mideye, fourth supralabial at level below anterior half of eye, and subcaudals aligned in single row in *S. leonardovaldesi*). *Sphaerodactylus guanajae* has a mean of 8.8 Toe IV subdigital lamellae and 7.4 Finger IV subdigital lamellae (versus means of 9.8 and 8.7, respectively, in *S. leonardovaldesi*). *Sphaerodactylus millepunctatus* has 42–57 dorsal scales between the levels of the axilla and groin (versus 58–69 in *S. leonardovaldesi*). *Sphaerodactylus poindexteri* has a female HL/SVL ratio of 0.272–0.273 (versus 0.210–0.260 in *S. leonardovaldesi*). Also, *S. continentalis*, *S. millepunctatus*, and *S. poindexteri* lack a short pale line above each pelvis (versus that pale pelvic line frequently present, curving shortly inward on base of tail in *S. leonardovaldesi*). Also, *S. guanajae* usually has the pelvic stripe completely crossing the top of the base of the tail to connect with its counterpart on the other

Plate 22.   *Sphaerodactylus leonardovaldesi.* FMNH 281531. Islas de la Bahía: Roatán Island, Palmetto Bay.

side (versus each pale pelvic line widely separated from counterpart in *S. leonardovaldesi*). *Sphaerodactylus exsul*, of the West Indian *S. notatus* species group, has 22–30 dorsal scales between the levels of the axilla and groin (versus 58–69 dorsal scales in *S. leonardovaldesi*). *Sphaerodactylus glaucus* (not placed in any species group) has smooth dorsal scales (versus keeled dorsals in *S. leonardovaldesi*). *Sphaerodactylus alphus* and *S. rosaurae*, of the predominately West Indian *S. scaber* species group, have 1–3 middorsal rows of granular scales sharply and distinctively differentiated from the larger surrounding dorsal scales (versus all middorsals of similar size in *S. leonardovaldesi*).

Illustrations (Plate 22).—McCranie et al., 2005 (adult, only those from Roatán; as *S. millepunctatus*); McCranie and Hedges, 2012 (adult); Stafford and Meyer, 1999 (adult; as *S. millepunctatus*).

Remarks.—*Sphaerodactylus leonardovaldesi*, of the island of Roatán, was considered conspecific with *S. millepunctatus* until McCranie and Hedges (2012) proposed the Roatán population was a distinct species. *Sphaerodactylus continentalis, S. dun-*

ni, *S. guanajae, S. leonardovaldesi, S. millepunctatus*, and *S. poindexteri* are the Honduran members of the *S. millepunctatus* species group (McCranie and Hedges, 2012, 2013a; also see Remarks for *S. dunni*), and molecular results from all of those species suggests each is distinct.

Natural History Comments.—*Sphaerodactylus leonardovaldesi* is known from near sea level to 30 m elevation in the Lowland Moist Forest formation. This predominately diurnal species is probably active throughout the year, as it was found on each trip to Roatán (January, February, May, June, and from September to November), including both the rainy and dry seasons. Active individuals were on top of leaf litter during both the morning and afternoon, and it was also found under leaf litter, palm fronds, rocks, brush piles, and other debris on the ground. One was active at night on a wall of a building and was feeding on insects attracted to an electric light. Wilson and Hahn (1973) reported it active in leaf litter in oak forest and under palm fronds at the base of coconut palms in July and August. Hudson (1981) reported the species was found in brush piles (month of collection

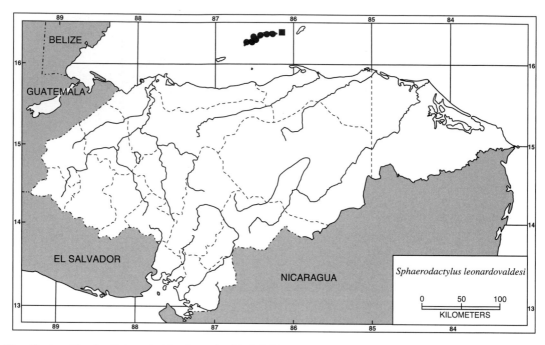

Map 19. Localities for *Sphaerodactylus leonardovaldesi*. Solid circles denote specimens examined and the solid square represents an accepted record.

not given). Grismer et al. (2001) found them moving in the shade in leaf litter, under and in rotten tree trunks, and under rocks, logs, and fallen palm fronds in July. Nothing has been reported on reproduction in this species, but like other species of *Sphaerodactylus*, females probably deposit multiple clutches of a single egg, often in communal nests. Feeding data has also not been reported for this species, but it likely feeds on small insects, including ants.

Etymology.—The name *leonardovaldesi* is a patronym honoring Leonardo Valdés Orellana, a Honduran citizen who helped collect the type series of the species.

Specimens Examined (64 + 1 C&S [7]; Map 19).—**ISLAS DE LA BAHÍA**: Isla de Roatán, Camp Bay, FMNH 282791, USNM 579990–93; Isla de Roatán, near Coxen Hole, FMNH 34541; Isla de Roatán, near Diamond Rock, USNM 570180; Isla de Roatán, about 3.2 km W of French Harbor, LSUMZ 22390, 22392, UMMZ 152733

(C&S; formerly LSUMZ 22391); Isla de Roatán, near French Harbor, CM 64525; Isla de Roatán, Gibson Bight, LSUMZ 33796–97; Isla de Roatán, near Oak Ridge, UTA R-10723–32; Isla de Roatán, Palmetto Bay, FMNH 281531, 282785–88, USNM 579987–89; Isla de Roatán, 1 km E of Pollytilly Bight, FMNH 282789–90; Isla de Roatán, Port Royal, USNM 578832–36; Isla de Roatán, near Port Royal Harbor, LSUMZ 33801, 33806–08, MCZ R150935, TCWC 52427–28; Isla de Roatán, 3.2 km W of Roatán, CM 64527; Isla de Roatán, Roatán, LSUMZ 22338–40, UF 28489; Isla de Roatán, Rocky Point, USNM 570181–82; Isla de Roatán, near Sandy Bay, KU 203127, 203132–33; Isla de Roatán, West End, USNM 578832; Isla de Roatán, West End Point, USNM 570183–88; Isla de Roatán, West End Town, USNM 570189; "Isla de Roatán," USNM 570190.

Other Records (Map 19).—**ISLAS DE LA BAHÍA**: Isla Barbareta, LSUHC 3695,

LSUPC 4782–84 (photographs) (Grismer et al., 2001); Isla de Roatán, Roatán, UF 28489 (Harris and Kluge, 1984, now apparently missing); Isla Morat, LSUHC 3710–11, LSUPC (photographs) 4785–87 (Grismer et al., 2001).

Sphaerodactylus millepunctatus Hallowell, 1861

Sphaeriodactylus [sic] *millepunctatus* Hallowell, 1861: 480 (two syntypes, USNM 6057a,b, now lost, neotype UMMZ 173053, designated by Harris and Kluge, 1984: 18; type locality by neotype selection: "NICARAGUA: Río San Juan; Isla Mancarrón of the Solentiname Archipiélago, 11°10′N, 85°02′W").

Sphaerodactylus millepunctatus: Cope, 1862a: 499; Harris and Kluge, 1984: 17 (in part); O'Shea, 1989: 16; Köhler, 2000: 52; Köhler, McCranie, and Nicholson, 2000: 425; Nicholson et al., 2000: 29; Wilson et al., 2001: 137 (in part); Castañeda, 2002: 15; McCranie, Castañeda, and Nicholson, 2002: 25; Savage, 2002, plate 295 (see Remarks); Köhler, 2003a: 76; McCranie et al., 2006: 110 (in part); Wilson and Townsend, 2006: 105 (in part); Köhler, 2008: 80; McCranie and Hedges, 2012: 76; McCranie and Hedges, 2013a: 41; Solís et al., 2014: 132; McCranie, 2015a: 366.

Sphaerodactylus continentalis: Cruz Díaz, 1978: 24; O'Shea, 1986: 33.

Geographic Distribution.—*Sphaerodactylus millepunctatus* occurs at low and moderate elevations from northeastern Honduras to northern Costa Rica on the Atlantic versant and in southwestern Nicaragua and northwestern Costa Rica on the Pacific versant. It also occurs on Isla Maíz Grande, Nicaragua. In Honduras, this species occurs at low elevations in the Mosquitia region in the northeastern portion of the country.

Description.—The following is based on eight males (FMNH 282793–96; USNM 570174, 570179, 570192, 579985) and ten females (FMNH 282792; UF 150302, 150304, 150310–11; USNM 570172–73, 570176, 570179, 573094). *Sphaerodactylus millepunctatus* is a tiny lizard (maximum recorded SVL 28 mm [UF 150302, a female]); dorsal surface of head covered by small scales, at least some scales on snout and in parietal region weakly keeled; 9–13 (9.6 ± 1.2) snout scales in 14; 4 supralabials and 3 infralabials to level below anterior half of eye; superciliary spine at level varying from anterior to mideye to about mideye; moveable eyelid absent; pupil circular; rostral with median cleft and small posterior notch; 2 enlarged supranasals, anterior supranasal separated by 1–2 (most often 2) smaller internasals; dorsal surface of body covered by slightly larger scales than those of head, scales essentially homogeneous, keeled, juxtaposed to imbricate, 42–57 (51.7 ± 4.8) between levels of axilla and groin; ventral scales larger than dorsal scales, smooth, flat, imbricate, 32–44 (35.8 ± 3.4) between levels of axilla and groin; 59–72 (65.4 ± 4.0) scales around midbody in 17; 9–12 (9.8 ± 0.8) narrow subdigital lamellae on Digit IV of hind limb, 7–9 (8.1 ± 0.7) on Digit IV of forelimb in nine, 34–37 (35.0 ± 1.1) combined Digit IV lamellae on fore- and hind limb in nine; retractile claw located laterally between enlarged terminal pad and several smaller scales; webbing absent between digits; subcaudal scales aligned in single median row; femoral and precloacal pores absent in both sexes; male escutcheon slightly hypertrophied, somewhat bell-shaped with subfemoral extension, extension sometimes reaching level of knee, escutcheon extending slightly onto groin region; pair of distinctively swollen granules at each corner of cloacal region in males; SVL 22.5–27.3 (25.2 ± 1.7) mm in males, 23.0–28.0 (25.2 ± 1.7) mm in females; HW/SVL 0.11–0.13 in males, 0.10–0.12 in females; HL/SVL 0.21–0.27 in males, 0.21–0.25 in females; HD/SVL 0.09–0.10 in males, 0.08–0.10 in eight females; TAL/SVL 1.02–1.10 in three males, 0.65–1.04 in three females; SEL/EEL 1.04–1.24 in males, 1.10–1.28 in females.

Color in life of an adult female (USNM 570191): dorsal surfaces of head, body, and tail pale brown with dark brown lines and streaking present on head plus dark brown mottling and spots on all three surfaces, spots larger than 1 scale in size on body; sacral spot and spot at base of tail Dark Brownish Olive (129), those dark spots separated by white spot; dorsolateral pinkish brown stripe extending from axilla region onto base of tail; dorsal surfaces of fore- and hind limb pale brown with yellowish brown knee spot, yellowish brown band on shank, and 2 yellowish brown bands on forelimb; ventral surface of body pale orange-yellow; ventral surface of head pale gray; iris pale orange-brown with pale yellow rim around pupil.

Color in alcohol: dorsal surfaces of head, body, and tail brown; dark brown suprascapular and suprapelvic patches vary from absent to distinct in adults; dorsal surface of body usually with darker brown spots (larger than 1 scale in size) and/or mottling; posterior end of tail usually with 2 black rings in adults; dorsolateral cream stripe extending from axilla region onto base of tail in some; dorsal and lateral surfaces of head with distinct, dark brown lines; dorsal surfaces of fore- and hind limb brown with 1–2 cream bands on forelimb, cream spot on knee, and cream band on shank; supraocular scales dark brown; ventral surface of head usually lined or spotted with dark brown; ventral surface of body cream with brown flecking; subcaudal surface similar in color to adjacent venter of body.

Diagnosis/Similar Species.—The combination of lacking moveable eyelids, having the dorsal surface of the head covered with granular scales, having each claw displaced laterally by expanded terminal subdigital lamella, and having a superciliary spine distinguishes *Sphaerodactylus millepunctatus* from all other Honduran lizards, except the remaining *Sphaerodactylus. Sphaerodactylus millepunctatus* differs from all other Honduran species of the genus in both molecular and morphological analyses. Also, *S. continentalis* has slightly smaller dorsal scales with 59–76 between the levels of the axilla and groin (versus 42–57 in *S. millepunctatus*). *Sphaerodactylus dunni* has the superciliary spine located posterior to the level of the mideye, has the third supralabial at the level below the anterior half of the eye, and has alternating subcaudals (versus superciliary spine at level varying from mideye to anterior of mideye, fourth supralabial at level below anterior half of eye, and subcaudals aligned in single row in *S. millepunctatus*). *Sphaerodactylus guanajae* has a distinct dorsal short pale pelvic stripe frequently crossing the base of tail to connect with its counterpart on the other side, has a mean number of 8.8 subdigital lamellae on Toe IV, and has 29–35 combined Toe IV subdigital lamellae (versus no such short pelvic stripe, mean of 9.8 subdigital scales on Toe IV, and 34–37 combined Toe IV subdigital scales in *S. millepunctatus*). *Sphaerodactylus leonardovaldesi* has a short dorsal pale pelvic line that frequently curves inward shortly on the base of the tail and 58–69 dorsal scales between the levels of the axilla and groin (versus no such line and 42–57 dorsals in *S. millepunctatus*). *Sphaerodactylus poindexteri* females have a HL/SVL ratio of 0.272–0.273 and have 63–72 dorsal scales between the levels of the axilla and groin (versus HL/SVL ratio 0.214–0.247 and 42–57 dorsals in *S. millepunctatus*). *Sphaerodactylus exsul*, of the West Indian *S. notatus* species group, has 22–30 dorsal scales between the levels of the axilla and groin (versus 42–57 dorsals in *S. millepunctatus*). *Sphaerodactylus glaucus* (not placed in any species group) has smooth dorsal scales (versus keeled dorsals in *S. millepunctatus*). *Sphaerodactylus alphus* and *S. rosaurae*, of the predominately

Plate 23. *Sphaerodactylus millepunctatus.* FMNH 282793, adult male, SVL = 27 mm. Gracias a Dios: confluence of ríos Sikiatingni and Warunta.

West Indian *S. scaber* species group, have 1–3 middorsal rows of granular scales sharply and distinctively differentiated from the larger surrounding dorsal scales (versus all middorsals of similar size in *S. millepunctatus*).

Illustrations (Plate 23).—Guyer and Donnelly, 2005 (adult); Harris and Kluge, 1984 (snout, supracaudal, and subcaudal scales); Köhler, 2000 (adult), 2001b (adult), 2003a (adult), 2008 (adult); McCranie and Hedges, 2012 (adult); McCranie et al., 2006 (adult); Savage, 2002 (adult [from Matamoros only], adult and juvenile dorsal pattern, escutcheon scale pattern [fig. 10.26b only; reprinted from Harris and Kluge, 1984]); Taylor, 1956b (escutcheon scale pattern).

Remarks.—Harris and Kluge (1984) provided a systematic revision of *Sphaerodactylus millepunctatus*, which also included the then unrecognized *S. continentalis, S. guanajae,* and *S. leonardovaldesi.* The Honduran specimen illustrated in Savage (2002, plate 295) and stated to be from "Cortés: Matamoros, about 10 m" should read "Olancho: Matamoros, 150 m."

Natural History Comments.—*Sphaerodactylus millepunctatus* is known from near sea level to 190 m elevation in the Lowland Moist Forest formation. This predominately diurnal species is probably active throughout the year, with known months of collection including from May to July and September to November. The species appears most active for about the first 1–4 hours after sunrise. *Sphaerodactylus millepunctatus* can be especially common under palm frond debris of collapsed roofs of *champas.* Individuals are also commonly seen climbing or descending poles supporting the palm frond roofs of standing *champas.* It also is active on leaf litter surface and can be seen darting under leaf litter and logs well away from human habitation, as well as in edificarian situations, including inside houses and under piles of yard debris. Another was active at dawn and darted into roots of a small plant on a steep riverbank in an attempt to escape. Searching for that tiny lizard uncovered a subadult of the snake *Bothrops asper* (Garman, 1884b: 124). Nothing has been reported on reproduction in this

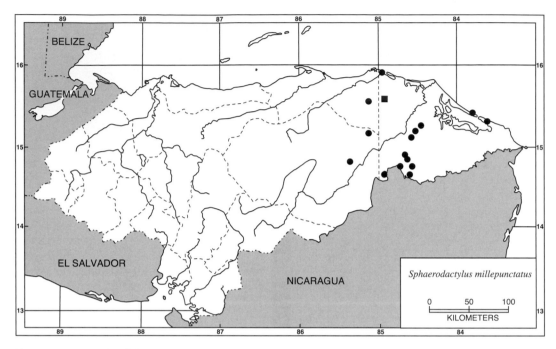

Map 20.   Localities for *Sphaerodactylus millepunctatus*. Solid circles denote specimens examined and the solid square represents an accepted record.

species, but like other species of *Sphaero-dactylus*, females probably deposit multiple clutches of a single egg, often in communal nests. Numerous eggs and their shells of *S. millepunctatus* were found under palm fronds of collapsed *champa* roofs. Those eggs and shells were found from May to July and in November. Feeding data has also not been reported for this species, but it likely feeds on small insects, including ants.

Etymology.—The name *millepunctatus* is derived from the Latin words *mille* (a thousand) and *punctum* (spot) and the Latin suffix *-atus* (provided with). The name alludes to the spotted pattern of the syntypes.

Specimens Examined (31 [0]; Map 20).—**COLÓN**: near Barranco, UMMZ 58408. **GRACIAS A DIOS**: Awasbila, USNM 570171; Bachi Kiamp, FMNH 282792, USNM 579985–86; Bodega de Río Tapal-wás, USNM 570172–73; Caño Awalwás, UF 140810, 150310–11; Cauquira, UF 150304;

Concho Kiamp, FN 257059 (still in Honduras because of permit problems); Dos Bocas, FMNH 282793–96; Leimus (Río Warunta), USNM 573094; Palacios, BMNH 1985.1280; Rus Rus, UF 150302, USNM 570174–76; Sadyk Kiamp, FN 257001 (still in Honduras because of permit problems); San San Hil Kiamp, USNM 570177–78; Tapalwás, USNM 570179; Yahurabila, USNM 573095. **OLANCHO**: Matamoros, SMF 79852–53, USNM 570191; confluence of ríos Sausa and Wampú, USNM 570192.

Other Records (Map 20).—**GRACIAS A DIOS**: Baltiltuk, UNAH 5505, 5515 (Cruz Díaz, 1978; specimens now lost).

Sphaerodactylus poindexteri McCranie and Hedges, 2013a

Sphaerodactylus millepunctatus: Köhler, 1996c: 25; Köhler, 1998b: 141; Monzel, 1998: 156 (in part); Powell, 2003: 36; McCranie et al., 2005: 80 (in part).

Spaeodactylus [sic] *millepunctatus*: Lundberg, 2000: 8.

Sphaerodactylus continentalis: McCranie and Hedges, 2012: 72 (in part).

Sphaerodactylus poindexteri McCranie and Hedges, 2013a: 42 (holotype, FMNH 283685; type locality: "near Trade Winds on the east coast of Isla de Utila, 16.102567°, –86.883117°, Islas de la Bahía, Honduras, 3 m elev."); McCranie and Valdés Orellana, 2014: 45; Solís et al., 2014: 132; McCranie, 2015a: 366.

Geographic Distribution.—*Sphaerodactylus poindexteri* occurs only on Utila Island in the Bay Islands, Honduras.

Description.—The following is based on three females (FMNH 283682, 283684–85). *Sphaerodactylus poindexteri* is a tiny lizard (maximum recorded SVL 30 mm [FMNH 283684]); dorsal surface of head covered by small scales, at least some scales on snout and in parietal region weakly keeled; 10–11 (10.3 ± 0.6) snout scales; 4 supralabials and 3 infralabials to level below anterior half of eye; superciliary spine at level slightly anterior to mideye; moveable eyelid absent; pupil circular; rostral with median cleft and small posterior notch; 2 supranasals, anterior largest, anterior separated medially from counterpart on other side by 1–2 smaller internasals; dorsal surface of body covered by slightly larger scales than those of head, scales essentially homogeneous, keeled, juxtaposed to imbricate, 63–72 (68.0 ± 4.6) between levels of axilla and groin; ventral scales larger than dorsal scales, smooth, flat, imbricate, 34–44 (39.7 ± 5.1) between levels of axilla and groin; 58–63 (60.5 ± 3.5) scales around midbody in two; 9–11 (9.5 ± 0.8) narrow subdigital lamellae on Digit IV of hind limb, 7–8 (7.2 ± 0.4) on Digit IV of forelimb, 32–37 (33.7 ± 2.9) combined lamellae on Digit IV of fore- and hind limb; retractile claw located laterally between enlarged terminal lamella and several smaller scales; webbing absent between digits; subcaudal scales aligned in single median row; femoral and precloacal pores absent in both sexes; SVL 22.0–29.6 (25.2 ± 4.0) mm; HW/SVL 0.11–0.12; HL/SVL 0.27; HD/SVL 0.08–0.09; TAL/SVL 0.90 in one; SEL/EEL 1.03–1.04.

Color in life of the female holotype (FMNH 283685; Plate 24) was described by McCranie and Hedges (2013a: 43): "dorsal ground color of body Ground Cinnamon (239) with scattered Vandyke Brown (221) small spots; dorsal surface of head Ground Cinnamon with Vandyke Brown postorbital stripe and scattered spots; Vandyke Brown occipital spot present; paired Vandyke Brown nuchal blotches present, blotches outlined by Tawny Olive (223D); dorsal surfaces of limbs Ground Cinnamon with golden brown spot on knee and band on shank; dorsal surface of basal part of tail (note that when color in life was recorded and the specimen was photographed, the tail was complete for about two-thirds of its original length; tail currently broken and lost at base) Ground Cinnamon with Vandyke Brown mottling, remainder of dorsal surface of tail to broken point Vinaceous Pink (221C); venter of head and body pale brown with scattered brown flecking; subcaudal surface Vinaceous Pink distal to pale brown base; iris with golden yellow ring around Vandyke Brown pupil."

Color in alcohol of the holotype (FMNH 283685) was described by McCranie and Hedges (2013a: 43–44): "dorsal surfaces brown without scattered darker brown scales on body; snout medium brown with indistinct brown postnasal stripe extending nearly to orbit; supraocular scales dark brown, top of head posterior to that point medium brown with indistinct reticulated pattern of brown lines; postocular stripe dark brown; paired dark brown nuchal blotches present, separated medially by five pale brown scales; dorsal surfaces of limbs similar in color to that of dorsal surface of body, but with distinct pale brown dorsal spot on knee and pale brown dorsal band on

Plate 24.   *Sphaerodactylus poindexteri.* FMNH 283685, adult female, SVL = 24 mm. Islas de la Bahía: Utila Island, east coast near Tradewinds.

shank; supralabials pale brown with dark brown flecking; mental and infralabials pale brown, without distinct markings; gular region pale brown with brown flecking; belly cream with brown flecking on scale edges; no pale pelvic line evident." Those authors also said (p. 44) "one paratype (FMNH 283682) has dark brown blotches on the nape, but they are smaller than those in the holotype, whereas dark nape blotches are absent in the other paratype (FMNH 283684). One paratype (FMNH 283684) has widely separated dark brown dorsal spots, most of which are confined to one scale; also, that specimen has distinct dark brown lines on the posterior end of the head and in the nuchal region."

Diagnosis/Similar Species.—The combination of lacking moveable eyelids, having the dorsal surface of the head covered with granular scales, having the claws displaced laterally by terminal expanded subdigital lamella, and having a superciliary spine distinguishes *Sphaerodactylus poindexteri* from all other Honduran lizards, except the other *Sphaerodactylus. Sphaerodactylus*

poindexteri differs from all other members of the genus using molecular techniques, as well as by morphological data. Also, *S. continentalis* differs in having 64–80 scales around midbody and usually having dark brown dorsal spots larger than a single scale in size (versus 58–63 scales around midbody and dark dorsal spots confined to single scale per spot in *S. poindexteri*). *Sphaerodactylus dunni* has the superciliary spine located posterior to the level of the mideye, has the third supralabial at a level below the anterior half of the eye, and has alternating subcaudals (versus superciliary spine at level varying from mideye to anterior of mideye, fourth supralabial at level below anterior half of eye, and subcaudals in single row in *S. poindexteri*). *Sphaerodactylus guanajae* has a distinct short pale pelvic stripe that frequently dorsally crosses the base of tail to connect with its counterpart on the other side (versus no such pelvic stripe in *S. poindexteri*). *Sphaerodactylus leonardovaldesi* has a short dorsal pale pelvic line (versus no pale pelvic line in *S. poindexteri*). *Sphaerodactylus millepuncta-*

tus has 42–57 dorsal scales between the levels of the axilla and groin (versus 63–72 in *S. poindexteri*). *Sphaerodactylus exsul*, of the *S. notatus* West Indian species group, has the third supralabial at a level below the anterior half of the eye and has 22–30 dorsal scales between the levels of the axilla and groin (versus usually fourth supralabial at level below anterior half of eye and 63–72 dorsal scales in *S. poindexteri*). *Sphaerodactylus glaucus* has smooth dorsal scales (versus keeled dorsals in *S. poindexteri*). *Sphaerodactylus alphus* and *S. rosaurae*, of the predominately West Indian *S. scaber* species group, have 1–3 middorsal rows of granular scales sharply and distinctively differentiated from the larger surrounding dorsal scales (versus all middorsal scales of similar size in *S. poindexteri*).

Illustrations (Plate 24).—Köhler, 1998b (adult; as *S. millepunctatus*), 1999b (subcaudal scales; as *S. millepunctatus*), 2001b (subcaudal and subdigital scales; as *S. millepunctatus* and *Sphaerodactylus*, respectively), 2003a (subdigital scales; as *Sphaerodactylus*), 2008 (subdigital scales; as *Sphaerodactylus*), McCranie and Hedges, 2013a (adult); McCranie et al., 2005 (adult [from Utila Island only], subdigital scales; as *S. millepunctatus* and *Sphaerodactylus*, respectively); 2006 (subdigital scales; as *S. millepunctatus*); Powell, 2003 (adult; as *S. millepunctatus*).

Remarks.—McCranie and Hedges (2013a) reported results of a molecular analysis of the *Sphaerodactylus millepunctatus* population from Utila Island, which demonstrated that population represented a distinct cryptic species. The *S. poindexteri* clade formed a sister clade to a lineage containing *S. continentalis*, *S. guanajae*, *S. leonardovaldes*i, and *S. millepunctatus* (McCranie and Hedges, 2013a). Those species in that clade, plus *S. dunni*, are the Honduran members of the *S. millepunctatus* species group.

Despite the very few specimens of *Sphaerodactylus poindexteri* in museum collections, the species is quite common under the correct conditions on Utila Island. Apparently no one has tried to collect a series of these geckos.

Natural History Comments.—*Sphaerodactylus poindexteri* is known from near sea level to 10 m elevation in the Lowland Moist Forest. The type series was collected in September by raking through Sea Grape (*Cocoloba uvifera*) and hardwood leaf litter at two nearby sites during one morning. One of those sites is located inland from the beach and an adjacent area of uplifted coral rock. The Sea Grape trees grow adjacent to the rocky area. The lizard's darting habits make them difficult to capture without damaging the extremely fragile skin. Köhler (1996c: 26) stated that this species "inhabits leaf litter." *Sphaerodactylus poindexteri* is likely active throughout the year. Nothing has been reported on reproduction in this species, but like other species of *Sphaerodactylus*, females probably deposit multiple clutches of a single egg, often in communal nests. Feeding data has also not been reported for this species, but it likely feeds on small insects, including ants.

Etymology.—The specific name *poindexteri* is a patronym honoring James A. Poindexter of the USNM Support Center in Suitland, Maryland. James has been, and continues to be, extremely helpful over the years in providing copies of important, difficult-to-obtain literature and photographs for my work on the amphibians and reptiles of Honduras.

Specimens Examined (7 [1]; Map 21).—**ISLAS DE LA BAHÍA**: Isla de Utila, Don Quickset Pond, SMF 779981; Isla de Utila, trail to Rock Harbor, SMF 77997; Isla de Utila, east coast near Trade Winds, FMNH 283682, 283684–85; Isla de Utila, 3 km N of Utila Town, SMF 77118–19.

Other records.—**ISLAS DE LA BAHÍA:** Isla de Utila, 3 km N of Utila Town, UNAH (Köhler, 1996c).

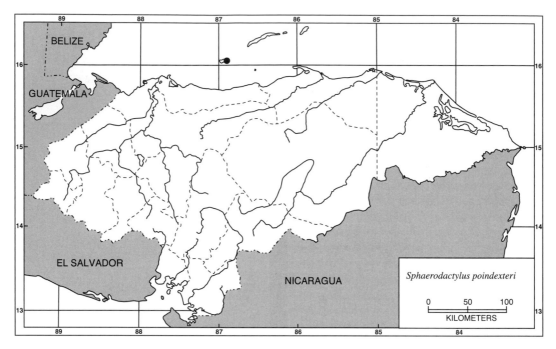

Map 21.   Localities for *Sphaerodactylus poindexteri*. Solid circle denotes specimens examined from nearby localities.

Sphaerodactylus rosaurae Parker, 1940

Sphaerodactylus rosaurae Parker, 1940: 264 (holotype, BMNH 1938.10.4.1; type locality: "Helene Island [= Santa Elena, Isla de Roatán], Bay Islands, Honduras"); M. A. Smith, 1941: 34; Wermuth, 1965: 172; Meyer, 1969: 210; Peters and Donoso-Barros, 1970: 253; Meyer and Wilson, 1973: 13; Wilson and Hahn, 1973: 106 (in part); Schwartz, 1975: 15 (in part); MacLean et al., 1977: 4; Schwartz and Garrido, 1981: 20 (in part); Köhler, 1994a: 4; Kluge, 1995: 23; Kluge and Nussbaum, 1995: 20; Köhler, 1995b: 102; Köhler, 1998d: 374, 379, 382 (in part); Monzel, 1998: 156 (in part); Köhler, 2000: 52; Lundberg, 2000: 3; Grismer et al., 2001: 135 (in part); Wilson et al., 2001: 137 (in part); Köhler, 2003a: 78; Powell, 2003: 36; Wilson and McCranie, 2003: 59 (in part); McCranie et al., 2005: 82 (in part); Wilson and Townsend, 2006: 105 (in part); Köhler, 2008: 80; Frazier et al., 2011: 391; Daza and Bauer, 2012: 24; McCranie and Hedges, 2013a: 50; McCranie and Valdés Orellana, 2014: 45; Solís et al., 2014: 132; McCranie, 2015a: 366.

Sphaeodactylus [sic] *rosaurae*: Lundberg, 2001: 26.

Geographic Distribution.—*Sphaerodactylus rosaurae* is known to occur on the islands of Barbareta, Morat, Roatán, and Utila and on Cayo Cochino Menor, all in the Bay Islands, Honduras.

Description.—The following is based on 12 males (FMNH 283665, 283667, USNM 520261, 570199, 570201, 570205–06, 570209–13) and 14 females (FMNH 283662, 283667, 283669–70; KU 203128–29; USNM 520259, 520262, 570196, 570198, 570200, 570204, 570207–08). *Sphaerodactylus rosaurae* is a small lizard (maximum recorded SVL 39 mm [USNM 570196, a female]); dorsal surface of head covered by small to granular scales, scales conical in parietal region; 8–12 (10.1 ± 1.0) snout scales; 3–4 supralabials to level below anterior portion of eye; 4 infralabials to level below mideye; superciliary spine at level of about mideye; moveable eyelid absent; pupil circular; rostral with median cleft and small posterior notch; enlarged supranasal single, separated from counterpart on other side by 1–2 (rarely 2) smaller internasals; dorsal surface of body with median zone of granules (1–3 granules wide) and larger, keeled, slightly imbricate to juxtaposed scales lateral to that granule zone, 21–30 (26.0 ± 2.4) scales along innermost row of larger scales between levels of axilla and groin; ventral scales slightly smaller than largest dorsal scales, smooth, flat, imbricate, 24–35 (30.7 ± 2.4) between levels of axilla and groin; usually with zone of lateral scales noticeably smaller than ventral or dorsal scales; 30–53 (46.8 ± 5.1) scales around midbody; 10–14 (11.3 ± 1.5) narrow subdigital lamellae on Digit IV of hind limb, 8–11 (9.5 ± 0.6) on Digit IV of forelimb, 39–45 (41.8 ± 1.8) combined Digit IV lamellae in 23; retractile claw located laterally between enlarged terminal lamella and several smaller scales; webbing absent between digits; subcaudal scales aligned in single median row; femoral and precloacal pores absent in both sexes; male escutcheon slightly hypertrophied, about 12–13 scales long and about 10–15 scales wide; SVL 33.3–38.1 (35.7 ± 1.6) mm in males, 31.5–38.5 (35.1 ± 2.1) mm in females; HW/SVL 0.12–0.14 in males, 0.11–0.13 in females; HD/SVL 0.09–0.12 in males, 0.09–0.11 in females; HL/SVL 0.23–0.25 in both sexes; TAL/SVL 0.67–1.06 in seven males, 0.67–0.99 in seven females; SEL/EEL 1.12–1.45 in males, 1.10–1.54 in females.

Color in life of an adult male (USNM 570201): dorsal surface of body Olive (30) with scattered black spots; dorsal surface of head rusty brown; dorsal surfaces of fore- and hind limb Olive with black spots; dorsal surface of tail pale brown with dark brown mottling; lateral surface of head anterior to eyes pale brown, except supralabials with alternating dark brown and cream bars; cream, dark brown outlined postsubocular band present, band extends onto throat region; area around ear opening rust brown with yellowish brown mottling; ventral surface of body gray; subcaudal surface cream with dark brown reticulations; toe pads white; iris dark brown with pale brown rim.

Color in alcohol: males have scales of dorsal surfaces of body and tail pale olive brown without distinct markings, females have those dorsal surfaces pale olive brown to pale brown, with numerous dark brown spots; males have dorsal surface of head pale brown with dark brown spots, except in largest specimens, females have dorsal surface of head pale brown, usually with dark brown mottling and striping; ventral surfaces of head and body cream, lightly flecked with brown in both sexes; chin and throat of females with brown mottling and lines, those of males with reduced pattern; subcaudal surface of both sexes more heavily flecked with brown than posterior portion of body. Juveniles of both sexes have the dorsal surfaces of the body and tail pale brown with dark brown crossbands.

Diagnosis/Similar Species.—The combination of lacking moveable eyelids, having the dorsal surface of the head covered with small to granular scales, having the claws displaced laterally by expanded terminal subdigital lamella, and having a superciliary spine distinguishes *Sphaerodactylus rosaurae* from all other Honduran lizards, except the remaining *Sphaerodactylus*. *Sphaerodactylus rosaurae* differs from those species of *Sphaerodactylus*, except *S. alphus*, by having 1–3 middorsal rows of granular scales sharply and distinctively differentiated from the larger surrounding dorsal scales. *Sphaerodactylus alphus* has a white occipital spot and is known to reach a SVL of 41.2 mm (versus no white occipital spot and maximum known SVL 38.5 mm in *S. rosaurae*). Molecular studies also recovered

Plate 25.   *Sphaerodactylus rosaurae.* FMNH 282674, adult male. Islas de la Bahía: Roatán Island, near Pollytilly Bight.

those two species as separate evolutionary species.

Illustrations (Figs. 21, 34; Plates 25, 26).—Köhler, 1998d (adult), 2000 (adult), 2003a (adult, dorsal scales), 2008 (adult, dorsal scales); McCranie et al., 2005 (adult, middorsal scales); Wilson and Hahn, 1973 (adult dorsal pattern variation [except Guanaja specimen], juvenile dorsal pattern).

Remarks.—Wilson and Hahn (1973: 109) stated that "it is entirely possible that *S. rosaurae* is conspecific with *S. copei*" Steindachner (1867: 18) of the West Indian *S. scaber* species group. Schwartz (1975)

Plate 26.   *Sphaerodactylus rosaurae.* FMNH 282675, juvenile. Islas de la Bahía: Roatán Island, near Pollytilly Bight.

and Schwartz and Garrido (1981) concluded that *S. rosaurae* was a species distinct from *S. copei*, but was a member of their proposed *S. copei* (= *S. scaber*) species group. Molecular and morphological studies of Honduran Bay Island *S. rosaurae* and the newly described *S. alphus* confirmed validity of the Bay Island species and their close relationships with *S. copei* of the *S. scaber* species group (McCranie and Hedges, 2013a).

All LSUMZ specimens of *Sphaerodactylus rosaurae* (and *S. alphus*) listed by Wilson and Hahn (1973) lost their tags and were essentially destroyed when the alcohol in their jar evaporated (see list in Other Records). As of January 2009, only 27 soft, tagless specimens, some in several pieces, remained of the 32 original specimens in that jar.

Natural History Comments.—*Sphaerodactylus rosaurae* is known from near sea level to 20 m elevation in the Lowland Moist Forest formation. This diurnal species is probably active throughout the year. Wilson and Hahn (1973: 109) reported that specimens (apparently also including *S. alphus*) were collected from "stilt rootlets of coconut palm stumps, under rotten palm logs, inside hollow standing thorn palms, under palm fronds, in the axils of fronds on coconut palms, in crevices of rock coral, in abandoned roofs of huts, and under rocks." Subsequent observations confirm that this species can be found in just about any terrestrial situation, as well as on trunks of mangrove trees, and at any time of the year. However, the species appears to be most easily observed in and under dead Sea Grape leaves (*Cocoloba uvifera*) and under piles of dead coconut palm fronds (*Cocos* sp.). Nothing has been reported on reproduction, but like other species of *Sphaerodactylus*, females probably deposit multiple clutches of a single egg, often in communal nests. Also, nothing has been reported on diet in *S. rosaurae*, but it likely feeds on small insects, including ants. *Sphaerodactylus rosaurae* was for a long time known from Utila Island by few specimens, but my two most recent trips to Utila (2012) targeting this and another species of *Sphaerodactylus*, revealed it to be a common species in Sea Grape leaf litter and under coconut palm debris on the east coast. The reasons for these discrepancies are not understood.

Etymology.—The name *rosaurae* is for the yacht Rosaura, owned by Lord Moyne, the vessel used on the cruise on which the holotype was collected. Moyne (1938) also wrote a book that included some of his adventures on that cruise.

Specimens Examined (80, 2 C&S [28]; Map 22).—**ISLAS DE LA BAHÍA**: Isla Barbareta, near southwest shore, USNM 520258–62; Isla de Roatán, Coxen Hole, FMNH 34542; Isla de Roatán, near Diamond Rock, USNM 570196–97; Isla de Roatán, Flowers Bay, USNM 570198; Isla de Roatán, between Flowers Bay and West End Point, USNM 570199; Isla de Roatán, about 3.2 km W of French Harbor, UF 28557 (C&S), 28558–59, UMMZ 152732 (C&S; formerly LSUMZ 22385); Isla de Roatán, 2.5 km N of, 3 km E of Oak Ridge, UF 91320; Isla de Roatán, E of Oak Ridge, MCZ R150938–42; Isla de Roatán, near Oak Ridge, TCWC 52422–25, UTA R-10700–06; Isla de Roatán, Palmetto Bay, FMNH 283665, 283667, 283669–70; Isla de Roatán, 1 km E of Pollytilly Bight, FMNH 282674–77; Isla de Roatán, Port Royal, USNM 578838; Isla de Roatán, about 4.8 km W of Roatán, UF 28536; Isla de Roatán, about 3.2 km W of Roatán, CM 57184; Isla de Roatán, near Roatán, CM 64523, UF 28488, 28496; Isla de Roatán, Rocky Point, USNM 570200; Isla de Roatán, near Sandy Bay, USNM 570201–02; Isla de Roatán, Sandy Bay, KU 203128–31; Isla de Roatán, W of Santa Elena near mangroves dividing Isla de Roatán, MCZ R150936–37; Isla de Roatán, near Santa Elena, TCWC 52426; Isla de Roatán, Santa Elena, BMNH 1938.10.4.1, UMMZ 142648 (2), UTA R-10699; Isla de

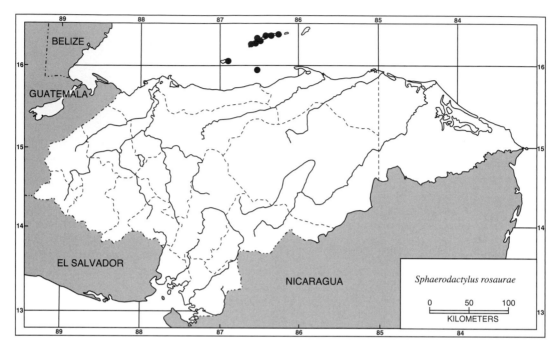

Map 22.   Localities for *Sphaerodactylus rosaurae*. Solid circles denote specimens examined.

Roatán, 6.6 km E of West End Point, MVZ 263859; Isla de Roatán, West End Point, USNM 570203–10; Isla de Roatán, West End Town, USNM 570211–13; "Isla de Roatán," UF 149595, USNM 570214, UTA R-55247–51; Isla de Utila, east coast near Trade Winds, FMNH 283662, 283667, 283683; Isla de Utila, 2 km N of Utila, SMF 81207.

Other Records.—**ISLAS DE LA BA-HÍA**: Cayo Cochino Pequeña, UNAH (2); Isla de Roatán, about 3.2 km W of French Harbor, LSUMZ 22385–89 (see Remarks for all LSUMZ specimens), UF 28557 (C&S); Isla de Roatán, Gibson Bight, LSUMZ 33794–95; Isla de Roatán, near Port Royal Harbor, LSUMZ 33802–05; Isla de Roatán, 0.5–1.0 km N of Roatán, LSUMZ 21939–41; Isla de Roatán, about 4.8 km W of Roatán, LSUMZ 22352–62; Isla de Roatán, Sandy Bay, LSUMZ 33798–800; Isla de Utila, Utila LSUMZ 22297–98;

Isla Morat, LSUHC 3712, LSUPC 4789–93 (Grismer et al., 2001).

Infraorder Neoanguimorpha Vidal and Hedges, 2009

The infraorder Neoanguimorpha contains the superfamilies Anguioidea Gray (1825: 201; as Anguidae), Helodermatoidea Gray (1837: 132; as Helodermidae), and Xeno-sauroidea Cope (1867:3 22; as Xenosaur-idae; Conrad, 2008: 113 erroneously gave 1886 as the date of that Cope publication). Of these three superfamilies, only the Anguioidea is represented in Honduras.

Superfamily Anguioidea Gray, 1825

Anguioidea includes the families Angui-dae, Anniellidae Cope (1865c: 230), and Diploglossidae (see Vidal and Hedges, 2009: 133). There are no members of the Anniel-lidae in Honduras, but the remaining two families occur in the country. Externally, the Honduran members of this superfamily can be distinguished from all remaining Hon-

Figure 42.   Uniform, striated, cycloid dorsal scales. *Diploglossus bivittatus* USNM 563521 from Chupocay, Intibucá.

duran lizards by the combination of having moveable eyelids and the dorsal surface of the head with enlarged scales or plates, including 2 pairs of scales (anterior and posterior internasals) between the rostral scale and the first unpaired plate. As far as is known, all Honduran members of this superfamily are viviparous.

Remarks.—Males and females of Honduran *Abronia* are similar in size to each other (but very few specimens are known), males of *Mesaspis* are significantly larger than females, and too few male specimens of Honduran *Diploglossus* are known to understand sexual size differences.

KEY TO HONDURAN FAMILIES OF THE
SUPERFAMILY ANGUIOIDEA

1A. Body covered by uniform, striated cycloid scales (Fig. 42); distinct ventrolateral folds absent . . . .
. . . . Diploglossidae (p. 149)
1B. Body covered with rectangular-shaped, weakly keeled scales on dorsum (Fig. 43), or if body with oblong scales, scales strongly keeled (Fig. 44); distinct ventrolateral folds present (Fig. 45). . . . . . .
. . . . Anguidae (p. 135)

CLAVE PARA LAS FAMILIAS HONDUREÑAS DE LA
SUPERFAMILIA ANGUIOIDEA

1A. Cuerpo cubierto con escamas uniformes en tamaño, cicloideas y estriadas (Fig. 42); un pliegue ventrolateral distintivo ausente. . .
. . . . Diploglossidae (p. 149)
1B. Cuerpo cubierto con escamas en forma rectangular y debilmente quilladas en el dorso (Fig. 43), o si el cuerpo está cubierto con escamas oblongas, las escamas están fuertemente quilladas (Fig. 44), un pliegue ventrolateral distintivo presente (Fig. 45) . . . .
. . . . Anguidae (p. 135)

Family Anguidae Gray, 1825

This family of lizards is distributed in the Western Hemisphere in largely disjunct populations from the eastern, central, and western USA southward through Mexico (including Baja California) to western Panama. In the Eastern Hemisphere, it occurs

Figure 43.   Rectangular-shaped, weakly keeled dorsal scales. *Abronia salvadorensis*. KU 195560 from Zacate Blanco, Intibucá.

in western Eurasia and northwestern Africa, southeastern Asia, and the islands of the Sunda Shelf. Ten genera containing at least 65 named species are included in this family. Two genera containing three species are known from Honduras.

KEY TO HONDURAN GENERA OF THE FAMILY ANGUIDAE

1A. Distinct ventrolateral fold present (Fig. 45); dorsal scales oblong-shaped, distinctly keeled middor-

Figure 44.   Oblong, strongly keeled dorsal scales. *Mesaspis moreletii*. USNM 578861 from La Esperanza, Francisco Morazán.

Figure 45.　Ventrolateral fold present on each side, with small and/or granular scales present above each fold. *Mesaspis moreletii.* USNM 578861 from La Esperanza, Francisco Morazán.

sally, keels on each scale in contact with, or nearly in contact with keel on following scale (Fig. 44) . . . . . .
. . . . *Mesaspis* (p. 144)

1B.　Ventrolateral fold weak or absent (at least in Honduran species), but ventrolateral smaller or granular scales present, also some skin visible between some of those smaller scales (Fig. 46); dorsal scales rectangular-shaped, also some scales weakly keeled (Fig. 43)
. . . . *Abronia* (p. 137)

CLAVE PARA LOS GÉNEROS HONDUREÑOS DE LA FAMILIA ANGUIDAE

1A.　Un distintivo pliegue ventrolateral presente (Fig. 45); escamas dorsales oblongas y distivamente quilladas mediodorsalmente, quillas en cada escama en contacto o casi en contacto con la quilla en la próxima escama (Fig. 44) . . . *Mesaspis* (p. 144)

1B.　Un pliegue ventrolateral débil o ausente o (al menos en las especies de Honduras) con escamas ventro-

laterales granulares o pequeñas presentes, también un poco de piel visible entre las escamas pequeñas (Fig. 46); escamas dorsales en forma rectangular, algunas escamas débilmente quilladas (Fig. 43)
. . . . *Abronia* (p. 137)

Genus *Abronia* Gray, 1838b

Abronia Gray, 1838b: 389 (type species: *Gerrhonotus deppii* Wiegmann, 1828: col. 379, by subsequent designation of Tihen, 1949: 587).

Geographic Distribution and Content.— This genus occurs in disjunct populations from southwestern Tamaulipas, Mexico, to western Honduras, and northern El Salvador. Twenty-nine named species are recognized, two of which are known from Honduras.

Remarks.—Phylogenetic analyses of the genus *Abronia* were provided by Campbell and Frost (1993) and Campbell et al. (1998). However, the Pyron et al. (2013) study, using more species, recovered a paraphyletic *Abronia*.

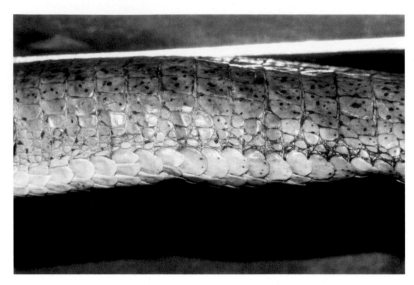

Figure 46.   Ventrolateral fold weak or absent on each side, but smaller or granular scales present, with some skin visible between some of those smaller scales. *Abronia salvadorensis*. KU 195560 from Zacate Blanco, Intibucá.

Scale terminology used herein follows Campbell and Frost (1993).

Etymology.—The name *Abronia* is formed from the Greek word *habros* (pretty, graceful, dainty, tender, delicate) and Greek suffix *-ia* (pertaining to), and alludes to "the generally graceful and attractive appearance of these lizards" (Campbell and Frost, 1993: 99).

KEY TO HONDURAN SPECIES OF THE GENUS *ABRONIA*

1A. Five occipital scales present (Fig. 47); dorsal surfaces of body and tail with pale crossbars present, except pale crossbars faint in adults .....
.... *montecristoi* (p. 138)

1B. One to 4, usually 3, occipital scales present (Fig. 48); dorsal surfaces of body and tail pale brown with darker brown crossbars in life and preservative.... *salvadorensis* (p. 142)

CLAVE PARA LAS ESPECIES HONDUREÑAS DEL GÉNERO *ABRONIA*

1A. Con 5 escamas occipitales (Fig. 47); superficies dorsal del cuerpo y la cola con bandas transversales pálidas presentes, excepto borrosas en adultos ....
.... *montecristoi* (p. 138)

1B. De 1 a 4, usualmente 3, escamas occipitales (Fig. 48); superficies dorsal del cuerpo y la cola parda pálida, con bandas transversales oscuras en vida y en alcohol .....
.... *salvadorensis* (p. 142)

Abronia montecristoi Hidalgo, 1983

Abronia montecristoi Hidalgo, 1983: 6 (holotype, KU 184046; type locality: "Hacienda Montecristo, Metapán, Cordillera de Alotepeque-Metapán, Departamento de Santa Ana, El Salvador, 2250 meters"); McCranie and Wilson, 1999: 127; Köhler, 2003a: 57; Wilson and McCranie, 2003: 59; Wilson and McCranie, 2004b: 43; Köhler et al., 2005: 95; McCranie, 2005: 20; Köhler, 2008: 61; Solís et al., 2014: 129; McCranie, 2015a: 366.

Geographic Distribution.—*Abronia montecristoi* occurs at moderate and intermedi-

Figure 47.   Five occipital scales present (outlined). *Abronia montecristoi*. USNM 520001 from Quebrada Grande, Copán.

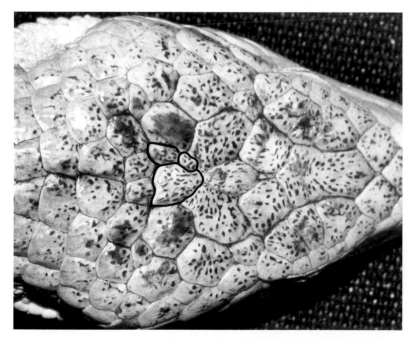

Figure 48.   Less than five occipital scales present (outlined). *Abronia salvadorensis*. USNM 520002 from Zacate Blanco, Intibucá.

Plate 27.   *Abronia montecristoi.* USNM 520001, adult female, SVL = 87 mm. Copán: Quebrada Grande.

ate elevations of northwestern El Salvador on the Pacific versant and western Honduras on the Atlantic versant. It is known in Honduras only from a single locality in northwestern Copán.

Description.—The following is based on one female (USNM 520001). *Abronia montecristoi* is a moderately large lizard (maximum recorded SVL 90 mm [Hidalgo, 1983]; 87 mm in only Honduran specimen [USNM 520001]); dorsal head scales enlarged, smooth; 4 internasal scales, with posterior pair about 2 times larger than anterior pair; supranasal scales not expanded; 1 median frontonasal scale, separated from single frontal scale by paired prefrontal scales; prefrontal scales not contacting anterior superciliary scale; 1 canthal scale; 5 medial supraoculars; parietal scales not contacting median supraocular; 5 occipital scales; posterolateral head scales not convex nor knobby; supra-auricular scales not protuberant; anterior superciliary in broad contact with cantholoreal on both sides; 1 loreal; 10–11 supralabials with antepenultimate one contacting subocular and postocular scale series; moveable eyelid present; pupil circular; 4 anterior temporals, lower 2 contacting postoculars, lowermost separated from posteriormost scale of subocular series; 1 row of preauricular scales above ear opening, 2 rows of much smaller scales along lower third of ear opening; postmental larger than mental, not divided; 4 pairs of enlarged chinshields; 10 infralabials, most not elongated; 6 transverse rows of nuchal scales; dorsal body scales large, rectangular-shaped, each scale usually rather weakly keeled or rugose, 32 along dorsal midline between occipital and base of tail, in 14 crossrows between weakly developed ventrolateral folds at midbody; ventral body scales nearly rectangular, flat, smooth, imbricate, 50 along midline between posterior pair of chinshields and cloacal scale, in 12 crossrows at midbody, in 10 crossrows posteriorly, lateral row expanded on posterior portion of body; femoral pores absent; 5 digits on all limbs; 20 subdigital lamellae per side on Digit IV of hind limb; SVL 86.5 mm; HW/SVL 0.15; HL/SVL 0.21; TAL/SVL 1.73.

Color in life (adult female, USNM 520001; Plate 27): "dorsal surface of body Antique Brown (37) with narrow, paler, irregular cross-bands colored Cinnamon

(39) dorsally, grading to Buff-Yellow (53) laterally; top of head Cinnamon (39), most scales on top of head with set of small black punctations; side of head Cinnamon (39) without black punctations; forelimbs Cinnamon (39); hind limbs Tawny (38); dorsal and lateral surfaces of tail Antique Brown (37) with Cinnamon (39) cross-bands; ventral surfaces of head and body Cinnamon (39); iris pale greenish silver with black reticulations" (McCranie and Wilson, 1999: 127).

Color in alcohol: all dorsal surfaces brown, with posterior edges of scales on body and tail slightly paler brown; pale gray, irregular crossbands present on body; dorsal surface of tail with paler brown crossbands anteriorly; ventral surface of head cream colored anteriorly, changing to gray in throat region; belly gray.

Diagnosis/Similar Species.—The combination of having moveable eyelids, the dorsal surface of the head with enlarged scales including 2 pairs of internasals between the rostral and first unpaired scale, and the body covered with large rectangular-shaped scales distinguishes *Abronia montecristoi* from all remaining Honduran lizards, except the other anguids. Species of *Diploglossus* have striated, cycloid, imbricate body scales and lack a ventrolateral fold (versus rectangular-shaped, platelike dorsal body scales and weak ventrolateral fold or small ventrolaterally placed scales present in *A. montecristoi*). *Mesaspis moreletii* has several rows of small and/or granular scales above a distinct ventrolateral fold and has oblong-shaped, distinctly keeled dorsal body scales (versus ventrolateral fold weakly developed and without distinct granules and/or small scales above fold and platelike, rectangular-shaped, weakly keeled or rugose dorsal scales in *A. montecristoi*). *Abronia salvadorensis* has 1–4 (usually 3) occipital scales and has the dorsal surface of the body brown with darker brown crossbands (versus 5 occipital scales and body brown with paler brown crossbands in *A. montecristoi*).

Illustrations (Fig. 47; Plate 27).—Campbell and Frost, 1993 (head scales and pattern, adult); Hidalgo, 1983 (head scales); Köhler, 1996e (head), 2000 (head scales), 2003a (adult), 2008 (adult); Köhler et al., 2005 (adult); Wilson and McCranie, 2004a (adult).

Remarks.—*Abronia montecristoi* is apparently most closely related to the Honduran endemic *A. salvadorensis*. Good (1988: 94–95) and Casas-Andreu and Smith (1990: 323) placed *A. montecristoi* and *A. salvadorensis* in the *A. aurita* species group. Campbell and Frost (1993: 99) placed *A. montecristoi* in the new subgenus *Abaculabronia* and *A. salvadorensis* by itself in the new subgenus *Lissabronia* (p. 100; those authors placed *A. aurita* Cope, 1869: 306, in the also new subgenus *Auriculabronia*). The morphological analysis by Campbell et al. (1998: 232) demonstrated *A. montecristoi* to be the sister species of *A. salvadorensis*; thus, subsequently Campbell et al. (1998) transferred *A. montecristoi* to the rediagnosed subgenus *Lissabronia* (pp. 231–232). Campbell et al. (1998: 231) also placed their new species, *A. frosti*, in the subgenus *Lissabronia*.

Natural History Comments.—*Abronia montecristoi* is known only from 1,370 m elevation in the Lower Montane Wet Forest formation. The single Honduran specimen was collected in July as it was active on the ground beside a hollow log during the day. Nothing has been reported on reproduction or food habits of this species, but like other species of *Abronia*, it likely bears live young and feeds at least on orthopterans.

Campbell and Frost (1993: 55) postulated that *Abronia montecristoi* "may already be extinct" because "no additional specimens have been collected in over 30 years" at its type locality, and at that time its only known locality. Subsequent collection of a single specimen of this species in Copán, Honduras, in 1996 offers hope for the continued existence of this species. The species might also occur in several, currently forested

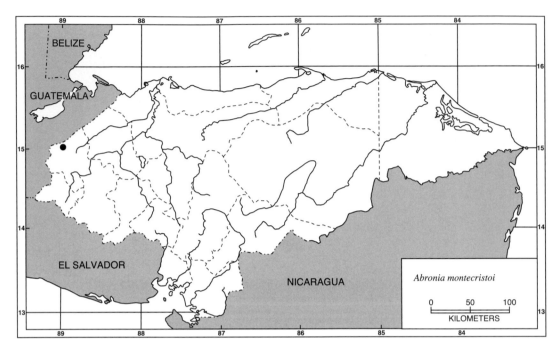

Map 23. Locality for *Abronia montecristoi*. Solid circle denotes single specimen examined, and only specimen known from Honduras.

mountain ranges between the type locality and the Copán locality, although those few remaining forested areas are generally above 2,000 m elevation.

Etymology.—The name *montecristoi* refers to the species' type locality, Hacienda Montecristo.

Specimens Examined (1 [0]; Map 23).— **COPÁN**: Quebrada Grande, USNM 520001.

Abronia salvadorensis Hidalgo, 1983

Abronia salvadorensis Hidalgo, 1983: 1 (holotype, KU 184047; type locality: "Cantón Palo Blanco, 10 km NE Perquín, Cordillera de Nahuaterique, Departamento de Morazán, El Salvador, 1900 meters" [= Palo Blanco, Montaña La Sierra, Departamento de La Paz, Honduras]; see Remarks); Good, 1988: 95; Campbell and Frost, 1993: 43; Campbell et al., 1998: 229; McCranie and Wilson, 1999: 127; Wil-

son et al., 2001: 134; Köhler, 2003a: 60; Wilson and McCranie, 2003: 59; Wilson and McCranie, 2004b: 43; Köhler, 2008: 61; Solís et al., 2014: 129; McCranie, 2015a: 366.

Abronia montecristoi: Wilson, Porras, and McCranie, 1986: 4.

Geographic Distribution.—*Abronia salvadorensis* occurs at intermediate elevations in the Montaña La Sierra and Cordillera de Opalaca in the departments of La Paz and Intibucá, respectively, on both sides of the Continental Divide in southwestern Honduras (see Remarks).

Description.—The following is based on three males (KU 195560–61; UTA R-26108) and two females (KU 184087; USNM 520002). *Abronia salvadorensis* is a moderately large lizard (maximum recorded SVL 111 mm [KU 195560, USNM 520002, a male and female, respectively]); dorsal head scales enlarged, smooth; 4 internasal scales, with posterior pair about 3 times larger than anterior pair; supranasal scales not expanded; 1 median frontonasal scale, separated from single frontal scale by paired prefrontal scales; prefrontal scales not contacting anterior superciliary scale; 1 canthal scale; 5 medial supraoculars; parietal scales not contacting median supraocular; 3 occipital scales; posterolateral head scales not convex or knobby; supra-auricular scales not protuberant; anterior superciliary in broad contact with cantholoreal; 2 loreals; 10–11 supralabials with antepenultimate one contacting subocular and postocular scale series; moveable eyelid present; pupil circular; 4 anterior temporals, lower 2 contacting postoculars, lowermost separated from posterior subocular series; 1 row of preauricular scales above ear opening; postmental larger than mental, not divided; 3 pairs of enlarged chinshields; 8–9 infralabials, posteriormost not elongated; 6 transverse rows of nuchal scales; dorsal body scales large, rectangular-shaped, each scale usually rather weakly keeled or rugose, 31–34 (33.2 ± 1.3) along dorsal midline between occipital and base of tail, in 14–15 crossrows between weakly developed ventrolateral fold at midbody; ventral body scales nearly rectangular, flat, smooth, imbricate, 50–55 (52.6 ± 1.9) along midline between posterior pair of chinshields and cloacal scale, in 12–14 crossrows at midbody, in 11–12 crossrows posteriorly, lateral row slightly expanded on posterior portion of body; femoral pores absent; 5 digits on all limbs; 17–21 (19.1 ± 1.0) subdigital scales per side on Digit IV of hind limb; SVL 94–111 (100.0 ± 9.5) mm in males, 78–111 (94.5) in females; HW/SVL 0.15–0.18 in males, 0.15 in both females; HL/SVL 0.22–0.23 in males, 0.21–0.22 in females; TAL/SVL 1.58–1.60 in two males, 0.93–1.53 in females.

Color in life of KU 195560 (an adult male): "dorsum pale brown with indistinct brown cross-bands; head horn color; chin white; venter dirty white; palms of hands and [soles of] feet yellowish-tan" (Wilson, Porras, and McCranie, 1986: 5; as *Abronia montecristoi*).

Color in alcohol: dorsal surface of body brown, with indistinct darker brown crossbands; dorsal surface of head pale brown to brown, with a few dark brown spots; dorsal surface of tail brown with indistinct darker brown crossbands anteriorly, crossbands not obvious on distal half; ventral surface of head nearly white; ventral surface of body with grayish tinge; subcaudal surface darker gray than belly, especially on distal half.

Diagnosis/Similar Species.—The combination of having moveable eyelids, the dorsal surface of the head with enlarged scales, including 2 pairs of internasal scales between the rostral and first unpaired scale, and the body covered with platelike, rectangular-shaped scales distinguishes *Abronia salvadorensis* from all remaining Honduran lizards, except some other anguids. Species of *Diploglossus* have striated, cycloid, imbricate body scales and lack a ventrolateral fold (versus rectangular-shaped, platelike body scales and a weak ventrolateral fold or small ventrolaterally placed scales present in *A. salvadorensis*). *Mesaspis moreletii* has several rows of small scales above a distinct ventrolateral fold and has oblong-shaped, strongly keeled dorsal body scales (versus ventrolateral fold weakly developed, without distinctively granular and/or small scales above fold and platelike, rectangular-shaped, weakly keeled or rugose dorsals in *A. salvadorensis*). *Abronia montecristoi* has 5 occipital scales and pale brown crossbands on the body and tail (versus 1–4, usually 3, occipital scales and dorsal surfaces of body and tail brown with darker brown crossbands in *A. salvadorensis*).

Illustrations (Figs. 7, 43, 46, 48; Plate 28).—Campbell and Frost, 1993 (head scales, adult); Hidalgo, 1983 (adult); Köhler, 2003a (adult), 2008 (adult); Wilson and McCranie, 2004a (adult).

Remarks.—*Abronia salvadorensis* is apparently most closely related to *A. montecristoi* (see Remarks for latter species).

Plate 28.   *Abronia salvadorensis.* USNM 520002, adult male, SVL = 111 mm. Intibucá: Zacate Blanco.

The type locality of *A. salvadorensis* is in an area long disputed as being either El Salvadoran or Honduran territory by the respective governments, but a binding 11 September 1992 International Court of Justice decision placed the area within Honduras (see Anonymous, 2002; also see McCranie [p. 567], *In* McCranie and Wilson [2002]).

The key provided by Hidalgo (1983) to distinguish *Abronia montecristoi* and *A. salvadorensis* will not consistently distinguish those two species from each other. The single known Honduran *A. montecristoi* will key out to *A. salvadorensis* using that key.

Natural History Comments.—*Abronia salvadorensis* is known from 2,020 to 2,125 m elevation in the Lower Montane Moist Forest formation. Specimens of this diurnal species were active about 4 m above the ground in a tree with loose bark, on the ground in a cornfield, and on the ground at the edge of a dirt road. The specimen in the tree was exposed to the late afternoon sun and slowly retreated beneath tree bark when discovered. Specimens were collected in June and August. Nothing has been reported on reproduction or diet of this species, but like other *Abronia*, it likely bears live young and feeds at least on orthopterans.

Etymology.—The name *salvadorensis* refers to the country El Salvador, where the type locality was thought to be located; however, that locality is now considered to be in Honduras (see Remarks).

Specimens Examined (5 [0]; Map 24).— **INTIBUCÁ**: Zacate Blanco, KU 195560, USNM 520002. **LA PAZ**: Cantón Palo Blanco, KU 184047; about 5 km S of Santa Elena, KU 195561. "HONDURAS": UTA R-26108.

Genus *Mesaspis* Cope, 1877

Mesaspis Cope, 1877: 96 (type species: *Gerrhonotus moreletii* Bocourt, 1871: 102, by subsequent designation of Dunn and Dunn, 1940: 73).

Geographic Distribution and Content.—This genus occurs in disjunct montane habitats from west-central Veracruz and central Guerrero, Mexico, to extreme western Panama. Six named species are recognized, one of which occurs in Honduras.

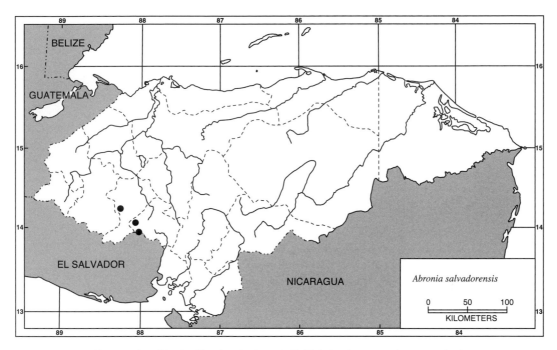

Map 24.   Localities for *Abronia salvadorensis*. Solid circles denote specimens examined.

Remarks.—The molecular study presented by Pyron et al. (2013, fig. 14) recovered a nonmonophyletic *Mesaspis* with respect to *Abronia*.

Scale terminology for *Mesaspis* used herein follows that of Good (1988).

Etymology.—The name *Mesaspis* is formed from the Greek *mesos* (middle) and *aspis* (viper) and probably alludes to the superficial resemblance of these Middle American lizards to a small snake.

Mesaspis moreletii (Bocourt, 1871)

Gerrhonotus moreletii Bocourt, 1871: 102 (eight syntypes, MNHN 1188, 1188A–B, 1267, 1267A, 1268, 2005, ZMB 7225 [see Good et al., 1993: 48]; type locality: "le Peten, ainsi que les forêts de pins de la haute Vera-Paz [Guatémala], à une altitude de 1,440 mètres").

Mesaspis moreletii: Cope, 1877: 96; Wilson et al., 1991: 70; Köhler, 2000: 44; Vesely and Köhler, 2001: 186; Köhler, 2003a: 66; Lundberg, 2003: 26; Wilson and McCranie, 2004b: 43; Wilson and McCranie, 2004c: 24; Castañeda, 2006: 29; Townsend, 2006a: 34; Townsend and Wilson, 2006: 245; Townsend et al., 2006: 35; Townsend et al., 2007: 10; Köhler, 2008: 70; Townsend and Wilson, 2009: 68; Gutsche, 2012: 71; Solís et al., 2014: 129; McCranie, 2015a: 366; Solís et al., 2015: 558.

Gerrhonotus moreletii salvadorensis: Dunn and Emlen, 1932: 28.

Barisia moreleti: Meyer, 1969: 198; Meyer and Wilson, 1973: 9; Wilson et al., 1979b: 62.

Mesaspis moreleti: Good, 1988: 84; Caceres, 1993: 114; Cruz et al., 1993: 28; Espinal, 1993, table 3; Anonymous, 1994: 116; Espinal et al., 2001: 106; Wilson et al., 2001: 135; Townsend and Wilson, 2008: 138.

Mesaspis moreletti [sic]: Townsend et al., 2007: 5.

Geographic Distribution.—Mesaspis

moreletii occurs in disjunct populations at intermediate and high elevations (and rarely upper moderate elevations) from central Chiapas, Mexico, to northwestern Nicaragua on the Atlantic versant and from southeastern Chiapas, Mexico, to northwestern El Salvador and southwestern Honduras on the Pacific versant. In Honduras, this species is known from scattered montane localities from the western border eastward to about central Olancho.

Description.—The following is based on ten males (ANSP 30619, 30621, 22220; KU 194262, 194264–65; SMF 78420; USNM 570239, 570243, 579543) and ten females (KU 192326, 192328, 194261, 194263, 194266, 200586; USNM 549353, 570235, 570238, 570242). *Mesaspis moreletii* is a moderately large lizard (maximum recorded SVL 97 mm [USNM 570243, a male]); dorsal head scales enlarged, smooth, with 2 pairs of internasal scales, 1 frontonasal scale (usually larger than combined sizes of all internasals), usually 2 prefrontal scales (rarely absent), prefrontal scales in contact medially or separated by frontonasal-frontal contact, and 1 frontal scale; postrostral scale absent; frontal scale contacting interparietal scale; 5 medial supraoculars; 2–3 (usually 3) lateral supraoculars; 3 loreals, postnasal sometimes fused to anterior loreal, anterior loreal otherwise located above postnasal, middle largest, posterior loreal contacting a supralabial; anterior superciliary scale-prefrontal contact present or absent; 1 subocular; 1–2 (rarely 2) postmental scales; 6–8 (rarely 8) supralabials and 4–6 infralabials to point below eye; moveable eyelid present; pupil circular; dorsal body scales oblong-shaped, imbricate, strongly keeled, 48–56 (51.7 ± 2.0) along dorsal midline from posterior to occipital to base of tail, with keels on some scales touching keel on following scale; 18 or 20 (usually 20) dorsal scales across midbody, with 12 crossrows above hind limb; 10 rows of nuchal dorsal scales; distinct ventrolateral fold present, zone of granular scales located above fold;

ventral body scales more or less oblong-shaped, but most scales have truncate posterior ends, smooth, imbricate, in 12 transverse rows at midbody, 8–10 transverse rows at level of forelimb; femoral pores absent; 5 digits on all limbs; 13–18 (15.8 ± 1.1) subdigital scales on 39 sides of Digit IV of hind limb; SVL 73–97 (86.6 ± 7.1) mm in males, 68–89 (78.5 ± 6.8) mm in females; HW/SVL 0.11–0.17 in males, 0.12–0.16 in females; HL/SVL 0.18–0.23 in males, 0.18–0.22 in females; TAL/SVL 1.56–1.75 in three males, 1.42–1.81 in five females.

Color in life of an adult female (USNM 570235; Plate 29): dorsal surface of body Cinnamon-Brown (33) with Sepia (119) blotches; lateral field of body Tawny (38) with Sepia blotches; dorsal surface of head Cinnamon-Brown; lateral surface of head Tawny with Sepia markings below and posterior to eye; ventral surface of head cream with bronze tinge, that of body Cinnamon-Rufous (40); subcaudal surface Cinnamon-Rufous at base, becoming grayish brown at about anterior third of length; iris Tawny.

Color in alcohol: dorsal surface of head brown to dark brown; dorsal surface of body with median brown to dark brown swath, with some darker brown spotting in those with paler swath; body swath continuing onto dorsal surface of tail; lateral surface of head brown to dark brown, with white to gray spots; lateral surface of body sometimes pale brown with scattered dark brown spots, but usually spotted with white to gray and brown; ventral surfaces of head and body cream with gray cast.

Diagnosis/Similar Species.—The combination of having moveable eyelids, having a distinct ventrolateral fold with small and/or granular scales above the fold, having keeled oblong-shaped dorsal body scales, having smooth oblong-shaped ventral scales, and having 2 pairs of internasal scales distinguishes *Mesaspis moreletii* from all remaining Honduran lizards.

Plate 29. *Mesaspis moreletii*. USNM 570235, adult female, SVL = 84 mm. Intibucá: near El Rodeo.

Illustrations (Figs. 44, 45; Plate 29).— Álvarez del Toro, 1983 (adult; as *Barisia*); A. H. A. Duméril et al., 1870–1909b (adult, head scales; as *Gerrhonotus* Wiegmann, 1828: col. 379); Good, 1988 (adult); Günther, 1885, *In* Günther, 1885–1902 (adult; as *Gerrhonotus*); Gutsche, 2012 (adult); Köhler, 2000 (adult), 2003a (adult), 2008 (adult); Köhler et al., 2005 (adult, head scales); Lundberg, 2003 (adult); Pianka and Vitt, 2003 (adult); Schmidt, 1928a (head scales; as *G. salvadorensis* Schmidt, 1928a: 196); Solís et al., 2015 (adult); Sunyer and Köhler, 2007 (adult); Townsend and Wilson, 2006 (adult), 2008 (adult); Vesely and Köhler, 2001 (adult, head scales).

Remarks.—*Mesaspis moreletii* is a member of the *M. moreletii* species group (Good, 1988: 86) and was recovered as the sister species to *M. monticola* (Cope, 1877: 97) of Costa Rica and western Panama (Good, 1989: 231), based on patterns of allozyme variation.

Natural History Comments.—*Mesaspis moreletii* is known from 1,450 to 2,530 m elevation in the Lower Montane Wet Forest and Lower Montane Moist Forest formations. This species is terrestrial and diurnal.

It is found crawling deliberately in sunny areas on the forest floor and in pastures with ground debris and inactive under logs and other ground debris. When active, its movements are usually rather deliberate and is usually easily captured if in an open area. The species appears to be active throughout much of the year, as long as sunlight is available. In Guatemala, this species is reported to give birth in May and June (Stuart, 1948, 1951). Fitch (1970) reported adult females with late embryos were collected in February and that 15 gravid females contained 4–9 embryos (countries of origin not given). Vesely and Köhler (2001) reported that nine dissected adult females from El Salvador contained 4–10 follicles and one collected in March or May contained eight well-developed embryos. Cooper and Habegger (2001) studied prey discrimination in captive specimens based on material from "Honduras" purchased from commercial animal dealers. Those captives ate crickets and tenebrionid beetle larva, but refused all vegetable matter offered. Thus, Cooper and Habegger (2001: 87) concluded that *M. moreletii* "presumably consumes a wide range of

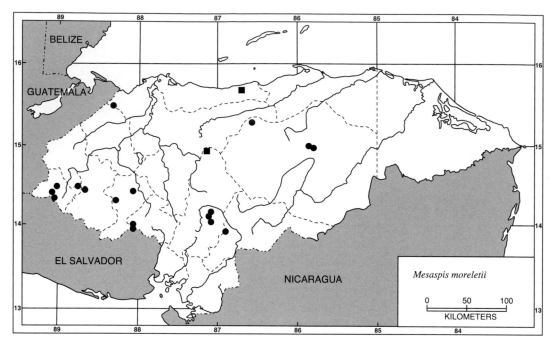

Map 25.   Localities for *Mesaspis moreletii*. Solid circles denote specimens examined and solid squares represent accepted records.

arthropods and other small animals" and that its diet "is largely restricted to animal prey."

Etymology.—The specific name *moreletii* is a patronym honoring Pierre Marie Arthur Morelet, a French naturalist, who collected in Guatemala and Mexico during the 19th century.

Specimens Examined (118, 2 skeletons [28]; Map 25).—**CORTÉS**: Bosque Enano, UF 144727; Cantiles Camp, UF 144734–35, 147634–36; Cerro Jilinco, UF 147632–33; El Cusuco, KU 200587–88, USNM 549353–54. **EL PARAÍSO**: Monserrat, MCZ R49958–59 (both skeletons). **FRANCISCO MORAZÁN**: Cerro La Tigra NNE of El Hatillo, KU 200591–93, LSUMZ 24412, SMF 78419–20; La Esperanza, USNM 578861; Parque Nacional La Tigra at Jutiapa, UTA R-53230–33; Parque Nacional La Tigra near San Juancito, KU 192326–28; San Juancito Mountains, AMNH 70398–99, ANSP 22213–22, UF 124827, UMMZ

112329; 11.3 km SW of San Juancito, LACM 47757–58, LSUMZ 21440; Sendero La Cascada, USNM 549355. **INTIBUCÁ**: near El Rodeo, USNM 570235–37; Zacate Blanco, KU 194258–61, LSUMZ 38826–27, USNM 570238. **LA PAZ**: Cantón Sabaneta, KU 184365; about 5 km S of Santa Elena, KU 194262–66. **LEMPIRA**: E slope of Cerro Celaque, KU 200586. **OCOTEPE-QUE**: El Chagüitón, KU 200589–90, USNM 570239–40; 3.6 km S of El Portillo de Ocotepeque, MVZ 262868, 263365; near El Portillo de Ocotepeque, UF 166355–69; El Portillo de Ocotepeque, LACM 47759–69, LSUMZ 33686, USNM 570241–42; 20.1 km ENE of Nueva Ocotepeque, UF 124828–30; ENE of Nueva Ocotepeque, ANSP 30617–21; road between Nueva Ocotepeque and La Labor, UTA R-46866–67; Sumpul, SMF 77956–58, USNM 570243. **OLANCHO**: Cerro La Picucha, USNM 579542; Parque Nacional La Mura-lla Centro de Visitantes, USNM 342267–68,

FN 256983 (still in Honduras because of permit problems); between Quebrada de Agua and Cerro La Picucha, USNM 579543. "HONDURAS": UTA R-46918–19.

Other Records (Map 25).—**ATLÁNTIDA**: Parque Nacional Pico Bonito (UTADC-8621). **CORTÉS**: Sendero de Cantiles (Townsend and Wilson, 2008). **FRANCISCO MORAZÁN**: Cataguana (Townsend et al., 2007; Townsend and Wilson, 2009). **LEMPIRA**: Cerro Celaque, UNAH 2045, 2397, 2838–40, 2851–52 (Cruz et al., 1993).

Family Diploglossidae Cope, 1865c

Members of this family occur in largely disjunct populations from Puebla and Oaxaca, Mexico, southward to Ecuador, north-central Argentina, Uruguay, and southeastern Brazil. It also occurs in the West Indies on Jamaica, the Cayman Islands, Cuba, Hispaniola, Puerto Rico, Navassa, and Montserrat. Two genera (but see Remarks) and about 45 named species are included in this family, with one genus and three named species known from Honduras (but see Species of Probable Occurrence).

Genus *Diploglossus* Wiegmann, 1834b

Diploglossus Wiegmann, 1834b: 36 (type species: *Diploglossus fasciatus* Wiegmann, 1834b: 36 [= *Tiliqua fasciatus* Gray, 1831: 71, *In* Gray, 1830–1831], by subsequent designation of Fitzinger, 1843: 23).

Geographic Distribution and Content.— Mainland members of this genus occur from Puebla, Mexico, to Bolivia and northeastern Brazil on the Atlantic versant and from southwestern Guatemala to northwestern Costa Rica, in west-central Panama and western Colombia and Ecuador (including Malpelo Island) on the Pacific versant. This genus also occurs on the Greater Antillean islands of Cuba, Hispaniola, Jamaica, Puerto Rico, Navassa, the Cayman Islands, and the Lesser Antillean island of Montserrat. Thirty-three named species are recognized, with three known from Honduras (but see Species of Probable Occurrence).

Remarks.—There has been considerable disagreement in the literature concerning the validity of the genera *Celestus* and *Diploglossus*. Savage and Lips (1994) and Savage et al. (2008) provided a thorough review of that literature and recognized *Celestus* as a valid genus distinct from *Diploglossus*. On the other hand, Myers (1973) reviewed the literature regarding the validity of *Celestus* and *Diploglossus* and regarded *Diploglossus* as the valid genus name. However, *Celestus* and *Diploglossus*, as recognized by those authors, were recovered as paraphyletic in the Pyron et al. (2013) study. Pyron et al. (2013: 12) stated "*Diploglossus* and *Celestus* are strongly supported as paraphyletic with respect to each other and to *Ophiodes*." The study of Gauthier (1982: 37) had also concluded that both *Celestus* and *Diploglossus* are "defined by plesiomorphic characters and thus paraphyletic." I use *Diploglossus* over that of *Celestus* for the species occurring in Honduras until their relationships are better understood. *Diploglossus* is also the older of the two names.

I use the head scale terminology suggested by Savage et al. (2008) in the descriptions below.

Etymology.—The name *Diploglossus* is derived from the Greek words *di-* (prefix meaning "two, double"), *plico* (fold), and *glossa* (tongue) The reference is to the apex of the bifid tongue being retractile into the basal portion, which forms a sheath.

Key to Honduran Species of the Genus *Diploglossus*

1A. Claws almost completely hidden by scaly sheath, only claw tip visible .... *bilobatus**
1B. Claws fully exposed................. 2

Figure 49.   Pale dorsolateral stripe present on each side. *Diploglossus scansorius*. USNM 335049 from NNE of La Fortuna, Yoro.

2A. Fourteen to 17 subdigital lamellae on Digit IV on hind limb . . . .
. . . . *bivittatus* (p. 152)

2B. Twenty or more subdigital lamellae on Digit IV on hind limb 3

3A. Pale dorsolateral stripes present (Fig. 49), stripes usually most distinct on head and anterior third of body; no distinct alternating pale and dark vertical bars or well defined ocelli on neck and flanks
. . . . *scansorius* (p. 159)

3B. Pale dorsolateral stripes absent; distinct alternating pale and dark vertical bars or well-defined ocelli present on neck and flanks 4

4A. A single pair of prefrontal scales present (Fig. 50); flank pattern of ocelli (Fig. 51) . . . . . *montanus* (p. 156)

4B. Prefrontal scales absent (Fig. 52); flank pattern of distinct alternating pale and dark vertical bars (Fig. 53)
. . . . *rozellae*°

CLAVE PARA LAS ESPECIES HONDUREÑAS DEL GÉNERO *DIPLOGLOSSUS*

1A. Garras casi completemente escondidos por una vaina, solamente la

punta de la garra es visible. . . *bilobatus*°

1B. Garras completamente visibles 2

2A. De 14 a 17 lamelas subdigitales en el Dígito IV en las extremidades posteriores. . . . . *bivittatus* (p. 152)

2B. Más de 20 lamelas subdigitales en el Dígito IV en las extremidades posteriores. 3

3A. Rayas dorsolaterales pálida presentes (Fig. 49), rayas usualmente más distinguibles en la cabeza y en el tercio anterior del cuerpo; sin barras pálidas y oscuras verticales que se alternan u ocelos bien desarrollados presentes sobre el cuello y costados. . *scansorius* (p. 159)

3B. Rayas dorsolaterales pálida ausentes; barras distintivas verticales alternas u ocelos bien desarrollados sobre el cuello y costados 4

4A. Un solo par de escamas prefrontales presentes (Fig. 50); patrón de los costados con ocelos (Fig. 51). . . . . *montanus* (p. 156)

4B. Escamas prefrontales ausentes (Fig. 52); patrón de los costado con barras verticales pálidas y oscuras alternadas (Fig. 53). . *rozellae*°

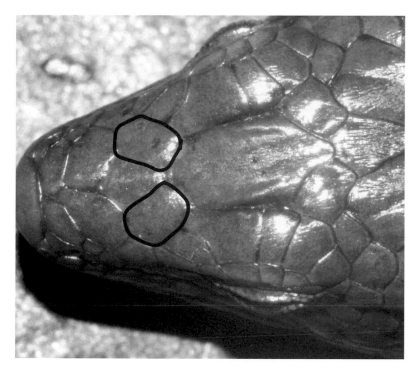

Figure 50.   Single pair of prefrontal scales present (outlined). *Diploglossus montanus*. UF 144903 from Cantiles Camp, Cortés.

Figure 51.   Ocellated flank pattern present on each side. *Diploglossus montanus*. UTA R-9443 from S of Santa Elena, Cortés. Photograph by Louie Porras.

Figure 52.   Prefrontal scales absent. *Diploglossus rozellae.** USNM 496640 from Belize. Compare with Fig. 50.

Diploglossus bivittatus Boulenger, 1895

Diploglossus bivittatus Boulenger, 1895: 732 (holotype, BMNH 1946.29.37 [see Wilson, Porras, and McCranie, 1986: 6]; type locality: "Hacienda [Santa] Rosa de Jericho [= Jericó], Nicaragua [Departamento de Matagalpa], 3250 feet"); Campbell and Camarillo R., 1994: 204 (in part); McCranie, 2015a: 367.

Celestus bivittatus: Wilson, Porras, and McCranie, 1986: 5; Villa and Wilson, 1988: 423.1; Savage and Lips, 1994: 834 (in part); McCranie and Wilson, 1996: 259; Wilson et al., 2001: 135; Wilson and McCranie, 2004b: 43; Wilson and Townsend, 2007: 145; McCranie and Valdés Orellana, 2011a: 240; Solís et al., 2014: 130.

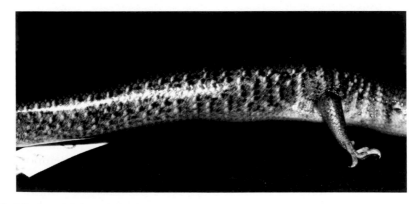

Figure 53.   Flank pattern of distinct alternating pale and dark vertical bars on each side. *Diploglossus rozellae.** USNM 496640 from Belize.

Geographic Distribution.—*Diploglossus bivittatus* occurs at moderate and intermediate elevations in eastern Guatemala on the Atlantic versant and from southwestern Honduras to northwestern Nicaragua on both versants. It is known in Honduras from several localities on both sides of the Continental Divide in the southwestern portion of the country.

Description.—The following is based on nine males (KU 194665, 194668, 194679; USNM 335050–52, 335054–55, 563522) and five females (KU 194666–67; USNM 563521, 563523, 573885). *Diploglossus bivittatus* is a moderately large lizard (maximum recorded SVL 104 mm [USNM 563522, KU 194667, a male and female, respectively]); dorsal head scales enlarged, smooth, with 1 pair of anterior internasal scales, 1 pair of posterior internasal scales, 1 frontonasal scale, 0 prefrontal scales, 1 frontal scale, paired frontoparietal scales, 1 interparietal scale, 1 pair of parietal scales, and 1 interoccipital scale; 5 medial supraoculars, fifth smallest, second contacting frontonasal scale or not; 3 loreals, second largest, usually fused with second canthal scale, third loreal not divided; 9–12 (usually 11 or 12) supralabials, with seventh and eighth or eighth and ninth (rarely sixth and seventh) at level below eye; suture between first and second supralabials at level about equal to anterior edge of nostril opening; 3–5 (usually 4) postoculars, arranged in a series and juxtaposed with subocular scales; moveable eyelid present; pupil circular; 7–10 (usually 9 or 10) infralabials; dorsal body scales cycloid, imbricate, strongly striated, 67–79 (75.2 ± 3.5) along dorsal midline between posterior edge of occipital and level above posterior surface of thigh, 50–58 (54.0 ± 2.3) between levels of axilla and groin; 14–20 (18.0 ± 1.7) dorsal scales between level above axilla and occipital; 29–31 (30.2 ± 0.8) scales around midbody; no ventrolateral fold; ventral body scales cycloid, weakly striated, imbricate, 78–89 (81.6 ± 3.0) along midline between first pair

of chinshields and cloacal scale, 46–53 (49.4 ± 2.2) between levels of axilla and groin; 8–10 scales along anterior edge of cloacal opening; femoral pores absent; 5 digits on fore- and hind limb; 12–14 (13.0 ± 0.7) subdigital scales on Digit III of forelimb, 14–18 (15.9 ± 1.0) on Digit IV of hind limb; SVL 62–104 (87.7 ± 14.7) mm in males, 72–104 (91.0 ± 12.9) mm in females; HW/SVL 0.11–0.13 in males, 0.11–0.12 in females; HL/SVL 0.14–0.17 in males, 0.13–0.17 in females; TAL/SVL 0.61–1.45 in eight males, 0.82–1.31 in four females; AGL/SVL 0.58–0.62 in males, 0.55–0.65 in females.

Color in life of an adult female (USNM 573885; Plate 30): middorsum Burnt Umber (22) with obscure paler brown stripes; dorsolateral stripes Pale Pinkish Buff (121D), dorsolateral stripes extend forward along each upper eyelid to unite as single stripe across snout; dorsolateral stripes extend onto anterior portion of tail where merging to form Pale Pinkish Buff dorsal surface of tail; lateral surface of body Sepia (119) with pale brown mottling; lateral surface of head Sepia, except supralabials with pale brown lower edges or halves; ventral surfaces of head and body pale brown; subcaudal surface pale brown with orange tinge; iris Sepia. Wilson, Porras, and McCranie (1986: 6) described color in life of an adult male (KU 194665): "middorsal area dark brown with each scale darker medially, paler laterally; dorsolateral stripes bright golden pink grading to dull copper at about midbody, fusing on base of tail to become color of tail; lateral area anterior to forelimb insertion dark brown, posterior to that point lateral stripe invaded by small pale spots from venter; venter pale yellow with copper patina; limbs brown with bronze patina, upper forelimbs with few small drab spots; head dark brown medially, laterally with bright copper stripes; temporal area dark brown; preorbital area pale brown; supralabials with cream-colored markings; area between ear opening and anterior limb insertion cream mottled with brown; chin

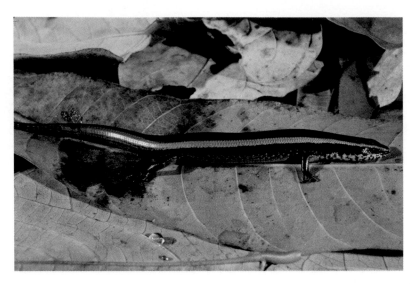

Plate 30.　*Diploglossus bivittatus*. USNM 573885, adult female, SVL = 72 mm. Ocotepeque: El Portillo de Cerro Negro.

pale yellow." Variation in adult color and pattern for five other adults was "relatively minor, primarily involving the overall intensity in brown shading and gradual change of color in the dorsolateral striping" (Wilson, Porras, and McCranie, 1986: 6). Wilson, Porras, and McCranie (1986: 6) described color in life of a series of juveniles (KU 194669–78): "dorsum of body dark chocolate; dorsolateral stripes gold on head grading to golden bronze posteriorly; tail brilliant red-orange; limbs dark reddish brown; labials and lateral neck area pale chartreuse; posterior portion of venter orangish green grading to brilliant reddish orange on tail just anterior to vent; underside of limbs brick red."

Color in alcohol: middorsal surfaces of body and head grayish brown to dark brown, with darker brown spots on most scales; dorsolateral stripe slightly paler grayish brown, brown, or pale brown (in juveniles and one adult), beginning on head and extending to posterior end of body, those pale stripes prominent in all juveniles and in one adult from Ocotepeque; lateral areas slightly darker brown than middorsum, heavily mottled with darker brown in most adults, slightly paler brown than middorsum, heavily mottled with darker brown in one adult; ventral surfaces of head and body cream with grayish brown spots or mottling on most scales in many adults, cream with gray flecks on most scales in one adult from Ocotepeque.

Diagnosis/Similar Species.—The cycloid and imbricate body scales, with the dorsal scales strongly striated in combination with having moveable eyelids and having 2 pairs of internasal scales distinguishes *Diploglossus bivittatus* from all other Honduran lizards, except *D. montanus* and *D. scansorius*. *Diploglossus montanus* has 1 pair of prefrontal scales, has 21–24 subdigital scales on Digit IV of the hind limb, has a pattern of ocelli on the lateral surface of the body, and lacks pale dorsolateral stripes (versus prefrontal scales absent, 14–18 subdigital scales on Digit IV of hind limb, pale brown mottling present on lateral side of body, and pale dorsolateral stripes present in *D. bivittatus*). *Diploglossus scansorius* has 21–22 subdigital scales on Digit IV of the hind limb and has 15–16 such scales on Digit III of the forelimb (versus 14–18 subdigital lamellae on Digit IV of hind limb and 12–14

Plate 31. *Diploglossus bivittatus*. USNM 573886, juvenile. Ocotepeque: El Portillo de Cerro Negro.

lamellae on Digit III of forelimb in *D. bivittatus*).

Illustrations (Fig. 42; Plates 30, 31).— Boulenger, 1895 (juvenile); Campbell and Camarillo R., 1994 (head scales); Köhler, 2000 (adult; as *Celestus*), 2003a (adult; as *Celestus*), 2008 (adult; as *Celestus*); Villa and Wilson, 1988 (juvenile; as *Celestus*); Villa et al., 1988 (adult, juvenile; as *Celestus*); Wilson and McCranie, 2004a, (adult; as *Celestus*); Wilson, Porras, and McCranie, 1986 (adult, juvenile; as *Celestus*).

Remarks.—*Diploglossus bivittatus* appears to be most closely related to *D. scansorius* and *D. atitlanensis* (P. W. Smith, 1950: 195, *In* Smith and Taylor, 1950b; see McCranie and Wilson, 1996). Campbell and Camarillo R. (1994) resurrected *D. atitlanensis* from the synonymy of *D. bivittatus*, where Wilson, Porras, and McCranie (1986) had placed it. Campbell and Camarillo R. (1994), in making that decision, failed to specifically discuss any characters they used to resurrect that species from that synonymy. A comparison of their diagnosis (p. 203) to my data herein for *D. bivittatus* reveals overlap between these two nominal forms in

all characters except maximum known SVL (113 mm in *D. atitlanensis* versus 104 mm in *D. bivittatus*). However, *D. atitlanensis* is known from very few specimens (only two examined by Campbell and Camarillo R., 1994). Campbell and Camarillo R. (1994) also said *D. atitlanensis* is arboreal (versus primarily terrestrial in *D. bivittatus*). Molecular analyses are needed to clarify the systematic relationships of these two nominal forms. Villa and Wilson (1988) briefly redescribed *D. bivittatus* and reviewed the literature; however, their concept of *D. bivittatus* also included *D. atitlanensis*.

Two *Diploglossus bivittatus* (one adult female and one juvenile) were taken at a higher elevation (2,100 m) in broadleaf cloud forest, whereas the remaining Honduran specimens were from lower elevations (1,330–1,980 m) and open pine forest. The high-elevation adult differs from all other known adults in retaining the juvenile coloration and having a less mottled venter, but there are no known morphological differences between the two populations.

Natural History Comments.—*Diploglossus bivittatus* is known from 1,330 to 2,100

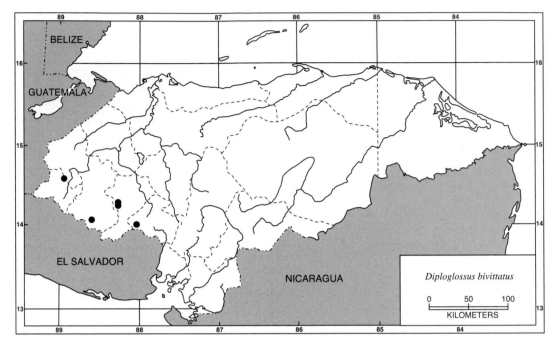

Map 26.   Localities for *Diploglossus bivittatus*. Solid circles denote specimens examined.

m elevation in the Premontane Moist Forest and Lower Montane Moist Forest formations. Most specimens were collected under pine logs or in rotten pine stumps, and it appears to be largely terrestrial, at least in pine forests. Two others were in rock crevices on a steep bank above a trail and another escaped by quickly climbing the embankment after being exposed by turning a rock. Specimens of this diurnal species were collected in January, May, July, August, and September. Campbell and Camarillo R. (1994) reported the Guatemalan specimens of this species were found in dry pine-oak forest. A Honduran female gave birth to ten young on 1 March (Wilson, Porras, and McCranie, 1986). Nothing has been reported on the food habits of this species.

Etymology.—The name *bivittatus* is formed from the Latin prefix *bi*- (two, twice) and Latin *vittatus* (decorated or bound with a ribbon), and alludes to the paired dorsolateral stripes (most prominent in juveniles) of this species.

Specimens Examined (26 [0]; Map 26).— **INTIBUCÁ**: Chupocay, USNM 563521–22; 18.1 km NW of La Esperanza, USNM 335050–55; 14.4 km WNW of La Esperanza, KU 194665; 11.3 km WNW of La Esperanza, KU 194666–78. **LA PAZ**: El Chilador, USNM 563523. **LEMPIRA**: 3 km N of Gualcince, KU 194679. **OCOTEPEQUE**: El Portillo de Cerro Negro, USNM 573885–86.

Diploglossus montanus (Schmidt, 1933)

Celestus montanus Schmidt, 1933: 21 (holotype, FMNH 5066; type locality: "mountains west of San Pedro, Honduras [the Sierra de Merendon sic = Sierra de Omoa]. Altitude 4500 feet" [see Remarks]); M. A. Smith, 1933b: 34; Marx, 1958: 455; Meyer, 1969: 198; Peters and Donoso-Barros, 1970: 90; Meyer and Wilson, 1973: 9; Strahm and Schwartz, 1977: 71; Wilson, Porras, and McCranie, 1986: 10; Campbell and Vannini, 1989: 16; Savage and Lips, 1994: 834; McCranie and Wilson, 1996: 263; Wilson et al., 2001: 135; Köhler, 2003a: 61; Wilson and McCranie, 2003: 59; Wilson and McCranie, 2004b: 43; Townsend, Hughes et al., 2005: 67; Townsend, 2006a: 34; Townsend, 2006b: 834.1; Townsend et al., 2006: 31; Wilson and Townsend, 2006: 104; Köhler, 2008: 64; Townsend and Wilson, 2008: 134; Solís et al., 2014: 130.

Celestus [sp.]: Schmidt, 1942: 29.

Diploglossus montanus: Wermuth, 1969: 10; Campbell and Camarillo R., 1994: 204; McCranie, 2015a: 367.

Geographic Distribution.—*Diploglossus montanus* occurs at moderate and intermediate elevations on the Atlantic versant in northwestern Honduras and in adjacent eastern Guatemala. In Honduras, this species is known only from a few montane localities in Cortés.

Description.—The following is based on four females (FMNH 5066; UF 142324, 144903; UTA R-9443; adult males not available). *Diploglossus montanus* is a moderately large lizard (maximum recorded SVL 94 mm [UTA R-9443]); dorsal head scales enlarged, smooth, with 1 pair of anterior internasal scales, paired posterior internasal scales, 1 frontonasal scale flanked on each side by prefrontal scale, 1 frontal scale, paired frontoparietal scales, 1 interparietal scale, 1 pair of parietal scales, and 1 interoccipital scale; 5 medial supraoculars, fifth smallest, second usually not contacting frontonasal scale (on 5 of 8 sides); 3 loreals, second largest, fused with second canthal, third loreal not divided; 10–11 supralabials, with seventh and/or eighth at level below mideye; suture between first and second supralabials at level slightly anterior to edge of nostril opening; 3–4 postoculars, arranged in a series and juxtaposed with subocular scales; moveable eyelid present; pupil circular; 10–11 infralabials, with seventh and/or eighth at level below mid-

eye; dorsal body scales cycloid, imbricate, strongly striated, 67–72 (69.5 ± 2.4) along dorsal midline posterior to occipital to level above posterior surface of thigh, 44–52 (48.3 ± 3.5) between levels of axilla and groin; 16–19 (17.3 ± 1.3) dorsal scales between level of axilla and occipital; 32–34 (33.0 ± 0.7) scales around midbody; no ventrolateral fold; ventral body scales cycloid, weakly striated, imbricate, 67–75 (71.0 ± 3.4) along midline between first pair of chinshields and cloacal scale, 42–49 (45.0 ± 3.2) between levels of axilla and groin; 9 scales along anterior edge of cloacal opening; femoral pores absent; 5 digits on fore- and hind limb; 16–19 (17.6 ± 1.4) subdigital scales on Digit III of forelimb, 21–24 (23.1 ± 1.4) on Digit IV of hind limb; SVL 34.2–94.0 (69.4 ± 25.9) mm; HW/SVL 0.11–0.13; HL/SVL 0.17–0.25; TAL/SVL 0.92–1.70 in two; AGL/SVL 0.52–0.57.

Wilson, Porras, and McCranie (1986: 10) recorded color in life of UTA R-9443 (adult female; Plate 32): "dorsum of body uniform brownish olive; sides brownish olive with numerous black-outlined pale lime green ocelli; venter pale yellowish green; head grayish blue." Schmidt (1933: 22) recorded the color in life of the female holotype (FMNH 5066) as: "Olive green above, lighter on the sides, yellowish green beneath; small black spots on the back and sides."

Color in alcohol: dorsal surface of head brownish gray; dorsal surface of body with broad, brownish gray middorsal swath; lateral surface of body slightly paler brownish gray than middorsal swath, with numerous pale gray ocellated spots surrounded, at least in part, by dark brown; dorsal surface of tail brownish gray; ventral surfaces of head, body, and tail pale gray, especially on posterior two-thirds of body.

Diagnosis/Similar Species.—The cycloid and imbricate body scales, with the dorsal scales strongly striated, in combination with having moveable eyelids and 2 pairs of internasal scales, distinguishes *Diploglossus*

Plate 32.   *Diploglossus montanus.* UTA R-9443, adult female, SVL 94 mm. Cortés: near Santa Elena. Photograph by Louis Porras.

montanus from all other Honduran lizards, except *D. bivittatus* and *D. scansorius*. *Diploglossus bivittatus* and *D. scansorius* lack prefrontal scales, lack pale ocelli on the lateral surface of the body and have pale dorsolateral stripes (versus 1 pair of prefrontal scales, pale ocelli present on lateral surface of body, and no pale dorsolateral stripes in *D. montanus*). Additionally, *D. bivittatus* has 14–18 subdigital scales on Digit IV of the hind limb and *D. scansorius* has 21–22 such scales (versus 23–25 scales on Digit IV of hind limb in *D. montanus*).

Illustrations (Figs. 50, 51; Plate 32).— Campbell and Camarillo R., 1994 (head scales); Köhler, 2008 (adult; as *Celestus*); Strahm and Schwartz, 1977 (middorsal osteoderm; as *Celestus*); Townsend, 2006b (adult, juvenile; as *Celestus*); Townsend and Wilson, 2008 (adult; as *Celestus*); Townsend et al., 2006 (adult; as *Celestus*); Wilson, Porras, and McCranie, 1986 (adult; as *Celestus*).

Remarks.—By using the information in Schmidt (1942) in conjunction with copies of Schmidt's field notes, I believe I have been able to pinpoint the type locality of *Diploglossus montanus* to along the Quebrada del Infierno on the eastern slope of Cerro de La Virtúd, about 6.5 airline km W of San Pedro Sula, Cortés (also see McCranie, 2002: 28, 562, *In* McCranie and Wilson, 2002; McCranie, 2011a: 627). Campbell and Camarillo R. (1994) concluded that this species belonged to a species group that included *D. rozellae* (Smith, 1942: 372; as *Celestus*), *D. atitlanensis*, and *D. bivittatus* (their *D. bivittatus* included the subsequently described *D. scansorius*). Townsend (2006b) provided an overview of the species' morphology and a literature review.

Natural History Comments.—*Diploglossus montanus* is known from 915 to 1,780 m elevation in the Premontane Wet Forest and Lower Montane Wet Forest formations. The holotype was in the palm-thatched roof of a small *champa* in April (Schmidt, 1933). A juvenile was active among rocks near a stream in July (Townsend, Hughes et al., 2005). Nothing else has been reported on natural history of this diurnal species other than two were "collected on a hill" (Wilson, Porras, and McCranie, 1986: 10); however, those two specimens were from a Honduran

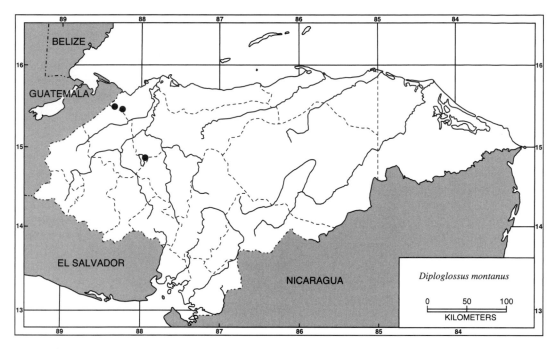

Map 27.   Localities for *Diploglossus montanus*. Solid circles denote specimens examined.

animal dealer who was sometimes unreliable with his locality data. The holotype of this viviparous species is a gravid female collected in April (Schmidt, 1933). No information has been published on diet in *D. montanus*.

Etymology.—The name *montanus* is Latin (pertaining to mountains) and alludes to the type locality of the species being in a mountainous area.

Specimens Examined (5 [1]; Map 27).— **CORTÉS**: Cantiles Camp, UF 144903; Guanales Camp, UF 142324; mountains W of San Pedro Sula, FMNH 5066; S of Santa Elena, LSUMZ 36659, UTA R-9443.

Diploglossus scansorius (McCranie and Wilson, 1996)

Celestus bivittatus: Wilson et al., 1991: 70; Savage and Lips, 1994: 834 (in part).
Diploglossus bivittatus: Campbell and Camarillo R., 1994: 201 (in part).

Celestus sp.: Wilson and McCranie, 1994b: 148; Wilson et al., 2001: 134 (as *C. scansorius* elsewhere in that work).
Celestus scansorius McCranie and Wilson, 1996: 260 (holotype, USNM 335049; type locality: "2.5 airline km NNE La Fortuna [15°25′N, 87°19′W], 1550 m elevation, Cordillera Nombre de Dios, Departamento de Yoro, Honduras"); Köhler, 2003a: 62; Wilson and McCranie, 2003: 59; Wilson and McCranie, 2004b: 43; Wilson and Townsend, 2007: 145; Köhler, 2008: 65; Townsend et al., 2010: 12; Townsend et al., 2012: 109; Solís et al., 2014: 130.
Diploglossus scansorius: McCranie, 2015a: 367.

Geographic Distribution.—*Diploglossus scansorius* occurs at intermediate elevations in the western portion of the Cordillera Nombre de Dios and the Montaña Macuzal in Yoro, Honduras.

Plate 33. *Diploglossus scansorius.* USNM 335049, adult female, SVL = 111 mm. Yoro: 2.5 airline km NNE of La Fortuna.

Description.—The following is based on two females (FMNH 236386; USNM 335049). *Diploglossus scansorius* is a moderately large lizard (maximum recorded SVL 111 mm [FMNH 236386; USNM 335049]); dorsal head scales enlarged, smooth, with paired anterior internasal scales, paired posterior internasal scales, 1 frontonasal scale, 0 prefrontal scales, 1 frontal scale, paired frontoparietal scales, 1 interparietal scale, paired parietal scales, and 1 interoccipital scale; 5 medial supraoculars, fifth smallest, second not contacting frontonasal; 3 loreals, second largest, not fused with second canthal scale, third loreal not divided; 11–12 supralabials, with seventh and eighth or eighth and ninth at level below eye; suture between first and second supralabials at level about equal to anterior edge of nostril opening; 4–5 postoculars arranged in a series and juxtaposed with subocular scales; moveable eyelid present; pupil circular; 9–10 infralabials; dorsal body scales cycloid, imbricate, strongly striated, 74 (in both) along dorsal midline between posterior edge of occipital to level above thigh, 50–51 (50.5) between levels of axilla and groin; 18–19 (18.5) dorsal scales between level of axilla and occipital; 29–31 (30.0) scales around midbody; no ventrolateral fold; ventral body scales cycloid, weakly striated, imbricate, 76–78 (77.0) along midline between first pair of chinshields and cloacal scale, 47 (in both) between levels of axilla and groin; 11–12 scales along anterior edge of cloacal opening; femoral pores absent; 5 digits on all limbs; 15–16 (15.3 ± 0.5) subdigital scales on Digit III of forelimb, 21–22 (21.5 ± 0.6) on Digit IV of hind limb; SVL 111 mm; HW/SVL 0.13; HL/SVL 0.16; TAL/SVL 0.60–0.87 (regenerated in USNM 335049 and tip missing in FMNH 236386); AGL/SVL 0.58–0.60.

Color in life of the adult female holotype (USNM 335049; Plate 33) was described by McCranie and Wilson (1996: 262): "dorsum Verona Brown middorsally (color 223B) with fragmented dark brown stripe down each row; dorsolateral stripe Verona Brown with metallic sheen; middorsum of head coppery brown; head stripes coppery tan; lateral surface of head pale coppery brown anterior to eye, coppery brown posterior to eye; labial surface cream with bronze sheen; lateral surfaces of body mottled creamy bronze, coppery brown, and very dark brown; dorsal surfaces of limbs coppery brown with indistinct very dark brown spotting; dorsal surface of tail same as that of body; ventral surfaces of head, body, and tail pale golden yellow with pale gray scale edges; iris dark brown with gold flecking." Color in life of the adult female paratype (FMNH 236386) was described by McCranie and Wilson (1996: 262): "dorsum coppery brown with very dark brown streaks on each scale; dorsolateral stripe coppery brown; middorsum of head coppery brown with very dark brown markings; supralabials brown above, pale yellow below; dark brown temporal band grading into darker colored lateral band; lateral band on body with small coppery brown spots; dorsal surfaces of limbs dark brown; dorsal surfaces of tail coppery brown with very dark brown streaking; ventral surfaces of head, body, limbs, and tail greenish yellow with pale gray scale edges."

Color in alcohol for the adult female holotype (USNM 335049) was provided by McCranie and Wilson (1996: 262): "dorsum pale brown with bluish cast and diffuse, fragmented stripe down middle of each scale row; dorsolateral stripe pale brown with bluish cast; middorsum of head pale bluish gray, smudged with dark brown; labials cream with dark brown smudging; lateral surfaces of body mottled with pale blue and dark brown; dorsal surface of forelimb mottled with pale blue and dark brown; dorsal surface of hind limb dark gray-brown; dorsal surface of original portion of tail as for body; ventral surfaces of head, body, and tail pale gray except preanal scales which are cream." These authors also stated "the paratype essentially is colored the same as the holotype save for being somewhat more faded due to longer emersion in preservative" (p. 262).

Diagnosis/Similar Species.—The cycloid and imbricate body scales, with the dorsal ones strongly striated in combination with having moveable eyelids and 2 pairs of internasal scales, distinguishes *Diploglossus scansorius* from all other Honduran lizards, except *D. bivittatus* and *D. montanus*. *Diploglossus bivittatus* has 14–18 subdigital scales on Digit IV of the hind limb and 12–14 subdigital scales on Digit III of the forelimb (versus 21–22 and 15–16 subdigital lamellae, respectively, in *D. scansorius*). *Diploglossus montanus* has a single pair of prefrontal scales, has 23–25 subdigital scales on Digit IV of the hind limb, has a pattern of ocelli on the lateral surface of the body, and lacks pale dorsolateral stripes (versus prefrontal scales absent, 21–22 subdigital scales on Digit IV of hind limb, pale ocelli absent on lateral surface of body, and pale dorsolateral stripes present in *D. scansorius*).

Illustrations (Fig. 49; Plate 33).—Köhler, 2003a (adult; as *Celestus*), 2008 (adult; as *Celestus*); McCranie and Wilson, 1996 (adult; as *Celestus*).

Remarks.—*Diploglossus scansorius* appears to be most closely related to *D. bivittatus*. The known ranges of these two species are separated by about 130 km of territory that is too dry, too warm and open, and, the vast majority of which is also, too low in elevation to support populations of those two species.

Natural History Comments.—*Diploglossus scansorius* is known from 1,550 to 1,590 m elevation in the Lower Montane Wet Forest formation and peripherally in the Premontane Moist Forest formation. One was underneath bark of a tall pine tree stump about 2 m above the ground. When the portion of bark underneath which the lizard was hidden was removed, the lizard fell to the ground but quickly climbed the opposite side of the tree stump. The other lizard was lying in a heavily shaded spot on a horizontal tree trunk growing from a vertical cliff in disturbed broadleaf forest. The lizard was about 2.5 m above the forest floor and appeared to be asleep, as it made no effort

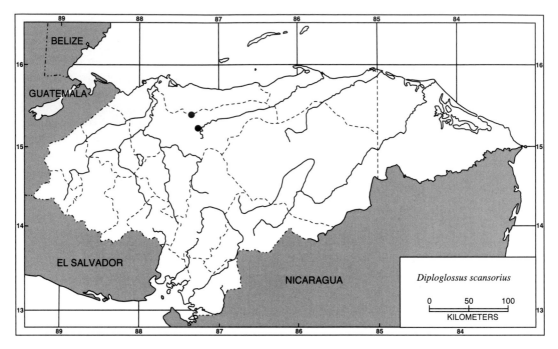

Map 28.   Localities for *Diploglossus scansorius*. Solid circles denote specimens examined.

to escape while I worked my way through the undergrowth to position myself beneath the lizard in order to capture it. Thus, the limited data suggests *D. scansorius* is arboreal. Both specimens were collected in July. Nothing has been reported on reproduction or diet of this species, but like other species of *Diploglossus*, it likely bears live young.

Etymology.—The name *scansorius* is Latin (of climbing) and alludes to the arboreal habits of the two known specimens.

Specimens Examined (2 [0]; Map 28).— **YORO**: 2.5 airline km NNE of La Fortuna, USNM 335049; Montaña Macuzal, FMNH 236386.

Infraorder Neoiguania Vidal and Hedges, 2009

Vidal and Hedges (2009: 134) coined the name Neoiguania, which equals the Pleurodonta of Cope (1865b: 181, 1865c: 226). That name is an equivalent to an infraorder

name and contains the superfamily Iguanoidea (Bell, 1825: 206; as Iguanidae).

Superfamily Iguanoidea Bell, 1825

Bell (1825: 206) coined the family name Iguanidae for a marine iguana occurring on the Galapagos Islands. The Principle of Coordination of the Code (ICZN, 1999) dictates that Bell is also author of this superfamily name Iguanoidea. Honduran families in this superfamily are the Corytophanidae, Dactyloidae, Iguanidae, Leiocephalidae, Phrynosomatidae, and Polychrotidae. Externally, the Honduran members of this superfamily can be distinguished from all remaining Honduran lizards by the combination of having moveable eyelids, the dorsal surface of the head with small to enlarged scales (never covered with granular scales), the snout scales small and irregular, and the dorsal surface of the body without large cycloid or rectangular scales that are similar in size and shape to those of the

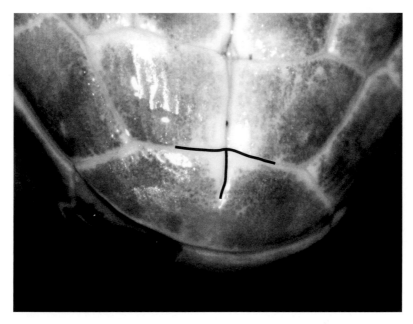

Figure 54.   Mental scale partially divided (outlined). *Polychrus gutturosus*. UF 137404 from Bodega de Río Tapalwás, Gracias a Dios.

venter. Most Honduran members of the Iguanoidea are oviparous, with two notable exceptions. *Corytophanes percarinatus* of the Corytophanidae and lizards in the *Sceloporus malachiticus* species complex of the Phrynosomatidae give birth to living young. McCranie and Köhler (2015) treated the Honduran species of the Dactyloidae, with the exception of one species reported from the country too late to be included in a complete species account in that book. An account of that species (*Norops wermuthi*) is included herein in the Dactyloidae. The report of that species, *N. wermuthi*, from Honduras, is a bit tainted as no specimen of that species from Honduras was deposited in any museum, despite a museum number being given in the report. Also, subsequent to the McCranie and Köhler (2015) review of the Dactyloidae, Köhler, Townsend, and Petersen (2016) proposed a new taxonomy for the *N. tropidonotus* species complex. No separate species accounts for those species

in that new taxonomic suggestion are included in this work because of the late publication of that work.

Remarks.—The stem name Iguanidae of the superfamily Iguanoidea apparently originated with the Oppel (1811b) nonavailable name Iguanoides.

KEY TO HONDURAN FAMILIES OF THE
SUPERFAMILY IGUANOIDEA

1A. Mental scale partially divided (Fig. 54) 2
1B. Mental scale entire (Fig. 55) 3
2A. Femoral pores present in both sexes (Figs. 56, 57) . . . .
. . . . Polychrotidae (p. 331)
2B. Femoral pores absent in both sexes . . . . Dactyloidae (p. 218)
3A. Scales on dorsal surface of body relatively large, strongly keeled, strongly imbricate, and often spiny at posterior tips (Fig. 58); body dorsally compressed (Fig. 59) 4

Figure 55.   Mental scale entire. *Laemanctus waltersi.* USNM 580367 from Tela, Atlántida.

3B. Scales on dorsal surface of body relatively small, smooth to weakly keeled, slightly imbricate to juxta-posed (Fig. 60); body not dorsally compressed (Fig. 60) 5

4A. Supraocular scales smooth, large, in only one regular row (Fig. 61); femoral pores absent in both sexes .... Leiocephalidae (p. 288)

4B. Supraocular scales keeled, in more than 1 irregular row (Fig. 62);

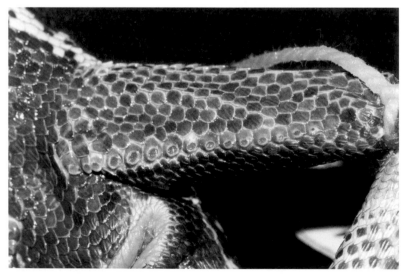

Figure 56.   Femoral pores present in both sexes. *Polychrus gutturosus.* USNM 563277 (male) from Wakling Tingni Kiamp, Gracias a Dios. Males have femoral pores more distinct than females.

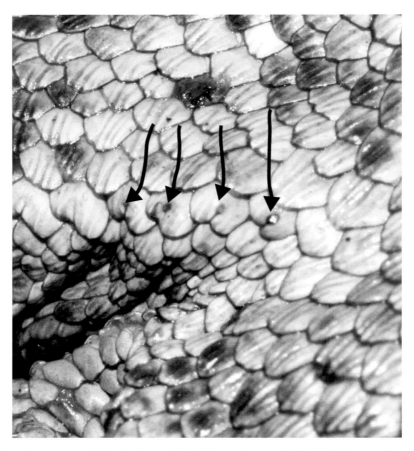

Figure 57. Femoral pores present in females (arrows) of *Polychrus gutturosus*. USNM 570298 from near Rus Rus, Gracias a Dios. Many femoral pores indistinct in female *P. gutturosus*.

femoral pores present in both sexes (Figs. 56, 57), but femoral pores best developed in males; femoral pore series relatively short in one species..... Phrynosomatidae (p. 291)

5A. Distinctive parietal crests present, although those crests reduced to knobs in subadults and females of one genus; no distinct caudal whorls and enlarged scale ventral to tympanum....... Corytophanidae (p. 168)

5B. Parietal crests absent; distinct caudal whorls present or, if absent, large scale present ventral to tympanum........... Iguanidae (p. 225)

CLAVE PARA LAS FAMILIAS HONDUREÑAS DE LA SUPERFAMILIA IGUANOIDEA

1A. Escama mentonal parcialmente dividida (Fig. 54) 2

1B. Escama mentonal completa (Fig. 55) 3

2A. Poros femorales presentes en ambos sexos (Figs. 56, 57) .... Polychrotidae (p. 331)

2B. Poros femorales ausentes en ambos sexos........... Dactyloidae (p. 218)

3A. Escamas en el dorso relativamente grandes, fuertemente quilladas, muy imbricadas y frecuentemente con una espina en el extremo posterior de la escama (Fig. 58);

Figure 58.   Dorsal scales relatively large, strongly keeled, and imbricate. *Sceloporus variabilis*. USNM 580378 from Punta Novillo, Isla Zacate Grande, Valle.

cuerpo deprimido (Fig. 59) 4
3B. Escamas en el dorso relativamente pequeñas, lisas o débilmente quilladas, desde ligeramente imbricadas hasta yuxtapuestas (Fig. 60); el cuerpo no está deprimido (Fig. 60). . 5

4A. Escamas supraoculares grandes, lisas y en una sola hilera, en organización regular (Fig. 61); poros femorales ausentes en ambos sexos . . . . Leiocephalidae (p. 288)
4B. Escamas supraoculares quilladas en al menos dos hileras irregulares

Figure 59.   Body dorsally compressed. *Sceloporus* cf. *hondurensis* (probably undescribed). FN 256155 (still in my collection for further study) from Quebrada Las Cuevas Joconales, Santa Bárbara.

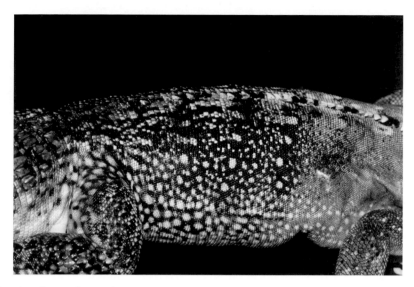

Figure 60.   Dorsal scales small, smooth to weakly keeled, and body not dorsally compressed. *Ctenosaura quinquecarinata.* USNM 581879 from Las Pilas, Choluteca.

(Fig. 62); poros femorales presentes en ambos sexos (Figs. 56, 57), pero más dessarrollado en machos; la línea de poros femorales relativamente corta en una especies ... Phrynosomatidae (p. 291)

5A. Crestas parietales presentes y distinguibles, aunque esas crestas pueden estar reducidas a protuberancias en subadultos y en las hembras de un género; sin anillos distinguibles en forma de espinas

Figure 61.   Supraocular scales smooth with one row of large scales regular in arrangement. *Leiocephalus varius.* USNM 494807 from Isla Grande, Swan Islands, Gracias a Dios (introduced population).

Figure 62.   Supraocular scales keeled, in more than one irregular row. *Sceloporus variabilis.* USNM 580777 from Isla Tigrito, Valle.

sobre la cola; escama agrandada subtimpánica ausente . . . . Corytophanidae (p. 168)

5B. Crestas parietales ausentes; anillos distinguibles de escamas agrandadas en forma de espinas sobre la cola presentes, o si están ausentes, hay una escama agrandada subtimpánica . . . . Iguanidae (p. 225)

Family Corytophanidae Fitzinger, 1843

This family (*sensu* Frost and Etheridge, 1989: 34; Frost et al., 2001: 13) of lizards is limited in distribution to the Western Hemisphere, where it ranges from Tamaulipas and Jalisco, Mexico, southward through Central America into South America as far south as west-central Venezuela east of the Andes and western Ecuador west of the Andes. One, possibly two, species have been introduced and established in southern Florida, USA. Three genera comprising 11 named species are included in this family. All three genera and nine species occur in Honduras. Males of *Basiliscus* are significantly larger than are females, but there is no significant size differences between males and females of *Corytophanes* and *Laemanctus*.

KEY TO HONDURAN GENERA OF THE FAMILY CORYTOPHANIDAE

1A. Cephalic crest flat-topped (Fig. 63) . . . . *Laemanctus* (p. 199)

1B. Cephalic crest with raised parietal ridges or laterally compressed flap of skin 2

2A. Cephalic crest formed by laterally compressed flap of skin lacking raised parietal ridges (Figs. 64, 65), cephalic crests reduced to knobs on posterior portion of head in subadults and adult females; lateral flap of scales present on outer edge of digits III–V on hind limbs (Fig. 66); male middorsal crest composed of strongly raised flap of skin with spines (Fig. 67). . . . . *Basiliscus* (p. 172)

2B. Cephalic crest with raised parietal ridges that unite posteriorly (Fig. 68); cephalic crests well developed and similar shaped in both sexes;

Figure 63.   Flat-topped cephalic crest with smooth posterior margin. *Laemanctus longipes.* USNM 570297 from Río Amarillo, Copán.

lateral flap of scales absent on outer edges of digits on hind limbs; middorsal crest of both sexes similar, composed of raised and similar-shaped scales (Fig. 69) . . . . *Corytophanes* (p. 185)

CLAVE PARA LOS GÉNEROS HONDUREÑOS DE LA FAMILIA CORYTOPHANIDAE

1A. La superficie dorsal de la cresta de la cabeza plana (Fig. 63) . . . . *Laemanctus* (p. 199)

Figure 64.   Male cephalic crest formed by laterally compressed, single lobe (flap of skin). *Basiliscus vittatus.* FMNH 283753 from Utila Island, Islas de la Bahía. Best developed in males.

Figure 65.   Bilobed cephalic crest of a male *Basiliscus plumifrons*. USNM 572085 from Crique Yulpruan, Gracias a Dios. Photograph by James A. Poindexter.

Figure 66.   Lateral flap of scales present on outer three toes on each hind limb. *Basiliscus vittatus*. USNM 580365 from Agua Chiquito, Atlántida.

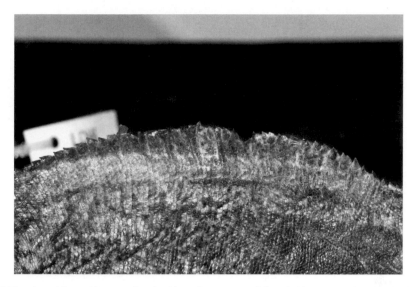

Figure 67.   Middorsal crest formed by strongly raised laterally compressed flap of skin connected to raised spines. *Basiliscus vittatus*. UF 144601 from Rus Rus, Gracias a Dios.

1B. La cresta en la superficie dorsal de la cabeza comprimida, formanda una aleta o con crestas parietales abultadas lateralmente 2

2A. Cresta en la superficie dorsal de la cabeza en la zona parietal con una aleta de piel comprimida, la cresta puede estar reducida a una protu-berancia en subadultos y en hembras adultas (Figs. 64, 65); piel lateral de los dedos III–V en las extremidades posteriores, formando un fleco de escamas (Fig. 66); cresta mediodorsal de piel fuertemente abultada y espinosa en los machos (Fig. 67)... *Basiliscus* (p. 172)

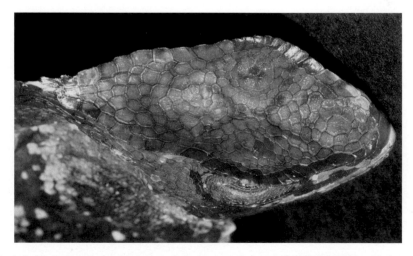

Figure 68.   Cephalic crest with raised parietal ridges that unite posteriorly and dorsal head scales smooth to weakly rugose. *Corytophanes cristatus*. USNM 580366 from near Mezapita, Atlántida.

Figure 69.   Male middorsal crest formed by raised, laterally compressed scales, with nuchal crest continuous with middorsal crest. *Corytophanes cristatus*. UF 166339 from Los Pinos, Cortés.

2B. Cresta en la superficie dorsal de la cabeza abultado a cada lado de la zona parietal, se unen posteriormente (Fig. 68); crestas cefálicas bien desarrolladas, similares en ambos sexos; piel lateral de los dedos III–V en las extremidades posteriores no formando un fleco de escamas; borde de la cresta mediodorsal formanda por escamas agrandadas, de igual en forma en ambos sexos (Fig. 69)...........
.... *Corytophanes* (p. 185)

Genus *Basiliscus* Laurenti, 1768

Basiliscus Laurenti, 1768: 50 (type species: *Basiliscus americanus* Laurenti, 1768: 50 [= *Lacerta basiliscus* Linnaeus, 1758: 206], by monotypy as well as tautonymy).

Geographic Distribution and Content.—This genus ranges from northwestern Jalisco and southern Tamaulipas, Mexico, southward through Central America to western Ecuador and west-central Venezuela. One species has been introduced and established into southern Florida, USA, and another species was introduced and might be established in southern Florida, USA. Four species are currently recognized, two of which occur in Honduras.

Etymology.—The name *Basiliscus* is derived from the Greek *basiliskos* (a small king) and alludes to the parietal crest in the type species that is emblematic of a crown.

KEY TO HONDURAN SPECIES OF THE GENUS
BASILISCUS

1A. Ventral scales keeled, imbricate (Fig. 70); dorsal color some shade of brown in life; distinct yellowish or cream postorbital stripe extending onto dorsolateral portion of body, becoming indistinct posteriorly; adult cephalic crest single lobed (Fig. 64)....... *vittatus* (p. 178)
1B. Ventral scales smooth, nonimbricate (Fig. 71); dorsal color some shade of green in life; no yellow or cream postorbital stripe present; adult male cephalic crest bilobed (Fig. 65).......... *plumifrons* (p. 174)

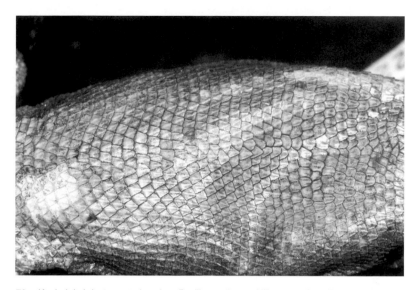

Figure 70.   Keeled, imbricate ventral scales. *Basiliscus vittatus.* UF 144601 from Rus Rus, Gracias a Dios.

CLAVE PARA LAS ESPECIES HONDUREÑAS DEL
GÉNERO *BASILISCUS*

1A. Escamas ventrales quilladas, imbricadas (Fig. 70); color de dorso con
algún tono de pardo en vida; franja
distinguible de color amarillo o
crema extendiéndose desde la región postorbital hasta porción dorsolateral del cuerpo, volviéndose
indistinguible en el parte posterior
del cuerpo; cresta cefálica en los
adultos con un solo lóbulo (Fig. 64)
.... *vittatus* (p. 178)

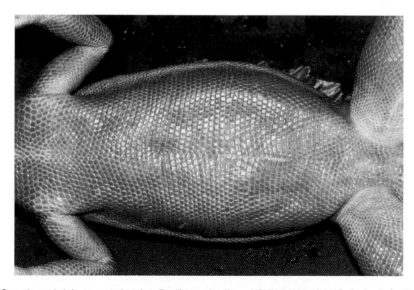

Figure 71.   Smooth, nonimbricate ventral scales. *Basiliscus plumifrons.* USNM 563291 from Caño Awalwás, Gracias a Dios.

1B. Escamas ventrales lisas, no imbricadas (Fig. 71); color de dorso algún tono de verde en vida; sin una franja pálida postorbital que se extiende hacia la parte posterior del cuerpo; cresta cefálica bilobulada en los machos adultos (Fig. 65) .... *plumifrons* (p. 174)

Basiliscus plumifrons Cope, 1875

Basiliscus plumifrons Cope, 1875: 125 (six syntypes, UIMNH 40735, USNM 32622–26 [see Cochran, 1961: 94; Smith et al., 1964: 43]; type locality: "Sipurio" [Costa Rica]); Maturana, 1962: 32; Meyer, 1969: 239; Meyer and Wilson, 1973: 21; Cruz Díaz, 1978: 27; Böhme, 1988: 34; Köhler, 1999e: 49; Köhler, 2000: 56; Köhler, McCranie, and Nicholson, 2000: 425; Nicholson et al., 2000: 29; Wilson et al., 2001: 134; McCranie et al., 2006: 112; Wilson and Townsend, 2006: 105; McCranie, 2007a: 176, Solís et al., 2014: 129; McCranie, 2015a: 367.

Geographic Distribution.—*Basiliscus plumifrons* occurs at low and moderate elevations from northeastern Honduras to northwestern Panama on the Atlantic versant and in southeastern Costa Rica on the Pacific versant. This species was introduced into southern Florida, USA, but it is not certain if it is established there. In Honduras, this species is known from low elevations in the northeastern portion of the country.

Description.—The following is based on ten males (USNM 337558, 549370, 549372, 549376, 563227, 563239, 563291–92, 570284–85) and nine females (FMNH 282570; USNM 549375, 563229, 563234–38, 563244). *Basiliscus plumifrons* is a large corytophanine (maximum recorded SVL 207 mm [USNM 337558, a male]) with a long tail and long limbs; male cephalic crest beginning just posterior to interparietal scale, laterally compressed, bilobed in

males, anterior lobe small and narrow, posterior lobe large, extending posteriorly to about midlength of neck; female cephalic crest smaller, less developed than male crest, bilobed only in largest females, anterior lobe located just posterior to interparietal scale, posterior lobe located on posterior portion of head; subadult and smaller adult female crests consist of small knob on posterior portion of head; scales of cephalic crest of adults enlarged, smooth; dorsal head scales in frontal region keeled, rugose; 2–4 (usually 3–4) scales separating supraorbital semicircles; 1–3 (usually 2–3) scales between supraorbital semicircles and interparietal; parietal eye distinct; interparietal scale larger than surrounding scales; tympanum distinct, higher than long; nasal scale single, nostril located more-or-less centrally in scale, nostril opening directed laterally or slightly posterolaterally; moveable eyelid present; pupil circular; 6–8 (usually 7) supralabials and infralabials; 2–4 chinshields in contact with infralabials; gular fold complete, continuous with antehumeral fold; gular scales elongate, smooth; body laterally compressed; dorsal and lateral body scales imbricate, smooth to weakly keeled; lateral body scales smaller than dorsal body scales; adult males with strongly raised middorsal crest extending from level anterior to forelimb to about midlevel of hind limb, middorsal crest arched anteriorly and posteriorly, with small gap of slightly raised dorsal crest present between body crest and caudal crest; male caudal crest strongly raised on anterior third of tail, arched anteriorly and posteriorly; male middorsal and caudal crests supported by bony rays connected by laterally compressed flap of skin, dorsal borders serrated because tissue between rays form concave border; adult female middorsal and caudal crests much lower and less developed than adult male crests, otherwise similar in structure to male crests; middorsal and caudal crests of subadults resemble those of adult females, but even less developed;

Plate 34. *Basiliscus plumifrons*. USNM 337558, adult male, SVL = 207 mm. Olancho: Caño El Cajón.

scales on nape region slightly smaller than dorsal body scales; ventral scales larger than dorsal scales, nonimbricate, smooth; 54–68 (59.2 ± 4.2) midventral scales between levels of axilla and groin in 16; 131–165 (144.8 ± 10.2) scales around midbody in 16; caudal autotomy absent; caudal scales keeled, although weakly, anteriorly; femoral and precloacal pores absent; scale arrangement on toes agree with patterns B or D of Maturana (1962), including variation in those patterns discussed by him; lateral flap of scales present on outer edge of at least 3 outer digits of hind limb, lateral flap on Digit IV of hind limb, when folded, flaps cover more than half lateral height of digit; 31–38 (34.9 ± 1.6) scales in lateral flap of Digit IV of hind limb on 32 sides; 36–44 (39.0 ± 2.2) subdigital scales on Digit IV of hind limb on 32 sides; subdigital scales keeled, with nonkeratinized knobs on anterior section of most scales; SVL 136–207 (178.6 ± 24.0) mm in males, 128–184 (152.6 ± 17.4) mm in females; TAL/SVL 2.42–3.05 in males, 2.24–2.93 in seven females; casque L/SVL 0.31–0.50 in nine males, 0.17–0.37 in seven females; HL/SVL 0.22–0.31 in nine

males, 0.23–0.27 in seven females; SHL/SVL 0.31–0.34 in nine males, 0.31–0.35 in seven females.

Color in life of an adult male (USNM 337558; Plate 34): dorsal surface of body Parrot Green (260) with 2 rows of ivory white spots on lateral portion; head Parrot Green, lips with grayish tinge and scattered white spots; anterior lobe of cephalic crest Parrot Green, outlined with white to bluish white border; posterior lobe of cephalic crest Parrot Green, outlined with white border and patterned with scattered bluish white spots; middorsal crest Yellow-Green (58) with scattered bluish white spotting and black bandlike markings outlined with Robin's Egg Blue (93); caudal crest Chartreuse (158) with interrupted Robin's Egg Blue border; tail posterior to crest Parrot Green, with broad dark gray crossbands; ventral surface of head a mixture of Yellow-Green and Robin's Egg Blue mottling; ventral surfaces of body, limbs, and tail Yellow-Green; iris Yellow Ocher (123C).

Color in alcohol: dorsal surface of body greenish olive to gray, with or without 4–5 darker, indistinct, narrow bands dorsally,

Plate 35.   *Basiliscus plumifrons*. FN 25703 (still in Honduras because of permit problems), adult female. Gracias a Dios: Bachi Kiamp.

bands (when present) extending laterally to about midheight of body; pale green spots sometimes present on body below crest and above hind limb; dorsal surface of head and cephalic crest gray to black; dorsal surface of tail gray with darker gray crossbands posterior to caudal crest; middorsal and caudal crests similar in color to body and tail, paler green streaks and spots sometimes present in male crests; ventral surfaces of head, body, and tail pale green to dark gray, subcaudal surface posterior to caudal crest usually with dark brown or gray crossbands.

Diagnosis/Similar Species.—*Basiliscus plumifrons* can be distinguished from all other Honduran lizards, except *B. vittatus*, by having a lateral flap of scales on the outer edge of at least the outer 3 digits on the hind limb. *Basiliscus vittatus* has keeled, imbricate ventral scales, has a single lobed cephalic crest in adult males, has a brown dorsal color in life, has a yellow or cream postorbital stripe, and has a less elevated male middorsal crest (versus smooth, non-imbricate ventral scales, bilobed cephalic crest in adult males, green dorsum in life without pale postorbital stripes, and male

middorsal crest greatly elevated in *B. plumifrons*).

Illustrations (Figs. 65, 71; Plates 34, 35).—Freiberg, 1972 (adult); Guyer and Donnelly, 2005 (adult); Kober, 2012 (adult, subadult); Köhler, 1993c (adult, juvenile, hatchling), 1999b (adult), 1999d (adult, juvenile), 2000 (adult), 2001b (adult), 2003a (adult, male head crest shape), 2008 (adult, male head crest shape); Köhler, McCranie, and Nicholson, 2000 (adult); Kundert, 1974 (adult); Maturana, 1962 (male head crest shape, chin scales, scale pattern of Phalanx III of Toe IV on hind foot); McCranie, 2007a (adult); McCranie et al., 2006 (adult, cephalic crest); Pianka and Vitt, 2003 (adult); Pough et al., 2003 (adult), 2015 (adult); Savage, 2002 (adult, head crest shape, dorsal head scales); W. Schmidt and Henkel, 1995 (adult); Trutnau, 1986 (adult); Vitt and Caldwell, 2014 (adult); Wirth, 2012a (adult, juvenile), 2012b (adult).

Remarks.—Lang (1989), in his monograph on "basiliscine [= corytophanine] iguanians," did not include Honduras within the range of *Basiliscus plumifrons*, even though the species had previously been reported from the country (Meyer and Wilson, 1973). Lang's (1989) phylogenetic analysis showed *B. plumifrons* to form a clade with *B. basiliscus* (Linnaeus, 1758: 206). Savage (2002) wrote that Costa Rican adult male *B. plumifrons* have a red iris in life; however, the two color photographs in his work (plates 236–237) show Costa Rican males with a yellow iris. All Honduran adult males I saw in life (most released or not captured), like the juveniles and adult females, had a yellow iris.

Adult *Basiliscus plumifrons* are voracious biters when captured and seem to carry infectious bacteria in their mouths. Both times I was bit hard by an adult *B. plumifrons*, the site of the bite became infected, with one of those bites taking over a week to heal, even after applying antibiotic cream to the site regularly.

Natural History Comments.—*Basiliscus plumifrons* is known from 40 to 225 m elevation in the Lowland Moist Forest formation. This diurnal species has been collected in January and February, from May to August, and in November, and thus is probably active throughout the year. Adults and juveniles were commonly seen sleeping at night on vegetation above streams and rivers in primary forest. Adults and juveniles were active during sunny days both on the ground and in trees along rivers and streams through primary and slightly disturbed forest (although adults are largely arboreal). Juveniles were frequently seen from boats on partially emergent logs near the water edge along large rivers. No *B. plumifrons* were seen away from the vicinities of rivers and streams. Vaughan et al. (2007) also noted the riparian habitats of *B. plumifrons* on a cacao farm in Costa Rica, but Sajdak et al. (1980, also in Costa Rica) reported a forest edge habitat away from water for this species. Lattanzio and La-Duke (2012) studied the ecology of a Costa Rican population of *B. plumifrons* and found juveniles were most frequently seen in more open and grassier habitats closer to water than were adults. Juvenile activity also declined significantly during the dry season, whereas adult activity did not decline significantly during the dry season, but timing of peak activity shifted to mostly early morning and late afternoon periods. Females deposit 4–17 eggs per clutch with much of the reproduction centered during the rainy season (May to September, December; literature summation by Savage, 2002: 431). Hirth (1963a) found small shrimp and crabs, a small bat, and vegetation in the form of seeds, fruits, and leaves in stomachs of Costa Rican specimens. Only adults contained vegetable matter. Cover (1986) reported a frog consumed by a subadult *B. plumifrons* in Costa Rica. Kober (2012) reported small fish (guppies) in the diet of captive individuals, and also said it forages in water. I have never seen *B. plumifrons* in the field foraging in water, but Hirth's (1963a) record of small shrimp in the diet supports foraging in aquatic habitats.

Etymology.—The name *plumifrons* is derived from the Latin *pluma* (soft feather) and *frons* (forehead, forepart) and alludes to the "accessory anterior portion" of the head crest as described by Cope (1875: 126).

Specimens Examined (47, 5 skeletons [1]; Map 29).—**GRACIAS A DIOS**: Awasbila, USNM 563231–34, 563292; Bachi Kiamp, USNM 573975, FN 257033 (still in Honduras because of permit problems); Bachi Tingni, FMNH 282570; Bodega de Río Tapalwás, UF 144599–600, USNM 563241–42, 570281; Caño Awalwás, UF 144598, USNM 563227–30, 563235, 563291; Crique Yulpruan, USNM 570282–83, 570285; Hiltara Kiamp, USNM 563236; Kalila Plapan Tingni, USNM 563237; Kaska Tingni, USNM 563240; Kipla Tingni Kiamp, USNM 563238–39, 573101–02; Río Coco, USNM 24523; Urus Tingni Kiamp, USNM 563243–44; Warunta Tingni Kiamp, USNM 565408, 570284. **OLAN-CHO**: Caño El Cajón, USNM 337558, 563245; Matamoros, SMF 80827, USNM

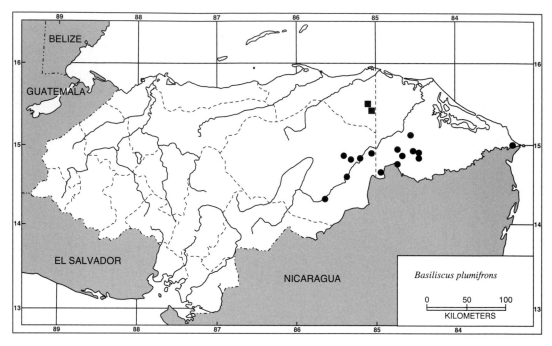

Map 29.   Localities for *Basiliscus plumifrons*. Solid circles denote specimens examined and solid squares those for accepted records.

549373; Quebrada El Guásimo, SMF 80836, USNM 549370; Quebrada El Mono, USNM 549374; Qururia, USNM 549375–76; Río Kosmako, USNM 549371–72, 563246. "HONDURAS": UF 67526 (skeleton), 67612 (skeleton), 68101 (skeleton), 68305 (skeleton), 68311 (skeleton).

Other Records (Map 29).—**COLÓN**: Empalme Río Chilmeca, UNAH 5435, Cañon Subterráneo del Río Plátano, UNAH 5473 (Cruz Díaz, 1978, specimens now lost). "HONDURAS": ZFMK 41935 (Böhme, 1988).

Basiliscus vittatus Wiegmann, 1828

Basiliscus vittatus Wiegmann, 1828: col. 373 (holotype not stated, but Lang, 1989: 121, listed ZMB 549–51 as syntypes; type locality not stated, but "Mexico" implied from title of publication containing description); Werner, 1896: 346; Dunn and Emlen, 1932: 28; Lynn, 1944: 190; Brattstrom and Howell, 1954: 117; Maturana, 1962: 32; Meyer, 1966: 175; Meyer, 1969: 239; Hahn, 1971: 111; Meyer and Wilson, 1973: 22 (in part); Wilson and Hahn, 1973: 113; Cruz Díaz, 1978: 27; Wilson et al., 1979a: 25; Hudson, 1981: 377; O'Shea, 1986: 38; Lang, 1989: 122; Cruz et al., 1993: 28; Espinal, 1993, table 3; Wilson and Cruz Díaz, 1993: 16; Köhler, 1994a: 4, 8; Köhler, 1994b: 12; Köhler, 1995b: 97, 102; Köhler, 1996d: 20; Cruz Díaz, 1998: 29, *In* Bermingham et al., 1998; Köhler, 1998d: 375, 382; Monzel, 1998: 157; Wilson and McCranie, 1998: 15; Köhler, 1999a: 214; Köhler, 1999d: 56; Köhler, McCranie, and Nicholson, 2000: 425; Lundberg, 2000: 4; Nicholson et al., 2000: 29; Espinal et al., 2001: 106; Grismer et al., 2001: 134; Lundberg, 2001: 23; Wilson et al., 2001: 134; Castañeda, 2002: 38; Lundberg, 2002a: 7; Lundberg, 2002b: 7; McCranie, Castañeda, and Nicholson, 2002: 25;

Köhler, 2003a: 122; Powell, 2003: 36; Gutsche, 2005c: 317; McCranie and Castañeda, 2005: 14; McCranie et al., 2005: 86; Castañeda and Marineros, 2006: 3.8; Lovich et al. 2006: 15; McCranie et al., 2006: 113; Townsend, 2006a: 34; Wilson and Townsend, 2006: 105; Townsend et al., 2007: 10; Wilson and Townsend, 2007: 145; Köhler, 2008: 134; Townsend and Wilson, 2010b: 692; McCranie, 2011a: 326; McCranie and Solís, 2013: 242; McCranie and Valdés Orellana, 2014: 45; Solís et al., 2014: 129; McCranie, 2015a: 367.

Basiliscus [sp.]: Schmidt, 1924: 86.
Basiliscus basiliscus: Barbour, 1928: 59.
Anolis allisoni: Meyer and Wilson, 1973: 15 (in part).

Geographic Distribution.—*Basiliscus vittatus* occurs at low and moderate elevations from central Veracruz, Mexico, to northwestern Colombia on the Atlantic versant and on the Pacific versant from northwestern Jalisco, Mexico, to west-central Nicaragua and from central Panama to northwestern Colombia. *Basiliscus vittatus* has also been introduced and established in southern Florida, USA. This species is widely distributed through much of Honduras, including the Bay Islands.

Description.—The following is based on ten males (UF 144601–02, 144606–07; USNM 549385, 549399–400, 563249, 563258, 563264) and ten females (KU 67209, 203150; UF 124036; USNM 549380–81, 549386, 549391, 549394, 563256, 563265). *Basiliscus vittatus* is a moderately large corytophanine (maximum recorded SVL 147 mm [USNM 549385, a male]) with a long tail and long limbs; male cephalic crest beginning just posterior to interparietal scale, laterally compressed, triangular in shape, extending posteriorly to about midlength of neck; female cephalic crest smaller, less well developed than male crest, located on posterior portion of head (subadult male and subadult female crests

consist of a small knob on posterior portion of head); scales of male cephalic crest enlarged, smooth, those of female cephalic crest small, smooth; dorsal head scales in frontal region keeled, rugose; 1–4 (usually 2–3) scales separating supraorbital semicircles; 1–3 (usually 1–2) scales between supraorbital semicircles and interparietal; parietal eye distinct; interparietal scale larger than surrounding scales; tympanum distinct, higher than long; nasal scale single, nostril located more-or-less centrally in scale, nostril opening directed laterally or slightly posterolaterally; moveable eyelid present; pupil circular; 6–8 (usually 7–8) supralabials; 7–9 (usually 7–8) infralabials; 1–3 chinshields in contact with infralabials; gular fold complete, continuous with antehumeral fold; gular scales elongate, smooth to weakly keeled; body laterally compressed; dorsal and lateral body scales imbricate, keeled; lateral body scales smaller than dorsal body scales; adult males with raised middorsal crest extending from level above shoulder region to about midlevel of hind limb, middorsal crest supported by bony rays, male rays better developed than females, bony rays connected by laterally compressed flap of skin, crest continuing onto tail as low serrated ridge; female and subadult middorsal crest consisting of row of serrated scales, serrated scales continue onto tail; scales on nape region smaller than dorsal body scales; ventral scales larger than dorsal scales, imbricate, keeled, some scales mucronate; 39–51 (43.7 ± 3.6) midventral scales between levels of axilla and groin in 17; 95–126 (109.5 ± 7.6) scales around midbody in 17; caudal autotomy absent; caudal scales keeled; femoral and precloacal pores absent; scale arrangement on toe patterns A or C of Maturana (1962), including variation in those patterns discussed by him; lateral flap of scales present on outer edge of outer 3 digits of hind limb, lateral flap on Digit IV of hind limb, when folded, covering less than half of lateral height of digit; 33–40 (36.0 ± 2.1) scales in lateral flap of Digit IV of hind limb on 34 sides; 31–38 (35.1 ± 2.0) subdigital scales on Digit IV of hind limb on 34 sides; subdigital scales keeled, with dark, keratinized knobs on anterior section of most scales; SVL 103–147 (128.6 ± 11.6) mm in males, 102–134 (111.5 ± 10.1) mm in females; TAL/SVL 2.28–3.05 in seven males, 2.73–3.21 in females; casque L/SVL 0.41–0.56 in seven males, 0.16–0.36 in females; HL/SVL 0.22–0.33 in seven males, 0.23–0.25 in females; SHL/SVL 0.32–0.36 in seven males, 0.30–0.36 in females.

Color in life of an adult male (USNM 549385): dorsal surface of head brown with orange cast; dorsal surface of body gray-brown with dark brown blotches, blotches flecked with rust brown; dorsolateral stripe pale gray-brown; venter tan with orange patina; limbs and tail gray-brown dorsally, paler brown ventrally. Color in life of an adult female (USNM 563265): dorsum Clay Color (26) with Drab-Gray (119D) flecks and Olive-Brown (28) crossbands; dorsolateral stripe Grayish Horn Color (91); ventrolateral area blotched with Pale Horn Color, Grayish Horn Color, and Drab-Gray; top of head Olive-Brown; side of head (postocular or temporal area) Brownish Olive (29), bounded above and below by Olive-Brown stripes; Cream Color (54) stripe passes from in front of eye obliquely to below eye to angle of mouth; supralabials and infralabials Olive-Yellow (52); chin mottled Pale Horn Color and Grayish Horn Color; ventral surfaces of body predominately Pale Horn Color with Grayish Horn Color blotching; ventral surfaces of fore- and hind limb Clay Color with Pale Horn Color flecking; dorsal and ventral surfaces of tail Grayish Horn Color with Pale Horn Color narrow bands. Color in life of another adult female (USNM 549386): dorsal surface of head dark olive green; dorsal surface of body mottled olive green and gray, with dark brown crossbands; dorsal surface of forelimb olive brown; dorsal surface of hind limb mottled olive green and dark olive green; lips pale gray; dorsal surface of tail gray with dark gray crossbands; chin pale pink-orange; belly gray with pale pink-orange cast.

Color in alcohol: dorsal surfaces of head and body of adult males brown to dark brown, with distinct pale brown dorsolateral stripe on each side extending from posterior edge of head to about midlength of body, another pale brown stripe extends from below each eye to region of forelimb; male head and dorsal crests usually slightly paler brown than dorsal ground color; dorsal surfaces of head and body of adult females pale brown to brown, with distinct darker brown crossbars or blotches present from immediately posterior to head to above hind limb, those crossbars or blotches divided by brown dorsolateral stripe on each side, stripe extending from above tympanum to above hind limb; pale brown or cream stripe usually also extending from below each eye to about region of forelimb in females; ventral surfaces of head, body, and tail of adult males pale brown to brown, with indistinct darker brown crossbands present on subcaudal surface; ventral surfaces of head, body, and tail of adult females dirty white to pale brown, dark brown to black flecks or small spots frequently present on lateral portion of belly and on ventral surfaces of limbs and tail.

Diagnosis/Similar Species.—*Basiliscus vittatus* can be distinguished from all other Honduran lizards, except *B. plumifrons*, by having a lateral flap of scales on the outer edge of the outer 3 digits on the hind limbs. *Basiliscus plumifrons* has smooth, nonimbricate ventral scales, has a bilobed cephalic crest in adult males, has a green dorsal color in life, and lacks a pale postorbital stripe (versus keeled, imbricate ventrals, single-lobed crest, and brown dorsum with yellowish or cream postorbital stripe in *B. vittatus*).

Illustrations (Figs. 64, 66, 67, 70; Plates 36, 37).—Álvarez del Toro, 1983 (adult); Calderón-Mandujano et al., 2008 (adult);

Plate 36.   *Basiliscus vittatus*. UF 144601, adult male, SVL = 133 mm. Gracias a Dios: Rus Rus.

Campbell, 1998 (adult); Conant and Collins, 1998 (adult); Davis and Dixon, 1961 (ontogenetic changes in male head crest); Günther, 1890, *In* Günther, 1885–1902 (adult); Guyer and Donnelly, 2005 (adult); Köhler, 1993c (adult, juvenile), 1999d (adult, juvenile), 2000 (adult, chin scales), 2001b (adult, chin scales), 2003a (adult, chin scales, male head crest shape), 2008 (adult, subadult, chin scales, male head crest shape); Köhler et al., 2005 (adult); Lee, 1996 (adult, head scales, subdigital lamellae), 2000 (adult, head scales, subdigital lamellae); Lundberg, 2001 (adult); Ma-

Plate 37.   *Basiliscus vittatus*. FMNH 283752, adult female. Islas de la Bahía: Roatán Island, Palmetto Bay.

turana, 1962 (male head crest shape, chin scales, scale pattern of Phalanx III of Digit IV of hind foot); McCranie et al., 2005 (adult, subadult, head crest), 2006 (adult, cephalic crest); Meshaka et al., 2004 (subadult); Powell, 2003 (subadult); Savage, 2002 (adult, male head crest shape); W. Schmidt and Henkel, 1995 (juvenile); Stafford and Meyer, 1999 (adult); Villa et al., 1988 (adult, juvenile); Wirth, 2012a (adult, head), 2012b (adult).

Remarks.—Lang (1989: 121) listed ZMB 549–51 as syntypes of *Basiliscus vittatus.* Stuart (1963: 66) had stated ZMB 549 was the "Type" of *B. vittatus.* According to Article 74 of the Code (ICZN, 1999), Stuart's act does not seem sufficient to have designated that specimen as the lectotype of *B. vittatus* Wiegmann.

Gray (1852: 439) described *Cristasaura mitrella* (a synonym of *Basiliscus vittatus*) based on material collected by a Mr. Dyson from "Honduras." However, according to Schmidt (1941: 503), the Dyson material in the BMNH collection originated from British Honduras (= Belize). Boulenger (1885b: 110) and Günther (1885: 55, *In* Günther, 1885–1902) also reported on the Dyson specimens of *B. vittatus* stated to be from "Honduras." To add to this confusion, Smith and Taylor (1950a: 319) restricted the type locality of Gray's *C. mitrella* to "Honduras," the incorrect country of origin.

Lang (1989) reviewed the morphology of this species and provided a phylogenetic analysis that showed *B. vittatus* (along with *B. galeritus* A. M. C. Duméril, 1851: 61, *In* A. M. C. Duméril and Duméril, 1851) as basal to the *B. plumifrons–B. basiliscus* clade. Savage (2002) pointed out errors in the distribution map of *B. vittatus* in Lang's (1989) revision. Significant differences in extent of development in the height of the dorsal crest of males occur in this geographically widely distributed species, with males from some Mexican populations having distinctly higher dorsal body crests than do Honduran males. More than a single species

appears to be represented under the name *B. vittatus.*

Natural History Comments.—*Basiliscus vittatus* is known from sea level to 1,400 m elevation in the Lowland Moist Forest, Lowland Dry Forest, Lowland Arid Forest, Premontane Wet Forest, Premontane Moist Forest, and Premontane Dry Forest formations. This diurnal species has been collected in every month of the year and occurs in highly disturbed areas, as well as in primary forest. *Basiliscus vittatus* is common around streams, rivers, and temporary and permanent ponds, both while active during daytime and while sleeping on vegetation at night. It can also be found at considerable distances from water. When active, adults are usually seen in vegetation, whereas juveniles are primarily terrestrial. Meyer (1966) reported two Honduran females collected in June contained six and seven eggs. Elsewhere, clutch size has been reported to be 2–18 eggs (Köhler, 1997f) or 3–18 eggs (Köhler, 1999d), but other authors have reported 3–12 eggs per clutch (Fitch, 1970; Lee, 1996, and references cited in those works). Females can deposit at least four clutches per year throughout an extended breeding season (Fitch, 1973a). Savage (2002), in a literature summary, said the main reproductive period in Costa Rica begins from mid-February to March and continues to October. Much original data on reproduction for a site in Costa Rica was given by Hirth (1963b). Gutsche (2005c) reported a large adult male *B. vittatus* feeding on a newly hatched *Ctenosaura bakeri* on Isla de Utila. Savage (2002), in a literature summation, wrote that young *B. vittatus* are exclusively carnivorous, feeding principally on insects and spiders, with those from coastal localities also eating terrestrial crustaceans. Adults eat "a wide variety of animal prey and considerable amounts of plant material (grasses, seeds, fruits, and stems)" (p. 432). Hirth (1963b) provided an extensive list and discussion of food items at a Costa Rican site. Meshaka et

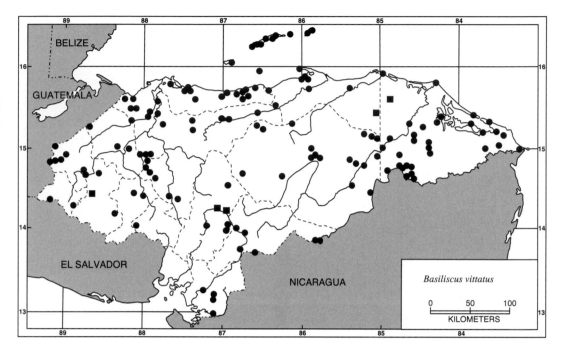

Map 30. Localities for *Basiliscus vittatus*. Solid circles denote specimens examined and solid squares those for accepted records.

al. (2004) recorded beetles, roaches, ants, true bugs (Hemiptera), and ficus fruits from stomachs of an introduced population in southern Florida, USA. A juvenile *B. vittatus* (USNM 573111) was removed from the stomach of a Honduran specimen (USNM 573263) of the snake *Leptodeira rhombifera* Günther (1872: 32). Telford (1977) reported saurian malaria in specimens of this species from "Honduras."

Etymology.—The name *vittatus* is Latin meaning "decorated or bound with a ribbon" and probably alludes to either the pale longitudinal facial or body stripe.

Specimens Examined (871, 1 skeleton [146]; Map 30).—**ATLÁNTIDA**: Agua Chiguito, USNM 580365; Carmelina, USNM 62969; Corozal, LACM 48361, LSUMZ 21651; Dacota, UMMZ 58368; Estación Forestal CURLA, USNM 508441, 563247, 578720; La Ceiba, CM 29006–07, LACM 48359, TU 19424, USNM 55243, 55246, 117604–06; Lancetilla, AMNH 70438–44,

CM 41224, MCZ R29607, 49996, 165724–25, UMMZ 69539 (12), 70325 (9), 71906, 74040 (3); near Pico Bonito Lodge, USNM 573974; Quebrada de Oro, UTA R-41223–24, USNM 509526 (skeleton), 508437–40, 549377–79; Río Viejo, UTA R-41222; Salado Barra, MCZ R191114–16; near Santa Ana, USNM 579546; 32.2 km ESE of Tela, LACM 48367; Tela, MCZ R21116, 21751–56, 27313–14, 27316, 27318–25, 29332, 165568–625, UMMZ 58367, 62520 (5), 70326. **CHOLUTECA**: Choluteca, MSUM 4648; El Despoblado, CAS 152970; El Faro, UNAH (1); La Fortuna, SDSNH 72717–23. **COLÓN**: Balfate, AMNH 58610, 58628–29; Cerro Calentura, LSUMZ 22461–63; Guaimoreto, LSUMZ 10287–91; Los Planes, CM 29368; Puerto Castillo, LSUMZ 22474; Río Guaraska, BMNH 1985.1145; Río Negro, ANSP 28122; Salamá, USNM 242444–513, 242516–69, 242571–72, 242574–75, 242577–81, 242586–600; 0.5–3.0 km W of Trujillo, KU 101435, LSUMZ

21652–54; about 3.2 km E of Trujillo, LSUMZ 22492; between Trujillo and Santa Fé, CM 64636; Trujillo, CM 64635, 64637–39. **COMAYAGUA**: 3 km W of Comayagua, TCWC 23813; about 8 km S of La Misión, MCZ R49976; Las Mesas, UTA R-41239–43; 1.6 km N of Pito Solo, LSUMZ 28530–31, 28533–36, 52501; Pito Solo, CM 64633, FMNH 5206–08, LSUMZ 29624; Siguatepeque, FMNH 5212. **COPÁN**: 12.9 km ENE of Copán, LACM 48362; 6 km NE of Copán, AMNH 156063; 1 km W of Copán, USNM 563248; Copán, FMNH 28533 (12), TCWC 23630, UMMZ 83034 (42); La Playona, USNM 549380–82, 563249; Río Amarillo, USNM 579545; Río Higuito, ANSP 22979–81, 26660; about 7 km ESE of San Antonio, LACM 72092; 8.0 km SE of Santa Rosa de Copán, LACM 48363. **CORTÉS**: Agua Azul, AMNH 70354, FMNH 282562–63, LSUMZ 52499–500, MCZ R49962–63, USNM 243308–32; Cofradía, LSUMZ 51429–30; 1.6 km NW of El Jaral, LACM 48364, LSUMZ 11642–46; El Paraíso, UF 144677; Hacienda Santa Ana, FMNH 5195–205; 7 km SW of La Lima, KU 67209; 5 km SW of La Lima, KU 67210; La Lima, BYU 22554–58, 22567–70, LACM 48366; Lago de Yojoa, MSUM 4645–47; Laguna Ticamaya, FMNH 5209–11; Los Pinos, UF 166336, USNM 573103; Quebrada Agua Buena, FMNH 283751; San Isidro, UF 166337; 3.2 km W of San Pedro Sula, LACM 48368; San Pedro Sula, LSUMZ 70592–94, MCZ R29392, USNM 24378; Santa Teresa, USNM 573890; 2.9 km E of Tegucigalpita, USNM 549383; 6 km ENE of Villanueva, CM 64632, LSUMZ 51428; 7.2 km ENE of Villanueva, LACM 48365. **EL PARAÍSO**: Arenales, LACM 16838–44; Boca Español, USNM 563250–51; 25.4 km W of Danlí, LACM 45069; between El Paraíso and Las Manos, UTA R-41233; near Ojo de Agua, AMNH 70350; 1 km S of Orealí, USNM 580773–74. **FRANCISCO MORAZÁN**: 5.0 km ENE of El Zamorano, KU 103240–41; El Zamorano, AMNH 70352–53, LACM 39771, MCZ

R49762, 49983, 165726–29; Hacienda San Diego, AMNH 69074–75, TCWC 19192–95; Río Yeguare, AMNH 70347–49, MCZ R48676–77, 49906, UMMZ 94042 (2); Tegucigalpa, AMNH 69085, BYU 18270–76, 18823, 18828–30, MCZ R49929. **GRACIAS A DIOS**: Awasbila, USNM 563252–53; Bachi Kiamp, USNM 573888, 579550; Barra Patuca, USNM 20297–303; Canco, UF 144607; Cauquira, UF 144606, 144611, USNM 565411; Crique Canco, USNM 563254; Crique Curamaira, USNM 563255; Crique Yulpruan, USNM 570286; Dos Bocas, FMNH 282564; Kakamuklaya, USNM 573104; Kaska Tingni, USNM 563256; Krahkra, USNM 563258–59; Krausirpe, LSUMZ 28477–79, 28483, 28490; Kyras, LACM 16837; about 15 km S of Mocorón, UTA R-42644; about 3 km S of Mocorón, UTA R-46184; Mocorón, UTA R-42637–43, 42645, 46179–83, 53512–20; Palacios, BMNH 1985.1292; Puerto Lempira, USNM 573106; confluence of Quebrada Waskista and Río Wampú, USNM 549384; Río Coco, USNM 24512–22; Rus Rus, UF 144601–04, USNM 563260–63; Sachin Tingni Kiamp, USNM 563524; Samil, USNM 573889; Sisinbila, USNM 579547; Tánsin, LACM 48357–58, LSUMZ 21643, USNM 573107; Tikiraya, UF 144605; Usus Paman, USNM 573887; Wampusirpe, LSUMZ 52498; Warunta, USNM 565410, FN 256938 (still in Honduras because of permit problems); Yahurabila, USNM 573108–11. **INTIBUCÁ**: 5.5 km S of Jesús de Otoro, LSUMZ 33688; La Rodadora, USNM 565409. **ISLAS DE LA BAHÍA**: Cayo Cochino Mayor, East End Village, USNM 549385–86; Cayo Cochino Menor, USNM 570531–32; Isla de Barbareta, USNM 520270–71; Isla de Guanaja, SE shore opposite Guanaja, LSUMZ 21646, UF 28573; Isla de Guanaja, La Playa Hotel, LSUMZ 21647; Isla de Guanaja, Savannah Bight, USNM 520272; "Isla de Guanaja," BMNH 1938.10.4.82, KU 101434–35; Isla de Roatán, near Coxen Hole, FMNH 34543 (8), 34556 (7), 34557–60, 34583–85, 34587–

92, 34597–601; Isla de Roatán, Diamond Rock, UTA R-55223–27; Isla de Roatán, near Oak Ridge, MCZ R150952–54, TCWC 52417, UTA R-10660–73; Isla de Roatán, Oak Ridge, CM 27597, FMNH 53830; Isla de Roatán, Palmetto Bay, FMNH 283752; Isla de Roatán, Port Royal Harbor, LSUMZ 33780, TCWC 52418; Isla de Roatán, about 3.2 km W of Roatán, LSUMZ 29625, 51423; Isla de Roatán, about 1.6 km N of Roatán, LSUMZ 51416–22; Isla de Roatán, 0.5–1.0 km NNE of Roatán, TCWC 21952–53; Isla de Roatán, 0.5–1.0 km N of Roatán, LSUMZ 21644–45; Isla de Roatán, near Roatán, UF 28461–62; Isla de Roatán, Roatán, LSUMZ 22318; Isla de Roatán, near Sandy Bay, USNM 563264; Isla de Roatán, Sandy Bay, KU 203149–52, LSUMZ 33779; Isla de Roatán, 6.6 km E of West End Point, MVZ 263855; "Isla de Roatán," MCZ R26756–58, UF 149593–94, UTA R-56369; Isla de Utila, east coast near Trade Winds, FMNH 283753; Isla de Utila, near Utila, LSUMZ 51424; Isla de Utila, Utila, LSUMZ 22269–71, UF 28389–91, 28406–14, 28452–53; "Isla de Utila," CM 29002, 29014, USNM 28626. **LA PAZ**: Marcala, CM 64634, LSUMZ 51425. **LEMPIRA**: El Rodeito, USNM 573105; Erandique, LSUMZ 51427. **OCOTEPEQUE**: 3 km S of Nueva Ocotepeque, TCWC 23732–34; 6 km NE of San Marcos de Ocotepeque, UF 124035–38. **OLANCHO**: Babilonia, USNM 579544; 5.9 km WSW of Campamento, LSUMZ 22265, 51431–32; Casamacoa, USNM 549387–88; about 9.7 km E of Catacamas, LACM 45132; 0.5 km WNW of Catacamas, LACM 48351–52; 19.3–22.5 km W of Catacamas, LACM 45166; 4.5 km SE of Catacamas, LACM 48353–56; Cuaca, USNM 579548, UTA R-53189; El Díctamo, USNM 342269; El Vallecito, USNM 342270; 22.3 km WSW of Juticalpa, LSUMZ 51426; Kauroahuika, USNM 563257; Matamoros, SMF 80826, USNM 549392–93; Quebrada El Guásimo, USNM 549389–91; Quebrada El Mono, USNM 549394; confluence of Quebrada Siksatara and Río Wampú, USNM 549396–97; confluence of ríos Sausa and Wampú, USNM 549395; confluence of ríos Yanguay and Wampú, USNM 549398–99; confluence of ríos Yapuwás and Patuca, LSUMZ 28471–72, 28480–82; Yapuwás, USNM 549400. **SANTA BÁRBARA**: Cerro Negro, USNM 565412, 570287; La Canadá, UTA R-41225–26; Quebrada Las Cuevas, USNM 580364. **YORO**: 0.5 km N of Coyoles, LACM 48360; 5 km E of Coyoles, LSUMZ 21650; Coyoles, LSUMZ 21648–49; El Negrito, MCZ R29391; 17 km NW of El Progreso, LSUMZ 24602; San Francisco, MVZ 52411–13; 3.5 km S of San Lorenzo Arriba, USNM 579549; 4.7 km ESE of San Lorenzo Arriba, USNM 563265; Subirana Valley, MCZ R38837–38, UMMZ 77839 (2). "HONDURAS": AMNH 36474, 50845, 70351, 46986–87, ANSP 8152, 8157, 28123, 33121–31, USNM 17804, 24337, 56792.

Other Records (Map 30).—**COLÓN**: Cañon Subterráneo del Río Plátano, UNAH 5469 (Cruz Díaz, 1978, specimen now lost). **FRANCISCO MORAZÁN**: Cantarranas, San Juancito (Dunn and Emlen, 1932). **GRACIAS A DIOS**: Baltiltuk, UNAH 5484 (Cruz Díaz, 1978, specimen now lost). **ISLAS DE LA BAHÍA**: Chachaute, Cayos Cochinos (Lundberg, 2002a); Isla de Barbareta, LSUHC 3693–94, LSUPC 4717–24 (Grismer et al., 2001); Isla de Utila, Iron Bound (Gutsche, 2005c). **LEMPIRA**: Villa Verde, UNAH 2012, 2088–90 (Cruz et al., 1993). "HONDURAS": (Werner, 1896).

Genus *Corytophanes* Boie, 1827

Corytophanes Boie, 1827: col. 290, *In* Schlegel, 1827 (type species: *Agama cristata* Merrem, 1820: 50, by monotypy).

Geographic Distribution and Content.— This genus ranges from southeastern San Luis Potosí and the Isthmus of Tehuantepec region, Oaxaca, Mexico, to northwestern

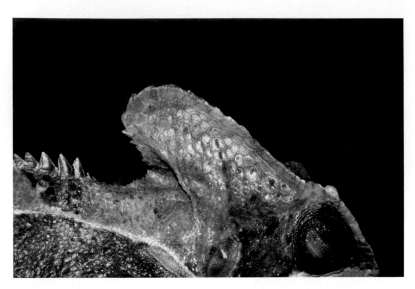

Figure 72.   Nuchal crest not continuous with middorsal crest. *Corytophanes hernandesii.* USNM 520003 from Las Rosas, Santa Bárbara.

Colombia. Three named species are recognized, all of which occur in Honduras.

Remarks.—Wagler (1830a: 151) emended Boie's *Corytophanes* to *Corythophanes*. However, Wagler's name is an unjustified emendation, even though it is technically the correct spelling and has frequently been used in the literature for all three species of the genus (Article 42; ICZN, 1999: 47).

Etymology.—The name *Corytophanes* is derived from the Greek *korythos* (helmet) and *phaneros* (visible, evident) and alludes to the prominent cephalic crest.

KEY TO HONDURAN SPECIES OF THE GENUS
CORYTOPHANES

1A. Nuchal crest not continuous with middorsal crest (Fig. 72)........
.... *hernandesii* (p. 193)
1B. Nuchal crest continuous with middorsal crest (Fig. 69)............... 2
2A. Dorsal head scales smooth to weakly rugose (Fig. 68); prominent lateral spine in squamosal area absent .... *cristatus* (p. 186)

2B. Dorsal head scales keeled to strongly rugose (Fig. 73); prominent lateral spine in squamosal area present (Fig. 74)...........
.... *percarinatus* (p. 196)

CLAVE PARA LAS ESPECIES HONDUREÑAS DEL
GÉNERO *CORYTOPHANES*

1A. Cresta nucal no continua con la cresta mediodorsal (Fig. 72) .....
.... *hernandesii* (p. 193)
1B. Cresta nucal continua con la cresta mediodorsal (Fig. 69)............... 2
2A. Escamas dorsales de la cabeza lisas a ligeramente rugosas (Fig. 68); espina en la región escamosal ausente .... *cristatus* (p. 186)
2B. Escamas dorsales de la cabeza quilladas hasta fuertemente rugosas (Fig. 73); espina en la región escamosal prominente (Fig. 74)..
.... *percarinatus* (p. 196)

Corytophanes cristatus (Merrem, 1820)

Agama cristata Merrem, 1820: 50 (holotype originally in the Seba collection [based

Figure 73.   Dorsal head scales keeled to strongly rugose. *Corytophanes percarinatus*. USNM 520004 from Guarín, Ocotepeque.

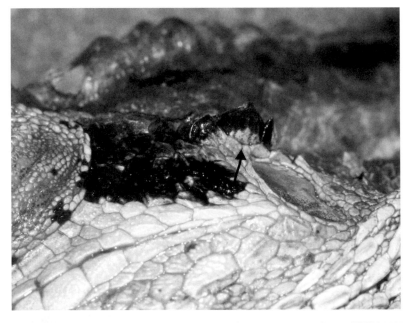

Figure 74.   Lateral squamosal spine present (arrow) on each side. *Corytophanes percarinatus*. USNM 520004 from Guarín, Ocotepeque.

on plate 94, fig. 4 in Vol. 1 of Seba, 1734]; type locality: "Ceylona" [in error]).

Corytophanes cristatus: Boie 1827: col. 290 (by inference), *In* Schlegel, 1827; Meyer, 1969: 242; Meyer and Wilson, 1973: 23; Cruz Díaz, 1978: 28; de Queiroz, 1982: 310; O'Shea, 1986: 38; Lang, 1989: 136; Köhler, 1996g: 65; Köhler, 1999d: 66; Köhler, 2000: 56; Sasa and Salvador Monrós, 2000: 361; Espinal et al., 2001: 106; Wilson et al., 2001: 135; Castañeda, 2002: 39; McCranie, Castañeda, and Nicholson, 2002: 25; Köhler, 2003a: 126; Townsend et al., 2004a: 788.1; Townsend et al., 2004b: 789.1; McCranie and Castañeda, 2005: 14; Townsend, Aldrich et al., 2005: 346; McCranie et al., 2006: 113; Townsend, 2006a: 34; Wilson and Townsend, 2006: 105; McCranie, 2007a: 175; Köhler, 2008: 138; Townsend et al., 2012: 100; McCranie and Solís, 2013: 242; Solís et al., 2014: 129; McCranie, 2015a: 367.

Corythophanes cristatus: Werner, 1896: 346; Dunn and Emlen, 1932: 28.

Geographic Distribution.—*Corytophanes cristatus* occurs at low and moderate elevations from Tabasco, the Yucatán Peninsula, and Chiapas, Mexico, to northwestern Colombia on the Atlantic versant (including Isla de Maíz Grande, Nicaragua) and from west-central Costa Rica to central Panama on the Pacific versant (see Remarks). It also occurs marginally on the Pacific versant in northwestern Costa Rica. In Honduras, this species is widespread in the northern and eastern portions of the country.

Description.—The following is based on ten males (LACM 47848; USNM 549404–06, 549412, 549414, 559544, 559547, 559551–52) and ten females (UF 137644; USNM 549403, 549409–10, 549413, 559545, 559549–50, 559554, 559557). *Corytophanes cristatus* is a moderately large corytophanine (maximum recorded SVL 124 mm [USNM 549404, a male]) with a long tail and long limbs; cephalic casque triangular, projecting posteriorly past head; dorsal head scales vary from smooth to weakly rugose; canthus raised into sharp ridge, ridge forming raised shelf above eye, ridge continues posteriorly to form distinct raised casque; ridges forming posterior edge of casque unite, and after about a 15–30 mm distance, tapering downward onto neck, casque continuous with long, well-developed nuchal crest; scales on upper posterior margin of casque and those on edge of downwardly tapered portion enlarged, triangular-shaped; deep frontal and parietal depressions present; parietal eye distinct to indistinct, in scale similar in size to surrounding scales or in scale slightly larger than most surrounding scales; tympanum distinct, higher than long; squamosal spine absent above tympanum, although an enlarged scale usually present, enlarged scale varies from strongly keeled to rounded above; short series of enlarged keeled scales absent from orbit to upper edge of tympanum; nasal scale single, nostril located more-or-less centrally in scale, nostril opening directed posterolaterally; moveable eyelid present; pupil circular; 11–15 supralabials; 10–13 infralabials; gular fold complete, continuous with antehumeral fold; gular scales elongate, strongly keeled, each scale with single keel, medial row with distinctively enlarged, strongly serrate scales; body laterally compressed; dorsal body scales imbricate, usually smooth, although occasional dorsal scales keeled, also occasional dorsal scales much larger than surrounding scales; lateral body scales imbricate, usually smooth, although occasional lateral scales keeled, most lateral scales smaller than dorsal scales; middorsal scale row enlarged, forming serrated dorsal crest with triangular-shaped scales extending from shoulder region to base of tail, crest more prominent anteriorly; middorsal crest continuous with nuchal crest; 28–34 (32.5 ± 1.8) scales in vertebral row between levels of anterior insertions of forelimb and groin; serrated row of scales forming distinct to indistinct ventrolateral fold on body; dorsolateral fold present on body, extending from shoulder region to usually about midlength of body; scales on nape region smaller than those of body; ventral scales large, subequal in size, imbricate, strongly keeled, varying from having rounded to broadly rounded posterior ends, some ventral scales also mucronate; 32–41 (35.5 ± 2.4) midventral scales between levels of axilla and groin; 96–130 (108.3 ± 8.3) scales around midbody; caudal autotomy absent; caudal scales keeled, although some lateral scales near base smooth; femoral and precloacal pores absent; 23–29 (26.6 ± 1.6) keeled subdigital scales on Digit IV of hind limb; SVL 95.1–123.8 (108.7 ± 9.1) mm in males, 90.7–123.0 (108.5 ± 10.9) mm in females; TAL/SVL 1.84–2.45 in males, 1.96–2.22 in females; casque L/SVL 0.48–0.55 in males, 0.45–0.49 in females; HL/SVL 0.25–0.29 in both sexes; SHL/SVL 0.30–0.35 in males, 0.29–0.32 in females.

Color in life of an adult male (USNM 559558): dorsal surface of body mottled Dark Grayish Brown (20), Olive-Gray (42), and Pale Horn Color (92); Flesh Ocher (132D) blotching at base of tail; tail banded Fawn Color (25) and Smoke Gray (44); head Lime Green (59) with brown X-shaped mark in center and associated brown mottling between superciliary crests; side of head Olive-Yellow (52) with area around orbit gray-brown; crest Cream Color (54) with brown mottling; chin tan; belly Pale Horn Color; iris Orange-Rufous (132C). Color in life of another adult male (USNM 549414): general color of dorsal surfaces pale olive green; head with darker green smudging above, side of head with darker green and paler green smudging; casque olive green with rust red smudging at base, smudging extending across neck to base of arm; gular region paler brown than dorsal surface of head, a kind of horn color, as is belly, ventral surfaces of fore- and hind

limb, and tail; vague paler green crossbands on dorsal surfaces of fore- and hind limb; dorsal surface of tail with vague rust wash along length; iris ocher with fine dark brown reticulations. Color in life of a juvenile (USNM 549401): dorsal surface of body pale olive green with about 4 olive brown oblique crossbars, prominent black spot on body below base of crest, another smaller spot located about midway along length of body; lateral surfaces of head and head crest olive yellow, with series of black lines radiating outward from orbit, and gray smudging on crest; dorsal area of head pale olive green; dorsal surface of forelimb banded pale olive green and olive brown, with prominent ivory white band at elbow; dorsal surface of hind limb pale olive green with prominent black spot at knee, midway along shank, and at base of each toe; gular region white to pale yellow, with dark streaks; belly pale olive cream with scattered dark brown spots; dorsal surface of tail banded pale olive green and gray-brown; iris copper.

Color in alcohol: dorsal surface of head brown, with or without darker brown reticulations; dorsal and lateral surfaces of body brown to dark brown, darker brown crossbands usually present on body; dorsal surfaces of fore- and hind limb brown, with or without distinct darker brown crossbars; lateral surface of head casque brown, with or without darker brown reticulations; chin and throat regions pale brown; belly pale brown with darker brown mottling or reticulations sometimes present laterally; dorsal surface of tail same color anteriorly as posterior end of body, usually with distinct darker brown crossbands extending length of tail, crossbands more evident anteriorly; subcaudal surface slightly paler brown than dorsal surface of tail, crossbands usually not evident on subcaudal surface.

Diagnosis/Similar Species.—*Corytophanes cristatus* can be distinguished from all other Honduran lizards, except *C. hernandesii* and *C. percarinatus*, by having

Plate 38.   *Corytophanes cristatus*. Gracias a Dios: Bachi Kiamp. FN 257016 (still in Honduras because of permit problems).

a cephalic crest formed by raised parietal ridges that unite posteriorly. *Corytophanes hernandesii* has a short, low nuchal crest that is not continuous with the middorsal crest and has a squamosal spine (versus long nuchal crest continuous with middorsal crest present and squamosal spine absent in *C. cristatus*). *Corytophanes percarinatus* has keeled to strongly rugose dorsal head scales and has a squamosal spine (versus smooth to weakly rugose dorsal head scales and no squamosal spine in *C. cristatus*).

Illustrations (Figs. 68, 69; Plate 38).— Calderón-Mandujano et al., 2008 (adult); Campbell, 1998 (adult); Gravenhorst, 1833 (adult, head scales; as *Corythophanes*); Gutsche, 2007 (adult); Guyer and Donnelly, 2005 (adult); Köhler, 1999b (adult), 1999d (adult), 2000 (adult), 2001b (adult), 2003a (adult, head crest shape), 2008 (adult, head crest shape); Kundert, 1974 (adult); Lang, 1989 (head scales); Lee, 1996 (adult, head scales), 2000 (adult, head scales); McCranie et al., 2006 (adult, cephalic crest); Pianka and Vitt, 2003 (adult); Savage, 2002 (adult); W. Schmidt and Henkel, 1995 (adult; as *C. cristatus* and *C. hernandezi*); Stafford and Meyer, 1999 (adult); Sunyer, Nicholson et al., 2013 (head); Taylor, 1956b (adult); Townsend et al., 2004a (adult), 2004b (adult); Townsend, Aldrich et al. 2005 (juvenile); Villa et al., 1988 (adult); Werner, 1896 (head scales).

Remarks.—Schwenk et al. (1982: 493) analyzed the karyotype of a specimen of *Corytophanes cristatus* purchased from an animal dealer that was "thought to have originated from the north coast of Honduras" (the major and longer running animal dealer was located in San Pedro Sula, Cortes, and a rather short-lived dealer was in Sambo Creek, Atlántida). Lang's (1989) phylogenetic analysis demonstrated *C. cristatus* and *C. percarinatus* to form a clade, with *C. hernandesii* as a sister species to that clade. Townsend et al. (2004b) presented an overview of the species' morphology and a literature review.

Smith and Taylor (1950a: 349, 1950b: 69) restricted the type locality of *Corytophanes cristatus* to "Orizaba, Veracruz [Mexico]." Their apparent source for that restriction was Cope (1866a: 195), who reported *Corythaeolus cristatus* (equals *Corytophanes cristatus*, according to Smith and Smith, 1976: L-B-51) from "Orizava [Veracruz], Mexico." However, the identification of Cope's specimen is questionable, as he included after *Corythaeolus cristatus* "(*Thysanodactylus*, Gray, *Dracontura*, Hallow)." *Thysanodactylus* Gray (1845b: 193) is a junior synonym of *Basiliscus* and "*Dracontura*, Hallow" apparently was a *lapsus* by Cope for *Draconura* Hallowell (1861: 482). *Dracontura* is also a junior synonym of *Basiliscus* (see Boulenger, 1885b: 109). Additional evidence supporting Cope's specimen being a *Basiliscus* is that no specimens of *Corytophanes cristatus* from Veracruz, Mexico, are in the USNM collection where Cope (1866a) mentions most of his specimens were deposited. In fact, no specimens of *C. cristatus* from Veracruz are in any U.S. museum, and the species apparently is not known to occur in that state (also see Pelcastre Villafuerte and Flores-Villela, 1992, who did not list the species from Veracruz). Smith and Taylor (1950b: 69) inaccurately stated (apparently based on above discussed issue) that *C. cristatus* occurred on the "Atlantic slopes from central Veracruz to Costa Rica." Similar range statements inaccurately including central Veracruz occur in much of the recent literature (e.g., Lang, 1989; Lee, 1996, 2000; Campbell, 1998; Köhler, 1999d, 2001b, 2003a, 2008; Savage, 2002). I explained this situation in Townsend et al. (2004b: 789.2). The Köhler (2008) statement including central Veracruz in the distribution of *C. cristatus* came after the results of that *Corythaeolus* research was published. None of those authors just mentioned who plotted localities on distribution maps plotted a corresponding locality for *C. cristatus* from Veracruz.

Agama cristata Mocquard (1905: 288), an African species of Agamidae Gray (1827a: 57) is a junior primary homonym of *Agama cristata* Merrem (1820: 50 = *Corytophanes cristatus*). That issue is deserving of discussion by someone familiar with *Agama cristata* of Mocquard.

Natural History Comments.—*Corytophanes cristatus* is known from near sea level to 1,300 m elevation in the Lowland Moist Forest and Premontane Wet Forest formations. This diurnal species was collected from May to November and in January and February, and thus is likely active throughout the year. Specimens were usually found on tree trunks during the day, although several times adult females were on the ground near trees. This is a sit-and-wait predator; thus, active individuals are rarely seen moving, instead relying on camouflage to avoid detection. The species can also be found sleeping on tree trunks at night. However, two females on the ground during the day ran slowly toward the nearest tree, rather than relying on camouflage; other females on the ground during the day made no effort to escape and appeared to be sleeping or waiting for prey to pass. A recently hatched specimen was sleeping at dusk in a small cave through which a small stream flowed, whereas other hatchlings were found sleeping on leafy vegetation at night. Females are reported to deposit 4–11 eggs in a clutch in shallow excavations they make in the forest floor, with breeding and egg deposition probably occurring throughout much of the year (Fitch, 1970; Lee, 1996; Köhler, 1997f, 1999d, and references cited in those works; also see list of reproduction references in Townsend et al., 2004b). A "Honduran" female (UTA R-32565) contained seven eggs when imported from an animal dealer. One Honduran adult (USNM 573113) regurgitated a moth caterpillar. Savage (2002: 433), in a literature summation, said individuals of *C. cristatus* "prey on large, generally slow-moving arthropods, primarily lepidopterous and coleopterous larvae and orthopterans, but they also feed on small lizards of the genus *Norops*." Sasa and Salvador Monrós (2000) reported on their examination of stomach contents of museum specimens and reported a wide range of prey items, including insects, especially adult coleopterans, lepi-

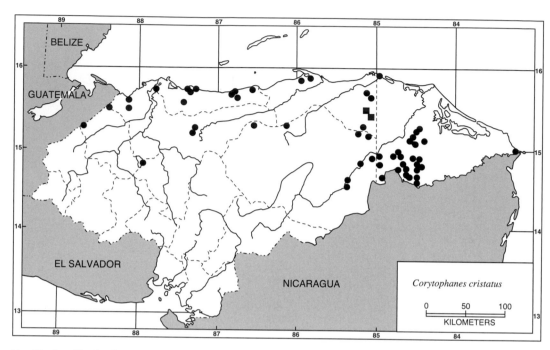

Map 31. Localities for *Corytophanes cristatus*. Solid circles denote specimens examined and solid squares those for accepted records.

dopteran larvae, and adult grasshoppers. Those authors also found scales of a small snake in the stomach of one Costa Rican individual. A Keel-billed Toucan (*Ramphastos sulfuratus* Lesson) in the Mosquitia of northeastern Honduras was seen eating an adult *C. cristatus* (USNM 570288) that the bird dropped, upon being harassed. Townsend, Aldrich et al. (2005) reported a juvenile with a myxomycete sporulating on its body. *Corytophanes cristatus* appears to require primary forest or old second growth forest near primary forest for its survival.

Etymology.—The name *cristatus* is Latin meaning "tufted, crested" and alludes to the distinctive crests in this species.

Specimens Examined (129, 9 skeletons, 1 egg lot [21]; Map 31).—**ATLÁNTIDA**: Colorado District, UMMZ 58370; Estación Forestal CURLA, USNM 578722; Guaymas District, UMMZ 58372, 58381; La Ceiba, MCZ R32035; Lancetilla, ANSP 28155, MCZ R29393; 3 km SE Mezapita, USNM 580366; Piedra Pintada, LACM 47847–48; Quebrada de Oro, USNM 343746, 549401; Río Mezapita, USNM 578723; Tela, MCZ R21153, 31483; near Tela, USNM 101279. **COLÓN**: Cerro Calentura, LSUMZ 27742, USNM 549402; Quebrada Machín, USNM 549403–05; Río Guaraska, BMNH 1985.1146, 1985.1263; 24.1 km E of Trujillo, USNM 38379. **CORTÉS**: El Paraíso, UF 144746–47; Los Pinos, UF 166338–39, USNM 565418, 573115; Montaña Santa Ana, MCZ R32036; near San Pedro Sula, UF 85408; Santa Teresa, USNM 573895–98. **GRACIAS A DIOS**: Awasbila, USNM 559544–45, 559548, 559557; Bachi Kiamp, USNM 565417, 573892, FN 257016–17 (both still in Honduras because of permit problems); Bodega de Río Tapalwás, UF 137168, 137644–46, USNM 559550–56, 559596 (skin and skeleton), 563269–70; Caño Awalwás, UF 137639–42, USNM 559546–47, 559558–59; Cerro Wahatingni, UF 137647; Concho Kiamp, FN 257053 (still in Honduras because of permit problems); Crique Yulpruan, USNM 570288; near Cueva de Leimus, USNM 579551–53; Dos Bocas, FMNH 282568; Hiltara Kiamp, USNM 563266; Kakamuklaya, USNM 565413, 573112; Kipla Tingni Kiamp, USNM 573113–14; Leimus (Río Warunta), FMNH 282569, USNM 565415; Mocorón, UTA R-53187–88; Rawa Kiamp, USNM 573893; Río Coco, USNM 24524; between Río Patuca and Río Coco, MCZ R38840–41; Río Sutawala, USNM 549406; near confluence of ríos Tapalwás and Rus Rus, USNM 559549; Rus Rus, UF 137643; Sachin Tingni Kiamp, USNM 563267, 563525; Sadyk Kiamp, FMNH 282569, USNM 565416, 573891, 579554; San San Hil, USNM 563268, 564061; near San San Hil Kiamp, USNM 564060, 570289; Sisinbila, USNM 579554; Urus Tingni Kiamp, USNM 563271–73, 564058–59; Wakling Tingni Kiamp, USNM 563274; Warunta Tingni Kiamp, USNM 563275, 565414. **OLAN-CHO**: Cuaca, UTA R-53224; Quebrada El Guásimo, USNM 549407; Quebrada de Las Marías, USNM 549409; confluence of Quebrada Siksatara and Río Wampú, USNM 549411–13; Qururia, USNM 549410; Río Kosmako, USNM 549408; Subterráneo, USNM 549414; Terrero Blanco, USNM 342271. **SANTA BÁRBARA**: La Cafetalera, USNM 573894. **YORO**: Montañas de Mataderos, FMNH 21794–95, MCZ R38842–44; Portillo Grande, FMNH 34754; Subirana Valley, MCZ R38839. "HONDURAS": BYU 42593–94, UF 56325 (skeleton), 57011 (skeleton), 57739 (skeleton), 60670 (skeleton), 67516 (skeleton), 68499 (skeleton), 69072 (skeleton), UMMZ 148960 (skeleton), UTA R-19526, 32565 (two vials containing 7 eggs), 35607.

Other Records (Map 31).—**COLÓN**: Empalme Río Chilmeca, UNAH 5466, Cañon Subterráneo del Río Plátano, UNAH 5434 (Cruz Díaz, 1978, specimens now lost). **GRACIAS A DIOS**: Palacios (O'Shea, 1986). "HONDURAS": (Werner, 1896); SDSNH 64543 (skeleton).

Corytophanes hernandesii (Wiegmann, 1831)

Chamaeleopsis hernandesii Deppe, 1830: 3 (*nomen nudum*; see Remarks).
Chamaeleopsis hernandesii Wiegmann, 1831: col. 298 (syntypes, ZMB 545–46 [see Remarks]; type locality not stated).
Corytophanes hernandesii: P. W. Smith and Burger, 1950: 166; McCranie et al., 2004: 790.1; McCranie et al., 2006: 216; Wilson and Townsend, 2006: 105; Solís et al., 2014: 129; McCranie, 2015a: 367; Ramos Galdamez et al., 2016: 1041.
Corytophanes hernandezii: McCranie and Espinal, 1998: 174; Köhler, 1999d: 71.

Geographic Distribution.—*Corytophanes hernandesii* occurs at low and moderate elevations on the Atlantic versant from extreme southeastern San Luis Potosí, Mexico, to northwestern Honduras (exclusive of most of the western portion of the Yucatán Peninsula). It is known in Honduras only from a few localities in Cortés and Santa Bárbara.

Description.—The following is based on one male (USNM 520003) and one female (ANSP 30509). *Corytophanes hernandesii* is a moderately small corytophanine (maximum recorded SVL 89 mm [USNM 520003, a male]) with a long tail and long limbs; cephalic casque triangular, projecting posteriorly past head; dorsal head scales vary from smooth to rugose in frontal region, scales of supraocular and parietal regions weakly keeled; canthus raised into sharp ridge, ridge forming raised shelf above eye, ridge continues posteriorly to form distinct raised casque; ridges forming posterior edges of casque unite and almost immediately curve downward onto neck, continuous with short, low nuchal crest; scales on upper posterior margin of casque enlarged, triangular-shaped, those on edge of downwardly curved portion small, serrated; deep frontal and parietal depressions present; parietal eye indistinct, in scale similar in size to surrounding scales or in scale slightly larger than most surrounding scales; tympanum distinct, higher than long; prominent squamosal spine present above tympanum; short series of enlarged keeled scales present from orbit to upper edge of tympanum; nasal scale single, nostril located more-or-less centrally in scale, nostril opening directed posterolaterally; moveable eyelid present; pupil circular; 11–13 supralabials; 8–10 infralabials; gular fold complete, continuous with antehumeral fold; gular scales elongate, strongly keeled, each scale with single keel, medial row with slightly enlarged, slightly serrate scales; body laterally compressed; dorsal body scales imbricate, usually smooth, although occasional dorsal scales keeled, also occasional dorsal scales much larger than surrounding scales; lateral body scales imbricate, usually smooth, although occasional lateral scales keeled, most lateral scales smaller than dorsal scales; middorsal scale row enlarged, forming serrated dorsal crest with triangular-shaped scales extending from shoulder region to base of tail, crest more prominent anteriorly; middorsal crest not continuous with low nuchal crest; 26–27 (26.5) scales in vertebral row between levels of anterior insertion of forelimb and groin; serrated row of scales forming indistinct ventrolateral fold on body; dorsolateral fold absent on body, although serrated row of scales forming distinct fold extending from lower edge of tympanum, curving upward to above level of forelimb insertion, and then downward to shoulder region just posterior to level of axilla; scales on nape region smaller than dorsal scales of body; ventral scales large, imbricate, strongly keeled, usually with rounded posterior ends, subequal in size; 31 (in both specimens) midventral scales between levels of axilla and groin; 75–78 (76.5) scales around midbody; caudal autotomy absent; caudal scales keeled; femoral and precloacal pores absent; 21–23 (22.0 ± 0.8) keeled subdigital scales on Digit IV of

hind limb; SVL 89.3 mm in male, 86.5 mm in female; TAL/SVL 2.80 in male, 2.66 in female; casque L/SVL 0.47 in male, 0.43 in female; HL/SVL 0.26 in male, 0.27 in female; SHL/SVL 0.39 in both.

Color in life of an adult male (USNM 520003; Plate 39): dorsal surface of body Light Drab (119C) with Hair Brown (119A) crossbars, except anteriormost crossbar Vandyke Brown (121); head casque, nuchal crest, and body immediately below and posterior to nuchal crest Drab-Gray (119D); Vandyke Brown blotch extending from eyelid across eye to anterior edge of tympanum, blotch bordered below by narrow white line; Vandyke Brown blotch beginning just posterior to tympanum, extending laterally on body to about midlength, blotch becoming slightly paler brown along lower edge, anterior and upper edges outlined with narrow dirty white line; dorsal surface of tail Hair Brown with narrow Light Drab crossbands anteriorly, crossbands fading out at point about one-third length of tail; ventral surface of body Light Drab; iris Prout's Brown (121A).

Color in alcohol: dorsal surface of head brown, with or without darker brown reticulations; dorsal and lateral surfaces of body brown to dark brown; dorsal surfaces of fore- and hind limb brown, with or without distinct darker brown crossbars; lateral surface of head casque brown with varying amounts of darker brown small spots; distinct dark brown to indistinct brown band extending from upper eyelid, across eye, to anterior edge of tympanum; dark band bordered below by white line in USNM 520003; USNM 520003 also has a large dark brown shoulder blotch on each side extending from near posterior edge of tympanum to about midlength of body, blotch extending dorsally to about ninth scale row below middorsal crest; shoulder blotch outlined with white line anteriorly, dorsally, and on upper portion posteriorly; chin and throat regions pale brown, darker brown reticulations present in ANSP 30509;

Plate 39.   *Corytophanes hernandesii*. USNM 520003, adult male, SVL = 89 mm. Santa Bárbara: Las Rosas.

belly pale brown; dorsal surface of tail same color anteriorly as posterior end of body, without distinct markings; subcaudal surface slightly paler brown than dorsal surface of tail.

Diagnosis/Similar Species.—*Corytophanes hernandesii* can be distinguished from all other Honduran lizards, except *C. cristatus* and *C. percarinatus*, by having a cephalic crest formed by raised parietal ridges that unite posteriorly. *Corytophanes cristatus* and *C. percarinatus* have well-developed nuchal crests that are continuous with the middorsal crest (versus low, short nuchal crest not continuous with middorsal crest in *C. hernandesii*). *Corytophanes cristatus* also lacks a squamosal spine (present in *C. hernandesii*).

Illustrations (Fig. 72; Plate 39).—Álvarez del Toro, 1983 (adult); Álvarez Solórzano and González Escamilla, 1987 (adult); Campbell, 1998 (adult); A. H. A. Duméril et al., 1870–1909b (head; as *Corythophanes mexicanus* Bocourt, 1874: 122, *In* A. H. A. Duméril et al., 1870–1909a); Freiberg, 1972 (adult); Gravenhorst, 1833 (head, body; as *Chamaeleopsis* Wiegmann, 1831); Köhler,

1999d (adult, hatchling), 2000 (adult), 2003a (adult, head crest shape), 2008 (adult, head crest shape); Lee, 1996 (adult, head scales), 2000 (adult, head scales); Lemos-Espinal and Dixon, 2013 (adult); McCranie et al., 2004 (adult); Ramos Galdamez et al., 2016 (adult); Stafford and Meyer, 1999 (adult); Townsend et al., 2004a (adult); Wiegmann, 1834b (adult; as *Chamaeleopsis*).

Remarks.—Smith (1971: 75) credited Deppe (1830: 3; [erroneously stated by Smith as "(Deppe, 1930)"]) as the author of the name *Chamaeleopsis hernandesii*. Gray (1845b: 194), Boulenger (1885b: 103), Smith and Taylor (1950b: 68), Stuart (1963: 67), and McCoy, (1970: 101, *In* Peters and Donoso-Barros, 1970), considered Wiegmann (1831: 45, *In* Gray, 1830–1831) as the author of the specific name *Chamaeleopsis hernandesii*. Wiegmann (1831 [a different publication than that *In* Gray]: col. 298) published a more detailed description of the species than the brief description in Gray (1830–1831). However, there is some uncertainty as to which of the two publications appeared first. I follow Bauer and Adler (2001: 326) in citing Wiegmann

(1831) as the original description of *Corytophanes hernandesii*.

Taylor (1969: iv) said there were three syntypes under the numbers ZMB 545–46. Lang (1989: 135) stated there were three syntypes catalogued under ZMB 545. However, there are only two syntypes (ZMB 545, 546) of *Chamaeleopsis hernandesii* Wiegmann, although the number 3 is written in front of the genus on the original label for ZMB 545 (R. Günther, in litteris, 8 March 2001). The original entries in the ZMB catalogue also indicate that only two types (ZMB 545, 546) were originally registered (R. Günther, in litteris). Stuart (1963: 67) stated that ZMB 545 was the "Type." According to Article 74 of the Code in ICZN (1999), that act does not seem sufficient to have designated ZMB 545 as the lectotype of *Chamaeleopsis hernandesii* Wiegmann.

Lang (1989) reviewed the morphology of *Corytophanes hernandesii* and provided a phylogenetic analysis showing the species to be sister to a clade containing *C. cristatus* and *C. percarinatus*. McCranie et al. (2004) presented an overview of the species' morphology and reviewed its literature. °Hernández (1849, published posthumously) used the name *Chamaeleo mexicanus*, a junior synonym of *Corytophanes hernandesii*, but the °Hernández (1849) book is extremely rare and almost never cited, and I could not locate a copy in any form. The specimen of *C. hernandesii* discussed by Ramos Galdamez et al. (2016) as being the first actual record of this species from Honduras was actually collected in Guatemala.

Natural History Comments.—*Corytophanes hernandesii* is known from about 150 to about 1,000 m elevation in the Lowland Moist Forest and Premontane Wet Forest formations. This diurnal species was collected in March and October. One Honduran specimen was kept alive for about 6 months and readily fed on grasshoppers (M. R. Espinal, personal communication, 1997). Nothing else has been recorded regarding its natural history in Honduras, other than one was sleeping on a low tree branch at night. Females deposit three to seven eggs in a clutch at least from May to July (Köhler, 1997f, 1999d), with the eggs being placed in a shallow excavation they make on the forest floor (Lee, 1996: 198, and references cited therein). McCranie et al. (2004) also provided other references on reproduction in this species. Sasa and Salvador Monrós (2000) studied stomach contents of museum specimens and reported a wide range of prey items, including insects, especially adult coleopterans, lepidopteran larvae, and adult grasshoppers.

Etymology.—The species name is a patronym honoring Francisco Hernández, a Spanish explorer-naturalist who traveled extensively in Mexico from 1570 to 1577 (Lee, 1996).

Specimens Examined (2 [0]; Map 32).— **CORTÉS**: 1 km W of San Pedro Sula, ANSP 30509. **SANTA BÁRBARA**: Las Rosas, USNM 520003.

Other Record (Map 32).—**COPÁN**: El Bijao (Ramos Galdamez et al., 2016). **SANTA BÁRBARA**: Totoca, CM 158759 (Ramos Galdamez et al., 2016).

Corytophanes percarinatus A. H. A. Duméril, 1856

Corytophanes percarinatus A. H. A. Duméril, 1856: 518 (holotype, MNHN 2117 [see Guibé, 1954: 40]; type locality: "Ascuintla, dans l'Amér. centrale, à 30 lieues de Guatemala" [also see Townsend et al., 2004c: 791.2]); McCranie and Wilson, 1998: 174; Köhler, 1999d: 73; Wilson and McCranie, 2003: 59; Townsend et al., 2004c: 791.1; Cruz et al., 2006: 3.6; Wilson and Townsend, 2007: 145; Solís et al., 2014: 129; McCranie, 2015a: 367.

Geographic Distribution.—*Corytophanes percarinatus* occurs at low (rarely), moder-

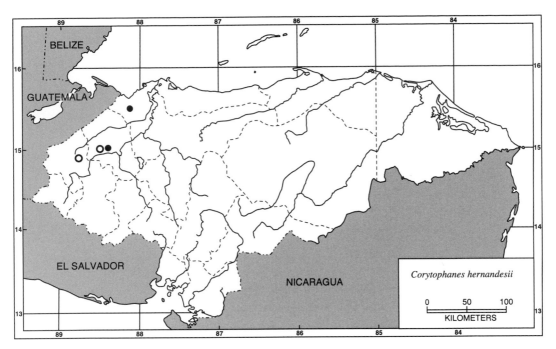

Map 32.   Localities for *Corytophanes hernandesii*. Solid circles denote specimens examined and open circles those of accepted records.

ate, and intermediate elevations from southeastern Chiapas, Mexico, to northwestern El Salvador and extreme southwestern Honduras on the Pacific versant. It also occurs on the Atlantic versant of eastern Chiapas, Mexico, and central Guatemala. Many of the populations are apparently disjunct. In Honduras, this species is known only from a few localities in Ocotepeque near the border with El Salvador.

Description.—The following is based on one male (USNM 520004) and one female (USNM 570290). *Corytophanes percarinatus* is a moderately small corytophanine (maximum recorded SVL 102 mm [USNM 570290, a female]) with a long tail and long limbs; cephalic casque triangular, projecting posteriorly past head; dorsal head scales keeled to strongly rugose or striated; canthus raised into sharp ridge, ridge forming raised shelf above eye, ridge continues posteriorly to form distinct raised casque; ridges forming posterior edge of casque unite and curve downward onto neck, continuous with long, well-developed nuchal crest; scales on upper posterior margin of casque and those on edge of downwardly curved portion enlarged, triangular-shaped; deep frontal and parietal depressions present; parietal eye indistinct, in scale similar in size to surrounding scales; tympanum distinct, higher than long; prominent squamosal spine present above tympanum; short series of enlarged keeled scales present from orbit to upper edge of tympanum; nasal scale single, nostril located more-or-less centrally in scale, nostril opening directed posterolaterally; moveable eyelid present; pupil circular; 13–14 supralabials; 10–13 infralabials; gular fold complete, continuous with antehumeral fold; gular scales elongate, strongly keeled, each scale with single keel, medial row with distinctively enlarged, strongly serrated scales; body laterally compressed; dorsal body scales imbricate, usually smooth,

Plate 40.   *Corytophanes percarinatus*. USNM 520004, adult male, SVL = 90 mm. Ocotepeque: Guarín.

although occasional dorsal scales keeled, also occasional dorsal scales much larger than surrounding scales; lateral body scales imbricate, usually smooth, although occasional lateral scales keeled, most lateral scales smaller than dorsal scales; middorsal scale row enlarged, forming serrated dorsal crest with triangular-shaped scales extending from shoulder region to base of tail, crest more prominent anteriorly; middorsal crest continuous with nuchal crest; 20–30 (25.0) scales in vertebral row between levels of anterior insertion of forelimb and groin; dorsolateral fold extending from shoulder region to about midlength of body; serrated row of scales forming indistinct ventrolateral fold present on body; scales on nape region smaller than dorsal body scales; ventral scales large, imbricate, strongly keeled, usually with rounded posterior ends, subequal in size; 26–28 (27.0) midventral scales between levels of axilla and groin; 94–99 (96.5) scales around midbody; caudal autotomy absent; caudal scales keeled; femoral and precloacal pores absent; 24–25 (24.8 ± 0.5) keeled subdigital scales on Digit IV of hind limb; SVL 90.4 mm in male, 102.0 mm in female; TAL/SVL 2.41 in male,

2.11 in female; casque L/SVL 0.45 in male, 0.40 in female; HL/SVL 0.25 in male, 0.23 in female; SHL/SVL 0.36 in male, 0.40 in female. Six newborns (USNM 570291–96) have SVLs of 34.1–37.7 (35.9 ± 1.4) mm (preserved same day as birth and measurements taken 1 week after preservation).

Color in life of an adult male (USNM 520004; Plate 40): dorsal surface of body Brownish Olive (29) with narrow Straw Yellow (56) crossbands on dorsolateral portion of body, breaking into Brownish Olive reticulations on Straw Yellow ground color laterally; dorsal surface of head Olive-Brown (28); lateral surface of head Olive-Yellow (52) with Sepia (219) blotch posterior to eye; dorsal surfaces of limbs and tail banded with Brownish Olive and Sulphur Yellow (57); ventral surface of body Sulphur Yellow; iris Raw Sienna (136) with dark brown flecking and narrow yellowish brown line around pupil; gular region with Olive-Yellow scales and pale blue skin.

Color in alcohol: dorsal surface of body dark brown with indistinct, incomplete white crosslines, white reticulations or elongated lines also present laterally; dorsal surface of head brown with occasional darker brown spots; dorsal surfaces of limbs dark brown with indistinct white crossbars; lateral surface of head casque dark brown; chin, throat, and belly pale brown; dorsal surface of tail same color anteriorly as posterior end of body, without distinct markings; subcaudal surface slightly paler brown than dorsal surface of tail, with paler brown crossbars anteriorly.

Diagnosis/Similar Species.—*Corytophanes percarinatus* can be distinguished from all other Honduran lizards, except *C. cristatus* and *C. hernandesii*, by having a cephalic crest formed by raised parietal ridges that unite posteriorly. *Corytophanes cristatus* has smooth to weakly rugose dorsal head scales and lacks a prominent squamosal spine above the tympanum (versus keeled to strongly rugose dorsal head scales and prominent squamosal spine present in *C. percarinatus*). *Corytophanes hernandesii* has a short, low nuchal crest that is not continuous with the middorsal crest (versus middorsal crest continuous with well-developed nuchal crest in *C. percarinatus*).

Illustrations (Figs. 73, 74; Plate 40).— Álvarez del Toro, 1983 (adult); Álvarez Solórzano and González Escamilla, 1987 (adult); A. H. A. Duméril, 1856 (adult); Köhler, 1999d (adult), 1999h (adult), 2000 (adult), 2003a (adult, head crest shape), 2008 (adult, head crest shape); Köhler et al., 2005 (adult); Townsend et al., 2004a (adult), 2004c (adult); Wilson and McCranie, 2004a (adult).

Remarks.—Lang (1989) reviewed the morphology of *Corytophanes percarinatus* and provided a phylogenetic analysis that demonstrated it to be sister to *C. cristatus*. Townsend et al. (2004c) provided an overview of the species' morphology and reviewed its literature.

Natural History Comments.—*Corytophanes percarinatus* is known from 1,350 to 1,700 m elevation in the Premontane Wet Forest and Lower Montane Moist Forest formations. One was collected in August under a large rock pile on the ground in disturbed gallery forest about 3 m from a small stream. The forest away from that gallery forest was completely denuded but was not formerly pine oak forest as indicated by Wilson and Townsend (2007). Another was sleeping at night in a tree on a coffee farm and gave birth to six young in the collecting bag the following day (23 June). Köhler (1999d) also provided reproduction notes on this viviparous lizard in El Salvador. Townsend et al. (2004c) provided literature references for reproduction in this species. Sasa and Salvador Monrós (2000) studied stomach contents of museum specimens and found a wide range of prey items, including insects, especially adult coleopterans, lepidopteran larvae, and adult grasshoppers. Two *C. percarinatus* also contained the remains of a snail and an earthworm.

Etymology.—The name *percarinatus* is derived from the Latin words *per* (very) and *carinatus* (keeled), and alludes to the keeled or rugose dorsal head scales.

Specimens Examined (8 [0]; Map 33).— **OCOTEPEQUE**: El Mojanal, USNM 570290–96; Guarín, USNM 520004.

Other Records.—**OCOTEPEQUE**: Las Hojas (Cruz [Díaz] et al., 2006).

Genus *Laemanctus* Wiegmann, 1834b

Laemanctus Wiegmann, 1834b: 16 (in part) (type species: *Laemanctus longipes* Wiegmann, 1834b: 46, by subsequent designation of Fitzinger, 1843: 16).

Geographic Distribution and Content.— This genus ranges from central Tamaulipas and central Oaxaca, Mexico, to central Nicaragua. Four named species, one described as a new species and one elevated from a subspecies in this work, are recognized, all of which are reported from Honduras.

Remarks.—The two most recent revisions of *Laemanctus* are McCoy (1968) and Lang (1989). Both recognized two named species, *L. longipes* Wiegmann (1834b: 46) and *L.*

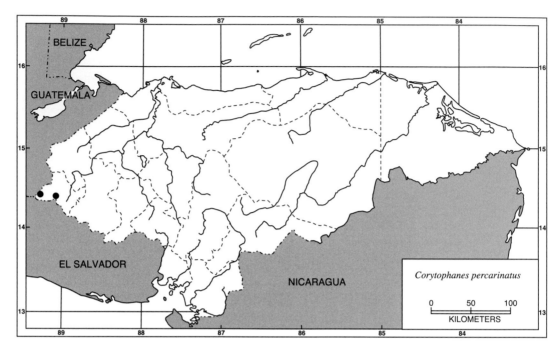

Map 33. Localities for *Corytophanes percarinatus*. Solid circles denote specimens examined.

serratus Cope (1865a: 176). The former was said to contain three nominal subspecies, the latter two nominal subspecies. *Laemanctus longipes* and *L. serratus* are easily separated into two groups by the size and shape of their postrostral scales.

The Honduran specimens of the *Laemanctus longipes* group are themselves readily separated into two morphological groups. One group (those from Atlántida and Cortés; = *L. waltersi*) lacks a gular fold or has an incomplete gular fold, has larger body scales (30–39, $x = 33.7 \pm 2.6$, $n = 10$, scales around midbody; 24–34, $x = 29.5 \pm 3.0$, $n = 10$, scales in vertebral row between levels of anterior portion of forelimb insertion and groin), and lacks a group of smaller lateral scales. Those in the second *L. longipes* group (Copán and Olancho specimens; = *L. longipes*) have a complete gular fold, smaller body scales (43–47, $x = 45.3 \pm 2.1$, $n = 4$, scales around midbody; 34–41, $x = 37.8 \pm 3.3$, $n = 4$, scales in the vertebral

row between the levels of the anterior portion of the forelimb insertion and groin), and have narrow patches, or isolated, smaller lateral scales present. Those four characters are shown in the figures listed in the following *Laemanctus* key. The two populations appear to be diagnosable as separate species based on the consistent morphological differences and non-overlapping geographical distributions. There appears to be no intergradation in morphological characters between those two nominal forms as postulated by McCoy (1968: 670). As a result, I consider *Laemanctus waltersi* Schmidt to represent a species distinct from *L. l. longipes* Wiegmann (1834b) and *L. l. deborrei*. A molecular and morphological study of the *L. longipes* complex is needed, but there seems no doubt that *L. waltersi* represents a distinct evolutionary species.

A specimen (USNM 572041) of *Laemanctus* from Tegucigalpa, Francisco Morazán, belongs to the *L. serratus* group with its postrostral scales being similar in size and arrangement to those of *L. serratus*. The Tegucigalpa specimen also does not appear to represent any species or subspecies of *L. serratus* recognized by McCoy (1968) or Lang (1989). At the time I had that specimen in my collection, I thought it was introduced to Tegucigalpa. Subsequently, in the most recent 4 years, several other specimens of that *Laemanctus* have been collected in the area (all in the UNAH collection); thus, it now seems certain the population is native, especially as it does not represent any species discussed by McCoy or Lang. Another contributing factor to my native population belief is that I saw a lizard that appeared to be *L. serratus* in a gallery forest of the same river valley where Tegucigalpa is located and only some 25 km E of Tegucigalpa in October 2008 (see McCranie and Köhler, 2004b: 796.2). A close examination of the USNM specimen from Tegucigalpa in November 2015 indicates it is an undescribed species more closely related to *L. serratus* than to *L. longipes* or *L. waltersi*. It agrees with *L. serratus* in having enlarged anterior dorsal head scales (postrostrals), having a higher number of scales around midbody, and having a distinct body color pattern. At the same time, USNM 572041 differs from *L. serratus* by lacking a serrated casque and serrated middorsal crest scales, thus distinguishing it from *L. serratus,* which has a serrated casque and serrated middorsal scales. The Tegucigalpa specimen also has a complete gular fold, thus further differentiating it from *L. waltersi*. This new species of the *L. serratus* group is described in this work.

Etymology.—The name *Laemanctus* is formed from the Greek *laimos* (throat) and *anctus* (press together) said to be in reference "to the tapered crowns of the marginal teeth" by Lang (1989: 141).

KEY TO HONDURAN SPECIES OF THE GENUS *LAEMANCTUS*

1A. Dorsal head scales from rostral to at least mideye (postrostrals) relatively enlarged, numbering 20–22, and rather regular in size and arrangement (Figs. 75, 76) 2
1B. All dorsal head scales relatively small, rather irregular in arrangement (Fig. 77); postrostrals numbering 36–55 3
2A. Cephalic casque with enlarged serrated, posteriorly or laterally directed, scales along posterior margin (Fig. 78); low, serrated middorsal crest scales present (Fig. 79) . . . . *serratus* (p. 213)
2B. Cephalic casque margin relatively smooth, without enlarged serrated scales, but some slightly conical (directed dorsally) along posterior margin (Fig. 80); serrated middorsal crest scales absent (Fig. 81) . . . . **sp. nov.** (p. 204)
3A. Gular fold absent (Fig. 82) or incomplete; group of smaller lateral scales absent (Fig. 83) . . . . *waltersi* (p. 216)
3B. Gular fold complete (Fig. 84); patches of smaller or isolated smaller scales present laterally on body (Fig. 85) . . . . . . *longipes* (p. 208)

CLAVE PARA LAS ESPECIES HONDUREÑAS DEL GÉNERO *LAEMANCTUS*

1A. Superficie dorsal de la cabeza desde la rostral hasta los ojos (postrostrales) con escamas relativamente agrandadas, casi regulares en tamaño y organización, y en número de 20–22 (Figs. 75, 76) . . 2
1B. Superficie dorsal de la cabeza desde del rostral hasta los ojos (postrostrales) con escamas relativamente pequeñas, bastante irregulares en tamaño y organización, y en

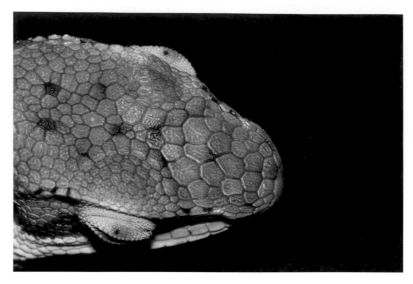

Figure 75. Dorsal head scales between rostral and at least mideye relatively enlarged (postrostrals), in rather regular arrangement. *Laemanctus julioi* USNM 572041 from Tegucigalpa, Francisco Morazán.

número 36–55 (Fig. 77) 3

2A. Cresta dorsal de la cabeza con escamas agrandadas aserradas sobre el margen posterior (Fig. 78); cresta mediodorsal en forma ase-

rradas poco prominente (Fig. 79) . . . . *serratus* (p. 213)

2B. Cresta dorsal de la cabeza con escamas relativamente lisas, sin escamas muy agrandadas y aserradas, pero ligeramente cónica (en

Figure 76. Dorsal head scales between rostral and at least mideye relatively enlarged (postrostrals), in rather regular arrangement. *Laemanctus serratus.* USNM 84550 from "Honduras." Both *L. julioi* and *L. serratus* are included to show similarities with each other and differences from the *L. longipes* group.

Figure 77.　Dorsal head scales small, rather irregular in arrangement. *Laemanctus longipes*. USNM 570297, from Río Amarillo, Copán.

dirección dorsal) a lo largo del margen posterior (Fig. 80); sin cresta mediodorsal presente (Fig. 81) .... **sp. nov.** (p. 204)

3A. Pliegue gular ausente (Fig. 82) o incompleto; escamas lateralmente pequeñas ausente en el cuerpo (Fig. 83)............ *waltersi* (p. 216)

3B. Pliegue gular completo (Fig. 84); escamas pequeñas presentes lateralmente en el cuerpo, en parches o aisladas (Fig. 85).. *longipes* (p. 208)

Figure 78.　Enlarged serrated scales present along posterior margin of cephalic crest, scales directed posteriorly or laterally. *Laemanctus serratus*. USNM 84550 from "Honduras."

Figure 79. Serrated middorsal crest present. *Laemanctus serratus.* USNM 84550 from "Honduras."

Laemanctus julioi sp. nov. McCranie, herein

Laemanctus julioi **sp. nov.** McCranie, 2018: herein (holotype, USNM 572041; type locality: "Residencial Plaza, 14°03.870′ N, 87°11.376′ W, 1000 m elev., Tegu-

cigalpa, Distrito Central, Francisco Morazán, Honduras").

Geographic Distribution.—*Laemanctus julioi* sp. nov. is known only from moderate elevations on the Pacific versant in south-central Honduras in the vicinity of its type

Figure 80. Flat-topped cephalic crest with smooth posterior margin, but some edge scales conical-shaped. *Laemanctus julioi* USNM 572041 from Tegucigalpa, Francisco Morazán.

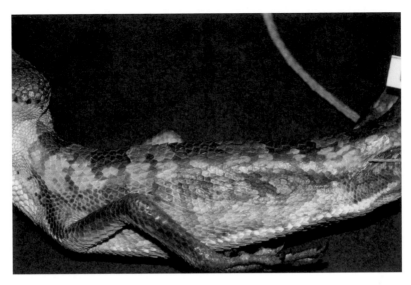

Figure 81. Serrated middorsal crest absent and pale dorsolateral longitudinal broad stripe present on each side (4–5 scales wide anteriorly). *Laemanctus julioi* USNM 572041 from Tegucigalpa, Francisco Morazán.

locality and probably in the Río Choluteca Valley about 25 km E of the type locality (see Remarks).

Description of holotype—The following is based on the female holotype (USNM 572041). *Laemanctus julioi* sp. nov. is a moderately large corytophanine (maximum recorded SVL 115 mm); mostly flat-topped cephalic casque present (slightly elevated medially on posterior portion); dorsal head scales strongly rugose; anterior dorsal head scales (postrostrals) much larger than posterior dorsal head scales, scales between rostral and orbits approximately bilaterally

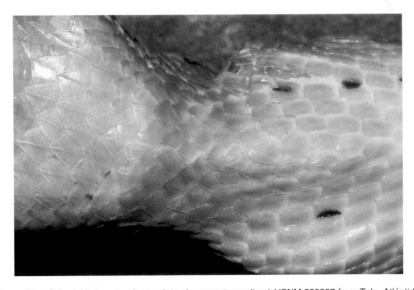

Figure 82. Gular fold absent or incomplete. *Laemanctus waltersi.* USNM 580367 from Tela, Atlántida.

Figure 83.   Smaller lateral scales absent. *Laemanctus waltersi*. USNM 580367 from Tela, Atlántida.

symmetrical, 20 postrostral scales in series, including canthal scales and centrally located azygous scale; posterior edge of casque with only slightly conical scales; 26 scales around posterior edge of casque (between posterior most superciliary on each side); parietal eye distinct, in scale similar in size to surrounding scales; tympanum distinct; nasal scale single, nostril located more-or-less centrally in scale, nostril opening directed posterolaterally; moveable eyelid present; pupil circular; 11 supralabials; 10–12 infralabials; gular fold complete, 3 rows of small scales contained in fold; antehumeral fold indistinct; gular scales weakly keeled, each scale with single keel; body

Figure 84.   Complete gular fold present. *Laemanctus longipes*. USNM 570297 from Río Amarillo, Copán.

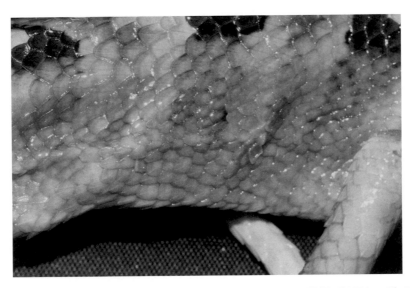

Figure 85. Some smaller lateral scales present on each side. *Laemanctus longipes*. USNM 570297 from Río Amarillo, Copán.

laterally compressed; dorsal body scales relatively small, keeled; scattered smaller lateral scales absent; middorsal scales not enlarged, not forming serrated dorsal crest; 37 scales in vertebral row between levels of anterior insertion of forelimb and groin; scales on nape region smaller than those of body; ventral scales large, imbricate, strongly keeled, usually pointed posteriorly, subequal in size; 34 midventral scales between levels of axilla and groin; 53 scales around midbody; caudal autotomy absent; caudal scales strongly keeled; femoral and precloacal pores absent; 31–33 (32.0) subdigital scales on Digit IV of hind limb; subdigital scales with keratinized knobs on anterior section of each scale; SVL 115.0 mm; TAL/SVL 3.82; casque L/SVL 0.29; HL/SVL 0.23; SHL/SVL 0.29.

Color in alcohol of holotype (USNM 572041): dorsal surface of body pale brown; middorsal brown longitudinal broad stripe extending from nape onto base of tail; middorsal stripe about 4 scales wide at widest point; dirty white, broad dorsolateral stripe about 7 scales wide at widest point present below broad, brown middorsal stripe; about 7 distinct to indistinct brown crossbars below dorsolateral pale broad stripe; head casque pale brown dorsally, without distinctive markings; few brown to black spots or short lines present on lateral surface of head casque; narrow white stripe extending from below orbit to just anterior to forelimb; pale brown to cream narrow ventrolateral stripe (1 scale wide) extending from axilla to groin; dorsal surfaces of fore- and hind limb brown, with distinct pale brown crossbands on hind limb; chin, throat, and belly cream to very pale gray; tail brown on all surfaces.

Diagnosis/Similar Species.—The prominent flat-topped cephalic crest will distinguish *Laemanctus julioi* from all other Honduran lizards, except the remaining *Laemanctus*. *Laemanctus longipes* and *L. waltersi* have relatively small anterior dorsal head scales (postrostrals) with 37–55 postrostral scales between the rostral and anterior edge of the eye, have 36–47 scales around midbody, and lack a pale colored dorsolateral broad stripe (versus 20 postrostral scales, 53 scales around midbody, and broad pale dorsolateral stripe present in *L. julioi*). *Laemanctus waltersi* also lacks a gular fold (versus complete gular fold in *L.*

Plate 41.   *Laemanctus julioi* USNM 572041, adult female, SVL = 115 mm. Francisco Morazán: Tegucigalpa, Residencial Plaza. Photograph by Leonardo Valdés Orellana.

julioi). *Laemanctus serratus* has a fringe of enlarged, erect, flattened, serrated scales on the posterior edge of the cephalic crest and has a serrated middorsal crest (versus no fringe of enlarged triangular scales on posterior edge of cephalic crest and no serrated middorsal crest in *L. julioi*).

Illustrations (Figs. 75, 80, 81; Plate 41).— None previously published.

Remarks.—I saw, but could not capture, what appeared to be this species (it also had a pale-colored dorsolateral broad stripe) in the Río Choluteca Valley near Ojo de Agua, El Paraíso, about 25 km E of Tegucigalpa in August 1998 (see McCranie and Köhler, 2004b: 796.2). Hurricane Mitch destroyed that gallery forest in October 1998.

Natural History Comments.—*Laemanctus julioi* is believed to occur from 650 to 1,000 m elevation in the Premontane Dry Forest formation. This is a diurnal species. The holotype was active on an outside wall of a house in July at 11:00 a.m. in the capital city of Honduras. One was seen on the ground along a dirt road in August (see Remarks). See Natural History Comments of its apparent closest relative *L. serratus* for notes on that species reproduction and feeding behaviors.

Etymology.—The specific name *julioi* is a patronym for Julio Enrique Mérida, the current curator of the Museum of Natural History at the UNAH in Tegucigalpa.

Specimens Examined (1, the holotype [0]; Map 34).—**FRANCISCO MORAZÁN:** Tegucigalpa, USNM 572041.

Other Records (Map 34).—**EL PARAÍSO:** near Ojo de Agua (personal sight record).

Laemanctus longipes Wiegmann, 1834b

Laemanctus longipes Wiegmann, 1834b: 46 (holotype, ZMB 494 [see Lang, 1989: 143]; type locality: "prope Jalapam" [Jalapa, Veracruz, Mexico]); Meyer, 1969: 252 (in part); Meyer and Wilson, 1973: 26 (in part); Köhler, 1999d: 78 (in part); Espinal et al., 2001: 106; Wilson et al., 2001: 135 (in part); McCranie, 2005: 20; Castañeda and Marineros, 2006: 3.8; McCranie et al., 2006: 216 (in part); Wilson and Townsend, 2006: 105 (in part); Wilson and Townsend, 2007: 145; Solís et al.,

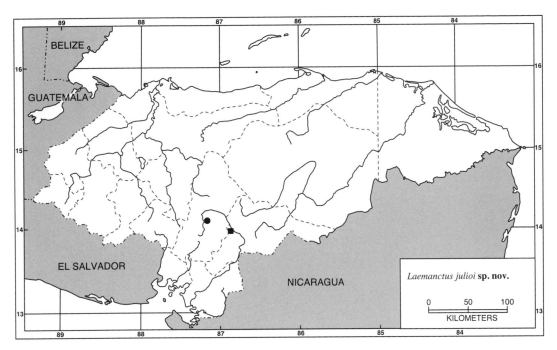

Map 34.   Localities for *Laemanctus julioi*. Solid circle denotes a specimen examined and the solid square represents a personal sight record (see text).

2014: 129 (in part); McCranie, 2015a: 367.

Laemanctus longipes deborrei: McCoy, 1968: 668; McCranie and Köhler, 2004a: 795.1.

Geographic Distribution.—*Laemanctus longipes* occurs at low and moderate elevations on the Atlantic versant from southern Veracruz, Mexico, to western Honduras, with an apparently isolated population in central Nicaragua (the systematics of that isolated population needs to be studied). In Honduras, this species is known from several interior localities from western Copán to northwestern Olancho.

Description.—The following is based on one male (USNM 283032) and three females (USNM 342272, 549415, 570297). *Laemanctus longipes* is a moderately large corytophanine (maximum recorded SVL 150 mm [McCoy, 1968]; SVL 137 mm in largest Honduran specimen [USNM 382272, a female]) with an extremely long tail and long limbs; flat-topped cephalic casque present; dorsal head scales strongly rugose to slightly carinate; anterior dorsal head scales similar in size, or slightly larger than posterior dorsal head scales; 37–55 (45.8 ± 7.4) postrostral scales present anterior to level of anterior edge of eye (including canthal scales); posterior edge of casque lacking fringe of enlarged, flattened, triangular scales; 30–35 (32.0 ± 2.2) scales around posterior edge of casque (between posterior most superciliary on each side); parietal eye distinct to indistinct, in small scale similar in size to surrounding scales, or in scale slightly larger than surrounding scales; tympanum distinct; nasal scale single, nostril located more-or-less centrally in scale, nostril opening directed posterolaterally; moveable eyelid present; pupil circular; 10–12 supralabials; 9–12 infralabials; gular fold complete, well developed, 2–4 rows of small scales contained in fold; antehumeral

Plate 42.   *Laemanctus longipes.* USNM 549415, adult female, SVL = 74 mm. Copán: between Laguna del Cerro and Quebrada Grande.

fold absent; gular scales keeled, each scale with 1–3 keels; body laterally compressed; dorsal body scales large, keeled, imbricate; group of smaller lateral scales, or isolated smaller scales present; middorsal scales not enlarged, not forming serrated dorsal crest; 34–41 (37.8 ± 3.3) scales in vertebral row between levels of anterior insertion of forelimb and groin; scales on nape region smaller than those of body; ventral scales large, imbricate, strongly keeled, usually pointed posteriorly, subequal in size; 30–33 (31.7 ± 1.5) midventral scales between levels of axilla and groin; 43–47 (45.3 ± 2.1) scales around midbody; caudal autotomy absent; caudal scales distinctively keeled; femoral and precloacal pores absent; 32–38 (34.9 ± 2.3) subdigital scales on Digit IV of hind limb; subdigital scales with keratinized knobs on anterior section of each scale; SVL 119.1 mm in male, 73.9–137.0 (102.6 ± 31.9) mm in females; TAL/SVL 4.01 in male, 2.92–3.56 in females; casque L/SVL 0.34 in male, 0.32–0.35 in females; HL/SVL 0.23 in male, 0.23–0.25 in females; SHL/SVL 0.29 in male, 0.28–0.29 in females.

Color in life of an adult female (USNM 549415; Plate 42): dorsal surface of body lime green with pale green-gray middorsal stripe about 4 scales wide, dorsum crossed by 4 crossbars, crossbars brown with dark brown anterior and posterior edging where crossing middorsal stripe and dark lime green edging lateral to that point; narrow, interrupted white line extending from axilla to groin, separating dorsal coloration from pale lime green ventral coloration; forelimb lime green; hind limb lime green with brown smudging and pale lime green cross-bands dorsally; top of head casque tan-gray with small black scattered punctations; posterior edge of casque outlined with black line, line breaking up anteriorly; side of head pale lime green with narrow white stripe beginning below orbit, continuing along upper lip, below tympanum, and along side of throat to just posterior to base of anterior edge of forearm, irregular bronze, black-flecked band above that stripe beginning posterior to orbit and continuing across tympanum to neck; tail lime green at base with alternating pale green-gray and brown blotches, grading to tan with brown crossbands on remainder of tail.

Color in alcohol: dorsal surfaces of head, body, and limbs purple, pale purple, or pale brown; middorsal pattern variable, series of dark brown blotches (usually with small paler central area) extending length of back usually present, dark brown blotches usually present on neck and anterior body region, brown blotches followed by indistinct blotches that are purple or paler brown than dorsal ground color, or length of dorsum with distinct to indistinct blotches paler shade than dorsal ground color; head casque with varying amounts of brown to black spots along posterior edge and varying amounts of brown to black spotting or reticulation on dorsal surface; varying amounts of brown to black spots also present on lateral surface of head and between tympanum and forelimb; narrow white to cream stripe usually extending from below orbit onto ventrolateral surface of forelimb; white to cream narrow ventrolateral stripe (about 1 scale wide) extending from axilla to groin; chin and throat regions pale purple to cream; belly pale purple, pale brown, or cream; dorsal surface of tail same color anteriorly as posterior end of body, with or without indistinct brown or paler brown markings, tail generally becoming browner posteriorly with slightly more distinct brown markings; subcaudal surface similar in color to dorsal surface of tail, but generally paler shade.

Diagnosis/Similar Species.—The prominent flat-topped cephalic crest will distinguish *Laemanctus longipes* from all other Honduran lizards, except the remaining species of *Laemanctus*. *Laemanctus julioi* and *L. serratus* have enlarged postrostral scales, rather regular in size, with 20–22 scales between the rostral and the level of the eye (versus 36–55 smaller postrostrals, irregular in size, in *L. longipes*). *Laemanctus julioi* also has 53 scales around midbody (versus 43–47 scales around midbody in *L. longipes*). *Laemanctus serratus* also has a

fringe of enlarged serrated, erect, flattened, triangular-shaped scales on the posterior edge of the cephalic crest and has a serrated middorsal crest (versus fringe of enlarged triangular scales and serrated middorsal crest absent in *L. longipes*). *Laemanctus waltersi* lacks a gular fold or has an incomplete gular fold, has 30–39 scales around midbody, and has 24–34 scales in the vertebral row between the levels of the anterior portion of the forelimb insertion and groin (versus gular fold complete, 43–47 scales around midbody, and 34–41 scales in vertebral row between levels of forelimb insertion and groin in *L. longipes*).

Illustrations (Figs. 63, 77, 84, 85; Plate 42).—Álvarez del Toro, 1983 (adult); Boulenger, 1877 (adult; as *L. deborrei*); Campbell, 1998 (adult); Hartdegen, 1998 (adult); Köhler, 1999d (adult, head casque scales), 2000 (adult), 2001b (adult), 2003a (adult), 2008 (adult); Lee, 1996 (adult, head scales), 2000 (adult, head scales); McCoy, 1968 (head casque scales); McCranie and Köhler, 2004a (adult); Stafford, 1994 (adult); Stafford and Mallory, 2002 (adult); Stafford and Meyer, 1999 (adult); Villa et al., 1988 (hatchling).

Remarks.—McCoy (1968) and Lang (1989) reviewed the morphology of *Laemanctus longipes*, whereas McCranie and Köhler (2004a) provided an overview of the species' morphology and a literature review. All three of those reviews included both *L. longipes* and *L. waltersi* as the same species (*L. longipes*). Data recovered for this book indicates that Honduran populations of *L. longipes* agree better with the *L. longipes deborrei* data given by McCoy (1968) than they do with more northerly *L. l. longipes*. Lang (1989) was not clear on any diagnostic characters to define *L. l. deborrei*. Those McCoy data also suggest that *L. deborrei* might represent a valid species and at the same time leaves little doubt that *L. waltersi* represents a distinct species.

Natural History Comments.—*Laemanctus longipes* is known from 600 to 1,200 m

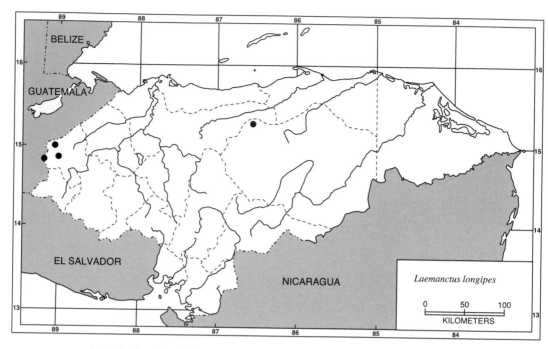

Map 35.   Localities for *Laemanctus longipes*. Solid circles denote specimens examined.

elevation in the Premontane Wet Forest formation and peripherally in the Premontane Moist Forest formation. The species probably occurs near water in Premontane Moist Forest. Known months of collection of this diurnal species are April to June and September to October. One specimen I saw was in vegetation above a trail. This has proven to be a difficult species to find in Honduras, as it appears to be a canopy inhabitant. *Laemanctus longipes* and its relatives need primary forest to old secondary forest for their survival. Females deposit three to five eggs per clutch with egg deposition known to occur in April and June (Lee, 1996; Köhler, 1999d; Stafford and Mallory, 2002, and references cited in those works). McCranie and Köhler (2004a) also provided a list of literature on reproduction in this species complex. Apparently nothing has been published on diet in natural populations of the *L. longipes* complex, but Hartdegen (1998) reported

adult and subadult crickets were taken in captivity, and Köhler (1999d) reported insects in diet of captives.

Etymology.—The name *longipes* is formed from the Latin *longus* (long) and *pes* (foot), in reference to the long feet of this species.

Specimens Examined (7 [1]; Map 35).— **COPÁN**: Copán, UMMZ 83032; between Laguna del Cerro and Quebrada Grande, USNM 549415; Río Amarillo, USNM 570297. **OLANCHO**: Terrero Blanco, USNM 342272. "HONDURAS": UF 83590, 90413–14.

Other Records.—UF 43115 (skeleton), 50251 (skeleton), 50470 (skeleton), 56401 (skeleton), 57597 (skeleton), 62082 (skeleton), 62561 (skeleton), 124598 (3 eggs; all of these skeletons and eggs were examined but lack locality data; thus, it is impossible to know which species of the *L. longipes* complex they represent).

Laemanctus serratus Cope, 1865a

Laemanctus serratus Cope, 1865a: 176
(holotype, RMNH 2845 [see McCoy,
1968: 673]; type locality: "Orizaba
Valley, Mexico"); Wilson and McCranie,
1998: 16; McCranie and Köhler, 2004b:
796.2; Townsend and Wilson, 2010b:
692; Solís et al., 2014: 129; McCranie,
2015a: 367.
Laemanctus longipes: Meyer, 1969: 252 (in
part); Meyer and Wilson, 1973: 26 (in
part).
Laemanctus serratus alticoronatus: Lang,
1989: 145.

Geographic Distribution.—*Laemanctus
serratus* occurs at low and moderate elevations on the Atlantic versant from central
Tamaulipas to the outer portion of the
Yucatán Peninsula, Mexico. Apparently
isolated populations occur in northwestern
Honduras (see Remarks). There are also
two 19th century records from the Pacific
versant of central Oaxaca, Mexico.

Description.—The following is based on
three females (USNM 83436, 84026,
84550). *Laemanctus serratus* is a moderately large corytophanine (maximum recorded
SVL 117 mm [USNM 83436]); mostly flat-
topped cephalic casque present (slightly
elevated medially on posterior portion);
dorsal head scales strongly rugose; anterior
dorsal head scales (postrostrals) much larger
than posterior dorsal head scales, scales
between rostral and orbit approximately
bilaterally symmetrical, 21–22 (21.3 ± 0.6)
postrostral scales in more-or-less regular
rows, including canthal scales and centrally
located azygous scale; posterior edge of
casque with projecting fringe of enlarged,
erect, flattened, triangular scales; 21–24
(22.7 ± 1.5) scales around posterior edge
of casque (between posterior most superciliary on each side); parietal eye distinct, in
scale smaller than surrounding scales, or
scale similar in size to surrounding scales, or
in large scale that is noticeably larger than
surrounding scales; tympanum distinct;

nasal scale single, nostril located more-or-
less centrally in scale, nostril opening
directed posterolaterally; moveable eyelid
present; pupil circular; 10–11 supralabials;
9–11 infralabials; gular fold complete, 3–4
rows of small scales contained in fold;
antehumeral fold distinct to indistinct,
continuous with gular fold; gular scales
keeled, each scale usually with single keel,
few scales with 2 keels; body laterally
compressed; dorsal body scales relatively
small, keeled; group of smaller lateral scales
absent; middorsal scales enlarged, forming
serrated dorsal crest; 30–33 (31.7 ± 1.5)
scales in vertebral row between levels of
anterior insertion of forelimb and groin;
scales on nape region smaller than those of
body; ventral scales large, imbricate, strongly keeled, usually rounded posteriorly, subequal in size; 34–42 (37.0 ± 4.4) midventral
scales between levels of axilla and groin; 58
scales around midbody in all three; caudal
autotomy absent; caudal scales strongly
keeled; femoral and precloacal pores absent;
31–38 (33.7 ± 3.1) subdigital scales on Digit
IV of hind limb; subdigital scales with
keratinized knobs on anterior section of
each scale; SVL 107.7–117.0 (113.2 ± 4.9)
mm; TAL/SVL 3.36–3.44; casque L/SVL
0.34–0.37; HL/SVL 0.25–0.26; SHL/SVL
0.30–0.34.

Color in alcohol: dorsal surface of body
pale green or pale brown; middorsal pattern
consisting of about 6–7 distinct to indistinct
brown crossbars, crossbars extend laterally
to lower half of dorsolateral stripe; middorsal crest with pale green to pale brown
scales, some middorsal scales tipped with
brown; indistinct pale brown dorsolateral
stripe about 2–5 scales broad, extending
between levels of axilla and groin; head
casque pale green to brown, without distinctive markings; a few brown to black
spots or short lines present on lateral
surface of head; distinct brown blotch or
broad stripe present on nape region anterior
to forelimb; pale brown dorsolateral broad
stripe usually present, but stripe can be

Plate 43.   *Laemanctus serratus.* UTA R-33236. Mexico: Yucatán Peninsula. Photograph by Eric N. Smith.

indistinct, about 4 scales wide on shoulder region; narrow white to cream stripe extending from below orbit to just anterior to forelimb; pale brown to cream narrow ventrolateral stripe (2–3 scales wide) extending from axilla to groin; dorsal surfaces of fore- and hind limb pale green to pale brown, with or without indistinct pale brown crossbands; chin and throat regions pale green to pale brown; belly pale green to pale brown, but slightly darker shade than chin and throat; dorsal surface of tail similar in color as body, with or without indistinct brown markings; subcaudal surface similar in color to dorsal surface of tail, but slightly paler green or brown.

Diagnosis/Similar Species.—The prominent flat-topped cephalic crest will distinguish *Laemanctus serratus* from all other Honduran lizards, except the remaining species of *Laemanctus.* All three remaining species of *Laemanctus* lack a fringe of enlarged triangular scales on the posterior edge of the cephalic crest and also lack a serrated middorsal crest (versus fringe of enlarged, erect, flattened, triangular scales present on posterior edge of cephalic crest

and serrated middorsal crest present in *L. serratus*). *Laemanctus longipes* and *L. waltersi* also have 36–45 postrostral scales (versus 21–22 in *L. serratus*).

Illustrations (Figs. 76, 78, 79; Plate 43).— Álvarez del Toro, 1983 (adult); Álvarez Solórzano and González Escamilla, 1987 (adult); Calderón-Mandujano et al., 2008 (adult); Campbell, 1998 (adult); Freiberg, 1972 (adult); Hartdegen, 1998 (head); Köhler, 1999d (adult, head casque scales), 2000 (adult), 2003a (adult), 2008 (adult); Lee, 1996 (adult, head scales, precloacal and thigh regions scales), 2000 (adult, head scales, precloacal and thigh scales); Lemos-Espinal, 2015 (adult); Lemos-Espinal and Dixon, 2013 (adult, head); McCoy, 1968 (head casque scales); McCranie and Köhler, 2004b (adult); Peters, 1948 (adult); W. Schmidt and Henkel, 1995 (adult); Stafford and Meyer, 1999 (adult); Villa et al., 1988 (adult).

Remarks.—McCoy (1968) and Lang (1989) reviewed the morphology of *Laemanctus serratus.* Lang (1989) included four symbols in northwestern Honduras on his distribution map of *L. serratus* but did not provide a list of specimens examined or locality data for those records. Lang, without providing supportive evidence, considered the Honduran specimens to represent *L. s. alticoronatus* Cope (1866a: 192), which otherwise occurs on the Yucatán Peninsula according to McCoy (1968) and Lang (1989). The Honduran specimens of *L. serratus* I examined lack specific locality data, except for one that was said to be from Tela, Atlántida. Those specimens were presented to the National Zoological Park in Washington, DC, USA, and then deposited in the USNM collection upon their deaths in 1931. A specimen of *L. waltersi* (USNM 83434) from "Honduras" was also presented to the National Zoological Park, and then to the USNM upon its death, also in 1931. *Laemanctus waltersi* is known to occur only in scattered low elevation localities in Atlántida and Cortés, Honduras. Thus, there seems to be no reason to doubt that those USNM *L. serratus* were collected in Honduras. Given that *L. serratus* occurs in subhumid habitats elsewhere in its range, it is likely that the Honduran specimens were collected in the subhumid Sula Plain of the ríos Chamelecón and Ulúa in the department of Cortés.

McCoy (1968) recognized two subspecies of *Laemanctus serratus*, *L. s. serratus* and *L. s. alticoronatus*. Pérez-Higareda and Vogt (1985: 140) described a third subspecies, *L. s. mccoyi*. Pérez-Higareda and Vogt apparently did not examine any specimens of *L. serratus* other than their type series of *L. s. mccoyi*. Their definitions of *L. s. serratus* and *L. s. alticoronatus* were taken verbatim from McCoy (1968), although their table 1 contains an error in mean number of scales around midbody for *L. s. serratus*. Lang (1989) also used the data of McCoy (1968) and Pérez-Higareda and Vogt (1985) to define those three subspecies of *L. serratus*. McCoy (1968) used four characters (number of scales around midbody, number of scales around posterior edge of head casque, number of subdigital scales on Digit

IV of hind limb, and presence or absence of an azygous scale among the enlarged anterior dorsal head scales) to distinguish between the subspecies *L. s. serratus* and *L. s. alticoronatus*. Pérez-Higareda and Vogt (1985) used those same four characters to define *L. s. mccoyi*. McCoy (1968) did not explain his methods for counting subdigital scales but, on p. 670, gave a total of 41/41 for the holotype (FMNH 5213) of *L. waltersi*. However, examination of FMNH 5213 revealed that in order to achieve a similar count, one has to count not only the subdigital scales on all phalanges, but also those scales in 1 row covering the metatarsal to the level of the point of insertion of Toe V. Assuming that McCoy (1968) was consistent in his methods of counting the subdigital lamellae in all specimens of *Laemanctus*, then his counts also included the scales covering the metatarsals on Toe IV in *L. serratus*. Pérez-Higareda and Vogt (1985) recorded lower numbers of subdigital scales for *L. s. mccoyi* than McCoy (1968) did for *L. s. serratus* and *L. s. alticoronatus*. Thus, it is likely that Pérez-Higareda and Vogt (1985) did not include scales covering the metatarsal in their counts for *L. s. mccoyi*, as was apparently done by McCoy (1968) in his counts for *L. serratus*. Assuming this to be the case, then there is considerable overlap between the three described subspecies of *L. serratus* in all four characters used to define these subspecies. Thus, those subspecies are established based on average values, some of which are inaccurate, for each of these characters and cannot be recognized as valid. Molecular and morphological data from additional *Laemanctus* specimens are needed to help resolve the systematics of *L. serratus* populations. Until that time, I herein place *L. s. mccoyi* Pérez-Higareda and Vogt in the synonymy of *L. serratus* Cope.

McCranie and Köhler (2004b) provided an overview of the species' morphology and reviewed its literature.

Natural History Comments.—The distribution of *Laemanctus serratus* in Honduras is unclear (see Remarks) with all three specimens from northern Honduras carrying the locality data of only "Honduras." Females deposit a clutch of three to seven eggs at least between May and July (Lee, 1996; Köhler, 1999d; and references cited in those two works). McCranie and Köhler (2004b) also listed references on reproduction in this species. Peters (1948: 2) reported a specimen from Tamaulipas, Mexico, contains seven eggs (month of collection not given). That same specimen also contained remains of insects in its stomach, apparently an orthopteran, and that it also fed on caterpillars in captivity (Peters, 1948: 2), and Martin (1958: 58) reported a snail shell, remains of an "*Anolis*," and a variety of arthropods, "mainly beetles and Orthoptera," in stomachs of other Mexican specimens.

Etymology.—The specific name *serratus* is Latin (toothed like a saw) and was used in reference to the serrated posterior edge of the casque.

Specimens Examined (3, 1 skeleton [2]).—"HONDURAS": USNM 83435 (skeleton), 83436, 84026, 84550.

Redescription of *Laemanctus waltersi* Schmidt, 1933

Laemanctus waltersi Schmidt, 1933: 20 (holotype, FMNH 5213; type locality: "Lake Ticamaya, east of San Pedro, between the Chamelecon [sic] and Ulua [sic] Rivers, Honduras"); M. A. Smith, 1933b: 32; Marx, 1958: 461; McCranie, 2015a: 367.

Laemanctus longipes waltersi: McCoy, 1968: 670; McCoy, 1970: 153, *In* Peters and Donoso-Barros, 1970; Lang, 1989: 143; McCranie and Köhler, 2004a: 795.1.

Laemanctus longipes: Meyer, 1969: 252 (in part); Meyer and Wilson, 1973: 26 (in part); Wilson and McCranie, 1998: 16; Köhler, 1999d: 78 (in part); Wilson et al., 2001: 135 (in part); McCranie et al., 2006: 216 (in part); Townsend, 2006a: 34; Wilson and Townsend, 2006: 105 (in part); Townsend and Wilson, 2010b: 692; Townsend et al., 2012: 100; Solís et al., 2014: 129 (in part).

Geographic Distribution.—*Laemanctus waltersi* occurs at low elevations on the Atlantic versant in northwestern Honduras in the departments of Atlántida and Cortés.

Description.—The following is based on two males (LSUMZ 33662; MCZ R29334), seven females (FMNH 5213; KU 187739; LACM 114240; MCZ R32034; UF 144748; USNM 580367; UTA R-6948), and one juvenile (MCZ R27907; measurements do not include the juvenile). *Laemanctus waltersi* is a moderately large corytophanine (maximum recorded SVL 144 mm [UF 144748, a female]) with an extremely long tail and long limbs; flat-topped cephalic casque present; dorsal head scales strongly rugose to slightly carinate; anterior dorsal head scales similar in size to, or slightly larger than, posterior dorsal head scales, irregular in shape; 36–45 (41.4 ± 3.5) postrostral scales present anterior to level of anterior edge of orbit (including canthal scales); posterior edge of casque lacking fringe of enlarged, flattened, triangular scales; 21–26 (23.2 ± 1.8) scales around posterior edge of casque (between posteriormost superciliary on each side); parietal eye distinct to indistinct, in scale similar in size to surrounding scales, or in scale slightly larger than surrounding scales; tympanum distinct; nasal scale single, nostril located more-or-less centrally in scale, nostril opening directed posterolaterally; moveable eyelid present; pupil circular; 9–11 supralabials; 8–11 infralabials; gular fold varies from absent to incomplete, 2–4 rows of small scales contained in fold when partially present; antehumeral fold absent; gular scales keeled, each scale with 1–3 keels; body laterally compressed; dorsal body scales large, keeled, imbricate; group of smaller lateral scales, or isolated smaller lateral scales, absent; middorsal scales not enlarged, not forming serrated dorsal crest; 24–34 (29.5 ± 3.0) scales in vertebral row between levels of anterior insertion of forelimb and groin; scales on nape region smaller than those of body; ventral scales large, imbricate, strongly keeled, usually pointed posteriorly, subequal in size; 24–38 (28.2 ± 4.3) midventral scales between levels of axilla and groin; 30–39 (33.7 ± 2.6) scales around midbody; caudal autotomy absent; caudal scales distinctively keeled; femoral and precloacal pores absent; 29–35 (32.5 ± 2.2) subdigital scales on Digit IV of hind limb; subdigital scales with keratinized knobs on anterior section of each scale; SVL 86.7–110.4 (98.6) mm in males, 110.0–144.0 (118.9 ± 12.0) mm in females; TAL/SVL 3.83–3.95 in males, 3.18–4.07 in females; casque L/SVL 0.32–0.33 in males, 0.29–0.32 in females; HL/SVL 0.23–0.24 in males, 0.20–0.25 in females; SHL/SVL 0.30–0.31 in males, 0.29–0.31 in females.

Color in life of an uncollected adult from CURLA (description from image taken in the field by Alexander Gutsche): dorsal surface of body Paris Green (63), with yellow tinge ventrolaterally and four Buff (24) vertebral blotches; head casque Buff with greenish tinge laterally and on supra-oculars; dorsal and lateral surfaces of tail Leaf Green (146) with series of Buff small blotches centered medially; side of head Yellow-Green (58); dorsal surfaces of fore- and hind limb Leaf Green; series of white ventrolateral spots present on body; pupil Tawny (38).

Color in alcohol: dorsal surfaces of head, body, and limbs purple, pale purple, or pale brown; middorsal pattern variable, series of dark brown blotches (usually with small paler central area) extending length of back usually present, dark brown blotches usually present on neck and anterior body region, brown blotches followed by indistinct paler brown or purple blotches, paler than dorsal ground color, or length of dorsum with distinct to indistinct blotches that are paler brown or paler purple than dorsal ground color; head casque with varying amounts of brown to black spots along posterior edge and varying amounts of brown to black spotting or reticulation on dorsal surface; varying amounts of brown to black spots also present on lateral surface of head and between tympanum and forelimb; narrow white to cream stripe usually extending from below orbit onto ventrolateral surface of forelimb; white to cream narrow ventro-lateral stripe (about 1 scale wide) extending from axilla to groin; chin and throat regions pale purple to cream; belly pale purple, pale brown, or cream; dorsal surface of tail same color anteriorly as posterior end of body, with or without indistinct brown or paler brown markings, tail generally becoming browner posteriorly with slightly more distinct brown markings; subcaudal surface similar in color to dorsal surface of tail, but generally paler shade of brown.

Diagnosis/Similar Species.—The prominent flat-topped cephalic crest will distinguish *Laemanctus waltersi* from all other Honduran lizards, except the other species of *Laemanctus*. *Laemanctus longipes* has a complete gular fold, has 43–47 scales around the midbody, has 34–41 scales in the vertebral row between levels of the forelimb insertion and groin, and has smaller lateral scales (versus gular fold absent, 30–39 scales around midbody, 24–34 scales in vertebral row, and no smaller lateral scales present in *L. waltersi*). *Laemanctus julioi* has 20 rather regularly arranged postrostral scales on top of the head in front of the levels of the eyes, has 53 scales around midbody, and has a distinct gular fold (versus 36–45 postrostral scales, 30–39 scales around midbody, and no gular fold in *L. waltersi*). *Laemanctus serratus* has 21–22 postrostrals, a fringe of enlarged triangular scales on the posterior edge of the cephalic crest, and has a serrated middorsal crest (versus 36–45 postrostral scales present, fringe of enlarged triangular

Plate 44. *Laemanctus waltersi*. Atlántida: Río Santiago. Photograph by Juan R. Collart.

cephalic scales absent, and serrated mid-dorsal crest scales absent in *L. waltersi*).

Illustrations (Figs. 55, 82, 83; Plate 44).—Apparently none previously published.

Remarks.—See Remarks for the genus *Laemanctus* for information on *L. waltersi*. This nominal form was originally described as a species, but treated as a subspecies of *L. longipes* in McCoy (1968) and Lang (1989).

Natural History Comments.—*Laemanctus waltersi* is known from near sea level to 300 m elevation in the Lowland Moist Forest formation and peripherally in the Lowland Dry Forest formation. The species probably occurs near rivers in Lowland Dry Forest. Known months of collection are June, July, and December. This arboreal and probably canopy-inhabiting species appears to need primary or old second growth forest for its survival. Apparently nothing has been published on reproduction or diet of wild *L. waltersi*, however some of the literature regarding captive *L. longipes* listed in McCranie and Köhler (2004a) could represent *L. waltersi*.

Etymology.—The patronym *waltersi* honors Leon L. Walters, the taxidermist at the FMNH, who accompanied Schmidt when the holotype was collected.

Specimens Examined (20, 1 C&S [4]; Map 36).—**ATLÁNTIDA**: Lancetilla, MCZ R32034; Tela, MCZ R27907, 29334, USNM 580367. **CORTÉS**: El Paraíso, UF 144748; Finca Fé, LSUMZ 33662; Laguna Ticamaya, FMNH 5213; near San Pedro Sula, KU 187739, LACM 11240, UTA R-6948–49, 6950 (C&S). "HONDURAS": BYU 42592, FMNH 206087, UF 124597, USNM 83434, UTA R-6948–49, 18052–53, 40730.

Other Records (Map 36).—**ATLÁNTIDA**: CURLA forest station (photograph); Jilamito Nuevo (Townsend, 2006a); near Pico Bonito Lodge (photograph); Río Santiago (photograph).

Family Dactyloidae Fitzinger, 1843.

This family of anole lizards is restricted to the Western Hemisphere where it occurs in the southeastern USA, portions of Mexico, southward into South America, the Greater and Lesser Antilles, and some other Caribbean and Pacific islands. Eight genera (Nicholson et al., 2012, 2014) and at least 400 species occur in this family. The number of genera is extremely controver-

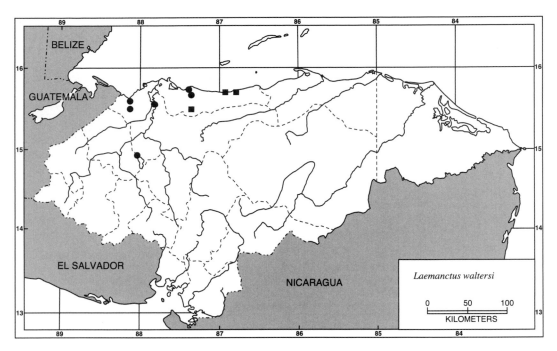

Map 36.   Localities for *Laemanctus waltersi*. Solid circles denote specimens examined and solid squares represent accepted records.

sial. No recent author disagrees that the anoles are a monophyletic unit. For that reason, the family name Dactyloidae was elevated (see T. M. Townsend et al., 2011) to recognize that monophyly. Now, and to get in step with other workers, it seems time for all workers on anoles to recognize monophyletic groups within the Dactyloidae unit as separate genera (i.e., *Norops*) when using the Linnaean taxonomic system, as has been shown by several recent phylogenetic analyses.

Poe (2013) strongly criticized the Nicholson et al. (2012) eight genera results for the anoles. Nicholson et al. (2014) argued point by point to the Poe (2013) criticisms, and as of 31 December 2016, Poe has not responded. Pincheira-Donoso et al. (2013: 1) realized that *Anolis* (*sensu stricto*), with its about 400 species, would be far and away "the most species-rich among amniote vertebrates on earth." Following the latest authorities (Nicholson et al., 2014), I

continue to treat *Anolis* (one species in Honduras) and *Norops* (41 species in Honduras, see below) as valid genera, as well as other monophyletic lineages (genera) recovered by Nicholson et al. (2012, 2014).

McCranie and Köhler (2015) treated the systematics, distribution, and conservation of the 39 named anole species known to occur in Honduras at the time of that work. McCranie and Köhler (2015) included those species in two genera, *Anolis* Daudin (1802: 50; the single species in Honduras treated by McCranie and Köhler, 2015) and *Norops* Wagler (1830a: 149; 38 of 41 species in Honduras treated by McCranie and Köhler, 2015). Sunyer, García-Roa, and Townsend (2013) added *N. wermuthi* Köhler and Obermeier (1998: 129) to the fauna of Honduras after it was too late to be included in McCranie and Köhler (2015). *Norops wermuthi* is treated fully herein from only a less than ideal literature review (see superfamily Iguanoidea). Köhler, Townsend, and Petersen (2016) described two new species of the *N. tropidonotus* species group, both of which occur in Honduras. These proposed changes are documented below, but no species accounts are included herein. Some recent publications including Honduran specimens are listed in the species list below. Other recent publications after McCranie and Köhler (2015) that include Honduran specimens or species that occur in Honduras are: phylogenies based on tissue analyses (Hertz et al., 2013; Helmus et al., 2014); morphologically based reviews (J. J. Köhler et al., 2015; G. Köhler, Townsend, and Petersen, 2016); and dorsal and ventral scale illustrations of *N. biporcatus*, *N. capito*, and the *N. sagrei* complex (Wegener et al., 2014).

Remarks.—Wagler (1830a: 149) used the name Dactyloae, 13 years before Fitzinger did, but it appears that Wagler did not intend that name for a group of anoles. A list of the 41 species of Dactyloidae known from Honduras as recognized herein is included below. A few annotations are also included for information published subsequent to the McCranie and Köhler (2015) work.

Anolis Daudin, 1802: 50 (one species)

Anolis allisoni Barbour, 1928:58.

Norops Wagler, 1830a: 149 (41 species, 20 endemic species, 1 introduced species).

Norops amplisquamosus McCranie, Wilson, and Williams, 1992: 209.
Norops beckeri (Boulenger, 1882: 921).
Norops bicaorum Köhler, 1996d: 21.
Norops biporcatus (Wiegmann 1834b: 47). Solís et al. (2015) reported this species from Santa Bárbara based on a digital image (UTADC-8622).
Norops capito (W. Peters, 1863b: 142).
Norops carpenteri (Echelle, Echelle, and Fitch, 1971: 355).
Norops crassulus (Cope, 1865a: 173). More than one species is likely represented among the *N. crassulus* complex in

Honduras (McCranie, 2015a; McCranie and Köhler, 2015). Indeed, several manuscripts in preparation will document those beliefs.
Norops cupreus (Hallowell, 1861: 481).
Norops cusuco McCranie, Köhler, and Wilson, 2000: 214.
Norops heteropholidotus (Mertens, 1952a: 89).
Norops johnmeyeri (Wilson and McCranie, 1982: 133).
Norops kreutzi McCranie, Köhler, and Wilson, 2000: 218.
Norops laeviventris (Wiegmann, 1834b: 47). Espinal, Solís, O'Reilly, and Valle (2014) reported a record from Choluteca. A recent visit (May 2015) to that locality proved the species to be easily found there, both during sunny afternoons and as animals were sleeping at night, usually on thin tree branches and in bromeliads in the vicinity of 3 m or more above the ground. One of the two nights spent at that site was very windy and chilly, with most sleeping sites moving with the wind; however, numerous *N. laeviventris* were still seen.
Norops lemurinus (Cope, 1861a: 213).
Norops limifrons (Cope, 1862c: 178).
Norops loveridgei (Schmidt, 1936: 47).
Norops mccraniei Köhler, Townsend, and Petersen, 2016: 11. Part of the former *N. tropidonotus* species complex. Köhler, Townsend, and Petersen (2016) no longer consider *N. tropidonotus* (W. Peters, 1863b: 135) to occur in Honduras.
Norops morazani (Townsend and Wilson, 2009: 63).
Norops muralla Köhler, McCranie, and Wilson, 1999: 285.
Norops nelsoni (Barbour, 1914: 287). Solís et al. (2014), without comment, continued to list this species as *N. sagrei*. Significant color differences in male dewlap and female head pattern occur between the Big and Little Swan Island populations (see McCranie et al., 2017). The significance of those color patterns remains unknown. McCranie et al.

(2017) also discussed some behavioral differences between those two island populations.

Norops ocelloscapularis Köhler, McCranie, and Wilson, 2001: 248.

Norops oxylophus (Cope, 1875: 123). Solís et al. (2014) listed this species as *N. lionotus* (Cope, 1861a: 210) without comment. *Norops lionotus* occurs in Panama to the south of *N. oxylophus* (as noted in McCranie and Köhler, 2015).

Norops petersii (Bocourt, 1873: 79, *In* A. H. A. Duméril et al., 1870–1909a).

Norops pijolense McCranie, Wilson, and Williams, 1993: 393. Several authors have used the unjustified emendation *N. pijolensis*, beginning with Köhler (2000: 63). McCranie and Köhler (2015) briefly discussed that point.

Norops purpurgularis McCranie, Cruz, and Holm, 1993: 386.

Norops quaggulus (Cope, 1885: 391).

Norops roatanensis Köhler and McCranie, 2001: 240.

Norops rodriguezii (Bocourt, 1873: 62, *In* A. H. A. Duméril et al., 1870–1909a).

Norops rubribarbaris Köhler, McCranie, and Wilson, 1999: 280.

Norops sagrei (Cocteau, 1837: 149, *In* A. M. C. Duméril and Bibron, 1837). Solís et al. (2014) failed to credit Cocteau with the name *N. sagrei*, although that fact has been well documented in the recent literature (i.e., Nicholson et al., 2012). Systematic studies combining molecular and morphological data of the *N. sagrei* complex populations in Central America, including Caribbean island populations, are really needed. Espinal, Solís, O'Reilly, Marineros, and Vega (2014) reported *N. sagrei* from Santa Bárbara based on a digital record. Terán-Juárez et al. (2015) reported this species was introduced at two localities in Tamaulipas, Mexico. Those authors also mentioned introductions of this species in other states of Mexico.

Norops sminthus (Dunn and Emlen, 1932: 26).

Norops uniformis (Cope, 1885: 392).

Norops unilobatus (Köhler and Vesely, 2010: 217). Köhler et al. (2014; as *Anolis*) included photographs of adults, male and female dewlaps, and head scales of Mexican specimens.

Norops utilensis Köhler, 1996c: 24.

Norops wampuensis McCranie and Köhler, 2001: 228.

Norops wellbornae (Ahl, 1939: 246). McCranie and Gutsche (2016) provided information on this species on the Golfo de Fonseca islands in Valle.

Norops wermuthi Köhler and Obermeier, 1998: 129 (see below).

Norops wilsoni Köhler, Townsend, and Petersen, 2016: 27. Part of the former *N. tropidonotus*, and a Honduran endemic.

Norops yoroensis McCranie, Nicholson, and Köhler, 2002: 466.

Norops zeus Köhler and McCranie, 2001: 236. Townsend et al. (2012) commented that molecular data indicates *N. zeus* is a complex of more than one species.

Norops wermuthi Köhler and Obermeier, 1998.

Norops wermuthi Köhler and Obermeier, 1998: 129 (holotype, SMF 77323; type locality: "Nicaragua, Departamento Jinotega, road from Matagalpa to Jinotega at km 146"); Sunyer, García-Roa, and Townsend, 2013: 104; Solís et al., 2014: 130; McCranie, 2015a: 368; McCranie and Köhler, 2015: 292.

Geographic Distribution.—*Norops wermuthi* occurs at moderate and intermediate elevations of extreme southeastern Honduras (apparently) and in north-central Nicaragua. In Honduras, this species is reported from one intermediate elevation locality in El Paraíso some mere 3 m from the boundary with Nicaragua (see Remarks).

Description.—Since there are no known Honduran specimens of this nominal species in any museum collection, the following is based on a summation of the species in Sunyer et al. (2008), with some additional notes from the type description of *Norops wermuthi* in Köhler and Obermeier (1998). Unfortunately, some data in Sunyer et al. 2008, table 1) do not match the values given in their variation section. *Norops wermuthi* is a medium-sized anole (SVL 51 mm in largest Nicaraguan male, 56 mm in two Nicaraguan females): snout scales slightly bulging to keeled; dorsal head scales smooth, slightly tuberculate, or keeled in internasal, prefrontal, and frontal regions; frontal depression present; parietal depression shallow; 4–8 (5.9 ± 0.8) postrostrals; anterior nasal divided, lower section contacting rostral and first supralabial; 4–8 (6.3 ± 0.6) internasals; canthal ridge distinct; supraorbital semicircles well defined; 0–3 scales separating supraorbital semicircles and 0–3 scales separating supraorbital semicircles and interparietal at narrowest points; interparietal well defined, slightly to distinctly enlarged relative to adjacent scales, surrounded by scales of moderate size, longer than wide, smaller than ear opening; about 4–8 (total number) distinctly enlarged, keeled or wrinkled supraocular scales; 1–2 enlarged supraoculars in broad contact with supraorbital semicircles; 2–3 strongly keeled elongate superciliaries; usually 3 (rarely 4) enlarged canthal scales per side; 5–8 scales separate second canthals; 6–11 scales separate posterior canthals; loreal region slightly concave, 16–35 (24.2 ± 4.9) mostly strongly keeled loreal scales in maximum of 4–6 horizontal rows; 5–8 supralabials and 6–8 infralabials to level below mideye; suboculars weakly to strongly keeled, usually 2 suboculars in broad contact with supralabials; ear opening vertically oval; scales anterior to ear opening not granular, slightly larger than those posterior to ear opening; 4–6 postmentals, outer pair usually largest; keeled granular scales present on chin and throat; male dewlap relatively small, extending to level of axilla or slightly posteriorly; male dewlap

with 4–5 horizontal gorgetal scale rows, total of 14–18 enlarged gorgetal scales; sternal scales in 4–6 rows, with 3–7 scales per row; female dewlap smaller than that of male, extending to about level of shoulder in front of axilla; low nuchal crest present in males, no dorsal ridge; about 8–14 middorsal scale rows distinctly enlarged, keeled, small scales irregularly interspersed among enlarged dorsal scales; dorsal scales lateral to middorsal series abruptly larger than granular lateral scales; flank scales heterogeneous, solitary enlarged keeled scales scattered among laterals; 34–59 (47.3 ± 6.8) dorsal scales along vertebral midline between levels of axilla and groin; 23–36 dorsal scales along vertebral midline contained in 1 head length; ventral scales on midsection about same size as largest dorsal scales; ventral body scales smooth to weakly keeled, flat, imbricate; 35–48 (42.5 ± 3.4) ventral scales along midventral line between levels of axilla and groin; 22–30 (26.2 ± 2.6) ventral scales contained in 1 head length; 97–122 (108.5 ± 5.9) scales around midbody; tubelike axillary pocket absent; pair of greatly enlarged postcloacal scales present in males; tail slightly to distinctly compressed, TH/TW 1.05–1.40; all subcaudal scales keeled; middorsal caudal scales keeled, homogeneous, although indistinct division in segments discernable; dorsal medial caudal scale row enlarged, keeled, not forming crest; scales on anterior surface of antebrachium distinctly keeled, unicarinate; 19–29 (24.2 ± 1.7) subdigital lamellae on phalanges II–IV of Digit IV of hind limb (both hind limbs combined), 8–10 (9.1 ± 0.6) lamellae on Phalanx I of Digit IV of hind limb; TAL/SVL 2.08–2.53 in males, 2.02–2.35 in females; HL/SVL 0.25–0.27 in males, 0.25–0.29 in females; SHL/SVL 0.23–0.30 in males, 0.22–0.28 in females; longest toe of adpressed hind limb usually reaching to about eye.

Color in life and in preservative of the adult male holotype (SMF 77323 from Jinotega, Nicaragua was described as follows by Köhler and Obermeier (1998: 130): "the dewlap color was Flame Scarlet (Color 15) with white scales. ... Ground color in life and after 12 months in preservative, Drab-Gray (Color 119D). A dark Drab (Color 119B) interorbital bar is present. Four dorsolateral Dark Drab (Color 119B) triangular markings open laterally to Light Drab (Color 119C) lateral coloration. The third (left) and fourth (right) ultimate dorsolateral body markings are each connected to a small Dark Drab (Color 119B) vertebral spot. Venter uniform pale brown (paler than Light Drab Color 119C). Legs with irregular transverse dorsal Dark Drab (Color 119B) crossbars. Original part of tail with 7 Dark Drab (Color 119B) crossbands, which are narrowest dorsally."

Hemipenis: the everted hemipenis of SMF 78604 is a moderately large, bilobed organ; asulcate processus divided, with medial flap; sulcus spermaticus bordered by well-developed sulcal lips, opening into broad concave area at base of apex; lobes strongly calyculate; truncus with transverse folds.

Diagnosis/Similar Species.—*Norops wermuthi* can be distinguished from all other Honduran lizards, except *Polychrus gutturosus* and the remaining species of anoles, by having a partially divided mental scale and a well-developed male dewlap. *Polychrus gutturosus* has femoral pores, has Digit IV of the hind limb about the same length as Digit III, and lacks tail autotomy (versus femoral pores absent, Digit IV of hind limb much longer than Digit III, and tail autotomy present in *N. wermuthi*). *Norops wermuthi* is distinguished from all other Honduran *Norops*, except *N. heteropholidotus*, *N. muralla*, and *N. sminthus*, by the combination of having about 8–14 distinctly enlarged middorsal scale rows, having heterogeneous lateral scales, having smooth to mostly weakly keeled and imbricate ventral scales, having a red male dewlap in life, and having a pair of greatly enlarged postcloacal scales in males. *Norops heteropholidotus* differs by having smooth midventral scales and the medial dorsal scales uniform in size, without interspersed small scales (versus weakly keeled and small midventral scales, and small scales irregularly interspersed among enlarged medial dorsal scales in *N. wermuthi*). *Norops muralla* differs by having completely smooth midventral scales (versus midventral scales weakly keeled in *N. wermuthi*). *Norops sminthus* differs by having enlarged median dorsal scales that are relatively regularly arranged (versus small scales irregularly interspersed among enlarged medial dorsal scales in *N. wermuthi*). *Norops amplisquamosus* shares most of those characters listed above with *N. wermuthi* but differs most notably in having an orange-yellow male dewlap in life (male dewlap red in *N. wermuthi*).

Illustrations (Plates 45, 46).—Köhler, 1999b (adult), 2000 (adult, dorsal and lateral body scales, chin scales, male dewlap), 2001b (male dewlap), 2003a (adult, dorsal and lateral body scales, chin scales, male dewlap), 2008 (adult, male dewlap, dorsal and lateral body scales, chin scales); Köhler and Obermeier, 1998 (adult, head, hemipenis, male dewlap); Sunyer, García-Roa, and Townsend, 2013 (adult, male and female dewlap; as *Anolis*).

Remarks.—*Norops wermuthi* was reported from Honduras around the same time as the cut-off date (31 June 2013) for substantial additions to the manuscript reviewing the anoles of Honduras (McCranie and Köhler, 2015). *Norops wermuthi* is a member of the *N. crassulus* species group. Also, Sunyer, García-Roa, and Townsend (2013) said the Honduran specimen was an adult female that was collected only "about three meters from the border" with Nicaragua.

Sunyer et al. (2008) presented a study of the external morphology and hemipenis of *Norops wermuthi* (as *Anolis*).

Natural History Comments.—*Norops wermuthi* is known only from 1,800 m elevation in the Lower Montane Moist Forest formation (in Honduras). According

Plate 45.   *Norops wermuthi.* Nicaragua: Jinotega, Cerro Kilambé. Photograph by Javier Sunyer.

to Sunyer, García-Roa, and Townsend (2013: 104), the Honduran individual was collected in July "at about 19:00 h." That individual, as well as a series from the Nicaraguan side of the border at that locality, were sleeping on low vegetation at night. A daytime search at the same locality by those authors did not reveal additional

Plate 46.   *Norops wermuthi.* Adult male dewlap. Nicaragua: Jinotega, Cerro Kilambé. Photograph by Javier Sunyer.

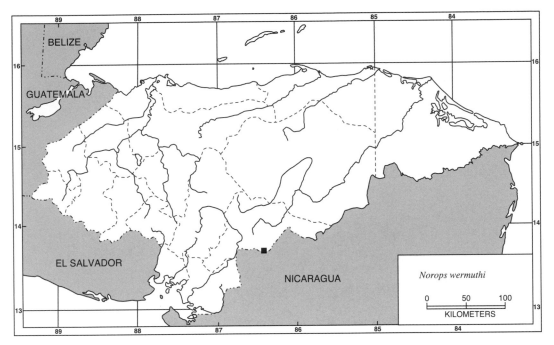

Map 37.   Locality for *Norops wermuthi*. Solid square denotes a purported Honduran record (see text).

individuals (Sunyer, García-Roa, and Townsend, 2013: 104). Köhler and Obermeier (1998) reported specimens in the type series were active on the ground during the day and sleeping on low vegetation at night (Sunyer, García-Roa, and Townsend, 2013, reported similar data for subsequently collected specimens). The known elevational range of *N. wermuthi* throughout its geographical distribution is 1,000 to 1802 m. Nothing has been published on reproduction or diet in *N. wermuthi*.

Etymology.—The patronym *wermuthi* honors the late Heinz Wermuth of Freiberg, Germany, himself a former herpetologist.

Specimens Examined (0 [0]).—None available.

Other Records (Map 37).—**EL PARAÍSO:** Cerro Jesús (Sunyer, García-Roa, and Townsend, 2013).

Family Iguanidae Bell, 1825.

This family (*sensu* Frost and Etheridge, 1989; Frost et al., 2001) of lizards ranges in the Western Hemisphere from southeastern California, southern Nevada, and southwestern and south-central Utah, USA, southward to the southern tip of the Baja California Peninsula (excepting the northwestern sector), the coastlands of western Mexico, and Tamaulipas, Mexico, thence through Central America into South America as far as western Ecuador west of the Andes and southeastern Brazil east of the Andes. The family also occurs on the Greater and Lesser Antilles, the Galapagos Islands of Ecuador, the Swan Islands and the Bay Islands of Honduras, and on several Pacific islands of Mexico to northwestern Colombia. In the Eastern Hemisphere, it is also distributed on the Fiji and Tonga Islands in the southwestern Pacific Ocean. Eight genera containing about 43 named species are included, with two genera comprising seven species occurring in

Figure 86.   Caudal whorls of spiny scales present, with single intercalary scale row. *Ctenosaura flavidorsalis.* USNM 581850 from Potrerillos, La Paz.

Honduras. Males of all Honduran species of Iguanidae are significantly larger than are females.

KEY TO HONDURAN GENERA OF THE FAMILY IGUANIDAE

1A. Tail bearing whorls of enlarged spiny scales, separated by 1 or 2 rows of slightly smaller intercalary scales (Figs. 86–91); no enlarged circular scale present on side of head ventral to tympanum . . . .
. . . . *Ctenosaura* (p. 226)
1B. Tail with scales more-or-less uniform in size, not in conspicuous whorls (Fig. 92); enlarged circular scale present on side of head ventral to tympanum (Fig. 93) . . .
. . . . *Iguana* (p. 281)

CLAVE PARA LOS GÉNEROS HONDUREÑOS DE LA FAMILIA IGUANIDAE

1A. La cola con verticilos distintivos de escamas agrandadas y espinosas presentes, separados por una o dos hileras de escamas intercalares

más pequeñas (Figs. 86–91); sin una escama subtimpánica agrandada presente . . . . . . *Ctenosaura* (p. 226)
1B. Las escamas de la cola son más o menos iguales en tamaño, sin anillos distinguibles (Fig. 92); con una escama subtimpánica agrandada presente (Fig. 93) . . . . *Iguana* (p. 281)

Genus *Ctenosaura* Wiegmann, 1828

Ctenosaura Wiegmann, 1828: col. 371 (type species: *Ctenosaura cycluroides* Wiegmann, 1828: col. 371 [= *Lacerta acanthura* Shaw, 1802: 216], by subsequent designation of Fitzinger, 1843: 16 and by monotypy).

Geographic Distribution and Content.— This genus ranges from central Tamaulipas and north-central Sonora, Mexico, southward to central Panama. *Ctenosaura* also occurs from east-central Baja California, Mexico, to the southern tip of that peninsula, as well as on many islands associated with it. Species of *Ctenosaura* likewise occur on many islands along both coasts of Middle America and are also introduced

Figure 87.   Caudal whorls of spiny scales present, with single intercalary scale row. *Ctenosaura melanosterna*. USNM 573903 from near mouth of Río Lorenzo, Yoro.

and established in southern Florida and southern Texas, USA, several Bahama Islands, and the Colombian islands of Malpelo, Providencia, and San Andrés. At least 18 named species are included in this genus, six of which occur in Honduras.

Remarks.—The six Honduran species of *Ctenosaura* are easily differentiated from each other based on external morphology. However, the molecular data produced thus far (i.e., Gutsche and Köhler, 2008; Pasachnik, Echternacht, and Fitzpatrick, 2010; F.

Figure 88.   Caudal whorls of spiny scales present, with single intercalary scale row. *Ctenosaura quinquecarinata*. USNM 581884 from El Rodeo, El Paraíso.

Figure 89.   Caudal whorls of spiny scales present, with two intercalary scale rows, at least laterally. *Ctenosaura bakeri.* USNM 26137 from Utila Island, Islas de la Bahía.

Köhler, unpublished data, based on a significant number of vouchers of all nominal Honduran species of *Ctenosaura* and collected from various parts of the country) demonstrate weak agreement with the morphologically based species. The Honduran species of *C. palearis* and *C. quinquecarinata* species groups are more distinct morphologically from each other than they are genetically. Köhler, Schroth,

Figure 90.   Caudal whorls of spiny scales present, with two intercalary scale rows (at least partially). *Ctenosaura oedirhina.* USNM 573905 from near Flowers Bay, Roatán Island, Islas de la Bahía.

Figure 91. Caudal whorls of spiny scales present, with two intercalary scale rows. *Ctenosaura similis*. USNM 580390 from Pájaros Island, Valle.

and Streit (2000) presented a phylogenetic analysis of the genus *Ctenosaura* and proposed recognition of three monophyletic subgroups as subgenera. See the Remarks for each species account of *Ctenosaura* for its placement in one of those subgenera.

The species of the *Ctenosaura palearis* group were placed on Convention on International Trade in Endangered Species of Wild Fauna and Flora (CITES) Appendix II in 2010. Three species of that group (*C. bakeri*, *C. melanosterna*, and *C. oedirhina*) occur in Honduras. Unfortunately, the CITES petitioners (see Pasachnik and Ariano, 2010) *made a serious error when they misidentified the previously exported*

Figure 92. Caudal whorls absent. *Iguana iguana*. UNAH from El Faro, Choluteca.

Figure 93.   Much enlarged scale present ventral to tympanum. *Iguana iguana.* UNAH from El Faro, Choluteca.

(emphasis mine) Honduran *Ctenosaura* as *C. melanosterna.* That *Ctenosaura* being exported was actually *C. quinquecarinata* (subgenus *Enyaliosaurus*) and not *C. melanosterna* (subgenus *Loganiosaura*). Additionally, no species of *Ctenosaura* were exported from Honduras for at least 5 years before the CITES petition was accepted. Documentation of that misidentification comes from my examination of the animal dealer specimen (BYU 39667) used by Sites et al. (1996; as *C. palearis;* but given as 34667) and Gutsche and Köhler (2008) in their respective phylogenetic studies. Gutsche and Köhler (2008) even wrote that the identification of that animal dealer BYU specimen was erroneous because it clustered with *C. flavidorsalis* in their study. The Gutsche and Köhler (2008) study had demonstrated that the BYU specimen formed a subclade with *C. flavidorsalis,* a *Enyaliosaurus* group member, and not to any of the *C. palearis* group species (*C. bakeri, C. oedirhina*) used in that same study. Those two studies were available to the CITES petitioners and should have alerted them to the potential problems regarding the correct identification of the

Ctenosaura formerly exported. Further documentation of the misidentification of the animal trade specimens comes from the animal dealer himself. That former animal dealer told me El Madreal, Choluteca, was the locality where his *Ctenosaura* were collected. Subsequent collection of *C. quinquecarinata* from El Madreal for tissue analyses demonstrated the El Madreal population formed a subclade within *C. quinquecarinata* along with the animal dealer BYU specimen from that locality (F. Köhler, Gutsche, and McCranie, unpublished data). Thus, a little research on the part of the CITES petitioners should have avoided their inaccurate conclusions. Additionally, A. Gutsche (personal communication to author, 24 November 2015), in response to my inquiry about European keeping of *Ctenosaura,* wrote: "there is no big interest among the European pet keepers to keep these iguanas. Keeping is too expensive and [they] are mostly too big and not really attractive. Consequently, they have no market relevance in the pet trade. There is no known private keeper of *C. oedirhina* in Europe, a few individuals of *C. bakeri* are only kept in Zoo Frankfurt and

Zoo Rotterdam, *C. melanosterna* is only kept in low numbers and only by very few people." Gutsche (personal communication to author, 24 November 2015) and his German colleague, with much experience on the European pet trade, were asked to review the original CITES proposal for placing *C. bakeri*, *C. melanosterna*, and *C. oedirhina* on CITES. They "were of the opinion that the CITES proposal was so bad and provided no serious data about trading of said species of *Ctenosaura* that it should be denied" for acceptance, but their suggestion was ignored by the CITES personnel. Gutsche (personal communication 31 January 2016) also wrote about a recent meeting with a well-known German reptile keeper where they talked about private keeping of *C. melanosterna* and other *Ctenosaura* in Europe. Their conclusion was "*C. melanosterna* and relatives continue to have no market relevance in Europe." The unfortunate misconception of international pet trade in *C. melanosterna* has subsequently been repeated in the literature (i.e., Pasachnik, Echternacht, and Fitzpatrick, 2011; Pasachnik, Montgomery, Ruyle et al., 2012; Pasachnik, Dannof-Burg et al., 2014; Montgomery et al., 2015.

Hasbún and Köhler (2009: 201) wrote: the smaller species of *Ctenosaura* "have always been relatively rare compared to the larger, sympatric *C. similis*." Also, two inaccurate International Union for Conservation of Nature (IUCN) reports regarding *C. flavidorsalis* (Köhler, 2004b) and *C. oedirhina* (Pasachnik, Ariani-Sánchez et al., 2010; changed to different authors and released in 2017, but with similar inaccurate assesment) classified those two species as declining, largely in part to habitat destruction. The biggest problem behind those inaccurate reports is those authors did not do substantial fieldwork searching for those lizards, instead only searching in select localities. Subsequently, Naccarato et al. (2015), citing Köhler (2004b), added to that inaccurate information and wrote that *C.* *flavidorsalis* was endangered; nothing could be further from the truth.

In the year 2011, I returned to Roatán Island and began to search areas I had previously searched for *Ctenosaura oedirhina*. I was surprised to discover that *C. oedirhina* had greatly expanded its distribution and contained robust populations where they were not seen in the 1980s. That expansion was a result of the destruction of the forests, thus creating more open areas for this sun-loving species to live and reproduce. Therefore, that 2011 trip demonstrated that the 2010 and 2017 IUCN reports were inaccurate (also see the *C. oedirhina* account below). That same year, I was doing general fieldwork in a subhumid area of El Paraíso. *Ctenosaura quinquecarinata*, another of those small species, contained robust populations at that site and was easily seen. Those recent discoveries challenged all that I had recently read about all of those smaller species of *Ctenosaura* being rarely seen, or in serious decline. Those discoveries increased my desire to search the subhumid zones in southern Honduras, which I had previously neglected, where three of those species of smaller *Ctenosaura* had been reported. Thus, I prepared a list of the known Honduran localities for *C. flavidorsalis*, *C. praeocularis*, and *C. quinquecarinata* (frequently misidentified as *C. palearis*) from the literature and from museum holdings. In April 2012, I began targeting those three species based on those locality records, including the type locality of *C. praeocularis* and a locality near the *C. flavidorsalis* type locality, which contained more rocky slopes than did its actual type locality. I was shocked by how easily those three species of *Ctenosaura* were to find, because all information in recent literature said those species were rare and are severly declining. Four *C. praeocularis* were collected, and others seen, on a 1-hour walk with a longtime resident of the area of its type locality on the first and only day spent there. Next we visited the vicinity of the *C. flavidorsalis* type locality and drove into some low mountains to the west of the flat terrain of the type locality. After a short drive, we came to an area with many boulders along the road. I stopped at the first house along the road to question the owner about the presence of these lizards. He quickly pointed to a boulder in front of his house where a beautiful adult male *C. flavidorsalis* was sitting. The remainder of that afternoon and the following morning resulted in eight captures (sufficient for my project) and numerous other *C. flavidorsalis* sightings from the vicinity of that house. Other similar-looking mountains, that also surely contain those *Ctenosaura*, were visible from where we worked.

On a subsequent trip in 2013, we drove to southern La Paz near the border with El Salvador. Köhler et al. (2005) had provided a locality for *Ctenosaura flavidorsalis* in northern El Salvador near the border with La Paz, Honduras. After a full rainy day of driving on a bad mountain road and a short drive the following day, we reached an area that looked promising for *Ctenosaura*. We stopped to talk with a group of three men walking to work in their fields. They told us a lizard fitting our description lived in a rocky hillside some 5 minutes from where we were talking. That afternoon we captured four *C. flavidorsalis* and toward nighttime, we drove about an hour away and the first man we talked with in that area also knew the *Ctenosaura* well and the next morning, as promised, a helper arrived to show us a *Ctenosaura* locality. The next 2 days showed that that species of *Ctenosaura* was extremely common in that area.

Earlier in July 2004, I had visited the El Madreal, Choluteca, site discussed above to collect *C. quinquecarinata*. Several were collected on the afternoon we were there. In 2012, with the renewed interest, we returned to El Madreal and also visited several sites in the area. With the help of local residents, we collected several *C. quinquecarinata* at every site we worked. Thus, at every site in Choluteca, El Paraíso, Francisco Morazán, and La Paz, where we targeted these smaller species of *Ctenosaura*, we were able to collect several individuals on our first day, and almost always during the first or second hour at each site we visited.

It seems obvious that personal fieldwork targeting certain species of *Ctenosaura* would reveal more reliable results than would be obtained by the previously used method of surveying residents of an area where *Ctenosaura* occur. Stephen et al. (2011) reported on their methods of surveying status, trade, and exploitation of Central American iguanas (also see Pasachnik et al., 2014). Their methods consisted of visiting "local markets, captive breeding facilities, NGO's, academic institutions, and government authorities" (p. 5). They also interviewed local people (see photographs on their p. 24), including women unlikely to spend much time in *Ctenosaura* habitat, about their knowledge of species of *Ctenosaura*, including their knowledge of the international pet trade. The Stephen et al. (2011; also see Pasachnik et al., 2014) methods are the preferred ones used by researchers investigating those types of data for most reptile species. Whereas those methods might be the best ones for some reptile species and can be beneficial for some groups (i.e., marine turtles), they can also result in misleading and inaccurate conclusions.

Those misleading and inaccurate second-hand results are likely the basis for most of the inaccurate statements in the literature regarding the population status of those smaller species of *Ctenosaura*.

Etymology.—The name *Ctenosaura* is formed from the Greek *ktenos* (comb) and *saura* (lizard), in reference to the serrated dorsal crests of adult males of the type species.

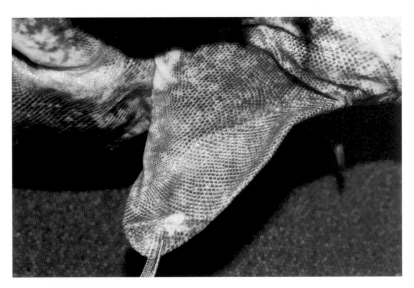

Figure 94. Pendulous male dewlap present. *Ctenosaura melanosterna.* USNM 573324 from at mouth of Río San Lorenzo, Yoro.

KEY TO HONDURAN SPECIES OF THE GENUS
CTENOSAURA

1A. Only a single row of intercalary scales present between caudal whorls (Figs. 86–88) 2

1B. Two rows of intercalary scales present between caudal whorls, at least dorsolaterally (Figs. 89–91) 4

2A. Chest black in adult males; pendulous male dewlap present (Fig. 94); <45 enlarged, strongly laterally compressed middorsal crest scales (Fig. 95); ≤10 enlarged middorsal crest scales in 1 head length .....
.... *melanosterna* (p. 251)

2B. Chest pale brown to brown or grayish brown; pendulous dewlap absent, gular fold present (Fig. 96); >45 enlarged middorsal crest scales (rarely as low as 39 in *C. flavidorsalis*), laterally compressed or not; >10 enlarged middorsal crest scales in 1 head length 3

3A. Enlarged middorsal crest scales not distinctively laterally compressed, barely raised, crest scales

<2 mm high in nuchal area in adult males (Fig. 97) ....
.... *flavidorsalis* (p. 246)

3B. Enlarged middorsal crest scales distinctively laterally compressed in males, crest scales >2 mm high in nuchal area in adult males (Fig. 98) .... *quinquecarinata* (p. 263)

4A. Scales on dorsal and anterodorsal surfaces of shanks not greatly enlarged and only slightly keeled (Fig. 99) .... *similis* (p. 273)

4B. Scales on dorsal and/or anterodorsal surfaces of shanks enlarged, some heavily keeled (Figs. 100–104) 5

5A. Pendulous male dewlap present (Fig. 94); enlarged middorsal crest scales distinctively laterally compressed, >4 mm high in nuchal area in adult males (Fig. 105); 38–42 enlarged middorsal crest scales .... *bakeri* (p. 240)

5B. Pendulous dewlap absent, only large gular extension present in adults of both sexes (Fig. 106);

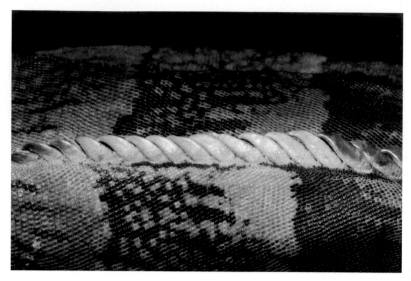

Figure 95.   Male middorsal crest scales laterally compressed. *Ctenosaura melanosterna*. USNM 573324 from mouth of Río San Lorenzo, Yoro.

enlarged middorsal crest scales only somewhat laterally compressed, <2 mm high in nuchal area in adult males (Fig. 107); 58–83 enlarged middorsal crest scales .... *oedirhina* (p. 258)

CLAVE PARA LAS ESPECIES HONDUREÑAS DEL GÉNERO *CTENOSAURA*

1A. Solamente una hilera de escamas intercalares entre cada verticilo de la cola (Figs. 86–88) 2

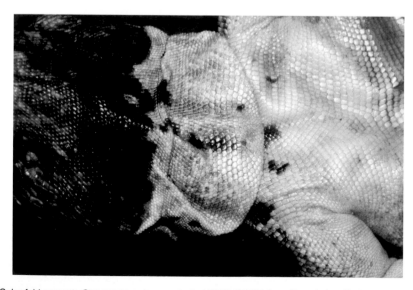

Figure 96.   Gular fold present. *Ctenosaura quinquecarinata*. USNM 581890 from Caserío Los Encinitos, Francisco Morazán.

Figure 97. Male middorsal crest scales not distinctly laterally compressed. *Ctenosaura flavidorsalis.* USNM 581850 from Potrerillos, La Paz.

1B. Dos hileras de escamas intercalares entre cada verticilo de escamas de la cola, al menos dorsolateralmente (Figs. 89–91) 4
2A. El pecho en machos adultos es de color negro; abanico gular presente en machos adultos (Fig. 94); menos de 45 escamas mediodorsales agrandadas y fuertemente comprimidas lateralmente de la cresta dorsal (Fig. 95); 10 o menos escamas agrandadas en la hilera

Figure 98.   Male middorsal crest scales laterally compressed. *Ctenosaura quinquecarinata.* USNM 581890 from Caserío Los Encinitos, Francisco Morazán.

Figure 99.   Dorsal surfaces of each shank and thigh without much enlarged and only relatively slightly keeled scales. *Ctenosaura similis.* USNM 580371 from Las Almejas Island, Valle.

mediodorsal presente en una longitud igual al tamaño de la cabeza
.... *melanosterna* (p. 251)
2B. El pecho del color pardo o pardo-
grisaceo; abanico gular ausente,
con un pliegue gular (Fig. 96);

más de 45 escamas mediodorsales
agrandadas presentes de la cresta
dorsal (raramente menos hasta 39
en *C. flavidorsalis*), que pueden
estar, o no, fuertemente comprimidas lateralmente; más de 10

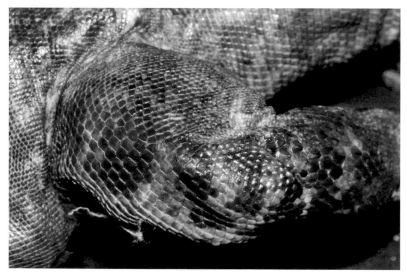

Figure 100.   Dorsal surfaces of each shank slightly enlarged, with somewhat spiny scales and those of thighs with enlarged, nonspiny scales. *Ctenosaura bakeri.* USNM 26137 from Utila Island, Islas de la Bahía.

Figure 101.   Dorsal surfaces of each shank and thigh with distinctly enlarged scales, those on shanks spiny. *Ctenosaura flavidorsalis*. USNM 581850 from Potrerillos, La Paz.

escamas agrandadas en la hilera mediodorsal presentes en una longitud igual al tamaño de la cabeza .. 3

3A. Las escamas mediodorsales agrandadas no están distinguiblemente comprimidas lateralmente, apenas

sobresalientes, escamas en la región nucal de los machos menos de 2 mm de alto (Fig. 97)..........
.... *flavidorsalis* (p. 246)

3B. Escamas mediodorsales distinguiblemente comprimidas lateral-

Figure 102.   Dorsal surfaces of each shank with enlarged strongly keeled scales and thigh scales not enlarged and only weakly keeled. *Ctenosaura melanosterna*. USNM 573903 from near mouth of Río Lorenzo, Yoro.

Figure 103.   Dorsal surfaces of each shank with much enlarged spiny scales and thighs with barely enlarged, slightly spiny scales. *Ctenosaura oedirhina*. USNM 573905 from near Flowers Bay, Roatán Island, Islas de la Bahía.

mente en los machos, escamas en la región nucal de los machos adultos más de 2 mm de alto (Fig. 98).... *quinquecarinata* (p. 263)

4A.   Escamas en las superficies dorsales y anterodorsales de las pantorillas

no están muy agrandadas y son debilmente quilladas (Fig. 99) ...
.... *similis* (p. 273)

4B.   Escamas en las superficies dorsales y/o de los anterodorsales de las pantorillas muy agrandadas y algu-

Figure 104.   Dorsal surfaces of each shank with much enlarged and spiny scales and thighs with only slightly enlarged, some slightly spiny scales. *Ctenosaura quinquecarinata*. USNM 581884 from El Rodeo, El Paraíso.

Figure 105.   Male middorsal crest scales distinctly laterally compressed. *Ctenosaura bakeri*. Utila Island Islas de la Bahía (released).

nas fuertemente quilladas (Fig. 100–104) 5

5A. Abanico gular de macho presente (Fig. 94); escamas de la cresta mediodorsales distinguiblemente comprimidas lateralmente, escamas en la región nucal de los machos adultos más de 4 mm de alto (Fig. 105); 38–42 escamas agrandadas en la cresta mediodorsales .... *bakeri* (p. 240)

5B. Abanico gular ausente en los machos, solamente hay un pliegue gular grande presente en ambos

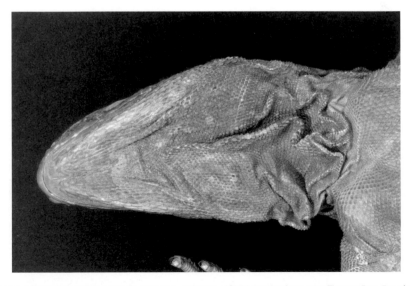

Figure 106.   Large gular extension present. *Ctenosaura oedirhina*. USNM 573905 from near Flowers Bay, Roatán Island, Isla de la Bahía. Photograph by James A. Poindexter.

Figure 107.   Male middorsal crest scales only somewhat laterally compressed. *Ctenosaura oedirhina* from West Bay, Roatán Island, Isla de la Bahía (released).

sexos (Fig. 106); escamas medio-dorsales no está distintivamente comprimidas lateralmente, escamas en la región nucal de los machos menos de 2 mm de alto (Fig. 107); 58–83 escamas de la cresta medi-odorsales agrandadas . . . .
. . . . *oedirhina* (p. 258)

Ctenosaura bakeri Stejneger, 1901

Ctenosaura bakeri Stejneger, 1901: 467 (holotype, USNM 26317; type locality: "Utilla [sic] Island, Honduras"); Bou-lenger, 1902: 19; Werner, 1904: 27; Bailey, 1928: 38; Barbour, 1928: 56 (in part); Meyer, 1969: 243 (in part); Peters and Donoso-Barros, 1970: 105 (in part); Wilson and Hahn, 1973: 114 (in part); MacLean et al., 1977: 4 (in part); Etheridge, 1982: 17 (in part); de Quei-roz, 1987a: 901; de Queiroz, 1987b: 176 (in part); de Queiroz, 1990a: 465.1; Köhler, 1993a: 70; Köhler, 1994a: 6; Köhler, 1994b: 12; Köhler, 1994c: 51; Köhler, 1994e: 688; Bender, 1995: 10; Köhler, 1995b: 94; Köhler, 1995d: 16; Köhler, 1995e: 9; Köhler, 1995f: 8;

Köhler, 1995h: 8; Köhler, 1995i: 43; Köhler, 1995k: 73; Köhler, 1996d: 20; Köhler, 1996f: 181; Buckley and Axtell, 1997: 148; Köhler, 1997a: 73; Köhler, 1997b: 10; Köhler, 1997d: 10; Köhler, 1997e: 20; Gravendyck et al., 1998: 35 (in part); van den Heuvel and Leenders, 1998: 90; Köhler, 1998a: 44; Köhler, 1998c: 417; Köhler, 1998d: 375; Köhler, 1998f: 228; Köhler, 1998g: 49; Köhler, 1998h: 52; Köhler, 1998i: 121; Monzel, 1998: 158; Köhler, 1999c:36; Köhler, 1999f:77; Köhler, 1999g:7; Köhler, 2000:75; Köhler, Schroth, and Streit, 2000: 190; Lundberg, 2000: 3; Wilson et al., 2001: 135; Gutman, 2002: 59; Köhler, 2002: 129; Binns, 2003: 30; Gees, 2003: 30; Gutman, 2003: 32; Gutsche, 2003: 29; Köhler, 2003a: 128; Wilson and McCranie, 2003: 59; Dirk-sen, 2004: 21; Hollingsworth, 2004: 30; Köhler, 2004a: 206; Gutsche, 2005a: 6; Gutsche, 2005c: 317; Gutsche, 2005d: 143; Gutsche, 2005e: 239; McCranie et al., 2005: 90; Dirksen and Gutsche, 2006: 51; Gutsche, 2006: 109; Pasach-nik, 2006: 268; Wilson and Townsend, 2006: 105; Diener, 2007: 59; Schulte,

2007b: 6; Gutsche and Köhler, 2008: 253; Köhler, 2008: 141; Gutsche and Streich, 2009: 105; Pasachnik et al., 2009: 1250; Pasachnik and Ariano, 2010: 136; Pasachnik, Echternacht, and Fitzpatrick, 2010: 1769; Schulte and Köhler, 2010: 141; S. Hallmen, 2011: 12; M. Hallmen and Hallmen, 2011: 11; Faulkner et al., 2012: 332; Gutsche et al., 2012: 157; Pasachnik, Montgomery, Martinez et al., 2012: 391; Pyron et al., 2013, fig. 17; Wilson et al., 2013: 66; Binn, 2014: 29; McCranie and Valdés Orellana, 2014: 45; Solís et al., 2014: 131; McCranie, 2015a: 369; Naccarato et al., 2015: 234.

Enyaliosaurus bakeri: Cochran, 1961: 105; Meyer and Wilson, 1973: 24 (in part).

Geographic Distribution.—*Ctenosaura bakeri* occurs only on Utila Island in the Bay Islands of Honduras.

Description.—The following is based on one male (USNM 26317) and two females (LSUMZ 22293; USNM 25324). *Ctenosaura bakeri* is a huge lizard (maximum recorded SVL 315 mm [Gutsche, 2003; males significantly larger than females]) with spiny caudal whorls; snout region scales equal in size to scales in frontal region, smooth, flat; snout acutely rounded in dorsal aspect, sloping downward in profile; 5–8 postrostrals; 6–7 internasals; 2 canthals, anterior equal in size to posterior; 7–9 scales between each second canthal; 1–3 scales (minimum) between supraorbital semicircles; interparietal scale larger than surrounding scales, parietal eye visible; 2 scales between interparietal and supraorbital semicircles; nasal single, nostril opening posterior to center of scale, directed laterally; 1–3 scales between nasal and rostral; moveable eyelid present; pupil circular; 10–11 supralabials; 9–10 infralabials; second to fourth subocular directly below eye, elongate; 1–3 preoculars; 1–3 scale rows separating suboculars from supralabials; 4 postmentals; pendulous dewlap present; gular fold absent; dorsal body scales gran-

ular on nape, becoming slightly larger posteriorly, those between level of forelimb to about three-quarters length of body slightly conical, those on posterior quarter of body broadly keeled with spiny distal ends; middorsal crest scales distinctively laterally compressed, greatly raised, especially in males (crest scales >4 mm high in nuchal area), scales longest on nape and shoulder regions, 38–42 (40.3 ± 2.1) enlarged middorsal crest scales between nuchal and sacral areas; 5–8 (5.3 ± 2.5) smaller scales between first differentiated middorsal crest scale and posterior end of head; 10–12 (10.7 ± 1.2) scales between ultimate differentiated middorsal crest scale and first caudal whorl; ventral scales flat, not keeled, not imbricate, larger than dorsal scales; dorsal and anterodorsal surfaces of shank with enlarged, somewhat spiny scales; dorsal surface of thigh with enlarged, non-spiny scales; 32–36 (33.7 ± 1.6) subdigital scales on Digit IV of hind limb; 9 femoral pores in male, 7–9 (8.0 ± 0.8) in females, 18 total femoral pores in male, 15–17 (16.0) in females; femoral pores much larger in males than in females; 21–28 caudal whorls, all separated by 2 intercalary scale rows (at least dorsolaterally), 2 intercalary scale rows complete from whorls 7 to 13; inner paramedian caudal scales of each whorl with smaller keels than outer paramedian caudal scales; 5–7 smooth paramedian intercalary scales in uninterrupted row on each side below caudal whorls 6–7; strongly keeled median intercalary scale present between all caudal whorls; reduction from 3 to 2 paramedian scales occurs at caudal whorls 12–15; reduction from 2 to 1 paramedian scales occurs at caudal whorls 21–22; SVL 230 mm in male, 133–200 (166.5) mm in females; HW/SVL 0.24 in male, 0.16 in one female; HL/SVL 0.23 in male, 0.19–0.22 in females; SL/SVL 0.11 in male, 0.10 in females; SHL/SVL 0.21 in male, 0.20–0.22 in females; TAW/TAH 0.88 in male, 0.81–0.96 in females; TAL/SVL

Plate 47.   *Ctenosaura bakeri.* Adult male. Islas de la Bahía: Utila Island (released). From the north coast breeding grounds.

1.33–1.35 in females (tail incomplete in male).

Color in life of an adult male (released; Plate 47): dorsal and lateral surfaces of head Blackish Neutral Gray (82) with a few pale gray scales; dorsal and lateral surfaces of body Plumbeous (78) with some pale gray scales; crest scales dirty white with some scales tipped with Jet Black (89); dorsal surface of forelimb Blackish Neutral Gray with some pale gray scales; dorsal surface of hind limb dark brownish black; dorsal surface of tail Medium Neutral Gray (84) with some Plumbeous crossbands; iris Jet Black. Color in life of an adult female (released; Plate 48): dorsal and lateral

Plate 48.   *Ctenosaura bakeri.* Adult female. Islas de la Bahía: Utila Island (released). From the north coast breeding grounds.

surfaces of head Blackish Neutral Gray (82) with a few pale gray scales; dorsal and lateral surfaces of body and forelimb mottled Plumbeous (78) and Light Neutral Gray (85); dorsal surface of hind limb mottled Blackish Neutral Gray, Medium Plumbeous (87) and brown; dorsal surface of tail as in male described above, except darker crossbands more evident; iris Jet Black (89).

Color in alcohol: dorsal ground color of head and body gray without distinct markings; middorsal crest scales generally same color as adjacent dorsal scales; lateral surface of body similar in color to middorsum; dorsal surface of tail brown with suggestion of darker brown bands; ventral surface of head grayish brown with pale brown mottling, that of body grayish brown with pale brown mottling; subcaudal surface pale brown with some indication of slightly darker brown bands.

Diagnosis/Similar Species.—*Ctenosaura bakeri* is distinguished from all other Honduran lizards, except the remaining *Ctenosaura*, by having distinct caudal whorls in combination with a large to huge size. *Ctenosaura flavidorsalis* and *C. quinquecarinata* lack a pendulous male dewlap, have distinctively enlarged scales on the dorsal surface of the thigh, and have only a single row of intercalary scales, at least between caudal whorls 4–20 (versus male dewlap present, no distinctively enlarged scales on dorsal surface of thigh, and 2 rows of intercalary scales in *C. bakeri*). *Ctenosaura melanosterna* has a black chest in adult males, has black forelimbs dorsally in adult males, and has only 1 intercalary scale row between caudal whorls 4–20 (versus those surfaces pale brown to brown and 2 intercalary scale rows between caudal whorls in *C. bakeri*). *Ctenosaura oedirhina* has 58–83 somewhat compressed middorsal crest scales and lacks a pendulous male dewlap (versus 38–42 middorsal crest scales between nuchal and sacral areas, crest scales distinctively laterally compressed in

males, and pendulous male dewlap present in *C. bakeri*). *Ctenosaura similis* has the scales on the dorsal and anterodorsal surfaces of the shank only slightly enlarged and only weakly keeled, has 51–89 enlarged middorsal scales, and lacks a pendulous male dewlap (versus scales on anterodorsal surface of shank distinctively enlarged and spiny, 38–42 middorsal crest scales, and pendulous male dewlap present in *C. bakeri*).

Illustrations (Figs. 89, 100, 105; Plates 47, 48).—Binn, 2014 (adult); Binns, 2003 (adult); Diener, 2007 (adult); Dirksen, 2004 (subadult); Dirksen and Gutsche, 2006 (two adults attempting to feed on a juvenile *Iguana iguana*); Gees, 2003 (adult); Gutman, 2002 (adult, subadult, juvenile), 2003 (adult); Gutsche, 2003 (adult), 2005d (adult), 2006 (adult); S. Hallmen, 2011 (juvenile); M. Hallmen and Hallmen, 2011 (adult, juvenile); van den Heuvel and Leenders, 1998 (adult); Köhler, 1994b (adult), 1994c (adult; on front cover), 1995b (adult), 1995d (adult), 1995e (adult), 1995i (adult, subadult), 1995k (adult, juvenile), 1997a (juvenile), 1997d (adult), 1998a (adult, hatchling), 1998c (adult, subadult, juvenile), 1998f (adult, juvenile, hatchling), 1998g (adult, hatchling), 1998h (adult, juvenile, hatchling), 1998i (adult), 1999c (adult), 1999f (adult, juvenile), 1999g (adult), 2000 (adult), 2002 (adult, juvenile, hatchling), 2003a (adult), 2004a (adult), 2008 (adult); Lundberg, 2000 (subadult); McCranie et al., 2005 (adult, head scales, caudal scales, dorsal scales); Pasachnik, 2006 (adult); Pasachnik and Ariano, 2010 (adult); Schulte, 2007b (adult, juvenile); Wartenberg, 2013 (adult, juvenile).

Remarks.—Based on a phylogenetic analysis of the genus *Ctenosaura*, Köhler, Schroth, and Streit (2000) recommended placing *C. bakeri* in the subgenus *Loganiosaura* (p. 187) along with *C. melanosterna*, *C. oedirhina*, and the Guatemalan *C. palearis* Stejneger. Gutsche and Köhler (2008) studied mitochondrial and nuclear

DNA in the Honduran segment of *Loganiosaura* and recovered a clade formed by those three species (their fig. 2). De Queiroz (1990a) provided a redescription/diagnosis of that species along with a literature review. Pasachnik, Montgomery, Martinez et al. (2012) studied body size and demography of *Ctenosaura bakeri*.

This species is being captive bred with eggs from gravid, wild-caught females incubated at the Iguana Station on Utila with those females released at the point of capture. Some of those offspring are "head-started" and then released into habitat where the adults were collected. Also, juvenile *Ctenosaura bakeri* from legally collected adults (pre-CITES) were bred in Germany and released onto Utila in November 1999.

Regarding *Ctenosaura bakeri* and international trade, Pasachnik and Ariano (2010: 138) wrote, "illegal exportation for the pet trade has not yet been documented for this species, but is *thought* (emphasis mine) to be occurring." Subsequently, Pasachnik, Montgomery, Martinez et al. (2012: 391) wrote, *C. bakeri* was placed on CITES "due to the presence of this species [*C. bakeri*] and closely related species in the pet trade" and cite Pasachnik and Ariano (2010) as the source for that new information. Thus, that new and inaccurate information regarding *C. bakeri* and the internation pet trade is not based on scientifically sound techniques, and will certainly be repeated in future literature.

Natural History Comments.—*Ctenosaura bakeri* is known only from near sea level in the Lowland Moist Forest formation. This species is diurnal and frequently arboreal. It occurs at varying heights in mangrove trees, with its hiding places usually in hollows in those trees. Gutsche (2005d) called the species a mangrove dweller, and Gutsche and Streich (2009) also mentioned records of *C. bakeri* from various other habitats on Utila, usually close to mangrove habitat. The species has also been reported from terres-

trial situations, such as grasses and rock formations (Wilson and Hahn, 1973; Buckley and Axtell, 1997). Pasachnik (2006: 269) also concluded that this species "can be found in nearly all areas of the island [Utila], with the sole exception of the savannah." Since female *C. bakeri* migrate some distances for egg deposition, it would be expected that individuals would occasionally be found in various habitats they usually do not frequent outside of egg deposition season. I saw numerous juveniles and subadults along a dirt road passing next to a denuded mangrove swamp on the east end of Utila. Gutsche and Streich (2009) estimated the mangrove swamp habitat of the species comprises about 1,091 ha with population densities of 35–78 and 72–114 adults/ha within three mangrove areas. Gutsche and Streich (2009) also found no evidence of an unbalanced demographic structure, with adults, juveniles, and hatchlings all present. Schulte and Köhler (2010) studied microhabitat selection and found that hatchlings preferred the fringes of the Red Mangrove (*Rhizophora mangle*) zone and perched on the ground or at low heights. Juveniles also preferred the Red Mangrove zone but perched at a mean height of 3.6 m, whereas adults preferred the White Mangrove (*Laguncularia racemosa*) zone with males perching at a mean height of 4.9 m and females perching at a mean height of 3.8 m. Schulte (2007b) also provided some natural history data on this species. Gutsche (2003, 2005a, 2005d, 2006) and Köhler (1995b) presented a wealth of information on the biology of this species. Females deposit clutches of 6–16 eggs (wild individuals, but captives have deposited up to 19 eggs; Gutsche, personal communication 24 November 2015) from mid-March through early May, with hatching occurring from mid-June to early August. Gutsche (2003: 28) also found that "with recent warming trends, a shift of the entire reproduction season to as much as two weeks earlier has been recorded." Females move to sandy areas of beach margin forest and coconut groves to deposit their eggs, with nesting sites restricted to only 109 ha (Gutsche and Streich, 2009). Local people eat the meat and eggs of this species, but Gutsche and Streich (2009) identified the loss of habitat and nesting sites caused by unrestricted, illegal development as the biggest threat facing *C. bakeri*. Dirksen and Gutsche (2006) reported this species feeding on a juvenile *Iguana iguana* in captivity, and M. Hallmen and Hallmen (2011) reported a gecko (*Hemidactylus frenatus*) and small crabs being consumed in captivity. Gutsche (2005c) reported on predation of this species by a snake (*Boa imperator*), a lizard (*Basiliscus vittatus*), a Turkey Vulture (*Cathartes aura* [Linnaeus]), and a group of Great-tailed Grackles (*Quiscalus mexicanus* [Gmelin]). Diener (2007) speculated that the snakes *Leptophis mexicanus* A. M. C. Duméril, Bibron, and Duméril (1854: 536) and *Oxybelis aeneus* (Wagler, 1824: 12) might eat juvenile *C. bakeri*, but that seems usually unlikely to occur often given the slender habitus of those two species. Gutsche et al. (2012) reported on ectoparasites and Faulkner et al. (2012) reported on endoparasites infecting this Honduran endemic species. Gravendyck et al. (1998) reported reoviruses and paramyxoviruses in *C. bakeri* on Utila.

Gutsche (2005a) studied diet of *Ctenosaura bakeri* over a period of 2 years. The following data is a personal communication from him. He collected and examined 364 *C. bakeri* samples of droppings that he returned to Germany legally and discovered the species to feed predominantly on mangrove parts (69%; leaves, flowers, and fruits), with crabs (4.4%) and insects and spiders (0.8%) also regularly consumed. Adult *C. bakeri* are normally herbivorous, but also feed on invertebrates throughout its life, but with a clear ontogenetic shift in the amount of invertebrates taken. Ninety-three percent of juvenile droppings contained some animal parts, 76.7% of subadult

droppings contained some animal parts, and 79.7% of adult droppings contained some animal parts. Juveniles eat more insects and spiders (5.3% of identifiable food) than crabs (0.9%), subadult droppings contained 3.9% of identifiable insect and spider remains versus 2.5% of crab remains, whereas adult droppings contained 0.4% of identifiable insect remains versus 4.7% of crab remains. Both sexes of *C. bakeri* showed a preference for Black Mangrove (*Avicennia germinans*) parts with 58.8% of identifiable remains in droppings containing Black Mangrove parts compared with 8.4% of Red Mangrove (*Rhizophora mangle*) and only 1.4% of White Mangrove (*Laguncularia racemosa*). Gutsche's studies also revealed that the raw protein dry matter was higher in Black Mangrove (9.4%) than in Red Mangrove (6.5%) and White Mangrove (4.6%).

Köhler and Blinn (2000) reported a case of natural hybridization between *Ctenosaura bakeri* and *C. similis* on Isla de Utila and included a photograph of the hybrid. Gutsche and Köhler (2004) reported on a suspected hybrid female *C. bakeri* × *C. similis* that deposited fertile eggs, the offspring of which were morphologically similar to *C. similis*. Gutsche and Köhler (2004) also included a photograph of the adult female suspected hybrid. Gutman (2003), Schulte (2007a), and Wartenberg (2013) also included photographs of other hybrid *C. bakeri* × *C. similis*. Gutsche and Köhler (2008), using mitochondrial and nuclear DNA sequence data (collected before CITES listing), also recovered evidence that those two species can hybridize on Isla de Utila, but Pasachnik et al. (2009: 1252) concluded that the level of gene flow between the two species among natural populations on Utila "is far too low to present a threat to the distinctiveness of *C. bakeri*."

Etymology.—The specific name *bakeri* is a patronym for Frank Baker, the then superintendent of the National Zoological

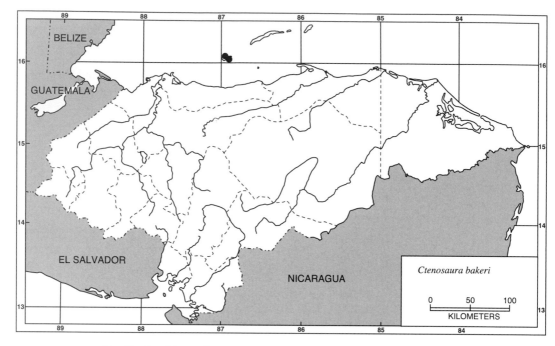

Map 38.   Localities for *Ctenosaura bakeri*. Solid circles denote specimens examined.

Park in Washington, D.C. The holotype lived in that zoological park for some time before dying and being placed in the USNM collection (also see *Laemanctus serratus* account).

Specimens Examined (14 [6]; Map 38).— **ISLAS DE LA BAHÍA**: Isla de Utila, Pumpkin Hill, UNAH 5294–95; Isla de Utila, Utila, LSUMZ 22275, 22293, UF 28437, 28471; "Isla de Utila," SMF 77093–94, 77498, 77705–06, 77925, USNM 25324, 26317.

Other Records.—**ISLAS DE LA BAHÍA**: Isla de Utila, Iron Bound, USNM Herp Image 2717; "Isla de Utila," UNAH 3613–15 (Köhler, Schroth, and Streit, 2000), Ralph W. Axtell 6550–69 (Buckley and Axtell, 1997).

Ctenosaura flavidorsalis Köhler and Klemmer, 1994

Enyaliosaurus quinquecarinatus: Meyer and Wilson, 1973: 25 (in part); Iverson,

1980: 93 (in part); Gicca, 1983: 329.1 (in part).
Ctenosaura quinquecarinata: Etheridge, 1982: 21 (in part); Köhler, 1993a: 105 (in part); Wilson and McCranie, 1998: 15.
Ctenosaura flavidorsalis Köhler and Klemmer, 1994: 197 (holotype, SMF 75845; type locality: "1 km südl. La Paz [750 m ü. N. N.; 14°16′, 87°40′; Dpto. La Paz, Honduras]"); Köhler, 1995a: 11; Köhler, 1995d: 19; Köhler, 1995g: 21; Köhler, 1995i: 27; Köhler, 1995j: 619; Köhler, 1995l: 5; Köhler, 1999a: 214; Köhler, 2000: 77; Köhler, Schroth, and Streit, 2000: 190; Hasbún and Köhler, 2001: 254; Hasbún et al., 2001: 60; Köhler and Hasbún, 2001: 266; Wilson et al., 2001: 134; Köhler, 2002: 130; Köhler, 2003a: 128; Hollingsworth, 2004: 31; Köhler, 2004a: 210; Köhler, 2004b: 4; Hasbún et al., 2005: 3099; Köhler, 2008: 141; Townsend and Wilson, 2010b: 692; Solís et al., 2014: 131;

McCranie, 2015a: 369; Naccarato et al., 2015: 234.

Geographic Distribution.—*Ctenosaura flavidorsalis* occurs at low and moderate elevations on the Pacific versant from extreme east-central Guatemala to eastern El Salvador and southwestern Honduras. It also occurs on the Atlantic versant in central Honduras. In Honduras, this species is known from the western portion of the Comayagua Valley and low mountains to the west in the department of La Paz and from several localities in Intibucá and La Paz near the El Salvador border.

Description.—The following is based on eight males (USNM 573899, 581850–51, 581854, 581856, 581859–60, 581862) and ten females (USNM 573900, 581849, 581852–53, 581855, 581858, 581861, 581863–65). *Ctenosaura flavidorsalis* is a large lizard (maximum recorded SVL 170 mm [Köhler, 2008]; 160 mm SVL in largest Honduran specimen [USNM 581854, a male] with spiny caudal whorls; snout region scales smaller than, or equal in size to scales in frontal region, snout scales smooth, flat; snout acutely rounded in dorsal aspect, sloping downward in profile; 5–6 (usually 6) postrostrals; 5–6 (usually 6) internasals; 2 canthals, anterior smaller than, or equal in size to posterior; 5–8 scales between second canthals; 1–2 (usually 1) scales (minimum) between supraorbital semicircles; interparietal scale larger than surrounding scales, parietal eye visible; 1–2 scales between interparietal and supraorbital semicircles; nasal single, nostril opening in center of scale, directed laterally; 1–2 (usually 2) scales between nasal and rostral; moveable eyelid present; pupil circular; 9–10 (usually 9) supralabials; 9–11 (usually 9 or 10) infralabials; second subocular directly below eye, elongate; 1–2 (rarely 2) preoculars; 1 scale row separating suboculars from supralabials; 4 postmentals; dewlap absent; gular fold distinct, barely rounded (nearly straight) posteriorly across throat; dorsal body scales granular on nape, becoming slightly larger posteriorly, those between level of forelimb to midbody slightly conical, those posterior to midbody broadly keeled with spiny distal ends; middorsal crest scales not distinctively laterally compressed, barely raised, forming low crest (crest scales <2 mm high in both sexes), largest on nape and shoulder regions, 39–70 (55.3 ± 8.0) enlarged middorsal crest scales; 6–27 (12.8 ± 6.4) granular scales between first differentiated middorsal crest scale and posterior end of head; 12–38 (25.1 ± 6.5) scales between last differentiated middorsal crest scale and first caudal whorl; 16–22 (19.2 ± 1.7) middorsal crest scales in 1 head length; ventral scales flat, not keeled, not imbricate, larger than dorsal scales; dorsal and anterodorsal surfaces of shank with distinctively enlarged, spiny scales and dorsal surface of thigh with distinctly enlarged scales; 26–31 (28.6 ± 1.2) subdigital scales on Digit IV of hind limb; 7–10 (8.2 ± 0.8) femoral pores on each side in males, 6–9 (8.0 ± 0.7) in females; 15–18 (16.3 ± 1.1) total femoral pores in males, 13–17 (16.0 ± 1.2) in females; femoral pores much larger in males than in females; 18–19 caudal whorls, all separated by 1 intercalary scale row, caudal whorls abruptly ending on distal third of tail, whorls replaced by regular small, strongly keeled scales; inner paramedian caudal scales of each whorl with smaller keels than outer paramedian caudal scales; median and lateral series of scales of caudal whorl enlarged; 5–8 smooth paramedian intercalary scales in uninterrupted row on each side between caudal whorls 6–7; strongly keeled median intercalary scale absent between caudal whorls 2–13, small keeled intercalary scale present between distal whorls; reduction from 3 to 2 paramedian scales occurs at caudal whorls 8–14; reduction from 2 to 1 paramedian scales occurs at caudal whorls 12–18; SVL 110–160 (134.7 ± 18.0) mm in males, 109–142 (123.5 ± 10.1) mm in females; HW/SVL 0.17–0.20 in males, 0.16–0.20 in

Plate 49.   *Ctenosaura flavidorsalis.* USNM 581850, adult male, SVL = 145 mm. La Paz: Potrerillos.

females; HL/SVL 0.20–0.22 in males, 0.19–0.23 in females; SL/SVL 0.08–0.11 in males, 0.08–0.10 in females; SHL/SVL 0.21–0.24 in males, 0.18–0.23 in females; TAW/TAH 1.11–1.44 in males, 1.11–1.43 in females; TAL/SVL 1.07–1.50 in six males, 0.89–1.51 in five females. A recently hatched individual (USNM 581857) had a SVL of 53.5 mm.

Color in life of an adult male (USNM 581854; southern La Paz): top of head Raw Umber (color 280 in Köhler, 2012); side of head Natal Brown (49) with pale brown mottling on scales anterior to eye, on supralabials, and on infralabials; side of head below tympanum Light Yellow Ocher (13) overlaid with Medium Chrome Orange (75), that combination of colors extending posteriorly to near axilla region; neck and body with 7 Raw Umber crossbands; scales along middorsal area of body varying from Lime Green (116), Chrome Orange, and Light Yellow Ocher; middorsal crest scales varying from Raw Umber to Chrome Orange; sides of body between dark crossbands brown with yellow-brown and green-brown spots; ventral surface of head and chest Dark Yellow Buff (54) with Chrome Orange spotting; belly Cream Color (12)

between Raw Umber crossbands; dorsal surface of forelimb Natal Brown with Buff (5) scales, that of hind limb Natal Brown with dirty white scales on anterior and posterior surfaces, dorsal surface of hind limb also with scattered dirty white scales and occasional Lime Green pointed spines; dorsal surface of tail with alternating Tawny Olive (17) and Clay Color (18) caudal whorls, both sets with some paler brown and darker brown crest scales; subcaudal surface pale brown with numerous dark brown scales; iris Mars Brown (25). Another adult male (USNM 581855; southern La Paz) was recorded as being similar in color to that of USNM 581854. An adult female (USNM 581856; southern La Paz) was also recorded as being similar in color to that of USNM 581854, except some scales on middorsum of body and top of forelimb Pistachio (color 102 in Köhler, 2012). Color in life of an adult male (USNM 581850; Plate 49; near type locality): dorsal surface of head Fuscous (21) with some orange-brown scales; rostral and postrostrals orange-brown; nape Dark Drab (119B) with two orange-brown blotches, gray-brown mottling, and some pale brown scales;

anterior half of dorsal surface of body Dark Drab with orange-brown blotches, posterior half Dark Drab with yellow-brown blotches; side of head Dark Drab with Antique Brown (37) and Tawny (38) scales surrounding eye, including suboculars; supralabials orange-brown with dark brown mottling; side of nape Clay Color (123B) with Dark Drab crossbands; side of body with Clay color swath around area of forelimb insertion, also with pale brown spots and extensions of Dark Drab crossbands; side of remainder of body mottled pale brown and dark brown, also with extensions of Dark Drab crossbands; top of tail medium brown with yellow-brown enlarged scales, becoming banded with yellow-brown and medium brown on distal third; chin with Tawny scales on pale brown ground color, throat pale brown with medium brown mottling; ventral surface of body pale brown with lateral extensions of dark dorsal crossbands; subcaudal surface pale brown; ventral surfaces of fore- and hind limb pale brown with dark brown spots on hind limb; iris golden brown with pale gold rim around pupil.

Color in alcohol: dorsal ground color of head and body brown with distinct pale brown to bluish brown spots on body; 3–7 dark brown crossbands on body, varying from indistinct to distinct; middorsal crest scales generally same color as adjacent dorsal scales; lateral surface of body generally slightly paler brown than dorsal surface, with 3–4 dark brown crossbands; dorsal surface of tail brown with suggestion of paler brown crossbands distally; ventral surface of head brown with darker brown lineate mottling, that of body pale brown with 3–4 dark brown bands or bars (extensions of lateral bands) not meeting medially; subcaudal surface pale brown with some indication of slightly darker brown bands distally.

Diagnosis/Similar Species.—*Ctenosaura flavidorsalis* can be distinguished from all remaining Honduran lizards, except the other *Ctenosaura*, by having distinct caudal whorls in combination with a large size. *Ctenosaura bakeri* has strongly compressed and raised middorsal crest scales, has a pendulous male dewlap, has 2 rows of intercalary scales, and lacks enlarged scales on the dorsal surface of the thigh (versus middorsal crest scales not distinctively compressed and barely raised, male dewlap absent, single row of intercalary scales, and enlarged scales on dorsal surface of thigh in *C. flavidorsalis*). *Ctenosaura melanosterna* has 27–42 distinctly compressed and raised middorsal crest scales (>4 mm high in nuchal area in males), has a series of granular to compressed scales between most enlarged middorsal crest scales, and has a pendulous male dewlap (versus 39–70 [rarely <45] not compressed and low middorsal crest scales, no granular to compressed scales between most enlarged middorsal crest scales, and no dewlap in *C. flavidorsalis*). *Ctenosaura oedirhina* has the dorsal surface of the thigh with only slightly enlarged scales and has 2 rows of intercalary scales (versus dorsal surface of thigh with distinctively enlarged scales and 1 row of intercalary scales in *C. flavidorsalis*). *Ctenosaura quinquecarinata* has the middorsal crest scales distinctively laterally compressed in males (versus middorsal crest scales not distinctively laterally compressed in both sexes in *C. flavidorsalis*). *Ctenosaura similis* has the scales on the dorsal surfaces of the shank and thigh not enlarged and only weakly keeled and has 2 intercalary scale rows between the anterior caudal whorls (versus scales on dorsal surfaces of shank and thigh distinctively enlarged and spiny with 1 intercalary scale row between anterior caudal whorls in *C. flavidorsalis*).

Illustrations (Figs. 86, 97, 101; Plate 49).—Hasbún et al., 2001 (adult, juvenile); Köhler, 1995a (adult), 1995d (adult), 1995g (adult; on cover), 1995i (adult), 1995j (adult), 1999a (adult), 2000 (adult, juvenile), 2002 (adult, juvenile, dorsal crest, femoral pore region), 2003a (adult, middorsal scales), 2004a (subadult), 2008 (adult, middorsal scales), 2016 (adult); Köhler and Klemmer, 1994 (adult, juvenile, head, posterior portion of body and adjacent tail, dorsal scales); Köhler et al., 2005 (adult); Sprackland, 1999 (adult); Townsend and Wilson, 2010b (subadult).

Remarks.—*Ctenosaura flavidorsalis* is morphologically and genetically similar to *C. quinquecarinata.* As noted above, *C. flavidorsalis* differs from *C. quinquecarinata* in lacking compressed middorsal crest scales in adult males (those present in *C. quinquecarinata*). *Ctenosaura flavidorsalis* also tends to have larger and more spiny scales on the dorsal surface of the thigh and more spiny intercalary scales on the caudal whorls than does *C. quinquecarinata.* According to Hasbún et al. (2005), those more spiny scales are an adaptation to its terrestrial and rock crevice habitats, whereas *C. quinquecarinata* is usually an arboreal, tree hole inhabitant (frequently Jícaro [*Crescentia alata*] and Nance [*Brysonima crassifolia*] trees).

Based on a phylogenetic analysis of *Ctenosaura*, Köhler, Schroth, and Streit (2000) recommended placing *C. flavidorsalis* in the subgenus *Enyaliosaurus* (p. 188) along with *C. alfredschmidti* Köhler (1995m: 5), *C. clarki* Bailey (1928: 44), *C. defensor* (Cope, 1866b: 124), and *C. quinquecarinata* (Gray, 1842: 59). Only the last-mentioned species occurs in Honduras. The subsequently described *C. oaxacana* Köhler and Hasbún (2001: 260) would also belong to this subgenus. Hasbún et al. (2005) provided a phylogenetic analysis of the *C. quinquecarinata* group that demonstrated *C. flavidorsalis*, as presently understood, might consist of more than one species. The review of *C. quinquecarinata* (as *Enyaliosaurus*) by Gicca (1983) also included *C. flavidorsalis* (the populations from El Salvador and south-central Honduras).

Ctenosaura flavidorsalis from southern La Paz near the frontier with El Salvador have distinct orange on the dorsal surfaces, whereas those from the vicinity of its type locality and the Intibucá population have yellow on those dorsal surfaces.

Naccarato et al. (2015) cited an IUCN list (Köhler, 2004b) that classified *Ctenosaura flavidorsalis* as endangered. That endangered statement is not true! Recent fieldwork by colleagues and me demonstrate the species to be extremely common in every rocky area we searched in southern Honduras and should hold true across the border in El Salvador (one of my field sites is about 2 km from that border). That short-term fieldwork looking for previously unknown populations of *C. flavidorsalis* and *C. quinquecarinata* proved both species to be more widespread and among the most common lizards on mainland Honduras, even rivaling *C. similis.* Also, those low mountains and rocky terrain, where those species were found, are considerably more widespread than those along the few scattered dirt roads we traveled. The Justification, Population, and Threats sections of the IUCN web document on *C. flavidorsalis* contains mostly erroneous and speculative statements, none of which were documented by extensive fieldwork. The author of that web document was apparently misled by his choice of field site, the *C. flavidorsalis* type locality and vicinity not being in prime *C. flavidorsalis* habitat. Instead of the flat valley area of the type locality, the species is considerably more common in the low, rocky mountains just west of the type locality, where it is well known to the locals as Rúmia.

Natural History Comments.—*Ctenosaura flavidorsalis* is known from 370 to 920 m elevation in the Lowland Dry Forest and Premontane Dry Forest formations. This species was collected on rocky outcrops, on the ground, and rarely in adjacent oak trees (*Quercus* sp.) with holes in April, June, and October. When disturbed, individuals quickly retreat to crevices in boulders or extremely rarely into tree holes near boulders. Adults and juveniles were quite common in June 2013 in southern La Paz near the El Salvador border. Numerous individuals could have been collected if so desired. I was able to see four juveniles and a single adult at the same time and without binoculars by waiting about 10 min during midday for lizards to reappear on sunny lookout points, mostly on the tops of boulders. The area involved was in second growth thorn forest in a hilly region with numerous boulders and small rocks. Juveniles were especially abundant among old, man-made rock walls used as fencerows in that region (La Estancia and San Antonio, La Paz). The type series was collected in April in rocky areas and tree holes (Köhler and Klemmer, 1994) in more-or-less flat areas, and thus not in primary *C. flavidorsalis* habitat. Captive females are known to deposit 6–13 eggs in February and March (Köhler, 2002, and references cited therein). Apparently nothing has been published on diet in free-living *C. flavidorsalis*. Köhler (2002) reported captives feeding on a variety of vegetative matter and on insects.

The rocky areas where we have targeted *Ctenosaura flavidorsalis* (La Estancia, Potrerillos, and San Antonio) are all in La Paz and within 100 m of a dirt road, and all contain what appear to be healthy populations. Those three sites all contain adults and numerous juveniles. *Ctenosaura flavidorsalis* is restricted to hot areas, with numerous rocks and boulders providing retreat sites. Those rocky areas are of no use to humans for crop fields and are mainly used for cattle. The word cattle usually brings negative thoughts concerning wildlife, but in the case of sun-loving *Ctenosaura*, the cattle appear to be beneficial in keeping those areas open. Cattle feed on the grass and sprouting shrubs. At La Estancia, it was obvious that many more *C. flavidorsalis* were present in grazed areas compared with adjacent sites with more dense secondary growth not currently used as grazing areas.

Etymology.—The specific name *flavidorsalis* is derived from the Latin *flavus* (yellow) and *dorsalis* (of the back), and refers to the yellow pigment on the back of the specimens in the type series.

Specimens Examined (30 [0]; Map 39).— **INTIBUCÁ**: Santa Lucía, SMF 79126–28, USNM 573899–900. **LA PAZ**: La Estancia, USNM 581857–65; 1 km S of La Paz, SMF 75845, 75910, 77084, 80897; near La Paz, LACM 72088–89; Potrerillos, USNM 581849–53; San Antonio, UNAH (1), USNM 581854–57.

Ctenosaura melanosterna Buckley and Axtell, 1997

Enyaliosaurus palearis: Echternacht, 1968: 151; Peters and Donoso-Barros, 1970: 116 (in part); Meyer and Wilson, 1973: 24; Iverson, 1980: 93 (in part).

Ctenosaura palearis: Meyer, 1969: 243; Etheridge, 1982: 20 (in part); de Queiroz, 1987a: 900; Buckley and Axtell, 1990: 491.1; Köhler, 1993a: 99; Wilson and Cruz Díaz, 1993: 17; Köhler, 1995c: 329; Köhler, 1995d: 17; Köhler, 1995i: 53 (in part); Köhler, 1996g: 65; Köhler and Vesely, 1996: 23; van den Heuvel, 1997: 155; Cruz Díaz, 1998: 29, *In* Bermingham et al., 1998; Wilson and McCranie, 1998: 15; Köhler, Schroth, and Streit, 2000: 191 (in part); Wilson et al., 2001: 135.

Ctenosaura quinquecarinata: Etheridge, 1982: 21; Köhler, 1993a: 105.

Enyaliosaurus quinquecarinatus: Gicca, 1983: 329.1 (in part).

Ctenosaura melanosterna Buckley and Axtell, 1997: 139 (holotype, KU 101441; type locality: "2 km south of Coyoles Central, Department of Yoro, Honduras"); Monzel, 1998: 158; Köhler, 2000: 77; Köhler, 2002: 130; Lundberg, 2002a: 7; van den Heuvel, 2003: 34; Köhler, 2003a: 128; Wilson and McCranie, 2003: 59; Hollingsworth, 2004: 33; McCranie et al., 2005: 92; Pasachnik, 2006: 267; Reed et al., 2006: 84; Wilson and Townsend, 2006: 105;

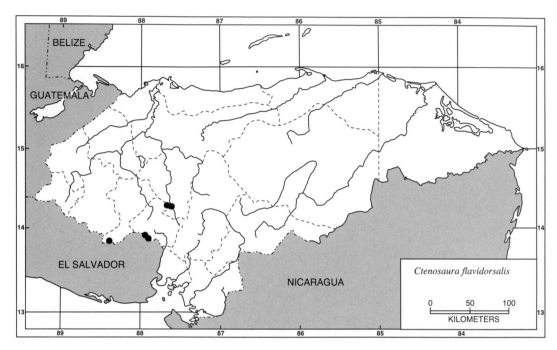

Map 39. Localities for *Ctenosaura flavidorsalis*. Solid circles denote specimens examined.

Gutsche and Köhler, 2008: 246; Köhler, 2008: 140; Pasachnik and Ariano, 2010: 137; Pasachnik, Echternacht, and Fitzpatrick, 2010: 1769; Townsend and Wilson, 2010b: 692; Pyron et al., 2013, fig. 17; Wilson et al., 2013: 66; McCranie and Valdés Orellana, 2014: 45; Solís et al., 2014: 131; McCranie, 2015a: 369; Montgomery et al., 2015: 281; Naccarato et al., 2015: 234.

Geographic Distribution.—*Ctenosaura melanosterna* occurs at low elevations on the Atlantic versant in the middle portion of the Río Aguán Valley in north-central Honduras and on the Cayos Cochinos off the coast of north-central Honduras.

Description.—The following is based on six males (LACM 48428; LSUMZ 21487; USNM 573324, 573902–04) and three females (KU 101441; USNM 322784, 573901). *Ctenosaura melanosterna* is a huge lizard (maximum recorded SVL 320 mm [Buckley and Axtell, 1997; males significant-

ly larger than females]) with spiny caudal whorls; snout region scales (at least those in prefrontal area) larger than, or same size as scales in frontal region, smooth, flat; snout acutely to broadly rounded in dorsal aspect, sloping downward in profile; 5–6 postrostrals; 5–6 internasals; 2 canthals, anterior smaller than, or same size as posterior; 4–7 scales between second canthals; 1 scale (minimum) between supraorbital semicircles; interparietal scale slightly larger than surrounding scales, parietal eye visible; 1–2 scales between interparietal and supraorbital semicircles; nasal single, nostril opening in center of scale, directed laterally; 2 scales between nasal and rostral; moveable eyelid present; pupil circular; 8–11 supralabials; 9–12 infralabials; second or third subocular directly below eye elongate; 1–2 preoculars; 2–3 scale rows separating suboculars from supralabials; 4 postmentals; pendulous male dewlap present; gular fold absent; dorsal body scales granular from nape to about

Plate 50.   *Ctenosaura melanosterna.* USNM 573324, adult male, SVL = 184 mm. Yoro: near mouth of Río San Lorenzo.

midbody, slightly increasing in size from about midbody posteriorly, keeled only in sacral region; middorsal crest scales strongly compressed (especially in adult males), forming high crest (crest scales >4 mm high in males), highest on nape and shoulder regions, 27–42 (35.7 ± 6.1) enlarged middorsal crest scales between nuchal and sacral areas; 8–10 (9.0 ± 0.6) middorsal crest scales in 1 head length; 2–12 (6.3 ± 3.5) granular scales between first differentiated middorsal crest scale and posterior end of head; 11–21 (15.4 ± 3.8) scales between ultimate differentiated middorsal crest scale and first caudal whorl; ventral scales flat, not keeled, some slightly imbricate, larger than dorsal scales; dorsal and anterior surfaces of shank with much enlarged, strongly keeled scales; dorsal surface of thigh without enlarged scales, thigh scales weakly keeled; 26–32 (29.6 ± 1.8) subdigital scales on 17 sides of Digit IV of hind limb; 6–9 (7.5 ± 0.9) femoral pores on each side in males, 3–11 (6.8 ± 3.4) in females, 12–17 (15.0 ± 1.8) total femoral pores in males, 7–22 (13.7 ± 7.6) in females; femoral pores much larger in males than in females; 23–36 caudal whorls, at least

whorls 4–20 separated by 1 intercalary scale row; caudal whorls indistinct on about distal third of tail; inner paramedian caudal scales of each whorl with smaller keels than outer paramedian caudal scales; strongly keeled median intercalary scale present on at least whorls 4–20; 4–6 smooth paramedian intercalary scales in uninterrupted row on each side between caudal whorls 6 and 7; median and lateral series of scales in caudal whorls 4–20 enlarged; 2 paramedian scales occur on anterior caudal whorl; reduction from 2 to 1 paramedian scales occurs at caudal whorls 11–18; SVL 115–195 (144.7 ± 35.4) mm in males, 99–134 (115.9 ± 17.7) mm in females; HW/SVL 0.15–0.16 in both sexes; HL/SVL 0.20–0.21 in males, 0.20–0.24 in females; SL/SVL 0.09–0.11 in males, 0.09–0.10 in females; SHL/SVL 0.20–0.23 in both sexes; TAW/TAH 1.00–1.09 in males, 1.03–1.10 in females; TAL/SVL 1.28–1.66 in three males, 1.92 in one female.

Color in life of an adult male (USNM 573324; Plate 50): dorsal surfaces of neck and body to slightly posterior to anterior limb insertion Olive-Brown (28), followed by Drab-Gray (119D) quadrangular blotch bounded by Sepia (119) border, posterior portion of border broadens laterally to become a crossband connecting to broad Sepia chest band; dorsal dark crossband followed by narrow Drab-Gray band and then by Sepia and Drab-Gray mottled crossband; remainder of dorsal surface of body Light Drab (119C) with longitudinal Sepia streaks suggesting about 3 more crossbands; each crossband bounded anteriorly by set of diffuse Pale Horn Color (92) spots; dorsal crest scales Buff (124) on tips and brownish gray at base, followed by 7 ivory white scales that are gray at their base, followed by 3 largely Jet Black (89) scales, finally followed by series of ivory white scales; dorsal surface of tail crossed by alternating Smoke Gray (45) and Pale Horn Color bands anteriorly, with pale and darker bands becoming more distinguishable posteriorly until end of tail where bands are alternately Pale Horn Color and Hair Brown (119A); dorsal surface of forelimb Sepia with scattered Pale Horn Color small spots, spots especially prevalent on bands, except at base of upper arm which is mottled olive gray and Pale Horn Color; dorsal surface of hind limb Light Drab with scattered Pale Horn Color and Sepia spots and streaks, except enlarged shank and foot scales mottled pale brown and pale washed out lime green; ventral surface of body divided primarily into Sepia chest band and Pale Pinkish Buff (121D) posteriorly; ventral surface of forelimb mottled pale tan, cream, and dark brown; ventral surface of hind limb Pale Pinkish Buff mottled with small cream spots; subcaudal surface Pale Pinkish Buff, except gray-brown distally; dorsal surface of head Brownish Olive (29), except for pale lime green pineal scale; side of head mottled Olive (30), pale orange, and Pale Horn Color; chin Trogon Yellow (153) mottled with olive gray; dewlap olive gray; iris Flesh Ocher (132D) with gold rim around pupil. Color in life of a subadult female (USNM 573901): dorsal surface of neck dark gray with Buff (24) mottling; dorsal surface of body Smoke Gray (44) with

Blackish Neutral Gray (82) crossbands, crossbands with longitudinal Jet Black (89) streaking, crossbands fading posteriorly to point where only dark streaking indicates extent of crossbands; dorsal surface of tail with alternating pale orange and pale olive bands anteriorly, bands alternating Grayish Horn Color (91) and Pale Horn Color (92) posteriorly; dorsal surface of forelimb mostly Jet Black with scattered white dots largely confined to hands, base of forelimb also mottled yellowish gray and black; dorsal surface of hind limb mottled gray, brownish gray, and pale turquoise blue; anterior of venter brownish gray, grading to dark brown at about midbody; ventral surface of posterior portion of body dirty yellow; underside of forelimb mottled with dark brown and brown; underside of hind limb gray-tan with cream flecking; subcaudal surface dirty cream with increasing evidence of dark crossbands posteriorly; top of head gray with cream mottling; side of head gray with pale ocher temporal stripe; chin cream with gray smudging; dewlap olive gray; dorsal crest scales mostly ivory white, except for region of first dorsal crossband where crest scales predominately black; iris Flesh Ocher (132D) with gold rim around pupil. Color in life of a juvenile (USNM 322785) was described by Wilson and Cruz Díaz (1993: 17; as *C. palearis*): "dorsum pale gray with a pale orange cast and dark gray cross-bands; head brownish gray; front limbs dark gray with obscure paler cross-bands; hind limbs mottled pale and dark gray; tail tan with gray cross-bands; chin cream with gray streaking; venter gray anteriorly, pale gray posteriorly, the two areas separated by a black chestband."

Color in alcohol: dorsal ground color of head and body brown with 2–3 dark brown to black crossbands on nape and anterior two-thirds of body, 2 bands posterior to forelimb extending onto lateral portion of body, with first band just posterior to forelimb and forming part of black chest band; posterior third of body brown with dark brown lineate pattern; dorsal surface of tail brown anteriorly, becoming paler brown with darker brown crossbands on distal half; dorsal surface of forelimb black; ventral surface of head anterior to dewlap cream with dark brown lines; dewlap mostly gray; ventral surface of nape posterior to dewlap mottled with black, black mottling connecting with black band just posterior to forelimb to form black chest in adult males; belly posterior to male black chest brown, with indication of extension of second dorsal and lateral band posterior to forelimb that nearly crosses belly; subcaudal surface pale brown with brown crossbands on distal half.

Diagnosis/Similar Species.—*Ctenosaura melanosterna* is distinguished from all other Honduran lizards, except the remaining *Ctenosaura*, by having distinct caudal whorls in combination with a large size. *Ctenosaura melanosterna* is the only Honduran *Ctenosaura* to have a black pendulous male dewlap and a black chest region in males. *Ctenosaura bakeri* also has 2 intercalary scale rows between the caudal whorls (versus only single intercalary scale row present between at least caudal whorls 4–20 in *C. melanosterna*). *Ctenosaura flavidorsalis* also lacks strongly compressed and high middorsal crest scales with those scales <2 mm high, lacks a pendulous male dewlap, has 39–70 (rarely <45) enlarged middorsal crest scales, and distinctively enlarged dorsal scales on the thigh (versus 27–42 enlarged, compressed, and high middorsal crest scales, pendulous male dewlap present, and distinctively enlarged thigh scales absent in *C. melanosterna*). *Ctenosaura oedirhina* also has 2 rows of intercalary scales, although the second row is sometimes located only dorsolaterally between the anterior whorls, has 58–83 enlarged middorsal crest scales, and lacks a pendulous male dewlap (versus only single intercalary scale row, 27–42 middorsal crest scales, and pendulous male dewlap present in *C. melanosterna*). *Ctenosaura quinquecarinata* also has distinctively enlarged scales on the dorsal surface of the thigh, has 50–70 enlarged middorsal crest scales, and lacks a pendulous male dewlap (versus thigh scales not enlarged, 27–42 middorsal scales, and pendulous male dewlap present in *C. melanosterna*). *Ctenosaura similis* also has the scales on the dorsal and anterodorsal surfaces of the shank only slightly enlarged and only weakly keeled, has 51–89 enlarged middorsal crest scales, and has 2 intercalary scale rows between the caudal whorls (versus scales on dorsal and anterodorsal surfaces of shank distinctively enlarged and spiny, 27–42 middorsal crest scales, and single intercalary scale row between caudal whorls 4–20 in *C. melanosterna*).

Illustrations (Figs. 87, 94, 95, 102; Plate 50).—Binns, 2007 (adult); Boonman, 2000 (adult); Braun, 1993 (adult; as *C. palearis*); Buckley and Axtell, 1990 (subadult; as *C. palearis*), 1997 (adult, head); Burghardt and Rand, 1982 (adult; as *C. quinquecarinata*); van den Heuvel, 1997 (adult; as *C. palearis*), 2003 (adult, juvenile); Köhler, 1995c (adult, head scales; as *C. palearis*), 1995d (adult; as *C. palearis*), 1995i (head scales; as *C. palearis*), 2000 (subadult), 2002 (adult), 2003a (juvenile), 2008 (juvenile); Köhler and Vesely, 1996 (adult; as *C. palearis*; Honduran specimen only); Lundberg, 2002a (adult); McCranie et al., 2005 (adult, caudal scales); Mora, 2010 (adult); Pasachnik and Ariano, 2010 (adult).

Remarks.—Köhler, Schroth, and Streit (2000: 187), based on a phylogenetic analysis of the genus *Ctenosaura*, placed *C. melanosterna* in the subgenus *Loganiosaura* (along with *C. bakeri*, *C. oedirhina*, and the Guatemalan *C. palearis*). Gutsche and Köhler (2008), using mitochondrial and nuclear DNA sequence data, also recovered strong evidence of a close relationship of that clade. The review of *C. palearis* by Buckley and Axtell (1990) and that of *Enyaliosaurus quinquecarinatus* by Gicca (1983) also included the subsequently described *C. melanosterna*.

Pasachnik et al. (2011) and Pasachnik, Montgomery, Ruyle et al. (2012) studied molecular and morphological characters of *Ctenosaura melanosterna* and found evolutionary significant units between those populations on the Cayos Cochinos and those of the Aguán Valley of mainland Honduras.

Ctenosaura melanosterna was recently placed on the CITES List, Category II because of perceived illegal collecting for the international pet trade. See the Remarks section for *Ctenosaura* for a discussion showing that the species formerly exported from Honduras for the international pet trade was actually *C. quinquecarinata* and not *C. melanosterna*. That was a serious error on the part of the CITES petitioners. The principal threat to survival of this arboreal species is the destruction of its tree-hole habitat in the Río Aguán Valley, and until that habitat devastation issue is addressed, the placing of this, and any other species, on CITES is akin to placing a band aid on a finger to stop the bleeding on a toe. It also falsely suggests that an effort to protect *C. melanosterna* has been taken, but in reality no productive effort to protect these animals has been accomplished. No amount of protection is provided by its misplacement on CITES and the discussions on IUCN lists will not keep *C. melanosterna* from soon being extirpated from that valley.

Natural History Comments.—*Ctenosaura melanosterna* is known from near sea level to 300 m elevation in the Lowland Moist Forest and Lowland Arid Forest formations. Wilson and Cruz Díaz (1993: 17) stated "On [Cayo] Cochino Grande, lizards were seen in the vicinity of a half-constructed house. One specimen was collected from beneath a pile of concrete blocks, the other two from the walls of the buildings. One specimen escaped by climbing a tree. On [Cayo] Cochino Pequeño, a young ctenosaur was seen in hill forest and another two on the isthmus behind Bonkes Nose Point. An adult ctenosaur was seen on a steep rocky cliff face along the shore of the island." Their specimens were seen in March. Buckley and Axtell (1997: 148) wrote, "We observed territorial behavior in very young *C. melanosterna* (under 100 mm SVL). Up to six individuals were observed living together on a wooden shed at La Ensenada, Cayo Cochino Grande. The largest of those showed aggressive territorial behavior toward smaller individuals, driving them from the highest positions. Adults were not seen together in the wild, and appear less likely than young to be seen in close proximity to each other. We saw an adult female about 10 m up in a tree on Cayo Cochino Pequeno [sic]." Reed et al. (2006) reported natural predation of an adult *C. melanosterna* by an adult *Boa imperator* on Cayo Cochino Menor. Echternacht (1968: 152) stated that mainland juveniles "were collected on and around the branches of a large fallen tree," and an adult "was collected from about 20 m up in a tree." Echternacht's specimens were collected in July. Other mainland specimens were also collected in July, with adults being found in trees and juveniles on the ground (Meyer, 1969). Adults and subadults were subsequently found in hollow fence posts and hollow trees in the Río Aguán Valley in April and June 2014. Both adults and juveniles were seen in mornings, although more were seen at midday and in the afternoons. Those seen during midday were in shady areas beneath trees. Two adult females deposited 11 and 23 eggs in captivity in April and June, respectively (Buckley and Axtell, 1997). Köhler (2002) reported females in captivity deposited 11–41 eggs per clutch. Braun (1993) and van den Heuvel (1997, 2003) also reported on captive breeding. Buckley and Axtell (1997) wrote residents of the Cayos Cochinos reported egg deposition from April to May on that island. Boonman (2000) stated captives fed on various vegetables and on grasshoppers. Montgomery et al. (2015: 280) reported the Cayos Cochinos

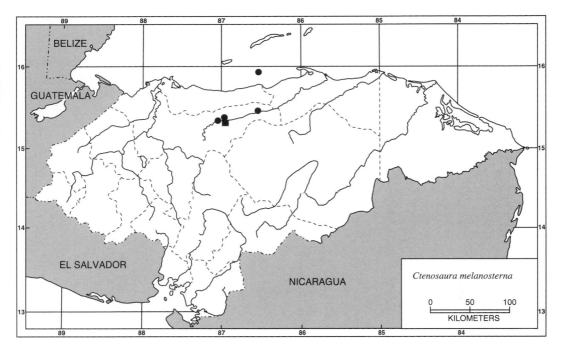

Map 40.　Localities for *Ctenosaura melanosterna*. Solid circles denote specimens examined and the solid square represents an accepted record.

populations are "an omnivore that feeds on a variety of vegetation, invertebrates, vertebrates, and food scraps." Montgomery et al. (2015) also thought the undocumented increasing *Iguana* population on Cayo Cochino Menor might be a new threat to *C. melanosterna*. Pasachnik et al. (2014) reported significant human consumption of this species in the Aguán Valley (that conclusion was based solely on interviews with people living in that valley, and thus not on actual observation) of the mainland, in stark contrast to my observations, where my responders admitted to only rarely eating *Ctenosaura* meat, if at all (although second-hand information). My responders were from the western portion of the valley in the area around San Lorenzo.

Etymology.—The specific name *melanosterna* is derived from the Greek *melanos* (black) and *sternon* (chest), and refers "to the black chest and forelegs possessed by

members of this species" (Buckley and Axtell, 1997: 139). Those black areas seem to occur only, or are at least stronger, in adult males.

Specimens Examined (34 [16]; Map 40).—**ISLAS DE LA BAHÍA**: Cayo Cochino Mayor, La Ensenada, USNM 322784–86; Cayo Cochino Menor, USNM 570537. **YORO**: 2 km S of Coyoles, KU 101439–41; 0.5 km N of Coyoles, LACM 48425–27, LSUMZ 21482–86; Poligono Air Force Base, UNAH (2); Rancho San Lorenzo, LACM 48428–30, LSUMZ 21487–88; near mouth of Río San Lorenzo, USNM 573324, 573901–03; San Patricio, SMF 75913, USNM 573904. "HONDURAS": SMF 75915, 77319, 77500, 78057–59.

Other Records (Map 40).—**ISLAS DE LA BAHÍA**: Cayo Cochino Mayor, La Ensenada, Ralph W. Axtell 6573–74, 6575 (skeleton), 6576, 6577 (skeleton), 6578–95 (Buckley and Axtell, 1997); Cayo Cochino

Mayor, Pelican Point (van den Heuvel, 2003). **YORO**: Agua Caliente (= Poligono Air Force Base near San Patricio), Arenal (Pasachnik, 2006). "HONDURAS": ZFMK 48424–25 (Köhler, Schroth, and Streit, 2000).

Ctenosaura oedirhina de Queiroz, 1987a

Ctenosaura bakeri: Barbour, 1928: 56 (in part; see Remarks); Meyer, 1969: 243 (in part); Peters and Donoso-Barros, 1970: 105 (in part); Wilson and Hahn, 1973: 114 (in part); MacLean et al., 1977: 4 (in part); Iverson, 1980: 93 (in part); Hudson, 1981: 377; Etheridge, 1982: 18 (in part); Oldham and Smith, 1983: 74; de Queiroz, 1987b: 176 (in part); Gravendyck et al., 1998: 35 (in part).
Enyaliosaurus bakeri: Meyer and Wilson, 1973: 24 (in part).
Ctenosaura oedirhina de Queiroz, 1987a: 892 (holotype, UF 28532; type locality: "approx. 4.8 km [converted from 3 miles] west of Roatán on the path to Flowers Bay, Isla de Roatán, Departamento de las Islas de la Bahía, Honduras, near sea level [less than 20 m elevation]"); de Queiroz, 1990b: 466.1; Köhler, 1993a: 91; Rossman and Good, 1993: 9; Köhler, 1994a: 2; Köhler, 1995b: 100; Köhler, 1995d: 16; Köhler, 1995i: 51; Buckley and Axtell, 1997: 148; Köhler, 1998d: 377; Köhler and Rittmann, 1998: 6; Monzel, 1998: 158; Köhler, 2000: 76; Köhler, Schroth, and Streit, 2000: 190; Grismer et al., 2001: 134, 135; Kaiser et al., 2001a: 253; Lundberg, 2001: 22; Wilson et al., 2001: 135; Köhler, 2002: 131; Köhler, 2003a: 128; Wilson and McCranie, 2003: 59; Hollingsworth, 2004: 33; McCranie et al., 2005: 94; Pasachnik, 2006: 268; Wilson and Townsend, 2006: 105; Rittmann, 2007: 33; Gutsche and Köhler, 2008: 253; Köhler, 2008: 141; Pasachnik and Ariano, 2010: 138; Pasachnik, Echternacht, and Fitzpatrick, 2010: 1769; Gandola and Hendry,

2011: 428; Goldberg et al., 2011: 600; Hendry and Gandola, 2011: 273; Pasachnik, 2011a: 107; Pasachnik, 2011c: 600; Pasachnik and Chavarria, 2011: 429; Pyron et al., 2013: Fig. 17; Wilson et al., 2013: 66; McCranie and Valdés Orellana, 2014: 45; Solís et al., 2014: 131; McCranie, 2015a: 369; Naccarato et al., 2015: 234.

Geographic Distribution.—*Ctenosaura oedirhina* occurs on Isla de Roatán and nearby Barbareta and some smaller satellite islands of the Bay Islands, Honduras.

Description.—The following is based on four males (CM 64681; FMNH 53831; LSUMZ 22367; UTA R-36574) and five females (KU 203153; UF 28532–33; USNM 573905–06). *Ctenosaura oedirhina* is a huge lizard (maximum recorded SVL 270 mm [UTA R-36574; males significantly larger than females; but see Remarks]) with spiny caudal whorls; snout region scales (at least those in prefrontal area) larger than dorsal head scales posterior to that point, smooth, slightly convex; snout acutely rounded in dorsal aspect, sloping downward in profile; 4–6 postrostrals; 4–6 internasals; 2 canthals, anterior largest; 5–7 scales between second canthals; 2 scales (minimum) between supraorbital semicircles; interparietal scale slightly larger than surrounding scales, parietal eye visible; 1–2 scales between interparietal and supraorbital semicircles; nasal single, nostril opening in center of scale, directed laterally; 1–2 scales between nasal and rostral; moveable eyelid present; pupil circular; 8–11 supralabials; 8–9 infralabials; second to fourth elongated subocular (usually second) directly below eye; 1–3 preoculars; 2–3 scale rows separating suboculars from supralabials; 2–4 postmentals; dewlap absent, but conspicuous compressed flap of skin present in gular area in adults; gular fold distinct, rounded posteriorly; dorsal body scales granular on nape, keeled from above forelimb posteriorly, those on body in front of sacral region largest; middorsal crest scales somewhat compressed in males, forming low crest (crest scales <2 mm high), largest on nape and shoulder regions, 58–83 (68.9 ± 8.2) enlarged middorsal crest scales; 5–16 (11.0 ± 3.6) granular scales between first differentiated middorsal crest scale and posterior end of head; 0–19 (10.9 ± 7.2) scales between ultimate differentiated middorsal crest scale and first caudal whorl in ten; ventral scales flat, not keeled, some slightly imbricate, larger than dorsal scales; dorsal and anterodorsal surfaces of shank with much enlarged, strongly keeled scales; dorsal surface of thigh with at best only slightly enlarged scales; 30–34 (31.5 ± 1.5) subdigital scales on Digit IV of hind limb; 5–13 (10.0 ± 3.1) femoral pores on each side in males, 7–16 (10.7 ± 2.9) in females; 13–26 (20.5 ± 5.4) total femoral pores in males, 16–30 (21.6 ± 6.1) in females; femoral pores much larger in males than in females; 22–30 caudal whorls, all separated by 2 intercalary scale rows, although whorls 1–9 can have second intercalary scale row present only laterally and with small scales; intercalary scales increase in size posteriorly until distinction between intercalary and whorl scales disappears on about distal third of tail; 5–7 smooth paramedian intercalary scales in uninterrupted row between caudal whorls 6–7; inner paramedian caudal scales of each whorl with smaller keels than outer paramedian caudal scales; strongly keeled median intercalary scale present between all caudal whorls; median and lateral series of scales in caudal whorls enlarged; reduction from 3 to 2 paramedian scales occurs at caudal whorls 6–7; reduction from 2 to 1 paramedian scales occurs at caudal whorls 17–25; SVL 180–270 (219.5 ± 41.0) mm in males 113–238 (184.3 ± 55.6) mm in females; HW/SVL 0.15–0.24 in males, 0.15–0.17 in females; HL/SVL 0.21–0.25 in males, 0.20–0.21 in females; SL/SVL 0.11–0.12 in males, 0.09–0.10 in females; SHL/SVL 0.23–0.26 in males, 0.20–0.23 in females; TAW/TAH 1.00–1.25 in males,

0.84–1.01 in females; TAL/SVL 1.59–1.79 in three males, 1.48–1.85 in two females.

Color in life of an adult female (USNM 573905): dorsal and lateral surfaces of head Blackish Neutral Gray (82) with scattered pale gray and brown scales; dorsal and lateral surfaces of body Blackish Neutral Gray with pale rusty brown large blotches and mottling; dorsal surfaces of fore- and hind limb Blackish Neutral Gray; dorsal surface of tail Blackish Neutral Gray with scattered pale brown mottling; ventral surfaces of head and body brown with scattered dark brown scales on body; subcaudal surface pale brown; iris reddish brown above, remainder of iris black. Color in life of a young adult female (USNM 573906): top and side of head Blackish Neutral Gray (82) with pale brown scales and brown mottling; nuchal area and anterior third of body Cinnamon-Brown (33) with black and white mottling; dorsal and lateral surfaces of remainder of body black with brown and white mottling; dorsal surfaces of fore- and hind limb black with pale brown and pale gray scales; dorsal surface of tail medium brown with dark brown crossbands, some scales also black; iris reddish brown.

Color in alcohol: dorsal ground color of head and body dark brown with large cream to orange spots or blotches on body, blotches largest near dorsal midline from above forelimb posteriorly to about two-thirds length of body, adult pattern suggestive of 3–4 dorsal bands; subadults and juveniles with only scattered pale brown spots and darker brown bands most prominent along vertebral area; middorsal crest scales same color as adjacent dorsal scales; dorsal surface of tail dark brown with suggestion of paler brown bands anteriorly, becoming paler brown on about distal half; ventral surfaces of head and body dark brown with some scattered pale brown scales on belly, but those surfaces of subadults and juveniles brown with scattered pale brown spots on belly; subcaudal

Plate 51. *Ctenosaura oedirhina.* Islas de la Bahía: Roatán Island, West Bay (released).

surface of adults, subadults, and juveniles paler brown than belly, becoming even paler brown on distal two-thirds.

Diagnosis/Similar Species.—*Ctenosaura oedirhina* can be distinguished from all other Honduran lizards, except the remaining *Ctenosaura*, by having distinct caudal whorls in combination with its large size. *Ctenosaura bakeri* has a pendulous male dewlap and has 38–42 distinctively laterally compressed middorsal crest scales that fold laterally in adult males between the nuchal and sacral areas (versus no dewlap and 58–83 only somewhat compressed middorsal crest scales in *C. oedirhina*). *Ctenosaura flavidorsalis* and *C. quinquecarinata* have a single intercalary scale row between at least caudal whorls 4–10 and have the dorsal surface of the thigh with distinctively enlarged scales (versus 2 intercalary scale rows between at least caudal whorls 10–30 and dorsal surface of thigh with at best only slightly enlarged scales in *C. oedirhina*). *Ctenosaura melanosterna* has a black chest in adult males, has only a single row of intercalary scales between at least caudal whorls 4–20, has 27–42 greatly compressed

and raised middorsal crest scales that are >4 mm high in the nuchal area in adult males, and has a pendulous male dewlap (versus brown chest, 2 intercalary scale rows between caudal whorls 10–30, 58–83 low and noncompressed middorsal crest scales, and pendulous dewlap absent in *C. oedirhina*). *Ctenosaura similis* has the scales on the anterodorsal surface of the shank not enlarged and only weakly keeled (versus scales on anterodorsal surface of shank greatly enlarged and spiny in *C. oedirhina*).

Illustrations (Figs. 90, 103, 106, 107; Plate 51).—de Queiroz, 1987a (adult, head), 1990b (adult, head); Gandola and Hendry, 2011 (adult feeding on fruit); Köhler, 1993a (adult), 1994d (adult), 1995b (adult, juvenile), 1995d (adult), 1995i (adult), 1998d (adult), 2000 (adult), 2002 (adult, juvenile), 2003a (adult), 2008 (adult); Köhler and Rittmann, 1998 (adult, juvenile, hatchling); Lundberg, 2001 (subadult); McCranie et al., 2005 (adult, subadult, shank scales, caudal scales); Pasachnik, 2006 (adult), 2011a (adult); Pasachnik and Ariano, 2010 (adult); Pasachnik and Chavarria, 2011 (adult feeding on hatchling *Trachemys*); Rittmann, 2007 (adult, juvenile).

Remarks.—Barbour (1928: 57) did not observe *Ctenosaura* on Roatán Island, but discussed having been told that *Ctenosaura* "were very rare on Roatán." Meyer and Wilson (1973; as *Enyaliosaurus bakeri*) and Wilson and Hahn (1973; as *C. bakeri*) were the first to publish fieldwork-based museum records of *Ctenosaura* from Roatán. Bailey (1928: 38) had earlier inaccurately speculated that *Ctenosaura bakeri* "may occur on Bonacca [= Guanaja Island] and Ruatan [= Roatán] Island."

De Queiroz (1987a) performed a morphological analysis of *Ctenosaura oedirhina* and recognized its distinctness from *C. bakeri*, with which it had previously been confused. De Queiroz (1990b) presented an overview of its morphology and reviewed literature on this species. Köhler, Schroth, and Streit (2000) presented a phylogenetic analysis of the genus *Ctenosaura* and recommended placing *C. oedirhina* in the subgenus *Loganiosaura* (along with *C. bakeri, C. melanosterna,* and the Guatemalan *C. palearis*). Gutsche and Köhler (2008), using mitochondrial and nuclear DNA sequence data, also recovered strong evidence of close relationship of that clade (their fig. 2). *Ctenosaura oedirhina* reaches a considerably larger size than what usually has been recorded in the literature. Adults of both sexes that appeared much larger than the maximum recorded SVL for the species were seen around the house of one of the former private owners of Barbareta Island in February 1990 (juveniles to adults were numerous on that property at the time of my visit). Köhler (1995b) estimated a SVL of 350 mm for some individuals in that population. Those *Ctenosaura* were afforded protection by the most recent former landowner, and the previous landowner also did not allow hunting of *Ctenosaura*; thus, adults were apparently able to reach a larger size than those populations under hunting pressure on Roatán. The current protection

status of the Barbareta population is not known, but fieldwork on Roatán in 2011–2012 revealed *C. oedirhina* to be expanding its population and to be significantly more easily seen than it was in the late 1980s to early 1990s.

Natural History Comments.—*Ctenosaura oedirhina* is known from near sea level to 20 m elevation in the Lowland Moist Forest formation. This diurnal species is both arboreal and terrestrial and is probably active throughout the year. It is found on the ground in sandy areas, on the ground in rocky areas, in trees, on buildings, and on rock cliffs. As just noted, this species is currently more widespread and significantly easier to see now on Roatán than it was 25 years ago when I made my first trip (see Remarks for *Ctenosaura*) to Roatán (in 1989) at the beginning of the real estate boom (McCranie et al., 2005). One of the prime objectives of my first Roatán trip was to photograph *Ctenosaura oedirhina*. Thus, numerous residents on the western half of the island were questioned about the presence of *Ctenosaura* on the island. The general consensus then was that the *Ctenosaura* occurred only in a few isolated ironshore formations (rocky, uplifted fossilized fringe reefs; also see Wilson and Hahn, 1973). Indeed, during my 5 days on the island in 1989, I saw only one *C. oedirhina* and that was in the ironshore formation just west of Flowers Bay. On that trip, I spent a long day walking along the beach and scrub vegetation east of what was then a tiny Sandy Bay, without seeing a single *Ctenosaura*. Wilson and Hahn (1973) also made collections around Sandy Bay without collecting any *Ctenosaura*. Today, similar work would reveal numerous *C. oedirhina*, but with the majority being juveniles. With the onslaught of the real estate boom during the late 1980s, much of the remaining hardwood forests of Roatán were developed (= destroyed), thus benefiting *C. oedirhina* by opening new habitat for this sun-loving species. During my most recent trip to

Roatán (September 2012), I stopped briefly to target collecting a different reptile species at four localities along the dirt road between Sandy Bay and Palmetto Bay. Juveniles of *C. oedirhina* were common at all four localities, and at least one adult *C. oedirhina* was seen at three of those four localities. Also, about a dozen juvenile *C. oedirhina* were seen basking along the side of that road the morning of 19 September 2012. Additionally, during the most recent 5–10 years, I usually stay at a hotel in Sandy Bay when I am on Roatán. One cannot eat breakfast there on a sunny morning without seeing several juvenile and subadult *C. oedirhina* in trees outside their restaurant. Adults of both sexes are also easily seen on the small hotel grounds during sunny afternoons. My impression is that *C. oedirhina* is the most common lizard species in any open situation along the north coast of Roatán between Sandy Bay and at least Palmetto Bay. As a result, the species is now easily seen in numerous places away from the ironshore formations. Further support for my observed changes in the occurrence of *C. oedirhina* on Roatán comes from the statement in Barbour (1928: 57) that the *Ctenosaura* "were very rare on Ruatan [= Roatán] and mostly found on the wild and inaccessible north coast," although Barbour was notorious for his personal wealth that allowed him to purchase most of his specimens through his "opportunity to conduct some of his field work from the comfort of a yacht" (Henderson and Powell, 2004: 298). The inaccessible north coast mentioned by Barbour most likely alluded to several ironshore formations and sandy areas with coconut palms in the vicinity of what is now Sandy Bay.

Adult *Ctenosaura oedirhina* are hunted by some of the human population on the Roatán. But all species of *Ctenosaura*, at least in Honduras, are resilient and maintain populations at sites that have been under hunting pressure for many years. They seem to breed at young ages and at relatively smaller sizes than the adults typically targeted for their meat. Even a cursory trip across Roatán researching the status of *C. oedirhina* would reveal numerous common and widespread *C. oedirhina* populations on the island. The real concern for the future of *C. oedirhina* on Roatán is the very recently introduced population of *C. similis*. That introduction deserves a dedicated effort to eradicate *C. similis* on Roatán. Several juvenile *C. similis* were also seen in September 2012 along the road near Sandy Bay mentioned above.

Kaiser et al. (2001a) reported warning vocalization for this species in mangrove swamp on the eastern end of Roatán. Captive females are known to deposit seven to nine eggs per clutch in January, July, and September (Köhler, 2002, and references cited therein). Rittmann (2007) also reported on captive breeding in this species with females depositing 7–12 eggs per clutch. Köhler (1995b) wrote that captive specimens eat hibiscus flowers, fruits (apple, banana, mango, and melon), and insects. Hendry and Gandola (2011) reported that an adult *C. oedirhina* on Roatán captured and ate a juvenile *Iguana iguana*, and Pasachnik and Chavarria (2011) reported an adult feeding on a hatchling *Trachemys venusta*. Gandola and Hendry (2011) observed an adult feeding on the fruit of a Hog Plum (*Spondias mombin*). An adult that I kept in captivity relished hibiscus flowers and also killed and ate a white mouse that was offered it; therefore, the species will probably eat any small animal it can overpower. Pasachnik (2011c) reported a case of partial limb regeneration. Goldberg et al. (2011) reported finding endoparasites in this Honduran endemic. Gravendyck et al. (1998) reported reoviruses and paramyxoviruses in *C. oedirhina* on Roatán (as *C. bakeri*).

Etymology.—The specific name *oedirhina* is formed from the Greek *oedos* (or *oidos*; a swelling) and *rhinos* (nose), and refers "to the profile of the [swollen] snout

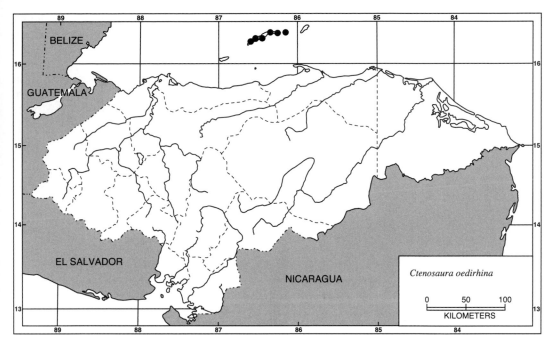

Map 41.   Localities for *Ctenosaura oedirhina*. Solid circles denote specimens examined.

compared with those of other *Ctenosaura*" (de Queiroz, 1987a: 892).

Specimens Examined (19, 1 skin and skeleton [12]; Map 41).—**ISLAS DE LA BAHÍA**: Isla Barbareta, USNM 573906–07; Isla de Roatán, near Flowers Bay, USNM 573905; Isla de Roatán, near French Harbor, UF 28553; Isla de Roatán, French Harbor, FMNH 53831; Isla de Roatán, Indian Cay, SMF 78036; Isla de Roatán, Indian Cay, SMF 78036; Isla de Roatán, Oak Ridge, UTA R-36574; Isla de Roatán, about 4.8 km W of Roatán, CM 64681, LSUMZ 22367, 22369, UF 28530 (skin and skeleton), 28531–33; Isla de Roatán, near Roatán, LSUMZ 22399; Isla de Roatán, Sandy Bay, KU 203153; Isla de Roatán, Santa Elena, BMNH 1938.10.4.82; Isla de Roatán, West End Point, CM 64682–83.

Other Records.—**ISLAS DE LA BA-HÍA**: Isla Barbareta, Big Pigeon Cay LSUPC 5805 (Grismer et al., 2001); Isla de Roatán, Gumbolimbo Park and Reserve

at West Bay (Pasachnik, 2011c); Isla de Roatán, about 4.8 km W of Roatán, LSUMZ 22368, 22370–71 (Meyer and Wilson, 1973; "on loan to Jaime Villa" as of February 2009); "Isla de Roatán," Ralph W. Axtell 6570–72 (Buckley and Axtell, 1997).

Ctenosaura quinquecarinata (Gray, 1842)

Cyclura quinquecarinata Gray, 1842: 59 (holotype, BMNH 1946.8.30.48 [formerly BMNH 1841.3.5.61; Etheridge, 1982: 20; Hasbún and Köhler, 2001: 247]; type locality: "Demerara?"

Ctenosaura quinquecarinata: Bocourt, 1882: 47; Hollingsworth, 1998: 175; McCranie, 2015a: 369.

Ctenosaura palearis: Sites et al., 1996: 1090.

Ctenosaura flavidorsalis: Hasbún et al., 2005: 3099; Gutsche and Köhler, 2008: 247, 253.

Ctenosaura "palearis": Gutsche and Köhler, 2008: 247, 253.

Ctenosaura praeocularis Hasbún and Köhler, 2009: 197 (holotype, SMF 79520; type locality: Cerro Las Mesitas, 10 km east of Sabanagrande toward Nueva Armenia, Montegrande, Departamento Francisco, Morazán, Honduras, 800 m, 13°46.43′N, 86°11.83′W″); McCranie et al., 2014: 100; Solís et al., 2014: 131.

Geographic Distribution.—*Ctenosaura quinquecarinata* is known from low and moderate elevations from southern Honduras to northwestern Costa Rica. In Honduras, this species is restricted to subhumid areas on the Pacific versant of the southern portion of the country.

Description.—The following is based on 15 males (USNM 573908–09, 581866–67, 581870, 581874, 581876, 581878, 581881, 581884, 581886–90) and 15 females (CAS 152981; USNM 573910–11, 581868–69, 581871–73, 581875, 581877, 581880, 581882–83, 581885, 581891). *Ctenosaura quinquecarinata* is a large lizard (maximum recorded SVL 195 mm [Köhler and Hasbún, 2001; males reach a significantly larger size than do females]) with spiny caudal whorls; snout region scales smaller than to equal in size to scales in frontal region, scales smooth, flat; snout acutely to broadly rounded in dorsal aspect, sloping downward in profile; 4–7 (most often 6) postrostrals; 5–7 internasals; 2 canthals, anterior smaller than, or equal in size to posterior; 4–8 scales between second canthals; 1–2 (usually 1) scales (minimum) between supraorbital semicircles; interparietal scale larger than surrounding scales, parietal eye visible, but sometimes indistinct; 1–2 (usually 2) scales between interparietal and supraorbital semicircles; nasal single, nostril opening in center of scale, or slightly posterior to center of scale, directed laterally; 1–2 (usually 2) scales between nasal and rostral; moveable eyelid present; pupil circular; 7–10 (most often 9) supralabials; 7–10 (most often 8 or 9) infralabials;

second, third, or fourth elongated subocular (usually second) directly below eye; 1–3 (usually 1 or 2) preoculars; 0–2 (usually 1) scale rows separating suboculars from supralabials; 2–4 (usually 2) postmentals; pendulous dewlap absent; gular fold distinct, barely rounded (nearly straight) posteriorly across throat; dorsal body scales granular on nape, becoming slightly larger posteriorly, those between level of forelimb to midbody slightly conical, those posterior to midbody broadly keeled with spiny distal ends; middorsal crest scales laterally compressed in males, but only barely compressed in females, forming low crest (crest scales >2 mm high in both sexes, highest in males), largest on nape and shoulder regions, 50–70 (58.5 ± 5.1) enlarged middorsal crest scales; 13–21 (17.5 ± 1.7) middorsal crest scales in one HL; 2–3 short gaps present or absent in middorsal crest scales in thoracic region; 2–14 (6.8 ± 3.1) granular scales between first differentiated middorsal crest scale and posterior end of head; 9–30 (16.8 ± 5.4) scales between ultimate differentiated middorsal crest scale and first caudal whorl; ventral scales flat, not keeled, not imbricate, larger than dorsal scales; dorsal and anterodorsal surfaces of shank with much enlarged, spiny scales and dorsal surface of thigh with somewhat enlarged, but not spiny, scales; 27–36 (30.9 ± 2.0) subdigital scales on 60 sides of Digit IV of hind limb; 5–8 (6.5 ± 0.5) femoral pores on each side in males, 5–8 (6.4 ± 0.8) in females; 11–15 (12.9 ± 1.2) total femoral pores in males, 10–15 (12.9 ± 1.4) in females; femoral pores larger in adult males than in females; 17–24 caudal whorls, all whorls separated by 1 intercalary scale row; caudal whorls abruptly ending on distal third of tail, whorls replaced by regular small, strongly keeled scales; paramedian scales anterior to reduction from 3 to 2 keeled; inner paramedian caudal scales of each whorl with smaller keels than outer paramedian caudal scales, although many paramedian caudal scales of each whorl

Plate 52.   *Ctenosaura quinquecarinata*. USNM 581890, adult male, SVL = 136 mm. Francisco Morazán: Caserío Los Encinitos.

barely spinose in some; 4–9 smooth paramedian intercalary scales in uninterrupted row on each side between caudal whorls 6–7; strongly keeled median intercalary scale usually absent between caudal whorls 2–15, strongly keeled median intercalary scale occasionally present on caudal whorls 1–8; small, keeled intercalary scale present on distal whorls; reduction from 3 to 2 paramedian scales occurs between caudal whorls 7–13; reduction from 2 to 1 paramedian scales occurs between caudal whorls 12–20; SVL 110–159 (135.5 ± 18.0) mm in males, 107–145 (129.7 ± 9.4) mm in females; HW/SVL 0.17–0.20 in males, 0.15–0.18 in females; HL/SVL 0.20–0.24 in males, 0.19–0.21 in females; SL/SVL 0.09–0.10 in males, 0.08–0.10 in females; SHL/SVL 0.21–0.25 in males, 0.19–0.23 in females; TAW/TAH 1.10–1.38 in males, 1.12–1.39 in females; TAL/SVL 1.00–1.70 in seven males, 1.08–1.43 in seven females.

Color in life of an adult male (USNM 581890; Plate 52): dorsal surface of body Light Drab (119C) with Sepia (119) and pale brown spotting dorsolaterally and Bunting Green (150) and Sepia spotting and mottling middorsally; 4 Sepia crossbands on body; crest scales on nape alternating between Sepia and pale gray, changing to Sepia and pale green on body; scales on caudal whorls alternating between Lime Green (159) and Sepia crossbands, scales dark brown and pale greenish brown distal to caudal whorls; top and side of head dark brown with some brown mottling dorsally; Sepia crossband on chin below levels of tympanum and posterior angle of jaw, chin mottled dark and pale brown anterior to that point; throat Chamois (123D) with dark brown scales and larger dark brown spots; midventer Pale Pinkish Buff (121D) with Sepia extensions of crossbars laterally; ventral surface of forelimb Chamois, that of hind limb paler brown, as is subcaudal surface; iris Chestnut (32) with gold rim around pupil. Color in life of a second adult male (USNM 573909): dorsal surfaces of head and anterior portion of body Olive-Brown (28) with Jet Black (89) blotches and crossbars; posterior portion of body black with Olive-Brown and greenish yellow spots, Jet Black crossbars extend laterally on Olive-Brown background; dorsal surface of tail black with pale green and Olive-Brown mottling; dorsal surface of forelimb Olive-Brown with pale greenish yellow mottling; dorsal surface of hind limb black with pale brown mottling; ventral surface of head dirty white with brown mottling; belly pale yellow medially, pale brown laterally, with extension of dorsal and lateral crossbars continuing onto lateral portion nearly to midventer; subcaudal surface pale brown; iris Amber (36). Color in life of another adult male (USNM 581874): anterior third of dorsal surface of body Drab (27) with Sepia (219) mottling and pale brown spotting laterally and with pale green and pale brown spotting mid-dorsally; crest scales pale brown with some Sepia color on nape and anterior third of body, becoming pale green or Sepia on posterior two-thirds of body; top and side of head Dark Drab (119B); dorsal surface of tail with pale green enlarged spiny scales, remainder of tail with pale brown scales with some dark brown mottling; chin and throat mottled dark and pale brown, belly pale brown with some dark brown spotting laterally; ventral surfaces of fore- and hind limb pale brown, as is that of tail; iris golden brown with gold rim. Color in life of a fourth adult male (USNM 581866): dorsal surfaces similar in color to that of USNM 581874, except 4 distinct Sepia (119) cross-bands present on body; top of tail with alternating Sepia and pale green crossbands; belly Cinnamon (39), remainder of ventral color similar to that of USNM 581874 (an adult male, see above). Color in life of another adult male (USNM 581884): dorsal surfaces of head and anterior portion of body Olive-Brown (28) with Jet Black (89) blotches and crossbars; posterior portion of body black with Olive-Brown and greenish yellow spots, Jet Black crossbars extend laterally on Olive-Brown background; dorsal surface of tail black with pale green and Olive-Brown mottling; dorsal surface of forelimb Olive-Brown with pale greenish yellow mottling; dorsal surface of hind limb

black with pale brown mottling; ventral surface of head dirty white with brown mottling; belly pale yellow medially, pale brown laterally with extension of dorsal and lateral crossbars continuing onto lateral portion nearly to midventer; subcaudal surface pale brown; iris Amber (36). Color in life of an adult female (USNM 581891): dorsal surface of body similar in color to that of USNM 581890 (an adult male, see above), except less spotting and dark crossbars present, with ragged edges around those markings; dorsal and lateral surfaces of head Drab (27); caudal whorls with alternating Drab and dark brown cross-bands; chin and throat Grayish Horn Color (91) with pale brown mottling; belly pale brown with dark brown spots laterally, formed by extensions of incomplete cross-bands; ventral surfaces of fore- and hind limb pale brown; subcaudal surface Chamois (123D), without markings; iris Chestnut (32) with gold rim around pupil. Color in life of another adult female (USNM 581875): dorsal surfaces of head and ante-rior third of body Dark Drab (119B), Dark Drab with Sepia (219) and pale brown mottling on posterior two-thirds of body; dorsal surface of tail pale brown with dark brown spotting, some hint of pale green on enlarged median and lateral caudal scales; remainder of color similar to that of USNM 581874 (an adult male, see above). Color in life of a third adult female (USNM 573910): dorsal surface of body Brownish Olive (29) with pale greenish yellow mottling anterior-ly and black mottling posteriorly; vague black crossbars also present; dorsal surface of tail Brownish Olive with black mottling; ventral surface of head dirty white with brown mottling; belly dirty white without brown crossbars; subcaudal surface dirty white; iris Amber (36). Hasbún and Köhler (2009: 199–200) described color in life of the male holotype of the synonym *C. praeocularis* (SMF 79520): "Dorsal ground color Sepia (119) with Yellow Green (58) and Sulphur Yellow (57) scales; neck region with Sky Blue (168c) blotches and Opaline Green (162d) scales posteriorly; head dorsally Vandyke Brown (121); temporal region Vandyke Brown (121) checkered with Prouts [sic] Brown (121a); loreal region checkered with Hair Brown (119a) and Light Drab (119c); rostrum Sky Blue (66); chin Vandyke Brown (121) anteriorly and pale Pinkish Buff (121d) and dirty white posteriorly; chest Sulphur Yellow (57); venter Salmon Color (6); lateral bands Sepia (119); upper arm Sepia (119), dorsally scattered with Sulphur Yellow (57) and Yellow Green (58) scales; ventral surface of forefoot Cream Color (54); dorsal rear foot Sepia (119) scattered with dirty white scales; ventral side of hind feet dirty white diffused with Salmon Color (6); dorsal surface of tail Sepia (119) with a suggestion of Lime Green (80) and Glaucus (159) on bands; ventral surface of tail dirty white; iris Raw Umber (223) grading into Tawny Olive (223d) with a fine Chamois ring (123d) around pupil."

Color in alcohol: dorsal ground color of head and anterior third of body brown to dark brown; dorsal surface of posterior two-thirds of body dark brown, usually with numerous pale brown spots; middorsal crest scales cream to pale brown on nuchal area and anterior third of body, becoming pale brown to dark brown on posterior two-thirds of body, some larger males also have scattered bluish brown middorsal crest scales posteriorly on body; lateral surface of body brown to dark brown, usually with numerous pale brown, bluish brown, or cream spots, especially on posterior two-thirds of body, sometimes also with 3–4 indistinct brown crossbars; dorsal surface of forelimb brown with some gray, pale brown, or bluish brown scales, that of hind limb dark brown with pale brown spots; dorsal surface of tail brown with dark brown mottling on some paramedian scales, posterior portion after termination of caudal whorls uniform brown, or with suggestion of paler brown dorsal crossbands; ventral

surface of head varies from pale brown with some lineate dark brown mottling to dark brown with varying amounts of paler brown mottling, to ventral surface of head almost entirely dark brown; ventral surface of body pale brown medially, mottled pale brown, cream, and dark brown laterally, occasional specimens also have 3–4 dark brown lateral bars (or indications thereof) that do not meet medially; subcaudal surface pale brown with some scales of caudal whorls and those posterior to caudal whorls dark brown edged, occasionally indication of slightly darker brown crossbands present distally; ventral surfaces of fore- and hind limb pale brown, dark brown spots also present on hind limb; palm, sole, and digits rusty brown.

Diagnosis/Similar Species.—*Ctenosaura quinquecarinata* is distinguished from all remaining Honduran lizards, except the other *Ctenosaura*, by having distinct caudal whorls in combination with a large size. *Ctenosaura bakeri* has 38–42 enlarged middorsal crest scales, has 2 intercalary scale rows between the caudal whorls, has a pendulous male dewlap, and lacks enlarged scales on the dorsal surface of the thigh (versus 50–70 middorsal crest scales, 1 intercalary scale row between caudal whorls, no pendulous male dewlap, and distinctively enlarged scales present on dorsal surface of thigh in *C. quinquecarinata*). *Ctenosaura flavidorsalis* has only somewhat laterally compressed middorsal crest scales <2 mm high and usually has substantial yellow or orange pigment on the dorsal surface of the body (versus distinctively laterally compressed middorsal scales >2 mm high in nuchal area in males and usually green dorsal pigment in *C. quinquecarinata*). *Ctenosaura melanosterna* has 27–42 enlarged middorsal crest scales >4 mm high in nuchal area in males, has a black chest in adult males, and has a pendulous black male dewlap (versus 50–70 middorsal crest scales <4 mm high in nuchal area in males, a brownish cream chest, and no pendulous male dewlap in *C. quinquecarinata*). *Ctenosaura oedirhina* has 2 intercalary scale rows between at least caudal whorls 10–30 and has the dorsal surface of the thigh without enlarged scales (versus single intercalary scale row and dorsal surface of thigh with distinctively enlarged scales in *C. quinquecarinata*). *Ctenosaura similis* has the scales on the anterodorsal surface of the shank not enlarged and only weakly keeled, and has 2 intercalary scale rows between the anterior caudal whorls (versus scales on anterodorsal surface of shank greatly enlarged and spiny and single intercalary scale row between anterior caudal whorls in *C. quinquecarinata*).

Illustrations (Figs. 60, 88, 96, 98, 104; Plate 52).—Hasbún and Köhler, 2009 (adult, head scales; as *C. praeocularis*); Huy, 2008 (adult, subadult, juvenile); McCranie et al., 2014 (subadult; as *C. praeocularis*); Villa et al., 1988 (adult).

Remarks.—Hasbún and Köhler (2009) described *Ctenosaura praeocularis* based on two populations, the type locality in the southern portion of Francisco Morazán, Honduras, and the other about 35 km to the south-southeast in Choluteca, Honduras. Hasbún and Köhler (2009: 201) considered *C. praeocularis* to be somewhat intermediate between *C. flavidorsalis* (type locality in La Paz, Honduras, about 85 km NW of the *C. praeocularis* type locality) and *C. quinquecarinata* (restricted type locality "Nicaragua and Costa Rica") because genetically it was only 1.2% divergent from *C. flavidorsalis*, compared with 2.6% divergent from *C. quinquecarinata*. However, "in general appearance and in the studied morphological characters, *C. praeocularis* is much more similar to *C. quinquecarinata*" than it is to *C. flavidorsalis* (p. 201). Hasbún and Köhler (2009) considered *C. quinquecarinata* to occur on the subhumid Pacific versant of Nicaragua adjacent to the Honduran departments of Choluteca and El Paraíso southward to northwestern Costa Rica. The closest locality for *C. quinquecar-*

inata reported by Hasbún and Köhler (2009) is near San Francisco del Norte, Estelí, Nicaragua, which lies about 80 km southeast of the *C. praeocularis* type locality, but only about 45 km southeast of the Choluteca locality for *C. quinquecarinata* reported by Hasbún and Köhler (2009). Thus, three closely related *Ctenosaura* nominal forms, *C. flavidorsalis*, *C. praeocularis*, and *C. quinquecarinata*, are all reported to occur in a relatively small geographic region of southern Honduras and adjacent western Nicaragua.

I have been aware for several years of two other *Ctenosaura quinquecarinata*–like populations at El Madreal and El Banquito, both in Choluteca, Honduras. The latter locality lies less than 30 km northwest of the San Francisco del Norte, Nicaraguan locality identified as *C. quinquecarinata* by Hasbún and Köhler (2009). Tissues from the El Madreal population have been available for years (see Sites et al., 1996; Gutsche and Köhler, 2008; also see discussion in *Ctenosaura* account above), although those authors were unaware of the locality for those tissues (I alerted Sites to the El Madreal locality during the mid 2000s). In November 2011, a population of *C. quinquecarinata*–like lizards was discovered in the vicinities of El Rodeo and Orealí in El Paraíso, Honduras, about 45 km east-northeast of the *C. praeocularis* type locality. Tissue samples were collected from the El Paraíso populations. That discovery (also see *Ctenosaura* account) spurred a colleague and me to undertake a study of the smaller species of *Ctenosaura* in southern Honduras. Collections were made of those *Ctenosaura* from the vicinities of the type localities of *C. flavidorsalis* and *C. praeocularis* and of the *C. quinquecarinata*–like lizards from several localities in Choluteca. Morphological study of these newly collected *Ctenosaura* demonstrated much variation in the preocular character largely used to diagnose *C. praeocularis* from *C. quinquecarinata*. Because of that variation and of the continuous distribution between the Francisco Morazán *C. praeocularis* through El Paraíso, Choluteca, to northwestern Nicaragua, I consider *C. praeocularis* Hasbún and Köhler a junior synonym of *C. quinquecarinata* Gray (see below and McCranie, 2015a: 369). Molecular data involving those recently collected specimens and other previously collected Honduran *Ctenosaura* show little molecular differences between morphologically well defined species (personal unpublished data; also see Gutsche and Köhler, 2008).

The taxonomy and distribution, the type specimen, and the type locality of *Ctenosaura quinquecarinata* have all been highly confused in the literature. Gray (1842: 59) gave the type locality for his *Cyclura quinquecarinata* as "Demerara?" (repeated by Gray, 1845b: 192). However, Demerara is in the South American country of Guyana (see Russell and Bauer, 2002: 753.8). Cope (1870: 161), in reference to information from Sumichrast, concluded the region of the Isthmus of Tehuantepec, Mexico, "is the undoubted home" of *C. quinquecarinata* (as *Cyclura*). Bocourt (1874: 138, *In* A. H. A. Duméril et al. 1870–1909a) repeated that Cope assumption about Tehuantepec as the type locality for this species. Sumichrast (1873: 259) reported *C.* (as *Cyclura*) *quinquecarinata* as abundant on the Isthmus de Tehuantepec and later (Sumichrast, 1880: 176) listed this species as occurring in the Tehuantepec area. Bocourt (1882: 47) listed this species in a table without mentioning any locality, only referring to Gray's (1845b) *Enyaliosaurus quinquecarinatus*. Boulenger (1885b: 198) listed the holotype of *C. quinquecarinata* as in the BMNH, but gave no locality data for that holotype, and listed the Isthmus of Tehuantepec as the range of this species based on two other BMNH specimens he also assigned to *C. quinquecarinata*.

Günther (1890: 58, *In* Günther, 1885–1902) listed a specimen of *Ctenosaura quinquecarinata* from "Honduras." That

locality was actually based on the holotype of *Cyclura quinquecarinata* in the BMNH (BMNH 1946.8.30.48). Bailey (1928: 43) examined that holotype and noted, "it is a stuffed skin without any locality or collector's label." Because of the significant number of specimens he identified as *C. quinquecarinata* from Tehuantepec in the BMNH, Bailey (1928: 43) also concluded that *C. quinquecarinata* "is perhaps confined to the Isthmus of Tehuantepec." Bailcy (1928: 43) thus restricted the type locality of *C. quinquecarinata* to "Tehuantepec, Oaxaca, Mexico." Smith (1987: xxx [Roman numeral pagination]), without examining the "Honduran" specimen, considered the specimen to be *C. palearis*, which does not occur in Honduras, but has been confused with a related Honduran species, *C. melanosterna*, just like the people responsible for the CITES proposal confused *C. quinquecarinata* with *C. melanosterna*. Hasbún and Köhler (2001: 248) took that error one step further and called the specimen *C. melanosterna*, without realizing it was the holotype of *C. quinquecarinata* that they had earlier restricted to the Costa Rica-Nicaragua locality (see below).

Hasbún and Köhler (2001: 247) examined the holotype of *Ctenosaura quinquecarinata* and wrote "the current specimen label of BMNH 1946.8.30.48 reads 'Honduras'" and concluded the label was presumably added to the specimen sometime after Bailey (1928) examined it. It is most likely the new label was added to the specimen when it was recatalogued with the current BMNH number. Apparently, the person responsible for the recataloging used Günther's (1890, *In* Günther, 1885–1902) reference to "Honduras" for the specimen. Hasbún and Köhler (2001: 254) restricted the type locality of *C. quinquecarinata* "to the southern portion of the distribution of *C. quinquecarinata* in Costa Rica and Nicaragua," again without realizing it was the same specimen they called *C. melanosterna* from "Honduras" and considered Bailey's (1928:

43) type locality restriction of this species to Tehuantepec, Oaxaca, Mexico, "as invalid" (but see below). Thus, the confusion surrounding this BMNH specimen by Hasbún and Köhler (2001) adds to the confusion already surrounding the *C. quinquecarinata* holotype. Schmidt (1941: 503) suggested that all BMNH specimens in Günther (1885–1902) carrying the locality data "Honduras" should be considered as having originated from British Honduras (= Belize). As just noted, Smith (1987: xxx [Roman numeral pagination]) concluded that Günther's (1890, *In* Günther, 1885–1902) specimen from "Honduras" (actually the holotype of *Cyclura* [= *Ctenosaura*] *quinquecarinata*) was "*Enyaliosaurus palearis*" (= *Ctenosaura palearis* Stejneger, 1899: 381), a species unknown from Belize or Honduras.

Gicca (1983: 329.1) reported *Ctenosaura quinquecarinata* from the departments of La Paz and Yoro, Honduras. Gicca's (1983) source for these records appears to be Etheridge (1982: 21), who stated "in Honduras from near La Paz and Yaro [sic]" in the geographic distribution statements for *C. quinquecarinata*. The *C. quinquecarinata* in Yoro is *C. melanosterna* and those from La Paz are *C. flavidorsalis*.

The collection of additional specimens from the vicinity of the type locality of *Ctenosaura praeocularis* Hasbún and Köhler (2009: 197) and numerous additional specimens from Choluteca and El Paraíso, Honduras, reveal there is substantial variation in the number of preocular scales that was used to diagnose *C. praeocularis* from *C. quinquecarinata*. The only remaining morphological characters Hasbún and Köhler (2009) found to distinguish *C. praeocularis* from *C. quinquecarinata* were two trivial osteological characters that would not likely stand examination of more specimens for those characters. Close morphological examination of the Francisco Morazán, Choluteca, and El Paraíso *C. praeocularis* leaves no doubt they are conspecific with each other and with *C. quinquecarinata* (following the contention of Hasbún and Köhler, 2009) of western Nicaragua and northwestern Costa Rica. There are no known morphological characters, and the molecular data is weak (F. Köhler personal communication), to distinguish Honduran *C. praeocularis* from *C. quinquecarinata* specimens from adjacent Nicaragua and Costa Rica. Also, there are no vegetational or physical barriers between the distributions of those two nominal forms, and their ranges are continuous across the borders between those two countries. Some of our new localities are only 20 km from some of those of Nicaraguan *Ctenosaura* examined and identified as *C. quinquecarinata* by Hasbún and Köhler (2009). The photographs in Huy (2008) of Nicaraguan *C. quinquecarinata* also demonstrate they are conspecific with *C. quinquecarinata* from adjacent southeastern Honduras. Hollingsworth (1998) had correctly identified a *Ctenosaura* specimen from El Banquito, Choluteca as *C. quinquecarinata* and McCranie (2015a: 369) suggested synonymizing *C. praeocularis* with *C. quinquecarinata*. The low molecular data differences of those *C. quinquecarinata* group species in Honduras (*C. flavidorsalis*, *C. quinquecarinata*) agree with the low molecular differences between other morphology very distinct species (i.e., *C. bakeri*, *C. oedirhina*) in Honduras (see Sites et al., 1996: 1090; Gutsche and Köhler, 2008: 253; Hasbún and Köhler, 2009: 197).

Ctenosaura quinquecarinata differs from the very closely related *C. flavidorsalis* in having the male middorsal crest scales compressed and slightly raised, whereas those scales are not compressed and barely raised in *C. flavidorsalis*. *Ctenosaura quinquecarinata* also has some green on the dorsal surfaces, whereas those areas are generally yellow or orange in *C. flavidorsalis*. *Ctenosaura quinquecarinata* also tends to have smaller and less spiny scales on the dorsal surface of the thigh and less spiny intercalary scales on the caudal whorls than does *C. flavidorsalis. Ctenosaura quinque-carinata* is also more arboreal and utilizes tree holes as retreats, whereas *C. flavidorsalis* is more terrestrial and usually uses rock crevices as retreats. Those characters and tree-hole habitats are also identical between topotypic *C. praeocularis* and nearby *C. quinquecarinata* populations in El Paraíso, Choluteca, and in northwestern Nicaragua. Thus, there are no morphological characters nor habitat differences to distinguish or separate *C. praeocularis* from the geographically continuous and otherwise identical *C. quinquecarinata.*

Sites et al. (1996) recorded *Ctenosaura palearis* from "Honduras" (based on a specimen [BYU 39667; given as BYU 34667 by Sites et al., 1996: 1090] purchased from an animal dealer) in a study based on morphological and mitochondrial DNA sequence data. Gutsche and Köhler (2008) concluded that identification of the same specimen was erroneous, because it clustered with *C. flavidorsalis* in their study (also see molecular data in Sites et al., 1996). Examination of morphological characters of BYU 39667 (tissue data for that specimen are available from GenBank under accession number U66229) revealed it to be *C. quinquecarinata* and with a strong resemblance to the El Madreal populations. An animal collector for a former Honduran animal dealer lived at El Madreal, Choluteca, until recently, a locality where *C. quinquecarinata* remains common (at least until 2012); thus, the BYU specimen is likely from the vicinity of that locality (see Remarks for *Ctenosaura* and *C. melanosterna*). Additionally, the BYU 39667 DNA sequence data are nearly identical to those of several specimens collected by me in recent years from El Madreal.

Hasbún and Köhler (2001: 254) stated "an independent molecular genetic study of the *quinquecarinata-flavidorsalis* group of the genus *Ctenosaura* including the holotype of *Ctenosaura quinquecarinata*

(Hasbún in preparation) provided further evidence that BMNH 1946.8.30.4 did not originate from Oaxaca, Mexico." Those molecular data were part of the data Hasbún and Köhler (2001) used to restrict the *C. quinquecarinata* holotype "to the southern portion of the distribution ... in Costa Rica and Nicaragua." Hasbún and Köhler (2001) did not present any molecular data in that work, nor are there any molecular data on that specimen in Hasbún and Köhler (2009), nor Hasbún et al. (2001, 2005). Additionally, apparently no molecular data on the *C. quinquecarinata* holotype has been filed with GenBank. The "Hasbún in preparation" mentioned by Hasbún and Köhler (2001) is the Hasbún (2001) unpublished dissertation. Hasbún (2001: 211) included a phylogram based on ND4 mitochondrial DNA haplotypes of the *C. quinquecarinata* holotype in a cluster with other *Ctenosaura* samples from Nicaragua and Costa Rica. Hasbún (2001: 152) stated the *C. quinquecarinata* holotype sequence data was identical to a haplotype from Estelí, Nicaragua. It is unfortunate the sequence data for that *C. quinquecarinata* specimen was apparently not sent to GenBank, nor provided in Hasbún (2001), so it could be used in future phylogenetic studies.

Natural History Comments.—*Ctenosaura quinquecarinata* is known from 95 to 1,000 m elevation in the Lowland Dry Forest, Lowland Arid Forest, and Premontane Dry Forest formations. This is a mostly arboreal and strictly diurnal species that is usually found in oak (*Quercus* sp.), Jícaro (*Crescentia alata*), and Nance (*Brysonima crassifolia*) trees with hollows in them. Most lizards were first seen in open sunny areas. Lizards were also seen on hollow fence posts and in man-made rock walls used as fencerows. Several were also found under logs near trees with holes where they retreated upon being exposed. Individuals also usually retreat into nearby tree holes when approached. Months of collection

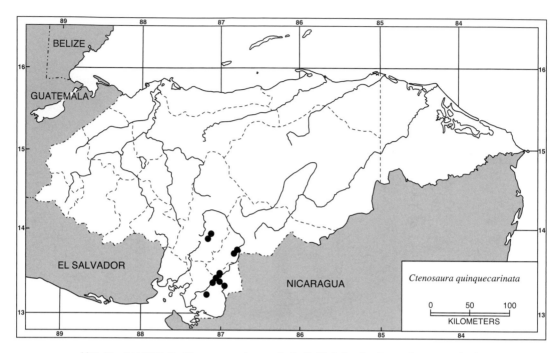

Map 42. Localities for *Ctenosaura quinquecarinata*. Solid circles denote specimens examined.

were January, April, July, August, November, and December; thus, the species is active year-round during sunny conditions in the hot zones where it occurs. It is apparently confined to places that contain nearby and suitable tree holes available for retreat. The species remains very common, even in the vicinity of El Madreal, Choluteca, the home of an animal collector for years (see Remarks). I was once in El Madreal when an animal buyer for the above-mentioned dealer, who told me of the El Madreal locality, visited the village to purchase about 30 *C. quinquecarinata* being held for him (along with boxes of tarantulas). Thus, that animal dealer was the source of numerous *C. quinquecarinata* that formerly appeared on price lists of at least one U.S. animal dealer as *C. palearis* (see Remarks for *Ctenosaura*). Both adults and juveniles were present at all localities where we did concentrated searches for this species, thus indicating healthy populations

are well represented. Several people living in the area where this species occurs indicated that they sometimes eat its meat. Hasbún and Köhler (2009) noted three females collected in December contained 8–13 oviductal eggs. Hasbún and Köhler (2009: 201) also reported *C. quinquecarinata* feeds "mainly on young leaves and insects." Villa and Scott (1967: 475) reported 5 of 10 stomachs of this species from Nicaragua "contained three Coleoptera, one Hemiptera (Siricidae), three leaves and several pieces of fruit."

Etymology.—The name *quinquecarinata* is formed from the Latin *quinque* (five) and *carinatus* (keeled). The meaning is unclear as are meanings of many names erected by Gray (see Pyron et al., 2014).

Specimens examined (48 [0]; Map 42).— **CHOLUTECA**: Apalcinagua, USNM 581866–72; El Banquito, CAS 152981; El Madreal, BYU 39667–68, USNM 573909–12, 581873; El Potrero, USNM 573908,

581874–76; Las Pilas, USNM 581877–79; 1 km E of Orocuina, SMF 79516–17. **EL PARAÍSO**: El Rodeo, USNM 581880–86, UNAH 5658; near Orealí, UNAH 5677; Orealí, USNM 581887. **FRANCISCO MORAZÁN**: La Varazón, Caserío Los Encinitos, SMF 80868–69, 82474, USNM 581888–91; La Varazón, Caserío Zacahuato, SMF 80867; Monte Grande, SMF 79518–20, 80866, 82475. "HONDURAS": UTA R-25386.

Other Records.—**CHOLUTECA**: El Madreal, UU 66229 (Gutsche and Köhler, 2008).

Ctenosaura similis (Gray, 1831)

Iguana (*Ctenosaura*) *similis* Gray, 1831: 38, *In* Gray, 1830–1831 (holotype formerly in the private collection of Thomas Bell of London, England, now presumed lost [see Bailey, 1928: 35]; type locality not stated; see Remarks).
Ctenosaura completa: Günther, 1890: 58, *In* Günther, 1885–1902.
Ctenosaura similis: Bailey, 1928: 32; Dunn and Emlen, 1932: 28; Meyer, 1966: 175; Meyer, 1969: 246; Peters and Donoso-Barros, 1970: 105 (in part); Meyer and Wilson, 1973: 23; Wilson and Hahn, 1973: 115; Gundy and Wurst, 1976: 116; Wilson et al., 1979a: 25; Iverson, 1980: 93; Etheridge, 1982: 21; O'Shea, 1986: 39; de Queiroz, 1987a: 901; de Queiroz, 1987b: 176; Köhler, 1994b: 12; de Queiroz, 1995: 22; Köhler, 1995a: 12; Köhler, 1995i: 59; Köhler, 1996d: 20; Köhler, 1996g: 65; Köhler and Streit, 1996: 42; Gravendyck et al., 1998: 35; Hollingsworth, 1998: 175; Köhler, 1998d: 375, 382; Köhler, 1998e: 17; Monzel, 1998: 159; Wilson and McCranie, 1998: 15; Köhler, 1999a: 214; Köhler, 2000: 76; Köhler, Schroth, and Streit, 2000: 191; Lundberg, 2000: 4; Wilson et al., 2001: 135; Köhler, 2002: 133; Lundberg, 2002b: 7; Köhler, 2003a: 130; McCranie et al., 2005: 96; Lovich et al., 2006: 15; McCranie et al.,

2006: 115; Pasachnik, 2006: 266; Townsend, 2006a: 35; Wilson and Townsend, 2006: 105; Diener, 2007: 59; Gutsche, 2007: 40; Townsend et al., 2007: 10; Wilson and Townsend, 2007: 145; Gutsche and Köhler, 2008: 253; Köhler, 2008: 142; Pasachnik et al., 2009: 1250; Townsend and Wilson, 2010b: 692; McCranie and Valdés Orellana, 2014: 45; McCranie et al., 2014: 100; Solís et al., 2014: 131; McCranie, 2015a: 369; Naccarato et al., 2015: 234; McCranie and Gutsche, 2016: 871.
Ctenosaura bakeri: Peters and Donoso-Barros, 1970: 105 (in part); MacLean et al., 1977: 4 (in part); Iverson, 1980: 93 (in part).

Geographic Distribution.—*Ctenosaura similis* occurs at low and moderate elevations on the Atlantic versant from central Tabasco, Mexico, to northeastern Honduras, in adjacent coastal northwestern Nicaragua, and in western Nicaragua, eastern Costa Rica, and central Panama. It occurs on the Pacific versant from southeastern Oaxaca, Mexico, to west-central Panama. It is also found on numerous islands in the Caribbean Sea, some of which are likely human introductions. This species has also been introduced and established in southern Florida, USA, the Colombian island of Malpelo, in north-central Veracruz, Mexico, a few islands in the Bahamas, and on islas de Guanaja and Roatán (the latter on purpose by the owner of a small resort on an offshore cay) in the Bay Islands, Honduras. In Honduras, this species is widespread in open habitats on the mainland and it also occurs on numerous islands in the Golfo de Fonseca, as well as the Bay Islands just mentioned.

Description.—The following is based on ten males (CM 27611; LSUMZ 34039; USNM 573918, 573920, 573924, 573931, 573934; ZMB 73629–31) and 11 females (CM 41232; MCZ R21101, 21745; USNM 573326, 573919, 573921, 573925, 573927,

573929, 573932, 573936). *Ctenosaura similis* is a huge lizard (maximum recorded SVL 489 mm [Köhler, 2008; males significantly larger than females], 302 mm SVL in largest measured Honduran specimen [CM 27611, a male]) with spiny caudal whorls; snout region scales similar in size to those in frontal region, scales smooth, flat; snout acutely rounded in dorsal aspect, sloping downward in profile posteriorly, but bluntly rounded at tip; 4–7 postrostrals; 5–8 internasals; 2 canthals, posterior larger than anterior, or both similarly sized; 5–9 scales between second canthals; 2 scales (minimum) medially between supraorbital semicircles; interparietal slightly larger than surrounding scales, parietal eye visible; 2–3 scales between interparietal and supraorbital semicircles; nasal single, nostril opening in center of scale, directed posterolaterally; 1–2 (usually 2) scales between nasal and rostral; moveable eyelid present; pupil circular; 11–13 supralabials; 10–13 infralabials; second, third, or fourth subocular directly below eye, elongated (usually second); 1–3 (most often 2) preoculars; 2–3 scale rows separating suboculars from supralabials; 3–4 (usually 4) postmentals; pendulous dewlap absent; gular fold distinct, rounded posteriorly across throat; dorsal body scales granular on nape, becoming slightly larger and flatter to about midbody, keeled on posterior half of body; 51–89 (74.0 ± 8.1) enlarged middorsal crest scales, crest scales strongly compressed in adult males, forming high crest (crest scales >5 mm high), middorsal crest scales not as strongly compressed in adult females, forming low crest (crest scales <5 mm high); middorsal crest scales largest on nape in both sexes; 15–22 (19.4 ± 2.0) middorsal crest scales in one HL in 14; 3–18 (8.2 ± 3.4) granular scales between first differentiated middorsal crest scale and posterior end of head; no scales between ultimate differentiated middorsal crest scale and first caudal whorl; ventral scales flat, not keeled, larger than dorsal scales; dorsal and anterior surfaces of shank and thigh without enlarged, strongly keeled or pointed scales; 29–37 (32.0 ± 1.8) subdigital scales on 40 sides of Digit IV of hind limb; 5–8 (6.4 ± 1.0) femoral pores on each side in males, 2–7 (4.2 ± 1.3) in females; 11–15 (12.8 ± 1.8) total femoral pores in males, 4–14 (8.5 ± 2.7) in females; femoral pores larger in males than in females; 18–26 caudal whorls, all separated by 2 intercalary scale rows, caudal whorls becoming indistinct on about distal third of tail; inner paramedian caudal scales of each whorl with keels similar in size to those on outer paramedian caudal scales; 17–22 (18.7 ± 1.7) paramedian intercalary scales in uninterrupted rows on each side between caudal whorls 6–7 on 26 sides, some smooth, some keeled; strongly keeled median intercalary scale present between all caudal whorls; reduction from 3 to 2 paramedian scales occurs at caudal whorls 9–12; reduction from 2 to 1 paramedian scales occurs at caudal whorls 15–24; SVL 122–302 (203.3 ± 53.3) mm in males, 155–251 (180.7 ± 27.3) mm in females; HW/SVL 0.14–0.18 in both sexes; HL/SVL 0.22–0.27 in males, 0.21–0.23 in females; SL/SVL 0.10–0.12 in males, 0.09–0.10 in females; SHL/SVL 0.22–0.27 in males, 0.21–0.26 in females; TAW/TAH 0.72–1.03 in nine males, 0.81–1.00 in ten females; TAL/SVL 1.5–2.4 in four males, 1.5–2.2 in nine females.

Color in life (Plates 53, 54) of an adult male (USNM 573920; not photographed): dorsal surface of head Fuscous (21), that of neck slightly paler brown than that of head; dorsal surface of body between 3 Dark Grayish Brown (20) crossbands a mixture of reddish brown, pale brown, and dark brown; dorsal spines creamy white, except those above crossbands dark brown; dorsal surfaces of fore- and hind limb dark brown with some pale brown scales; gular area reddish brown with pale brown scales on interior perimeter; belly pale brown with black bands nearly crossing belly; iris Amber (36). Color in life (Plate 54) of an adult

Plate 53. *Ctenosaura similis*. USNM 573325, adult male, SVL = 238 mm. Gracias a Dios: Tánsin.

female (from 4.7 km ESE of San Lorenzo Arriba, Yoro, but later ruined): dorsal surfaces of head and neck Smoke Gray (45) with Flesh Ocher (132D) tinge middorsally; dorsal surface of body Smoke Gray with Flesh Ocher tinge middorsally, and with Sepia (119) edged Grayish Horn Color (91) crossbands; dorsal surfaces of fore- and hind limb mottled Pale Horn Color (92) and Sepia; dorsal surface of tail with alternating Drab (27) and Olive-Brown (28) bands; ventral surface of body mottled dirty orange and pale olive gray; underside of forelimb dirty pale olive yellow; underside of hind limb pale olive with scattered olive dots; side of head mottled with pale ocher and

Plate 54. *Ctenosaura similis*. Adult female (discarded). Yoro: near San Lorenzo Arriba.

dark brown; chin mottled pale ocher, pale olive, and dark olive; iris rust red with gold rim around pupil. Color in life of another adult female (USNM 573919): dorsal surface of head grayish black; dorsal surface of body adjacent to head grayish brown, followed by Tawny Olive (223D) ground color with 3 Jet Black (89) crossbands medially and Robin Rufous (340) laterally, crossbands outlined with Sepia (119); dirty white spotting also present on posterior third of body, especially laterally; dorsal surface of tail banded pale brown and dark grayish brown; dorsal surfaces of fore- and hind limb Jet Black with numerous white scales; belly a mixture of pale brown with orange tinge and pale brown with scattered dirty white and black scales; black of dorsal crossbands extends well onto belly on both sides; iris Raw Sienna (136).

Color in alcohol: dorsal surface of head grayish brown to dark brown, with pale brown mottling; dorsal surface of nape generally pale brown with dark brown mottling, blotches, and/or spots; dorsal surface of body posterior to forelimb generally pale brown with 3–4 dark brown to black bands extending onto lateral surface of body; dark bands vary in intensity near dorsal midline from nearly solid color to having pale brown centers, dark crossbands becoming less distinct laterally, varying from black with pale brown spots or mottling to only having dark brown edges; dorsal crossbands becoming less distinct on posterior portion of body, generally replaced by pattern of dark brown with paler brown mottling, blotches, and/or spots; middorsal crest scales same color as adjacent body; lateral surface of posterior of body generally mottled dark brown and pale brown or dark brown with pale brown spots; dorsal surface of tail generally with alternating pale brown and dark brown crossbands, tail crossbands becoming more distinct on distal half; ventral surface of head varies from brown with dark brown streaks or mottling to dark brown with pale

brown mottling and/or spots; ventral surface of body generally pale brown with dark brown mottling, blotches, and/or spots ventrolaterally, ventrolateral extensions of dorsal and lateral crossbands sometimes present, those extensions sometimes nearly solid dark brown or black and distinct; subcaudal surface usually with dark brown mottling anteriorly, becoming progressively paler brown on distal half, usually with distinct darker brown crossbands.

Diagnosis/Similar Species.—*Ctenosaura similis* is distinguished from all remaining Honduran lizards, except the other *Ctenosaura*, by having distinct caudal whorls in combination with its large to huge size. *Ctenosaura similis* differs from all remaining Honduran *Ctenosaura* in lacking greatly enlarged scales on the anterodorsal surface of the shank.

Illustrations (Figs. 91, 99; Plates 53–55).—Álvarez del Toro, 1983 (adult); Álvarez Solórzano and González Escamilla, 1987 (adult); Binns, 2003 (adult); Burghardt and Rand, 1982 (adult); Calderón-Mandujano et al., 2008 (adult); Campbell, 1998 (adult); Diener, 2007 (adult, juvenile, juvenile being swallowed by the snake *Oxybelis aeneus*); Dion and Porras, 2014 (adult); Freiberg, 1972 (adult); García-Padilla et al., 2015 (juvenile); Günther, 1890, *In* Günther, 1885–1902 (adult, caudal scales, leg scales; as *C. completa*); Gutman, 2003 (juvenile); Gutsche, 2007 (subadult); Hollingsworth, 1998 (caudal scales); Köhler, 1991 (adult, subadult), 1993a (adult, subadult, juvenile), 1995a (adult), 1995d (adult), 1995i (adult, juvenile, postmentals, head scales, hind limb scales, caudal scales, subdigital scales), 1998e (adult), 1999b (caudal scales), 2000 (adult, juvenile, postmentals, hind limb scales, caudal scales, subdigital scales), 2001b (adult, head scales, caudal scales), 2002 (adult, juvenile, postmentals, hind limb scales, caudal scales), 2003a (adult, juvenile, lateral head scales, postmentals, hind limb scales, caudal scales), 2004a (adult), 2008 (adult, juvenile,

Plate 55. *Ctenosaura similis*. UNAH, green juvenile. Valle: Exposición Island.

lateral head scales, postmentals, hind limb scales, caudal scales); Köhler and Streit, 1996 (adult, juvenile, head scales, caudal scales); Köhler, Schmidt et al., 1999 (subadult); Köhler et al., 2005 (adult); Lee, 1996 (adult, caudal scales), 2000 (adult, juvenile, caudal scales); Lundberg, 2000 (adult), 2002b (juvenile); McCranie et al., 2005 (adult, subadult, shank scales, dorsal scales, caudal scales), 2006 (adult, caudal scales); Meshaka et al., 2004 (adult); Mora, 2010 (adult); Mora et al., 2015 (adults cannibalizing juveniles); Pasachnik, 2006 (adult, juvenile); Pasachnik and Corneil, 2011 (adult); Powell et al., 2012 (caudal whorls, intercalary scales); Savage, 2002 (adult); W. Schmidt and Henkel, 1995 (adult); Stafford and Meyer, 1999 (adult, hatchling); Sunyer, Nicholson et al., 2013 (juvenile); Wartenberg, 2013 (adult).

Remarks.—Köhler, Schroth, and Streit (2000) presented one of the several phylogenetic analyses of the genus *Ctenosaura* and recommended placing *C. similis* in the subgenus *Ctenosaura* (p. 187). Bailey (1928: 32) restricted the type locality of *C. similis* to "Tela, Honduras, Central America," which was accepted by some authors (i.e.,

Schwartz and Thomas, 1975: 111; Etheridge, 1982: 21; Schwartz and Henderson, 1988: 120, 1991: 387). De Queiroz (1995: 23) realized Bailey's (1928) type locality restriction was inappropriate. The records of this species from "Honduras" in Boulenger (1885b; as *C. acanthura* [Shaw, 1802: 216] and Günther [1890, *In* Günther, 1885–1902; as *C. completa* Bocourt, 1874: 145, *In* A. H. A. Duméril et al., 1870–1909a]) represent British Honduras (= Belize) specimens (see Schmidt, 1941: 503).

A thorough systematic study of the wide-ranging *Ctenosaura similis* is lacking. However, Gutsche and Köhler (2008), using mitochondrial and nuclear DNA, found strong support for recent gene flow between Honduran Caribbean island populations (Utila and Guanaja) and Honduran northern mainland populations. Those results apparently point to a recent man-made introduction of *C. similis* on Guanaja, which has been separated from the mainland much longer than has Utila Island. Molecular studies of other island populations might indicate that other human-dispersed populations occur (i.e., the islands of Providencia and San Andrés, Colombia;

Barbour and Shreve [1934: 197] described the Providencia population as *C. s. multipunctata*). Bailey (1928: 33) had speculated that the Providencia specimen "was in all probability carried to the island from the neighboring mainland by some fishing or turtling schooner." Savage (2002) wrote that the geographic distribution of this species is usually erroneously stated and then discussed its distribution in lower Central America.

Molecular data recovered by Naccarato et al. (2015) identified the source of an introduced population of *Ctenosaura similis* in Florida, USA, to be Honduras. The results of those authors also suggested that Pasachnik et al. (2009) misidentified some of their data sources of *C. similis* based on GenBank molecular data. Naccarato et al. (2015: 234) also implied that "morphological identification" is difficult between *C flavidorsalis* and *C. similis*, which is not true for workers familiar with their subject animals.

Fitch et al. (1982) discussed illegal trade in Honduran specimens of this species and *Iguana iguana* during the 1970s. Wartenberg (2013) mentioned, and included a photograph of, a person offering living *Ctenosaura similis* for sale along the road between Tegucigalpa and Choluteca in the south of Honduras. I have also witnessed for numerous years people selling live *C. similis* along that same road (and several others) every time I pass by, but never *C. quinquecarinata*, which might occur in Jícaro forest localities in the nearby low mountains.

Natural History Comments.—*Ctenosaura similis* is known from near sea level to 1,300 m elevation in the Lowland Moist Forest, Lowland Dry Forest, Lowland Arid Forest, and Premontane Dry Forest formations and peripherally in the Premontane Moist Forest formation. This diurnal, terrestrial to arboreal species is active on any sunny day throughout the year. It can be seen on or in rocky outcrops, rock walls used as fence-rows, fallen logs, trees (especially those with hollows in them), fence posts, piles of lumber, and walls of abandoned and inhabited buildings. It also is active on the ground in grassy fields, pastures, and mangrove flats. Sunyer, Nicholson et al. (2013) also reported juvenile *C. similis* sleeping at night on low vegetation on Isla del Maíz Pequeño, Nicaragua. I have found, over the last 20+ years, juveniles and subadults commonly sleeping at night in vegetation above streams or dry streambeds, in addition to well away from watercourses, especially in the southern portion of the country. That sleeping exposed at night might be a widespread trait of *C. similis* but is not a trait normally used by other ctenosaurs in the country. Juvenile *C. similis* are incredibly common in extreme southern Honduras, probably more so than any other active diurnal lizard anywhere else on the Honduran mainland (rivaled only by juvenile *C. flavidorsalis* and *C. quinquecarinata* in their prime habitat). Females from western Nicaragua have been reported to deposit 12–88 eggs per clutch, with egg deposition probably occurring from February to April (Fitch and Henderson, 1978). Eggs are deposited in burrows (Wiewandt, 1982). Van Devender (1982) studied Costa Rican populations and found juveniles feed primarily on insects, with larger animals progressively feeding more on plant matter such as fruit, leaves, and flowers. That author also noted that *C. similis* is an opportunistic feeder that also will eat juvenile *Iguana* and birds. Henderson (1973) and Fitch and Henderson (1978), plus references cited in those works, also demonstrated that *C. similis* is an opportunistic feeder that will eat leaves, flowers, fruits, seeds, invertebrates, and small vertebrates, including rodents, birds, bats, frogs, lizards, and lizard eggs, with young and smaller ctenosaurs more insectivorous than larger animals. Krysko et al. (2009) recorded similar results in their study of diet in two introduced populations in Florida, USA, and Meshaka et al. (2004) also reported individuals of a Florida population feeding on discarded meat from sandwiches. Mora (2010) recorded insects in the stomachs of 90% of the juveniles he sampled in Costa Rica, with plant material consumed increasing with aging of the lizards. Plant material consumed consisted of leaves, but flowers and fruits were also eaten (see Mora, 2010, for an extensive review of feeding habits of his subject animals). Pasachnik and Corneil (2011) reported a Costa Rican individual feeding on a dead toad (*Rhinella horribilis* (Wiegmann, 1833: col. 654); see Acevedo et al., 2016), and those authors also provided a literature review regarding diet in this species. Subsequently, Anderson and Enge (2012) reported this species feeding on a deer carcass (*Odocoileus virginianus* [Zimmermann]), Stroud and Krysko (2014) found an adult feeding on bay bean flowers (both in Florida, USA), and Dion and Porras (2014) reported a young adult feeding on an adult tarantula in Costa Rica. Mora et al. (2015) reported two occasions of adult *C. similis* feeding on conspecific juveniles in Costa Rica (also see references therein). Traveset (1990) studied *C. similis* scats in northwestern Costa Rica and concluded that individuals in that population are important seed dispersers, especially of *Acacia farnesiana*. Diener (2007) recorded a snake (*Oxybelis aeneus*) feeding on a juvenile *C. similis* on Utila Island. Telford (1977) reported saurian malaria in specimens of this species from "Honduras." Gravendyck et al. (1998) reported reoviruses and paramyxoviruses in *C. similis* on the Bay Islands. See the *C. bakeri* account for information on hybrids between *C. similis* and *C. bakeri*.

Ctenosaura similis has been recently introduced on Roatán in the Bay Islands, Honduras, where it might pose a serious threat to the endemic *C. oedirhina*. Juvenile *C. similis* were seen basking along with *C. oedirhina* juveniles along the dirt road between the large dump north of Coxen Hole and Palmetto Bay during the morning of 19 September 2012. *Ctenosaura similis* was originally intentionally introduced to a satellite island near the shore of the main island in the vicinity of Coxen Hole during the previous 6 years. Eradication of those *C. similis* on Roatán should be a top priority, to protect possible competition and interbreeding between that species and *C. oedirhina*.

Etymology.—The specific name *similis* is Latin meaning "like." The name probably refers to the resemblance of this species to some other *Ctenosaura* species recognized by Gray (1831, *In* Gray, 1830–1831).

Specimens Examined (354, 11 skeletons, 3 skulls [134]; Map 43).—**ATLÁNTIDA**: San Alejo, UF 20442; Tela, AMNH 65474–77, 156287, MCZ R21101–15, 21152, 22665, 22669, 21743–45, 21749–50, 167141–56, UMMZ 58373, 69536. **CHOLUTECA**: Cedeño, BYU 18288; 12.9 km E of Choluteca, USNM 147998; between Choluteca and San Lorenzo, AMNH 69625–27 (all skulls only); Choluteca, CAS 152984, MSUM 4649–53; El Despoblado, CAS 152959–63; El Faro, UNAH (1); El Madreal, USNM 573916; El Ojochal, USNM 573917; La Fortunita, SDSNH 72725; Ola, TCWC 20813 (skeleton); Pespire, FMNH 5191–93. **COLÓN**: Cerro Calentura, CM 64690, LSUMZ 22454; Laguna de Guaimoreto, LSUMZ 10266–67; Puerto Castilla, LSUMZ 22475–78; 0.5–1.5 km W of Trujillo, LACM 48416–17, LSUMZ 21659–61; 1.6–4.8 km E of Trujillo, LSUMZ 22445–47, 22495; 2 km E of Trujillo, LACM 48418, LSUMZ 21662. **COMAYAGUA**: 4.8 km SSE of Comayagua, LACM 48424; 1 km SW of Comayagua, UF 41538; Lo de Reina, UTA R-41271–73. **CORTÉS**: Amapa, LACM 67359; near Cofradía, USNM 128086–87; Hacienda Santa Ana, FMNH 5156–85; 9.7 km S of La Lima, CM 41226, 41229–32; 1.6 km W of La Lima, TCWC 19198; La Lima, BYU 22571–85, KU 299821, LACM 48419–22; Laguna Ticamaya, FMNH 5186; near Puer-

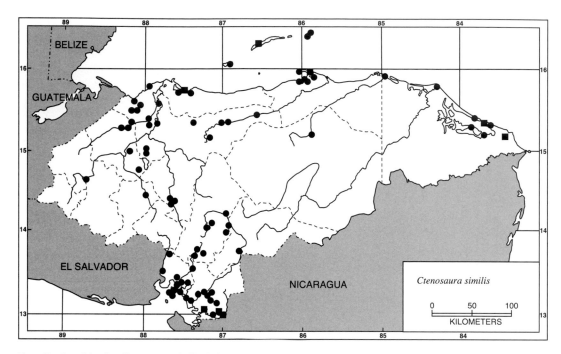

Map 43.   Localities for *Ctenosaura similis.* Solid circles denote specimens examined and solid squares represent accepted records.

to Cortés, USNM 69394; Puerto Cortés, FMNH 5190, TCWC 19196; San Luís, LSUMZ 33439; 3.2–4.8 km W of San Pedro Sula, LACM 48423, TCWC 19197; 9.7 km N of San Pedro Sula, UF 41536–37, 41562; W of San Pedro Sula, FMNH 5188–89, MCZ R29389–90; San Pedro Sula, LSUMZ 70512, USNM 24375–77. **EL PARAÍSO**: El Rodeo, UNAH 5657, 5662, 5664, 5678, USNM 579863. **FRANCISCO MORAZÁN**: Cantarranas, ANSP 28156; near Caserío Los Encinitos, UNAH (1); El Zamorano, AMNH 70342–46; La Venta, FMNH 5187; Los Corralitos, ZMB 73631; Río Yeguare, MCZ R49907–08; 21 km SSW of Sabanagrande, LSUMZ 24603; Tegucigalpa, MCZ R49927–28. **GRACIAS A DIOS**: Barra Patuca, USNM 20290–96; Cauquira, USNM 573929–34; Palacios, BMNH 1985.1264; Puerto Lempira, USNM 573326; Tánsin, LACM 48456, USNM 573325, 573935, ZMB 76328–30; Yahurabila, USNM 573936–37. **INTIBUCÁ**: 2.7 km

N of Jesús de Otoro, LSUMZ 33689, 34039; Valle de Otoro, SMF 77563. **ISLAS DE LA BAHÍA**: Isla de Guanaja, El Bight, USNM 573920–28; Isla de Guanaja, SE shore opposite Guanaja, LSUMZ 22406–09, UF 28575–77; Isla de Guanaja, La Playa Hotel, LACM 48412–14, 48457 (listed as 48458 by Meyer and Wilson, 1973), LSUMZ 21655; Isla de Guanaja, Northeast Bight, SMF 76000; Isla de Guanaja, 8.8 km W of Savannah Bight, TCWC 21954; Isla de Guanaja, near Savannah Bight, LSUMZ 22410–11, UF 28581–89, USNM 573913, 573919; "Isla de Guanaja," CM 27611, KU 101438, LSUMZ 10264–65; Isla de Utila, Pumpkin Hill, UNAH 5291; Isla de Utila, east coast near Trade Winds, USNM 581893; Isla de Utila, near Utila, LSUMZ 22276–77, 22287–92, 22299–302, UF 28392–95, 28415–29, 28445–51; "Isla de Utila," CM 64685; "Islas de la Bahía, Knob Hill," TCWC 66848–49. **LEMPIRA**: Copán-Lempira border, FMNH 40866.

OLANCHO: Las Trojas, UTA R-41274–76, 45184; San Esteban, LACM 138130. **SANTA BÁRBARA**: 33.0 km W of Chamelecón, LACM 48458; La Canadá, UTA R-41269–70; 4 km SW of Quimistán, KU 67222; near Quimistán, USNM 128094–97; 30 km SSE of Santa Bárbara, UF 41539. **VALLE**: near Amapala, USNM 579558–59; near Aramecina, FN 256975 (still in Honduras because of permit problems); Coyolitos, FN 256893–95 (still in Honduras because of permit problems); 7 km E of El Salvador border, MCZ R139421 (skeleton), 154741 (skeleton), 65388–89; Isla Comandante, USNM 580372; Isla Conejo, USNM 580369–70; Isla de Las Almejas, USNM 580371; Isla de Pájaros, USNM 580390; Isla Inglasera, USNM 580368; Isla Sirena, USNM 580391, 581892; Isla Zacate Grande, MSUM 4654–55, 4657–59, USNM 243391 (formerly MSUM 4656); 4.2 km W of and 27.5 km S of Jícaro Galán, TCWC 61765; La Laguna, SDSNH 72724; 10 km WNW of Nacaome, KU 67218–21; Playona Exposición, USNM 579556–57, 580813; Punta Novillo, USNM 580373; San Lorenzo, FMNH 5194. **YORO**: 0.5 km N of Coyoles, LACM 48415, LSUMZ 21656; El Progreso, MCZ R21742 (skeleton), UMMZ 58374; Rancho San Lorenzo, FMNH 5194, LSUMZ 21657–58; 4.7 km ESE of San Lorenzo Arriba, USNM 573915, 573918; 18.8 km N of Yoro-El Progreso road on road to El Ocotal, USNM 573914; near Yoro, MCZ R29401–03. "HONDURAS": ANSP 8189, BYU 39457, CM 27619, UF 63272 (skeleton), 67489 (skeleton), 67496 (skeleton), 67525 (skeleton), 67707 (skeleton), 67724 (skeleton), 69028 (skeleton), USNM 17799–803.

Other Records (Map 43).—**ATLÁNTIDA**: Cocolito, Miami, Puerto Carib, Tornabé (Pasachnik, 2006). **CHOLUTECA**: El Triunfo, La Bonanza (Pasachnik, 2006); vicinity of El Triunfo, vicinity of San Bernardo (Lovich et al., 2006). **COLÓN**: Casa Kiwi, Sambo Creek, San Antonio (Pasachnik, 2006). **CORTÉS**: El Paraíso

(Townsend, 2006a). **GRACIAS A DIOS**: Canco (personal sight record); Prumnitara (personal sight record). **ISLAS DE LA BAHÍA**: Isla de Roatán, W of Palmetto Bay (personal sight records of several juveniles, September 2012). **VALLE**: Amapala, ZMH 3546 (Bailey, 1928); Playa Blanca, San Juan Pali (Pasachnik, 2006). **YORO**: Agua Caliente (= Poligono Air Force Base near San Patricio), Arenal (Pasachnik, 2006). "HONDURAS": MTKD-D 3143, UNAH 1052, 1054–58, 1724, 1791, 1925, ZFMK 21490–92, 25607 (Köhler, Schroth, and Streit, 2000).

Genus *Iguana* Laurenti, 1768

Iguana Laurenti, 1768: 47 (type species: *Iguana tuberculata* Laurenti, 1768: 49 [= *Lacerta iguana* Linnaeus, 1758: 206], by subsequent designation of Stejneger, 1936: 136, and by tautonymy).

Geographic Distribution and Content.— This genus ranges from Sinaloa and extreme southern Tamaulipas, Mexico, southward to Paraguay and southeastern Brazil east of the Andes and extreme northwestern Peru west of the Andes. It also occurs on several Pacific coastal islands and on many islands in the Caribbean off the coast of Central and South America and from the Lesser Antilles to the Virgin Islands and Puerto Rico (some island populations represent human-aided introductions). One species is introduced into southern Florida and Hawaii, USA, the Dominican Republic, several islands in the Bahamas, Grand Cayman, Turks and Caicos, Fiji, and the Canary Islands. Two named species are recognized (but see Remarks), one of which occurs in Honduras.

Remarks.—Hollingsworth (2004) summarized the previously published studies addressing the intergeneric relationships of *Iguana* within the Iguanidae. Malone and Davis (2004), based on patterns of molecular variation and geographic distribution, found evidence that *Iguana iguana* as presently understood, possibly represents more than one species.

Etymology.—The family name *Iguana* might be formed from Igoana, a name used by "Les habitans de Saint-Domingue, d'après Hernandez" (A. M. C. Duméril and Bibron, 1837: 199).

Iguana iguana (Linnaeus, 1758)

Lacerta igvana Linnaeus, 1758: 206 (two syntypes, NHRM 114, ZMUU Linnaean collection 10 [see Avila-Pires, 1995: 40]; type locality: "Indiis" [restricted to "the confluence of the Cottica River and the Perica Creek, Surinam" by Hoogmoed, 1973: 44, who stated that the NHRM syntype probably came from that locality]). See Remarks.

Iguana nudicollis: Cope, 1870: 159.

Iguana delicatissima: Boulenger, 1885b: 191; Barbour, 1914: 297; Barbour, 1930b: 85; Barbour, 1935: 105; Barbour, 1937: 116.

Iguana tuberculata var. *rhinolopha*: Werner, 1896: 346.

Iguana iguana: Van Denburgh, 1898: 461 (by inference); Meyer, 1969: 249; Lazell, 1973: 7; Meyer and Wilson, 1973: 25; Wilson and Hahn, 1973: 116; Cruz Díaz, 1978: 28; Wilson et al., 1979a: 25; Iverson, 1980: 93; Hudson, 1981: 377; Etheridge, 1982: 31; Klein, 1982: 302 (in part); Morgan, 1985: 43; Schwartz and Henderson, 1988: 132; Schwartz and Henderson, 1991: 419; Köhler, 1996d: 20; Köhler, 1997b: 14; Köhler, 1997e: 20; Köhler, 1998d: 375, 382; Köhler, 1998e: 9; Monzel, 1998: 159; Wilson and McCranie, 1998: 16; Köhler, 1999a: 214; Köhler, McCranie, and Nicholson, 2000: 425; Lundberg, 2000: 8; Nicholson et al., 2000: 30; Grismer et al., 2001: 134; Wilson et al., 2001: 135; Castañeda, 2002: 39; Lundberg, 2002a: 5; Lundberg, 2002b: 7; McCranie, Castañeda, and Nicholson, 2002: 28; McCranie et al., 2005: 98; Castañeda and Marineros, 2006: 3.8; Lovich et al., 2006: 15; McCranie et al., 2006: 115; Pasachnik, 2006: 266; Townsend, 2006a: 35; Wilson and Townsend, 2006: 105; Townsend et al., 2007: 10; Wilson and Townsend, 2007: 145; Ferrari, 2008: A14, *In* Anonymous, 2008; Henderson and Powell, 2009: 134; Townsend and Wilson, 2010b: 692; Powell and Henderson, 2012: 92; Townsend et al., 2012: 100; McCranie, Valdés Orellana, and Gutsche, 2013: 288; McCranie and Valdés Orellana, 2014: 45; Solís et al., 2014: 131; McCranie, 2015a: 369; McCranie and Gutsche, 2016: 871; McCranie et al., 2017: 274.

Iguana iguana rhinolopha: Barbour, 1928: 56; Dunn and Emlen, 1932: 28; Dunn, 1934: 3; Barbour, 1935: 105; Barbour, 1937: 116; P. W. Smith, 1950: 55; Köhler, 1996a: 101; Gravendyck et al., 1998: 35; Köhler, 1998e: 13; Köhler, 2000: 80; Lundberg, 2000: 4; Lundberg, 2001: 24; Köhler, 2003a: 134; Ferrari, 2008: 14A, *In* Anonymous, 2008.

Iguana iguana iguana: Dunn, 1934: 3.

Iguana iguana var. *rhinolopha*: O'Shea, 1986: 39.

Iguana iguana rhinovola [sic]: Gravendyck et al., 1998: 35.

Geographic Distribution.—See the geographical distribution for *Iguana*, which is essentially the same as that of *Iguana iguana*. In Honduras, this species occurs across much of the mainland. It also occurs on the islands Barbareta, Guanaja, Roatán, Utila, Cayo Cochino Menor, the Swan Islands, and a few islands in the Golfo de Fonseca.

Description.—The following is based on ten males (CM 27617, 28998, 64718; USNM 142274–75, 145592–93, 163164–65, 573357) and three females (ANSP 11984; CM 29005; USNM 494799). *Iguana iguana* is a huge lizard (maximum recorded SVL 550 mm [Köhler, 2008; males significantly larger than females]; 381 mm SVL in largest Honduran specimen [USNM 145592, a male]) with a long tail; 3 snout scales raised, broadly pointed; snout broadly rounded in dorsal aspect and in profile; 6 postrostrals; 3 internasals; 3 canthals, anteriormost largest; 3 scales between second canthals; no medial scales between supraorbital semicircles; interparietal scale larger than surrounding scales, parietal eye distinct; 0 or 1 scale between interparietal and supraorbital semicircles; nasal single, nostril opening situated posteriorly in scale, directed laterally; nasal scale contacts rostral; moveable eyelid present; pupil circular; 7 supralabials and 9 infralabials to level below mideye; 6 suboculars, anteriormost largest; 2 scale rows separating subocular row from supralabials; large, circular scale located ventral to tympanum, circular scale/TYML 1.23–2.61 in males, 1.50–1.89 in females; tympanum usually higher than long, TYML/TYMH 0.76–1.07 in males, 0.69–0.87 in females; 3 postmentals; large, pendulous dewlap present, 9–15 (12.6 ± 1.9) large, pointed scales along anterior border in eight males; dorsal body scales granular with numerous enlarged, slightly pointed scales on nape, small, keeled middorsal scales present from level above axilla posteriorly; middorsal crest scales strongly compressed, forming high crest in males (crest scales >14 mm high), with crest scales highest in nuchal area; 47–58 (52.7 ± 3.8) strongly enlarged middorsal scales between nuchal and sacral areas in 12; 5–11 (8.4 ± 2.5) granular middorsal scales between first differentiated middorsal crest scale and posterior end of head in eight; no scales between ultimate differentiated middorsal crest scale and first undifferentiated caudal scale; ventral scales flat, not keeled, larger than dorsal scales; 29–41 (36.3 ± 3.3) subdigital scales on Digit IV of hind limb on 18 sides; 12–17 (14.2 ± 1.3) femoral pores per side in males, 12–14 (13.3 ± 0.8) in females; 26–33 (28.3 ± 2.5) total femoral pores in males, 25–28 (26.7 ± 1.5) in females; femoral pores larger in males than in females; caudal scales undifferentiated, formed by small scales of equal size, caudal

scales slightly increasing in size on about distal third of tail; SVL 217–381 (326.2 ± 55.5) mm in males, 211–267 (231.3 ± 31.0) mm in females; HW/SVL 0.10–0.13 in males, 0.10–0.12 in females; HL/SVL 0.17–0.19 in males, 0.15–0.16 in females; SL/SVL 0.07–0.10 in males, 0.07–0.09 in females; SHL/SVL 0.22–0.26 in males, 0.23–0.26 in females; TAL/SVL 1.77–2.87 in eight males, 2.28–2.75 in females.

Color in life of a young adult male (USNM 573357; Plate 56): dorsal surface of body Parrot Green (160) with black mottling; some tubercles on neck region black; dorsal crest scales vary from Parrot Green to brown; lateral vertical stripes on body bluish green; dorsal surface of head Parrot Green with some black spots anteriorly, snout tubercles black; dorsal surface of tail Parrot Green anteriorly, changing to Lime Green (159) at midlength, black crossbands also present on tail; dorsal surfaces of fore- and hind limb Parrot Green; ventral surfaces of head, body, and tail Yellow-Green (58); dewlap mottled Yellow-Green, black, and brown; iris brown with gold rim.

Color in alcohol: dorsal ground color of head and body uniform dark gray; middorsal crest scales gray, most with slightly pale brown bases; dorsal surface of tail dark gray anteriorly, becoming brown on distal third; ventral surfaces of head and body dark gray; pointed dewlap scales with pale brown to white tips; subcaudal surface dark gray anteriorly, becoming brown on distal third.

Diagnosis/Similar Species.—*Iguana iguana* differs from all remaining Honduran lizards in having a large circular subtympanic scale. It also has a huge adult size and a green coloration. Juvenile *Iguana* are also bright green, unlike the majority of the remaining Honduran lizards.

Illustrations (Figs. 92, 93; Plate 56).— Álvarez del Toro, 1983 (adult, juvenile); Álvarez Solórzano and González Escamilla, 1987 (adult, juvenile); Avila-Pires, 1995 (adult, head scales, foot scales); Barbour,

Plate 56.   *Iguana iguana*. USNM 573357, subadult male, SVL = 242 mm. Gracias a Dios: Urus Tingni Kiamp.

1926 (adult); Binns, 2003 (adult); Böhme, 1988 (hemipenis); Burghardt and Rand, 1982 (adult); Campbell, 1998 (juvenile); Dowling and Duellman, 1978 (hemipenis); Dunn, 1934 (head; as *I. i. iguana*); Duvernoy, 1836–1849 (adult, skull; as *Lacerta* Linnaeus, 1758: 200); Guyer and Donnelly, 2005 (adult head, juvenile head); Hoogmoed, 1973 (adult, juvenile head scales); Köhler, 1993b (adult, juvenile, hatchling), 1996a (adult), 1997c (adult), 1998e (adult, subadult, juvenile, head scales), 1999b (adult), 2000 (adult, subadult), 2001b (adult), 2003a (adult, subadult), 2008 (adult, juvenile); Köhler et al., 2005 (adult); Lazell, 1973 (head, some color pattern variation); Lee, 1996 (adult, juvenile, head scales), 2000 (adult, head scales); Lundberg, 2001 (adult), 2002a (juvenile); McCranie et al., 2005 (adult, subadult, head scales), 2006 (adult, head scales); Moyne, 1938 (adult); Murphy, 1997 (subadult); Pianka and Vitt, 2003 (juvenile); Savage, 2002 (adult, juvenile); W. Schmidt and Henkel, 1995 (subadult); Stafford, 1991 (adult); Stafford and Meyer, 1999 (adult, subadult); Sunyer, Nicholson et al., 2013 (juvenile head); Villa et al., 1988 (adult); Vitt and Caldwell, 2014 (adult).

Remarks.—Hoogmoed (1973: 152) restricted the type locality of *Lacerta iguana* to "the confluence of the Cottica River and the Perica Creek, Surinam," because he believed one of the two syntypes still extant came from that locality. Lazell (1973: 7) erroneously stated that no type specimens of *I. iguana* existed and then restricted the species' type locality to "the island of Terre de Haut, Les Illes des Saintes, Department de La Guadeloupe, French West Indies." Lazell's (1973: 7) only justification for that type locality restriction was this species "is extremely abundant on this island." Thus, Lazell's type locality restriction is invalid. Contrary to the statement in Etheridge (1982: 9), Lazell (1973) did not designate a neotype of *Lacerta iguana* Linnaeus. Lazell (1973) reviewed the morphological variation of this species in the Lesser Antilles. However, no similar thorough analysis exists for this species throughout the remainder of its geographic distribution. Malone and Davis (2004) presented evidence that possibly more than one species might be consumed within *I. iguana* as presently understood.

Fitch et al. (1982) discussed illegal trade in Honduran specimens of this species and *Ctenosaura similis* during the 1970s. Luxmoore et al. (1988) summarized data on threats to, and commercial exports of Honduran populations of this species. Seidel and Franz (1994) thought the population of *Iguana iguana* introduced on Grand Cayman Island might have originated from a Honduran population.

Molecular studies of *Iguana iguana* from the Swan Islands are needed to determine the geographic origin of those island populations. However, the placement of the species on CITES hampers any such work on the Swan Island *Iguana* that could be helpful toward conservation efforts. Swan Island *Iguana* are notable for their huge sizes and their habit of passively floating in shallow salt water near the island shores. I saw several Big Swan Island adult *Iguana* deliberately and casually entering the sea while I was sitting and observing several *Iguana* with binoculars. Dunn (1934: 3) reported two "subspecies" of *Iguana*, *I. i. iguana* and *I. i. rhinolopha* Wiegmann (1834b: 44) from the Swan Islands. Cope (1870: 159) reported *I. nudicollis* Merrem (1820: 48; Cope erroneously credited "Cuvier"), a synonym of *I. delicatissima* Laurenti (1768: 48) from the Swan Islands. Dunn (1934: 4) speculated that the Swan Island populations were introductions "(by whatever agency) from the nearest mainland." In 2007, while on Guanaja Island, I was told of a yacht owner bringing a sack full of *Iguana* from the Swan Islands and releasing them on Guanaja. A huge adult male hanging around a bar between Savannah Bight and Guanaja (town) was said to be one of those released iguanas.

Natural History Comments.—*Iguana iguana* is known from near sea level to 800 m elevation in the Lowland Moist Forest, Lowland Dry Forest, Lowland Arid Forest, Premontane Moist Forest, Premontane Dry Forest, and Lowland Dry Forest (West Indian Subregion) formations. This species is diurnal and mostly arboreal as an adult, with juveniles being largely terrestrial and/or climbing in low vegetation. Mainland adults are (or were in some cases) most commonly seen in trees or other vegetation along rivers and streams into which they frequently dive when approached. Adults are also sometimes seen in rocky areas overlooking rivers. Juveniles are frequently seen on banks, logs, or low vegetation overlooking rivers and streams. This species can also be seen at night sleeping in trees near rivers and streams. It can also occur in areas where no surface freshwater occurs (i.e., Swan Islands). Adult and subadult iguanas on the Swan Islands were seen to voluntarily enter the sea and seemingly to enjoy floating in the seawater near shore. That passive floating in seawater might explain some island colonization of *Iguana* in the past. The Swan Island population was seen on boulders jutting from the sea, on boulders connected to land, in grassy areas surrounding boulders, in trees, and on concrete walls of old and abandoned buildings in the interior of the large island. Several subadults also basked daily on the roof of the building inhabited by the Honduran naval personnel. Weigel (1973) mentioned iguanas on Isla Grande, in the Swan Islands, reaching 4–5 feet long, and their habit of jumping from cliffs into the sea when surprised. Montgomery et al. (2015: 280) said *Iguana* populations on Cayo Cochino Menor were expanding in size and appeared to be becoming a threat to the *Ctenosaura melanosterna* populations on the island, but they offered no supporting evidence. This species appears to be active under sunny conditions throughout the year. Klein (1982) studied reproduction in several populations from southern Honduras and concluded that egg deposition occurred in January and February with egg clutches varying from 19–54 eggs. Others were reported to deposit 10–86 eggs

(Köhler, 1997f) or 9–71 (Köhler, 1998e) in a single clutch per year and Wiewandt (1982 and references therein) recorded clutch sizes of 14–76 eggs. I have witnessed *I. iguana* depositing eggs in burrows they dig on the sandy beaches in April along the Río Rus Rus, Gracias a Dios, in northeastern Honduras. Those sandy river beaches are exposed during the dry season but are usually inundated during the extended rainy season. The nesting season of those populations is said by the local people to extend from February to March. Elsewhere, free-living iguanas have also been reported to deposit eggs in burrows and tunnels during the dry season (Harris, 1982b; Rand and Greene, 1982; Köhler, 2008; Henderson and Powell, 2009, and references cited in those works). Platt et al. (2010) reported *I. iguana* depositing eggs in crocodile nests in Belize. I have seen several nests of *Iguana* eggs in sandy beaches along small rivers in northeastern Honduras that were ruined by unseasonal high water from heavy rains. Free-living iguanas are primarily vegetarians as adults but also will eat insects, bird eggs, carrion, and mice in captivity (Van Devender, 1982; Köhler, 1998e; Henderson and Powell, 2009, and references cited in those works). Govender et al. (2012) also reported introduced *I. iguana* on Puerto Rico feeding on crabs (*Uca* sp.), and Anderson and Enge (2012) also reported this species feeding on a deer carcass (*Odocoileus virginianus*) in Florida, USA. Barrio-Amorós and Ojeda (2015) reported predation of an adult *Iguana* in Costa Rica by young tayras (*Eira barbara* [Linnaeus]). Telford (1977) reported saurian malaria in specimens of this species from "Honduras." Gravendyck et al. (1998) reported reoviruses and paramyxoviruses in *I. iguana* on the Bay Islands.

Some people in Honduras eat iguanas and their eggs. Adults and their eggs, were until recently, so abundant along the Río Rus Rus in March that people from Nicaragua regularly came there during that month to hunt them. However, Honduran soldiers were temporarily posted in Rus Rus in March 2004 to discourage that activity, but their effectiveness in slowing down the slaughter of iguanas seems to have been negligible. It appeared the soldiers made little or no effort to stop the influx of people, nor did they patrol the river beaches (personal observation) during a 2-week stay I made in the area in March 2004. Only the human destruction of those populations seems to have slowed down that human influx into the area. I have also witnessed people shooting basking iguanas with 22 rifles from boats along several rivers in northeastern Honduras. That killing for food has apparently seriously reduced the *Iguana* population along at least one river, the Río Warunta, as during 6 and 8 sunny days in a boat along that river in March 2011 and July 2015, respectively, did not result in the spotting of a single adult *Iguana*. The *Iguana* population on the Swan Islands is also hunted for a food source. In December 2012, we were given a ride on a Puerto Lempira lobster boat from Big Swan Island to Little Swan Island and saw several mesh sacks containing about 20 live iguanas on the boat (see McCranie et al., 2017). We also witnessed two of those iguanas being prepared for the day's meal. As lobster boats regularly visit the environs of the Swan Islands, that taking of iguanas is likely a regular occurrence that is ignored by the Naval Base personnel on Big Swan.

Etymology.—See the Etymology for the genus *Iguana*.

Specimens Examined (91, 4 skeletons [20]; Maps 5, 44).—**ATLÁNTIDA**: La Ceiba, CM 29005, INHS 4485; Tela, USNM 84249. **CHOLUTECA**: 26.4 km NNE of Cedeño, LSUMZ 33666; near Cedeño, UF 41559; Cedeño, BYU 18289–300; El Faro, UNAH (1). **COLÓN**: Barranco, ANSP 24190; Salamá, USNM 242601–05; 2 km E of Trujillo, LSUMZ 21507; about 12.9 km NE of Trujillo, LSUMZ 22502. **COPÁN**: Copán, FMNH 28531,

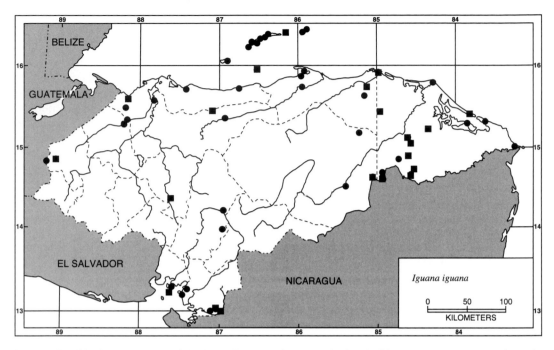

Map 44.   Localities for *Iguana iguana*. Solid circles denote specimens examined and solid squares represent accepted records. *Iguana iguana* also occurs on the Swan Islands (Map 5).

TCWC 23634, UMMZ 83033 (2). **CORTÉS**: near Cofradía, USNM 128085; Hacienda Santa Ana, FMNH 5153; Laguna Ticamaya, FMNH 5152, 5154–55. **FRANCISCO MORAZÁN**: Cantarranas, ANSP 24191–93; El Zamorano, MCZ R49937. **GRACIAS A DIOS**: Barra Patuca, USNM 20287–89; Caño Awawás, UNAH 5550; Cauquira, UNAH; Swan Islands, Isla Grande, USNM 142274–75, 145592–93, 494799–800; Swan Islands, no other data, ANSP 11984, FMNH 34674–76, MCZ R21695–98, 32243 (skeleton), 73905, USNM 76938, 163164–65; Río Coco, USNM 24508–11; Tánsin, LSUMZ 21506; Urus Tingni Kiamp, USNM 573357. **ISLAS DE LA BAHÍA**: "Isla de Guanaja," CM 27617; Isla de Roatán, near Coxen Hole, FMNH 34594–96; Isla de Roatán, Diamond Rock, UTA R-55222; Isla de Roatán, near Oak Ridge, UTA R-31234; Isla de Roatán, about 3 km N of Roatán, LACM 47849; Isla de Roatán, about 1.6 km N of Roatán, CM

64718; "Isla de Roatán," MCZ R32246–47 (skeletons); Isla de Utila, Utila, CM 28998, LSUMZ 22294. **OLANCHO**: Quebrada El Guásimo, UNAH (1); confluence of ríos Aner and Wampú, UNAH 5534–35. **SANTA BÁRBARA**: 12.3 km ENE of Quimistán, KU 67234–36. **VALLE**: Isla de La Vaca, UNAH 5653; Isla Zacate Grande, Coyolitos, UNAH (1). **YORO**: San Patricio, USNM 573323. "HONDURAS": ANSP 8087, 8089, BYU 39456, UF 99098 (skeleton), USNM 58667.

Other Records (Map 44).—**CHOLUTECA**: vicinity of El Triunfo (Lovich et al., 2006); La Bonanza (Pasachnik, 2006). **COLÓN**: Río Guaraska (O'Shea, 1986). **CORTÉS**: El Paraíso (Townsend, 2006a). **COMAYAGUA**: Comayagua Valley (Wilson and McCranie, 1998). **COPÁN**: Río Amarillo (Castañeda and Marineros, 2006). **GRACIAS A DIOS**: Bachi Kiamp, USNM Herp Image 2719; Caño Awalwás (personal sight record); Coco, USNM Herp Image

2718; Crique Wahatingni (personal sight record); Kakamuklaya (personal sight record); Palacios (O'Shea, 1986); Raudal Kiplatara, UNAH 5597 (Cruz Díaz, 1978, specimen now lost); Rawa Kiamp, USNM Herp Image 2720; Rus Rus (personal sight records); Urus Tingni Kiamp, USNM Herp Image 2721; Yahurabila (personal sight record). **ISLAS DE LA BAHÍA**: Cayo Cochino Menor (Lundberg, 2002a); Isla Barbareta (Grismer et al., 2001); Isla de Utila (Köhler, 1996d). **OLANCHO**: Quebrada Kuilma (personal sight record); ríos Patuca and Cuyamel (personal sight records). **VALLE**: Isla del Tigre (Lovich et al., 2006). **YORO**: San José de Texíguat (Townsend et al., 2012). "HONDURAS": (Werner, 1896).

Family Leiocephalidae Frost and Etheridge, 1989

This family (*sensu* Frost et al., 2001) of lizards occurs only in the Western Hemisphere, where it is found on the Cayman and Bahama Islands, Cuba, Hispaniola, and associated banks. It is also introduced and established in southern Florida, USA, and on the Swan Islands, Honduras. One genus comprising 25 named species is included in this family, with one introduced species occurring in Honduran territory (on Isla Grande in the Swan Islands).

Genus *Leiocephalus* Gray, 1827b

Leiocephalus Gray, 1827b: 207 (type species: *Leiocephalus carinatus* Gray, 1827b: 208, by monotypy).

Geographic Distribution and Content.— See statements for the family Leiocephalidae.

Remarks.—The method of counting scales follows Pregill (1992), with the exception of a few characters not used by him. Males of the single Honduran species are much larger than are females.

Etymology.—The generic name *Leiocephalus* is formed from the Greek *leios*

(smooth) and *kephalaios* (of the head) and refers to the smooth cephalic scales of the type species of the genus.

Leiocephalus varius Garman, 1887a

Liocephalus [sic] *varius* Garman, 1887a: 274 (syntypes, three under MCZ R6023 [see Barbour and Loveridge, 1929a: 296], one of which is now USNM 52405 [see Cochran, 1961: 124]; type locality not given, but "Grand Cayman" given in title and introduction to publication where description was made and "Grand Cayman Island" given by Garman, 1887b: 49).

Leiocephalus varius: Barbour, 1914: 300; Powell and Henderson, 2012: 92; McCranie, 2015a: 369; McCranie et al., 2017: 274.

Leiocephalus carinatus varius: Schwartz and Thomas, 1975: 128; MacLean et al., 1977: 4; Morgan, 1985: 43; Schwartz and Henderson, 1988: 135; Schwartz and Henderson, 1991: 424; Powell, 2004: 153; Henderson and Powell, 2009: 138.

Sceloporus spp.: Ferrari, 2008: A14, *In* Anonymous, 2008.

Leiocephalus carinatus: Solís et al., 2014: 131; Köhler, Rodríguez Bobadilla, and Hedges, 2016: 532.

Geographic Distribution.—*Leiocephalus varius* occurs on Grand Cayman Island. It is also introduced and established in Honduran territory on Isla Grande in the Swan Islands.

Description.—The following is based on four males (SMF 90447; USNM 494804, 494806, 494808) and ten females (MCZ R191132, 191158; SMF 90442, 90444; USNM 494801–03, 494805, 494807, 570244). *Leiocephalus varius* is a moderate-sized lizard (maximum recorded SVL 101 mm [USNM 494806, a male]) with a long tail and relatively short limbs; dorsal head scales smooth, except weak ridges usually present on posterior portions of parietals, with numerous tiny pits present in all scales; snout scale pattern Type II and parietal scale pattern Type III (Pregill, 1992); 2 internasal scales, both broadly contacting rostral and each other; 3 rows of scales between internasals and anterior frontals, including single pair of anterior frontal scales; single pair of posterior frontals; paired parietal scales, inner pair in contact medially posterior to interparietal; interparietal scale well defined; parietal eye distinct; lateral postparietal scales absent; 5–7 (usually 6; seventh when present, much smaller than others) supraocular scales in single row; cephalic scale ridge absent, except ridge weakly developed on posterior head scales; 3–7 (usually 4) loreals; fifth supralabial and infralabial below level of mideye; moveable eyelid present; pupil circular; 1–4 enlarged preauricular scales; 3–5 (usually 4) lorilabial scales anterior to enlarged subocular; 1 temporal scale enlarged or only slightly enlarged; elongated temporal scales absent; lateral neck scales keeled, undifferentiated; lateral neck fold moderate; antegular scale fringe absent; antehumeral fold moderately developed; body dorsally compressed; nuchal fold moderately convex; dorsal body scales strongly keeled, slightly mucronate, imbricate, in long rows, dorsal keels in line, arching inward posteriorly on body; 52–60 (56.6 ± 2.5) dorsal crest scales between occipital and level above cloaca; middorsal crest scales slightly raised; lateral fold absent on body; lateral body scales keeled, slightly mucronate, imbricate, subequal, only slightly smaller than dorsal scales; ventral scales smooth, imbricate, with multinotched posterior ends, 42–55 (47.4 ± 3.9) para-midventral scales between level of anterior edge of forelimb and cloacal scale; 38–49 (43.4 ± 3.2) scales around midbody; subdigital scales of median portions of digits not larger than those of remainder of digits; tricarinate scales of digits I–II of hind limb not greatly enlarged into combs, fringe weakly developed; 21–23 (22.3 ± 0.7) tricarinate subdigital scales on 27 sides of

Digit IV of hind limb; femoral pores absent; enlarged postcloacal scales present or absent in males; postcloacal escutcheons absent; tail slightly laterally compressed; SVL 82–101 (92.8 ± 8.5) mm in males, 61–89 (76.0 ± 10.1) mm in females; TAL/SVL 1.26–1.56 in three males, 1.26–1.60 in seven females; HL/SVL 0.25–0.26 in males, 0.22–0.27 in females; HW/SVL 0.18–0.20 in males, 0.17–0.19 in females; SHL/SVL 0.26–0.28 in males, 0.23–0.28 in females; S-OL/SVL 0.22 in all males, 0.21–0.24 in females.

Color in life of an adult female (MCZ R191122): middorsal ground color Pale Cinnamon (color 55 in Köhler, 2012) with Raw Umber (23) crossbars on nuchal area and anterior third of body and Citrine (119) crossbars on posterior two-thirds of body; lateral surface of body Pale Cinnamon with Glaucous (189) mottling and long lateral broad stripe; top of head Tawny Olive (17) with Raw Umber mottling; side of head below eye dirty white with small dark brown subocular spot; tail Pale Cinnamon with indistinct reddish brown lateral bars on about anterior half, posterior half becoming pale reddish brown with Raw Umber crossbands; belly Pale Cinnamon; venter of head pale dirty white; subcaudal surface Pale Cinnamon with brown mottling on anterior half, reddish brown with Raw Umber crossbands on posterior half; iris golden brown with incomplete pale yellow ring around pupil. Color of life of another adult female (MCZ R191123; Plate 57) was similar in all aspects to that recorded for MCZ R191122.

Color in alcohol: dorsal surfaces of body and head brown, without distinct markings other than dark brown middorsal crest scales forming narrow stripe; nuchal area dirty white with brown mottling suggesting lineate pattern; chin and throat cream with brown mottling forming lineate pattern; belly cream with brown flecking and dark brown line down middle; tail with indistinct brown and slightly darker brown crossband-

Plate 57. *Leiocephalus varius.* MCZ R191123, adult female. Gracias a Dios: Isla Grande, Swan Islands.

ing anteriorly, becoming strongly crossband-ed with pale and dark brown on distal third.

Diagnosis/Similar Species.—The combi-nation of the dorsally compressed body, strongly keeled dorsal scales with keels in longitudinal rows, scale rows in line and arching inward posteriorly, smooth dorsal head scales, and lack of femoral pores will distinguish *Leiocephalus varius* from all other Honduran lizards.

Illustrations (Fig. 61; Plate 57).—McCra-nie et al., 2017 (adult, including front cover); Powell and Henderson, 2003 (adult; as *L. carinatus varius*); Seidel and Franz, 1994 (adult; as *L. carinatus varius*).

Remarks.—This species was relatively recently introduced onto the Swan Islands (Hedges, personal communication regard-ing his molecular results of a Swan Island specimen, August 2016), where it was first collected in 1974. Schwartz and Thomas (1975: 128) allocated the Swan Island specimens to *Leiocephalus carinatus varius*, thus implying that they were introduced from Grand Cayman Island. All Honduran specimens have only 2 internasal scales, both of which are in broad contact with the

rostral scale and each other, in agreement with *L. c. varius* of Grand Cayman Island. Apparently all other subspecies of *L. carinatus* Gray (1827b: 208) have 3 inter-nasals (Pregill, 1992). Grant (1941: 39) provided a detailed description of an adult male and considered *L. varius* to be a species, in part, because of having only 2 internasal scales and by being isolated by deep seas from all other populations of *L. carinatus*. I also recognize *L. varius* as a distinct species because of its diagnostic, non-overlapping characters with the re-maining populations of the *L. carinatus* group and by its isolation of natural populations on Grand Cayman Island (Pow-ell and Henderson, 2012, listed the Swan Island population as *L. varius* based on my suggestion, but without acknowledgment; also see McCranie, 2015a: 369). Tissues for molecular analyses were collected on Big Swan Island in December 2012 and are in the MCZ tissue collection.

Seidel and Franz (1994: 421) erroneously said that Schwartz and Thomas (1975) reported *Leiocephalus carinatus granti* Rabb (1957: 109) from the Swan Islands.

Schwartz and Thomas (1975: 128) actually reported *L. c. varius* from those islands.

Dodd and Franz (1996) suggested that island populations of *Leiocephalus* were more vulnerable to extirpation or extinction by populations of introduced cats than were those of *Ameiva*. That suggestion was based on *Leiocephalus* being less wary than *Ameiva* and having the habit of sleeping at night under accessible leaf litter and rocks. By contrast, *Ameiva* are alert and quick in their movements and sleep at night in burrows. Despite the Dodd and Franz (1996) opinion, the *Ameiva* population on the Swan Islands was extirpated sometime before 1912 with introduced cats having been blamed for that event (Barbour, 1914: 214; also see McCranie and Gotte, 2014: 545). Dodd and Franz (1996) also suggested that cats might have decimated several populations of *Leiocephalus*, including the related *L. carinatus* on small islands in the Bahamas. Despite the presence of a long-haired feral cat population on Big Swan Island (still present in December 2012), *L. varius* remains abundant on that island. However, the entire Swan Island fauna would benefit if that feral cat population were eradicated. The cat eradication techniques discussed by Nogales et al. (2004) could be consulted as a guide.

Natural History Comments.—*Leiocephalus varius* is known from near sea level to 10 m elevation in the Lowland Dry Forest (West Indian Subregion) formation. Specimens were collected on Big Swan Island in February, April, May, and December. This diurnal species was abundant in virtually all open situations and forested areas where sunlight reaches the ground on Isla Grande in December 2012. It was seen from sandy grassy coastal areas to the wooded interior of the island (also see McCranie et al., 2017). Several *Leiocephalus* also freely moved in and out of the occupied buildings used by navy personnel and appeared to have little fear of humans, and at times, one had to step over a stationary lizard to avoid

stepping on it. Other individuals well away from those buildings were also easily approached and easily captured by hand. That *Leiocephalus* population on Big Swan does not frequently curl its tail as has been reported in other species (also McCranie, personal observation). Nothing else has been published on the natural history of *L. varius* on the Swan Islands, but Grant (1941) published detailed notes on this species on Grand Cayman. Grant (1941: 40) said that this diurnal species is "found in open rocky ground preferably near the beach, but [also] sometimes along stone walls or roads [and that] it spends much time on top of rocks in the sun." The species is also known from an elevated bluff habitat on Grand Cayman. It is apparently restricted to isolated colonies, with other favorable beach areas lacking populations (Grant, 1941). English (1912: 599) also said that it "lives on stony ground near the sea" on Grand Cayman. Nothing has been published on reproduction in this species other than "young of the species appear in early August, [but] no eggs were found" (Grant, 1941: 41). Grant (1941: 40) stated the species is "omnivorous, feeding on flowers, particularly of the 'bay-vine'" (*Ipomoea pescaprae*) and also on "insects and lizards."

Etymology.—The name *varius* is Latin meaning "different," apparently referring to the taxon "being different" from its closest allies.

Specimens Examined (36 [0]; Map 5).—**GRACIAS A DIOS**: Isla Grande, Swan Islands, MCZ R191072, 191122–23, 191131–35, 191137–39, 191155–64, SMF 90442–47, USNM 494801–08, 570244.

Family Phrynosomatidae Fitzinger, 1843.

This family (*sensu* Frost and Etheridge, 1989; Frost et al., 2001) of lizards occurs only in the Western Hemisphere, where it is distributed from extreme southern British Columbia, Alberta, and Saskatchewan, Canada, southward through most of the western U.S., excepting coastal Washington and Oregon, northern Idaho, western Montana, and the northern Great Plains, on through most of the eastern U.S. from southern Illinois, southern Indiana, southern Ohio, southern Pennsylvania, and southern New Jersey, excepting the Louisiana Delta and Mississippi River Basin regions and south-central and southwestern coastal Florida. From the U.S., it occurs through Mexico into Central America to eastern Honduras and in Costa Rica on the Atlantic versant and to west-central Panama on the Pacific versant. Nine genera containing about 145 named species belong to this family, with one genus containing five named species occurring in Honduras. Two of those species are named in this work and another, *Sceloporus schmidti*, is redescribed and resurrected from synonymy of *S. smaragdinus*. Other Honduran populations of the *Sceloporus malachiticus* species complex also have an unresolved specific status.

Genus *Sceloporus* Wiegmann, 1828

Sceloporus Wiegmann, 1828: col. 369 (type species: *Sceloporus torquatus* Wiegmann, 1828: col. 369, by subsequent designation of Wiegmann, 1834b: 18).

Geographic Distribution and Content.— This genus ranges from the U.S. Pacific northwest and extreme southern British Columbia, Canada, across most of the U.S. to New Jersey on the Atlantic versant, thence southward throughout most of the U.S., Mexico, and Central America to central Costa Rica and extreme western Panama. The genus contains about 100 named species, five of which occur in Honduras. Two of those are named herein; another is resurrected from synonymy, and around four other Honduran populations appear to need naming according to a recent molecular analysis (E. N. Smith, personal communication, 8 January 2016).

Remarks.—Werner (1896: 346) reported *Sceloporus aeneus* Wiegmann (1828: col.

370) and *S. serrifer* Cope (1866b: 124) from "Honduras," both without further comment. The Werner collection appears to have originated in Honduras and not Belize ("British Honduras"), as is the case in some other old literature. Both Werner species are extralimital to Honduras; thus, Werner's specimens represent misidentifications. The brief description given by Werner (1896) for his specimen of *S. serrifer* fits only that of the *S. malachiticus* species complex of the *S. formosus* group members in Honduras. The *S. aeneus* of Werner (1896) also likely represents a juvenile of the *S. malachiticus* species complex. Both of Werner's names, for convenience, are included in the synonymy of one species of *Sceloporus* I describe below.

Members of the genus *Sceloporus* have frequently been placed into two groups: the small-bodied and small-scaled radiation and the large-bodied and large-scaled radiation (see Sites et al., 1992, for a summation). The Honduran species *S. squamosus* and *S. variabilis* have been placed in two sections of the small-scaled group, whereas the remaining Honduran *Sceloporus* are of the *S. malachiticus* species complex of the *S. formosus* group and are in the large-scaled group. However, Wiens and Reeder (1997) demonstrated that the traditional division based on scale size is not supported by their phylogenetic analysis. Honduran female *Sceloporus* tend to be slightly larger than males.

Pérez-Ramos and Saldaña-de-la Riva (2008) broke the former *S. formosus* group into two species groups: the *S. salvini* group for four former members of the *S. formosus* group plus one species they themselves named. The *S. salvini* group occurs only in Mexico, Guatemala, and apparently El Salvador (but see comments for *S. acanthinus* below). The *S. formosus* group (*sensu stricto*) occurs from Mexico to Panama. All known Honduran species would remain in *S. formosus* group (*sensu stricto*).

The taxonomy of the *Sceloporus mala-chiticus*–like lizards in Central America is in need of much study, using both molecular and morphological data. The species of the *S. formosus* group are morphologically very conservative. The species *S. acanthinus* Bocourt (1873a: 1; misspelled as *S. acathinus*; see intended correct spelling by Bocourt, 1874: 180, *In* A. H. A. Duméril et al., 1870–1909a) and *S. salvini* Günther (1890: 68, *In* Günther, 1885–1902) were transferred to the poorly-defined *S. salvini* group by Pérez-Ramos and Saldaña-de La Riva (2008). The following species of various Central American populations remain in the also poorly-defined *S. formosus* group (*sensu stricto*): *S. lunaei* Bocourt (1873b: 1; although sometimes considered a synonym of *S. acanthinus* [of the *S. salvini* group!] E. N. Smith, personal communication, 8 January 2016); *S. malachiticus* Cope (1865a: 178); *S. smaragdinus* Bocourt (1873b: 1); and *S. taeniocnemis* Cope (1885: 399). Of those Central American species just listed, only *S. malachiticus* and *S. smaragdinus* need to be considered further in this work.

Bell et al. (2003; apparently also Pérez-Ramos and Saldaña-de La Riva, 2008) considered *Sceloporus schmidti* Jones (1927: 4) a junior synonym of *S. smaragdinus* Bocourt, apparently following H. M. Smith (1939). Cope (1875: 572, *In* Yarrow, 1875) created a junior primary homonym when he proposed the name *S. smaragdinus* for a green sceloporine occurring in Nevada and Utah, USA. Cope (1884: 15, 18, 21) also assigned the species name *S. smaragdinus* to several populations in the western U.S.. Cope's *S. smaragdinus* is a partial junior synonym of *S. occidentalis* Baird and Girard (1852: 175).

I agree with E. N. Smith (2001, unpublished dissertation) and an unpublished recent molecular analysis with additional species and populations (E. N. Smith, personal communication, 8 January 2016) and recognize *Sceloporus schmidti* as a valid species, instead of its usual placement in the

synonymy of *S. smaragdinus*. Therefore, a redescription of *S. schmidti* is provided below based on Honduran specimens from near its type locality. I also describe two new species for Honduran populations of the *S. malachiticus* species complex based largely on their unique color patterns combined with their dorsal head scale characters. Additionally, some remaining Honduran populations are discussed below. Further studies of those populations are needed using both molecular and morphological data. Such a study has been planned for about 10 years by E. N. Smith and me, but other projects have continued to delay that project. For those reasons, only two new species are described herein until our molecular and morphological data set with new and key sampling can be completed. Because work is incomplete on some populations, specimens with field numbers (FN, my personal collection) remaining in this work are being kept for the present time in the McCranie collection, awaiting a chance for further study. It is my thought that at least four additional species are recognizable among those Honduran populations, and the recent molecular analysis provided by E. N. Smith (E. N. Smith, personal communication, 8 January 2016) supports that opinion.

The confusing concept of the *Sceloporus malachiticus* species complex in Central America led Stuart (1971: 250) to say that the Cerro La Tigra, Francisco Morazán, Honduras, population was closer morphologically to Costa Rican populations than it was to the geographically much closer La Esperanza, Intibucá, Honduras, populations. The La Esperanza populations are described below as a new species.

Etymology.—The name *Sceloporus* is formed from the Greek words *skelos* (leg) and *poros* (hole, passage), in reference to the conspicuous femoral pores found in both sexes of the members of this genus.

Figure 108.   Dorsolateral pale stripe present on each side. *Sceloporus squamosus*. USNM 580787 from Potrerillos, La Paz.

KEY TO HONDURAN SPECIES OF THE GENUS
SCELOPORUS

1A. Dorsal pattern with dorsolateral pale stripe on each side (Fig. 108) .. 2

1B. Dorsal pattern lacking pale dorsolateral stripes 3

2A. Postfemoral dermal pocket present (Fig. 109); basal subcaudal scales mostly smooth and adjacent ventral scales moderately keeled (Fig. 110); total femoral pores 14–22; enlarged, paired postcloacal scales present in males (Fig. 110) ......
.... *variabilis* (p. 325)

2B. No postfemoral dermal pocket; basal subcaudal scales and adjacent ventral scales strongly keeled (Fig. 111); total femoral pores 6–12; enlarged postcloacal scales absent in males (Fig. 111)..............
.... *squamosus* (p. 321)

3A. Adult male abdominal semeions poorly separated, especially on anterior third of body in life and especially in alcohol (Fig. 112); no, or poorly defined black bars separating male abdominal semeions,

especially in alcohol; adult female venter grayish brown to black in alcohol (occurs in pine forests near La Esperanza, Intibucá) ....
.... **sp. nov. 1** (p. 298)

3B. Adult male abdominal semeions distinctly separated on belly in life and alcohol (Fig. 113), black borders of male abdominal semeions distinct; adult female venter usually pale brown or with greenish tinge in alcohol 4

4A. Anterior section of frontal most often transversally and/or longitudinally divided (Fig. 114); usually no contact between lateral frontonasals, frequently separated medially by smaller scales (Fig. 114), resulting in 5–6 scales between large canthals; dark gray medial ventral longitudinal line usually present, at least posteriorly in Honduran specimens (Fig. 115) (occurs in northwestern Honduras W of Río Ulúa) ..... *schmidti* (p. 311)

4B. Anterior section of frontal usually not divided (Fig. 116); contact usually present between paired

Figure 109.   Postfemoral dermal pocket present (arrow) on each side. *Sceloporus variabilis*. USNM 580776 from Río del Hombre, Francisco Morazán.

lateral frontonasals (Fig. 116), resulting in usually 3–4 scales between large canthals; dark gray medial ventral longitudinal line absent (occurs in eastern Honduras E of Lago de Yojoa region) . . . . **sp. nov. 2** (p. 304)

CLAVE PARA LAS ESPECIES HONDUREÑAS DEL GÉNERO *SCELOPORUS*

1A. Patrón dorsal con una banda dorsolateral longitudinal pálidas en cada lado (Fig. 108) 2

1B. Patrón dorsal sin bandas longitudinales pálidas 3

2A. Una bolsillo dérmico postfemoral presente (Fig. 109); escamas subcaudales en la base de la cola y las escamas ventrales adyacentes moderadamente quilladas (Fig. 110); 14–22 poros femorales en ambas piernas; escamas postcloacales agrandadas presentes en los machos (Fig. 110) . . . . *variabilis* (p. 325)

2B. Sin una bolsillo dérmico postfemoral; escamas subcaudales en la base de la cola y las escamas ventrales adyacente fuertemente quilladas (Fig. 111); 6–12 poros femorales en ambas piernas; escamas postcloacales agrandadas ausentes en los machos (Fig. 111) . . . . *squamosus* (p. 321)

3A. Machos adultos con parches abdominales muy pocos separados, especialmente en el tercis anterior del cuerpo en vida y particularmente en preservación (Fig. 112); poco delineadas o sin franjas negras en los parches ventrales; vientre de hembras adultas negro o gris-pardo en preservación (se

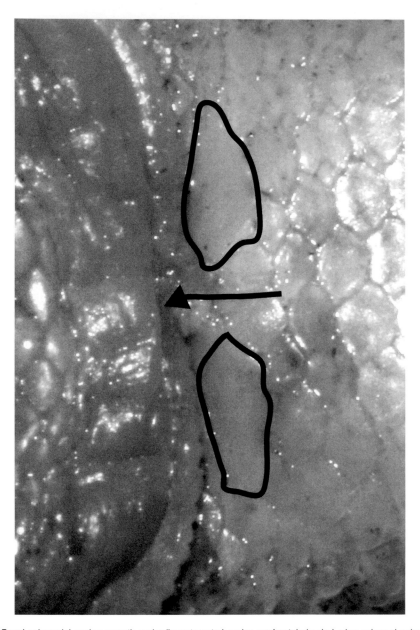

Figure 110.   Basal subcaudal scales smooth and adjacent ventral scales moderately keeled, plus enlarged paired postcloacal scales present in males (outlined), with arrow pointing to posterior edge of cloacal scale. *Sceloporus variabilis*. USNM 580777 from Tigrito Island, Valle.

encuentra solamente en los bosques del pino cerca de La Esperanza, Intibucá)............ .... **sp. nov. 1** (p. 298)

3B. Machos adultos con parches abdominales bien separados en vida y preservación (Fig. 113); franjas negras en bordes de parches bien

Figure 111.   Basal subcaudals and posterior ventrals strongly keeled, plus enlarged postcloacal scales absent. *Sceloporus squamosus.* USNM 578734 from del Tigre Island, Valle.

distinguibles; hembras adultas con un vientre usualmente pardo pálido, o con un tinte verdoso en preservación 4

4A. Sección anterior de la escama frontal muy frecuentemente dividida transversal y/o longitudinalmente (Fig. 114); usualmente sin contacto entre las frontonasales, frecuentemente separadas medialmente por escamas mas pequeñas (Fig. 114), con 5–6 escamas entre

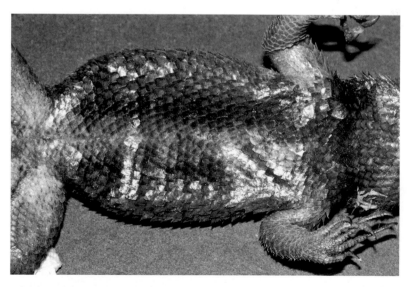

Figure 112.   Male abdominal semeions poorly separated. *Sceloporus esperanzae* sp. nov. USNM 589100 from NW of La Esperanza, Intibucá.

Figure 113.   Male abdominal semeions distinctly separated. Species D of *Sceloporus* "*hondurensis*" (probably not described). FN 11550 (still in my collection for further study) from Quebrada Machín, Colón.

las cantales grandes; línea longitudinal gris oscura usualmente presente medialmente en el vientre, y al menos posteriormente en espécimenes Hondureños (Fig. 115) (se encuentran en el noroeste Honduras oeste del Río Ulúa) ...
.... *schmidti* (p. 311)

4B. Sección anterior de la escama frontal usualmente no esta dividida (Fig. 116); usualmente contacto existe entre el par de laterales frontonasales (Fig. 116), con 3–4 escamas entre las cantales grandes; línea longitudinal gris oscura ausente medialmente en el vientre (se encuentran en el región central y este de Honduras este de la región del Lago de Yojoa) ....
.... **sp. nov. 2** (p. 304)

Sceloporus esperanzae sp. nov.
McCranie, herein

Sceloporus malachiticus: Meyer, 1969: 252 (in part); Meyer and Wilson, 1973: 26 (in part); Wilson et al., 2001: 136 (in part); Wilson and McCranie, 2004b: 43 (in part); Wilson and Townsend, 2007: 145 (in part); Solís et al., 2014: 131 (in part).

Sceloporus "*malachiticus*": Stuart, 1971: 238 (in part).

Sceloporus esperanzae **sp. nov.** McCranie, 2018, herein (holotype, USNM 589098, an adult male; type locality: La Soledád, Intibucá, Honduras, 14°18′N, 88°16′W, 1,760 m elevation, collected 21 June 2012 by James R. McCranie).

Geographic Distribution.—*Sceloporus esperanzae* occurs at moderate and intermediate elevations of the Atlantic and Pacific versant headwaters in the pine forests of Intibucá and La Paz surrounding the vicinity of La Esperanza, Intibucá. Broadleaf forests on peaks above those pine forests have other and apparently undescribed species.

Description.—The following is based on four males (UTA R-63123, 63125; USNM 589098, 589100) and nine females (UF 124600–01; UTA R-63120–22, 63124; USNM 589099, 589101–02). The holotype designated herein is an adult male (USNM 589098), whereas the remaining specimens just listed and used for the following

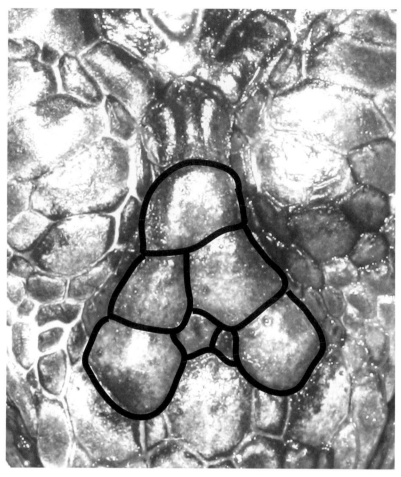

Figure 114.   Dorsal head scales showing anterior section of frontal longitudinally and transversely divided (outlined), can be only longitudinally or only transversely divided, or occasionally not divided, and usually without medial contact between lateral frontonasals (outlined), sometimes that contact made. *Sceloporus schmidti.* USNM 589133 from Cusuco, Cortés.

description (a composite description including the holotype; description of the holotype alone follows) are herein designated as paratypes. *Sceloporus esperanzae* sp. nov. is a large sceloporine (maximum recorded SVL 98 mm [UTA R-63121, a female]) with a long tail and relatively short limbs; dorsal head scales smooth to slightly rugose, except a few scales on snout keeled in some; 2–6 postrostral scales (most often 5), most wider than long, especially outermost ones; 3–4 (4 in all but 1) internasal scales in anterior row, generally longer than wide (occasionally wider than long); prenasal scales usually absent (1 present on 3 of 20 sides), with scales between nostril and rostral and contacting rostral considered a postrostral; 1 canthal, usually another small one forced above and medial to enlarged canthal, occasionally a well-developed (but small) second canthal present; 3 or 4 medial scales separating large canthal from counterpart on other side; lateral frontonasal scale contacting (in 7), or separated (in 6) from median frontonasal; paired prefrontal scales usually not separated by small azygous scale

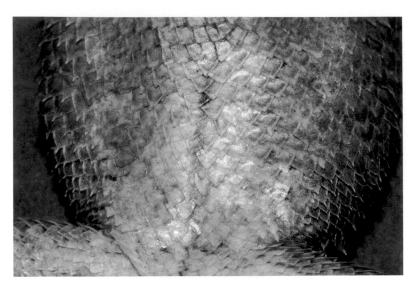

Figure 115.   Dark gray median ventral longitudinal line present, at least indistinctly and posteriorly. *Sceloporus schmidti*. USNM 589137 from Cerro Negro, Santa Bárbara.

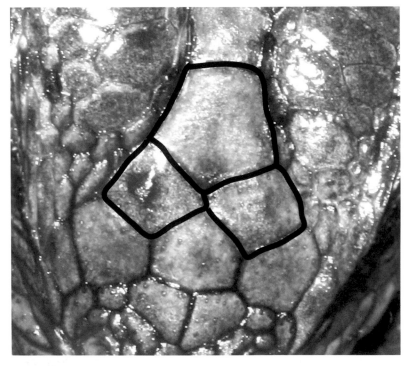

Figure 116.   Dorsal head scales showing anterior section of frontal not divided (outlined) and medial contact between lateral frontonasals (outlined). Species D of *Sceloporus* "*hondurensis*" (probably not described). FN 11350 (still in my collection for further study) from Quebrada de las Marías, Olancho.

(small azygous scale separating those paired prefrontals in 2 of 13); 2–4 (most often 2) frontal scales, with anterior one usually not divided (divided in 3 of 13); 1–2 (2 on 1 of 26 sides) frontoparietal scales almost always not contacting each other medially (contact made in 1 of 13 specimens); median frontoparietal absent in all but 1 of 13; frontal-interparietal contact present in 11 (small median frontoparietal preventing contact in 2); large interparietal scale completely separating each parietal scale from each other, interparietal much larger that each parietal; parietal eye distinct; 4–6 (4.6 ± 1.0) enlarged supraocular scales in inner row, almost always separated from parietal scales by complete row of small supraorbital semicircle scales (in all but 1 on 1 side); 1–5 (2.3 ± 1.0) supraorbital scales in outer row on 24 sides, smaller than scales in inner row; 5–6 (most often 6) superciliary scales, superciliaries 3–5, or 3–4 usually longest; first superciliary scale almost always contacting first preocular, second superciliary also rarely contacting first preocular; 2–3 scales between first superciliary and enlarged supraoculars in outer row; canthal (frontal) ridge usually well developed; 1–3 (usually 2) loreal scales, rarely 3 per side (an upper loreal present on 2 of 26 sides); 2–3 (rarely 3) rows of lorilabials; fourth or fifth supralabial and fifth or sixth (usually fifth) infralabial below level of mideye; moveable eyelid present; pupil circular; body dorsally compressed; dorsal body scales strongly keeled, mucronate, imbricate, relatively large, 39–43 (40.3 ± 2.2) in males, 34–40 (37.4 ± 1.9) in females between interparietal and level above posterior edge of hind limb; lateral body scales strongly keeled, mucronate, imbricate, only slightly smaller than dorsal scales; lateral nuchal scales not noticeably smaller than dorsal nuchal scales; ventral scales smooth, imbricate, with notched posterior ends; para-midventral scales between level of anterior edge of forelimb and cloacal scale 39–45 (41.5 ± 2.5) in males, 38–46 (43.0 ± 2.9) in females;

scales around midbody 40–48 (43.5 ± 3.4) in males, 41–47 (43.4 ± 2.1) in females; scales on posterior surface of thigh smaller than those on dorsal surface, but not granular; 13–16 (14.0 ± 0.9) femoral pores in males, 11–16 (13.6 ± 1.6) in females, each set separated by 7–9 (8.0 ± 0.8) scales in males, 8–12 (10.3 ± 1.2) scales in females; precloacal scales smooth, highly variable in number and clarity (with full or only partial division), ranging from about 8–13; paired enlarged postcloacal scales present in males, enlarged postcloacal scales absent in females; subcaudal scales smooth at base of tail; postfemoral dermal pocket absent; SVL 89–95 (91.8 ± 2.8) mm in males, 66–98 (78.4 ± 9.5) mm in females; TAL/SVL 1.29–1.36 in two males, 1.10–1.18 in four females; HL/SVL 0.21–0.25 in males, 0.20–0.24 in females; HW/SVL 0.21–0.23 in males, 0.19–0.22 in females; SHL/SVL 0.20–0.23 in males, 0.19–0.22 in females; S-OL/SVL 0.19–0.20 in males, 0.18–0.22 in females.

The adult male holotype (USNM 589098; Plate 58) has a SVL of 93 mm; an incomplete tail; a HL/SVL of 0.21; a SHL/SVL of 0.20; a S-OL/SVL of 0.19; 1 canthal with second small canthal forced above anterior portion of large canthal; paired prefrontals separated by small azygous scale; 4 internasals; anterior frontal divided; posterior frontal contacting interparietal scale; lateral frontonasal contacting medial frontonasal; no supraoculars contacting medial head scales; preocular not contacting first superciliary scale; 3–2 small scales between supraocular in outer row and first superciliary scale; 2 parietal scales plus 1 much larger interparietal scale; 38 dorsal scales; 45 ventral scales; 40 scales around midbody; 14–14 femoral pores, each set separated by 9 scales.

Color in life of adult male holotype (USNM 589098; Plate 58): scales of dorsal and lateral surfaces of body greenish yellow (changing to copper brown in life after placing in collecting bag) outlined by

Plate 58.   *Sceloporus esperanzae* USNM 589098, adult male holotype, SVL = 93 mm. Intibucá: La Soledád.

Vandyke Brown (221); scales of top and side of head and neck with blue-green centers, most scales also smudged with Vandyke Brown and paler brown; scales on top and side of tail Cyan (164) outlined by Vandyke Brown; throat (anterior to collar) and abdominal semeions Ultramarine (270) with Jet Black (89) collar crossing throat; abdominal semeions extending onto chest to contact throat collar; scapular band short, extending to level just above ear opening; abdominal semeions outlined medially distinctly by Jet Black on one side only, semeions indistinctly separated (except posterior quarter of venter) by dirty brown and black mottling; chin pale brown anterior to Ultramarine; dorsal surfaces of fore- and hind limb same color as dorsal surface and side of body; subcaudal surface Drab Gray (119D); iris bluish green.

Color in life of an adult male paratype (UTA R-63125): dorsal surface of head and nuchal region mottled with brown and iridescent pale green; dorsal surfaces of body and fore- and hind limb bright greenish yellow; dorsal surface of base of tail turquoise blue; tip of chin pale iridescent green; posterior portion of chin, throat,

and abdominal semeions brilliant cobalt blue. Color in life of an adult female paratype (UTA R-63121): dorsal surfaces and sides of body and fore- and hind limb pale greenish brown with Vandyke Brown (221) spots and blotches; scales of top and side of head brownish green, some with Vandyke Brown spots; some dark brown blotches on fore- and hind limb, especially latter; tail brown with dark brown blotches suggestive of crossbands; clear scapular band and throat collar absent; belly pale brown with some pale green mottling laterally and grayish green flecking on remainder of scales; throat and chin similar in color to that of belly, except infralabials pale green outlined; subcaudal surface pale brown with green tinge, especially along scale edges; iris bluish green.

Color in alcohol of adult male holotype (USNM 589098): all dorsal surfaces black; anterior part of chin dark gray, remainder of chin and throat black; scapular band indistinct, extending to just above level of ear opening; venter of body entirely black, except a few gray scales medially on posterior third of body; underside of forelimb gray to black; underside of hind limb gray with some brown scales; anterior third of subcaudal surface same as that of hind limb, then becoming dark gray until broken part.

Composite color in alcohol of paratypes: dorsal surface of body of males and females dark gray to black, no markings (spots or lines) noticeable; adult male scapular band short, indistinct, no extension above level of lower part of orbit; males with indistinct black throat collar, that collar usually absent in females; belly of males with black to dark gray abdominal semeions, each semeion not outlined medially with black (unlike color in life in some), semeions distinctly separated medially only on posterior quarter or third of venter, that separation narrow (1–2 scale rows wide) and by indistinct zone of brownish black scales; male abdominal semeions cross venter to unite with counterpart on other side on about anterior third of body, semeions extending onto lateral portion of throat to contact throat collar; venter of females grayish brown to dark gray, without dark midventral longitudinal line, with brown to gray central area; male throat anterior to collar dark gray to nearly jet black; chin of males dark gray to dark brown; chin and throat of females gray to bluish gray; ventral surfaces of fore- and hind limb and basal area of tail paler gray to brown than is belly, especially hind limb and tail.

Diagnosis/Similar Species.—The presence of strongly keeled and mucronate dorsal scales, a dorsally compressed (flattened) body, and femoral pores distinguishes *Sceloporus esperanzae* from all other Honduran lizards, except the other *Sceloporus*. *Sceloporus squamosus* and *S. variabilis* have pale brown to cream dorsolateral body stripes, and *S. variabilis* also has a postfemoral dermal pocket (versus dorsolateral pale stripes and dermal pocket absent in *S. esperanzae*). The two described (not including one named following this species) Honduran species of the *S. malachiticus* species complex are more difficult to distinguish from one another. Males of *S.*

schmidti have distinct separation of the male abdominal semeions in life and alcohol (versus no separation anteriorly to poorly separated by 1–2 scale rows on posterior third or quarter of body in *S. esperanzae*). *Sceloporus schmidti* also differs from *S. esperanzae* in having 5–6 scales between the large canthals and many Honduran specimens have a thin dark brown midventral line, at least posteriorly (versus 3–4 scales between large canthals, and dark midventral line absent in *S. esperanzae*).

Illustrations (Fig. 112; Plate 58).—None previously published.

Remarks.—Stuart (1971: 250; as *Sceloporus malachiticus*) studied a series of the herein described *S. esperanzae* and concluded the population was more closely related to *S. smaragdinus* and *S. acanthinus*, both occurring in Guatemala, than they were to Costa Rican *S. malachiticus* and a population from La Tigra, near Tegucigalpa, Honduras. *Sceloporus acanthinus* differs most obviously from *S. esperanzae* by having the nuchal collar either wide or narrow, complete or divided by as many as 5 dorsal scales (depending on source or authority for data), having well-defined paired abdominal semeions in males, and having 1–2 enlarged supraoculars contacting the parietal scales (versus no nuchal collar or well-defined paired semeions, and lacking enlarged supraoculars contacting parietal scales in *S. esperanzae*). *Sceloporus smaragdinus* of Guatemala has an extended scapular bar and 50–58 dorsals (versus no extended scapular band and 34–43 dorsals in *S. esperanzae*). The above data for *S. acanthinus* and *S. smaragdinus* were taken from Smith (1939) and Pérez-Ramos and Saldaña-de La Riva (2008).

Sceloporus esperanzae appears limited to pine forest habitats, as elevated peaks within that pine forest range where broadleaf forests occurred until recently (i.e., Zacate Blanco, Intibucá) are inhabited by a different apparently undescribed species listed below under an unknown category.

Natural History Comments.—*Sceloporus esperanzae* is known from 1,530 to 1,900 m elevation in the Lower Montane Moist Forest formation and peripherally in the Premontane Moist Forest formation. It is a diurnal lizard that regulates its body temperature by basking on elevated sites. It remains common at most localities in the pine forests around La Esperanza, as long as those forests are not burned on a regular basis. The presence of cattle at some sites within the pine forests appears beneficial to these lizards by keeping those grassy areas between the individual trees from becoming overgrown and then burned. This lizard is usually seen basking on fallen logs, lumber piles, and other elevated surfaces and is usually seen moving on pine needles on the ground. When threatened, individuals usually retreat rather deliberately to a nearby pine tree and climb the tree trunk on the opposite side of the observer. The species is easily seen (when ignored by the human population, which seems to be usual) in ornamental gardens (i.e., in restaurants in La Esperanza), as long as the associated structures have tin sheets or overlapping shingles on their roofs for retreat sites for the lizards in the relatively cold climate of that region. I have only visited sites where this lizard occurs during the rainy season from June to August, but the species is likely active throughout the year under sunny conditions, with the exceptions of cold weather that occurs at periods during the dry season (La Esperanza and surrounding areas are the coldest region of Honduras). No studies have been undertaken on diet or reproduction in *S. esperanzae*, but like other *S. malachiticus* complex lizards, it is surely ovoviviparous (Smith, 1939; Köhler and Heimes, 2002; personal observation) and insectivorous (Köhler and Heimes, 2002). Captives of several species of *Sceloporus* will also take vegetable and fruit matter (Köhler and Heimes, 2002). Hand-held newly captive individuals seem timid and rarely attempt to bite unless they are heavily provoked, and then their bites usually seem timid.

Etymology.—This species name *esperanzae* refers to La Esperanza, Intibucá, where this species occurs in the pine forests surrounding the town.

Specimens Examined (13 [5]; Map 45).— HOLOTYPE: **INTIBUCÁ**: La Soledád, USNM 589098. *PARATYPES* (12): **INTIBUCÁ**: 18.1 km NW of La Esperanza, USNM 589099; 17.5 km NW of La Esperanza, USNM 589100; 8.7 km NW of La Esperanza, USNM 589101; 10 km SE of La Esperanza, UF 124600–01; 7 km E of La Esperanza, USNM 589102, UTA R-63120; La Soledád, UTA R-63121–22; San Pedro La Loma, UTA R-63123–24. **LA PAZ**: 13.7 km N of Marcala, UTA R-63125.

Other Specimens Examined (9; Map 45)—**INTIBUCÁ**: 2.4 km NE of La Esperanza, LACM 72103–15; 4 km ENE of La Esperanza, UF 166382–83; 12.9 km ESE of La Esperanza, LACM 72116 (erroneously listed from two localities by Meyer and Wilson, 1973); vicinity of La Esperanza, LACM 45249; La Esperanza, USNM 589103; La Soledád, USNM 589104.

Other Records.—**INTIBUCÁ**: La Esperanza, Univ. So. California 3602–03, 3608–17, 3634, 3641 (Stuart, 1971).

Sceloporus hondurensis sp. nov.
McCranie, herein

Sceloporus serrifer. Werner, 1896: 346.
Sceloporus aeneus: Werner, 1896: 346.
Sceloporus malachiticus: Dunn and Emlen, 1932: 28; Meyer, 1969: 252 (in part); Meyer and Wilson, 1973: 26 (in part); Wilson et al., 1991: 69, 70 (in part); Caceres, 1993: 119; Espinal, 1993, table 3; Wiens, 1993: 294 (in part); Reeder and Wiens, 1996: 73; Köhler, 2000: 86; Espinal et al., 2001: 106; E. N. Smith, 2001: 148 (in part); Wilson et al., 2001: 136 (in part); Köhler and Heimes, 2002: 107 (in part); Lundberg, 2003: 26; Wilson and McCranie, 2004b: 43 (in part); McCranie and Castañeda,

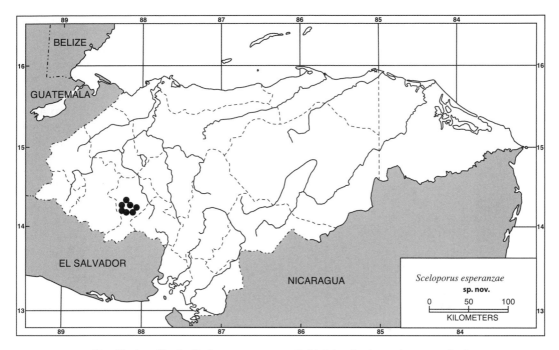

Map 45. Localities for *Sceloporus esperanzae*. Solid circles denote specimens examined.

2005: 14; Lovich et al., 2006: 13; Mahler and Kearney, 2006: 30; McCranie et al., 2006: 217 (in part); Wilson and Townsend, 2006: 105 (in part); Townsend et al., 2007: 10; Townsend and Wilson, 2009: 68; Lovich et al., 2010: 113; Gutsche, 2012: 70; Townsend et al., 2012: 100; McCranie and Solís, 2013: 242; Solís et al., 2014: 131 (in part).

Sceloporus formosus malachiticus: Smith, 1939: 46 (in part); Bogert, 1949: 418.
Sceloporus formosus: Bogert, 1959: 113.
Sceloporus "malachiticus": Stuart, 1971: 238 (in part).
Sceloporus hondurensis **sp. nov.** McCranie, 2018, herein (holotype, UTA R-63126, an adult male; see description below; type locality: from between the ríos Catacamas and Seco, Sierra de Agalta, 14°54′N, 85°55′W, 1,520 m elevation, Olancho, Honduras, collected 5 August 1986 by James R. McCranie).

Geographic Distribution.—*Sceloporus hondurensis* occurs at moderate and intermediate elevations on the Atlantic versant from the Cordillera Nombre de Dios and Montaña Pico Pijol eastward to the Sierra de Agalta and southward to the Sierra de Dipilto, which occurs on both sides of the Honduran and Nicaraguan border and the mountains of Choluteca in southeastern Honduras.

Description.—The following is based on 12 males (USNM 342363–64, 342367–69, 342372, 589128; UTA R-53704, 53709, 63126, 63128–29) and ten females (USNM 342361, 342365, 342370, 589105–06, 589122; UTA R-53705–06, 63127, 63130). The holotype designated herein is an adult male (UTA R-63126), whereas the remaining specimens, just listed and used for the description (including the holotype in a composite description; description of the holotype follows this description), are considered paratypes. *Sceloporus hondurensis* sp. nov. is a relatively large sceloporine (maximum recorded SVL 89 mm [UTA R-53706 mm, a female]) with a long tail and relatively short limbs; dorsal head scales smooth to slightly rugose, except a few snout scales keeled in some; 0–5 postrostral scales, most wider than long, especially outer ones; 3–6 (most often 4) internasal scales in anterior row, usually longer than wide; 1 prenasal scale occasionally present, usually not contacting rostral when present; 1 canthal with almost always another small one forced above and medial to anterior section of each enlarged canthal; 3–5 (most often 3, 4 in 1, 5 in 1) scales medially separating each large canthal; lateral frontonasal scales in contact medially on 27 of 38 sides; paired prefrontal scales in contact medially in 10, thus without small azygous scale present, or paired prefrontals separated by small azygous scale in 8; usually 2 frontal scales with anterior one usually not divided in 16, or occasionally transversely divided in 4; usually lateral frontoparietal scales in contact medially (on 27 sides), small medium frontoparietal occasionally present (on 11 sides); interparietal large, larger than either parietal scale; parietal scales not in contact medially; parietal eye distinct; 4–8 (5.3 ± 1.0) enlarged supraocular scales in inner row, almost always separated from parietal scales by complete row of semicircle scales; semicircular row rarely incomplete posteriorly, allowing contact of 1 supraocular with parietal scale only on 1 of 44 sides; 0–3 enlarged supraoculars in outer row; 5–6 (most often 5) superciliary scales, with 2–3 or 2–4 longest; first superciliary scale usually contacting first preocular; 1–4 scales between first superciliary and enlarged supraocular in outer row (when present); canthal ridge weakly to well developed; usually 2 loreals, loreals sometimes divided into 3 scales; 2–3 (most often 2) rows of lorilabials; fourth to sixth supralabial and fourth or fifth infralabial below level of mideye; moveable eyelid present; pupil circular; body dorsally compressed; dorsal body scales strongly keeled, mucronate, imbricate, relatively large, 34–39 (35.8 ± 2.1) in males, 34–39 (36.1 ± 1.6) in females, between interparietal and level above posterior edge of hind limb; lateral body scales strongly keeled, mucronate, imbricate, only slightly smaller than dorsal scales; lateral nuchal scales not noticeably smaller than dorsal nuchal scales; ventral scales smooth, imbricate, with notched posterior ends, para-midventral scales between level of anterior edge of forelimb and cloacal scale 37–46 (41.2 ± 2.7) in males, 39–46 (42.0 ± 2.3) in females; 35–43 (38.8 ± 2.6) scales around midbody in males, 35–43 (39.8 ± 1.0) in females; scales on posterior surface of thigh smaller than those on dorsal surface of thigh, but not granular; 13–17 (15.0 ± 1.2) femoral pores in males, each series separated medially by 5–10 (8.3 ± 1.4) scales; 12–15 (13.0 ± 0.8) femoral pores in females, each series separated medially by 8–11 (9.1 ± 1.0) scales; precloacal scales smooth, highly variable in number and clarity, in full or partial division, ranging from about 9–14; 1 enlarged postcloacal scale present per side in males, enlarged postcloacal scales absent in females; subcaudal scales smooth at base of tail; postfemoral dermal pocket absent; SVL 69–88 (81.4 ± 5.7) mm in 11 males, 56–89 (73.5 ± 11.2) mm in females; TAL/SVL 0.94–1.31 in six males, 0.97–1.11 in two females; HL/SVL 0.21–0.25 in all males and eight females; HW/SVL 0.20–0.22 in four males, 0.19–0.21 in seven females; SHL/SVL 0.21–0.23 in 11 males and eight females; S-OL/SVL 0.19–0.21 in 11 males, 0.18–0.22 in eight females.

The adult male holotype (UTA R-63126) has a SVL of 83 mm; an incomplete tail; a HL/SVL of 0.21; a SHL/SVL of 0.22; a S-OL/SVL of 0.20; 1 canthal per side with second small canthal forced above anterior portion of large canthal; paired prefrontals separated by azygous scale at only about posterior two-thirds of prefrontals; 4 internasals; anterior frontal not divided; posteri-

Plate 59.   *Sceloporus hondurensis* sp. nov. UNAH, adult male. Olancho: Quebrada del Agua. Photograph by Alex Gutsche.

or frontal not contacting interparietal; lateral frontonasal contacts medial frontonasal; no supraocular contacting medial head scales; preocular not contacting first superciliary; 1–2 small scales between supraocular in outer row and first superciliary; 2 parietal scales and 1 much larger interparietal; 34 dorsal scales; 39 ventral scales; 40 scales around midbody; 13 femoral pores per side, separated medially by 9 scales (no color in life available, but see below for color in preservative).

Color in life of an adult male (UNAH; Plate 59; based on digital photographs): dorsal surfaces of body and fore- and hind limb Yellow Green (color 103 in Köhler, 2012) with thin, dark olive brown crossbands on hind limb; unregenerated portion of tail Caribbean Blue (168); top of head Light Turquoise Green (146) with green brown smudging on interparietal scale forward to rostral; Jet Black (300) scapular extension reaching well above forelimb level; side of head Opaline Green (106); chin Salmon Color (83) on anterior half, Sky Blue (167) on posterior half, including throat; abdominal semeion Medium Blue (169), separated by dirty white, less than 1

scale wide on chest, 5–6 scales wide on posterior half of ventral surface of body; venter of tail dirty white with Caribbean Blue on ventrolateral edges, especially on anterior third. Color in life of another adult male (USNM 589118): middorsum Greenish Olive (49) with Citrine (51) scale centers; lateral portion of body Lime Green (159) with Citrine scale centers; mental area pale green, followed by Flesh Ocher (132D) anterior to throat, that color followed by Ultramarine (270) complete throat collar, each scale in Ultramarine throat collar outlined with black; nuchal collar Jet Black (89); abdominal semeions Ultramarine laterally with black borders medially; belly dirty white between black bordered abdominal patches.

Color in alcohol of adult male holotype (UTA R-63126): all dorsal surfaces black; mental region brown, remainder of chin and throat black; scapular band indistinct, extending only to just above level of ear opening; throat collar distinct, complete; entire chest region posterior to throat collar pale brown with large gap between throat collar and anterior extent of semeions; abdominal semeions dark gray, each bordered medially by distinct black bar 4–5 scale rows wide, those bars separated from each other medially by 2–6 scale rows of pale brown (midbelly); semeions extending from posterior to forelimb to groin area; underside of forelimb pale brown to gray; underside of hind limb mostly pale brown with some gray scales; base of subcaudal surface similar in color to that of hind limb (tail broken near base).

Composite color in alcohol of 21 paratypes: dorsal surface of body of males dark green to dark brown, without distinct darker colored spots or blotches; dorsal surface of body of females brown to dark green, usually with dorsal markings consisting of scattered black or dark brown spots or blotches; males with short, black scapular bands reaching level above forelimb, black collar extending across throat; females with brown to black indistinct scapular band extending to level of about ear opening, collar absent on throat of females; male gular region greenish gray to grayish brown; male chin varies from greenish gray to dark green, to grayish brown; belly of males with blue to dark blue abdominal semeions, semeions outlined with black border about 1–5 scale rows wide, separated medially for 2–5 scale rows at midlength by pale brown to dirty white with black mottling; male abdominal semeions separation variable, some semeions completely separated and not extending to lateral chest region, some completely separated and crossing lateral portion of chest to unite with throat collar, occasionally abdominal semeions cross chest just posterior to throat to unite with each other and with throat collar; belly of females varying from pale brown to dark gray, indistinct black outlined gray abdominal semeions present in some; no distinct dark ventral line.

Diagnosis/Similar Species.—The strongly keeled and mucronate dorsal scales and dorsally compressed body will distinguish *Sceloporus hondurensis* from all other Honduran lizards, except the other *Sceloporus. Sceloporus squamosus* and *S. varia-*

bilis have pale dorsolateral body stripes and *S. variabilis* also has a postfemoral dermal pocket (both characters lacking in *S. hondurensis*). The three named species of the *S. malachiticus* complex recognized (*S. esperanzae, S. hondurensis,* and *S. schmidti*) are more difficult to distinguish from one another. Males of *S. esperanzae* lack, or only rarely have a black bar bordering each semeion (only on 1 of 44 sides), lack distinct abdominal semeion separation or have those semeions poorly separated and only separated posteriorly on the body by 1–2 scale rows in life and alcohol (versus black bar bordering each abdominal semeion and distinct semeion separation in *S. hondurensis*). *Sceloporus schmidti* has a tendency to have more dorsal head scale fragmentation resulting in usually having the lateral frontonasals separated medially by a small azygous scale, having the anterior frontal usually divided transversely and/or longitudinally, having 5–6 scales between the canthals, and frequently having a dark midventral line, at least posteriorly in Honduran specimens (versus lateral frontonasals usually not separated by small scale[s], anterior frontal usually not divided, 3–4 scales between large canthals, and dark midventral line absent in *S. hondurensis*).

Illustrations (Plate 59).—Gutsche, 2012 (adult; as *S. malachiticus*); Köhler, 2000 (ventral pattern, fig. 187; as *S. malachiticus*), 2003a (adult from Nicaragua; as *S. malachiticus*); Köhler and Heimes, 2002 (adult, juvenile, figs. 104, 187; as *S. malachiticus*); Lundberg, 2003 (adult, ventral pattern, dorsal surfaces; as *S. malachiticus*).

Remarks.—*Sceloporus hondurensis* as just discussed, could represent two or more species (E. N. Smith, molecular data, personal communication, 8 January 2016), both of which are apparently more closely related to the southern *S. malachiticus* Cope than they are to *S. esperanzae* and *S. schmidti*. Hopefully, a planned project using both morphological and molecular data will elucidate species boundaries in this species complex within Honduras. The results of the E. N. Smith molecular work to date divided *S. hondurensis*, as recognized herein, into northern and southeastern clades that might represent separate species. Those from the vicinity of Cerro La Tigra, Francisco Morazán, southward form a subclade within the northern *S. hondurensis* clade. That northern clade occurs in the mountains of Atlántida and Yoro eastward to the Sierra de Agalta, Olancho, and southward to at least northern Francisco Morazán. It will take further study, including new tissues and adult specimens from more populations, to determine whether two or more species are involved.

Males from scattered localities in Honduras have a chin region of bright orange or reddish orange (see photograph in Gutsche, 2012, and Plate 59 herein). However, that color pattern is not consistent at any known Honduran locality. According to Wettstein (1934: 25), the orange-colored throat blotch does not occur in Costa Rican *S. malachiticus*.

Natural History Comments.—*Sceloporus hondurensis* is known from 650 to 2,300 m elevation in the Premontane Wet Forest, Lower Montane Wet Forest, and Lower Montane Moist Forest formations and peripherally in the Premontane Moist Forest formation. Like other species of *Sceloporus*, *S. hondurensis* is a diurnal, sunloving species. *Sceloporus hondurensis* is apparently a canopy species in broadleaf forests that quickly adapts to man-made structures and edge situations in larger clearings. It appears to be active throughout the year under sunny conditions, as the species has been collected in both the rainy and dry seasons (from March to December). No studies have been undertaken on diet or reproduction in *S. hondurensis*, but like other *S. malachiticus* complex lizards, it is surely ovoviviparous (Smith, 1939; Köhler and Heimes, 2002; personal observation) and insectivorous (Köhler and Heimes, 2002), but captives of several *Sceloporus*

species also will take vegetable and fruit matter (Köhler and Heimes, 2002).

Sceloporus hondurensis occurs up to at least 2,300 m elevation above the point where the elfin forest on Cerro La Picucha, Olancho, itself becomes less than one-third of a meter tall. That locality is along a windswept ridge, with fast-moving clouds overhead, that quickly shifts from sunny with warm rays to cloudy conditions with chilly, windy temperatures. An adult male and an adult female were sunning from the top of a rock crevice one afternoon at that high elevation, but quickly disappeared beneath the crevice when approached.

The vicinity of the *Sceloporus hondurensis* type locality has changed considerably from 1986 when the holotype was collected. In 1986, the only other primates my field party saw were spider monkeys (*Ateles geoffroyi* Kuhl) and an unusually curious troop of howler monkeys (*Alouatta palliata* [Gray]). Today, those entire forests between the ríos Catacamas and Seco are essentially devastated of their previous forest and fauna and now only contain numerous humans and their villages, cattle farms, pastures, dirty and/or dried streams, and various crop fields, all of which are now unfit for the former natural world, including those two passive primates that used to occur there.

Etymology.—The species name *hondurensis* refers to the country of Honduras, where this species appears to be largely confined. The name also represents the similarly ruined condition of the forests at the species' type locality and those of the current status of the vast majority of the Honduran forests 30 years later.

Specimens Examined (259, 2 skeletons [39]; Map 46).—*HOLOTYPE*: **OLANCHO**: between ríos Catacamas and Seco, UTA R-63126. *PARATYPES* (21): **OLANCHO**: Cuaca, UTA R-53704–06, 53709; between El Díctamo and Parque Nacional La Muralla Centro de Visitantes, USNM 342361, 342363–64; Montaña de Las Parras, USNM 342367; Montaña de Liquidambar, USNM

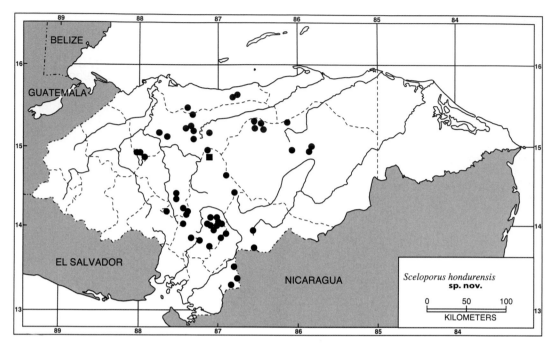

Map 46.   Localities for *Sceloporus hondurensis*. Solid circles denote specimens examined and the solid square an accepted record.

342365; Parque Nacional La Muralla Centro de Visitantes, USNM 342372; Quebrada El Pinol, USNM 342368; Quebrada Las Cantinas, USNM 342369–70; between ríos Catacamas and Seco, UTA R-63127; Río Cuaca, UTA R-63128; **YORO**: Cerro de Pajarillos, UTA R-63129; 2.5 airline km NNE of La Fortuna, USNM 589105–08; E slope of Pico Pijol, UTA R-63130.

Referred Specimens Examined.—**ATLÁN-TIDA**: S slope of Cerro Búfalo, KU 194320; La Liberación, USNM 578725–30; Quebrada de Oro, USNM 589109. **CHOLUTECA:** La Caguasca, FN 256950 (still in Honduras because of permit problems); Quebrada La Florida, SDSNH 72765; near San Marcos de Colón, CAS 152983. **COMAYAGUA:** Cerro El Volcán, UF 166371; Cerro La Granadilla, UF 166370, 166372; 38.6 km ESE of La Paz, TNHC 48853; Montaña de Comayagua, LSUMZ 88076–77; near Río Negro, KU 200565–66; 55.2 km NW of Tegucigalpa, TNHC 32150–51, 32153. **CORTÉS:** near

Agua Azul, AMNH 70298; near Peña Blanca, LSUMZ 88093. **EL PARAÍSO:** near Agua Fría, AMNH 70299–301; Cerro Moncerrato, USNM 589110–11; Las Manos, UTA R-41277–78; 30.6 km NW of Mandasta, KU 209315; Monserrat, AMNH 70303, MCZ R49952, 171089–94. **FRANCISCO MORAZÁN**: Cataguana, UF 156683–92, USNM 589112–13; Cerro Cantagallo, KU 209316, USNM 589114–15; Cerro La Tigra NNE of El Hatillo, CM 64830–31, KU 194321–23, LSUMZ 24202, 24215, SMF 80425, UF 124602–03, USNM 589116–19; near Cerro Uyuca, UTA R-41279; Cerro Uyuca, AMNH 70311–13, KU 103242–44, 200567–68, MCZ R49939–41, 171087–88; El Hatillo, CM 34055, LACM 72090, LSUMZ 24207, TCWC 64176; above El Rosario, SMF 80426; 9 km NW of El Zamorano, AMNH 150180–81; La Montañita, AMNH 70304, MCZ R49933 (listed as MCZ R49934 in Meyer and Wilson, 1973); 5 km W of Maraita, UF 124604; Montaña de Guaimaca, AMNH

70302; Montaña de la Sierra, UF 166384; Monte Crudo, AMNH 70316; near Ojojona, UTA R-46250, 46255; Quebrada Cataguana, UF 149439–41; 15 km N of Sabanagrande, AMNH 150176–79, LSUMZ 24604–06; 11.7 km SW of San Juancito, LSUMZ 21505; near San Juancito, UF 124605–07; San Juancito, ANSP 28117, 33132–37, CM 34056; San Juancito Mountains, AMNH 70305–10, 70314–15, ANSP 28117, 33132–37, CM 34056; Sendero de La Cascada, USNM 549367; Tatumbla, MCZ R49984; 41 km NW of Tegucigalpa, AMNH 150175, LSUMZ 24607; 16 km SE of Tegucigalpa, TCWC 26712–15; near Tegucigalpa, LSUMZ 24137; 4.8 km N of Valle de Ángeles, TNHC 48859–62; 5 km W of Zambrano, KU 67270. **OLANCHO**: Cerro Picucha, USNM 589120; Cuaca, UTA R-53212, 53707–08, 53710–12; El Aguacatal, UTA R-53213–20; between El Díctamo and Parque Nacional La Muralla Centro de Visitantes, USNM 342362; near Los Planes, USNM 342366; Parque Nacional La Muralla Centro de Visitantes, USNM 342371; Piedras Blancas, USNM 589121–22; Quebrada del Agua, USNM 589123–25; between ríos Catacamas and Seco, FMNH 236392, USNM 589126; Río Cuaca, USNM 589127; Terrero Blanco, USNM 342373. **YORO**: El Panal, UF 166378; 2.5 airline km NNE of La Fortuna, UF 166379, USNM 589128; Montaña La Ruidosa, KU 200574; Montaña Macuzal, KU 200572–73, UF 166373–77; Montañas de Mataderos, MCZ R38845, 171095; Portillo Grande, FMNH 21872–75, MCZ R38851–54; Subirana, MCZ R32283–92, UMMZ 71289 (10); Subirana Valley, FMNH 21793, MCZ R38846–50, 171096–97, UMMZ 77844 (2); "no other data," UTA R-46246–49. "HONDURAS": FMNH 31039 (skeleton), MCZ R9515–16, UF 72808 (skeleton).

Redescription of *Sceloporus schmidti* Jones, 1927

Sceloporus schmidti Jones, 1927: 4 (holotype, FMNH 5214, now lost [see Smith, 1939: 42; also A. Resetar, personal communication, 5 December 2013; lectotype designated herein UTA R-42636; see brief description herein for lectotype designation and its locality]; type locality: "Mountain Camp—west of San Pedro [Sula], 4,500 feet altitude, Honduras"); Flower, 1928: 50; Smith, 1936: 95; Townsend and Wilson, 2006: 245; Townsend and Wilson, 2008: 142; McCranie, 2015a: 369.
Sceloporus formosus smaragdinus: Smith, 1939: 42.
Sceloporus malachiticus: Meyer, 1969: 252 (in part); Meyer and Wilson, 1973: 26 (in part); Wilson et al., 1991: 70 (in part); Wiens, 1993: 294 (in part); Anonymous, 1994: 116; E. N. Smith, 2001: 147; Wilson et al., 2001: 136 (in part); Köhler and Heimes, 2002: 106–107 (in part); Wilson and McCranie, 2004b: 43 (in part); Wilson and McCranie, 2004c: 24; McCranie, 2005: 20; Castañeda and Marineros, 2006: 3.8; McCranie et al., 2006: 117 (in part); Townsend, 2006a: 35; Townsend et al., 2006: 35; Wilson and Townsend, 2006: 105 (in part); Solís et al., 2014: 131 (in part).
Sceloporus smaragdinus: Köhler and Heimes, 2002: 135 (in part); Bell et al., 2003: 153.

Geographic Distribution.—*Sceloporus schmidti* occurs at moderate and intermediate elevations on the Atlantic versant west of the Río Ulúa in the Sierra de Omoa west of San Pedro Sula and southwestward through the Sierra Espíritu Santo in Copán. It also occurs in adjacent Guatemala in those two sierras.

Description.—The following is based on four adult males (KU 200571, 203006; USNM 589132, 589136) and ten adult to subadult females (KU 194316, 200569–70, 200575–76; UTA R-4263; USNM 589130–31, 589133, 589137). *Sceloporus schmidti* is a large sceloporine (maximum recorded SVL 115 mm [USNM 589137, a female]) with a long tail and relatively short limbs; dorsal head scales have tendency for more fragmentation than do other Honduran populations of the *S. malachiticus* species complex, dorsal head scales smooth to slightly rugose, except few scales on snout keeled in some; 1–5 postrostral scales, most wider than long, especially outermost ones; 5–16 internasal scales, generally longer than wide, occasionally wider than long; 1 prenasal scale usually present per side, but scales between nostril and rostral (postrostral) occasionally present; 1 large canthal per side, anterior or second well-developed canthal present per side in some, always smaller than large one, and second (when present) partially forced above large canthal; 5 or 6 scales separate large canthals medially; lateral frontonasal scale usually not contacting median frontonasal; paired prefrontal scales usually medially separated by small azygous scale (in 9 of 14); 2–5 (most often 2) frontal scales, with anterior one most often transversely and/or longitudinally divided (divided in 8 of 14); frontoparietal scales usually not in contact medially (contact made in 3 of 14); median frontonasal absent in all but 1 of 14; frontal-interparietal contact present in seven, absent in seven; interparietal scale much larger than parietal scale on each side; interparietal scale completely separates each parietal from each other; parietal eye distinct; enlarged supraocular scales separated from parietals on 23 of 24 sides; 5–6 (most often 6) superciliary scales, superciliaries 3–5, 2–6, or 3–4 usually longest; first superciliary scale usually contacting first preocular, first preocular contact with upper loreal variable, present or absent; 4–5 enlarged supraoculars in inside row, 0–3 in outside row, those of outer row irregular in arrangement when present; 2–3 scales between first superciliary and large supraocular in outer row; canthal ridge weakly to well developed; 2–3 loreal scales; 2–3 (usually 3) rows of lorilabials; fourth to sixth supralabial and fifth or sixth infralabial below level of mideye; moveable eyelid present; pupil circular; body dorsally compressed; dorsal body scales strongly keeled, mucronate, imbricate, relatively large, 38–45 (40.5 ± 3.1) in males, 35–45 (40.5 ± 3.1) in females between interparietal and level above posterior edge of hind limb; lateral body scales strongly keeled, mucronate, imbricate, only slightly smaller than dorsal scales; lateral nuchal scales not noticeably smaller than dorsal nuchal scales; ventral scales smooth, imbricate, with notched posterior ends; para-midventral scales between level of anterior edge of forelimb and cloacal scale 40–44 (41.8 ± 1.7) in males, 41–47 (43.4 ± 1.7) in females; scales around midbody 37–44 (40.3 ± 3.0) in males, 37–42 (40.2 ± 2.0) in females; scales on posterior surface of thigh smaller than those on dorsal surface of thigh, but not granular; 11–14 (13.1 ± 1.2) femoral pores in males, 10–14 (12.4 ± 1.3) in females, each series separated by 8–10 (8.8 ± 1.0) scales in males, 8–12 (9.8 ± 1.4) scales in females; precloacal scales smooth, highly variable in number and clarity, in full or only partial division, ranging from about 8–12; 1 pair of enlarged postcloacal scales present in males, enlarged postcloacal scales absent in females; subcaudal scales smooth at base of tail; postfemoral dermal pocket absent; SVL 90–103 (96.8 ± 5.6) mm in males, 43–115 (71.2 ± 21.6) mm in females; tail incomplete in all males, TAL/SVL 1.22 in one female; HL/SVL 0.22–0.25 in males, 0.19–0.25 in nine females; HW/SVL 0.18–0.21 in males, 0.19–0.23 in females; SHL/SVL 0.20–0.22 in males, 0.21–0.24 in females; S-OL/SVL 0.18–0.21 in males, 0.19–0.21 in females.

The holotype (FMNH 5214) of *Sceloporus schmidti* was lost by the time Smith (1939) revised the genus; therefore, that specimen was apparently not examined by another worker studying *Sceloporus* or the Honduran herpetofauna, including Dunn and Emlen (1932) and Meyer and Wilson (1973). Thus, it seems appropriate to designate a lectotype of *S. schmidti* Jones. Jones (1927: 4) gave the type locality of *S.*

schmidti as Schmidt's mountain camp west of San Pedro Sula where Schmidt collected the holotype (see Schmidt, 1942: 29, fence lizard comment). By using topographical maps and copies of Schmidt's field notes, I have been able to determine (see McCranie, 2002: 28, 562, *In* McCranie and Wilson, 2002; McCranie, 2011a: 627) that Schmidt's locality is apparently near the Quebrada del Infierno on the eastern slope of Cerro La Virtúd. An adult female (UTA R-42636) collected at 12.7 km W of San Pedro Sula (15°.516′N, 88°.1322′W at 1,500 m elevation on 8 August 1997 by James R. McCranie) is very near Schmidt's type locality and is herein designated the lectotype of *Sceloporus schmidti* Jones. UTA R-42636 is essentially topotypic with the type locality and apparently lies along the ridge above the Quebrada del Infierno mountain camp of Schmidt.

The lectotype (UTA R-42636) has a SVL of 84 mm, a regenerated tail, a HL/SVL of 0.21, a SHL/SVL of 0.22, a S-OL/SVL of 0.19, 1 canthal with a second very small canthal forced above anterior portion of large canthal, paired prefrontal on each side separated medially by azygous scale, 38 dorsal scales, 42 ventral scales, 42 scales around midbody, and 12–13 femoral pores separated by 8 scales (no color notes in life).

Color in life of a subadult male (USNM 589131): dorsum of body and fore- and hind limb Citrine (51) with Pistachio (161) outlined dark olive green crossbands and spots; tail similar in color to that of body, but with bronze cast; top of head Blackish Neutral Gray (82) with Yellow-Green (58) spots; side of head Yellow-Green with dark brown spots and lines; chin Yellow-Green medially, 3–4 scale rows posteriorly grading to pale turquoise green; pale orange gular patch present; black scapular band short, not reaching above level of forelimb; venter of tail Salmon Color (6). Color in life of an adult female (USNM 589133): dorsal surface of head Shamrock Green (162B) with Sepia (119) elongated blotches; dorsal

surface of body Peacock Green (162C) with slight yellowish tinge and Sepia large spots and transverse bars; top of forelimb same color as that of body; dorsal surfaces of tail and hind limb pale greenish brown with Sepia crossbands; Sepia nuchal collar connected to Sepia nuchal blotch; ventral surface of head and chin bluish green, that of body golden brown with gray flecking; posterior surface of thigh pale greenish brown with Sepia blotches, blotches followed below by Sepia transverse bar, bar followed below by pale greenish brown bar, that bar followed below by Sepia bar, that bar followed below by golden brown ventrolateral bar; iris golden brown.

Color in alcohol: dorsal surface of body of males dark gray to black, no noticeable markings; males with indistinct darker black scapular band extending just above level of forelimb, or scapular band not visible; throat collar absent or indistinctly crossing throat in males, throat collar very indistinct to absent in females; dorsal surface of body of females dark green to dark gray, distinct to indistinct dark gray dorsal blotches or spots present; belly of males with black to dark gray abdominal patches that extend onto lateral chest to connect with black throat collar, patches also crossing chest just posterior to throat collar, patches not noticeably outlined medially with black, patches noticeably separated medially on posterior venter or about three-quarters of venter by 3–5 scale rows of pale brown scales; dark gray longitudinal midline present on posterior quarter of belly in males; belly of females pale brown to greenish brown, with dark gray to black longitudinal line medially, line frequently extending length of belly, sometimes distinct only on posterior portion of belly; male throat anterior to black collar (when present) dark blue to nearly jet black; chin of males dark gray to brown; chin and throat of females gray to bluish gray; ventral surfaces of fore- and hind limb and basal area of tail of both sexes paler gray to brown than belly, especially hind limb and tail.

Diagnosis/Similar Species.—The strongly keeled and mucronate dorsal scales, dorsally compressed body, and presence of femoral pores distinguishes *Sceloporus schmidti* from all remaining Honduran lizards, except the other *Sceloporus*. *Sceloporus squamosus* and *S. variabilis* have a pale brown or cream dorsolateral longitudinal body stripe and *S. variabilis* also has a postfemoral dermal pocket (versus pale stripe and dermal pocket absent in *S. schmidti*). The three described species of the *S. malachiticus* complex (*S. esperanzae, S. hondurensis*, and *S. schmidti*) are more difficult to distinguish from one another, but *S. schmidti* has a tendency for having more dorsal head scale fragmentation. *Sceloporus esperanzae and S. hondurensis* also differ from *S. schmidti* in having paired prefrontals that are usually not separated by a small azygous scale, having 3–4 scales separating the well-developed canthals medially, and lacks a dark brown midventral line (versus lateral frontonasals usually medially separated by small scale[s], anterior frontal usually divided, 5–6 scales between large canthals, and dark midventral line present, at least posteriorly in female *S. schmidti*). *Sceloporus esperanzae* males also lack black bars bordering the abdominal semeions, with those semeions poorly separated, or not separated in alcohol (versus black bars and well-developed semeions present in *S. schmidti*). The extralimital (Guatemala to Mexico) *S. acanthinus* differs in molecular analysis results, has a wide or narrow nuchal collar, has a single row of enlarged supraocular scales, and has contact between 1–2 enlarged supraoculars and parietal scales (versus only a short scapular band present, 2 rows of enlarged supraocular scales, and lacking contact between an enlarged supraocular and a parietal scale in *S. schmidti*; those data for *S. acanthinus* from Pérez-Ramos and Saldaña-de La Riva, 2008). The also extralimital (Guatemala and Mexico) *S.*

smaragdinus is distinguished from *S. schmidti* in molecular phylogenetic results, plus in having 50–58 dorsals, having 48 scales around the midbody, a single row of enlarged supraoculars, and having 15–17 femoral pores (versus 35–40 dorsals, 37–47 scaled around midbody, 2 somewhat irregular rows of supraoculars, and 10–14 femoral pores in *S. schmidti*; those data for *S. smaragdinus* from Pérez-Ramos and Saldaña-de La Riva, 2008).

Illustrations (Figs. 114, 115; Plate 60).— Köhler and Heimes, 2002 (adult; figs. 185, 186; as *S. malachiticus*); McCranie et al., 2006 (adult; as *S. malachiticus*); Townsend and Wilson, 2006 (adult), 2008 (adult).

Remarks.—*Sceloporus schmidti* was considered a junior synonym of *S. formosus smaragdinus* Bocourt (1873b: 1) by Smith (1939: 41) in his revision of the genus *Sceloporus*. Subsequently, Smith (1942) transferred *S. smaragdinus* to a subspecies of *S. malachiticus*, and Stuart (1970: 141) used "*malachiticus* like" for the lizards of this complex in Central America. *Sceloporus schmidti* was not mentioned by either Smith (1942) or Stuart (1970, 1971), because they apparently considered *S. schmidti* a synonym of *S. smaragdinus*, but Smith (1942: 356) did state that *S. m. smaragdinus* occupied the "central Plateau of Guatemala and its extensions." Honduran to Costa Rican populations are usually called *S. malachiticus* in the literature, but E. N. Smith (2001, in an unpublished dissertation) suggested elevating *S. schmidti* and other populations to species status. In this work, I consider *S. schmidti* to represent a valid species, apparently more closely related to the remaining Guatemalan species than it is to more eastern and southern *S. malachiticus* complex members. However, considerable variation occurs in the extent of the male abdominal semeions in alcohol in the two *S. schmidti* specimens examined in June 2015. *Sceloporus schmidti* is distinct from the remaining Honduran populations by a suite of dorsal head scale characters, mainly

Plate 60. *Sceloporus schmidti*. USNM 589132, adult male, SVL = 95 mm. Cortés: El Cusuco.

by a tendency to have more fragmentation among those scales, and in having a midventral dark line (see Description for *S. schmidti* above). The ambiguous data in the literature regarding the number of canthal scales for *S. smaragdinus* (Smith, 1939; Stuart, 1971; H. M. Smith et al., 2001) make it impossible to distinguish *S. schmidti* from *S. smaragdinus* based on data in the literature for that character.

Natural History Comments.—*Sceloporus schmidti* is known from 600 to about 2,240 m elevation in the Premontane Wet Forest and Lower Montane Wet Forest formations. Like other malachite *Sceloporus* occurring in broadleaf forests, *S. schmidti* is a canopy inhabitant that quickly becomes quite abundant closer to the ground whenever a clearing is made in the forest. This diurnal, sun-loving species quickly adapts to man-made structures. The walls and roofs of every building within its former forest habitat usually have their own population. It is also common in edge situations of larger clearings. *Sceloporus schmidti* appears to be active under sunny conditions throughout the year, as it has been seen during every visit into its range during both the rainy and dry seasons (April to June and in September). Likewise, as in all known reproductive data for *S. malachiticus* complex members, *S. schmidti* is probably ovoviparous (Smith, 1939; personal observation) and primarily insectivorous (Köhler and Heimes, 2002).

Etymology.—The name *schmidti* is a patronym honoring Karl Patterson Schmidt, the collector of the holotype. Schmidt was the first professional herpetologist to make amphibian and reptile collections in Honduras, and his quick success with finding new species in the lower cloud forest was instrumental in my desire to explore similar regions throughout Honduras.

Specimens Examined (34 [1, but that holotype listed by Meyer and Wilson lost by 1930s]; Map 47).—**COPÁN**: Cerro Los Dantos, KU 200576; La Leona, USNM 589129; Quebrada Grande, KU 200570–71, 200575, 203006, UF 166385–88, USNM 589130–31. **CORTÉS**: Bosque Enano, UF 144693; Cantiles Camp, UF 147628–30, 147639; El Cusuco, KU 194316, 200569, 200620, MVZ 263420, UF 144111–12, 166389, USNM 589132–34; Guanales, UF 144666; 12.7 km W of San Pedro Sula, UTA

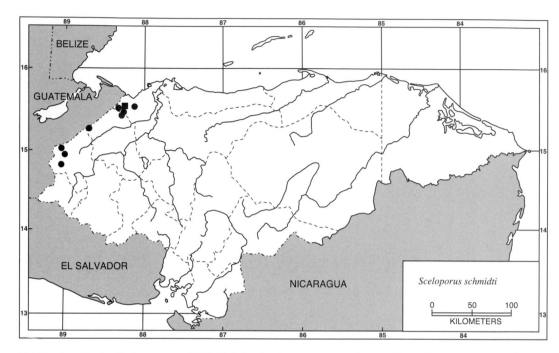

Map 47.   Localities for *Sceloporus schmidti*. Solid circles denote specimens examined and the solid square an accepted record.

R-42636 (*LECTOTYPE*). **SANTA BÁRBARA**: Cerro Negro, USNM 589135–37; La Fortuna Camp, UF 144694, 144715.

Other Records (Map 47).—**CORTÉS**: Finca Naranjito (specimen lost); W of San Pedro Sula, FMNH 5214 (holotype of *Sceloporus schmidti*, lost).

Comments on other Honduran *Sceloporus malachiticus* species complex members

Comments on various Honduran populations assigned to the *Sceloporus malachiticus* species complex are made here (see Map 48); according to molecular data (E. N. Smith, personal communication, 8 January 2016), most of those populations discussed below represent undescribed taxa. Those populations lacking names will not key out correctly in the identification keys presented above. Smith, E. N. (2001: 134–135) recognized four species as occurring in Honduras (only one of which, *S. schmidti*, had a name at that time, the remainder called *S. malachiticus*). Another species (*S.*

acanthinus) was noted to occur on the Guatemalan slopes of Cerro Montecristo adjacent to southwestern Honduras and thus should be expected on the Honduran side of that cerro. Above, I describe two new species and redescribe *S. schmidti*. Below, I discuss those remaining populations, plus others retrieved, as their own clade in the E. N. Smith recent molecular analysis, plus one population lacking molecular study.

Sceloporus acanthinus

Populations on the Honduran side of Cerro Montecristo should be the same species as on the Guatemalan and El Salvadoran sides, usually identified in the literature as *S. acanthinus*. Six specimens (FN 20044–45, 212557, 252206–07, 252220; two polygons on Map 48 in extreme SW Honduras) are available from the Honduran side of Cerro Montecristo within less than 10 km from the Guatemalan border (those from southeast of Chiquimula, Guatemala,

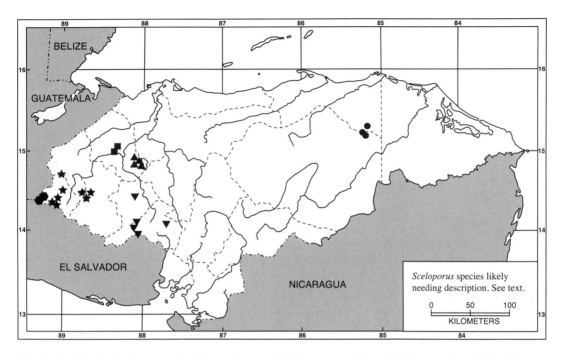

Map 48.   Localities for *Sceloporus malachiticus* species group that do not agree with any Honduran named species in that group. At least two or more of these populations probably need describing as new species. Solid symbols denote specimens examined. See text for explanation of various symbols.

listed under "ACANTHINUS_Montecristo" and those from "ACANTHINUS_El_Salvador" in E. N. Smith, 2001: 139–140). One Honduran Cerro Montecristo specimen (FN 252207) was phylogenetically studied by using molecular techniques by E. N. Smith, and those results placed it in a *S. acanthinus* clade including Cerro Montecristo specimens from El Salvador and Guatemala (E. N. Smith, personal communication, 8 January 2016). However, Honduran Montecristo specimens do not have the diagnostic characters said to define *S. acanthinus*, as provided by Smith (1939, 1942), Stuart (1971), H. M. Smith et al. (2001), and Pérez-Ramos and Saldaña-de La Riva (2008). Honduran specimens all lack a nuchal collar and a large supraocular contacting a parietal scale, and all have only a single canthal per side, usually without even a tiny second canthal present (versus nuchal collar present, 1–2 large supraocu-

lars contacting medium head scales, 2 pairs of canthals in *S. acanthinus*). The Honduran Montecristo specimens also have the first superciliary in contact with the first subocular, but not an upper loreal (character used by E. N. Smith, 2001). However, that character seems to be too variable to separate any Honduran populations as a distinct species. The single Honduran Montecristo male (FN 252220) does agree with *S. acanthinus* by having the abdominal semeions well separated and not extending onto the lateral chest region. Unfortunately, the extent of the male abdominal semeion pattern is extremely variable within and among many Honduran populations of malachite *Sceloporus* (one exception is *S. esperanzae*). Despite not agreeing with the diagnostic characters previously given for *S. acanthinus*, these Honduran Montecristo specimens also do not agree with the morphological characters of any of the remaining species or populations of the *S. malachiticus* species group. The six specimens available differ from all remaining Honduran malachite *Sceloporus* in having a single row of large supraoculars extending side to side and without any trace of a second row. Despite the molecular results placing it with *S. acanthinus*, these specimens do not agree with the morphological characters considered diagnostic of *S. acanthinus* as given in the literature. Clearly, the El Salvador and Guatemala populations occurring on Cerro Montecristo have not been properly studied. If they are similar morphologically to the Honduran Montecristo specimens, which they should be, those populations apparently represent an undescribed species. A study is planned to compare the specimens from Cerro Montecristo in all three countries. Additionally, some morphological characters given in the literature for Guatemalan *S. schmidti* do not agree with my data for the Honduran population (including those from the area of its type locality) of that nominal form (see the Redescription of *S. schmidti* above).

Another issue with *Sceloporus acanthinus* is its poorly documented geographical distribution. The most recent effort to understand the *S. acanthinus* distribution (H. M. Smith et al., 2001) left the shading off the map purported to show that distribution. Much of the published works concerning morphological data and distribution of *S. acanthinus* have been sloppy, inconsistent, and contradictive (even beginning with the misspelling of the name in the original description). For those reasons, the identification of the Guatemalan and El Salvadoran populations needs to be reinvestigated by performing a new morphological study. The current, noncongruent, character states given in the literature for El Salvadoran and Guatemalan Montecristo populations compared with my data from those Honduran populations are perplexing. Also, Pérez-Ramos and Saldaña-de La Riva (2008) divided *S. acanthinus* into two forms,

one with a wide, complete nuchal collar and the other with a narrow, complete nuchal collar. The Guatemalan and El Salvadoran Montecristo *S. acanthinus* specimens were not assigned to either group by Pérez-Ramos and Saldaña-de La Riva (2008). Those authors also made no mention of the divided nuchal collar stated by Stuart (1971) for some *S. acanthinus* he examined. Clearly, *S. acanthinus* remains poorly studied taxonomically, and those contradictive diagnostic morphological characters given for it in the literature are unsatisfactory. It is also possible that the Bocourt *S. acanthinus*, with its type locality on the "plateau of Guatemala" (Stuart, 1963: 70), might well not be conspecific with the populations called *S. acanthinus* on Cerro Montecristo in El Salvador and Guatemala, thus further questioning the previously published data regarding *S. acanthinus*.

Color in life of an adult male from the Honduran side of Cerro Montecristo (FN 252220): dorsal ground color Lime Green (159) with slightly darker green and dark brown mottling; abdominal semeions Cobalt Blue (168) with distinct black edges (2–3 scale rows wide); abdominal semeions separated medially by dirty white about 1–2 scale rows wide at midbody, 3–4 scale rows wide posteriorly, male abdominal semeions not crossing lateral chest region to connect with throat collar.

Illustrations.—None published of Honduran populations.

Natural History Comments.—The Honduran populations from Cerro Montecristo are known from 1,350 to 1,650 m elevation in the Premontane Wet Forest and Lower Montane Moist Forest formations. This *Sceloporus* was collected in June and August, and appears quite common as several were seen on each trip to the region.

Specimens Examined (6 [0]; Map 48)— **OCOTEPEQUE:** El Mojanal, FN 212557; Las Hojas, FN 20044–45, 252206–07, 252220.

Sceloporus populations likely representing undescribed species

The new molecular data from E. N. Smith suggested several populations from Honduras, as well as from Guatemala, represent undescribed species (E. N. Smith, personal communication, 8 January 2016). Those Honduran populations' collecting localities are shown on Map 48 and listed here. A, Montaña de Joconales, Santa Bárbara; B, Cerro Santa Bárbara, Santa Bárbara; C, Ocotepeque (non-Montecristo population) and Cerro Celaque, Lempira; D, a relatively lowland population from northeastern Olancho and southeastern Colón; E, an isolated broadleaf cloud forest group of populations from Intibucá and La Paz. Those montane broadleaf forest populations from Intibucá and La Paz were not studied using molecular techniques.

The E. N. Smith molecular data suggest four additional cryptic species, other than *S. esperanzae*, *S. hondurensis*, and *S. schmidti*, occur among these morphologically similar lizards and need to be described. Additional morphological work and additional molecular data on these populations is hampered by the lack of sufficient specimens from most populations. Below, I discuss what little has been published and what else is known regarding those populations of *S. "malachiticus"* likely representing undescribed species, all but one of which was used in the unpublished E. N. Smith molecular study.

Species A, Montaña de Joconales, Santa Bárbara species (squares on Map 48)

I was unable to find any outstanding distinct morphological characters to distinguish this molecular "species" from other Honduran populations of these morphologically conservative malachite lizards. It seems that all of their head scale characteristics and color patterns are shared by one or more of the other malachite named Honduran species, with the exception of the non–Cerro Montecristo Ocotepeque populations. A more thorough study of these populations is planned for the near future.

Illustrations (Fig. 59).—None previously published.

Natural History Comments.—This apparently undescribed species is known from 1,100 to 2,010 m elevation in the Premontane Wet Forest and Lower Montane Wet Forest formations. Until recent years, gangs of carjackers controlled all roads into this area, thus only two trips were made to this isolated mountain range. Those trips were in April and July during the dry and wet times of the year. This apparently undescribed species is typical in its habitats with those of other Honduran malachite *Sceloporus*. Four ready-for-birth neonates (FN 253468) were taken from FN 253462 on 18 April. That adult female was badly hampered in her movements by the presence of her young.

Specimens Examined (7, 4 neonates [0]; Map 48)—**SANTA BÁRBARA:** Las Quebradas, USNM 57824; Nuevo Joconales, FN 253461–63, 253468 (4 neonates); Quebrada de las Cuevas, FN 256155, 256172, 253496.

Species B, Cerro de Santa Bárbara species (triangles on Map 48)

Sceloporus malachiticus: Meyer, 1969: 252 (in part); Meyer and Wilson, 1973: 26 (in part); Townsend et al., 2007: 10; Solís et al., 2014: 131 (in part).

I was also unable to find any outstanding distinct morphological characters to distinguish this molecular species from other Honduran populations of the morphologically conservative malachite lizards. However, I had only one adult specimen (FN 213399) in my possession for this study.

Illustrations.—None published.

Remarks.—Cerro Santa Bárbara is a completely isolated mountain range surrounded by habitat unsuitable for *Sceloporus malachiticus* species complex members.

Natural History Comments.—This apparently undescribed species is known from

1,080 to 1,650 m elevation in the Premontane Wet Forest and Lower Montane Wet Forest formations. Months of collection were June to August during the rainy season. It is a typical *Sceloporus malachiticus* complex species in that it basks in the sun to achieve suitable body temperatures to search, or sit-and-wait, for prey. It was not seen during heavily clouded periods. This is the only *S. malachiticus* species complex member in the country I have seen sleeping partially exposed at night during some unpleasantly cold temperatures.

Specimens Examined (8 [3]; Map 48)— **SANTA BÁRBARA:** N slope of Cerro Santa Bárbara, FN 13891; E slope of Cerro Santa Bárbara, LACM 72100–02; 8.8 km SW of El Jaral, LACM 72099; near El Mochito, AMNH 70296–97; San José de Los Andes, FN 213399.

Species C, Ocotepeque (but not Cerro Montecristo) and Cerro Celaque, Lempira, species (stars on Map 48)

Sceloporus malachiticus: Meyer, 1969: 252 (in part); Meyer and Wilson, 1973: 26 (in part); Wilson et al., 1979b: 62; Cruz [Díaz] et al., 1993: 28; E. N. Smith, 2001: 148 (in part); Wilson et al., 2001: 136 (in part); Solís et al., 2014: 131 (in part).

The four specimens I examined carefully (FN 252237, 252240, 252252, 252273) of this molecular species (E. N. Smith, personal communication, 8 January 2016) are unique among all other Honduran malachite populations in having small, irregularly arranged anterior head scales (snout region). These four specimens have 20–27 (24.3 ± 3.0) scales between the rostral and frontonasal and canthal scales. All other Honduran populations examined to date have fewer than 20 scales in that region. If this character continues to hold true after examining more specimens from this group, then they are definitely deserving of species-level recognition. The specimen numbers in bold below were studied by E. N. Smith using molecular techniques.

Color in life of an adult male of this apparently undescribed species (FN 252252): dorsal ground color Lime Green (159) with slightly darker green and dark brown mottling; abdominal semeions Cobalt (68) without black borders; scales between abdominal semeions dirty white; throat between and in front of Jet Black (89) throat collar Spectrum Blue (69), changing to pale green on chin.

Illustrations.—None published.

Natural History Comments.—This apparently undescribed species is known from 1,400 to 2,530 m elevation in the Lower Montane Wet Forest and Premontane Moist Forest formations. It was collected from April to June and in August during both the dry and rainy seasons. Specimens were collected in primary broadleaf forest and in denuded areas. Otherwise, this form seems to have the normal *S. malachiticus* species complex–like habits of other Honduran populations.

Specimens Examined (34 [3]; Map 48).— **LEMPIRA:** E slope of Cerro Celaque, KU 200562; near Parque Nacional Celaque Centro de Visitantes, UTA R-**42633**. **OCOTEPEQUE:** Cerro El Pital, MVZ 40129, 40152–59; El Chagüitón, KU 194317–19; El Portillo de Cerro Negro, FN 252252, **252255**, 252273; El Portillo de Ocotepeque, LACM 72096–98, LSUMZ 33691–92; road between Nuevo Ocotepeque and La Labor, UTA R-46241–45, 46781–83; 1.5 km SE of Plan del Rancho, UTA R-**42635**; Quebrada La Quebradona, UTA R-42634; Sumpul, FN **252237**, 252240.

Other Records.—**LEMPIRA:** Cerro Celaque, UNAH 2018, 2370–71 (Cruz [Díaz] et al., 1993).

Species D, Lowland population from northeastern Olancho and southeastern Colón (circles on Map 48)

Sceloporus malachiticus: E. N. Smith, 2001: 146 (in part); McCranie et al., 2006: 217 (in part); Solís et al., 2014: 131 (in part).

I was unable to find any outstanding characters to distinguish this population from all other Honduran populations, although molecular study indicates it apparently needs to be named. However, those that were carefully examined always have a pair of enlarged scales bordering the anterior edge of the enlarged frontonasals. That same character also occurs in some *S. hondurensis*, but in those cases the two outside internasals are always elongated and rather large. That might be a defining character should it hold true after studying more specimens of various Honduran populations.

Illustrations.—(Figs. 113, 116)—None previously published.

Natural History Comments.—This apparently undescribed species is known from 540 to 660 m elevation in the Lowland Moist Forest and the Premontane Wet Forest formations. It was collected on both trips into the area in July and August during the rainy season. This species was collected in both nearly primary broadleaf forests and newly denuded broadleaf forest. All known localities for this form lie in the buffer and nuclear zones of the Río Plátano Biosphere Reserve. However, that occurrence in a world heritage site means nothing toward the survival of these lizards. My two trips to these areas were close to 15 years ago, and the forest devastation was already full speed ahead. Flying over the Biosphere Reserve during the last 3 years has revealed the forest devastation has now reached the very center of the reserve and has increased during the most recent 5 or so years.

Specimens Examined (9, 1 skeleton [0]; Map 48).—**COLÓN**: Quebrada Machín, **USNM 536492** (skin, skeleton), FN 11523 (skin only), 11526, 11550, 11648–51. **OLANCHO**: between La Llorona and confluence of ríos Lagarto and Wampú, FN 9690; Quebrada de Las Marías, FN 11350.

Species E, Intibucá and La Paz broadleaf populations (inverted triangles on Map 48)

Sceloporus malachiticus: Wilson et al., 1991: 70 (in part); Solís et al., 2014: 131 (in part).

I had only one adult available of these populations for this study; thus, I can make no morphological comments on these populations. Tissues are not available from any specimen in this group.

Illustrations.—None published.

Remarks.—The population assigned to this group from Intibucá occurs in former broadleaf forest sites (cloud forest) on peaks above the pine forest *Sceloporus esperanzae* occupies. However, the broadleaf forest specimens do not resemble the pine forest specimens, in having much different abdominal semeion patterns. Molecular studies of these populations are needed.

Natural History Comments.—This possibly undescribed species is known from 1,800 to 2,150 m elevation in the Lower Montane Moist Forest formation. The isolated mountaintops where this form occurs were largely deforested by 1982.

Specimens Examined (15 [0]; Map 48).—**INTIBUCÁ:** near El Rodeo, FN 212745–46; Montaña de Mixcure, FN 212589; Zacate Blanco, KU 194324–26, LSUMZ 38828. **LA PAZ:** Cantón El Zancudo, KU 184218–21; mountains S of San Pedro de Tutule, KU 200563–64; about 5 km S of Santa Elena, KU 194327–28.

Sceloporus squamosus Bocourt, 1874

Sceloporus squamosus Bocourt, 1874: 212, *In* A. H. A. Duméril et al., 1870–1909a (nine syntypes, MNHN 3180, 3180A–C, 3181, 3181A, 3182, 3182A, USNM 10964 [see Brygoo, 1989: 93–94; Cochran, 1961: 143; but not Bell et al., 2003: 156, who gave only a partial list of the syntypes]; type locality: "les environs de Guatemala et de l'Antigua, à une altitude de 1.500 mètres environ" and "le littoral du Guatemala sur le

Pacifique, a l'embouchure du Nagua-late"); Smith, 1939: 319; Bogert, 1949: 418; Stuart, 1954: 14; Meyer, 1969: 255; Hahn, 1971: 111; Meyer and Wilson, 1973: 27; Wilson et al., 1976: 179; Wilson et al., 1979a: 25; Wilson et al., 1991: 70; Eisenberg and Köhler, 1996: 11; Wiens and Reeder, 1997: 46, 47; Wilson and McCranie, 1998: 16; Köhler, 1999a: 214; Wilson et al., 2001: 136; Köhler and Heimes, 2002: 138; Köhler, 2003a: 143; Wilson and Town-send, 2007: 145; Köhler, 2008: 155; Townsend and Wilson, 2010b: 692; Solís et al., 2014: 131; McCranie, 2015a: 369; McCranie and Gutsche, 2016: 872.

Geographic Distribution.—*Sceloporus squamosus* occurs at low and moderate elevations on the Pacific versant from southeastern Chiapas, Mexico, to north-western Costa Rica. It is also found on the Atlantic versant in southeastern Guatemala and southwestern and north-central Honduras. In Honduras, this species occurs in largely disjunct populations in the western two-thirds of the country.

Description.—The following is based on ten males (KU 103245–47, 194329; LACM 72771, 72778; LSUMZ 33695; UF 124608; USNM 570245–46) and ten females (CAS 152968; KU 209317; LACM 72777; LSUMZ 24608–09, 33694, 36583–84; UF 124615; USNM 570247). *Sceloporus squamosus* is a relatively small sceloporine (maximum re-corded SVL 57 mm [Smith, 1939]; 52 mm SVL in largest Honduran specimen mea-sured [CAS 152968, a female]) with a long tail and relatively short limbs; dorsal head scales rugose; 2 postrostral scales, both wider than long; 2 internasal scales, both longer than wide; 1–2 canthals; lateral frontonasal scales separated from median frontonasal; paired prefrontal scales sepa-rated medially by smaller azygous scale; usually 4 frontal scales; 2 frontoparietal scales, in broad contact medially; interpari-etal large, roughly triangular-shaped; parie-

tal eye distinct; 4–5 large supraocular scales, separated from median head scales by usually complete row of 6–11 (8.7 ± 1.4) supraorbital semicircle scales, supraorbital semicircle row sometimes incomplete pos-teriorly; 5–6 superciliary scales, separated from large supraoculars by 1 complete and 1 incomplete row of small scales; frontal ridge distinct to absent; nasal scale single, sepa-rated from rostral; 1 row of lorilabials; fourth supralabial and third or fourth infralabial below level of mideye; moveable eyelid present; pupil circular; 1–3 loreals; 13–20 (16.1 ± 2.4) gular scales; body dorsally compressed; dorsal body scales strongly keeled, mucronate, imbricate, 27–35 (30.9 ± 2.2) between interparietal and level above posterior edge of hind limb; lateral body scales strongly keeled, mucro-nate, imbricate, smaller than dorsal scales; lateral nuchal scales smaller than dorsal nuchal scales; ventral scales smooth to weakly keeled, imbricate, with pointed posterior ends, 31–41 (34.5 ± 3.2) para-midventral scales between level of anterior edge of forelimb and cloacal scale; 32–47 (36.8 ± 4.1) scales around midbody; scales on posterior surface of thigh smaller than those on dorsal surface of thigh, but not granular; 4–6 (4.9 ± 0.6) femoral pores in males, 3–6 (4.3 ± 0.9) in females, each series separated by 9–13 (11.0 ± 1.3) scales in males, 8–14 (10.4 ± 1.6) scales in females; precloacal scales keeled in females; enlarged postcloacal scales absent in males; subcaudal scales strongly keeled at base of tail; postfemoral dermal pocket absent; SVL 40.5–51.9 (45.9 ± 4.4) mm in males, 39.7–52.1 (47.2 ± 4.1) mm in females; TAL/SVL 1.70–2.23 in eight males, 1.44–2.24 in eight females; HL/SVL 0.25–0.27 in males, 0.22–0.27 in females; HW/SVL 0.17–0.19 in males, 0.16–0.19 in females; SHL/SVL 0.23–0.28 in males, 0.22–0.27 in females; S-OL/SVL 0.22–0.24 in males, 0.21–0.24 in females.

Color in life of an adult male (USNM 580374; Plate 61): dorsal ground color

Plate 61.   *Sceloporus squamosus*. USNM 580374, adult male. Valle: Amapala, del Tigre Island, near communications tower.

Cinnamon (123A) with middorsal, black edged Brussels Brown (121B) blotches, blotches gradually becoming darker brown posteriorly until Prout's Brown (121A) on last 3 blotches; Prout's Brown lateral stripe present on body and base of tail; ventrolateral surface of body Brussels Brown; top of head Cinnamon with indistinct darker brown mottling; tail Brussels Brown with darker brown mottling, tail becoming dark brown at transition from lateral to ventral portion; belly Brussels Brown ventrolaterally, becoming paler brown medially.

Color in alcohol: dorsal surface of body medium brown to dark brown; pale brown paired dorsolateral stripes extending from level just posterior to ear opening onto base of tail; series of short, black, oblique stripes present between dorsolateral stripes and vertebral midline; ventral surfaces of head and body cream with varying amounts of gray to nearly black flecking; ventrolateral abdominal patches absent in both sexes.

Diagnosis/Similar Species.—The combination of having strongly keeled and mucronate dorsal scales, a dorsally compressed body, and the presence of femoral pores in both sexes distinguishes *Sceloporus squa-*mosus from all other Honduran lizards, except the remaining *Sceloporus*. Members of the *S. malachiticus* species complex lack pale dorsolateral stripes, and males usually have distinct ventrolateral abdominal patches (versus pale dorsolateral stripe present and males lack abdominal patches in *S. squamosus*). *Sceloporus variabilis* has a postfemoral dermal pocket, has notched ventral scales, has smooth precloacal scales in females, has enlarged postcloacal scales in males, has smooth to weakly keeled subcaudal scales at the base of the tail, and has ventrolateral abdominal patches in males (versus no postfemoral dermal pocket, ventral scales pointed posteriorly, keeled precloacal scales in females, no enlarged postcloacal scales in males, subcaudal scales at base of tail strongly keeled, and abdominal patches absent in *S. squamosus*).

Illustrations (Figs. 108, 111; Plate 61).— A. H. A. Duméril et al., 1870–1909b (adult, head scales, ear opening, dorsal scale, ventral scale); Eisenberg and Köhler, 1996 (adult); Fitch, 1973a (adult); Köhler, 1999a (adult), 2000 (adult), 2001b (adult), 2003a (adult, lateral head scales), 2008 (adult, lateral head scales); Köhler and Heimes,

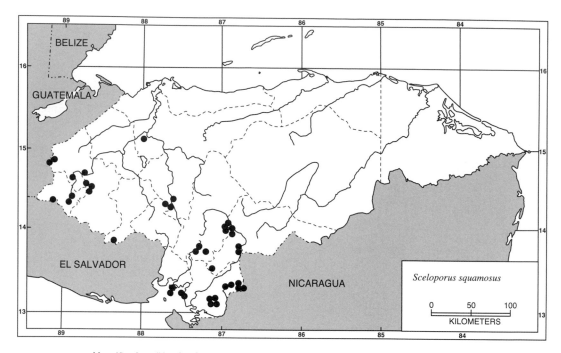

Map 49.   Localities for *Sceloporus squamosus*. Solid circles denote specimens examined.

2002 (adult); Köhler et al., 2005 (adult); Savage, 2002 (adult); Werning, 2002 (head and forebody); Wiens and Reeder, 1997 (dorsal and lateral head scales, scales surrounding ear opening).

Remarks.—Sceloporus squamosus belongs to the *S. siniferus* species group (Bell et al., 2003; Wiens et al., 2010).

Natural History Comments.—Sceloporus squamosus is known from near sea level to 1,470 m elevation in the Lowland Dry Forest, Lowland Arid Forest, Premontane Moist Forest, and Premontane Dry Forest formations. This diurnal species is both terrestrial and arboreal and is probably active throughout the year, as it has been found in March, April, July, November, and December. It is active on rock walls used as fencerows, on large rocks, on fence posts, and low on tree trunks. However, it is most often seen on the ground in the vicinity of the above-mentioned situations, especially where low shrubs are well represented.

Very little is known about reproduction in this species. Females from a population in La Paz, Honduras, are reported to deposit four or five eggs per clutch from June to July (Eisenberg and Köhler, 1996). Apparently nothing has been published on food habits of this species, but it likely feeds on a variety of arthropods, especially insects.

*Etymology.—*The name *squamosus* is Latin for scaly and refers to the relatively large dorsal scales of this species.

Specimens Examined (103, 1 C&S [25]; Map 49).—**CHOLUTECA**: 1.0 km N of Cedeño, LSUMZ 33694–95, 36584; 1.9 km SW of Comalí, SMBU 16943; El Banquito, CAS 152982; El Despoblado, CAS 152965–66, 152968–69; Finca Guayabal, SDSNH 72779–80; Finca Monterrey, USNM 579566; Finca Santa Fe, SDSNH 72766; La Fortunita, SDSNH 72774; La Isnaya, SDSNH 72770–73; La Pacaya, SDSNH 72768–69; Punta Ratón, KU 209317; Quebrada del Horno, SDSNH 72767; 1.4 km

SW of San Marcos de Colón, SMBU 16935. **COMAYAGUA**: 4.8 km SSE of Comayagua, LACM 47294. **COPÁN**: Cabañas, USNM 570245; near Copán, UF 124608; Copán, UMMZ 84274; 17 km NE of Cucuyagua, LSUMZ 24608; Río Higuito, ANSP 28119; 11.9 km E of Santa Rosa de Copán, USNM 570246. **CORTÉS**: Potrerillos, UNAH 5279. **EL PARAÍSO**: El Rodeo, USNM 580788–89; Monserrat, MCZ R49953; near Ojo de Agua, AMNH 70388; 1 km S of Orealí, USNM 580790; Orealí, UNAH (1); Soledád, KU 192321. **FRANCISCO MORAZÁN**: Caserío Los Encinitos, USNM 580782–83; near El Zamorano, AMNH 70386, LACM 39775, MCZ R49909 (2), 49938, 49985, UF 124613; 8 km ENE of El Zamorano, KU 103249 (C&S), 103250; 7 km E of El Zamorano, CM 64859; El Zamorano, KU 103245–48, LACM 39776, MCZ R49763 (5); Río Yeguare, MCZ 48679–82, UMMZ 94041 (3); 12 km SSW of Sabanagrande, LSUMZ 24609; 21 km SW of Sabanagrande, LSUMZ 24610; Villa San Francisco, AMNH 70385, 70387. **IN-TIBUCÁ**: Santa Lucía, SMF 79145. **LA PAZ**: 1 km E of La Paz, CM 64860; 15.0 km N of Marcala, USNM 570247; Potrerillos, FMNH 283601, USNM 580784–87. **LEM-PIRA**: 11.3 km NNW of Gracias, LACM 47293; 4 km N of Gracias, UF 124646; 4 km N of and 3 km W of Gracias, UF 124616–19; 5 km S of Gracias, UF 124615; Gracias, FMNH 28561, 40867. **OCOTEPEQUE**: 10 km N of La Labor, UF 124609–11; 5 km N of Nueva Ocotepeque, TCWC 23773; 1.5 km SE of San Marcos de Ocotepeque, UF 124614. **VALLE**: Isla del Tigre, near summit, USNM 580374–75; Isla del Tigre, on trail to summit, USNM 578733–35; Isla del Tigre, SDSNH 72775–78; Isla Zacate Grande, KU 194329, LSUMZ 36583.

Sceloporus variabilis Wiegmann, 1834b

Sceloporus variabilis Wiegmann, 1834b: 51 (seven syntypes, ZMB 650–653 [see Taylor, 1969: v; Bell et al., 2003: 162]; type locality not stated, but "Mexico" inferred from title of publication containing description); Dunn and Emlen, 1932: 28; Bogert, 1959: 110; Meyer, 1969: 258; Wilson and Meyer, 1969: 146; Hahn, 1971: 111; Meyer and Wilson, 1973: 27; Wilson et al., 1991: 70; Cruz [Díaz] et al., 1993: 28; Espinal, 1993, table 3; Köhler, 1996g: 66; Wilson and McCranie, 1998: 16; Köhler, 2000: 86; Espinal et al., 2001: 106; Wilson et al., 2001: 137; Castañeda, 2002: 40; McCranie, Castañeda, and Nicholson, 2002: 25; Castañeda and Marineros, 2006: 3.8; Lovich et al., 2006: 13; McCranie et al., 2006: 117; Townsend, 2006a: 35; Townsend et al., 2006: 32; Wilson and Townsend, 2006: 105; Townsend and Wilson, 2008: 144; Townsend and Wilson, 2010b: 692; Solís et al., 2014: 131; McCranie, 2015a: 369; McCranie and Gutsche, 2016: 873.

Sceloporus variabilis olloporus H. M. Smith, 1937: 11; Smith, 1939: 282; Bogert, 1949: 418; Peters, 1952: 38; Stuart, 1954: 15; Marx, 1958: 471; Meyer, 1966: 175; Smith et al., 1993: 113.

Sceloporus variabilis variabilis: Sites and Dixon, 1982: 16; Mather and Sites, 1985: 373.1.

Sceloporus olloporus: Mendoza-Quijano et al., 1998: 360; Wilson and Townsend, 2007: 145.

Geographic Distribution.—*Sceloporus variabilis* occurs at low, moderate, and lower portions of intermediate elevations on the Atlantic versant from southern Texas, USA, to northeastern Nicaragua and marginally in central Costa Rica and on the Pacific versant from the Isthmus of Tehuantepec, Mexico, to northwestern Costa Rica (but see Remarks). In Honduras, this species occurs across much of the country.

Description.—The following is based on ten males (USNM 570255, 570260, 570263, 570269, 570272, 570275–77, 570279–80) and ten females (USNM 570248–51, 570259, 570265, 570267, 570271, 570274, 570278). *Sceloporus variabilis* is a relatively moderate-sized sceloporine (maximum recorded SVL 76 mm [Smith et al., 1993]; 63 mm SVL in largest Honduran specimen measured [USNM 570280, a male]) with a long tail and relatively short limbs; dorsal head scales rugose; 2–4 (usually 4) postrostral scales; 2–3 (usually 2) internasal scales; 2 canthals; lateral frontonasal scales separated from median frontonasal; paired prefrontal scales separated medially by smaller azygous scale; anterior section of frontal scale usually divided, posterior section usually entire; 2–3 frontoparietal scales, usually in broad contact medially; interparietal large, roughly triangular-shaped, usually partially divided posteriorly and sometimes anterolaterally and/or laterally; parietal eye distinct; usually 5 large supraocular scales, separated from median head scales by usually complete row of 13–17 (14.7 ± 1.3) supraorbital semicircle scales; 5–6 (usually 6) superciliary scales, usually separated from large supraoculars by 2 incomplete rows of small scales; frontal ridge not distinct; nasal scale single, separated from rostral; 1 lorilabial row; third or fourth supralabial below level of mideye; usually third infralabial below level of mideye; moveable eyelid present; pupil circular; 1–2 (usually 1) loreals; 19–27 (22.7 ± 2.1) gular scales; body dorsally compressed; dorsal body scales strongly keeled, mucronate, imbricate, 45–52 (48.8 ± 2.1) between interparietal and level above posterior edge of hind limb; lateral body scales strongly keeled, mucronate, imbricate, smaller than dorsal scales; lateral nuchal scales smaller than dorsal nuchal scales; ventral scales smooth, imbricate, most with notched posterior ends; 56–67 (60.7 ± 2.7) para-midventral scales between level of anterior edge of forelimb and cloacal scale; 63–75 (71.0 ± 3.5) scales around midbody; scales on posterior surface of thigh granular; 7–11 (9.8 ± 1.2) femoral pores in males, 8–10 (9.1 ± 0.9) in females, each series separated by 17–24 (21.6 ± 2.2) scales in males, 19–23 (21.1 ± 1.9) in

females; precloacal scales smooth in females; paired, enlarged postcloacal scales in males; subcaudal scales smooth to weakly keeled at base of tail; postfemoral dermal pocket present; SVL 51.0–63.2 (58.9 ± 3.8) mm in males, 45.1–61.1 (53.6 ± 4.5) mm in females; TAL/SVL 1.20–1.53 in eight males, 1.00–1.27 in seven females; HL/SVL 0.24–0.26 in males, 0.23–0.27 in females; HW/SVL 0.19–0.21 in males, 0.18–0.21 in females; SHL/SVL 0.23–0.27 in males, 0.21–0.26 in females; S-OL/SVL 0.21–0.23 in males, 0.21–0.24 in females.

Color in life of an adult male (USNM 570280): middorsum Grayish Horn Color (91) with scattered small Turquoise Green (64) spots divided into rectangular area by indistinct tan, short crossbands, those on either side divided by narrow tan middorsal stripe and bounded laterally by broken Buff-Yellow (53) dorsolateral stripe; area below dorsolateral stripe Kingfisher Rufous (240) with scattered Turquoise Green spots; Jet Black (89) spot above axilla, spot edged anteriorly with Sulfur Yellow (157); dorsal surfaces of fore- and hind limb Hair Brown (119A) with slightly darker brown crossbands; dorsal surface of tail Drab-Gray (119D) with narrow Light Drab (119C) chevrons; dorsal surface of head Light Drab with scattered Sepia (119) spots and streaks; side of head with pale brownish gray postocular stripe, stripe grading into dorsolateral pale brown stripe, stripe also with rust brown along lower edge in temporal region; area around orbit Flesh Ocher (132D); chin Rose Pink (108D) with Smalt Blue (70) flecks; chest mottled pale copper and pale turquoise green; ventrolateral semeions Rose Pink, edged with Ultramarine Blue (170A) medially and Blue Black (90) anteriorly and posteriorly; undersides of fore- and hind limb gray-brown mottled with cream; iris dark brown. Color in life of another adult male (USNM 570273): dorsal surface of head brown; shoulder region black with central gold blotch; middorsum brown with dark brown blotches followed by smaller tan blotches on either side of pale brown middorsal stripe; dorsolateral stripe tan; lateral surface of body brown with tan spotting; dorsal surfaces of fore- and hind limb pale brown with dark brown crossbars; dorsal surface of base of tail rust brown, remainder of tail barred with pale rust brown and dark brown; black patch at groin; chin and chest pearl gray; abdominal semeions pale pink, semeions separated medially by pale blue. Color in life of an adult female (USNM 570250): dorsum Tawny Olive (223D) with Sepia (119) marks on either side of vertebral stripe; vertebral stripe slightly paler brown than dorsal ground color; dorsolateral stripes dirty white above forelimb, pale brown elsewhere; lateral surface of body Vandyke Brown (121) with pale brown and darker brown spotting; dorsal and lateral surfaces of head medium brown with dark brown flecking, except area below eye dirty white; area in front of forearm insertion orange; yellow vertebral bar above forearm connecting with dorsolateral stripe; belly pale brownish white with pale gray lateral abdominal semeions; iris orange-brown. Another adult female (USNM 570251) was similar in color to that of USNM 570250, except color in front of forearm Peach Red (94), Peach Red extending onto lateral surface of head and infralabial area. Color in life of another adult female (USNM 570274): dorsal surface of head brown; middorsum brown with dark brown blotches; dorsolateral stripe pale brown; area below pale dorsolateral stripe brown; dorsal surfaces of fore- and hind limb brown with dark brown crossbars; dorsal surface of tail barred brown and pale brown; lips with pale red-orange wash; chin pale gray; bright red-orange blotch on lateral gular region; belly pale bronze.

Color in alcohol: dorsal surface of body medium brown to dark brown; pale brown paired dorsolateral stripes extend from just posterior to eye onto base of tail; less distinct pale brown middorsal stripe usually evident, stripe extending from level above ear opening to posterior end of body; dorsolateral and middorsal stripes usually less distinct in specimens from lowland pine savanna of northeastern Honduras than those from remainder of country; a series of dark brown to nearly black spots present between dorsolateral and middorsal stripes; males with black band extending from axilla toward midventer, then extending posterior on belly on each side of midventral area, then arching to each groin; remainder of ventral surfaces of males cream with gray flecking, especially on chest in front of ventral bands; ventral surfaces of head and body of females cream with varying amounts of gray to nearly black flecking, many females have a midventral cream stripe separating flecked ventrolateral parts of belly.

Diagnosis/Similar Species.—The combination of having strongly keeled and mucronate dorsal scales, a dorsally compressed body, and the presence of femoral pores in both sexes distinguishes *Sceloporus variabilis* from all remaining Honduran lizards, except the other *Sceloporus*. *Sceloporus variabilis* is the only Honduran *Sceloporus* to have a postfemoral dermal pocket.

Illustrations (Figs. 58, 62, 109, 110; Plate 62).—Bartelt and Prassel, 2009 (adult, male abdominal semeions patch, juvenile; as *S. olloporus*); Fitch, 1973a (adult); Köhler, 2000 (adult), 2001b (adult), 2003a (adult), 2008 (adult); Lemos-Espinal and Dixon, 2013 (adult, male abdominal semeions patch); McCranie et al., 2006 (adult); Powell et al., 2012 (post femoral dermal pocket); Savage, 2002 (adult); Taylor, 1956b (adult; as *S. variabilis olloporus*); Townsend and Wilson, 2008 (adult); Villa et al., 1988 (adult); Werning, 2002 (adult; as *S. olloporus*).

Remarks.—Controversy exists in the literature on the systematics of *Sceloporus variabilis*. Based on patterns of molecular (isozyme) variation, Mendoza-Quijano et al. (1998) suggested elevating *Sceloporus variabilis marmoratus* Hallowell (1854b: 178;

Plate 62. *Sceloporus variabilis.* USNM 579563, adult female. Choluteca: Finca Monterrey.

described as a species) and *S. v. olloporus* H. M. Smith (1937: 11) to species level. However, Mendoza-Quijano et al. (1998) collected isozyme data for only one specimen they assigned to *S. olloporus* (from Chiapas, Mexico) and data from specimens from only one locality for *S. marmoratus*. The locality of their *S. marmoratus* is over 500 km distant from their closest molecular samples of *S. variabilis*. Complicating matters is that the locality of the *S. olloporus* specimen from which Mendoza-Quijano et al. (1998) extracted molecular data lies outside of the geographic distribution of *S. v. olloporus* as defined by H. M. Smith (1937, 1939). Mather and Sites (1985) presented an overview of the morphology of *S. variabilis* and briefly reviewed the literature (including various subspecies of *S. variabilis* now recognized by some workers as species; i.e., *S. teapensis* Günther, 1890: 75, *In* Günther, 1885–1902); *S. smithi* Hartweg and Oliver, 1937: 1; both recognized as species by Bell et al., 2003). A thorough review of the *S. variabilis* species complex using both molecular and morphological data is needed. Such a study would likely demonstrate that more than one

species is included in this complex, including in addition to those currently recognized as species.

Sceloporus variabilis belongs to the *S. variabilis* species group (Bell et al., 2003; Wiens et al., 2010).

Natural History Comments.—*Sceloporus variabilis* is known from near sea level to 1,760 m elevation in the Lowland Dry Forest, Lowland Arid Forest, Premontane Moist Forest, and Premontane Dry Forest formations and peripherally in the Lowland Moist Forest, Premontane Wet Forest, Lower Montane Wet Forest, and Lower Montane Moist Forest formations. The diurnal species occurs in lowland pine savanna in Lowland Moist Forest and can be especially common in moderate elevation pine forests (Premontane Moist Forest, if not frequently burned), and in open areas in Premontane Wet Forest and Lower Montane Wet Forest. It is usually active on the ground but will climb slightly above ground level and has been found in rocky areas, on rock walls used as fencerows, on lower parts of tree trunks, on logs, and on brush piles. Individuals on the ground usually retreat by climbing nearby trees or other nearby

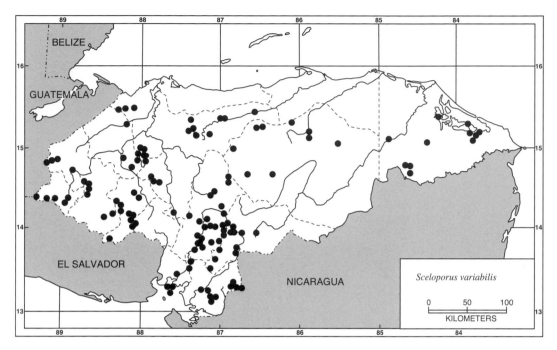

Map 50.   Localities for *Sceloporus variabilis*. Solid circles denote specimens examined.

objects such as those just mentioned. Daytime and nighttime retreats include under logs and rocks. The species is active under favorable conditions throughout the year, as it has been collected in every month. Bogert (1959) studied regulation of body temperature in this species near El Zamorano, Francisco Morazán. Savage (2002), in a literature review, noted that hatching occurs throughout the year in Costa Rica, with the highest reproduction occurring from July to November (= rainy season). Food is said to consist of small arthropods, especially insects (Savage, 2002, in a literature review). Also, Köhler and Fried (2012) reported an adult *S. variabilis* feeding on a juvenile *Aspidoscelis deppii* in Nicaragua.

Telford (1977) reported saurian malaria in specimens of this species from "Honduras."

Etymology.—The specific name *variabilis* is Latin for variable. Cope (1900: 400)

speculated that this taxon was "probably named on account of the wide difference in color between the males and females." Wiegmann (1834b) did allude to color differences between the sexes in his description of the species.

Specimens Examined (672 [159]; Map 50).—**CHOLUTECA**: Choluteca, KU 192322; El Despoblado, CAS 152964, 152967, 152971–78; El Jocote, SDSNH 72795, USNM 570252; El Madreal, USNM 570249–51, 580779; Finca Monterrey, USNM 579563–64; Finca Santa Clara, SDSNH 72789; La Fortunita, SDSNH 72790; La Isnaya, SDSNH 72782–83; La Libertad, SDSNH 72791–92; La Pacaya, SDSNH 72787–88; Moramulca, BYU 18239–41; Pespire, FMNH 5218; Quebrada del Horno, SDSNH 72793; Quebrada La Florida, SDSNH 72781; Tres Pilas, SDSNH 72784–86; 16 km W of Nicaraguan border, LSUMZ 51799. **COMAYAGUA**: 38.6 km SSE of La Paz, TNHC 48852; 43.4 km ESE of La Paz, TNHC 48851; 11.1 km NW of Siguatepeque, TNHC 32158; 13.7 km E of Siguatepeque, KU 67286–89; Siguatepeque, FMNH 5219–24, 5225 (5), UF 124630, 124632, 124639; 55.2 km NW of Tegucigalpa, TNHC 32159–63. **COPÁN**: 19.3 km ENE of Copán, LACM 47839–40; 6 km NE of Copán, AMNH 156477, LSUMZ 24611; Copán, AMNH 70318–21, 140274, FMNH 28514–15, UF 124620, 124628, 124641, UMMZ 84275; 13 km NE of Cucuyagua, LSUMZ 24612; Río Amarillo, USNM 570248. **CORTÉS**: 6 km N of Agua Azul, LSUMZ 24613; Agua Azul, AMNH 70326–27, LACM 61953, MCZ R49965, USNM 243334–35; Banaderos, UF 144109; 1.6 km W of El Jaral, LACM 47842, LSUMZ 24214; 1.6 km SE of El Jaral, LACM 47843; Los Pinos, UF 166390, 166392–93, USNM 570253, 573120–21, 578731–32; Parque Nacional Cusuco Centro de Visitantes, UF 144110; 37 km SW of Potrerillos, CM 64875, LSUMZ 30875, 52661; 4.8 km W of San Pedro Sula, LACM 47845–46; near Santa Elena, LSUMZ 38835. **EL PARAÍSO**: Agua Fría, AMNH 70317; 25.4 km W of Danlí, LACM 45066–68; El Rodeo, UNAH (3); 4.2 km NW of Güinope, KU 209318–19; 6 km SE of Los Limones, LSUMZ 24614; 30.6 km NW of Mandasta, KU 209320–22; Monserrat, MCZ R49954–57; Orealí, USNM 580780; Soledád, KU 192323–24. **FRANCISCO MORAZÁN**: Cantarranas, ANSP 28118, 33138–39; between Cantarranas and Talanga, UF 90202; 15.9 km SSW of Cofradía, LSUMZ 30872, 52659–60; 7 km E of El Zamorano, LSUMZ 52658; 8.0 km S of El Zamorano, AMNH 69641; 22 km NW of El Zamorano, AMNH 156485; El Zamorano, AMNH 70324–25, LACM 39774 (listed as 38774 by Meyer and Wilson, 1973), MCZ R49764 (4); 35.4 km W of Guaimaca, LACM 45095; 18.7 km WSW of Guaimaca, LSUMZ 22266, 24136; Guasculile, BYU 18234–38; Hacienda San Diego, AMNH 69069–73, 69076–79, TCWC 19199; La Venta, FMNH 5216–17; Los Corralitos, USNM 573118–19; 8 km W of

Maraita, UF 124636; 7 km W of Maraita, UF 124638; 10 km W of Maraita, UF 124622, 124631; Montaña de Guaimaca, AMNH 70322, 70328; 5 km N of Nueva Armenia, UF 124627; Ojojona, UTA R-46251–52; Río del Hombre, USNM 580775–76; Río Yeguare, LACM 39772–73, MCZ R48683–87, 49910–19, UMMZ 94038 (5), USNM 121122–24; 11–15 km N of Sabanagrande, AMNH 156478–81, LSUMZ 24615–17; 12 km SSW of Sabanagrande, LSUMZ 24618; S of San Antonio de Oriente, TU 19778 (now at LSUMZ, but not catalogued); San Francisco, AMNH 70323; 18.4 km NE of Talanga, LSUMZ 22267–68, 30873; 30.2 km N of Tegucigalpa, LACM 45242–43; Tegucigalpa, AMNH 69080–84, BYU 18831–35, FMNH 5215 (13), UTA R-41231; 2–6 km SW of Valle de Ángeles, AMNH 156482–84, LSUMZ 24619–20; 16 km SW of Valle de Ángeles, LSUMZ 24621; Zambrano, SMF 53687–93. **GRACIAS A DIOS**: Finca Nakunta, USNM 580781; Krausirpe, LSUMZ 52502; Mavita, USNM 570254–55, 589162; 2 km E of Mistruk, USNM 573116–17; Mocorón, UTA R-53179–86; Puerto Lempira, USNM 565423–24; Rus Rus, USNM 570256–57; Samil, USNM 573939; Sisinbila, USNM 579561; Tánsin, LSUMZ 21503–04; Walpatá, LACM 47836. **INTIBUCÁ**: 25.7 km NW of La Esperanza, USNM 570258–63; 18.1 km WNW of La Esperanza, UF 166396, USNM 570264–67; 0.9 km S of and 8.7 km NW of La Esperanza, USNM 570268; 16.1 km ESE of La Esperanza, USNM 570269; 15.0 km SE of La Esperanza, USNM 570270–71; 10 km SE of La Esperanza, UF 124637; about 15 km E of La Esperanza, KU 194331–33; La Rodadora, USNM 570272; La Soledád, FMNH 283629–30; 5–7 km SE of Masaguara, UF 124626; near San Miguelito, UF 124624, 124645; Santa Lucía, SMF 79116. **LA PAZ**: 13.7 km N of Marcala, USNM 570273–78; Marcala, CM 64876, UF 166391. **LEMPIRA**: Erandique, CM 64877–78, LSUMZ 30868–71, 52635–52, 52663; Gualcince, KU 194330; 11.3 km

NNW of Gracias, LACM 47841; 5 km S of Gracias, UF 124644; between Gracias and Villa Verde, KU 200579; Gracias, CM 64879, LSUMZ 30876–77, 52653–57, UF 124623, 124635; Las Culebras, UF 124640; Parque Nacional Celaque Centro de Visitantes, USNM 549368–69. **OCOTEPE-QUE**: near El Güisayote, FMNH 283630; 10 km N of La Labor, UF 89471, 124621, 124625, 124643; 1 km W of Nueva Ocotepeque, TCWC 23776; 12 km NNE of Nueva Ocotepeque, UF 124629; Nueva Ocotepeque, TCWC 23635–36; San Marcos de Ocotepeque, UF 124634. **OLANCHO**: 33.9 km E of Guaimaca, LSUMZ 30874; 18.7 km WSW of Juticalpa, LSUMZ 52662; Las Trojas, UTA R-41229–30; Montaña de Las Parras, USNM 342374–76; Montaña El Armado, USNM 342378; Pataste, MSUM 4660–78; Quebrada de Las Mesetas, USNM 342377; Río Cuaca, USNM 589161; 10.5 km S of San Esteban, KU 200577; San Esteban, UTA R-41228. **SANTA BÁRBARA**: 1.6 km NE of El Sauce, LACM 47844; 4.8 km E of Quimistán, LACM 47838; San Pedro Zacapa, TNHC 32619; ˙Santa Bárbara, MCZ R28195. **VALLE**: Coyolitos, FN 256882–83, 256886–87 (all still in Honduras because of permit problems); Isla del Tigre, near summit, USNM 580376–77; Isla Tigrito, USNM 580777–78; Isla Zacate Grande, KU 194334, LSUMZ 36582; 3.2 km W of Nacaome, LACM 47837; Playa Negra, SDSNH 72794; Punta Novillo, USNM 580378. **YORO**: 2 km S of Coyoles, KU 101442–43; Montaña La Ruidosa, KU 200578; Montañas de Mataderos, FMNH 21868–69, MCZ R38856 (2); Portillo Grande, FMNH 35152–53, MCZ R32269–78, 38855; between San Francisco and La Fortuna, UF 166394–95; San Francisco, MVZ 52414–15, USNM 570279; 3.5 km S of San Lorenzo Arriba, USNM 579562; 4.7 km ESE of San Lorenzo Arriba, USNM 565419–22, 570280; San Patricio, SMF 78422–23, USNM 579565; Subirana, MCZ R32279–82; Subirana Valley, BYU 36045–49, FMNH 21779 (+17 untagged speci-mens), 21780 (13), 21834–37, MCZ R38857–78, 38880–98 (+47 untagged specimens), UMMZ 77856 (= 22 untagged specimens), 77857 (28), 77858 (24), 77859 (20), 80459 (2).

Family Polychrotidae Fitzinger, 1843

This family (*sensu* T. M. Townsend et al., 2011) of lizards is limited in distribution to the Western Hemisphere, where it ranges from northwestern Honduras to western Peru and southeastern Brazil. It includes one genus and seven named species, one of which occurs in Honduras.

Genus *Polychrus* Cuvier, 1816

Polychrus Cuvier, 1816: 40 (type species: *Lacerta marmorata* Linnaeus, 1758: 208, by monotypy).

Geographic Distribution and Content.—See family Polychrotidae.

Etymology.—The generic name *Polychrus* is formed from the Greek *poly* (many, very) and *chros* (color of skin), and apparently alludes to the colorful pattern found in the type species of the genus.

Polychrus gutturosus Berthold, 1845

Polychrus gutturosus Berthold, 1845: 38 (holotype, ZFMK 21341 [see Böhme and Bischoff, 1984: 184; Böhme, 2010: 92]; type locality not stated ["Neu-Granada" implied from title of paper where name was proposed], but given as: "Provinz Popayan, etwa 2° N. B. und 301° L" by Berthold, 1846: 4 ["western Colombia, probably on the Pacific versant" according to Myers and Böhme, 1996: 17]); Meyer and Wilson, 1973: 26; Wilson et al., 2001: 136; McCranie et al., 2006: 128; Wilson and Townsend, 2006: 105; McCranie, 2007a: 175; Townsend and Wilson, 2010b: 692; Solís et al., 2014: 131; McCranie, 2015a: 370.
Polychrus gutterosus [sic]: Meyer, 1969: 252.

Geographic Distribution.—*Polychrus gutturosus* occurs at low and moderate elevations on the Atlantic versant from northwestern Honduras to northwestern Colombia and from west-central Costa Rica to western Ecuador on the Pacific versant. It is also found marginally on the Pacific versant in northwestern Costa Rica. In Honduras, this species is known from a few low elevation localities in the northern portion of the country.

Description.—The following is based on three males (UF 137431; USNM 563277, 579560) and four females (UF 137404; UMMZ 58369; USNM 563276, 570298). *Polychrus gutturosus* is a relatively large lizard (maximum recorded SVL 172 mm [USNM 570298, a female]) with an extremely long, nonautotomic tail (>3 times SVL); dorsal head scales in internasal and prefrontal areas multicarinate, those in frontal area smooth to rugose; most scales in parietal area smooth to rugose; frontal and parietal depressions present; 4–5 postrostrals; anterior nasal single, contacting rostral, second supralabial, and 2 loreals; 5 internasals (see Remarks); canthal ridge broadly rounded; scales comprising supraorbital semicircles smooth to rugose, largest scale in semicircles larger than largest supraocular scale; supraorbital semicircles poorly defined; 0–2 (usually 1) scales between supraorbital semicircles; 1–2 scales between supraorbital semicircles and interparietal; interparietal well defined, irregular in outline, longer than wide, smaller than ear opening; 3–15 enlarged, multicarinate supraocular scales in 2–3 rows; enlarged supraoculars varying from completely separated from supraorbital semicircles by 1 row of small scales to 1–5 enlarged supraoculars in broad contact with supraorbital semicircles; no elongate superciliaries; 2 enlarged canthals; 3–5 scales between second canthals; 3 rows of multicarinate loreal scales, 4–6 loreals; 5–6 supralabials and 4–5 infralabials to level below mideye; moveable eyelid present; pupil circular; suboculars multicarinate, in broad contact with supralabials; ear opening vertically oval; 2 large postmentals; gular scales smooth to faintly keeled; male dewlap large, extending to level of axilla; male dewlap with 2–3 horizontal gorgetal-sternal scale rows, 8 (mean number) scales per row; 1 anterior marginal pair in male dewlap ($n = 2$); female dewlap rudimentary; no nuchal crest or dorsal ridge; middorsal scales keeled, slightly larger than, or equal in size to homogeneous multicarinate lateral scales; 52–78 (63.3 ± 13.3) dorsal scales along vertebral midline between levels of axilla and groin in males, 59–67 (62.8 ± 3.4) in females; 20–31 (23.7 ± 5.5) dorsal scales along vertebral midline contained in 1 head length in males, 20–25 (23.0 ± 2.2) in females; ventral scales on midsection about same size as largest dorsal scales; ventral body scales keeled (multicarinate in adults), imbricate; 57–75 (67.0 ± 9.2) ventral scales along midventral line between levels of axilla and groin in males, 62–67 (64.5 ± 2.1) in females; 23–31 (27.0 ± 4.0) ventral scales contained in 1 head length in males, 24–28 (26.0 ± 1.8) in females; 66–87 (77.3 ± 7.7) scales around midbody; axillary pocket absent; 23–30 (27.7 ± 4.0) total femoral pores in males, 18–36 (26.8 ± 7.6) in females, pores much better developed in males; precloacal scales multicarinate; enlarged postcloacal scales absent in males; tail round; basal subcaudal scales multicarinate, remainder of caudal scales unicarinate; lateral caudal scales unicarinate, homogeneous; dorsal medial caudal scale row not enlarged, keeled, not forming crest; scales on lateral surface of antebrachium multicarinate; 25–32 (28.5 ± 2.5) subdigital lamellae on phalanges II–IV of Digit IV of hind limb, 8–12 (10.8 ± 2.4) lamellae on Phalanx I of Digit IV of hind limb; Digit IV of hind limb same length as, or only slightly longer than Digit III of hind limb; SVL 87–122 (109.7 ± 19.7) mm in males, 127–172 (150.3 ± 18.4) mm in females; TAL/SVL 3.00–3.36 in two males, 3.02–3.51 in females; HL/SVL 0.21–0.24 in males, 0.21–0.23 in females; SHL/SVL 0.18–

Plate 63.   *Polychrus gutturosus*. USNM 563277, adult male, SVL = 122 mm. Gracias a Dios: Wakling Tingni Kiamp.

0.19 in both sexes; SHL/HL 0.76–0.87 in males, 0.77–0.91 in females.

Color in life of an adult male (USNM 563277): dorsal surface of head Parrot Green (60); middorsum of body alternating Tawny Olive (223D) and reddish brown; lateral surface of body Shamrock Green (162B) with crossbands of various shades of reddish brown, white, and green; dorsal surfaces of fore- and hind limb Lime Green (159); dorsal surface of tail alternating grayish brown and brown; lateral surface of head varying from green to greenish white, with 1 white postorbital spot and 2 black postorbital stripes present; dewlap scales pale green, skin of dewlap blue with reddish brown mottling; belly Yellow-Green (58); subcaudal surface grayish brown with indistinct brown bands; eyelid green with black dorsal extensions of postorbital stripes; iris brown with gold rim. That adult (USNM 563277; Plate 63) exhibited rapid color changes in life with the various shades of green being replaced by dark brown, when handled. After being placed in the collecting bag, it would slowly return to its green color and pattern.

Color in alcohol: dorsal surfaces of head and body dark brown with slightly paler brown middorsal stripe; white postorbital stripe or blotch present; 3–5 pale brown or white crossbands present anteriorly on body; white lateroventral longitudinal stripe present on body, remainder of ventral surface dark brown (uniformly dark brown in males, mottled paler and darker brown in females); subcaudal surface dark brown.

Diagnosis/Similar Species.—Polychrus gutturosus is distinguished from all remaining Honduran lizards, except *Anolis* and *Norops*, by having a partially divided mental scale. *Anolis* and *Norops* lack femoral pores in both sexes, have tails capable of autotomy, and have Digit IV much longer than Digit III on the hind limb (versus femoral pores present in both sexes, nonautotomic tails, and digits III–IV on hind limb about same length in *P. gutturosus*).

Illustrations (Figs. 54, 56, 57; Plate 63).— Guyer and Donnelly, 2005 (adult, head); Koch et al., 2011 (adult, head and limb scales); Köhler, 2000 (adult), 2001b (adult), 2003a (adult, head scales), 2003b (adult), 2008 (adult, head scales); McCranie, 2007a (adult); McCranie et al., 2006 (adult); Savage, 2002 (adult); Taylor, 1956b (adult).

Remarks.—The systematics of *Polychrus gutturosus* remained one of the most poorly known of all Honduran nonendemic lizard species in the literature until Koch et al. (2011) reviewed its systematics based on 27 specimens from southeastern Nicaragua to Ecuador. Savage (2002: 445) plotted 20 localities on his distribution map of *P. gutturosus* in Costa Rica, but missed a chance to offer new systematic data on that significant number of specimens. Despite the study of Koch et al. (2011) and others before that, no author has applied consistent terminology for the dorsal head scales in *Polychrus*. Thus, we are left with vague descriptions such as "frontonasal broken into three parts" or "ten or more scales in the frontal area" (Taylor, 1956b: 158). For lack of standard terminology, I use post-rostrals for the dorsal snout scales bordering the rostral and use internasals for the 2 dorsal head scales following the postrostrals.

Natural History Comments.—*Polychrus gutturosus* is known from 10 to 190 m elevation in the Lowland Moist Forest formation. The specimen of this diurnal, canopy-dwelling species from Atlántida was taken on the trunk of a large tree immediately after it was felled (UMMZ catalogue). An adult female (UF 137404) with eight fully developed eggs was collected in September about 1 m above the ground among vines leading to the canopy of a tree. That female apparently was descending the tree to deposit her eggs in leaf litter on the ground. The eggs were deposited in the collecting bag the following day. Another adult female (USNM 570298) that contains well-developed eggs was collected on the forest floor in early November. Other adult specimens were sleeping at night in trees about 5 m above a stream and a river in May and November. A juvenile was in a tree near a river in October. Individuals of *P. gutturosus* are capable of rapid color change from bright green to brown. Unstressed specimens were green when first seen, but quickly changed to brown or dark brown when handled (also see

Roberts, 1997; also see comments on color in life above). Roberts (1997) reported a copulating pair of this species found in May in Costa Rica, and Taylor (1956b) reported a Costa Rican female contained nine (four eggs in right and five in left ovary) large eggs (month of collection not given). Koch et al. (2011) also reported that a preserved Costa Rican female contains six ovarian eggs. Taylor (1956b: 161) noted that "stomach contents consist chiefly of insect remains, with a meager portion of vegetable matter ingested, perhaps accidentally." Savage (2002: 446), in a literature review, said *P. gutturosus* "feeds primarily on relatively large arthropods but also eats leaves, flowers, fruits, and seeds."

Polychrus gutturosus is a canopy-dwelling species; thus, it depends on well-protected lowland broadleaf forest to maintain healthy populations. The accelerated deforestation of those, until recently, well-protected forests in northeastern Honduras (even in the Río Plátano Biosphere Reserve) is quickly eliminating those forests, along with this and many other species.

Etymology.—The name *gutturosus* is Latin meaning "with enlarged throat," and alludes to the extensible, saclike dewlap of males of this species.

Specimens Examined (10, 2 egg lots [1]; Map 51).—**ATLÁNTIDA**: Guaymas District, UMMZ 58369. **GRACIAS A DIOS**: Bodega de Río Tapalwás, UF 137404 (adult and 8 eggs); Kipla Tingni Kiamp, USNM 563276; Rus Rus, UF 137431; near Rus Rus, USNM 570298; Wakling Tingni Kiamp, USNM 563277; Warunta, USNM 579560. "HONDURAS": UF 124599, 137404 (2 eggs), UTA R-19528, 26107.

Infraorder Scincomorpha Camp, 1923

Hedges and Conn (2012) and Hedges (2014) revised the skink classification to recognize seven and nine families, respectively, in the Scincomorpha Camp (1923: 296, 313). The Hedges (2014) classification included three superfamilies for those nine families.

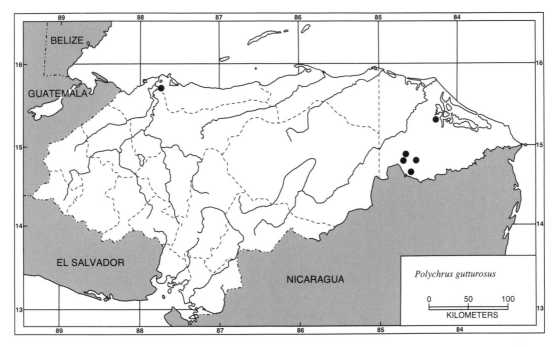

Map 51. Localities for *Polychrus gutturosus*. Solid circles denote specimens examined.

Two of those superfamilies, the Lygosomoidea Mittleman (1952: 3; as Lygosominae) and the Scincoidea Gray (1825: 201; as Sincidae, see Remarks below) occur in Honduras. The third superfamily, the Acontoidea Gray (1839b: 336; as Acontiadae), is restricted to the Old World. Hedges (2014) listed the genera and species of this infraorder.

KEY TO HONDURAN SUPERFAMILIES OF THE INFRAORDER SCINCOMORPHA

1A. Lower eyelid window present (Fig. 10; except window divided into 2–3 scales in one species (Fig. 117) .. .. Lygosomoidea (p. 335)

1B. Lower eyelid window absent, with 5–6 divided scales (Fig. 11) .... Scincoidea (p. 360)

CLAVE PARA LAS SUPERFAMILIAS HONDUREÑAS DEL INFRAORDEN SCINCOMORPHA

1A. El párpado inferior con un disco translucido sin dividir (Fig. 10),

excepto en una especie en que el disco está dividido en 2–3 escamas (Fig. 117)..... Lygosomoidea (p. 335)

1B. El párpado inferior sin un disco translucido, el párpado está dividido en 5–6 escamas (Fig. 11) .... Scincoidea (p. 360)

Superfamily Lygosomoidea Mittleman, 1952

This superfamily includes seven skink families, which includes all Scincomorpha families, except Acontidae and Scincidae. Two families of Scincomorpha occur in Honduras (Mabuyidae and Sphenomorphidae). Externally, the Honduran lygosomids are distinguished from all other Honduran lizards by the combination of having moveable eyelids, 5 digits on the forelimbs, the dorsal surface of the head with enlarged scales or plates that include fewer than 2 pairs of scales between the rostral and first unpaired plate, the dorsal and ventral surfaces of the body with large and smooth

Figure 117.   Two to three scales present in each lower eyelid window (outlined). *Scincella cherriei.* UNAH from near San José de Texíguat, Atlántida.

cycloid scales, and the lower eyelid with a translucent window, except that window divided into 2 or 3 scales in one species.

Both viviparous (Mabuyidae) and oviparous (Sphenomorphidae) families occur among the Honduran members of the Lygosomoidea.

Remarks.—Hedges and Conn (2012: 28) elevated the Lygosominae of Mittleman (1952: 5) to the superfamily Lygosomoidea for five skink families, two of which occur in Honduras.

Key to Honduran Families of the Superfamily Lygosomoidea

1A. Internasal (supranasal) scales absent, replaced by 1 large frontonasal scale that broadly contacts rostral scale (Fig. 118); a single frontoparietal scale present (Fig. 118) ...... Sphenomorphidae (p. 348)

1B. Paired internasal (supranasal) scales present (Fig. 119); paired frontoparietal scales present (Fig. 119) .... Mabuyidae (p. 338)

Clave para las Familias Hondureñas de la Superfamilia Lygosomoidea

1A. Escamas internasales (supranasales) ausentes, en su lugar hay una frontonasal (una escama grande); escama frontonasal en contacto amplio con la escama rostral (Fig. 118); solo una escama frontoparietal presente (Fig. 118) .... Sphenomorphidae (p. 348)

1B. Una par de escamas internasales (supranasales) presentes (Fig. 119); un par de escamas frontopa-

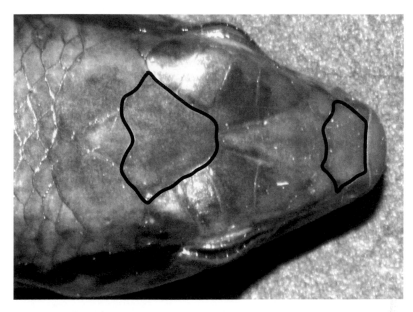

Figure 118.   Internasal absent, replaced by one large frontonasal scale (outlined) and a single, large frontoparietal scale (outlined). *Scincella cherriei.* USNM 581900 from near San José de Texíguat, Atlántida.

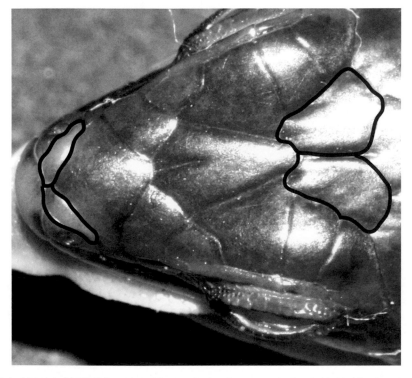

Figure 119.   Paired internasal (outlined) and paired frontoparietal scales present (outlined). *Marisora brachypoda.* USNM 589170 from Finca Monterrey, Choluteca.

rietales presentes (Fig. 119) . . . . .
. . . . Mabuyidae (p. 338)

Family Mabuyidae Mittleman, 1952.

In the Western Hemisphere, this family occurs from southern Jalisco and northeastern Hidalgo, Mexico, to southern Panama, in South America east of the Andes to central Argentina, Trinidad and Tobago, the Lesser Indies, and much of the Greater Antilles (with the exception of Cuba and the Bahama Islands). It also occurs on the Caribbean islands of Cozumel, the Honduran Bay Islands, Great Corn Island, Nicaragua, and San Andrés and Providencia, Colombia. In the Eastern Hemisphere, this family occurs in Asia, China, India, Sri Lanka, Africa, and Madagascar. Twenty genera and about 195 named species are recognized, with two species of a single genus known to occur in Honduras (but see Remarks for *Marisora brachypoda*).

Genus *Marisora* Hedges and Conn, 2012

Marisora Hedges and Conn, 2012: 119 (type species: *Mabuia unimarginata* Cope, 1862c: 187 by original designation).

Geographic Distribution and Content.— The distribution of this genus is from northeastern Hidalgo and southern Jalisco, Mexico, southward to southern Panama and then from Colombia to Venezuela along the north coast of South America, Trinidad and Tobago, Granada, the Grenadines, St. Vincent, and on several Caribbean islands close to the Central American mainland (Cozumel, Mexico, the Bay Islands, Honduras, and Great Corn Island, Nicaragua). It also occurs on some Pacific islands in the Golfo de Fonseca in southern Honduras. Seven named species are included in *Marisora*, two of which occur in Honduras (but see Remarks).

Remarks.—Until recently, the genus *Marisora* was included in *Mabuya* Fitzinger (1826: 23), a wide-ranging genus occurring in both hemispheres. Mausfield et al. (2002: 288–289) partitioned *Mabuya* (sensu lato) into four genera, with *Mabuya* (sensu stricto) confined to the Western Hemisphere. Whiting et al. (2006: 720) said that the Mausfield et al. (2002) suggestion was "premature." Whiting et al. also recovered four distinct clades of "*Mabuya*," including one for the South American species of *Mabuya* they studied. Miralles and Carranza (2010: 862) wrote that the phylogenetic relationships between Neotropical *Mabuya* recovered by Whiting et al. (2006) "were incorrect as a result of contamination problems." Miralles and Carranza (2010: 866) recovered a monophyletic *Mabuya* with "at least five distinct lineages" occurring in Amazonia alone. Subsequently, Hedges and Conn (2012) further divided the Western Hemisphere *Mabuya* into 16 genera (also see Hedges, 2014). Hedges and Conn (2012: 218) estimated, based on a molecular timetree, that the ancestors of *Marisora* dispersed to Central America from South America during the Miocene some 6.8 Ma. Recently, Pinto-Sánchez et al. (2015: 201) relegated *Marisora* to a synonym of *Mabuya* and considered seven named species as not valid, including four (*M. agilis* (Raddi, 1822: 62), *M. brachypoda* (Taylor, 1956b: 308), *M. heathi* Schmidt and Inger, 1951: 455, and *M. roatanae* Hedges and Conn, 2012: 132) relegated to the synonymy of *M. unimarginata* Cope (1862c: 186).

Despite the decision by Pinto-Sánchez et al. (2015), which lacked any comparison between *Marisora unimarginata* and *M. brachypoda* (also unsupported by morphological data), those species were used as part of their taxonomic decisions. I recognize *Marisora* as a valid genus and *M. brachypoda* and *M. roatanae* as valid species. Morphologically, all Honduran populations are short-limbed and thus do not represent the longer limbed *M. unimarginata*. After the *Marisora* species accounts below were written, additional *Marisora* tissues from

Mexico and Central America were used in a phylogenetic analysis performed by Hedges (2 September 2016, unpublished data). That study demonstrates that at least three species of *Marisora* are represented among Honduran populations (only two of which are discussed below). The mainland and Bay Islands populations are apparently *M. roatanae* (despite the morphological data separating those two populations based on number of scales around midbody), and the Golfo de Fonseca populations belong to a *M. brachypoda* clade and a clade representing an undescribed nominal form). Because of those conflicting data, no diagnostic key is provided below.

All scale counts used herein for the genus *Marisora* follow the methods of Hedges and Conn (2012), with few exceptions (Greer and Broadly, 2000: 3; Greer and Nussbaum, 2000: 616; Miralles, 2006: 2; but see Comments by Hedges and Conn, 2012: 13).

Etymology.—The generic name *Marisora* is derived from the Latin *maris* (sea) and *ora* (coast) and refers to the "distribution of this genus occurring predominately in low elevations near the coast (Caribbean, Atlantic, and Pacific), with relatively few inland and upland localities. Three of the seven species occur exclusively on islands" (Hedges and Conn, 2012: 120).

Marisora brachypoda (Taylor, 1956b)

Mabuia agilis: Werner, 1896: 347.
Mabuya agilis: Dunn and Emlen, 1932: 31.
Mabuya mabouya: Dunn, 1936: 544; Meyer, 1966: 176; Meyer, 1969: 264 (in part); Hahn, 1971: 111; Meyer and Wilson, 1973: 29 (in part); Wilson and Hahn, 1973: 116 (in part); Cruz Díaz, 1978: 29; Wilson et al., 1979a: 25; O'Shea, 1986: 41.
Mabuya mabouya mabouya: P. W. Smith, 1950: 55; Greer, 1970: 172.
Mabuya mabouya alliacea: Burger, 1952: 186.
Mabuya brachypodus Taylor, 1956b: 308 (holotype, KU 36258; type locality: "4

km. ESE of Los Angeles de Tilarán, Guanacaste" [Costa Rica]).
Mabuya unimarginata: Wilson et al., 1991: 70; Köhler, 1994a: 8; Köhler, 1995b: 97; Köhler, 1996d: 20; Köhler, 1998b: 141; Köhler, 1998d: 375; Monzel, 1998: 161 (in part); Wilson and McCranie, 1998: 16; Köhler, 2000: 90; Lundberg, 2000: 6; Wilson et al., 2001: 135 (in part); Castañeda, 2002: 15; McCranie, Castañeda, and Nicholson, 2002: 27; Goldberg and Bursey, 2003: 369; Honda et al., 2003: 80; Köhler, 2003a: 147; Powell, 2003: 36; McCranie et al., 2005: 114 (in part); Lovich et al., 2006: 14; McCranie et al., 2006: 129; Wilson and Townsend, 2006: 105 (in part); Wilson and Townsend, 2007: 145; Köhler, 2008: 159; Miralles, Chaparro, and Harvey, 2009: 68; Miralles, Fuenmayor, et al., 2009: 602; Miralles and Carranza, 2010: 861; Townsend and Wilson, 2010b: 692; McCranie, 2011a: 177, 360.
Marisora brachypoda: Hedges and Conn, 2012: 119, 244; McCranie and Valdés Orellana, 2014: 45 (in part); Solís et al., 2014: 131; McCranie, 2015a: 370; McCranie and Gutsche, 2016: 874.

Geographic Distribution.—*Marisora brachypoda* (*sensu lato*) occurs at low and moderate elevations and rarely at lower limits of intermediate elevations on the Atlantic versant from northeastern Hidalgo, Mexico, to about northeastern Nicaragua and from Jalisco, Mexico, to northwestern Costa Rica on the Pacific versant (see Remarks). It also occurs on Utila Island, Honduras, and the islands off the coast of Belize and Quintana Roo, Mexico. In Honduras, this species is widespread in open habitats throughout much of the mainland and on Utila Island in the Bay Islands. It also occurs on several islands in the Golfo de Fonseca (but see updated Remarks for *Marisora* above).

Description.—The following description is based on 25 unsexed adults (SMF 77097;

USNM 570299–302, 570311, 578839–40, 589167, 589191, 589193–97, 589169–75, 589199–201). *Marisora brachypoda* is a moderate-sized lizard (maximum recorded SVL 90 mm [USNM 578840]); dorsal head scales enlarged, smooth, platelike, with paired internasals (supranasals) usually in contact medially, a single frontonasal, paired prefrontals usually separated medially (contact made in 4 of 25), a single frontal, paired frontoparietals in contact medially, 4 supraoculars, and paired parietal scales almost always in contact posterior to single interparietal scale (no such contact in 1 of 25); frontal usually separated from first supraocular (occasionally narrow contact made between those 2 scales); frontal in broad contact with second supraocular; frontonasal contacting anterior loreal; 4 superciliary scales; nostril opening near posterior edge of nasal; 1 small postnasal; 2 loreals; 6–8 (usually 7) supralabials, usually fifth (rarely sixth) at level below eye; 1 post-supralabial; 2–3 (usually 2) and 2–4 (usually 3) preoculars and postoculars, respectively; 6–8 (usually 7) infralabials; moveable eyelid present; pupil circular; lower eyelid window undivided, transparent, with 1 scale row above window; 4 scales bordering upper edge of eyelid window; eyelid window L/SVL 0.010–0.024; 2–4 (almost always 4) superciliary scales; superciliary 2 L/SVL 0.005–0.015; 1 (rarely 2) primary temporal scales; 2–3 (usually 2) secondary temporals, with upper overlapping lower and parietal overlapping upper; 3 tertiary temporals; 2 (rarely 1) pretemporals; 1 large mental; 1 large postmental, postmental almost always contacting first and second infralabials; 2 pairs of chinshields, separated medially, contacting 1 or 2 infralabials; gular scales smooth, cycloid, imbricate, large; 1 row of 1 pair of enlarged primary nuchal scales; dorsal body scales smooth, cycloid, imbricate; 50–60 (55.2 ± 2.9) rows of dorsal scales in paravertebral row from end of parietal to level above cloacal scale; 28 or 30 scales around midbody; ventral body scales smooth, cycloid, imbricate, in 50–64 (57.6 ± 3.4) rows between mental and cloacal scale; dorsals + ventrals 103–123 (112.8 ± 5.5); precloacal scales similarly sized as those on remainder of venter; femoral and precloacal pores absent; limbs relatively short, forelimb + hind limb L/SVL 0.45–0.58; 13–17 (15.1 ± 1.2) subdigital scales on Digit IV of hind limb, 9–14 (12.0 ± 1.3) scales on Digit IV of forelimb (only 48 sides counted); 5 digits on forelimb; SVL 60.1–90.2 (70.5 ± 7.5) mm; TAL/SVL 1.12–1.94 in 11; HL/SVL 0.16–0.19; HW/SVL 0.11–0.13; SW/SVL 0.027–0.051; SHL/SVL 0.09–0.17 in 14.

Color in life of an adult (FMNH 236393; sex unknown): middorsal surface of body and top of head brown; lateral dark stripe dark brown; lateral pale stripe pale copper; area below lateral pale stripe dark brown, grading to pale bronze on belly; iris black. Color in life of another adult (USNM 570303; sex unknown): middorsal area of body and head Grayish Horn Color (91); lateral Sepia (119) stripe, bounded above by bronze stripe covering about adjacent halves of 2 scale rows, bounded below by iridescent pale green lateral stripe; iridescent pale green lateral stripe bounded below by narrow Sepia ventrolateral stripe, area below narrow Sepia ventrolateral stripe pale copper, grading to iridescent pale green of venter; dorsal surfaces of fore- and hind limb mottled Sepia and pale copper; iris Sepia; lateral dark stripe continues onto side of head across eye to nasal scale. Color in life of another unsexed adult (USNM 589178): dorsum of body Grayish Olive (43) with dark brown spotting on some scales suggesting incomplete linear pattern; top of head Grayish Olive with some dark brown flecking; dorsolateral stripe Sepia (119) on head to anterior third of body; lateral pale stripe Amber (36) with yellow tinge and Sepia spotting extending from snout to groin and again from posterior insertion of hind limb onto base of tail; dorsal surfaces of fore- and hind limb brown with Sepia spotting; venter of body pale brown with green tinge and incomplete dark brown lines, that of chin and throat similar in color, but with more cream than brown; subcaudal surface yellow-brown with distinct dark brown lines on anterior third of tail, becoming progressively darker brown on posterior two-thirds of tail; sole and palm dark brown; iris dark brown.

Color in alcohol: middorsal surfaces of head and body grayish brown to brown, frequently with indication of darker brown spots suggesting linear pattern; dark brown, rather broad lateral stripe extending from snout onto tail; indistinct, thin pale brown dorsolateral line bordering upper edge of dark lateral stripe sometimes present; white to pale brown narrow lateral stripe bordering lower edge of lateral dark stripe, pale lateral stripe extending from posterior most supralabial, passing below ear opening and above forelimb onto lower body to anterior portion of tail, pale lateral stripe usually involving adjacent edges or adjacent halves of 2 scale rows, rarely coursing down middle of 1 scale row; ventrolateral surface of body below pale lateral stripe same color as middorsum to slightly paler brown; ventral surfaces of head and body white or cream, with or without gray flecking around scale edges suggestive of longitudinal lines; ventral surface of anterior portion of tail with distinct dark gray lines; dorsal surface of unregenerated tail generally same color as middorsal surface of body; dorsal surfaces of fore- and hind limb dark brown to nearly black, with some pale brown mottling; palm and sole cream, paler than dark brown subdigital scales.

Diagnosis/Similar Species.—The moveable eyelid, a lower eyelid window, large smooth scales on the head, 5 digits on the forelimb, and smooth cycloid body scales distinguishes *Marisora brachypoda* (but see Remarks in *Marisora* account above) from all other Honduran lizards, except *Scincella* and *M. roatanae*. *Scincella* have broad contact between the frontal and first supra-

ocular, lack paired supranasals (internasals), lack a distinct lateral pale stripe, and have a maximum SVL of 68 mm (versus almost always no broad contact between frontal and first supraocular, paired supranasals present, distinct pale lateral stripe present, and maximum SVL of 90 mm in *M. brachypoda*). *Marisora roatanae* has 32 scales around midbody (versus 28–30 in *M. brachypoda*; but see updated Remarks in *Marisora* account above).

Illustrations. (Fig. 119; Plate 64)—Köhler, 1999b (head scales; as *Mabuya unimarginata*), 2000 (adult, ventral scales, head scales; as *Mabuya unimarginata*), 2001b (adult; as *Mabuya unimarginata*), 2003a (adult, head scales, ventral scales; as *Mabuya unimarginata*; except specimen from Bartola, Nicaragua), 2008 (adult, head scales, ventral scales; as *Mabuya unimarginata*; except specimen from Bartola, Nicaragua); Köhler et al., 2005 (adult; as *Mabuya unimarginata*); Lee, 1996 (adult, head scales; as *Mabuya unimarginata*), 2000 (adult, head scales; as *Mabuya unimarginata*); Lundberg, 2000 (adult; as *Mabuya unimarginata*); McCranie, Castañeda, and Nicholson, 2002 (adult; as *Mabuya unimarginata*); McCranie et al., 2005 (adult, head scales; as *Mabuya unimarginata*), 2006 (adult, head scales; as *Mabuya unimarginata*); Powell, 2003 (adult; as *Mabuya unimarginata*); Taylor, 1956b (adult; as *Mabuya*).

Remarks.—The phylogenetic analyses based on molecular and morphological data performed by Hedges and Conn (2012) demonstrated that *Marisora brachypoda*, as currently understood (Mexico to Costa Rica), apparently represents a complex of several species. A thorough systematic revision of this skink complex from Central America and Mexico is badly needed. See the *Marisora* account above for more recent data regarding the Honduran populations.

Natural History Comments.—*Marisora brachypoda* is known from near sea level to 1,510 m elevation in the Lowland Moist Forest, Lowland Dry Forest, Lowland Arid

Plate 64.   *Marisora brachypoda.* USNM 589194, SVL = 69.0 mm. Valle: del Tigre Island, near communications tower.

Forest, Premontane Wet Forest, Premontane Moist Forest, and Premontane Dry Forest formations. This diurnal species is probably active throughout the year, as it has been collected in the months of February, from March to August, and from October to December. However, the species appears to be more active during the rainy season. Individuals were crawling on the ground and climbing onto tree trunks, brush piles, fence posts, large rocks, rock walls, and buildings (inhabited and uninhabited) to bask. Inactive individuals were found under logs and other debris on the ground, stacked roofing tiles on the ground, under bark of standing trees, and in palm thatched roofing of *champas.* Two were under coconut debris on the ground in September but attempted to escape by climbing nearby palms or hardwood trees. Meyer (1969) noted finding a large series in banana bunches in a processing plant. Wilson and Hahn (1973: 117) reported this species as being "largely arboreal; specimens from Utila were collected on the sides of trees. One was found about 8 m up the trunk of a mango tree." However, those authors listed only one specimen from Utila.

Marisora brachypoda is viviparous and gives birth to 4–6 (Webb, 1958; an apparent literature summation and apparently from various Mexican populations), 4–9 (Lee, 1996, a literature review of Mexican and Guatemalan females), or 2–7 (Köhler, 2003a, apparently a literature review of populations from Mexico and Central America) neonates between at least May and July. One Honduran female (USNM 589174) captured on 28 November gave birth to three young (UNAH [1]; USNM 589175–76) the following day in the collecting bag. Luja (2006) reported that a female captured in late April in Quintana Roo, Mexico, had six fully formed young. Goldberg (2009d) found evidence of an extended breeding season in Costa Rica. However, the concept of *M. brachypoda* of those authors is a composite of more than one species. Apparently nothing has been published on diet in this species other than general statements that it feeds on invertebrates, especially insects and spiders (Lee, 1996). A subadult (USNM 559686) of the snake *Oxyrhopus petolarius* (Linnaeus, 1758: 225) and an adult of the snake *Oxybelis fulgidus* (Daudin, 1803b: 352;

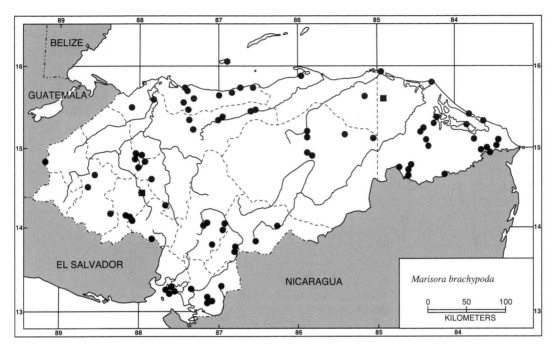

Map 52. Localities for *Marisora brachypoda*. Solid circles denote specimens examined and solid squares accepted records.

USNM 561969) each contained an adult *M. brachypoda* in their stomachs. Goldberg and Bursey (2003) reported endoparasites in several museum specimens from Honduras.

Etymology.—The name *brachypoda* is formed from the Greek *brachys* (short) and *podos* (foot), and alludes to the relatively short limbs in this species.

Specimens Examined (229, 1 skin, 1 C&S, 4 skeletons [59]; Map 52).—**ATLÁNTIDA**: Carmelina, USNM 62968; Corozal, LACM 47753–54; El Naranjal, USNM 589167; Estación Forestal CURLA, USNM 578839; Jilamito Nuevo, USNM 578840; 13 km E of La Ceiba, LACM 47755–56; La Ceiba, INHS 4488; Lancetilla, ANSP 28120, 33147, MCZ R29888; near Pico Bonito Lodge, USNM 589168; San José de Texíguat, USNM 589169; Tela, MCZ R21150, 21768, 27326–27. **CHOLUTECA**: 21.1 km W of Choluteca, UMMZ 123017; El Banquito, KU 200581; El Despoblado,

CAS 152979; Finca Monterrey, USNM 589170; La Fortunita, SDSNH 72728. **COLÓN**: Barranco, ANSP 28121; Trujillo, CM 65385–87, LSUMZ 22428. **COMAYAGUA**: Siguatepeque, FMNH 5063, UF 124824. **COPÁN**: 1 km S of Copán, USNM 570299–301; Copán, AMNH 70339–40, 140273, UMMZ 83029 (3). **CORTÉS**: Agua Azul, AMNH 70337, MCZ R49966–67 (+ 3 unnumbered), TCWC 19211–12; 1.6 km NW of El Jaral, LSUMZ 52317–18; El Jaral, FMNH 5062; Laguna Ticamaya, FMNH 5061; 1 km N of Los Pinos, USNM 573175; San Pedro Sula, FMNH 5060. **EL PARAÍSO**: El Rodeo, USNM 589172–73; Mapachín, USNM 589171; Orealí, UNAH (1), USNM 589174–77; Valle de Jamastrán, AMNH 70380. **FRANCISCO MORAZÁN**: El Picacho, USNM 570302; El Zamorano, AMNH 70338; 8.0 km W of Maraita, UF 143819; Río Yeguare, UMMZ 94040; near Tegucigalpa, BYU 18226; Tegucigalpa, FMNH 5064–65. **GRACIAS A DIOS**:

Awasbila, USNM 570303–04; Barra Patuca, USNM 20306–09; Cauquira, UF 150308; Dursuna, USNM 570305; Finca Nakunta, USNM 589178; SW end of Isla del Venado, USNM 573961; Kakamuklaya, USNM 573169; Kisalaya, LACM 16860; Krahkra, USNM 570306, 573958–60; Leimus (Río Warunta), USNM 589179–81, FN 257060 (still in Honduras because of permit problems); Mavita, USNM 589182–83; near Mocorón, UTA R-42650–51, 42653; Mocorón, UTA R-42652, 42654 (skin), 46175–78, 53521–24; Palacios, BMNH 1985.1293–94; Quiguastara, LACM 16859; Rus Rus, USNM 559560, 570307–10, 589184; Samil, USNM 573957, 589185; Swabila, UF 150307, 150312; Tánsin, LACM 47726–28, USNM 573171; Tikiraya, UF 150309, 150322, 150328; Usus Paman, USNM 573953–56; Warunta, USNM 589186–89, FN 256934 (still in Honduras because of permit problems); Yahurabila, USNM 573172–74. **INTIBUCÁ**: 15.0 km SE of La Esperanza, USNM 570311; 17.0 km N of Marcala, FMNH 236393. **ISLAS DE LA BAHÍA**: Isla de Utila, 2.5 km N of Utila, SMF 77097; Isla de Utila, Utila, LSUMZ 22309, SMF 79851; "Isla de Utila," CM 65381. **LA PAZ**: La Estancia, FN 256868 (still in Honduras because of permit problems); 13.7 km N of Marcala, USNM 570312–14; Potrerillos, FMNH 283593. **LEMPIRA**: El Rodeito, USNM 573170; Erandique, CM 65382–83; Gracias, CM 65384. **OLANCHO**: 1 km WNW of Catacamas, LACM 47720; 4.5 km SE of Catacamas, LACM 45165, 47721–25; 12.1 km E of Dulce Nombre de Culmí, LACM 45151; Las Trojas, UTA R-41227; confluence of Quebrada Siksatara and Río Wampú, USNM 570315; near Río Catacamas, USNM 589190; 10.5 km S of San Esteban, KU 200580. **SANTA BÁRBARA**: El Sauce, AMNH 70341; SW corner of Lago de Yojoa, USNM 589191. **VALLE**: near Amapala, FN 256926 (still in Honduras because of permit problems); Isla de Pájaros, USNM 589192; Isla del Tigre, summit, UNAH (1), USNM 589193; Isla del Tigre, near summit, UNAH (2); Isla Inglesera, UNAH (1); Isla Zacate Grande, KU 194267, LSUMZ 36578; Playa Negra, SDSNH 72727; Playona Exposición, USNM 589196; Punta El Molino, UNAH (1), USNM 589197; Punta Novillo, USNM 589198; near San Carlos, USNM 589194–95. **YORO**: 5 km E of Coyoles, LACM 47729; Coyoles, LACM 47730–52; Río San Lorenzo, USNM 589199; San Francisco, MVZ 52416; 5.5 km ESE of San Lorenzo Arriba, USNM 589200; 4.7 km ESE of San Lorenzo Arriba, USNM 570316; San Patricio, USNM 589201; Subirana Valley, FMNH 21784, 21785, 21786 (C&S), 21787, 21826–28, MCZ R32037–39, 32040 (skeleton), 38934, 38935 (skeleton), 38936, UMMZ 77848 (6). "HONDURAS": AMNH 37833, 46989, LSUMZ 56295; UF 42910 (skeleton), 43114 (skeleton).

Other Records (Map 52).—**GRACIAS A DIOS**: Baltiltuk, UNAH 5460 (Cruz Díaz, 1978, specimen now lost). **INTIBUCÁ**: Otoro Valley (Wilson and McCranie, 1998).

Marisora roatanae Hedges and Conn, 2012

Mabuya mabouya: Meyer, 1969: 264 (in part); Meyer and Wilson, 1973: 29 (in part); Wilson and Hahn, 1973: 116 (in part).

Mabuya unimarginata: Köhler, 1994a: 4; Köhler, 1995b: 102; Monzel, 1998: 161 (in part); Lundberg, 2001: 27; Wilson et al., 2001: 135 (in part); Lundberg, 2002b: 9; McCranie et al., 2005: 114 (in part); Wilson and Townsend, 2006: 105 (in part).

Marisora roatanae Hedges and Conn, 2012: 132 (holotype, TCWC 21955; type locality: "Jonesville, Isla de Roatán, Islas de la Bahía, Honduras, 3 m"); McCranie and Valdés Orellana, 2014: 45; Solís et al., 2014: 131; McCranie, 2015a: 370.

Geographic Distribution.—*Marisora roatanae* occurs on Guanaja and Roatán islands in the Bay Islands, Honduras (but see updated Remarks for *Marisora* account above).

Description.—The following is based on five unsexed subadults to adults (TCWC 21955; USNM 589204–07). *Marisora roatanae* is a moderate-sized lizard (maximum recorded SVL 90 mm [TCWC 21955]); dorsal head scales enlarged, smooth, plate-like, with paired internasals (supranasals) in contact medially, single frontonasal, paired prefrontals separated medially, single frontal, paired frontoparietals in contact medially, 4 supraoculars, and paired parietal scales in contact posterior to single interparietal scale; frontal separated from first supraocular; frontal in broad contact with second supraocular; frontonasal contacting anterior loreal; 4 superciliary scales; nostril opening near posterior edge of nasal; 1 small postnasal; 2 loreals; 7–8 (usually 7) supralabials, fifth at level below eye; 1 postsupralabial; 2 and 4–5 preoculars and postoculars, respectively; 7–8 infralabials; moveable eyelid present; pupil circular; lower eyelid window present, undivided, transparent, with 1 scale row above window, 4 scales bordering upper edge of window; 1 primary temporal; 2 secondary temporals, with upper overlapping lower and parietal overlapping upper; 3 tertiary temporals; 1 large mental; 1 large postmental, postmental contacting first infralabial; 2 pairs of chin-shields, separated medially, contacting 1 or 2 infralabials; gular scales smooth, cycloid, imbricate, large; 1 row of 1 pair of primary nuchal scales; dorsal body scales smooth, cycloid, imbricate; 54–58 (55.8 ± 1.5) rows of dorsal scales in paravertebral row from end of parietal to level above cloaca; 32 scales around midbody; ventral body scales smooth, cycloid, imbricate, in 59–67 (61.8 ± 3.6) rows between mental and cloacal scale; 114–125 (117.6 ± 4.8) dorsals + ventrals; precloacal scales similarly sized as those on remainder of venter; femoral and precloacal pores absent; limbs relatively short, forelimb + hind limb L/SVL 0.55–0.59; 15–18 (16.8 ± 1.2) subdigital scales on Digit IV of hind limb, 13–16 (14.8 ± 1.2) scales on Digit IV of forelimb; 5 digits on forelimb; SVL 52.8–90.2 (78.8 ± 15.0) mm; TAL/SVL 1.36–1.61 in two; HL/SVL 0.16–0.21; HW/SVL 0.12–0.13; SW/SVL 0.024–0.046; eyelid window L/SVL 0.010–0.014; superciliary 2 L/SVL 0.010–0.020.

Color in life of an adult female: (USNM 589205; Plate 65): middorsum from head to regenerated portion of tail Russet (34) with narrow Mikado Brown (121C) border below, border extending onto anterior portion of tail; Vandyke Brown (121) lateral band below Mikado Brown border, extending from snout onto anterior portion of tail; Buff (124) lateral stripe below Vandyke Brown lateral band, stripe extending from tip of snout across upper edges of supralabials, below tympanum, above forearm insertion along body to above hind limb insertion onto anterior portion of tail; indistinct Buff ventrolateral line present on body and anterior portion of tail, Buff line extending onto dorsolateral and ventrolateral edges of thigh; dorsal surfaces of fore- and hind limb mottled Vandyke Brown and Buff, mottled pattern extending to tips of toes; all ventral surfaces Pale Pinkish Buff (121D), except palm, sole, and digits Sepia (119); iris Vandyke Brown. Color in life of another adult (USNM 589203): middorsum Natal Brown (219A) with Sepia (119) spotting; Sepia lateral band beginning at posterior edge of rostral and extending just posterior to hind limb; Sepia lateral band bordered above by interrupted pale brown spots suggesting a line; Sepia lateral band bordered below by lateral white stripe beginning on snout and passing below tympanum and above forelimb to groin, stripe then extends from posterior to hind limb shortly onto tail; field below lateral white stripe mottled dark and pale brown; dorsal surface of tail similar in color to that of body, except with dark brown color

Plate 65. *Marisora roatanae.* USNM 589205, SVL = 82 mm. Islas de la Bahía: Roatán Island, Turquoise Bay.

forming short longitudinal lines; ventral surfaces of head, body, and anterior half of tail pale brown with gray lateral scale edges forming indistinct longitudinal lines, those lines less distinct on posterior third of body and anterior half of tail; posterior half of subcaudal surface brown; palm and sole dark brown; iris Sepia (219).

Color in alcohol of the adult female holotype (TCWC 21955) was described by Hedges and Conn (2012: 134): "dorsal ground color medium brown with relatively few dark brown spots, distributed in two dorsolateral zones on body, in discontinuous stripes on tail, and uniformly on limbs. Dark dorsolateral stripes absent. Dark lateral stripes present, dark brown, extending from loreal region past hind limbs. Pale middorsal stripe absent. Pale dorsolateral stripes present between bands of dorsolateral dark spots and dark lateral stripes. Two pale ventrolateral stripes present, whitish, extending from below eye to last third of body (upper stripe continues past hind limbs and lower stripe continues onto hind limbs), each bordered below by a dark line. Forelimbs and hind limbs with large dark spots. Ventral surface of body without pattern. Palmar and plantar surfaces dark brown." Color in alcohol of five recently collected adults (USNM 589203–07): middorsal surfaces of head and body brownish black with a few scattered black spots; indistinct, narrow pale brown dorsolateral stripe passing along upper edge of lateral black stripe from posterior to head to about level above cloacal opening; black, rather broad (2–3 scales high) lateral stripe extending from loreal region onto anterior portion of tail; distinct white lateral stripe bordering lower edge of lateral dark stripe, white stripe 1.0–1.5 scales high, extending from posterior most supralabial, passing below ear opening and above forelimb onto lower body and upper edge of hind limb onto anterior portion of tail; ventrolateral surface of body below white lateral stripe gray; dorsal surface of unregenerated tail same color as that of dorsal surface of body; dorsal surfaces of fore- and hind limb brownish black with brown mottling; ventral surfaces of head and body cream with some indistinct gray mottling and scale edges; subcaudal surface cream with less gray mottling and scale edges; palmar and plantar surfaces dark brown; subdigital scales on fore- and hind limb brownish black.

Diagnosis/Similar Species.—The moveable eyelid, a lower eyelid window, large smooth scales on the head, and smooth cycloid scales on the body will distinguish *Marisora roatanae* from all other Honduran lizards, except *M. brachypoda* and *Scincella*. Species of *Scincella* have broad contact between the frontal and first supraocular, lack a distinct pale lateral stripe, lack paired supranasals, and have a maximum SVL of about 68 mm (versus no broad contact between frontal and first supraocular, distinct pale lateral stripe present, paired supranasals present, and maximum SVL around 90 mm in *M. roatanae*). *Marisora brachypoda* has 28–30 scales around the midbody (versus 32 in *M. roatanae*; but see updated Remarks in *Marisora* account).

Illustrations (Figs. 9, 10; Plate 65).— Hedges and Conn, 2012 (adult, subadult, head).

Remarks.—Hedges and Conn (2012) concluded that the population from Isla de Roatán on the Bay Islands of Honduras represented a species (*Marisora roatanae*) distinct from those on the Honduran mainland. As of 2011, only two museum specimens of *Marisora* were known from Roatán, despite the species being quite common at some localities. I tentatively include the Guanaja population as *M. roatanae*, because the two recently collected Guanaja specimens have 32 scales around midbody like the Roatán population. However, unpublished molecular data (S. B. Hedges, personal communication, 2 September 2016) allies *Marisora* from the mainland of Honduras with the Roatán and Guanaja populations, despite differences in number of scales around midbody (30 in former, 32 in latter).

Natural History Comments.—*Marisora roatanae* is known from near sea level to 20 m elevation in the Lowland Moist Forest formation. Two were collected in November on the trunk of a coconut palm in the yard of an occupied house during a brief period of bright sunshine at 9:00 a.m. At least three others were seen at that time on the same palm. Shortly thereafter, the skies became cloudy and no other *Marisora* were seen. One other was sunning on top of a fallen log in secondary growth in September. Another was found dead on a dirt road in November. The above notes are from Roatán. On Guanaja, one was first uncovered under coconut palm debris on the ground and darting into surrounding grass. After about 10 minutes of searching, the lizard was captured as it climbed a nearby tree. Also on Guanaja, a second *Marisora* was captured after it was discovered under a concrete slab of an abandoned hotel. Both Guanaja specimens were taken a few days apart in September. This species, as does its mainland relatives, appears to tolerate humans and its buildings and is likely common on Roatán, despite the few museum specimens. Each of my two recent trips to Roatán and Guanaja, in which part of a day was spent targeting *Marisora*, were successful. An American living near Jonesville, Roatán, who had a copy of the Bay Island book (McCranie et al., 2005), told me those skinks, referring to a photograph of *Marisora* in that book, were common on his property but were mostly seen during the rainy season. Also, the person living in the house where the two were collected on the coconut palm told me that he usually only sees these skinks during the rainy season. Nothing has been reported on diet or reproduction in this species, but both are likely similar to *M. brachypoda*.

Etymology.—The specific name *roatanae* refers to its occurrence on Isla de Roatán.

Specimens Examined (9 [1]; Map 53).— **ISLAS DE LA BAHÍA**: Isla de Guanaja, Posada del Sol hotel ruins, USNM 589202; Isla de Guanaja, Savannah Bight, USNM 589203; "Isla de Guanaja," LSUMZ 21883; Isla de Roatán, Jonesville, TCWC 21955;

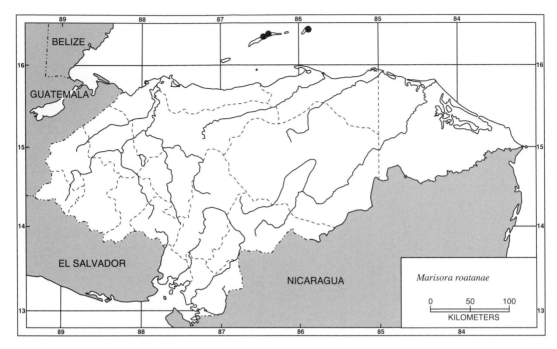

Map 53.   Localities for *Marisora roatanae*. Solid circles denote specimens examined.

Isla de Roatán, Oak Ridge, UTA R-55232; Isla de Roatán, 1 km E of Pollytilly Bight, USNM 589204; entrance to Turquoise Bay, USNM 589205–07.

Family Sphenomorphidae Welch, 1982

This family occurs in the Western Hemisphere from much of the eastern U.S., southward to Panama and in the Eastern Hemisphere in Australia, New Zealand, the Paupan Region, Melanesia, China, and southeastern Asia. Thirty-four genera and about 560 named species are recognized in this family, with three species placed in a single genus occurring in Honduras.

Genus *Scincella* Mittleman, 1950

Scincella Mittleman, 1950: 19 (type species: *Scincus lateralis* Say, 1823: 324, *In* James, 1823, by original designation).

Geographic Distribution and Content.— In the Western Hemisphere, this genus ranges from the southeastern U.S. to western Panama. It also occurs in eastern Asia in the Eastern Hemisphere. More than 35 named species are included in this genus, three of which occur in Honduras (see Remarks).

Remarks.—Honda et al. (2003: 77), based on mitochondrial DNA sequence data, suggested placing the Western Hemisphere members of *Sphenomorphus* Fitzinger (1843: 23) in the genus *Scincella*. A more recent molecular study of the *Sphenomorphus* group of skinks also recovered the *Scincella* clade for this group (Linkem et al., 2011: 1237). Hedges (2014: 332) listed the Panamanian *S. rarus* (Myers and Donnelly, 1991: 2) as a member of *Sphenomorphus*, but that species likely belongs to *Scincella* (see Linkem et al., 2011: 1237).

Etymology.—The name *Scincella* is formed from the Latin *scincus* (a kind of a lizard, a skink) and *-ellus* (little), in reference to the small size of the type species of the genus.

Key to Honduran Species of the Genus
Scincella

1A. Thirty to 34 scales around mid-body; adpressed fore- and hind limb overlapping in adults; 2–3 scales in lower eyelid window (Fig. 117)............ *cherriei* (p. 352)
1B. Twenty-four to 31 scales around midbody; adpressed fore- and hind limb separated by 3 or more lateral scales in adults; single, median scale in lower eyelid window (Fig. 10) 2
2A. Twenty-four to 26 scales around midbody; fewer than 65 dorsal scales in paravertebral row between parietal and level above cloaca............... *incerta* (p. 358)
2B. Twenty-seven to 31 scales around midbody; more than 65 dorsal scales in paravertebral row between parietal and level above cloaca............... *assata* (p. 349)

Clave para las Especies Hondureñas del
Género *Scincella*

1A. 30–34 escamas alrededor de la parte del media del cuerpo; cuando los miembros anteriores y posteriores se doblan sobre la parte media del cuerpo, estos se sobreponen en los adultos; el disco translucido del párpado inferior dividido en 2–3 escamas (Fig. 117)
.... *cherriei* (p. 352)
1B. 24–31 escamas alrededor de la parte del media del cuerpo; cuando los miembros anteriores y posteriores se doblan sobre la parte media del cuerpo, estos no se sobreponen en los adultos, están separados por tres o más escamas; el párpado inferior con un gran disco translucido sin dividir (Fig. 10) 2

2A. 24–26 escamas alrededor de la parte del media del cuerpo; menos de 65 escamas dorsales en la hilera paravertebral entre la escama parietal hasta al nivel superior de la cloaca............... *incerta* (p. 358)
2B. 27–31 escamas alrededor de la parte del media del cuerpo; más de 65 escamas dorsales en la hilera paravertebral y hasta al nivel superior de la cloaca............... *assata* (p. 349)

Scincella assata (Cope, 1865a)

Lampropholis assatus Cope, 1865a: 179 (holotype, ANSP 9465 [see Malnate, 1971: 355]; type locality: "Guatimala [sic] ... near the Volcano of Isalco [sic = Volcano de Izalco, Sonsonate, El Salvador]").
Scincella assata: Mittleman, 1950: 20 (combination inferred by use of trinomial *S. a. assata*); McCranie, 2015a: 370.
Sphenomorphus assatus: McCranie and Köhler, 1999: 111; Köhler, 2000: 94; Köhler et al., 2005: 136; Townsend and Wilson, 2010b: 692; Solís et al., 2014: 132.

Geographic Distribution.—*Scincella assata* occurs at low and moderate elevations on the Pacific versant from coastal Jalisco, Mexico, to eastern El Salvador and southwestern Honduras. In Honduras, this species is known only from a single low elevation locality in Intibucá.

Description.—The following is based on one unsexed subadult (SMF 78933). *Scincella assata* is a small lizard (maximum recorded SVL 53 mm [Leenders and Watkins-Colwell, 2004]; 26 mm SVL in Honduran specimen); dorsal head scales enlarged, smooth, platelike, with single frontonasal (internasals or supranasals absent) scale that broadly contacts rostral, frontonasal scale with rear margin concave, paired prefrontals separated medially, single frontal, single frontoparietal, 4 supraoculars,

Plate 66.   *Scincella assata*. SMF 78933, subadult, SVL = 26 mm. Intibucá: Santa Lucía.

and paired parietal scales in contact medially posterior to single interparietal scale; frontal in contact with first supraocular throughout length of that supraocular, frontal also in broad contact with second supraocular; 8–9 superciliary scales; nasal entire, nostril opening in center of scale; 2 loreals; 7 supralabials; 2 and 3 preoculars and postoculars, respectively; 6 infralabials; moveable eyelid present; single median window present in lower eyelid; pupil circular; 1+2 temporals, with upper secondary temporal overlapping lower secondary temporal and parietal overlapping upper secondary temporal; pretemporal single; mental single, large; postmental single, large, contacting first and second infralabials; gular scales smooth, cycloid, imbricate, large; medial pair of nuchal scales slightly enlarged; dorsal body scales smooth, cycloid, imbricate; 71 [≥67; Köhler, 2008] rows of dorsal scales from parietals to level of midlength of thigh in paravertebral row; 28 [27–31 throughout geographic distribution; Köhler, 2008] scales around midbody; ventral body scales smooth, cycloid, imbricate, in 42 rows between levels of axilla and groin; pair of enlarged precloacal scales present; femoral and precloacal pores absent; limbs relatively short, adpressed fore- and hind limb with about 3 lateral scales separating adpressed limbs; 14–15 (14.5) subdigital scales on Digit IV of hind limb, 10–11 (10.5) scales on Digit IV of forelimb; 5 digits on forelimb; SVL 26 mm; TAL/SVL 1.65; HL/SVL 0.19; HW/SVL 0.12; SHL/ SVL 0.10.

Color in life of a subadult (SMF 78933; Plate 66): dorsal surface of body Deep Vinaceous (4) with zig-zag dark brown lines; top of head Robin Rufous (340) with indistinct, incomplete dark brown lines; dorsal, lateral, and ventral surfaces of tail Spinel Red (108B) with dark brown dorsolateral and middorsal line on anterior third of tail; distinct dark brown dorsolateral stripe extending from posterior edge of eye onto dorsolateral anterior third of tail, that stripe most distinct on head, gradually thinning on body until disappearing on posterior two-thirds of tail; lateral surfaces of head and body a mixture of pink and Deep Vinaceous; ventral surface of head cream, that of body same as lateral surface of body.

Color in alcohol: dorsal surfaces of head and body brown with dark brown flecking; distinct, dark brown, generally straight edged dorsolateral stripe extending from snout onto tail; dark dorsolateral stripe not bordered by thin white line; lateral surface of body below dorsolateral stripe pale brown with slight brown flecking; lateral surface of head pale brown with dark brown flecking on supralabials; dorsal surface of tail brown with darker brown flecking anteriorly, becoming paler brown with less flecking on distal two-thirds of length; ventral surface of head cream with brown flecking on infralabials; ventral surface of body cream, that of tail pale brown.

Diagnosis/Similar Species.—The combination of having a moveable eyelid, a window in the lower eyelid, large scales on top of the head, the venter with large and smooth, imbricate cycloid scales, the supranasal scales replaced by a single frontonasal scale that broadly contacts the rostral, and a single frontoparietal scale distinguishes *Scincella assata* from all other Honduran lizards, except *S. cherriei* and *S. incerta*. *Scincella cherriei* has 30–34 scales around midbody, has 2–3 scales in the lower eyelid window, has longer limbs that usually overlap when the fore- and hind limb are adpressed against the body in adults, and has a brown tail in life (versus 27–31 scales around midbody, single median scale in lower eyelid window, limbs separated by more than 3 lateral scales when adpressed in adults, and unregenerated tail red or orange in life in *S. assata*). *Scincella incerta* has 66 or fewer dorsal scales in the paravertebral row, has 24–26 scales around midbody, and usually lacks a red or orange tail in life, except the regenerated tail is sometimes red or orange (versus 65 or more dorsal scales in paravertebral row, 27–31 scales around midbody, and unregenerated tail red or orange in life in *S. assata*).

Illustrations (Plate 66).—Álvarez del Toro, 1983 (adult); Castiglia et al., 2013 (adult); García and Ceballos, 1994 (drawing of adult; as *Sphenomorphus*); Köhler, 2000 (subadult; as *Sphenomorphus*), 2003a (adult; as *Sphenomorphus*), 2008 (adult as *Sphenomorphus*); Köhler et al., 2005 (adult; as *Sphenomorphus*); Ramírez-Bautista, 1994 (adult; figure caption erroneously listed as *Leptotyphlops humilis dugesi* [Bocourt, 1881: 81]).

Remarks.—The only Honduran specimen of *Scincella assata* was collected in October 1998. Recent and rather extensive collecting efforts on the Pacific versant of southern Honduras, including the islands in the Golfo de Fonseca, did not result in collection of additional specimens. Results of molecular analysis of Linkem et al. (2011: 1225, 1227) recovered a clade consisting of *S. assata* and *S. cherriei*; those two species were in turn sister to *S. lateralis* (Say, 1823: 324, *In* James, 1823) of the southeastern U.S., which was in turn sister to *S. reevesii* (Gray, 1839a: 292) of China. Linkem et al. (2011) did not study the two other Central American species of *Scincella* (*S. incerta* and *S. rarus*). Castiglia et al. (2013) provided information on the chromosomes of *S. assata* from Chiapas, Mexico.

Natural History Comments.—This species is known in Honduras only from 370 m elevation in the Lowland Dry Forest formation. The single Honduran specimen of *Scincella assata* was under a rock during midmorning along a rock fencerow in October in a cattle pasture. Köhler et al. (2005) reported this species was found in leaf litter and under a brush pile in El Salvador. Oliver (1937) reported two individuals were collected in rotten logs in coastal Colima, Mexico. García and Ceballos (1994) said it was found in decomposing tree trunks and under litter in tropical deciduous and semideciduous forest and in palm forests in coastal Jalisco, Mexico. Ramírez-Bautista (1994) reported *S. assata* living under dead leaves, trunks, and rocks in coastal Jalisco, Mexico. Nothing has been published on reproduction of this species other than reports in Álvarez del Toro

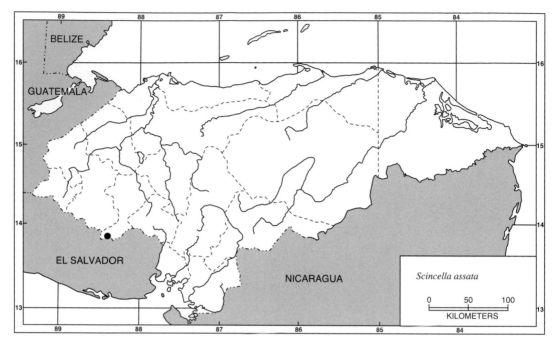

Map 54.   Locality for *Scincella assata*. Solid circle denotes single specimen examined. That single specimen is the only known Hounduran specimen of this species.

(1983) and Ramírez-Bautista (1994) that females deposit two to four or two to five eggs, principally during the rainy season. Apparently nothing has been published on diet of *S. assata* other than general statements in Álvarez del Toro (1983) and Ramírez-Bautista (1994) that it feeds on insects.

Etymology.—The name *assata* is formed from the Latin word *assis* (a copper coin) and the Latin suffix *-ata* (provided with, having the nature of, pertaining to). The name refers to the "fulvous" coloration described by Cope (1865a: 180) for the holotype.

Specimens Examined (1 [0]; Map 54).— **INTIBUCÁ**: Santa Lucía, SMF 78933.

Scincella cherriei (Cope, 1893)

Mocoa cherriei Cope, 1893: 340 (holotype, AMNH 9551 [see Myers and Donnelly, 1991: 6]; type locality: "Palmar" [Costa Rica]).

Leiolopisma assatum: Dunn and Emlen, 1932: 31.
Lygosoma assatum cherriei: Stuart, 1940: 13.
Scincella cherriei: Mittleman, 1950: 20 (combination inferred by use of trinomial *S. c. cherriei*); Meyer, 1969: 265 (in part); Wilson and Meyer, 1969: 146; Meyer and Wilson, 1973: 29 (in part); McCranie and Solís, 2013: 242; Espinal, Solís, O'Reilly, et al., 2014: 299; McCranie, 2014: 292; McCranie and Valdés Orellana, 2014: 45; McCranie, 2015a: 370.
Scincella assata assata: Brattstrom and Howell, 1954: 118.
Leiolopisma cherriei cherriei: Meyer, 1966: 175.
Leiolopisma cherriei: Greene, 1969: 55; J. F. Jackson, 1973: 309.
Leiolopisma cherrei [sic] *cherrei* [sic]: Greer, 1970: 172.
Sphenomorphus cherriei: Greer, 1974: 34; O'Shea, 1986: 41; Espinal, 1993, table

3; Wilson and McCranie, 1994a: 420; Wilson and McCranie, 1998: 17; Köhler, McCranie, and Nicholson, 2000: 425; Nicholson et al., 2000: 30; Espinal et al., 2001: 106; Wilson et al., 2001: 137; Castañeda, 2002: 15; Lundberg, 2002a: 6; McCranie et al., 2002: 27; Wilson and McCranie, 2004b: 43; Wilson and McCranie, 2004c: 24; McCranie, 2005: 20; McCranie and Castañeda, 2005: 15; McCranie et al., 2005: 116; Castañeda, 2006: 32; Castañeda and Marineros, 2006: 3.8; McCranie et al., 2006: 130; Townsend, 2006a: 35; Wilson and Townsend, 2006: 105; Wilson and Townsend, 2007: 145; Townsend and Wilson, 2008: 164; Townsend, 2009: 298; Townsend and Wilson, 2010b: 692; Townsend et al., 2012: 102; Townsend, Wilson, et al., 2013: 197; Solís et al., 2014: 132 (in part).

Geographic Distribution.—*Scincella cherriei* occurs at low, moderate, and occasionally intermediate elevations on the Atlantic versant from central Veracruz, Mexico, to extreme western Panama and on the Pacific versant in southeastern Honduras and from northwestern Costa Rica to extreme western Panama. Apparently isolated populations also occur on the outer end and central portions of the Yucatán Peninsula, Mexico. It is also found on Isla del Maíz Grande, Nicaragua. In Honduras, this species occurs throughout most of the mainland, except the extreme southern portion.

Description.—The following is based on 20 unsexed adults (USNM 570332–33, 570342, 570346, 570350–51, 570354–56, 570359–61, 570363–65, 570367, 570370–73). *Scincella cherriei* is a small lizard (maximum recorded SVL 68 mm [Savage, 2002, but no documenting data provided]; 59 mm SVL in largest Honduran specimen [USNM 573065]); dorsal head scales enlarged, smooth, platelike, with single frontonasal (internasals or supranasals absent) scale broadly contacting rostral scale, fron-

tonasal with rear margin straight, paired prefrontals separated medially, single frontal, single frontoparietal, 4 supraoculars, and paired parietal scales in contact medially posterior to single interparietal scale; frontal in contact with first supraocular throughout length of supraocular, frontal also in broad contact with second supraocular; 8–9 (rarely 9) superciliary scales; nasal entire, nostril opening in center of scale; 2 loreals; 7 supralabials; 2–3 (usually 2) and 3 preoculars and postoculars, respectively; 6–7 (usually 7) infralabials; moveable eyelid present; pupil circular; 2–3 scales in lower eyelid window; 1+2 temporals, with upper secondary temporal usually overlapping lower secondary temporal (occasionally lower secondary temporal overlapping upper secondary temporal; see Remarks) and parietal overlapping upper secondary temporal; 2 pretemporals; mental single, large; postmental single, large, contacting first and second infralabials; gular scales smooth, cycloid, imbricate, large; medial pair of nuchal scales varying from enlarged to not enlarged; dorsal body scales smooth, cycloid, imbricate; 57–65 (61.6 ± 2.1) rows of dorsal scales in paravertebral row from parietal to level of midthigh; 30–34 (31.1 ± 1.2) scales around midbody; ventral body scales smooth, cycloid, imbricate, in 39–45 (41.2 ± 2.1) rows between levels of axilla and groin; pair of enlarged precloacal scales present; femoral and precloacal pores absent; limbs relatively short, adpressed fore- and hind limb overlapping to about 3 lateral scales separating adpressed limbs in adults; 15–18 (16.7 ± 1.0, $n = 39$) subdigital scales on Digit IV of hind limb, 9–12 (10.3 ± 0.6) scales on Digit IV of forelimb; 5 digits on forelimb; SVL 47.0–58.9 (54.3 ± 3.4) mm; TAL/SVL 1.09–1.70 in ten; HL/SVL 0.18–0.22; HW/SVL 0.11–0.14; SHL/SVL 0.10–0.13.

Color in life of an adult female (USNM 570349): dorsum Dark Grayish Brown (20), bounded below by indistinct Sepia (119) lateral stripe; area below lateral stripe Hair

Brown (119A), grading to pale gold of venter; head Dark Grayish Brown above; Sepia postocular stripe present, confluent with lateral stripe on body; supralabials pale brown anteriorly, dark brown posteriorly; fore- and hind limb Dark Grayish Brown; tail Dark Grayish Brown dorsally, Pale Neutral Gray (86) ventrally. Color in life of another adult female (USNM 570325): dorsum dark brown; black stripe extending from nostril, above eye, and onto dorsolateral region of body, disappearing about third of distance along length of body; lateral surface of body gray-brown; dorsal surfaces of fore- and hind limb brown; tail brown; chin pale pink with gray edging on many scales; belly pale golden orange with gray edging on scales; subcaudal surface gray; iris black. Color in life of a third adult female (FMNH 252590): dorsal surfaces Tawny (38) with much Sepia (119) mottling; indistinct Cinnamon (39) dorsolateral stripe extending from posterior to eye onto tail; broad Sepia band extending below Cinnamon stripe from posterior to eye to above forelimb insertion, becoming mottled with Sepia and brown posterior to that point; top of head Tawny with Sepia flecking; Sepia stripe from anterior edge of eye to tip of snout; supralabials mottled with Sepia and pale brown; infralabials mottled with Sepia and cream; chin and throat pale brown with Sepia stripes along scale seams; neck and all of belly Tawny with dark brown longitudinal mottling along scale seams; top of tail Sepia with Tawny narrow crossbands, crossbands fading until disappearing at about midlength of tail; subcaudal surface pale brown with dark brown flecking; iris Tawny.

Color in alcohol: dorsal surfaces of head and body brown with dark brown spots or mottling present on body; dark brown, ragged edged, rather broad dorsolateral stripe extending from snout to anterior third to three-quarters of body length, dark stripe becoming less distinct posteriorly until disappearing; lateral surface of body below dark stripe pale brown to cream, with small brown spots; lateral surface of head pale brown to cream, with dark brown spots or markings on supralabials; dorsal surface of unregenerated tail generally same color as dorsal surface of body; ventral surface of head cream with dark brown to brown spots present on most scales; ventral surface of body cream; subcaudal surface of unregenerated tail cream with brown flecking laterally.

Diagnosis/Similar Species.—The combination of having moveable eyelids, large scales on top of the head, the venter with large, smooth, imbricate cycloid scales, having the supranasal (internasal) scales replaced by a single frontonasal scale, and having a single frontoparietal scale distinguishes *Scincella cherriei* from all remaining Honduran lizards, except the other *Scincella*. *Scincella assata* has 67–77 dorsal scales in the paravertebral row, has 27–31 scales around midbody, has a single median window in the lower eyelid, and usually has a red or orange tail in life when unregenerated (versus 57–65 dorsal scales in paravertebral row, 30–34 scales around midbody, 2–3 scales in lower eyelid window, and brown tail in life in *S. cherriei*). *Scincella incerta* has 24–26 scales around midbody, has a single median window in the lower eyelid, and has a distinct dorsolateral dark stripe bordered below by a thin white stripe, both of which extend at least to the posterior portion of the body and frequently continuous with similar tail stripe (versus 30–34 scales around midbody, 2–3 scales in lower eyelid window, and no distinct dorsolateral stripe extending length of body in *S. cherriei*).

Illustrations (Figs. 117, 118; Plate 67).—Álvarez del Toro, 1983 (adult); Campbell, 1998 (adult; as *Sphenomorphus*); Castiglia et al., 2013 (adult); García-Vázquez and Feria-Ortiz, 2006 (adult; as *S. cherriae* [sic]); Guyer and Donnelly, 2005 (adult; as *Sphenomorphus*); Köhler, 2000 (adult, head scales; as *Sphenomorphus*), 2001b (adult; as *Sphenomorphus*), 2003a (adult; as *Spheno-*

Plate 67. *Scincella cherriei*. FMNH 282588, adult male. Copán: San Isidro.

morphus), 2008 (adult; as *Sphenomorphus*); Lee, 1996 (adult, head scales; as *Sphenomorphus*), 2000 (adult, head scales; as *Sphenomorphus*); Lundberg, 2002a (adult; as *Sphenomorphus*); McCranie et al., 2005 (adult; as *Sphenomorphus*), 2006 (adult, head scales; as *Sphenomorphus*); Myers and Donnelly, 1991 (hemipenis; as *Sphenomorphus*); Savage, 2002 (adult; as *Sphenomorphus*); Stafford and Meyer, 1999 (adult; as *Sphenomorphus*); Taylor, 1956b (temporal scales; as *Leiolopisma* A. M. C. Duméril and Bibron, 1839: 742); Townsend and Wilson, 2008 (adult; as *Sphenomorphus*).

Remarks.—Myers and Donnelly (1991) provided a systematic review of *Scincella cherriei* based on Costa Rican and Panamanian specimens. The Honduran specimens for which I took comparable data agree well with the *S. cherriei* description provided by Myers and Donnelly (1991). Savage (2002) concluded that individual variation and sexual dimorphism in the Costa Rican specimens obscured most of the geographic races recognized by some previous workers but offered no data to back that claim. Castiglia et al. (2013) reported on the chromosomes of *S. cherriei* from Chiapas, Mexico.

Greer and Shea (2003) discussed the possible systematic importance of the secondary temporal overlap in sphenomorphine skinks. Honduran *Scincella cherriei* usually have the upper secondary temporal overlapping the lower secondary temporal, but the reverse is also occasionally true. Two of 20 specimens examined (USNM 570332, 570355) have different secondary temporal overlap patterns on each side of head, and two others (USNM 570360, 570373) have the lower secondary temporal overlapping the upper secondary temporal on both sides of the head.

See Remarks for *Scincella assata* for molecular comments.

Natural History Comments.—*Scincella cherriei* is known from near sea level to 1,860 m elevation in the Lowland Moist Forest, Lowland Dry Forest, Lowland Arid Forest, Premontane Wet Forest, Premontane Moist Forest, and Premontane Dry Forest formations and peripherally in the Lower Montane Wet Forest and Lower Montane Moist Forest formations. This diurnal species has been collected in every

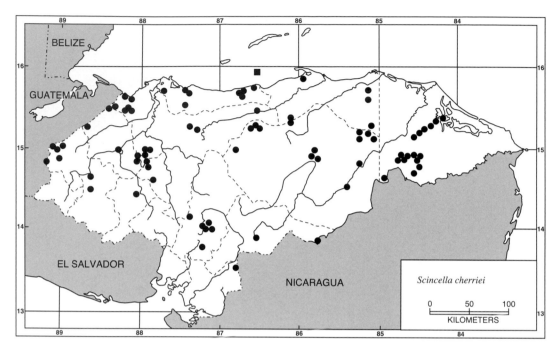

Map 55.   Localities for *Scincella cherriei*. Solid circles denote specimens examined and the solid square an accepted record.

month of the year, and it is usually found in areas with abundant leaf litter for foraging. Exposed individuals usually dart beneath leaf litter when they are aware of danger. It is also found under logs and other objects on the ground. Greene (1969) included two Honduran specimens in his note on reproduction in this species; however, Greene did not give locality data or specimen numbers, but they apparently were KU 67393–94. This species deposits one to five eggs at a time, with the breeding season probably occurring throughout the year in wetter climates, but more restricted in drier climates (Greene, 1969; Fitch, 1973a,b, 1983, 1985; Watling et al., 2005; Goldberg, 2008). Goldberg (2008) also found evidence of multiple clutches per season in this species. Food consists largely of small arthropods, with a wide variety of insects consumed, including larvae and pupae (Fitch, 1983). Fitch also recorded adults feeding on hatchling conspecifics.

Etymology.—The name *cherriei* is a patronym honoring George K. Cherrie, at that time "the well-known zoölogist of San José, who has added much to our knowledge of the life of Costa Rica" (Cope, 1893: 341).

Specimens Examined (235, 1 egg lot [54]; Map 55).—**ATLÁNTIDA**: mountains S of Corozal, LACM 47813–14, 47820, LSUMZ 21509–13; Estación Forestal CURLA, USNM 508443, 570324, 578843; Guaymas District, UMMZ 58383; 7.4 km SE of La Ceiba, USNM 570323; 14.5 km E of La Ceiba, LACM 47819; La Liberación, USNM 578848; Lancetilla, AMNH 70445–46, ANSP 20880, FMNH 21831, MCZ R29397–400, 32214–15, 38929, TCWC 19210, UMMZ 78083 (2), USNM 578845–47; Quebrada de Oro, KU 200582, USNM 508442, 570325–29; near Tela, MCZ R38930, 38931 (2 eggs); Tela, UMMZ 58384–85. **CHOLUTECA**: Las Mesas, CM 157705, FN 256954–55 (still in Honduras because of permit problems).

COLÓN: Cerro Calentura, LSUMZ 22453, USNM 570330; Quebrada Botaderos, UTA R-53279; Quebrada Machín, USNM 570331; Río Claura, UMMZ 58411; Tulito, BMNH 1985.1295–96. **COMAYAGUA**: east end of Lago de Yojoa, UF 87869; Siguatepeque, UF 124825. **COPÁN**: Copán, TCWC 23637, UIMNH 52516, UMMZ 83030–31; Laguna del Cerro, UF 142463, USNM 570332; below Quebrada Grande, SMF 79144; Río Amarillo, USNM 570333–34, 579571; San Isidro, FMNH 282588, 282590. **CORTÉS**: 6 km N of Agua Azul, AMNH 156734–35; Agua Azul, AMNH 70505–06; Buenos Aires, UF 144108; 1.6 km W of El Jaral, LACM 47815, LSUMZ 24213, 88072; 1.6 km SE of El Jaral, CM 65513, LACM 45299, 45349–51, 47816–18, 47822 (listed as LACM 47882 by Meyer and Wilson, 1973); El Paraíso, UF 144728–30, 144733; Finca Whisky River, LACM 108839; E side of Lago de Yojoa, KU 67393–94; Lago de Yojoa, MSUM 4704–05; Los Pinos, UF 166401, USNM 565494–96; Montaña Santa Ana, MCZ R32216; Naranjito, USNM 570335; near Peña Blanca, LSUMZ 88091; 3.2 km W of San Pedro Sula, LACM 47821; 1 km W of San Pedro Sula, ANSP 30518; W of San Pedro Sula, FMNH 5054–59; near Santa Elena, LSUMZ 38839–41, 38852; Santa Teresa, USNM 573968; Sendero Las Minas, SMF 78864; about 1 km SSE of Tegucigalpita, USNM 570336. **EL PARAÍSO**: Arenales, LACM 20476; Danlí, BYU 18197–98. **FRANCISCO MORAZÁN**: Cerro Uyuca, KU 200583, LSUMZ 24413; El Picacho, USNM 570337; Nueva Armenia, UNAH 4423; 9.9 km N of Tegucigalpa, LSUMZ 24180; Tegucigalpa, UTA R-41238, near Zambrano, USNM 579569. **GRACIAS A DIOS**: Bachi Kiamp, USNM 565493, 573963–64, FN 257003, 257012 (still in Honduras because of permit problems); Bodega de Río Tapalwás, UF 150325–26, USNM 570338–46; Caño Awalwás, UF 150323–24, 150327, 150330, USNM 570347–48; Cerro Wahatingni, UF 150329, USNM 570349; Cerros de Sabaní, USNM 570350–53; near Cueva de Leimus, USNM 579572; Hiltara Kiamp, USNM 570354; Kakamuklaya, USNM 573176; Kipla Tingni Kiamp, USNM 570355; Leimus (Río Warunta), FMNH 282589, USNM 565491; Mavita, USNM 579570; Sachin Tingni Kiamp, USNM 570356; Sadyk Kiamp, USNM 565492; Samil, USNM 573965; San San Hil, USNM 570357–62; San San Hil Kiamp, USNM 570363–65; Urus Tingni Kiamp, USNM 570366–70; Usus Paman, USNM 573962; Warunta, USNM 579573, FN 256933 (still in Honduras because of permit problems); Warunta Tingni Kiamp, USNM 565489–90, 570371–74. **INTIBUCÁ**: La Rodadora, USNM 580447. **LEMPIRA**: Villa Verde, FMNH 283734. **OLANCHO**: Caobita, SMF 80829–30; 4.5 km SE of Catacamas, LACM 45156–57, 47810; 6 km NW of Catacamas, UF 90203; Cerro de Enmedio, USNM 342379; Cuaca, UTA R-53278; between El Díctamo and Parque Nacional La Muralla Centro de Visitantes, USNM 342380; Montaña de Las Parras, USNM 342381–82; Piedra Blanca, USNM 579567–68; Quebrada de Las Escaleras, USNM 342383–84; Quebrada de Las Marías, USNM 570380–81; Quebrada El Guásimo, SMF 80828; Quebrada La Calentura, USNM 342385; Quebrada Las Cantinas, USNM 342386–87; confluence of ríos Aner and Wampú, USNM 570376–78; Río de Enmedio, USNM 342389; confluence of ríos Sausa and Wampú, USNM 570379; along Río Wampú between ríos Aner and Sausa, USNM 570382–83; confluence of ríos Yanguay and Wampú, USNM 570375; Terrero Blanco, USNM 342388. **SANTA BÁRBARA**: Buena Vista, USNM 579574; Cerro Negro, USNM 573177–78; Compaña Agrícola Paradise, USNM 578844; La Cafetalera, USNM 573966–67; W shore of Lago de Yojoa, CM 59125. **YORO**: 5 km E of Coyoles, LACM 47811–12; Montañas de Mataderos, MCZ R38928, UMMZ 77843; 2.5 km SE San José de Texíguat, USNM

581900; Subirana Valley, FMNH 21838–40, MCZ R38927, 38932, UMMZ 77850.

Other Records (Map 55).—**ISLAS DE LA BAHÍA:** Cayo Cochino Menor (Lundberg, 2002a).

Scincella incerta (Stuart, 1940)

Lygosoma incertum Stuart, 1940: 10 (holotype, FMNH 20307; type locality: "Volcán Tajumulco, Guatemala, at 5500 feet").

Scincella incertum: Mittleman, 1950: 20.

Scincella cherriei: Meyer, 1969: 265; Meyer and Wilson, 1973: 29 (in part).

Sphenomorphus incertus: Wilson and McCranie, 1994a: 419; Wilson et al., 2001: 137; Köhler, 2003a: 153; Wilson and McCranie, 2004b: 43; Townsend, 2005: 337; McCranie et al., 2006: 218; Townsend, 2006a: 35; Townsend et al., 2006: 32; Wilson and Townsend, 2006: 105; Köhler, 2008: 165; Townsend and Wilson, 2008: 168; Townsend, 2009: 298; Townsend et al., 2010: 12.

Scincella incerta: Linkem et al., 2011: 1237; McCranie, 2014: 292; Solís et al., 2014: 132; McCranie, 2015a: 370.

Geographic Distribution.—*Scincella incerta* occurs at moderate and intermediate elevations in disjunct populations on the Atlantic versant from central Guatemala to northwestern Honduras. It is also found on the Pacific slopes of western Guatemala and extreme southwestern Honduras. In Honduras, this species is known from isolated montane habitats in the western third of the country.

Description.—The following is based on four adult males (FMNH 283731; MCZ R38933; UF 144061; USNM 589147), three adult females (FMNH 283730, 283732; UF 14732), and four juveniles and subadults (UF 144731; USNM 330189, 570384–85). *Scincella incerta* is a small lizard (maximum recorded SVL 67 mm [MCZ R38933, a male]); dorsal head scales enlarged, smooth, platelike, with internasal (supranasal) replaced by single frontonasal scale broadly contacting rostral, frontonasal with straight rear margin, paired prefrontals separated medially, single frontal, single frontoparietal, 4 supraoculars, and paired parietal scales in contact medially posterior to single interparietal scale; frontal in contact with first supraocular throughout length of supraocular, frontal also in broad contact with second supraocular; 7–8 superciliary scales; nasal entire, nostril opening in center of scale; 1–2 (rarely 1) loreals; 6–7 supralabials; 2 and 3 preoculars and postoculars, respectively; 6–7 (usually 6) infralabials; moveable eyelid present; pupil circular; single medial window in each lower eyelid; 1+2 temporals, usually with lower secondary temporal overlapping upper secondary temporal (rarely upper secondary temporal overlapping lower secondary temporal) and parietal overlapping upper secondary temporal; 2–3 pretemporals; 1 large mental; 1 large postmental contacting first and second infralabials; gular scales smooth, cycloid, imbricate, large; medial pair of nuchal scales enlarged; dorsal body scales smooth, cycloid, imbricate; 54–64 (59.4 ± 3.2) dorsal scales in paravertebral row between parietal and level of midthigh; 24–26 (25.1 ± 1.0) scales around midbody; ventral body scales smooth, cycloid, imbricate, in 36–52 (44.7 ± 5.4) rows between levels of axilla and groin; pair of enlarged precloacal scales present; femoral and precloacal pores absent; limbs relatively short, 0–2 lateral scales separate adpressed fore- and hind limb in juveniles, 7–18 lateral scales separating adpressed limbs in adults; 11–14 (12.6 ± 1.0) subdigital scales on Digit IV of hind limb, 7–11 (9.0 ± 1.0) scales on Digit IV of forelimb in 12; 5 digits on forelimb; SVL 51.4–67.3 (56.1 ± 10.6) mm in males, 43.7–57.5 (50.5 ± 6.8) mm in females, 26.4–39.6 (30.2 ± 6.3) mm in juveniles and subadults; TAL/SVL 0.78–1.83 in three males, 1.26–1.48 in two females, 1.38 in one juvenile; HL/SVL 0.15–0.17 in males, 0.13–0.17 in females, 0.18–0.25 in juveniles and subadults; HW/SVL 0.10–0.11 in males, 0.10–0.12 in

Plate 68. *Scincella incerta*. FMNH 283730, adult female, SVL = 58 mm. Ocotepeque: Las Hoyas.

females, 0.12–0.15 in juveniles and sub-adults; SHL/SVL 0.09–0.11 in males, 0.08–0.10 in females, 0.09–0.11 in juveniles and subadults.

Color in life of an adult female (FMNH 283730; Plate 68): middorsum and top of head Natal Brown (219A) with Hair Brown (119A) flecking; middorsum bordered below by thin Cinnamon (39) line; top of tail Natal Brown with Hair Brown flecking, but without thin line below; Sepia (119) dorso-lateral line with Cinnamon flecking extending from posterior to eye to base of tail; lateral surface of body golden brown with Sepia mottling; side of head golden brown with Sepia spots on supralabials; chin pale brown with dark brown lineate flecking, throat and belly golden brown with dark brown around scale edges; subcaudal surface gray with some dark brown flecking; iris Sepia.

Color in alcohol: dorsal surfaces of head and body brown with darker brown spots or flecking on body; distinct, dark brown, generally straight edged dorsolateral stripe extending from snout onto tail; dark dorso-lateral stripe bordered below by thin white line throughout length and bordered above by similar pale stripe extending from posterior to eye onto about anterior third of body; lateral surface of body below dorsolateral stripe and line pale brown with brown flecking; lateral surface of head pale brown with dark brown spotting on supra-labials; dorsal surface of unregenerated tail generally same color as dorsal surface of body; ventral surface of head cream with brown spots on infralabial scales; ventral surface of body cream; subcaudal surface of unregenerated tail cream with brown fleck-ing laterally.

Diagnosis/Similar Species.—The combi-nation of having moveable eyelids, having a window in the lower eyelid, large scales on top of the head, the venter with large, smooth, imbricate cycloid scales, the supra-nasal scales replaced by a single frontonasal scale that broadly contacts the rostral, and having a single frontoparietal scale distin-guishes *Scincella incerta* from all remaining Honduran lizards, except the other *Scincel-la. Scincella assata* has 67–77 dorsal scales in the paravertebral row, has 27–31 scales around midbody, and usually has a red or orange tail in life when unregenerated (versus 53–66 dorsal scales in paravertebral row, 24–26 scales around midbody, and brown tail in life in *S. incerta*). *Scincella cherriei* has 30–34 scales around midbody, has longer limbs that overlap when the fore- and hind limb are adpressed against the body in adults, has 2–3 scales in the lower eyelid window, and has a dark dorsolateral stripe with ragged edges that fade out on the body (versus 24–26 scales around midbody, adpressed fore- and hind limb separated by 5+ lateral scales in adults, single median scale in lower eyelid, and dark brown dorsolateral stripe usually extending length of body in *S. incerta*).

Illustrations (Plate 68).—Townsend and Wilson, 2008 (adult; as *Sphenomorphus*).

Remarks.—Although first recognized as a distinct species in 1940, little information has been published concerning *Scincella incerta*. Whereas, the two other species of *Scincella* occurring in Honduras have been studied using molecular techniques, that is not the case with *S. incerta* (Linkem et al., 2011: 1237, although I did later send Linkem tissues of Honduran *S. incerta*). Although this species is easily distinguished morphologically from *S. cherriei*, Meyer and Wilson (1973) synonymized *S. incerta* with that species. Several distinct characters easily distinguish each species from the other.

Wilson and McCranie (2004c: 24) and Townsend and Wilson (2008: 313) reported SMF 78864 from Sendero Las Minas, Cusuco, Cortés, as *Scincella cherriei*. However, it is more likely the specimen actually represents *S. incerta*. Unfortunately, the specimen has been on loan for at least 7 years, and as of the end of December 2015, the specimen still had not been returned (G. Köhler, personal communication, 22 December 2015).

Natural History Comments.—*Scincella incerta* is known from 1,100 to 1,670 m elevation in the Premontane Wet Forest and Lower Montane Wet Forest formations. This diurnal species was crawling on the ground in disturbed areas in August, under dense mats of pine needles in June, and under a log in July. Townsend (2005) recorded one from a trail through mixed pine–broadleaf forest in March, and Townsend et al. (2006) reported one from a pitfall trap (month of collection not given). Nothing has been published on reproduction or feeding habits of this species, but both are probably similar to those of *S. cherriei*.

Etymology.—The name *incerta* is from the Latin *incertus* (doubtful) and alludes to the "extremely obscure" systematic relationships of this species to other Western Hemisphere members of this group of skinks (Stuart, 1940: 12).

Specimens Examined (11 [1]; Map 56).— **CORTÉS**: near Parque Nacional Cusuco Centro de Visitantes, UF 144061; Sendero Las Minas, UF 144732. **OCOTEPEQUE**: Las Hojas, FMNH 283730–32. **SANTA BÁRBARA**: La Fortuna Camp, UF 144731; Quebrada Las Cuevas, USNM 589147. **YORO**: Cerro de Pajarillos, USNM 570384–85; 2.5 airline km NNE of La Fortuna, USNM 330189; Portillo Grande, MCZ R38933.

Superfamily Scincoidea Gray, 1825

This superfamily includes only the Family Scincidae (see Vidal and Hedges, 2009; Hedges and Conn, 2012; Hedges, 2014). Externally, Honduran members of the Scincoidea can be distinguished from all other Honduran lizards by the combination of having moveable eyelids, the dorsal surface of the head with enlarged scales or plates that include fewer than 2 pairs of scales between the rostral and first unpaired plate, 5 digits on the forelimb, the dorsal and ventral surfaces of the body with large and smooth cycloid scales, and lacks a lower eyelid window. One Honduran species of the Scincoidea (*Plestiodon sumichrasti*) is known to be oviparous and the other Honduran species (*Mesoscincus managuae*) is presumed to also be oviparous.

Family Scincidae Gray, 1825

This family occurs in the Western Hemisphere from southern British Columbia,

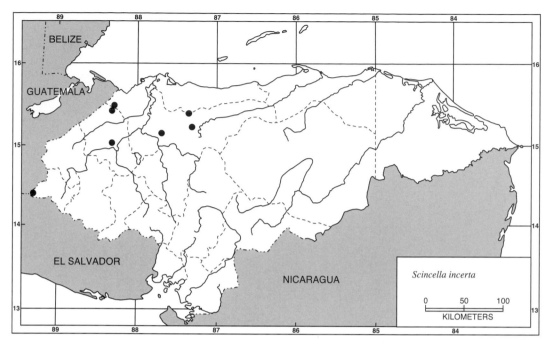

Map 56.   Localities for *Scincella incerta*. Solid circles denote specimens examined.

Canada, and most of the U.S. southward through Mexico and into Central America to northwestern Costa Rica, and on the Atlantic island of Bermuda. It occurs in the Eastern Hemisphere in Madagascar, from across northern and eastern Africa to southwestern Asia, and in southeastern Asia and associated islands. Thirty-three genera containing about 275 named species are included in this family, with two species in two genera known to occur in Honduras.

Remarks.—Brandley et al. (2010: 2–3) recovered the interesting results that the ancestors of the Bermuda Island endemic *Plestiodon longirostris* Cope (1861b: 313) descended from a lineage that dispersed overwater from the eastern North American mainland to Bermuda before that founder lineage became extinct on that mainland. That dispersal event was dated from 11.5 to 19.8 Ma by Honda et al. (2003) and Brandley et al. (2010). However, the proposed timing of that dispersal event predates the existence of Bermuda Island by well over 10 million years (Brandley et al., 2010: 2). Brandley et al. (2012: 178) stated that Bermuda is "just 1–2" million years old.

KEY TO HONDURAN GENERA OF THE FAMILY
SCINCIDAE

1A. Median row of greatly transversely expanded dorsal scales extending from shoulder region to base of tail (Fig. 120); no median pale stripe present, even in juveniles........
.... *Mesoscincus* (p. 363)

1B. No median row of greatly expanded scales (Fig. 121); median pale stripe extending length of body and onto tail, at least in juveniles and subadults, bifurcating on posterior portion of frontal . *Plestiodon* (p. 365)

CLAVE PARA LOS GÉNEROS HONDUREÑOS DE LA
FAMILIA SCINCIDAE

1A. Hilera de escamas vertebrales conspicuamente más anchas que las

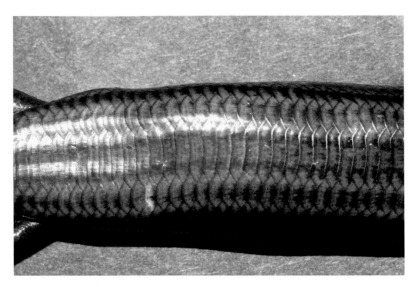

Figure 120. Enlarged middorsal scale row present. *Mesoscincus managuae*. USNM 580379 from del Tigre Island, Valle.

escamas de las hileras paravertebrales, extendiéndose desde la
región del hombro hasta la base
de la cola (Fig. 120); sin una raya
longitudinal pálida en la parte
media del cuerpo, incluso en juve-
niles . . . . *Mesoscincus* (p. 363)

1B. Sin una hilera de escamas vertebrales conspicuamente más anchas
que las escamas de las hileras
paravertebrales (Fig. 121); con
una raya pálida longitudinal presente en la parte media del cuerpo,
la raya del cuerpo extendiéndose a

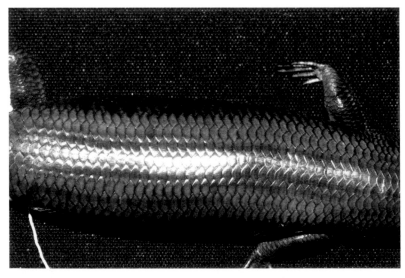

Figure 121. Enlarged middorsal scale row absent and five digits present on forelimbs. *Plestiodon sumichrasti*. USNM 578842 from Estación Forestal CURLA, Atlántida.

lo longitud del cuerpo de la cola (al menos en jóvenes y subadultos), esta raya se bifurca en la porción posterior de la escama frontal....
.... *Plestiodon* (p. 365)

Genus *Mesoscincus* Griffith, Ngo, and Murphy, 2000

Mesoscincus Griffith, Ngo, and Murphy, 2000: 10 (type species: *Eumeces schwartzei* Fischer, 1884: 3, by original designation).

Geographic Distribution and Content.— This genus occurs in disjunct populations from Michoacán, Mexico, to northwestern Costa Rica on the Pacific versant and the Yucatán Peninsula, Mexico, southward to north-central Guatemala and central Belize on the Atlantic versant. Three named species are included in the genus, one of which occurs in Honduras.

Remarks.—Taylor (1936), in his classic monograph on the systematics of skinks allied with *Eumeces* Wiegmann (1834b: 36), placed *E. altamirani* Dugès (1891: 485), *E. managuae* Dunn, and *E. schwartzei* Fischer (1884: 3) in the *E. schwartzei* species group (pp. 93–94). That species group of Taylor corresponds with today's concept of the genus *Mesoscincus*.

Etymology.—The generic name *Mesoscincus* "refers to the Middle American distribution of this group, and its current position as a middle group within the scincid cladogram" (Griffith et al., 2000: 10).

Mesoscincus managuae (Dunn, 1933a)

Eumeces managuae Dunn, 1933a: 67 (holotype, USNM 89474; type locality: "Managua, Nicaragua, Aviation Field"); Cruz [Díaz] et al., 1979: 26; Reeder, 1990: 467.1; Wilson and McCranie, 1998: 16; Wilson et al., 2001: 135; Lovich et al., 2006: 14; Wilson and Townsend, 2007: 145.

Mesoscincus managuae: Griffith et al., 2000: 10; Townsend and Wilson, 2010b: 692; Valdés Orellana et al., 2011b: 242; McCranie et al., 2014: 101; Solís et al., 2014: 132; McCranie, 2015a: 370; McCranie and Gutsche, 2016: 874.

Geographic Distribution.—*Mesoscincus managuae* occurs at low and moderate elevations on the Pacific versant from east-central Guatemala to northwestern Costa Rica and on the Atlantic versant near Lagos de Managua and Nicaragua in western Nicaragua. In Honduras, this species is known from several localities in the southern departments of Choluteca, El Paraíso, and Valle, including several islands in the Golfo de Fonseca.

Description.—The following is based on nine unsexed adults and subadults (USNM 565830–31, 570317, 579864, 580379–83). *Mesoscincus managuae* is a moderate-sized lizard (maximum recorded SVL 125 mm [Reeder, 1990]; 110 mm SVL in largest Honduran specimen [USNM 579864]); dorsal head scales enlarged, smooth, plate-like, paired internasals (supranasals) in contact medially, single frontonasal, usually 2 (rarely 3) prefrontals, single frontal, paired frontoparietals usually separated medially, 4 supraoculars, paired parietal scales not in contact medially, plus a single interparietal scale; frontal in broad contact with first and second supraoculars, frontal also narrowly contacting third supraocular on three sides; frontonasal contacting anterior loreal; 6–10 superciliary scales; nasal entire, nostril opening at posterior edge of scale; 1 small postnasal; 2 loreals; 7–8 (usually 7) supra-labials; 1 postsupralabial; 2–3 (usually 3) presuboculars and postsuboculars; 7 infrala-bials; moveable eyelid present; lower eyelid window absent, disc divided into 5–6 small scales; 1 small scale row above those scales; 4–5 scales bordering upper edge of eyelid disc; pupil circular; 1 primary temporal; 2 secondary temporals, upper overlapping lower and parietal overlapping upper; tertiary temporals absent; 2 pretemporals; 1 large mental; 1 large postmental contacting first and second infralabials; gular scales smooth, cycloid, imbricate, large; 7–13 (10.1 ± 2.0) rows of paired primary nuchal scales (all similar in size); dorsal body scales smooth, cycloid, imbricate, median row greatly enlarged; 50–57 (53.8 ± 2.3) rows of dorsal scales in paravertebral row from end of ultimate nuchal row to level above posterior margin of hind limb; 16–18 (16.9 ± 0.6) scales around midbody; ventral body scales smooth, cycloid, imbricate, in 45–51 (48.9 ± 1.7) rows between level of forelimb to cloacal opening; precloacal scales similarly sized as those on remainder of venter; femoral and precloacal pores absent; limbs relatively short, adpressed fore- and hind limb broadly separated; 11–15 (13.1 ± 1.1) subdigital scales on Digit IV of hind limb, 8–11 (9.7 ± 0.5) scales on Digit IV of forelimb on 16 sides (Finger IV damaged on one side and forelimb missing on one side); 5 digits on forelimb; SVL 55.7–110.0 (82.1 ± 21.3) mm; TAL/SVL 0.86–1.63 in six; HL/SVL 0.12–0.20; HW/SVL 0.09–0.13; SHL/SVL 0.06–0.13.

Color in life of an adult male (USNM 565830): dorsum of body and tail Buff (24) with Sepia (119) stripes, except middorsal stripe Sepia with Buff mottling; lateral surface of tail paler brown with Sepia stripes; subcaudal surface pale brown with dark brown mottling; side of head Sepia, except supralabials pale brown with Sepia bars; venter of head Buff with pinkish brown tinge, except infralabials pale brown with Sepia bars; iris Sepia. Color in life of an adult female (UNAH 5521) was described by Cruz [Díaz] et al. (1979: 26): "dorsal ground color yellowish-brown; two black paravertebral stripes beginning on the parietals and extending to the posterior limb insertion, thereafter continuing onto dorsum of tail, gradually breaking up into weak spots; two dorsolateral stripes on each side beginning just above ear opening and continuing to posterior limb insertion, thereafter fusing, breaking into inconspicuous spots, and continuing onto tail; venter immaculate white; each scale on top and sides of head with a well-defined black spot."

Color in alcohol: dorsal surface of body pale brown with 2 dark brown stripes beginning on first nuchal row and extending onto tail; additional 5 dark brown dorsolateral and lateral lines extending from posterior to head onto tail; dorsal surface of head pale brown with dark brown spots (some elongated); lateral surface of head pale brown with dark brown spots or lines along anterior and posterior edges of supralabials; temporal area pale brown with dark brown lineate pattern; tops of limb surfaces pale brown with dark brown stripes, those stripes less distinct ventrally; dorsal surface of tail same color as dorsal surface of body, but dark brown stripes less distinct; ventral surface of head pale brown, except dark brown along infralabial sutures; flecking on subcaudal surface pale brown.

Diagnosis/Similar Species.—The moveable eyelids, the large smooth scales on top of the head, the smooth cycloid dorsal body scales, a single pair of scales between the rostral and first unpaired scale (frontonasal), and the lack of a window in the lower eyelid distinguishes *Mesoscincus managuae* from all remaining Honduran lizards, except the other Scincidae. *Mesoscincus managuae* differs from those Scincidae in having the median row of dorsal body scales much larger than the adjacent dorsals (versus all dorsal scales similar-sized in other Scincidae).

Illustrations (Figs. 8, 11, 120; Plate 69).—Acevedo, 2006 (adult); Köhler, 2000 (adult, ventral scales; as *Eumeces*), 2001b (adult, ventral scales; as *Eumeces*), 2003a (adult, ventral scales), 2008 (adult, ventral scales); Köhler et al., 2005 (adult); Reeder, 1990 (head scales; as *Eumeces*); Savage, 2002 (adult; as *Eumeces*); Taylor, 1936 (adult, head scales; as *Eumeces*), 1955 (adult; as

Plate 69. *Mesoscincus managuae*. USNM 580379, SVL = 87 mm. Valle: del Tigre Island, near communications tower.

Eumeces), 1956b (adult; as *Eumeces*); Villa et al., 1988 (adult; as *Eumeces*).

Remarks.—Reeder (1990) provided a brief morphological description of *Mesoscincus managuae* and included a literature review. This species was first collected in Honduras in 1977 and until recently was known from the country based on only two specimens. However, *M. managuae* proved to be quite common on several islands in the Golfo de Fonseca in the Pacific Ocean.

Natural History Comments.—*Mesoscincus managuae* is known from near sea level to 920 m elevation in the Lowland Dry Forest, Lowland Arid Forest, and Premontane Moist Forest formations. This diurnal species was taken under logs, under rocks, in rock crevices, and basking on tree stumps. Another was under a pile of rocks near a house in a small town (Cruz Díaz et al., 1979). The species was collected in February, June, July, and December. Nothing has been reported on reproduction or diet in *M. managuae*.

Etymology.—The name *managuae* refers to the city of Managua, Nicaragua, the type locality of the species.

Specimens Examined (14 [0]; Map 57).—**CHOLUTECA**: El Banquito, USNM 570317; Punta Ratón, UNAH 5521. **EL PARAÍSO**: 1 km S of Orealí, USNM 579864. **VALLE**: Isla del Tigre, near summit, USNM 580379–80; Playona Exposición, USNM 565830–31; Punta Copalillo, USNM 580381; Punta Novillo, USNM 580382–83, FN 256915–17, 256925 (still in Honduras because of permit problems).

Other Records (Map 57).—**CHOLUTECA**: Cerro Guanacaure (UNAH; Lovich et al., 2006).

Genus *Plestiodon* A. M. C. Duméril and Bibron, 1839

Plestiodon A. M. C. Duméril and Bibron, 1839: 697 (type species: *Plestiodon quinquelineatum* A. M. C. Duméril and Bibron, 1839: 707 [= *Lacerta quinquelineata* Linnaeus, 1766: 366], by subsequent designation of Fitzinger, 1843: 22, but see Remarks).

Geographic Distribution and Content.—This genus ranges in the Western Hemisphere from southeastern British Columbia, Canada, and most of the U.S. (excluding the

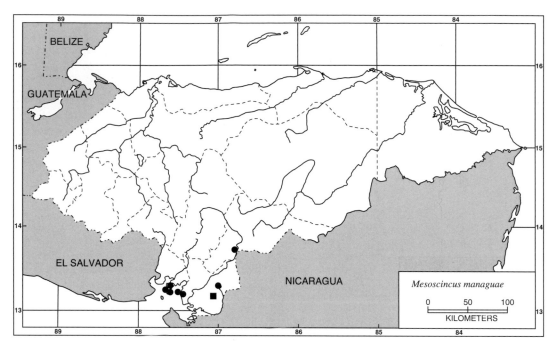

Map 57.   Localities for *Mesoscincus managuae*. Solid circles denote specimens examined and the solid square an accepted record.

extreme northwestern portion, the north-central region, and the extreme northeastern area) southward through much of Mexico to north-central Honduras. It also occurs on Bermuda Island in the Atlantic Ocean. In the Eastern Hemisphere, the genus occurs across eastern Eurasia. Forty-four named species are included in this genus, one of which occurs in Honduras.

Remarks.—Griffith et al. (2000) divided *Eumeces* into four genera. Those authors stated (p. 9) "The *schneiderii* species group would retain the name *Eumeces* Wiegmann, 1834, with the type-species *E. pavimentatus* (Geoffroy-Saint-Hilaire, 1829: 50)." Schmitz et al. (2004: 84) and Brandley et al. (2005: 388) noted that the Griffith et al. (2000) decision to try to retain *Eumeces* for that section was not justifiable. Smith (2005: 15) and Brandley et al. (2005: 388) demonstrated that *Plestiodon* was the oldest available name for this group of skinks (*Pariocela* section of Griffith et al., 2000;

also see Brandley et al., 2005). Smith (2005: 15) also noted that it is uncertain if the type species of *Plestiodon* (*Lacerta quinquelineata* Linnaeus, 1766: 366) applies to *P. fasciata* (Linnaeus, 1758: 209) or *P. laticeps* (Schneider, 1801: 189).

Etymology.—The generic name *Plestiodon* is derived from the Greek words *pleos* (full, full of) and *odous* (tooth). The name apparently alludes to the numerous teeth of these lizards.

Plestiodon sumichrasti Cope, 1867

Plistodon [= *Plestiodon*] *sumichrasti* Cope, 1867: 321 (holotype, USNM 6601; type locality: erroneously stated to be "Orizava" [said to be "en los encinales del Potrero cerca de Córdoba, á una altura de 590 metros" by Sumichrast, 1882: 40, who collected the holotype]).

Eumeces schmidti Dunn and Emlen, 1932: 30 (holotype, ANSP 19877; type locality: "Lancetilla, Honduras"); M. A. Smith, 1933a: 36; Malnate, 1971: 360.

Eumeces sumichrasti: Taylor, 1936: 178; Marx, 1958: 459; Meyer, 1969: 261; Peters and Donoso-Barros, 1970: 119; Meyer and Wilson, 1973: 28; Wilson et al., 2001: 135; McCranie and Castañeda, 2005: 14; McCranie et al., 2006: 218; Wilson and Townsend, 2006: 105.

Plestiodon sumichrasti: Liner and Casas-Andreu, 2008: 73; Townsend et al., 2012: 102; McCranie and Solís, 2013: 242; Solís et al., 2014: 132; McCranie, 2015a: 370.

Geographic Distribution.—*Plestiodon sumichrasti* occurs at low and moderate elevations on the Atlantic versant from central Veracruz, Mexico, to north-central Honduras. In Honduras, this species is known only from the department of Atlántida, although it surely occurs in suitable sites in Cortés.

Description.—The following is based on eight juveniles and subadults (ANSP 19877; LACM 47295; USNM 570318–22, 578841) and three unsexed adults (FMNH 13004; KU 194680; USNM 578842). *Plestiodon sumichrasti* is a moderate-sized lizard (maximum recorded SVL 96 mm [Taylor, 1936]; 93 mm SVL in largest Honduran specimen [USNM 578842]); dorsal head scales enlarged, smooth, platelike, with paired internasals in contact medially, single frontonasal, paired prefrontals usually separated medially, single frontal, paired frontoparietals in contact medially, 4 supraoculars, and paired parietals that are not in contact posterior to single interparietal scale; frontal in broad contact with first and second supraoculars, frontal also frequently in contact with third supraocular; frontonasal contacting anterior loreal or not, separated by small canthal when not in contact; 0–2 (usually 0) small canthals; 7–8 superciliary scales; nasal entire, nostril opening at posterior edge of scale; 1 small postnasal; 2 loreals; 6–8 supralabials; 2 postsupralabials (one above the other); 2–3 (usually 3) and 2–4 (rarely 2) preoculars and postoculars, respectively; 6–8 infralabials; moveable eyelid present; lower eyelid disc divided into several scales; 1 scale row above divided disc; 6–7 scales bordering upper edge of divided eyelid disc; pupil circular; 1 primary temporal; 2 secondary temporals, with upper overlapping lower and parietal overlapping upper; tertiary temporals absent; 2–3 (usually 2) pretemporals; 1 large mental; 1 large postmental, contacting first and second infralabials; gular scales smooth, cycloid, imbricate, large; 2 rows of nuchal scales, 2–3 (usually 2) scales in each row; dorsal body scales smooth, cycloid, imbricate; 48–52 (49.4 ± 1.5) dorsal scales in paravertebral row from ultimate nuchal row to level above posterior margin of hind limb in seven; 26–30 (29.0 ± 1.4) scales around midbody in ten; ventral body scales smooth, cycloid, imbricate, in 32–41 (37.1 ± 3.1) rows between level of forelimb and cloacal opening in ten; precloacal scales similarly sized as those on remainder of venter; femoral and precloacal pores absent; limbs relatively short, adpressed fore- and hind limb overlapping to barely touching; 14–18 (15.8 ± 1.3) subdigital scales on Digit IV of hind limb, 11–14 (12.3 ± 1.0) scales on Digit IV of forelimb; 5 digits on forelimb; SVL 32.6–56.0 (41.0 ± 7.9) mm in juveniles and subadults, 65.0–93.0 (77.9 ± 14.1) mm in adults; TAL/SVL 1.36–1.72 in three juveniles and subadults, 1.16–1.32 in adults; HL/SVL 0.20–0.25 in juveniles and subadults, 0.18–0.21 in adults; HW/SVL 0.14–0.15 in juveniles and subadults, 0.15–0.17 in adults; SHL/SVL 0.12–0.14 in juveniles and subadults, 0.14–0.15 in adults.

Color in life of a juvenile (USNM 570319): body Jet Black (89) with creamy white middorsal, dorsolateral, and lateral stripes, stripes extending onto base of tail; top of head Jet Black with Tawny (38) tinged stripes, stripes especially distinct on snout; dorsal surfaces of fore- and hind limb

Jet Black with creamy white stripes, except hand and foot lacking stripes; all of tail Cobalt (68) except Jet Black at base; belly and subcaudal surfaces Jet Black; chin and venter of neck region creamy white with black flecking, flecking especially dense just anterior to insertion of forelimb.

Color in alcohol of juveniles and sub-adults: dorsal surface of body dark brown to black, with middorsal white stripe extending well onto tail; white to pale brown dorso-lateral stripe extending from above eye onto tail; white to pale brown lateral stripe extending from tip of snout across supra-labials and passing across ear opening and above forelimb onto body, lateral stripe fading posteriorly; dorsal surface of head with middorsal white to pale brown stripe that bifurcates anteriorly on frontal, with both branches extending across each nostril opening to connect with lateral stripe on first supralabial; dorsal surface of unregen-erated tail same color as dorsal surface of body; ventral surface of head cream, that of chest becoming flecked with black, that of belly and subcaudal surface black. The above-described striped pattern is much faded in older subadults. Color in alcohol of adult male (KU 194680): dorsal surface of body brown with narrow dark brown spot at posterior edge of each dorsal scale; paired, slightly paler brown dorsolateral stripes poorly defined; slightly paler brown, poorly defined middorsal stripe present anteriorly on body; dorsal and lateral surfaces of head pale brown without paler stripes; dorsal surfaces of fore- and hind limb brown with narrow dark brown at posterior edge of each scale; dorsal surface of anterior portion of tail same color as posterior surface of body, but without pale stripes; dorsal surface of tail paler brown on distal half; ventral surface of head pale brown with dark brown small spots on ventrolateral scales and in gular region; belly pale brown with dark brown small spots on each scale anteriorly, dark brown spots increasing in size posteri-orly.

Diagnosis/Similar Species.—The move-able eyelids, the large smooth head scales, the smooth cycloid body scales, and a single pair of scales between the rostral and the first unpaired scale (frontonasal) on the dorsal surface of the head distinguishes *Plestiodon sumichrasti* from all remaining Honduran lizards, except the other Scinci-dae and Lygosomoidea. *Marisora brachy-poda* and *M. roatanae* lack broad contact between the frontal and first supraocular, lack a pale middorsal stripe, and have a window in the lower eyelid (versus broad contact between frontal and first supra-ocular, pale middorsal stripe present in juveniles and subadults, and lower eyelid window absent in *P. sumichrasti*). *Scincella* lack paired supernasal scales and two of the three species have a window in the lower eyelid (versus paired supernasals present and lower eyelid window absent in *P. sumichrasti*). *Mesoscincus managuae* has the medial row of dorsal scales greatly transversely enlarged from the level above the shoulders to the base of the tail and lacks pale dorsal and dorsolateral stripes (versus medial row of dorsal scales not larger than adjacent dorsals and pale dorsal and dorsolateral stripes present in juveniles and subadults of *P. sumichrasti*).

Illustrations (Fig. 121; Plates 70, 71).—Álvarez del Toro, 1983 (adult, subadult; as *Eumeces*); Calderón-Mandujano et al., 2008 (juvenile); Campbell, 1998 (adult, head scales; as *Eumeces*); Köhler, 2000 (juvenile; as *Eumeces*), 2003a (juvenile, dorsal pattern; as *Eumeces*), 2008 (juvenile, dorsal pattern; as *Eumeces*); Lee, 1996 (subadult, dorsal scales; as *Eumeces*), 2000 (subadult, dorsal scales; as *Eumeces*); Stafford and Meyer, 1999 (juvenile, head scales; as *Eumeces*); Taylor, 1936 (adult, subadult, head scales; as *Eumeces*); Townsend et al., 2012 (juvenile).

Remarks.—Taylor (1936: 178), in his revision of the genus *Eumeces*, placed *E. sumichrasti* as the sole member of the *E. sumichrasti* species group. Brandley et al. (2012: 182), in a phylogenetic analysis of the

Plate 70. *Plestiodon sumichrasti*. Adult. Izabal: Sierra de Caral. Photograph by Jonathan A. Campbell.

genus *Plestiodon*, recovered evidence suggesting *P. sumichrasti* is a member of the *P. brevirostris* species group.

Natural History Comments.—*Plestiodon sumichrasti* is known from 30 to 880 m elevation in the Lowland Moist Forest and Premontane Wet Forest formations. The 880-m-elevation locality is based only on a sight record from along the Quebrada de Oro, Atlántida. This diurnal species was active on tree trunks near the ground, on wooden walls of buildings, on a boulder above a stream, and on the ground. Individuals were seen in January and February and from May to August; thus, it is probably active throughout the year.

Plate 71. *Plestiodon sumichrasti*. USNM 570320, juvenile, SVL = 33 mm. Atlántida: Estación Forestal CURLA.

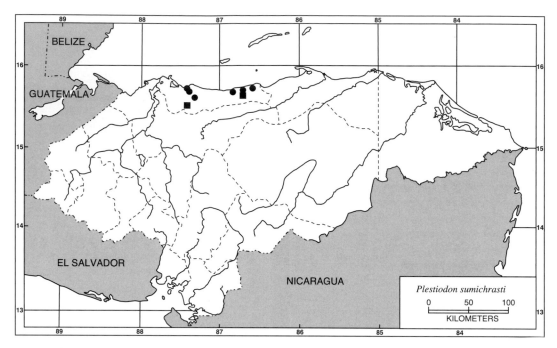

Map 58. Localities for *Plestiodon sumichrasti*. Solid circles denote specimens examined and solid squares accepted records.

Miller (1997) reported a clutch of 11 eggs of *P. sumichrasti* from Belize that was in a cavity in a felled tree that would have been about 20 m above the ground. The eggs and the apparently brooding adult female were found on 4 April, with the eggs hatching on 9 April. These lizards apparently feed primarily on insects (Lee, 1996), but small lizards have also been reported in its diet (Álvarez del Toro, 1983). Streicher et al. (2011) documented a tarantula (*Brachypelma* sp.) feeding on a juvenile *P. sumichrasti* in Guatemala.

Etymology.—The name *sumichrasti* is a patronym honoring François Sumichrast, a 19th century Swiss naturalist who collected the holotype while on an expedition sponsored by the Smithsonian Institution, Washington, D.C.

Specimens Examined (11 [2]; Map 58).—**ATLÁNTIDA**: mountains S of Corozal, LACM 47295; Estación Forestal CURLA, USNM 570319–22, 578841–42; 7.4 km SE of La Ceiba, USNM 570318; Lancetilla, ANSP 19877; 17 km ESE of Tela, KU 194680; Tela, FMNH 13004.

Other Record (Map 58).—**ATLÁNTIDA**: Quebrada de Oro (personal sight record); Refugio de Vida Silvestre Texíguat (Townsend et al., 2012).

Infraorder Teiformata Vidal and Hedges, 2005 (or Gymnoformata McCranie, *In* McCranie and Gutche, 2016; see Remarks)

Gymnophthalmidae Fitzinger (1826:11) is an older name than Teiidae Gray (1827a: 55), but in the interest of stability, I hesitantly use the infraorder name Teiformata Vidal and Hedges (2005:1005). This infraorder, regarding forms occurring in Honduras, includes the superfamily Gymnophthalmoidea and contains two families, Gymnophthalmidae, and Teiidae (see Goicoechea et al., 2016).

Remarks.—It has been generally overlooked that the family name Gymnophthalmidae Fitzinger (1826: 11; who also used

Gymnophthalmoidea) is an older name than Teiidae Gray (1827a: 55). Therefore, the Principle of Coordination (ICZN, 1999) would dictate the correct infraorder name should be formed from the family name Gymnophthalmidae, not from the name Teiidae. Goicoechea et al. (2016) realized Gymnophthalmoidea was the older name and thus correctly used that name as the correct superfamily name. McCranie (*In* McCranie and Gutsche, 2016: 867) suggested the new infraorder name Gymnoformata. However, one reviewer of an earlier version of this book manuscript objected to that infraorder name, citing Article 64 (choice of type genus) of ICZN (1999). I remain confident that Gymnoformata is the correct infraorder name for this group of lizards, especially since Goicoechea et al. (2016) used the superfamily name Gymnophthalmoidea over a superfamily name formed from Teiidae. After reading Article 64, as cited by that reviewer, I am further convinced that it also dictates the use of Gymnoformata as the correct infraorder name over a name formed from Teiidae, but for this work I hesitantly retain the name Teiformata.

Superfamily Gymnophthalmoidea Fitzinger, 1826

Fitzinger (1826:11) named Gymnophthalmoidea as a family and *Gymnophthalmus* Merrem (1820:74) was assigned as the type genus (see Goicoechea et al., 2016: 29, table 7). Three families, 65 genera, and 405 named species occur in this superfamily, with two families occurring in Honduras (Gymnophthalmidae and Teiidae). External characters that in combination differentiate the Honduran members of this superfamily from all other Honduran lizards are: (Gymnophthalmidae)-moveable eyelid absent, dorsal surface of head without paired scales between rostral scale and first unpaired scale, dorsal and ventral body scales enlarged, smooth, cycloid-shaped, and only 4 digits present on the

forelimb; or (Teiidae)-moveable eyelid present, dorsal surface of head with enlarged scales or plates including 1 or no pair of scales between rostral scale and first unpaired scale, ventral body scales large, smooth, rectangular-shaped, dorsal body scales granular, and 5 digits present on the forelimb. All Honduran species of this superfamily are oviparous.

KEY TO HONDURAN FAMILIES OF THE SUPERFAMILY GYMNOPHTHALMOIDEA

1A. Moveable eyelids absent (Fig. 122); dorsal and ventral body scales cycloid-shaped (Fig. 123); only 4 digits on forelimbs (Fig. 122) ....
.... Gymnophthalmidae (p. 371)

1B. Moveable eyelids present (Fig. 124); dorsal body scales granular (Fig. 125); ventral body scales rectangular-shaped (Fig. 126); 5 digits on forelimbs (Fig. 121) ....
.... Teiidae (p. 378)

CLAVE PARA LAS FAMILIAS HONDUREÑAS DE LA SUPERFAMILIA GYMNOPHTHALMOIDEA

1A. Párpados movibles ausentes (Fig. 122); escamas dorsales y ventrales cicloideas (Fig. 123); solamente cuatro dígitos presentes en las extremidades anteriores (Fig. 122)
.... Gymnophthalmidae (p. 371)

1B. Párpados movibles presentes (Fig. 124); escamas dorsales granulares (Fig. 125); escamas ventrales rectangulares (Fig. 126); cinco dígitos presentes en las extremidades anteriores (Fig. 121) ... Teiidae (p. 378)

Family Gymnophthalmidae Fitzinger, 1826

This family of lizards is restricted in distribution to the Neotropics, where it occurs from eastern Oaxaca, Mexico, and Belize southward to northwestern Peru west of the Andes and central Argentina east of the Andes. The Gymnophthalmidae also

Figure 122.   Moveable eyelids absent and four digits present on forelimbs. *Gymnophthalmus speciosus*. USNM 589158 from Palmetto Bay, Roatán Island, Islas de la Bahía.

Figure 123.   Cycloid dorsal scales present. *Gymnophthalmus speciosus*. USNM 579592 from Puerto Lempira, Gracias a Dios.

Figure 124.   Moveable eyelids present. *Ameiva fuliginosa*. USNM 14710 from Swan Islands, Gracias a Dios.

occurs on Trinidad and Tobago and other islands off the north coast of South America, on the Lesser Antilles, and on the Honduran Bay Islands. Forty-seven genera containing about 250 named species are recognized in this family, with one species occurring in Honduras (but see Species of Probable Occurrence).

Genus *Gymnophthalmus* Merrem, 1820

Gymnophthalmus Merrem, 1820: 74 (type species: *Gymnophthalmus quadrilineatus* Merrem, 1820: 74 [= *Lacerta 4-lineata* Linnaeus, 1766: 371, = *Lacerta lineata* Linnaeus, 1758: 209], by monotypy).

Figure 125.   Granular dorsal scales present. *Holcosus undulatus*. USNM 563572 from Copán, Copán.

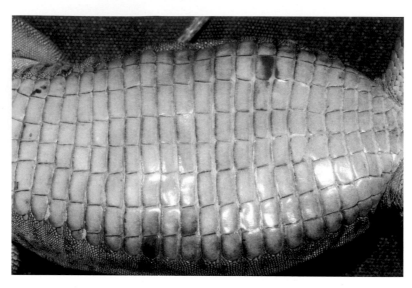

Figure 126.   Rectangular ventrals present. *Holcosus festivus.* USNM 578849 from Los Pinos, Cortés.

Geographic Distribution and Content.— Members of this genus range from northern Belize and eastern Oaxaca, Mexico, to Argentina and northern Brazil. They also occur on the Lesser Antilles, Trinidad and Tobago, and other islands off the north coast of South America, and the Bay Islands of Honduras. Seven named species are recognized (but see Remarks for *Gymnophthalmus speciosus*), one of which occurs in Honduras.

Etymology.—The generic name *Gymnophthalmus* is derived from the Greek *gymnos* (bare, naked) and *ophthalmos* (eye) and refers to absence of a moveable eyelid in this genus.

Gymnophthalmus speciosus (Hallowell, 1861)

Blepharactisis speciosa Hallowell, 1861: 484 (originally two syntypes, now lost [see Peters, 1967: 23]; type locality: "Nicaragua").

Gymnophthalmus speciosus: Stuart, 1939: 4; Echternacht, 1968: 152; Meyer, 1969: 280; Meyer and Wilson, 1973: 34; Wilson and Hahn, 1973: 119; Hahn and Wilson, 1976: 179; Wilson et al., 1979a: 25; O'Shea, 1986: 44; Cole et al., 1990: 27; Wilson et al., 1991: 69, 70; Köhler, 1998d: 380, 382; Monzel, 1998: 161; Wilson and McCranie, 1998: 16; Köhler, 2000: 102; Grismer et al., 2001: 134, 135; Lundberg, 2001: 27; Wilson et al., 2001: 135; Lundberg, 2002b: 9; Köhler, 2003a: 85; McCranie et al., 2005: 118; McCranie et al., 2006: 131; Wilson and Townsend, 2006: 105; Wilson and Townsend, 2007: 145; Köhler, 2008: 89; Townsend and Wilson, 2010b: 692; McCranie and Valdés Orellana, 2014: 47; Solís et al., 2014: 131; McCranie, 2015a: 370; McCranie and Gutsche, 2016: 867.

Gymnophthalmus "*speciosus*": Thomas, 1965: 152.

Gymnophthalus [sic] *speciosus*: Hudson, 1981: 377.

Geographic Distribution.—*Gymnophthalmus speciosus* occurs at low and moderate elevations on the Atlantic versant from northern Belize to Guyana and from eastern Oaxaca, Mexico, to Panama on the Pacific versant. It is also found on a satellite island of Trinidad and on several islands in the Bay Islands, Honduras (but see Remarks). In

Honduras, this species occurs in open habitats throughout much of the mainland in addition to Islas Barbareta, Guanaja, Morat, Roatán, and Utila in the Bay Islands. The species is also known from two islands in the Golfo de Fonseca in southern Honduras.

Description.—The following is based on ten males (AMNH 70362; USNM 520273, 563602, 573181, 573186, 579592; UTA R-10683, 10686–87, 10716) and ten females (KU 203154, 209314, 216146; USNM 520277, 563601, 563603–05, 565497, 570386). *Gymnophthalmus speciosus* is a small lizard (maximum recorded SVL 48 mm [USNM 520273, a male]); dorsal head scales enlarged, smooth, platelike, with single frontonasal scale bordering rostral, paired prefrontals (in contact medially), single frontal, paired dorsally expanded superciliaries, single large supraocular, paired parietal scales, and single interparietal scale; frontal separated from frontonasal; nasal entire, nostril opening in center of scale; prefrontal contacting single loreal; 1 frenocular; 4–5 supralabials to level of posterior edge of eye; lingual sheath absent; moveable eyelid absent; pupil circular; 3–5 infralabials to posterior edge of eye; gular scales smooth, cycloid, imbricate, large; dorsal body scales smooth, cycloid, imbricate, 13 or 15 around midbody (see Remarks); 31–36 (33.4 ± 1.6) rows of dorsal scales between interparietal and level of posterior margin of thigh, paravertebral rows larger than vertebral row; ventral body scales large, cycloid, imbricate; 8–12 (10.8 ± 1.3) total femoral and precloacal pores in males, those pores absent in females; enlarged dorsal brachial and antebrachial scales continuous with each other, scales nearly as wide as long, or as wide as long; 12–16 (13.9 ± 1.1) subdigital scales on Digit IV of hind limb on 38 sides; 4 digits on forelimb; SVL 36.6–48.0 (39.6 ± 3.8) mm in males, 30.0–46.7 (39.8 ± 4.2) mm in females; TAL/SVL 1.01–1.67 in seven males, 0.98–1.49 in five females; HL/SVL 0.18–0.21 in males, 0.16–0.20 in females; SHL/SVL 0.09–0.11 in males, 0.07–0.09 in females.

Color in life of an adult female (USNM 573179): dorsal surface of body Vandyke Brown (121) with Sepia (119) pigment forming dorsolateral incomplete stripes; dorsal surface of head Vandyke Brown; dorsal surface of tail reddish brown; lateral surfaces of body and head Sepia; ventral surface of head dirty white with Sepia spots on most scales; ventral surface of body dirty white with some scattered tiny Sepia spots; subcaudal surface Pratt's Rufous (140); iris Sepia.

Color in alcohol: dorsal surfaces of head and body pale brown to dark brown; pale brown to cream, narrow dorsolateral stripe bordered below by dark brown stripe in some (i.e., USNM 520273, 563601–03), but stripes absent in others (i.e., USNM 563604); dorsal surface of tail pale brown to pinkish brown for distal two-thirds in those with dorsal stripes, dorsal surface of tail dark brown in those with dark brown bodies; ventral surfaces of head and body vary from cream with gray mottling on most scales, to gray with grayish black mottling, to nearly uniformly black; subcaudal surface cream with gray spotting anteriorly in those with pale dorsal surface of tail, gray with grayish black mottling in those with dark dorsal surface of tail.

Diagnosis/Similar Species.—*Gymnophthalmus speciosus* is unique among all Honduran lizards in having only 4 digits on the forelimb. Additionally, the combination of lacking moveable eyelids, having the top of the head with large platelike scales, and having large, cycloid-shaped scales around the body also distinguishes *G. speciosus* from all remaining Honduran lizards.

Illustrations (Figs. 6, 122, 123; Plate 72).—Álvarez del Toro, 1983 (adult); Köhler, 1998d (adult), 2000 (adult), 2001b (adult), 2003a (adult, head scales), 2008 (adult, head scales); Köhler et al., 2005

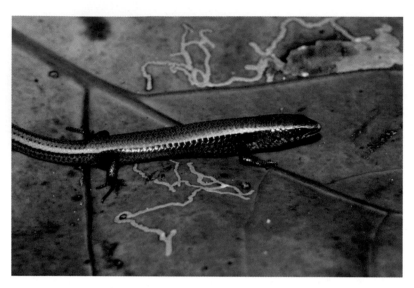

Plate 72.   *Gymnophthalmus speciosus.* FMNH 283587. Islas de la Bahía: Utila Island, east coast near Tradewinds.

(adult); Lee, 1996 (adult), 2000 (adult); Lundberg, 2002b (adult); McCranie and Valdés Orellana, 2014 (adult); McCranie et al., 2005 (adult, head scales), 2006 (adult, head scales); Presch, 1978 (hemipenis); Savage, 2002 (adult); Stafford and Meyer, 1999 (adult); Stuart, 1939 (head scales; as *G. birdi* and *G. speciosus*).

Remarks.—Kizirian and Cole (1999: 399) concluded that *Gymnophthalmus speciosus* "may not be an evolutionary entity but an artificial group of lineages that are morphologically similar but genetically diverse." Carvalho (1997: 162) had recently described one of those lineages as *G. vanzoi*, apparently while the Kizirian and Cole publication was under review or in press. Ugueto and Rivas (2010: 181) thought that the species called *G. speciosus* in Central America is not the same taxon occurring in Venezuela under that name. Because of taxonomic uncertainties involving this species, only illustrations of Mexican and Central American specimens are referenced in the illustrations section above. The geographic distribution of *G. speciosus* in South America remains largely unknown but has been reported in some literature to reach northern Brazil. Murphy (1997: 135) identified a population from a satellite island off the coast of Trinidad as *G. speciosus* but noted that Vanzolini and Carvalho (1991: 222) wrote "that the specimens that have been called *G. speciosus* in Venezuela certainly do not belong to this Central American species." Vanzolini and Carvalho (1991) found differences in scale counts and color pattern between two El Salvador *G. speciosus* and a series from Venezuela also identified as that species. Cole et al. (2013: 452) also made brief comments on the likelihood of more than one species "masquerading under the name *G. speciosus* within its broad range." Ugueto et al. (2013: 1076) inaccurately said the type locality of *G. speciosus* was in Honduras.

The Honduran mainland specimens of *Gymnophthalmus speciosus* have 13 scales around midbody, whereas only one of the Bay Island specimens have 13 scales around midbody, with the remaining eight having 15 such scales. *Gymnophthalmus speciosus* typically has 13 scales around midbody throughout its wide geographical distribution.

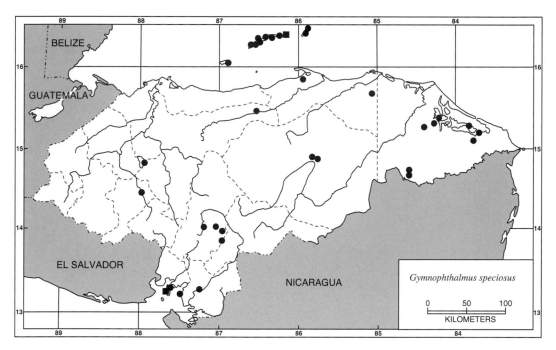

Map 59. Localities for *Gymnophthalmus speciosus*. Solid circles denote specimens examined and solid squares accepted records.

Natural History Comments.—*Gymnophthalmus speciosus* is known from near sea level to 1,320 m elevation in the Lowland Moist Forest, Lowland Dry Forest, Lowland Arid Forest, Premontane Moist Forest, and Premontane Dry Forest formations. This small and shy diurnal and terrestrial species was seen crawling on the ground and in leaf litter. It was also found under logs and rocks and by raking through leaf litter and other debris on the ground. Several also quickly retreated into root systems of small shrubs in otherwise open pastureland. When exposed from its hiding places, individuals usually flee immediately and disappear under nearby objects. It is likely active throughout the year, as it has been taken in nearly every month of the year. Females are much more frequently collected in Honduras than are males. Females of this species are reported to deposit one to four eggs from February through March in Panama (Telford, 1971) and two or three eggs in March in Chiapas,

Mexico (Álvarez del Toro, 1983), with females producing up to three egg clutches per year (Telford, 1971). Not much has been published on diet in this species other than general statements that it feeds on "small invertebrates, predominately insects" (Lee, 1996: 252) or that it feeds on "small insects, but also eats other arthropods" (Savage, 2002: 522, in a literature review).

Etymology.—The name *speciosus* is Latin meaning "showy, splendid." Hallowell (1861) gave no indication to what his name referred, but it probably refers to the handsome dorsal body and tail coloration.

Specimens Examined (103 [21]; Map 59).—**CHOLUTECA**: Choluteca, USNM 570386; Finca El Rubí, SDSNH 72796; Punta Ratón, KU 216146. **COLÓN**: Cerro Calentura, CM 90173–74; Las Champas, BMNH 1985.1291. **CORTÉS**: Santa Elena, LSUMZ 38853, USNM 563605. **EL PARAÍSO**: 30.6 km NW of Mandasta, KU 209314. **FRANCISCO MORAZÁN**: near El Zamo-

rano, LACM 39768, MCZ R48688; El Zamorano, AMNH 70361–63, 85477, FMNH
95981, MCZ R49776–80 (+3 untagged),
49997; Tegucigalpa, BYU 18203, LSUMZ
24129; near Valle de Ángeles, UTA R-17217.
GRACIAS A DIOS: Finca Nakunta, USNM
589149; Mavita, USNM 573182–85, 589150–
53; Puerto Lempira, USNM 565497, 573190,
579592, 589148, 589154; Rus Rus, USNM
563604, 573181, 573186, 589155; Samil,
USNM 573941; Tánsin, LACM 47298,
USNM 573187–89; Usus Paman, USNM
573940, 589156; Warunta, USNM 573179–
80, 589157. **INTIBUCÁ**: 3.0 km N of Jesús
de Otoro, LSUMZ 33664. **ISLAS DE LA
BAHÍA**: Isla de Guanaja, East End, FMNH
283806; Isla de Guanaja, La Playa Hotel,
LSUMZ 21494; Isla de Guanaja, Savannah
Bight, SMF 78136–37, USNM 520273; "Isla
de Guanaja," KU 101352; Isla de Roatán,
Barbarette, UTA R-10687; Isla de Roatán,
between Flowers Bay and West End Point,
USNM 563601–03; Isla de Roatán, Mudd
Hole, SMF 802662, 802265; Isla de Roatán,
W of Oak Ridge, UTA R-10716; Isla de
Roatán, near Oak Ridge, UTA R-10686; Isla
de Roatán, Palmetto Bay, USNM 589158;
Isla de Roatán, about 1.6 km N of Roatán,
CM 90175; Isla de Roatán, 0.5 km N of
Roatán, LSUMZ 21493; Isla de Roatán,
about 4.8 km W of Roatán, LSUMZ 22349;
Isla de Roatán, 1.2 km S, 0.4 km N of Sandy
Bay, KU 203154; Isla de Roatán, Sandy Bay,
KU 203155–59, LSUMZ 33778; Isla de
Roatán, Santa Elena, USNM 520277, UTA
R-10681–85; Isla de Roatán, West End
Point, SMF 79210–11, 81780–81, 81785; Isla
de Utila, east coast near Trade Winds,
FMNH 283587. **OLANCHO**: 4.5 km SE of
Catacamas, LACM 47296–97, LSUMZ
21492; Catacamas, LACM 45136. **VALLE**:
Isla Zacate Grande, LSUMZ 36579; Playona
Exposición, USNM 589159; Punta El Moli-
no, USNM 589160. **YORO**: 5 km E of
Coyoles, LACM 47299.
    Other Records (Map 59).—**ISLAS DE
LA BAHÍA**: Isla Barbareta, LSUHC 3689–
90, LSUPC 4868–87 (Grismer et al., 2001);

Isla Morat, LSUPC 5778–95 (Grismer et al.,
2001). **VALLE**: Isla Inglesera (personal
sight record).

Family Teiidae Gray, 1827a

This family of lizards is restricted in
distribution to the Western Hemisphere,
where it is found from Maryland westward
to Idaho and Oregon, USA, southward
through Mexico and Central America to
central Chile west of the Andes and central
Argentina east of the Andes. The family also
occurs on both Caribbean and Pacific
islands off the coast of Mexico and Central
America and on the Greater and Lesser
Antilles, and Trinidad and Tobago and other
islands off the north coast of South America.
Eighteen genera comprised of about 155
named species are included in this family.
Four genera containing six species occur (or
occurred) in Honduran territory.
    Remarks.—The ICZN (1985a: 130)
placed Teiidae Gray (1827a: 55) on the
Official List of Family-Group Names in
Zoology.
    Methods and terminology for morpho-
logical characters follow those used by
Harvey et al. (2012) unless otherwise noted.

KEY TO HONDURAN GENERA OF THE FAMILY
TEIIDAE

1A. Twelve transverse rows (rarely ten)
    of ventral scales at midbody; small
    scales separate enlarged scales on
    dorsolateral surfaces of brachium
    and antebrachium (Fig. 127).....
    .... *Ameiva* (p. 383)
1B. Eight transverse rows of ventral
    scales at midbody; scales on dorsal
    surfaces of brachium and antebra-
    chium not separated by small
    scales (Fig. 128) 2
2A. Four parietal scales plus 1 interpa-
    rietal scale present (Fig. 129); male
    cloacal spurs present (Fig. 130)..
    .... *Cnemidophorus* (p. 400)
2B. Two parietal scales plus 1 interpa-
    rietal scale present (Fig. 131); male

Figure 127.   Enlarged brachial scales on upper and lower arms separated by small scales. *Ameiva fuliginosa*. USNM 14710 from Swan Islands, Gracias a Dios.

cloacal spurs absent.............. 3
3A. Enlarged scales forming central platelike midgular patch present (Fig. 132).......... *Holcosus* (p. 409)
3B. Central platelike gular patch of enlarged scales absent (Fig. 133) .... *Aspidoscelis* (p. 387)

Clave para los Géneros Hondureños de la Familia Teiidae

1A. Doce hileras transversales (raramente 10) de escamas ventrales a mitad del cuerpo; escamas pequeñas separadas por escamas agran-

Figure 128.   Enlarged brachial scales of upper and lower arms not separated by small scales. *Holcosus festivus*. USNM 563570 from confluence of ríos Yanguay and Wampú, Olancho.

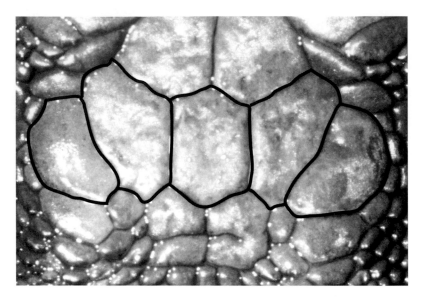

Figure 129.   Four parietal scales plus one interparietal scale present (all outlined). *Cnemidophorus ruatanus*. USNM 580928 from Agua Chiquito, Atlántida.

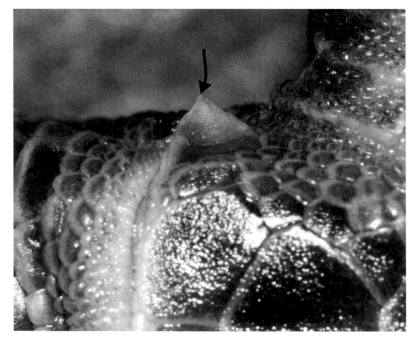

Figure 130.   Pair of male cloacal spurs present (arrow). *Cnemidophorus ruatanus*. USNM 580929 from Barra de Colorado, Atlántida.

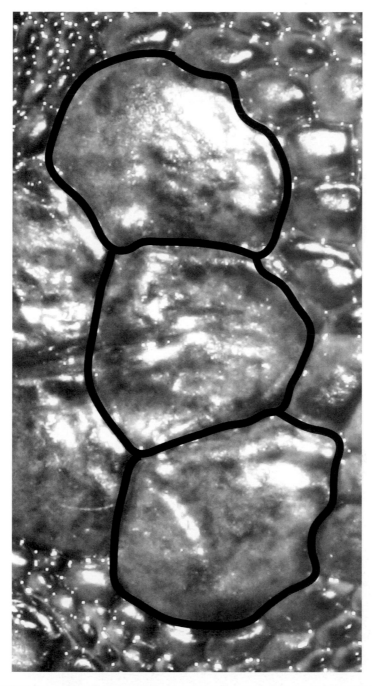

Figure 131.   Two parietal scales plus one interparietal scale present (all outlined). *Holcosus festivus*. USNM 578849 from Los Pinos, Cortés.

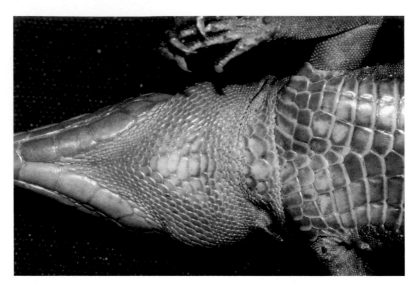

Figure 132. Enlarged midgular scales forming a central, platelike gular patch; gular patch somewhat regular in arrangement, without one scale much larger than others. *Holcosus undulatus.* USNM 578853 from NW of El Jaral, Cortés.

dadas en las superficies dorsales del brazo y antebrazo (Fig. 127) .... *Ameiva* (p. 383)

1B. Ocho hileras transversales de escamas ventrales a mitad del cuerpo; escamas agrandadas en las superficies dorsales del brazo y antebrazo

no están separadas por escamas pequeñas (Fig. 128) 2

2A. Cuatro escamas parietales más una escama interparietal presentes (Fig. 129); con una par de escamas "espuelas" cloacales presentes en

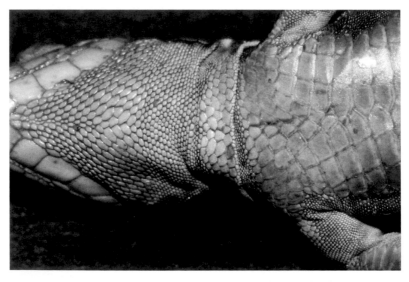

Figure 133. Central, platelike gular patch absent. *Aspidoscelis deppii.* USNM 580915 from El Rodeo, El Paraíso.

machos (Fig. 130) ....
.... *Cnemidophorus* (p. 400)
2B. Dos escamas parietales más una escama interparietal presentes (Fig. 131); sin una par de escamas "espuelas" cloacales en machos..... 3
3A. Escamas en la región mediogular agrandadas en forma de placa presentes (Fig. 132)..... *Holcosus* (p. 409)
3B. Escamas en la región mediogular no está agrandadas en forma de placa (Fig. 133). *Aspidoscelis* (p. 387)

Genus *Ameiva* Meyer, 1795

Ameiva Meyer, 1795: 27 (type species: *Ameiva americana* Meyer, 1795: 28 [= *Lacerta ameiva* Gmelin, 1789: 1070, = *Lacerta ameiva* Linnaeus, 1758: 202], by tautonymy; also see ICZN, 1985a: 130).

Geographic Distribution and Content.—On the mainland, this genus ranges from extreme southern Costa Rica southward to Bolivia east of the Andes. It also occurs on the Greater and Lesser Antilles, and numerous islands off the northern coast of South America and the eastern coast of southern Central America. Two species have been introduced and established in southern Florida, USA. Thirteen named species are recognized, one of which was formerly known from Honduran territory, but has now been extirpated.

Remarks.—Reeder et al. (2002: 20) presented evidence that *Ameiva* is paraphyletic, and Harvey et al. (2012) provided a phylogenetic analysis based on morphological data that resulted in the suggested separation of *Ameiva* into three genera (p. 74). The Pyron et al. (2013: 11, fig. 11) molecular phylogenetic analysis also recovered a paraphyletic *Ameiva*, as well as a monophyletic *Holcosus*, one of the genera separated from *Ameiva* by Harvey et al. (2012). Giugliano et al. (2013: 482) called the Harvey et al. (2012) broad taxonomic changes premature because it did not contain molecular data. Those authors should read Hurtado et al. (2014) and McMahan et al. (2015; but see Harrington et al., 2016) regarding evidence that molecular data is not superior to morphological data in determining squamate evolution, as Giugliano et al. (2013: 482) believe. The Harvey et al. (2012) work is extremely helpful with the identification of various species and lineages and also contains often-ignored detailed scale characteristics. Thus, the Harvey et al. (2012) work should not be dismissed contra the actually premature Giugliano et al. (2013; too few nominal forms sampled).

Etymology.—The exact derivation of *Ameiva* is unknown. "Duméril and Bibron (1839) remarked that the name had been used by earlier authors and was said to be a common name used somewhere in the New World" (Harvey et al., 2012: 84).

Ameiva fuliginosa (Cope, 1892a)

Tiaporus fuliginosus Cope, 1892a: 132 (lectotype, USNM 14710 [designated by McCranie and Gotte, 2014: 546]; type locality: "Swan Island in the Caribbean Sea"); Boulenger, 1893: 23; Boettger, 1893: 105; Barbour, 1914: 314; Barbour and Loveridge, 1929a: 351.
Tiaporus fulginosus [sic]: Barbour, 1914: 214; Barbour 1930a: 80.
Ameiva fuliginosa: Burt and Burt, 1931: 303; McCranie and Gotte, 2014: 546; McCranie, 2015a: 371; McCranie et al., 2017: 275.
Ameiva ameiva fuliginosa: Dunn and Saxe, 1950: 155; Cochran, 1961: 150; Tamsitt and Valdivieso, 1963: 135; Schwartz and Thomas, 1975: 49; MacLean et al., 1977: 4; Schwartz et al., 1978: 9; Morgan, 1985: 43; Schwartz and Henderson, 1988: 52; Henderson and Powell, 2009: 319; Harvey et al., 2012: 88; Powell and Henderson, 2012: 91.
Ameiva ameiva fulginosa [sic]: Schwartz and Henderson, 1991: 181; Ugueto and Harvey, 2012: 114.

Ameiva ameiva: Solís et al., 2014: 132.

Geographic Distribution.—*Ameiva fuliginosa* is known only from one of the Swan Islands, Honduras (extirpated), and Isla de Providencia (Old Providence Island), Colombia.

Description.—The following is based on four females (MCZ 20294; USNM 14710, 32119–20). *Ameiva fuliginosa* is a large lizard (maximum recorded SVL 105 mm [USNM 14710]); dorsal head scales enlarged, smooth, platelike, with paired anterior nasal scales dorsally expanded and in contact with each other by short median sutures, single frontonasal, paired prefrontals, single frontal (partially divided transversely in USNM 32120), paired frontoparietals, 4 parietal scales (although 1 or 2 parietal scales sometimes broken into additional smaller scales anteriorly) and interparietal scale broken transversely anteriorly into 1 or 2 other small scales in two of four, or an interparietal that is divided longitudinally anteriorly for about a third of its length in one, or a single interparietal in one; rostral not contacting frontonasal; short rostral groove present in three, absent in one (USNM 32119); posterior nasal scales not contacting prefrontal; first superciliary contacting prefrontal; frontal ridge absent; frontal-frontoparietal suture contacts third supraocular, varying from contacting near suture between supraoculars 2 and 3 to contacting at about midlength of supraocular 3; scales in frontoparietal region smooth, flat; no key-hole depression in frontoparietal region; interparietal scale narrower than, to about equal in width to adjacent parietal scales; longitudinal furrow absent in interparietal scale; occipitals forming small to enlarged row of scales bordering parietals and interparietal; 4–5 supraoculars (4 on five sides, 5 on three sides), with fourth scale (and fifth when present) smaller than first three; 21–23 (22.3 ± 1.0) occipital scales, median ones slightly larger than, or same size as first enlarged row of dorsal scales; 5–6 (5.9 ± 0.4) granular circumorbital scales, scales not reaching frontal-frontoparietal suture, granules extending to about level of middle of third supraocular; 14–17 (16.1 ± 1.4) lateral supraocular granules in 1.0–1.5 rows; nostril opening centered slightly anterior to suture between pre- and postnasal scales; nostril opening oval; 1 loreal (horizontally divided on one side in MCZ 20294); first subocular entire, separated from supralabials by second subocular; first and second supraoculars in broad contact; 6–8 (7.4 ± 0.7) superciliaries, second or third elongated; first subocular contacting first superciliary; subocular keel present; 4–5 suboculars; 5–7 (6.1 ± 0.4) supralabials; ventral margin of first supralabial curved to slightly "toothy" (see Harvey et al., 2012), first supralabial equal in length to second; lingual sheath present; moveable eyelid present; pupil reniform, usually with small median indentation along upper margin; 5–7 (5.9 ± 0.6) infralabials; supratemporals slightly enlarged, first supratemporal separated from parietal by several rows of small scales; scales in front of auditory meatus slightly enlarged; auricular flap absent; preauricular fold absent; anterior pair of chinshields contacting infralabials, in contact with each other medially; no intergular sulcus; 36–44 (40.8 ± 3.4) anterior gular scales; no central patch of enlarged gular scales; 15–21 (19.0 ± 2.8) posterior gular scales; intertympanic sulcus present; sharp transition from larger anterior gular scales to smaller posterior gular scales; mesoptychial scales slightly enlarged, bordered anteriorly by sharp transition from small scales; no serrated edge on gular fold; large dorsal platelike brachial and antebrachial scales separated by 8–9 transverse rows of small scales, those large dorsal foreleg scales 1.5–2.0+ times wider than long, with each set of enlarged scales smooth, extending well beyond center of arm; postaxial antebrachial scales distinctively enlarged; postaxial brachial scales granular; dorsal body scales conical, some weakly keeled, 169–178 (173.5 ± 4.2) scales around midbody between enlarged ventral plates; 316–335 (321.5 ± 9.0) middorsal scales between occipitals and first enlarged caudal scale; middorsal scales similar in size to lateral body scales; chest scales large, flat; pectoral sulcus absent; ventral body scales large, platelike, squarish, juxtaposed, smooth, in 32 longitudinal rows between gular fold and level of anterior margin of hind limb insertion in all; ventral plates usually in 12 (10 in one) transverse rows (at 15th longitudinal row) with outer scale about half size of other ventrals; precloacal plate formed by row of 5–8 enlarged scales, scales lateral to enlarged scales small or granular; scales immediately lateral to smaller (half sized) outside ventral plate progressively decreasing in size; 16–18 (17.0 ± 0.9) femoral-abdominal pores, with 8–12 (10.8 ± 1.9) scales medially separating each set from each other; 10–12 (11.0 ± 0.5) prefemoral scales; 17–18 (17.5 ± 0.6) subdigital scales on Digit IV of forelimb; subarticular scales of fingers III–IV homogeneous in size, entire; 39–43 (40.5 ± 1.7) subdigital scales on Digit IV of hind limb, distal ones smooth; no row of distinctly enlarged scales between toes IV–V; continuous row of small scales separating supradigital scales from subdigital scales; denticulate fringe absent along postaxial edge of outer toe; Toe V shortened, claw of Toe V not reaching skin between toes III–IV; scales on heel small; tibiotarsal shield and spur absent; caudal annuli complete; proximal subcaudal scales smooth; SVL 92–105 (100.7 ± 7.5) mm in three; TAL/SVL 1.92 in one; HL/SVL 0.23–0.25 in three; HW/SVL 0.13–0.14 in three; HW/HD 1.00–1.12 in three; HD/SVL 0.13–0.14 in three; SL/SVL 0.13 in three; SHL/SVL 0.22–0.26 in three; foot L/SVL 0.39–0.44 in three; hand L/SVL 0.13–0.16 in three.

Color in alcohol (based on USNM 32119; the specimen with the most pattern): dorsal surface of body with median gray swath with black, indistinct spots; black stripe lateral to median swath extending from posterior edge of head to level above hind limb and onto at least two-thirds length of tail, but with short interruptions near hind limb and base of tail; lateral surface of body with 2 gray stripes, with area above and below lateralmost stripe mottled or spotted with black; dorsal surface of head grayish black with some irregular gray stippling; ventral surfaces of head and body black. Another (USNM 32120) is essentially black and unpatterned, except for an indication of the dorsolateral darker black stripes. Plate 73 in this book indicates that the dark unpatterned phase might have been the typical pattern for this species on the Swan Islands.

Diagnosis/Similar Species.—The combination of having large platelike scales on top of the head and on the venter, moveable eyelids, femoral pores in both sexes, and granular dorsal body scales distinguishes *Ameiva fuliginosa* from all remaining Honduran lizards, except the other Teiidae. *Ameiva fuliginosa* is the only Honduran Teiidae usually with 12 (rarely 10) transverse rows of ventral plates and the only Honduran teiid with enlarged brachial scales on the anterior surface of the upper and lower forearm that are distinctively separated from each other by small scales.

Illustrations (Figs. 124, 127; Plate 73).— Cope, 1892a (head scales, forelimb and hind limb scales, lateral body scales; as *Tiaporus*; but see McCranie and Gotte, 2014: 546); McCranie and Gotte, 2014 (adult, head scales, brachial scales; adult also on cover of issue of journal).

Remarks.—McCranie and Gotte (2014) discussed the confused collecting history and taxonomic status of *Ameiva fuliginosa* on the Swan Islands of Honduras. McCranie and Gotte (2014) also compared the type series of this species with that of *A. panchlora* Barbour (1921a: 83) from Isla de Providencia, Colombia, and concluded that Dunn and Saxe (1950) were correct in

Plate 73.   *Ameiva fuliginosa.* Colombia: Providencia Island (released). Courtesy of Gabriel Ugueto.

placing *A. panchlora* in the synonymy of *A. fuliginosa*. McCranie and Gotte (2014) published a photograph of an adult *A. fuliginosa* from Isla de Providencia, Colombia, that demonstrated that population to be an essentially black lizard. *Ameiva fuliginosa* has frequently and erroneously been considered a "subspecies" of *A. ameiva* (Linnaeus, 1758: 202), but those two species are not particularly closely related among the species of *Ameiva* (see McCranie and Gotte, 2014).

Ugueto and Harvey (2012) provided a revision of the *Ameiva ameiva* species complex and concluded that at least four species were recognizable (their study was centered in Venezuela, but they also examined many Bolivian, Colombian, and Panamanian specimens). Those authors also concluded, "it is likely that … *A. a. fulginosa* [sic] represent … [a] full species" (p. 163). Harvey et al. (2012) placed *A. fuliginosa* in the *A. ameiva* species group, but McCranie and Gotte (2014) removed *A. fuliginosa* from that species group and

placed it as *incertae sedis* within the genus *Ameiva*. *Ameiva ameiva* and close relatives have the enlarged brachial and antebrachial scales continuous with each other, whereas *A. fuliginosa* has those large scales separated by smaller scales.

See Remarks for *Leiocephalus varius* for brief comments about the dangers introduced cats can have on island populations of *Ameiva*. Introduced cats have been blamed for the loss of the Swan Island *Ameiva* population. The current personnel of ICF would be wise to make an effort to eradicate the feral cat population on the Swan Islands. They could use Nogales et al. (2004) as an example on eliminating island cat populations.

Natural History Comments.—*Ameiva fuliginosa* is known only from near sea level in the Lowland Dry Forest (West Indian Subregion) formation, a formation occurring in Honduras only on some of the few Caribbean islands. Nothing is recorded about the natural history of this species from the Swan Islands, where it was apparently long ago extirpated (see McCranie and Gotte, 2014; also McCranie et al., 2017). Dunn and Saxe (1950: 156) reported that Isla de Providencia *A. fuliginosa* were "observed only on the ground." Nothing has been published on reproduction in this presumably oviparous species. Dunn and Saxe (1950) reported that on Isla de Providencia, *A. fuliginosa* ate insects and one stomach contained an anole.

Etymology.—The name *fuliginosa* is Latin for "sooty, painted black" and likely refers to the "lead-colored shade" and "lead color" mentioned by Cope (1892a: 132) for the type series.

Specimens Examined (4 [0]; Map 5).— **GRACIAS A DIOS**: Isla Grande, Swan Islands, MCZ R20294 (formerly part of USNM 14710, later numbered USNM 32118, afterwards sent to MCZ), USNM 14710, 32119–20.

Genus *Aspidoscelis* Fitzinger, 1843

Aspidoscelis Fitzinger, 1843: 20 (type species: "*Cnemidoph. sexlineatus* Wiegm."; see Remarks [= *Lacerta 6-lineata* Linnaeus, 1766: 364], by original designation).

Geographic Distribution and Content.— This genus ranges over much of the U.S. (exclusive of about the northern third) southward through much of Mexico and Central America to northeastern Nicaragua and west-central Costa Rica. One species has been introduced and established in southern Florida, USA. Forty-two named species are recognized, two of which occur in Honduras.

Remarks.—Fitzinger (1843:2 0) designated "*Cnemidoph. sexlineatus*. Wiegm." as the type species of *Aspidoscelis*. However, Wiegmann (1834b) did not use that combination, but in a footnote on p. 27, he used the combination "*Lacertam sexlineatam. Lin.*"

The phylogenetic analysis of *Cnemidophorus* Wagler (1830a: 154) and allies by

Reeder et al. (2002: 21) recognized a monophyletic clade for which they resurrected the name *Aspidoscelis*. Harvey et al. (2012), based on morphological characters, concluded *Aspidoscelis* to be a well-supported clade and wrote *Aspidoscelis* and *Holcosus* were the only teiids "with postanal plates in males" (p. 99). Pyron et al. (2013), based on molecular data, also recovered *Aspidoscelis* as a clade (p. 10, fig. 11). Goicoechea et al. (2016: 48), also using molecular data only, placed the two Honduran species into two species groups. Ashton (2003: 110; as *Cnemidophorus*) noted that all males examined of both species of *Aspidoscelis* occurring in Honduras have enlarged postcloacal scales, whereas all females examined of those two species lacked them, thus providing a way to determine the sex of specimens of those two species, even in juveniles.

Etymology.—"The name probably was derived from two Greek nouns, *aspido*, meaning 'shield', and *scelis*, meaning 'rib' or 'leg'. This seems appropriate, because it could refer to the large scales on the legs and has a meaning similar to that of *Cnemidophorus*: 'equipped with leggings'" (Reeder et al., 2002: 21–22).

KEY TO HONDURAN SPECIES OF THE GENUS *ASPIDOSCELIS*

1A. Supraoculars normally 3 (Fig. 134); postaxial antebrachial scales granular (Fig. 135); dorsal surface of body striped in adults; tail some shade of blue in life; SVL to 93 mm .... *deppii* (p. 389)

1B. Supraoculars normally 4 (Fig. 136); patch of enlarged postaxial antebrachial scales present (Fig. 137); dorsal and lateral surfaces of body spotted in adults; tail some shade of pink to orange in life; SVL to 139 mm .... *motaguae* (p. 395)

Figure 134.   Three supraocular scales present on each side. *Aspidoscelis deppii*. USNM 580384 from Inglesera Island, Valle.

Clave para las Especies Hondureñas del Género *Aspidoscelis*

1A. Normalmente tres escamas supra-oculares (Fig. 134); escamas post-axiales del antebrazo granulares (Fig. 135); superficie dorsal del cuerpo en adultos con líneas páli-das; la cola tiene algún tono de azul en vida; LHC hasta 93 mm . . . . *deppii* (p. 389)

1B. Normalmente cuatro escamas supraoculares (Fig. 136); escamas postaxiales del antebrazo agranda-das (Fig. 137); superficies dorsal y lateral del cuerpo en adultos con manchas pálidas; la cola tiene

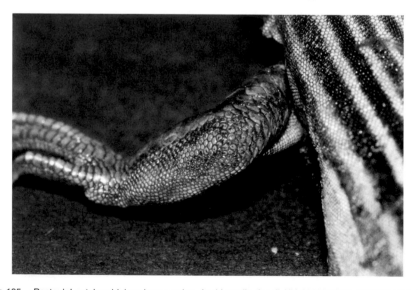

Figure 135.   Postaxial antebrachial scales granular. *Aspidoscelis deppii*. KU 101320 from near Coyoles, Yoro.

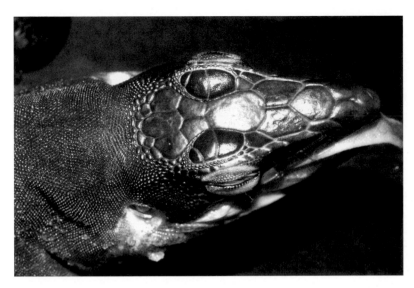

Figure 136.   Four supraoculars present on each side. *Aspidoscelis motaguae*. USNM 580919 from near Orealí, El Paraíso.

algún tono de rosado a anaranjado en vida; LHC hasta 139 mm.....
.... *motaguae* (p. 395)

Aspidoscelis deppii (Wiegmann, 1834b)

Cnemidophorus deppii Wiegmann, 1834b: 28 (holotype, ZMB 882 [*fide* Bauer and Günther, 1994: 270]; type locality not stated, but "Mexico" inferred from title of publication containing description); Köhler, 1996g: 65; Köhler, 1999a: 214.

Cnemidophorus deppii deppii: Burt, 1931: 56.

Cnemidophorus deppei [sic] *deppei* [sic]: Meyer, 1966: 176; Echternacht, 1968: 152; Villa, 1994: 33.

Figure 137.   Postaxial antebrachial scales enlarged. *Aspidoscelis motaguae*. USNM 580919 from near Orealí, El Paraíso.

Cnemidophorus deppii: Meyer, 1969: 274; Meyer and Wilson, 1973: 32; Wilson et al., 1976: 179; Wilson and McCranie, 1998: 15; Wilson et al., 2001: 135.

Aspidoscelis deppii: Reeder et al., 2002: 22; Lovich et al., 2006: 14; McCranie et al., 2006: 133; Wilson and Townsend, 2006: 105; Townsend et al., 2007: 10; Wilson and Townsend, 2007: 145; Townsend and Wilson, 2010b: 692; Solís et al., 2014: 132; McCranie, 2015a: 371; McCranie and Gutsche, 2016: 868.

Aspidoscelis motaguae: C. A. Smith and Krysko, 2007: 265 (in part).

Geographic Distribution.—*Aspidoscelis deppii* occurs at low and moderate elevations on the Pacific versant from Michoacán, Mexico, to west-central Costa Rica. The species also crosses the Isthmus of Tehuantepec to reach the Atlantic versant of northern Veracruz and western Campeche, Mexico. It also occurs disjunctly on the Atlantic Versant from southeastern Campeche, Mexico, to northeastern and southwestern Nicaragua. In Honduras, this species is found in open habitats on both versants of the country.

Description.—The following is based on 11 males (KU 101315, 101318–20, 101322–25; UF 150292; USNM 563578, 580915) and 11 females (KU 101313, 101317, 101326; UF 150293, 150301; USNM 563579–83, 580916). *Aspidoscelis deppii* is a moderate-sized lizard (maximum recorded SVL 93 mm [Duellman and Wellman, 1960]; 84 mm SVL in largest Honduran specimen measured [KU 101325, a male]); dorsal head scales enlarged, smooth, plate-like, paired anterior nasal scales dorsally expanded, in contact medially by short sutures, single frontonasal, paired prefrontals, single frontal, paired frontoparietals, paired parietal scales and single interparietal scale, however, longitudinal, incomplete furrow sometimes present in interparietal; rostral not contacting frontonasal; postnasal contacting prefrontal; prefrontal and first superciliary usually separated, occasionally in contact; frontal ridge absent; frontal-frontoparietal suture usually aligned near midlength of supraocular 3, or just anterior to that point; scales in frontoparietal region smooth, flat; interparietal varies from slightly narrower than, to slightly broader than flanking parietals; usually 3 supraoculars (occasionally a fourth smaller supraocular present posteriorly); first supraocular contacting second supraocular; median pair of occipital scales not greatly enlarged, but larger than first dorsal scale row; 11–21 (16.0 ± 3.3) occipitals in 21; supratemporals moderately enlarged, separated from parietals; short rostral groove usually present; nostril anterior to nasal suture, but contacting nasal suture; nostril opening oval; 1 row of lateral supraocular granules, 11–28 (16.2 ± 2.8) granules on 42 sides; anterior extent of circumorbital semicircles reaches between midlength and posterior third of supraocular 3; 3–7 (5.4 ± 1.0) circumorbitals on 41 sides; 5–7 (5.8 ± 0.8) superciliaries on 42 sides, usually with second or third (rarely first) elongated; 4–5 (rarely 5) suboculars, first usually entire, occasionally divided, lower edge (when entire) or lower part (when divided) sometimes contacting supralabials 3 and 4 (rarely contacting only supralabial 3), sometimes not contacting any supralabial; subocular keel present; enlarged scales located in front of auditory meatus; no auricular or preauricular flap or fold; 1 loreal; 6 supralabials; 6–7 (only occasionally with 7) infralabials; first supralabial curved ventrally, shorter than second; lingual sheath absent; moveable eyelid present; pupil reniform; 1 pair of chinshields, contacting infralabials, chinshields usually separated only at posterior edge (rarely not separated at all); intergular sulcus absent; 18–29 (22.7 ± 2.7) anterior gulars in 21; 10–20 (14.9 ± 2.8) posterior gulars in 21; gular patch of distinctively enlarged scales absent, although midgulars moderately enlarged compared with surrounding gulars; intertympanic sulcus absent; sharp transition from anterior gulars to smaller posterior gulars; mesoptychial scales moderately enlarged, differentiated, bordered anteriorly by sharp transition to small scales; edge of gular fold not serrated; dorsal scales conical, 226–270 (249.6 ± 12.9) middorsals between occipitals and first enlarged caudal scale in 21; 105–127 (118.1 ± 6.2) granules around midbody; 10–18 (13.7 ± 2.6) granules between paravertebral stripes at midbody in ten; middorsal scales subequal in size to lateral scales; chest scales large, flat; pectoral sulcus absent; ventral body scales large, platelike, squarish, juxtaposed, smooth, in 25–34 (29.5 ± 2.0) rows between gular fold and level of hind limb, in 8 transverse rows at midbody; scales immediately lateral to outside ventral plate small, granular; paired enlarged terminal scales forming precloacal plate, those enlarged scales flanked by single small subtriangular scale, that scale flanked by granular scales; paired precloacal plates usually larger than scale anterior to them; 4–6 (4.7 ± 0.6) precloacal scales in 21; precloacal spur and postcloacal button absent; pair of slightly enlarged postcloacal scales present in males, those enlarged scales absent in females; caudal annuli complete, tail lacking crest or dorsolateral row of serrated scales; proximal subcaudals smooth; large dorsal brachial and antebrachial scales not separated by smaller scales; enlarged dorsal scales on brachial and antebrachial surfaces 1.0–1.5 times as wide as long, both sets of enlarged scales smooth, extending well beyond center of arm; postaxial antebrachial scales granular; 12–16 (14.6 ± 1.4) subdigital scales on Finger IV on 41 sides; subarticular scales on fingers III–IV homogeneous in size, entire; 25–30 (27.6 ± 1.5) subdigital scales on 39 sides of Toe IV, distal ones smooth; no row of distinctively enlarged scales between toes IV–V; small scales separating supradigital scales from subdigital scales mostly restricted to phalangeal articulations; denticulate

fringe absent along postaxial edge of outer toe; Toe V reduced, claw reaching only to point of articulation of toes III–IV; 5–9 (7.1 ± 1.3) prefemoral scales in 21; heel with 3 moderately enlarged scales; tibiotarsal shield present; tibiotarsal spur absent; 35–42 (37.9 ± 2.1) total femoral-abdominal pores in males, 32–39 (36.2 ± 2.0) in females; no gap between femoral and abdominal pores; 4–5 (most often 4) medial scales separating each set of femoral and abdominal pores from each other; SVL 64.8–84.0 (76.1 ± 6.0) mm in males, 59.5–80.2 (67.8 ± 6.1) mm in females; TAL/SVL 1.78–2.36 in eight males, 1.86–2.23 in eight females; HL/SVL 0.23–0.26 in males, 0.21–0.25 in females; HW/SVL 0.14–0.16 in ten males, 0.11–0.14 in eight females; HD/SVL 0.13–0.15 in ten males, 0.11–0.14 in eight females; HW/HD 0.98–1.15 in ten males, 1.01–1.08 in eight females; SL/SVL 0.10–0.12 in ten males, 0.10–0.11 in eight females; SHL/SVL 0.17–0.22 in males, 0.18–0.24 in females; foot L/SVL 0.31–0.37 in ten males, 0.31–0.35 in eight females; hand L/SVL 0.12–0.15 in ten males, 0.11–0.15 in eight females.

Color in life of an adult male (UF 150292): dorsal ground color Jet Black (89) between middorsal stripe and each of 4 dorsolateral stripes, Pratt's Rufous (140) below that point; middorsal stripe Warm Buff (118), lateral stripes Sulfur Yellow (157), ventrolateral stripe pale turquoise blue; top of head Grayish Horn Color (91); venter pale turquoise; dorsal surfaces of fore- and hind limb Jet Black with Sulfur Yellow to tan spotting or lines; middle of subcaudal surface pale turquoise blue grading to Grayish Horn Color posteriorly.

Color in alcohol of adults: middorsal ground color dark brown to black, dorsolateral ground color pale brown to dark brown; 9–11 (usually 10), rather indistinct (especially in larger specimens) grayish brown to bluish white stripes, stripes beginning posterior to (dorsally), or on posterior part (laterally) of head, extending well onto tail,

Plate 74. *Aspidoscelis deppii*. USNM 580914. Yoro: San Patricio.

especially laterally; lowermost stripe extending across anterior surface of hind limb; lowermost stripe sometimes broken into dashes or spots anteriorly on body; no conspicuous pale spots on dorsal surface of body; dorsal surface of head pale brown to dark brown or grayish brown; ventral surfaces of males varying from dark gray to black in throat region and at midbelly to entirely dark gray to black under head, body, and forelimb, those ventral surfaces in females varying from bluish white to bluish gray; ventral surfaces of hind limb and tail varying from dirty white to cream. Color in alcohol of juveniles: middorsal ground color black, with ten distinct white to bluish white straight stripes beginning on posterior portion of head and extending length of body.

Diagnosis/Similar Species.—The combination of having large platelike scales on top of the head and on the venter, moveable eyelids, femoral pores in both sexes, and granular dorsal body scales distinguishes *Aspidoscelis deppii* from all remaining Honduran lizards, except the other Teiidae. The single Honduran species of *Ameiva* usually has 12 transverse ventrals at mid-

body, and always has small scales distinctively separating the platelike brachial and antebrachial dorsal foreleg scales (versus 8 ventrals and no small scales separating enlarged foreleg scales in *A. deppii*). The species of *Holcosus* have a midgular patch of enlarged scales (versus no patch of enlarged midgular scales in *A. deppii*). *Cnemidophorus ruatanus* has 5 parietal plus interparietal scales, usually has 4 supraocular scales, and has paired precloacal spurs in the males (versus 3 parietal plus interparietal scales, usually 3 supraocular scales, and no precloacal spur in *A. deppii*). *Aspidoscelis motaguae* is a larger species reaching 139 mm SVL, usually has 4 supraoculars, has a patch of enlarged postaxial antebrachial scales, has spotted dorsal and lateral surfaces of the body in adults, and has the tail usually some shade of pink to orange in life (versus maximum known SVL of 93 mm, usually 3 supraoculars, granular postaxial antebrachial scales, a striped dorsal surface of adult body, and turquoise blue to brown tail in life in *A. deppii*).

Illustrations (Figs. 133–135; Plate 74).— Alemán and Sunyer, 2014 (adult, juvenile); Álvarez del Toro, 1983 (adult; as *Cnemido-*

phorus); Duellman and Wellman, 1960 (schematic drawings of ontogenetic change in dorsal stripes; as *Cnemidophorus d. deppei*); A. H. A. Duméril et al., 1870–1909b (adult, head scales, forelimb scales, precloacal region scales; as *Cnemidophorus*); Fitch, 1973a (adult; as *Cnemidophorus*); Köhler, 1999b (head scales; as *Cnemidophorus*), 2000 (adult, ventral scales, head scales; as *Cnemidophorus*), 2001b (adult, ventral scales, head scales; as *Cnemidophorus*), 2003a (adult, ventral scales, head scales), 2008 (adult, ventral scales, head scales); Köhler et al., 2005 (adult, head scales); Lee, 1996 (head scales, scales in precloacal and thigh regions; as *Cnemidophorus*), 2000 (head scales, scales in precloacal and thigh regions; as *Cnemidophorus*); McCranie et al., 2006 (adult, head scales); Mertens, 1952b (adult; as *Cnemidophorus*); Sasa and Solórzano, 1995 (adult; as *Cnemidophorus*); Savage, 2002 (adult; as *Cnemidophorus*); Taylor, 1956b (adult; as *Cnemidophorus*); Villa et al., 1988 (adult; as *Cnemidophorus*); Wright, 1993 (adult; as *Cnemidophorus*).

Remarks.—Wiegmann (1834b: 28), in the original description of *Aspidoscelis deppii*, used the spelling *Cnemidophorus deppii* to honor Ferdinand Deppe. Some authors have tried to change Wiegmann's, apparently intentional (consistently spelled that way), original spelling of this species to *A. deppei*. Since Wiegmann's original spelling appears to be intentional, the original spelling of *deppii* should be retained as the correct spelling of this specific name (Article 32 of the Code; ICZN, 1999). Goicoechea et al. (2016: 48) continued the trend of using the nonoriginal spelling *A. deppei*.

Duellman and Wellman (1960) presented a systematic study of Guatemalan and Mexican specimens of *Aspidoscelis deppii* and recognized three subspecies, one of which *A. cozumela* (Gadow, 1906: 316) is now known to be parthenogenetic (see Fritts, 1969: 522). Savage (2002: 516) stated,

"there seems little reason to recognize races in this species, since all color patterns described by Duellman and Wellman (1960) occur in the Costa Rican samples." Unfortunately, Savage provided no data supporting his claim, nor did he comment on the taxonomic status of *A. d. cozumela* of Duellman and Wellman (1960: 35).

This species is in the *Aspidoscelis deppii* species group according to Reeder et al. (2002: 22) and Goicoechea et al. (2016: 48).

Natural History Comments.—*Aspidoscelis deppii* is known from near sea level to 900 m elevation in the Lowland Moist Forest, Lowland Dry Forest, Lowland Arid Forest, Premontane Moist Forest, and Premontane Dry Forest formations. The species occurs in open areas in Lowland Moist Forest such as coastal shrub and pine savanna in northeastern Honduras. The sight records for this species at Quebrada Kuilma along the Río Coco within an area dominated by broadleaf rainforest seemed peculiar at first. However, humans had earlier that year (2000) deforested that area of the Río Coco, and pockets of open pine savannah occur in the region away from the river, some surprisingly close to the Río Coco. One of those pine savannahs is the likely source for that *Aspidoscelis* population. Individuals of this terrestrial, diurnal, fast-moving, and seemingly nervous species move in and out of shade in grassy, sandy, or rocky areas after apparently reaching their preferred body temperature by basking in the sun. They are less active and appear to spend more time in shade during midday. Retreats include under rocks, logs, leaf litter, and in holes in the ground. It has been collected in every month of the year except January and thus is apparently active on sunny days throughout the year. Females from Nicaragua and Costa Rica have been reported to deposit one to four eggs per clutch (Fitch, 1973a, 1985; Vitt et al., 1993) and, like other species of *Aspidoscelis*, probably deposit two to six clutches per year. The long, severe dry season where

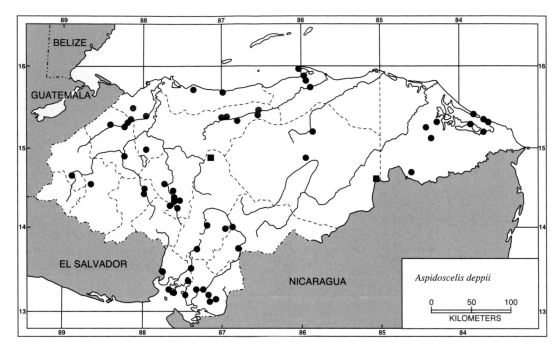

Map 60.   Localities for *Aspidoscelis deppii*. Solid circles denote specimens examined and solid squares accepted records.

many populations of *A. deppii* occur inhibits egg production (Fitch, 1985). Goldberg (2009b) recorded two oviductal eggs in one *A. deppii* from Costa Rica that was collected in July. Vitt et al. (1993: 2391), after studying a population of this species on the Pacific versant of Nicaragua, reported that, "forty-two types of prey were identified in stomachs, with termites, spiders, and various orthopterans accounting for most of the diet volumetrically." Alemán and Sunyer (2014) reported on a case of possible cannibalism in an adult in Nicaragua. One Honduran adult (MCZ R191166) was eaten in the collecting bag by an adult of the snake *Coniophanes imperialis* (Baird and Girard, 1859: 23, *In* Baird, 1859b; MCZ R191117) collected the same afternoon.

Etymology.—The name *deppii* is a patronym honoring Ferdinand Deppe, a German naturalist who collected the type series.

Specimens Examined (306 [140]; Map 60).—**ATLÁNTIDA**: Lancetilla, MCZ R29811 (listed as from Lago de Yojoa, Cortés, by Meyer and Wilson, 1973); Salado Barra, MCZ R191166. **CHOLUTECA**: 1.0 km N of Cedeño, LSUMZ 33698; Cedeño, BYU 18286; Choluteca, MSUM 4679–81; El Despoblado, CAS 152980; Finca La Libertad, SDSNH 72698–99; Finca Monterrey, USNM 580913; Ola, TCWC 19200; Pespire, FMNH 5101 (9), MCZ R49986 (2). **COLÓN**: near Bonito Oriental, TCWC 66857; Puerto Castilla, LSUMZ 22482–88; Río Aguán SE of Trujillo, TCWC 66850–56; 1.0 km E of Trujillo, LSUMZ 27736. **COMAYAGUA**: 6.4 km S of Comayagua, LACM 48037–40; 8 km N of Comayagua, UF 124822; 17.7 km NW of Comayagua, LACM 48041–43; near Comayagua, UF 124817; Las Mesas, UTA R-41260–63, 46022; about 7 km ESE of Villa San Antonio, LACM 72091; 1.4 km W of road to Villa San Antonio on road to La Paz, USNM 563577. **COPÁN**: near Cucuyagua, UF 124818. **CORTÉS**: 3 km WSW of

Cofradía, AMNH 157009–13, KU 67360–71; Cofradía, CM 65679–82, LSUMZ 50742–50, 88603; Hacienda Santa Ana, FMNH 5102–13, 5115–20, 5141–45, 5146 (15), 5147 (13); 9.7 km S of La Lima, CM 41227–28; San Luís, LSUMZ 72914; 3.2 km W of San Pedro Sula, LACM 48046–47; W of San Pedro Sula, FMNH 5121–40; San Pedro Sula, CM 8121, LSUMZ 72911–13, 72950–51, USNM 24367–70. **EL PARAÍSO**: El Rodeo, USNM 580915–16; near Ojo de Agua, AMNH 70329, 70331–32; Soledád, KU 192329–31. **FRANCISCO MORAZÁN**: El Zamorano, MCZ R49769–72, 49931, UF 124821; 21 km SSW of Sabanagrande, AMNH 157014, LSUMZ 24623; near Tegucigalpa, LSUMZ 24139. **GRACIAS A DIOS**: Cauquira, USNM 565510; Crique Ibantara, USNM 563578; Mocorón, UTA R-53744; Prumnitara, UF 150292–93, 150297, 150301; Puerto Lempira, USNM 565505–06, 573199; Tánsin, LACM 48023–25, LSUMZ 21628–31, USNM 573200–05, 573948–52; Usus Paman, USNM 573947; Warunta, USNM 565507–09, 573198, 580917, FN 256936, 256939 (still in Honduras because of permit problems). **INTIBUCÁ**: 3.0 km N of Jesús de Otoro, LSUMZ 33700; 5.5 km S of Jesús de Otoro, LSUMZ 33703; 8.0 km S of Jesús de Otoro, LACM 48048–50. **LA PAZ**: La Canadá, UF 127583–84; La Paz, UNAH 5227. **LEMPIRA**: near Gracias, UF 124819. **OLANCHO**: between Juticalpa and Catacamas, AMNH 70330; Las Trojas, UTA R-41257. **SANTA BÁRBARA**: 4 km S of Cofradía, LSUMZ 24624; 16 km WSW of La Flecha, CM 65678; 4 km SW of Quimistán, KU 67359; 4.8 km E of Quimistán, LACM 48044–45; about 10 km N of Santa Bárbara, LACM 45324–25; about 12 km N of Santa Bárbara, LACM 45326–32. **VALLE**: Isla Inglesera, USNM 580384–85; Isla Violín, USNM 580918; La Orilla, KU 220100; San Lorenzo, FMNH 5098–100. **YORO**: 2 km S of Coyoles, KU 101312–16, 101318–26; 6 km N of Coyoles, KU 101317; 0.5 km N of Coyoles, LACM 48026–32,

LSUMZ 21632–38; 5 km E of Coyoles, LACM 48033, LSUMZ 21639; Rancho San Lorenzo, LACM 48034–36, LSUMZ 21640–42; Río San Lorenzo, USNM 580912; 5.3 km W of road to San Juan, USNM 563579; near San Lorenzo Abajo, USNM 563580–82; 4.7 km ESE of San Lorenzo Arriba, USNM 563583, 565504; San Patricio, USNM 580914. "HONDURAS": ANSP 9608–15, UF 87653–56.

Other Records (Map 60).—**FRANCISCO MORAZÁN**: Marale (Townsend et al., 2007). **OLANCHO**: Quebrada Kuilma (personal sight record).

Aspidoscelis motaguae (Sackett, 1941)

Cnemidophorus sexlineatus: Werner, 1896: 347.

Cnemidophorus motaguae Sackett, 1941: 1 (holotype, ANSP 22143; type locality: "Motagua River Valley, about 10 kilometers northeast of Zacapa, Dept. of Zacapa, Guatemala"); Duellman and Zweifel, 1962: 190; Meyer, 1966: 176; Meyer, 1969: 277; Meyer and Wilson, 1973: 33; Wilson and McCranie, 1998: 15; Wilson et al., 2001: 135.

Cnemidophorus sacki motaguae: Stuart, 1954: 17.

Aspidoscelis motaguae: Reeder et al., 2002: 22; C. A. Smith and Krysko, 2007: 265 (in part); Wilson and Townsend, 2007: 145; Townsend and Wilson, 2010b: 692; Solís et al., 2014: 132; McCranie, 2015a: 371.

Geographic Distribution.—Aspidoscelis motaguae occurs at low and moderate elevations in disjunct populations on the Atlantic versant from eastern Chiapas, Mexico, to northwestern Honduras and on the Pacific versant from south-central Oaxaca, Mexico, to northwestern Nicaragua. This species apparently avoids the low, flat Pacific coastal plain, at least in extreme southern Honduras. It is also introduced and established in southern Florida, USA. In Honduras, *A. motaguae* is found in the central and southern portions of the country.

Description.—The following is based on 11 males (KU 103251–52, 200585; LSUMZ 21663, 21665–66, 21671–72; UF 127582; USNM 580920–21) and 13 females (FMNH 5094; KU 103253–54; LSUMZ 21664, 21673, 21676; UF 127581; USNM 563584–87, 580919, 580922). *Aspidoscelis motaguae* is a large lizard (maximum recorded SVL 145 mm [Duellman and Zweifel, 1962], SVL 139 mm in largest Honduran specimen [LSUMZ 21671, a male]); dorsal head scales enlarged, smooth, platelike, with dorsally expanded, paired anterior nasal scales in contact medially with short sutures, single frontonasal, paired prefrontals, single frontal, paired frontoparietals, paired parietal scales, and single interparietal scale, however, longitudinally incomplete furrow sometimes present in interparietal scale; rostral not contacting frontonasal; postnasal contacting prefrontal; prefrontal and first superciliary usually separated, occasionally in contact; frontal ridge absent; frontal-frontoparietal suture usually aligned near anterior edge of supraocular 3; scales in frontoparietal region smooth, flat; interparietal varies from slightly narrower than, to slightly broader than flanking parietals; usually 4 supraoculars (supraoculars 3 or 1, or both, sometimes divided); first supraocular contacting second supraocular, fourth supraocular usually smaller than first; median pair of occipital scales not greatly enlarged, but larger than first dorsal scale row; 13–24 (16.7 ± 2.7) occipitals in 15; supratemporals moderately enlarged, separated from parietals; rostral groove absent or short; nostril opening mostly anterior to nasal suture, but in contact with suture; nostril opening oval to subcircular; 1.5–2.0 rows of lateral supraocular granules, 15–32 (22.6 ± 5.1) on 30 sides; anterior extent of circumorbital semicircles reaching between anterior edge of, to middle of supraocular 3; 5–9 (6.7 ± 1.5) circumorbitals on 30 sides; 6–8 (6.7 ± 0.7)

superciliaries on 30 sides, usually third (rarely second) elongated; 4 suboculars, first divided, with lower part usually contacting fourth (rarely third) supralabial; subocular keel present; enlarged scales located in front of auditory meatus; no auricular or preauricular flap or fold; 1 loreal; 6 supralabials; 6–7 (rarely 7) infralabials; first supralabial curved ventrally, shorter than second; lingual sheath absent; moveable eyelid present; pupil reniform; 1 pair of chinshields, contacting infralabials, chinshields separated only at posterior edge; intergular sulcus absent; 17–29 (22.3 ± 3.7) anterior gulars in 15; 12–14 (13.0 ± 1.1) posterior gulars in 15; midgular enlarged patch absent, although midgulars moderately enlarged compared to remaining gulars; intertympanic sulcus absent; sharp transition from anterior gulars to smaller posterior gulars; mesoptychial scales moderately enlarged, differentiated, bordered anteriorly by sharp transition to small scales; edge of gular fold not serrated; dorsal scales conical, 216–261 (234.6 ± 14.6) middorsals between occipital and first enlarged and keeled caudal scale in 15; 117–160 (136.0 ± 12.5) granules around midbody; middorsal scales subequal in size to lateral scales; chest scales large, flat; pectoral sulcus absent; ventral body scales large, platelike, squarish, juxtaposed, smooth, in 28–37 (30.5 ± 2.1) rows between gular fold and level of hind limb, in 8 transverse rows at midbody; scales immediately lateral to outside ventral plate small or granular; paired enlarged terminal scales forming precloacal plate, those enlarged scales flanked by 1 slightly enlarged subtriangular scale, that scale flanked by small or granular scales; paired precloacal plates usually larger than scale anterior to them; 5–7 (6.1 ± 0.6) precloacal scales in 15; precloacal spurs and postcloacal buttons absent; pair of slightly enlarged postcloacal scales present in males, those enlarged postcloacal scales absent in females; caudal annuli complete, tail lacking crests or dorsolateral row of serrated scales; proximal subcaudals smooth; large dorsal brachial and antebrachial scales continuous, not separated by band of smaller scales; enlarged dorsal scales on brachial and antebrachial surfaces 1.0–1.5 times as wide as long, both sets of enlarged scales smooth, extending well beyond center of arm; postaxial antebrachial scales with patch of noticeably enlarged scales ventrolaterally; 13–16 (14.3 ± 0.7) subdigital scales on 32 sides of Finger IV; subarticular scales of fingers III–IV homogeneous in size, entire; 28–38 (31.8 ± 2.3) subdigital scales on 32 sides of Toe IV, distal ones smooth; no row of distinctly enlarged scales between toes IV–V; small scales separating supradigital scales from subdigital scales mostly restricted to phalangeal articulations; denticulate fringe absent along postaxial edge of outer toe; Toe V reduced, claw only reaching to point of articulation of toes III–IV; 8–12 (9.7 ± 1.2) prefemoral scales in 15; heel with 3 moderately enlarged scales; tibiotarsal shield present; tibiotarsal spur absent; 41–50 (46.0 ± 3.0) total femoral-abdominal pores in males, 41–49 (44.4 ± 2.7) in females; no gap between femoral and abdominal pores; 5–6 medial scales separating each set of femoral and abdominal pores from each other; SVL 98–139 (121.8 ± 13.0) mm in males, 105–134 (117.1 ± 9.4) mm in females; TAL/SVL 1.93–2.73 in nine males, 1.43–2.33 in nine females; HL/SVL 0.22–0.27 in males, 0.21–0.25 in females; HW/SVL 0.14–0.18 in four males, 0.12–0.13 in eight females; HD/SVL 0.14–0.16 in four males, 0.12–0.14 in eight females; HW/HD 0.96–1.07 in four males, 0.91–1.06 in eight females; SL/SVL 0.12–0.13 in four males, 0.11–0.12 in eight females; SHL/SVL 0.21–0.24 in males, 0.20–0.25 in females; foot L/SVL 0.32–0.38 in five males, 0.30–0.38 in eight females; hand L/SVL 0.12–0.16 in five males, 0.13–0.15 in eight females.

Color in life of an adult female (USNM 563587): dorsal ground color Maroon (31) with Spectrum Yellow (55) spots, spots decreasing in size middorsally, middorsal zone with greenish tinge; line of Olive-Yellow (52) closely set large spots extending laterally between points of fore- and hind limb insertions; area below that line of spots with another line of spots grading from Olive-Yellow laterally to Sky Blue (66) ventrally on Jet Black (89) background; front limb Grayish Horn Color (91) dorsally, dirty cream ventrally; hind limb Grayish Horn Color with small Straw Yellow (56) spots dorsally and pale greenish blue spots ventrally, except for dirty white femoral pore scales; dorsum of tail Grayish Horn Color grading to pale bluish green posteriorly; top of head Grayish Horn Color; chin dirty white; venter dirty white on chest, grading to pale greenish blue on belly.

Color in alcohol: dorsal ground color brown to dark brown; juveniles have 6 pale brown to white stripes beginning on posterior portion of head and extending onto base of tail; subadults with 4–6 pale brown to white stripes beginning posterior to (dorsally) or on posterior part of head (laterally) and extending onto base of tail, those stripes also visible in some adults up to about 115 mm SVL, other similar-sized adults and larger ones lack stripes; adults with stripes have some small cream spots on body, those larger ones without stripes have numerous, distinct cream spots on body and base of tail; about 27–35 cream to green-tinged flank spots, spots varying from large to small, small similarly colored spots also present on dorsal surface of thigh and on outer most row of ventral plates; throat and chest cream with scattered or somewhat dense dark gray spots; belly of females white to pale cream, some midbelly scales tinged with bluish gray or midventer can be largely black; belly of males largely black; posterior half of dorsal surface of tail paler brown than remainder of tail ground color.

Diagnosis/Similar Species.—The combination of having large platelike scales on top of the head and on the venter, moveable eyelids, femoral pores present in both sexes, and granular dorsal body scales distinguish-

Plate 75. *Aspidoscelis motaguae*. USNM 580919, adult female, SVL = 130 mm. El Paraíso: near Orealí.

es *Aspidoscelis motaguae* from all remaining Honduran lizards, except thc other Teiidae. The single Honduran species of *Ameiva* usually has 12 (rarely 10) transverse ventrals at midbody and has the brachial-antebrachial large scales separated by smaller scales (versus 8 transverse ventrals and enlarged brachial-antebrachial scales in contact in *A. motaguae*). *Holcosus* have a patch of enlarged midgular scales (versus no patch of enlarged midgulars in *A. motaguae*). *Cnemidophorus ruatanus* has 4 parietal scales plus 1 interparietal scale, has granular brachial scales on the postaxial surface of the upper arm, and has paired precloacal spurs in males (versus 2 parietal scales plus 1 interparietal scale, moderately enlarged postaxial brachial scales, and males lack precloacal spurs in *A. motaguae*). *Aspidoscelis deppii* almost always has 3 supraoculars, has granular scales ventrolaterally on the postaxial antebrachium, has a striped dorsal surface in adults, has the tail some shade of blue or brown in life, reaches only 93 mm SVL, and lacks distinct spots in adults (versus 4 supraoculars, patch of noticeably enlarged ventrolateral postaxial antebrachial scales, tail some shade of pink

to orange in life, reaches 139 mm SVL, and has distinctively spotted dorsal and lateral surfaces of body in *A. motaguae*).

Illustrations (Figs. 136, 137; Plate 75).— Álvarez del Toro, 1983 (adult; as *Cnemidophorus sackii* Wiegmann, 1834b: 27); Duellman and Zweifel, 1962 (ventral and ventrolateral scales on antebrachium; as *Cnemidophorus*); Köhler, 2000 (head scales; as *Cnemidophorus*), 2003a (adult, head scales), 2008 (adult, head scales); Köhler et al., 2005 (adult, head scales); Köhler, Salazar Saavedra et al., 2013 (adult); Mata-Silva et al., 2015 (adult); Mertens, 1952b (adult; as *Cnemidophorus s. bocourti* Boulenger, 1885b: 367); Meshaka et al., 2004 (adult; as *Cnemidophorus*); Sackett, 1941 (adult, head; as *Cnemidophorus*); C. A. Smith and Krysko, 2007 (head scales); Villa et al., 1988 (adult; as *Cnemidophorus*); Wright, 1993 (adult; as *Cnemidophorus*).

Remarks.—*Aspidoscelis motaguae* is a member of the *A. sexlineata* species group (Reeder et al., 2002: 22; Goicoechea et al., 2016: 48).

Natural History Comments.—*Aspidoscelis motaguae* is known from about 50 to 950 m elevation in the Lowland Dry Forest,

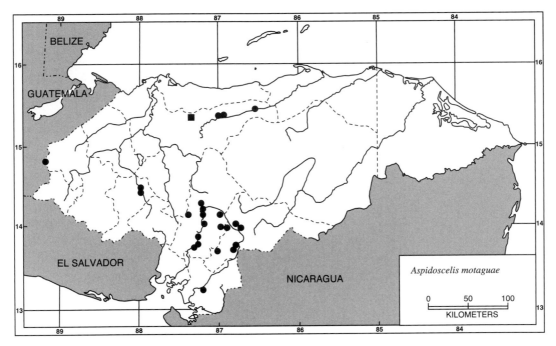

Map 61. Localities for *Aspidoscelis motaguae*. Solid circles denote specimens examined and the solid square an accepted record.

Lowland Arid Forest, Premontane Moist Forest, and Premontane Dry Forest formations. This sun-loving, terrestrial species is probably active throughout the year because it has been collected in March, April, July, November, and December during both the rainy and dry seasons. It occurs in open situations such as grassy, sandy, and rocky areas. Inactive individuals were under rocks, debris of a fallen house, and regular ground debris. It usually occurs in hilly areas but also can sometimes be seen in low, flat areas. This species tolerates disturbed areas and can still be seen in some areas of the capital city of Tegucigalpa. Meshaka et al. (2004) reported two females of Florida populations collected in late July contained three sets of previtellogenic follicles. Those authors also reported that population is an insectivore with food items "comprised mostly of beetles (Coleoptera) and their larvae, roaches (Dictyoptera), and ants (Hymenoptera)."

Etymology.—The name *motaguae* is derived from the Río Motagua Valley, Guatemala, plus the Latin suffix –*ae* (a derivation), denoting that the type locality lies in that valley.

Specimens Examined (203, 1 skeleton [97]; Map 61).—**CHOLUTECA**: El Madreal, USNM 563584–86, 580922. **COPÁN**: 0.5 km W of Copán, USNM 563587; Copán, UMMZ 83036. **EL PARAÍSO**: El Rodeo, USNM 580921; 15 km E of Guadalupe, LACM 39778; 6 km SE of Los Limones, AMNH 157378, LSUMZ 24628; 7 km S of Mansaragua, UF 127581–82; 1 km S of Orealí, USNM 580919–20; Orealí, USNM 589146. **FRANCISCO MORAZÁN**: 15.9 km SSW of Cofradía, CM 65727; 10.4 km N of Cofradía, CM 65728–29; Comayagüela, KU 200585; 12 km E of El Zamorano, LSUMZ 24632; 5 km ENE of El Zamorano, KU 103251–54; El Zamorano, AMNH 70460–65, 70467–69, 70474–99, MCZ R49773–75 (+35 unnumbered); La Venta,

FMNH 5094; 12 km SSW of Sabanagrande, AMNH 157371–77, LSUMZ 24629–30; 11 km N of Sabanagrande, LSUMZ 24631; 19.3 km N of Tegucigalpa, TCWC 19208–09; Tegucigalpa, AMNH 69086, BYU 16995–98, 18176–77, 18179, 18181–88, 18190, 18277–78, 18280–81, 18283–85, FMNH 5095 (17), LSUMZ 24140, MSUM 4682–89, UTA R-53299; Villa San Francisco, USNM 579540; Zambrano, UTA R-53191. **INTIBUCÁ**: 8.0 km S of Jesús de Otoro, LACM 48143–51; 3.0 km N of Jesús de Otoro, LSUMZ 33699; 5.5 km S of Jesús de Otoro, LSUMZ 33701–02. **YORO**: 0.5 km N of Coyoles, LACM 48129–40, 48142, 163969 (skeleton), LSUMZ 21663–76; Poligono, USNM 579541; Rancho San Lorenzo, LACM 48141.

Other Records (Map 61).—**YORO**: San Francisco (personal sight record). "HONDURAS": (Werner, 1896).

Genus *Cnemidophorus* Wagler, 1830a

Cnemidophorus Wagler, 1830a: 154 (type species: *Seps murinus* Laurenti, 1768: 63, by subsequent designation of Fitzinger, 1843: 20).

Geographic Distribution and Content.—This genus ranges from southern Belize to northeastern Nicaragua and from central Panama to northeastern Brazil east of the Andes and northwestern Colombia west of the Andes. It also occurs in the southern Lesser Antilles, on Trinidad and Tobago and numerous other islands off the coast of Venezuela, the Bay Islands of Honduras, and islands off the coast of Nicaragua. One species has also been introduced and established in southern Florida, USA. Nineteen named species are recognized, one of which occurs in Honduras.

Remarks.—Reeder et al. (2002: 14–20), using morphological and molecular data, recovered phylogenetic evidence that *Cnemidophorus*, as restricted by them, remained paraphyletic. Reeder et al. (2002: 21) found members of the *Cnemidophorus*

lemniscatus complex to be "more closely related to Central and South American '*Ameiva*' . . . than to members of the North American '*Cnemidophorus*'" clade. Harvey et al. (2012: 102–112), using morphological data, restricted *Cnemidophorus* to 14 named species (plus two unnamed species) in which males have a single pair of precloacal spurs. Those authors also recognized four species groups, one of which (the *C. lemniscatus* group) occurs in Honduras. The phylogenetic analysis based on molecular data by Pyron et al. (2013, fig. 11) recovered support for the monophyly of *Cnemidophorus* as restricted by Harvey et al. (2012), but they only tested four of the species included in the *C. lemniscatus* group by Harvey et al. (2012). Subsequently, McCranie and Hedges (2013c) described a new species (*C. duellmani*) from Panama in the *C. lemniscatus* group and suggested elevating four other nominal forms (*C. espeuti* Boulenger, 1885b: 362; *C. gaigei* Ruthven, 1915: 1; *C. ruatanus* Barbour, 1928: 60; *C. splendidus* Markezich et al., 1997: 46) to species level. Tissues of four of those five species were not available (only *C. ruatanus* available) to McCranie and Hedges (2013c).

Etymology.—*Cnemidophorus* is derived from the Greek *knemidotos* (with leggings) and *phoreus* (bearer, carrier). Wagler (1830a: 154) stated the name meant "*ocreis armatus*," which translates to equipped with protective armor for the shins. The name alludes to the several rows of large scales on the dorsal surface of the foreleg of the members of this genus.

Cnemidophorus ruatanus Barbour, 1928

Cnemidophorus lemniscatus ruatanus Barbour, 1928: 60 (holotype, MCZ R26759; type locality: "Coxen Hole, Ruatan [sic], Bay Islands of Honduras"); Flower, 1928: 53; Barbour and Loveridge, 1929a: 242; Rand, 1954: 260; Maslin and Secoy, 1986: 26; Wright, 1993: 79.

Cnemidophorus lemniscatus lemniscatus: Burt, 1931: 30; Lynn, 1944: 190; Dunn and Saxe, 1950: 157; P. W. Smith, 1950: 55; Rand, 1954: 260; Meyer, 1966: 176; Peters and Donoso-Barros, 1970: 94.

Cnemidophorus deppii deppii: Dunn and Emlen, 1932: 30.

Cnemidophorus lemniscatus: Echternacht, 1968: 152; Meyer, 1969: 277; Meyer and Wilson, 1973: 33; Wilson and Hahn, 1973: 117; Wilson et al., 1976: 179; Cruz Díaz, 1978: 29; O'Shea, 1986: 43; Wilson and Cruz Díaz, 1993: 16; Köhler, 1994a: 4; Köhler, 1995b: 102; Köhler, 1996d: 20; Cruz Díaz, 1998: 29, *In* Bermingham et al., 1998; Monzel, 1998: 161; Köhler, 2000: 95; Lundberg, 2000: 5; Lundberg, 2001: 27; Wilson et al., 2001: 134; Lundberg, 2002a: 7; Köhler, 2003a: 161; Powell, 2003: 36; Gutsche, 2005b: 14; McCranie and Castañeda, 2005: 15; McCranie et al., 2005: 120; McCranie et al., 2006: 134; Townsend, 2006a: 35; Wilson and Townsend, 2006: 106; Montgomery et al., 2007: 38; C. A. Smith and Krysko, 2007: 265; Köhler, 2008: 173; Butterfield et al., 2009: 47; Montgomery et al., 2011: 10; McCranie and Solís, 2013: 242.

Cnemidophorus ruatanus: McCranie and Hedges, 2013c: 304; McCranie and Valdés Orellana, 2014: 46; Solís et al., 2014: 132; McCranie, 2015a: 371.

Geographic Distribution.—*Cnemidophorus ruatanus* occurs at low elevations on the Atlantic versant from southern Belize (but see Remarks) and eastern Guatemala to northeastern Nicaragua, and on Roatán and Utila islands and the Cayos Cochinos in the Honduran Bay Islands. In Honduras, this species occurs along the northern coastal plain of the mainland (and inland along several rivers), as well as on the islands just mentioned.

Description.—The following is based on 21 males (FMNH 283560–61, 283563–64, 283568; KU 192621, 203160, 220101; USNM 563589–90, 563593–94, 563596, 563598–99, 570397, 570399, 570401, 573206, 573209, 580929) and 14 females (FMNH 283562, 283566; KU 101334, 101340, 192622; USNM 563588, 563592, 563595, 563597, 573207–08, 573211–12, 580928). *Cnemidophorus ruatanus* is a moderately large lizard (maximum recorded SVL 113 mm [FMNH 283561, a male]; dorsal head scales enlarged, smooth, plate-like, with dorsally expanded anterior nasal scales contacting each other medially with short sutures, single frontonasal, paired prefrontals, single frontal, paired frontoparietals, 2 pairs of parietal scales (occasionally 1 or 2 parietals partially divided) plus a single interparietal scale; rostral not contacting frontonasal; postnasal not contacting prefrontal; prefrontal and first superciliary usually in contact; frontal ridge absent; frontal-frontoparietal suture usually aligned near midlength of supraocular 3; scales in frontoparietal region smooth, flat; interparietal varies from slightly narrower than, to slightly broader than flanking parietals; 1–6 (2.8 ± 1.1) scales between fourth supraocular and innermost parietal; 1–6 (4.0 ± 1.0) scales between fourth supraocular and outermost parietal; usually 3–5 supraoculars, only occasionally a fifth smaller supraocular present posteriorly, rarely only 3 present; first supraocular contacting second supraocular; median pair of occipital scales distinctively enlarged, much larger than first dorsal scale row; 17–28 (21.4 ± 2.5) occipitals (occipitals of Harvey et al., 2012, plus scales bordering frontoparietals); supratemporals moderately enlarged, separated from parietals; short rostral groove present or absent; nostril centered in nasal suture; nostril opening oval; 1.0 to 1.5 rows of 21–52 (29.0 ± 6.4) lateral supraocular granules; anterior extent of circumorbital semicircles in single row reaches between midlength and posterior third of supraocular 3, that of double row not reaching seam between supraoculars 3–4; 4–9 (5.5 ±

1.5) circumorbitals (small scales bordering supraocular 4, except those on outer side); 5–6 (usually 6) superciliaries, usually with second (rarely third) elongated; 4–5 (rarely 5) suboculars, first usually entire (occasionally divided), lower edge (when entire) or lower part (when divided) usually contacting supralabial 3 (rarely contacting both supralabials 3 and 4); subocular keel present; patch of enlarged scales located in front of auditory meatus; no auricular or preauricular flap or fold; 1 loreal; 5–7 (rarely 5 or 7) supralabials and 5–6 (occasionally 6) infralabials; first supralabial straight ventrally, usually longer than second; lingual sheath absent; moveable eyelid present; pupil reniform; first pair of chinshields contacting infralabials, those chinshields separated from each other only at posterior edge; intergular sulcus absent; 18–26 (22.2 ± 2.2) anterior gulars; 9–18 (14.0 ± 2.5) posterior gulars; midgular patch of distinctively enlarged scales absent; intertympanic sulcus absent; sharp transition from anterior gulars to smaller posterior ones; mesoptychial scales moderately enlarged, bordered anteriorly by sharp transition to small scales; edge of gular fold not serrated; dorsal scales conical, 190–230 (212.0 ± 9.0) middorsals between occipitals and first enlarged caudal scale; 97–123 (106.9 ± 6.0) granules around midbody; middorsal scales subequal in size to lateral scales; chest scales large, flat; pectoral sulcus absent; ventral body scales large, platelike, squarish, juxtaposed, smooth, in 25–33 (28.9 ± 2.6) rows between gular fold and level of hind limb, in 8 transverse rows at midbody; scales immediately lateral to outside ventral plate small, granular; paired enlarged terminal scales forming precloacal plate with larger, single plate bordering paired enlarged terminal scales; 9–13 (10.6 ± 1.4) scales bordering all 3 enlarged plates in 55 (those listed above plus FMNH 283565, 283567; KU 101328–30, 101332–33, 101339, 101341–42, 101344–47, 101349–50; USNM 69397–400); males usually with 1 smaller subtrian-

gular scale flanking paired terminal plates and 1–2 smaller rounded scales between precloacal plate and small precloacal spur; 4–6 (4.8 ± 0.6) precloacal scales; postcloacal buttons absent; pair of slightly enlarged postcloacal scales forming spines present in males, those enlarged scales absent in females; caudal annuli complete, tail lacking crests or dorsolateral row of serrated scales; proximal subcaudals keeled; large dorsal brachial and antebrachial scales continuous, not separated by band of small scales; enlarged dorsal scales on brachial and antebrachial surfaces 1.0–1.5 times as wide as long, both sets of enlarged scales smooth, extending well beyond center of arm; postaxial brachial scales in continuous row with preaxial brachial scales; postaxial antebrachial scales slightly enlarged; 29–34 (31.4 ± 1.2) combined subdigital scales on Finger IV in 20; subarticular scales of fingers III–IV homogeneous in size, entire; 54–64 (59.7 ± 2.3) combined subdigital scales on both sides of Toe IV in 19, distal ones smooth; subarticular scales of toes III–IV divided, each scale smaller than remaining scales; no row of distinctively enlarged scales between toes IV–V; small scales separating supradigital scales from subdigital scales continuous, or nearly continuous; denticulate fringe absent along postaxial edge of outer toe; Toe V not reduced, claw extending beyond level of articulation of toes III–IV; 5–11 (7.4 ± 1.4) prefemoral scales on left side; heel without expanded scales; tibiotarsal shield present; tibiotarsal spur absent; 38–47 (42.4 ± 2.4) total femoral-abdominal pores in males, 33–46 (39.7 ± 3.5) in females; no gap between femoral and abdominal pores; 3–4 (most often 3) medial scales separating each femoral and abdominal pore series from each other; SVL 61.7–113.4 (79.6 ± 12.2) mm in males, 60.3–75.7 (67.0 ± 4.1) mm in females; TAL/SVL 1.60–2.79 in 13 males, 1.70–2.36 in eight females; HL/SVL 0.23–0.27 in males, 0.21–0.27 in females; HW/SVL 0.13–0.16 in males, 0.12–0.18 in females; HD/SVL 0.13–0.17 in males, 0.11–0.15 in females; HW/HD 0.93–1.13 in males, 1.04–1.23 in females; SL/SVL 0.09–0.11 in males, 0.09–0.12 in females; SHL/SVL 0.19–0.23 in males, 0.18–0.22 in females; foot L/SVL 0.29–0.37 in males, 0.30–0.46 in females; hand L/SVL 0.11–0.16 in males, 0.12–0.16 in females.

Color in life of an adult male (FMNH 283564): middorsal longitudinal broad stripe Ground Cinnamon (239) with two dark brown stripes in dorsolateral field of broad stripe; middorsal broad stripe bordered by Dark Brownish Olive (129) stripe (about 6 granules wide), that dark stripe bordered ventrally by narrow Chartreuse (158) dorsolateral stripe; lateral surface of body greenish yellow with scattered brownish green mottling; top of head Emerald Green (163) along outer edges, becoming brown with green tinge medially; anterior surfaces of fore- and hind limb Opaline Green (162D); ventral surfaces of head, throat, and chest Robin's Egg Blue (93); belly white with pale blue tinge; outer 2 ventral plates Emerald Green; ventral surfaces of limbs and tail pale blue, except palm and sole Robin's Egg Blue. Color in life of another adult male (KU 220101): dorsal surface of head tan; lateral surface of head yellow-green, grading to turquoise blue on lips; outer edge of supraoculars outlined by yellow-green line; middorsal broad stripe tan, flanked by dark brown stripe; lateral surfaces yellowish green on neck and near groin, pinkish tan between those points, with yellow to white spots; dorsal surface of forelimb olive green; dorsal surface of hind limb tan with white spots; dorsal surface of tail tan; chin turquoise blue; chest pale turquoise blue; belly green with bluish cast; ventral surface of forelimb white; ventral surface of hind limb turquoise blue. Color in life of an adult female (UF 150291): middorsal surface of body Drab-Gray (119D) with slightly paler gray paramiddorsal stripes; dorsolateral field Hair Brown (119A) bounded by pale tan stripes; lateral and ventrolateral fields Chartreuse (158) bounded by pale tan to white stripes; venter Light Sky Blue (116D) grading to pale bluish green laterally; top of head Grayish Horn Color (91); fore- and hind limb grayish brown with white spots dorsally, pale grayish blue ventrally; tail Drab-Gray dorsally, pale bluish green ventrally. Montgomery et al. (2007: 43) recorded additional color in life notes for a population on Cayo Cochino Menor and stated that males in that population had "a lesser intensity of green or blue coloration on the dorsal and lateral surfaces of the head and on the fore legs" than do those from the Honduran mainland and "on internet sites [= various species of the *C. lemniscatus* complex]." Those authors also stated that none of the Cayo Cochino males "had the intensity of blue coloration on the lateral surfaces of the head" as in the Utila male photographed by Gutsche (2005b). Montgomery et al. (2007) also reported nine pale longitudinal lines on the dorsal surface of *C. ruatanus* on Cayos Cochinos, but their count also included a barely visible paler line inside the middorsal broad stripe.

Color in alcohol: juveniles have 8 longitudinal white to dirty white stripes dorsally and laterally on a body that is dark brown laterally and slightly paler brown middorsally; adults and juveniles of both sexes have a complete, pale brown middorsal swath extending from posterior edge of head to base of tail; adult females have middorsal swath bordered laterally by thin paler brown to cream stripe, that in turn bordered below by a broad dark brown stripe that extends to posterior end of body, last mentioned stripe also bordered below by cream to pale brown thin stripe, with area below that thin stripe pale brown with or without indistinct cream stripe or cream to pale yellow small spots; adult males with middorsal swath with or without evidence of paler brown thin border stripe, but with a dark brown broader border stripe (as in females) extending to posterior end of body, with area below dark brown border stripe with cream to pale

Plate 76.   *Cnemidophorus ruatanus*. FMNH 283563, adult male, SVL = 82 mm. Cortés: Tegucigalpita.

yellow spots; ventral surfaces of head and body cream to white with pale bluish gray tinge, or entirely black, except throat region pale blue to pale gray in males. Adults of both sexes from the Bay Islands have pale blue ventral surfaces, whereas those from the mainland can have either pale blue or black ventral surfaces. Adults of both sexes from Isla de Roatán retain pale stripes to a larger size than do those from the Honduran mainland.

Diagnosis/Similar Species.—The combination of having large platelike scales on top of the head and on the venter, moveable eyelids, femoral pores in both sexes, and granular dorsal body scales distinguishes *Cnemidophorus ruatanus* from all remaining Honduran lizards, except the other Teiidae. *Cnemidophorus ruatanus* is the only Honduran teiid with enlarged precloacal spurs in males. Additionally, the *Ameiva* has enlarged brachial-antebrachial scales separated by small scales and usually has 12 (rarely 10) transverse rows of ventral scales at midbody (versus enlarged brachial-antebrachial scales in contact and 8 enlarged transverse ventrals in *C. ruatanus*). *Holcosus* also have a patch of distinctively

enlarged midgular scales (versus midgular scales not distinctively enlarged in *C. ruatanus*). *Aspidoscelis deppii* also has 2 parietal scales plus 1 interparietal scale and has 3 supraocular scales (versus 4 parietal scales plus 1 interparietal, and usually 4 supraocular scales in *C. ruatanus*), and *A. motaguae* also has slightly enlarged brachial scales on the posterior surface of the upper arm, has the tail usually some shade of pink to orange in life, and reaches 139 mm SVL (versus noticeably enlarged postantebrachial scales absent, tail gray, brown, or bluish green in life, and maximum known SVL 113 mm in *C. ruatanus*).

Illustrations (Figs. 129, 130; Plates 76–78).—Gutsche, 2005b (adult, hatchling; as *C. lemniscatus*); Köhler, 1996b (adult; as *C. lemniscatus*), 1999b (head scales; as *C. lemniscatus*), 2000 (adult, head scales; as *C. lemniscatus*), 2001b (head scales; as *C. lemniscatus*), 2003a (adult, head scales; as *C. lemniscatus*), 2008 (adult, head scales; as *C. lemniscatus*); Lee, 2000 (adult; as *C. lemniscatus*); Lundberg, 2000 (adult; as *C. lemniscatus*); McCranie and Hedges, 2013c (adult, cloacal scales, laterals and lateral ventrals); McCranie et al., 2005 (adult, head

Plate 77. *Cnemidophorus ruatanus.* FMNH 283568, adult male, SVL = 62 mm, Islas de la Bahía: Utila Island, east coast near Tradewinds.

scales, femoral pores and hind limb scales; as *C. lemniscatus*), 2006 (adult, head scales; as *C. lemniscatus*); Montgomery, Boback et al., 2011 (juvenile; as *C. lemniscatus*); Montgomery, Reed et al., 2007 (adult; as *C. lemniscatus*); Powell, 2003 (adult; as *C. lemniscatus*); Stafford and Meyer, 1999

(adult; as *C. lemniscatus*); Sunyer et al., 2009 (adult; as *C. lemniscatus*).

Remarks.—Barbour (1928: 60) described *Cnemidophorus lemniscatus ruatanus* based on a single specimen from Roatán Island. Barbour (1928) distinguished *C. l. ruatanus* from *C. l. lemniscatus* (Linnaeus, 1758: 209)

Plate 78. *Cnemidophorus ruatanus.* USNM 563595, adult female, SVL 71 mm. Islas de la Bahía: Roatán Island, Flowers Bay.

in having only "one scale in the angle between the outer parietal scute and the postorbital scute." Barbour (1928: 60) also distinguished *C. l. ruatanus* from *C. l. gaigei* Ruthven of northern Colombia in having "but a couple of elongate shields between the supraoculars and the postfrontals and parietals." Barbour (1928: 60) went on to say about his *C. l. ruatanus* that "a median postfrontal is present, and the frontals are narrower and extend further in an anterior direction than they do" in *C. l. lemniscatus* and *C. l. gaigei*. Burt (1931: 34–35) found that the characters used by Barbour were also shared by other populations of *C. l. lemniscatus* and, thus, synonymized *C. l. ruatanus* with his concept of *C. l. lemniscatus*. Rand (1954: 260) studied 47 specimens of *Cnemidophorus* from Roatán and 13 specimens from the Honduran mainland and suggested retaining the name *C. l. ruatanus* for the Roatán population, primarily because of retention of the juvenile dorsal pattern to a larger size in the Roatán series than those on the Honduran mainland. Thus, though not stated by Rand (1954), *C. l. lemniscatus* was inferred to be the form occurring on the Honduran mainland. McCranie and Hedges (2013c) performed a morphological and molecular study of the Honduran mainland and Bay Island populations of *Cnemidophorus* and resurrected *C. ruatanus* for those populations (p. 304), as well as those from adjacent Belize (probably human-made introduction), Guatemala, and Nicaragua.

Echternacht (1968: 152) compared a series of 24 Honduran mainland *C. ruatanus* with a series of *C. lemniscatus* from Panama and found the two series showed differences in four scale characters: number of transverse ventral scales; nature of contact between frontal and frontoparietal suture and granular scales separating supraorbital scales from median head scales; extent of the double row of granular scales between supraciliaries and supraorbitals; and number of dorsal granules around body.

Despite these differences, Echternacht (1968) did not make any taxonomic suggestions for the populations studied. McCranie and Hedges (2013c: 308) described that Panamanian population as a new species, *C. duellmani*. Examination of numerous *C. ruatanus*, a few *C. duellmani*, and some *Cnemidophorus* from northern South America confirms that the characters used by Barbour to distinguish *C. ruatanus* are not valid. Butterfield et al. (2009) studied scale data on 30 *C. ruatanus* from Utila Island and found some values that slightly expand scale number ranges for some counts given here in the description. Their data also confirmed that the Utila population reaches a smaller SVL than do the Roatán and Honduran mainland populations. Morphological data and color descriptions given by Butterfield et al. (2009: 47) for a Miami, Florida, area population of the *C. lemniscatus* complex indicate that the Honduran species, *C. ruatanus*, was not the source of the Florida populations. Butterfield et al. (2009: 53) speculated that the original source of those Florida *C. lemniscatus* complex populations might be Venezuela (except not the Peninsula de Paraguana population), Colombia, or one of the islands in the Caribbean Sea (but not a Honduran island).

Dunn and Emlen (1932: 30) reported *Cnemidophorus ruatanus* (as *C. l. lemniscatus*) from Lago de Yojoa based on a MCZ specimen. That is an unlikely locality for this species. Dunn and Emlen (1932: 30) also confused *Aspidoscelis deppii* with *C. ruatanus*; however, there are no extant specimens of either of those two species from Lago de Yojoa. Thus, the identity of the Dunn and Emlen specimen is problematic, although it might be the same MCZ specimen of *A. deppii* Meyer and Wilson (1973: 32) listed from Lago de Yojoa. However, that specimen is actually from Lancetilla, Atlántida. Schwartz and Thomas (1975: 110), MacLean et al. (1977: 4), and Morgan (1985: 43) listed *C. ruatanus* (as *C. l. lemniscatus*) from the Swan Islands, Honduras, apparently based on the erroneous Swan Island record of Cope (1892b: 30; as *C. espeutii* [sic]). As noted by Dunn and Saxe (1950: 159): "The material basis for Cope's record ... from Swan I ... has not been located." *Cnemidophorus* was not listed from the Swan Islands in the Schwartz and Henderson (1988, 1991) and Powell and Henderson (2012) checklists but was inaccurately said to occur on the Swan Islands by Auth (1994: 14) and Savage (2002: 517).

Cnemidophorus ruatanus is a member of the *C. lemniscatus* species group, as diagnosed by Harvey et al. (2012; morphological data) and Goicoechea et al. (2016; molecular data). McCranie and Hedges (2013c) inaccurately said *C. ruatanus* had round pupils.

Natural History Comments.—*Cnemidophorus ruatanus* is known from sea level to 400 m elevation in the Lowland Moist Forest and Lowland Dry Forest formations. This terrestrial, sun-loving species occurs in open sandy areas, such as beach vegetation and along rivers. It has the unusual habit of pausing to lift and shake one of its forelegs as it moves across the sand. It is active on sunny days throughout the year. Echternacht (1968) found *C. ruatanus* to inhabit the beach and coastal strand at La Ceiba and Trujillo, Honduras. Echternacht (1968) also reported *C. ruatanus* from along the Río Cangrejal in the environs of La Ceiba but said it seemed to be less common upstream. Currently, *C. ruatanus* does not occur along the beaches at La Ceiba but remains common along the Rio Cangrejal to at least the village of Río Viejo. The species also occurs for a short distance up the Río Viejo, a tributary of the Río Cangrejal. *Cnemidophorus ruatanus* has also dispersed up the Río Motagua in Guatemala and into Honduras along a tributary of the Río Motagua to at least as far as La Playona, Copán, an unusual dispersal event regarding the remaining known populations of this species. Montgomery et al. (2007: 43)

reported on the distribution and coastal zone habitat of this species on Cayo Cochino Menor, where it occurs "in a windswept coastal zone with dense low-lying vegetation interspersed with enough open spaces and exposed ground" (also see Montgomery et al. 2011). Gutsche (2005b) presented information on mating, egg deposition, and hatching for a population on Utila Island. Two clutches of four and five eggs (each likely from more than one female) were found buried on a sandy beach in March, and four eggs were incubated and hatched in 52 days (Gutsche, 2005b). Montgomery et al. (2011) recorded one or two eggs per clutch in seven females from Cayo Cochino Menor, with females capable of multiple clutches per year. Diet of this species on Cayo Cochino Menor includes amphipods, arachnids, insects, and plant matter (Montgomery et al., 2011).

Etymology.—The specific name *ruatanus* refers to Isla de Roatán (misspelled Ruatan by Barbour, 1928) plus the Latin suffix –*anus* (inhabiting, belonging to), referring to where the type specimen of this species was collected.

Specimens Examined (532 [130]; Map 62).—**ATLÁNTIDA**: Agua Chiquito, USNM 580928; Corozal, LACM 48067, LSUMZ 21693–95; Barra de Colorado, USNM 580929; 14.5 km E of La Ceiba, LACM 48070; 1 km W of La Ceiba, KU 101328–40; 2 km SE of La Ceiba, KU 101341; La Ceiba, CM 29008, INHS 4486, LACM 48051–62, LSUMZ 21677–88, USNM 62973–79, 117608; Lancetilla, AMNH 69640, 70449–54, 70455 (15), UMMZ 72412 (10); Piedra Pintada, LACM 48071; Punta Sal, USNM 580930–31; Río Cangrejal near La Ceiba, FMNH 283561; Río Cangrejal about 12 km SSE of La Ceiba, KU 101351; along Río Viejo, KU 200584, USNM 563588; San Marcos, USNM 570397–402; about 80 km ESE of Tela, FMNH 13007; Tela, AMNH 46917–19, 157317–41, MCZ R21117–49, 21757–67, 27570–75 (+12 untagged), UMMZ

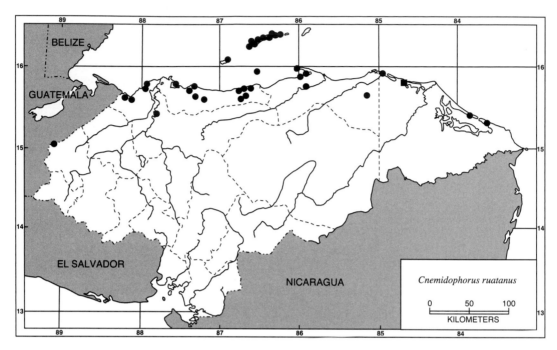

Map 62.   Localities for *Cnemidophorus ruatanus*. Solid circles denote specimens examined and the solid square represents an accepted record.

69537 (2), 62509 (6). **COLÓN**: Balfate, AMNH 58624–26; Barranco, ANSP 28124; Puerto Castilla, LSUMZ 22479–81; Salamá, USNM 242610, 242612–13, 242627–37; 1–3 km W of Trujillo, KU 101342–48, LACM 48068–69; between Trujillo and Santa Fé, CM 65704–18; Trujillo, CM 65719–25, KU 101349–50. **COPÁN**: La Playona, USNM 563589–90. **CORTÉS**: Cieneguita, BYU 22588–618; El Paraíso, UF 144706–07; Puerto Cortés, AMNH 37864–65, FMNH 5096, 213524–27, 213529–35, TCWC 19201–07, UIMNH 66642, UMMZ 79071, USNM 69395–401; about 0.5 km SSE of Tegucigalpita, SMF 79013–14; Tegucigalpita, FMNH 283560, 283563, 283565–67, 283569; USNM 563591–93. **GRACIAS A DIOS**: Cauquira, UF 150291, 150296, 150298–99; Laguna Bacalar, BMNH 1985.1286–90; Yahurabila, USNM 573206–12. **ISLAS DE LA BAHÍA**: Cayo Cochino Menor, near Bonkes Nose Point, KU 220101; Isla de Roatán, near Coxen Hole,

FMNH 34492–538; Isla de Roatán, Flowers Bay, USNM 563594–97, 563599–600; Isla de Roatán, about 3.2 km W of French Harbor, LSUMZ 22384, UF 28554–56; Isla de Roatán, W of Oak Ridge, UTA R-10677–80; Isla de Roatán, near Oak Ridge, MCZ R150947–49, TCWC 52419–21, UTA R-55245; Isla de Roatán, Oak Ridge, KU 192621–22; Isla de Roatán, Palmetto Bay, FMNH 283562, 283564; Isla de Roatán, between Port Royal Harbor and Calabash Bight, UTA R-10674–75; Isla de Roatán, Port Royal Harbor, LSUMZ 33781; Isla de Roatán, about 3.2 km W of Roatán, CM 65699–700, LSUMZ 29632–33, 29669, 46595, 46601–02, 50770; Isla de Roatán, 0.5–1.0 km W of Roatán, LACM 48063–64, LSUMZ 21689–90; Isla de Roatán, about 4.8 km N of Roatán, CM 65701–02, LSUMZ 29631, 46597, 46599–600, 50771; Isla de Roatán, about 1.6 km N of Roatán, LSUMZ 29634, 46596, 46598; Isla de Roatán, 0.5–4.0 km N of Roatán, LACM

48065–66, LSUMZ 21691–92, UF 28485; Isla de Roatán, about 4.8 km W of Roatán, UF 28509–13, 28534; Isla de Roatán, near Roatán, CM 65703, LSUMZ 29629–30, 46603; Isla de Roatán, Sandy Bay, KU 203160–62; Isla de Roatán, Santa Elena, UTA R-10676; Isla de Roatán, West End Town, USNM 563598; "Isla de Roatán," MVZ 160192; Isla de Utila, east coast near Trade Winds, FMNH 283568; Isla de Utila, Jake's Bight, SMF 77111; Isla de Utila, Pumpkin Hill, UNAH 5289–90; Isla de Utila, Utila, CM 28999–9001, LSUMZ 22278–86, UF 28366–88, 28430–36, 28444. **YORO**: 17 km NE of El Progreso, AMNH 157312–16, 157342, LSUMZ 24625–27. "HONDURAS": UF 76231, 90901–08, 99331, 99433, 99662.

Other Records (Map 62).—**GRACIAS A DIOS**: Barra Río Plátano, UNAH 5263, 5386 (Cruz Díaz, 1978, specimens now lost).

Genus *Holcosus* Cope, 1862b

Holcosus Cope, 1862b: 60 (type species: *Ameiva septemlineata* A. H. A. Duméril, 1851: 114, *In* A. M. C. Duméril and Duméril, 1851 [designated by Harvey et al., 2012: 123]).

Geographic Distribution and Content.— This genus ranges from Tamaulipas and Nayarit, Mexico, southward to north-central Colombia and western Ecuador. Ten named species are recognized, two of which occur in Honduras.

Remarks.—Harvey et al. (2012: 118) resurrected the genus *Holcosus* from the synonymy of *Ameiva* based on a phylogenetic study of morphological variation in the family Teiidae. The analysis performed by Harvey et al. (2012: 75) recovered *Holcosus* as a well-supported sister clade to *Aspidoscelis*. Pyron et al. (2013, fig. 11) used molecular data for three of the ten species of *Holcosus* recognized by Harvey et al. (2012) and also recovered a monophyletic *Holcosus* clade but recovered it as sister to

the *Cnemidophorus* clade of Harvey et al. (2012: 75; also see Goicoechea et al., 2016: 48). Ashton (2003: 110) noted that males of both species of *Holcosus* occurring in Honduras have enlarged postcloacal scales, whereas females of those two species lack them, thus providing a way to determine the sex of specimens of those two species, even in juveniles.

Etymology.—Cope (1862b) did not discuss the etymology of his new genus. Harvey et al. (2012: 122) speculated that it might have been formed from the Latin word *holcus* (a kind of grain) in probable "reference to the many cephalic shields that have fragmented into numerous small and, therefore grainlike keeled scales in the type species." The Latin *–us* (of, pertaining to) was amended to form *Holcosus*.

KEY TO HONDURAN SPECIES OF THE GENUS *HOLCOSUS*

1A. Enlarged midgular scales irregular in arrangement, with one much enlarged midgular (Fig. 138); pale vertebral stripe present in all but largest adults; head relatively deep, HW/HD >1.07 ...... *festivus* (p. 410)

1B. Enlarged midgular scales usually in regular longitudinal row(s), without one much enlarged midgular (Fig. 132); pale vertebral stripe always absent in juveniles and adults; head relatively shallow, HW/HD <1.15 ... *undulatus* (p. 416)

CLAVE PARA LAS ESPECIES HONDUREÑAS DEL GÉNERO *HOLCOSUS*

1A. Escamas agrandadas mediogulares organizadas irregularmente, con una escama mediogular muy agrandada (Fig. 138); con una raya vertebral pálida presente en todos, excepto especímenes adultos muy grandes; cabeza relativamente honda, anchura de la cabeza/altura de

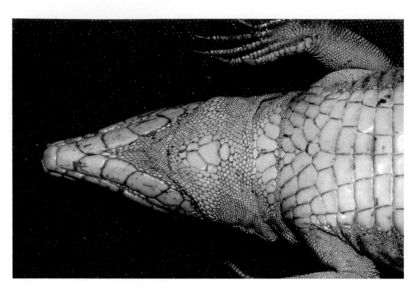

Figure 138.   Enlarged midgular scale patch present, irregular in arrangement, with one midgular scale much enlarged. *Holcosus festivus.* USNM 578849 from Los Pinos, Cortés.

la cabeza mayor a 1.07. . . . .
. . . . *festivus* (p. 410)

1B. Escamas agrandadas mediogulares organizadas en hileras, usualmente sin una escama mediogular muy agrandada (Fig. 132); raya vertebral pálida ausente en jóvenes y adultos; cabeza relativamente poco profunda, anchura de la cabeza/altura de la cabeza menor a 1.15
. . . . *undulatus* (p. 416)

Holcosus festivus (Lichtenstein and von Martens, 1856)

Cnemidophorus festivus Lichtenstein and von Martens, 1856: 13 [but see Remarks] (lectotype, ZMB 881a [designated by Echternacht, 1971: 26]; type locality: "Veragoa" [= Veraguas or Santiago de Veragua, Panama; see Echternacht, 1971: 26; Bauer and Günther, 1994: 270]).

Ameiva festiva: Barbour and Noble, 1915: 473; Dunn and Emlen, 1932: 30; Meyer, 1969: 265; Echternacht, 1971: 26; Meyer and Wilson, 1973: 30; Cruz Díaz, 1978: 28; O'Shea, 1986: 43; Espinal, 1993, table 3; Wilson and McCranie, 1998: 15; Köhler, McCranie, and Nicholson, 2000: 425; Nicholson et al., 2000: 30; Espinal et al., 2001: 106; Wilson et al., 2001: 134; Castañeda, 2002: 15; McCranie et al., 2002: 27; McCranie, 2005: 20; McCranie and Castañeda, 2005: 15; Castañeda and Marineros, 2006: 3.8; McCranie et al., 2006: 131; Townsend, 2006a: 35; Wilson and Townsend, 2006: 105; Goldberg, 2009a: 8; Townsend and Wilson, 2010b: 692.

Ameiva festiva edwardsi: Echternacht, 1968: 152.

Holcosus festivus: Harvey et al., 2012: 118; McCranie and Solís, 2013: 242; Solís et al., 2014: 133; McCranie, 2015a: 371; McCranie, 2016: 492.

Geographic Distribution.—*Holcosus festivus* occurs at low and moderate elevations on the Atlantic versant from Tabasco, Mexico, to north-central Colombia and on the Pacific versant marginally in northwestern Costa Rica and from south-central Costa Rica to northwestern and north-central Colombia. In Honduras, this species occurs in the northern half of the mainland.

Description.—The following is based on 13 males (USNM 563535, 563540–41, 563543–44, 563546, 563548, 563550–51, 563554, 563562, 5563570, 573943) and ten females (USNM 342390–92, 563536, 563539, 563558, 563560, 563565–66, 570392). *Holcosus festivus* is a moderate-sized lizard (maximum recorded SVL 131 mm [USNM 563570, a male]); dorsal head scales enlarged, smooth, platelike, with dorsally expanded anterior nasal scales that contact each other by short median sutures, single frontonasal, paired prefrontals, usually single frontal, frontal occasionally transversely divided into 2 scales, with posterior scale smaller than anterior when present, usually paired frontoparietals, paired parietal scales, but smaller scales sometimes separate parietals from frontoparietals, 1–2 additional smaller parietal scales usually present, and usually a single interparietal scale (interparietal sometimes divided into 2–4 scales; longitudinal, incomplete furrow also occasionally present); rostral not contacting frontonasal; postnasal contacting prefrontal or not; prefrontal and first superciliary usually separated, occasionally in contact; frontal ridge absent; frontal-frontoparietal suture usually aligned near posterior edge of supraocular 2, or anterior edge of supraocular 3; scales in frontoparietal region smooth, flat; interparietal varies from slightly narrower than, to about equal in size to adjacent parietal scales; 3–4 (3.2 ± 0.4) supraoculars (supraocular 4 small when present); first supraocular contacting second supraocular; 10–17 (13.6 ± 1.7) occipitals in 19, median ones at least slightly larger than first row of enlarged dorsal scales; supratemporals moderately enlarged, separated from parietals; short rostral groove present or absent; nostril located at posterior edge of nasal suture; nostril opening oval; 16–27 (21.9 ± 3.2) lateral supraocular granules in 1.0–1.5 rows on 38 sides; anterior extent of circumorbital semicircles usually reaching

about level of suture between second and third supraoculars, with single row sometimes extending further to reach posterior corner of first supraocular; 6–10 (7.7 ± 1.2) tiny circumorbital scales in 19; 5–6 (5.2 ± 0.4) superciliaries, second greatly elongated; 4 (rarely 5) suboculars, first usually entire, rarely divided into 2 scales, lower part broadly contacting supralabial 4, occasionally also narrowly contacting supralabial 3; subocular keel present; slightly enlarged scales located in front of auditory meatus; no auricular or preauricular flap or fold; 1 loreal; 6–7 (6.5 ± 0.5) supralabials; 5–6 (5.4 ± 0.6) infralabials; ventral edge of supralabial 1 varies from nearly straight to "toothy" (Harvey et al., 2012), also varies from slightly shorter than, to slightly longer than supralabial 2; lingual sheath present; moveable eyelid present; pupil reniform; 1 pair of chinshields, contacting infralabials, chinshields separated by small scales for entire length, or to nearly about two-thirds of length; intergular sulcus absent; 22–41 (27.8 ± 5.2) anterior gulars in 23 to point of first enlarged scale in midgular patch; 5–11 (7.4 ± 1.8) posterior gulars from ultimate enlarged scale in midgular patch in 23; midgular patch of 2–4 (usually 3) greatly enlarged, irregularly arranged (not in distinct row) scales present, with one enlarged scale much larger than others; intertympanic sulcus absent; sharp transition from anterior gulars to smaller posterior gulars; mesoptychial scales moderately enlarged, differentiated, with central scales greatly enlarged, bordered anteriorly by sharp transition to small scales; edge of gular fold not serrated; dorsal scales conical, not keeled, 239–272 (256.6 ± 6.8) granular middorsal scales between occipitals and first enlarged keeled scale at base of tail; 155–191 (175.1 ± 7.8) dorsal granular scales around midbody between enlarged ventrals; middorsal granules subequal in size to lateral granules; chest scales large, flat; pectoral sulcus absent; ventral body scales large, platelike, squarish, juxtaposed, smooth, in 25–28 (26.9 ± 0.9) longitudinal rows, in 8 transverse rows at 15th longitudinal row; 1–2 progressively smaller scales present or absent immediately lateral to outer ventral plate; 3–6 (3.7 ± 0.8) slightly enlarged terminal precloacal scales, those enlarged scales flanked by single row of small scales, that scale row flanked by granular scales; precloacal spur and postcloacal button absent; pair of enlarged postcloacal scales present in males; caudal annuli complete, tail lacking enlarged dorsal crests; dorsolateral row of serrated caudal scales present basally; proximal subcaudals smooth; enlarged dorsal brachial and antebrachial scales continuous, those enlarged scales 1.5–2.0 times as wide as long, both sets of enlarged scales smooth, large brachial and antebrachial scales extending well beyond center of arm; postaxial antebrachial scales not noticeably enlarged ventrolaterally; postaxial patch of enlarged scales present on brachial surface; subarticular scales of fingers III–IV homogeneous in size, entire; 13–17 (15.3 ± 0.9) subdigital scales on Digit IV of forelimb on 46 sides; 26–31 (28.5 ± 1.5) combined subdigital scales on Digit IV of hind limb, distal ones smooth; no row of distinctively enlarged scales between toes IV–V; small scales separating supradigital scales from subdigital scales restricted to phalangeal articulations; denticulate fringe present along proximal edge of outer toe; Toe V reduced, claw only reaching point of articulation of toes III–IV; 3–6 (3.7 ± 0.8) prefemoral scales on left hind limb; heel with moderately enlarged scales; tibiotarsal shield present; tibiotarsal spur absent; 35–42 (38.9 ± 2.6) total femoral-abdominal pores in males, 33–42 (36.9 ± 2.8) in females; no gap between femoral and abdominal pores; 5–8 (6.5 ± 0.8) medial scales between each femoral and abdominal pore set in both sexes combined; SVL 69.7–131.0 (96.4 ± 17.9) mm in males, 92.1–117.0 (101.2 ± 9.4) mm in females; TAL/SVL 1.68–2.44 in males, 1.85–2.37 in females; HL/SVL

0.24–0.28 in males, 0.22–0.28 in females; HW/SVL 0.14–0.18 in males, 0.13–0.15 in females; HD/SVL 0.10–0.15 in males, 0.12–0.14 in females; HW/HD 1.11–1.46 in males, 1.07–1.24 in females; SL/SVL 0.11–0.13 in males, 0.10–0.12 in females; SHL/SVL 0.24–0.30 in males, 0.22–0.28 in females; foot L/SVL 0.38–0.48 in males, 0.36–0.48 in females; hand L/SVL 0.12–0.17 in males, 0.13–0.18 in females.

Color in life of an adult male (USNM 563554): dorsal surface of head Clay Color (26), supraoculars edged medially with Sepia (119), scales posterior to frontal blotched with Sepia; middorsal body broad stripe Buff (24), becoming less distinct posteriorly; para-middorsal stripe Chestnut (32) with series of Sepia blotches connecting it to dorsolateral stripe; dorsolateral stripe Sepia, bordered by Trogon Yellow (153) dashes on either side; lateral region of body Chestnut with scattered Trogon Yellow spots and dashes; forelimb dark brown with pale blotching at base; hind limb olive brown with paler brown mottling; top of tail olive brown with dark brown mottling, increasingly overcast with blue-green posteriorly; venter Salmon Color (6); base of subcaudal surface Salmon Color, becoming increasingly blue-green posteriorly; chin horn color. Color in life of an adult female (USNM 563532): dorsal surface of head olive brown; dorsal surface of body copper on midregion, with lime green middorsal broad stripe extending from nape to base of tail, stripe gradually becoming fainter posteriorly; black dorsolateral stripe extending from posterior of eye to groin, stripe gradually becoming fragmented into spots surrounded by rust brown, narrow portions of some spots extend dorsally and less obviously ventrally, intruding into dorsal and lateral fields, respectively; lateral surface of body coppery green; dorsal surface of forelimb olive green with vague darker olive gray smudging; dorsal surface of hind limb olive green with dark olive gray reticulations; tail olive brown; infralabials

Plate 79. *Holcosus festivus*. Gracias a Dios: Bachi Kiamp. FN 257006 (still in Honduras because of permit problems).

ivory yellow; belly and ventral surfaces of fore- and hind limb pale pinkish orange; iris black.

Color in alcohol: dorsal surface of body with well-defined gray vertebral stripe, stripe extending well onto tail, vertebral stripe becoming less obvious or absent in large adults, especially males; dorsal surface of head with well-defined gray medial stripe in juveniles and subadults, head stripe disappearing in most adults; dorsal surface of body lateral to vertebral stripe dark brown, bordered below by white to gray, narrow, interrupted dorsolateral stripe, that stripe extending onto tail; black dorsolateral stripe below pale dorsolateral stripe visible in all but darkest adults, black stripe bordered below by white to gray, narrow, interrupted stripe; ventrolateral area of body dark brown with irregular gray spots and/or bars; ventral surface of body grayish white to nearly black.

Diagnosis/Similar Species.—The combination of having large platelike scales on top of the head and on the venter, moveable eyelids, femoral pores in both sexes, and granular dorsal body scales distinguishes *Holcosus festivus* from all remaining Hon-

duran lizards, except the other Teiidae. The presence of an enlarged midgular patch in *H. festivus* distinguishes it from those other Teiidae, except. *H. undulatus. Holcosus undulatus* has enlarged midgular scales that are usually in longitudinal rows, lacks a midgular patch scale much larger than the others, has a more shallow head with a HW/ HD of 0.98–1.14, and lacks a pale vertebral stripe, even in juveniles (versus enlarged midgular scales usually irregular in arrangement, 1 enlarged midgular scale much larger than others, deeper head with HW/ HD 1.07–1.46, and pale vertebral stripe present in juveniles and subadults in *H. festivus*).

Illustrations (Figs. 12, 13, 126, 128, 131, 138; Plate 79; all but Harvey et al., 2012, as *Ameiva*).—Campbell, 1998 (subadult); A. H. A. Duméril et al., 1870–1909b (adult, head scales, forelimb scales, anal plate region scales); Echternacht, 1971 (adult, subadult, head scales, precloacal and thigh region scales); Guyer and Donnelly, 2005 (adult, venter, juvenile); Harvey et al., 2012 (adult, ventral head scales, lateral body scales; cloacal region; postaxial antebrachial scales, subarticular scales, and heel scales);

Hödl, 1996 (young adult); Köhler, 1999b (adult, head scales), 2000 (juvenile, head scales), 2001b (adult, head scales), 2003a (adult, head scales), 2008 (adult, head scales); Lattanzio, 2014 (adult, juvenile); Lee, 1996 (subadult, head scales, precloacal and thigh regions scales), 2000 (subadult, head scales, precloacal and thigh regions scales); McCranie et al., 2006 (adult, head scales); Savage, 2002 (adult, juvenile, head scales); Stafford and Meyer, 1999 (subadult); Taylor, 1956b (adult).

Remarks.—Echternacht (1971) provided a thorough morphological review of *Holcosus festivus* and discussed four color pattern types. Several subspecies have been proposed for *H. festivus*, but Echternacht (1971) did not recognize any geographic subdivision or subspecies. Harvey et al. (2012: 124) placed *H. festivus* in the *H. undulatus* species group. Goicoechea et al. (2016: 48) did not recognize species groups in this genus.

See the Remarks for the Infraorder Cryptodira for a discussion about the conflicting opinions on the availability of the Lichtenstein and von Martens (1856) publication, in which *Holcosus festivus* was described. Also see Gutsche (2016) for more information on other names proposed by Lichtenstein and von Martens (1856) as being valid, available, and necessary to use.

Natural History Comments.—*Holcosus festivus* is known from near sea level to 1,400 m elevation in the Lowland Moist Forest and Premontane Wet Forest formations and peripherally in the Lowland Dry Forest formation. The Lowland Dry Forest records probably came from gallery forest. This usually diurnal and terrestrial species is most active during periods of bright sunshine when it is seen crawling on the broadleaf forest floor, usually in areas where sunlight reaches the ground. It also occurs in forest edge situations, including along rivers and edges of cleared forest. On two occasions, I saw two different *H. festivus* adults actively feeding on spiders at night at the mouth of burrows in a stream bank. It is also found inactive under logs and rocks during the night and on cloudy days. *Holcosus festivus* has been collected in every month of the year. Lattanzio (2014) studied escape behavior of a Costa Rican population. Costa Rican females deposit three or four clutches of one to four eggs a year and are reproductively active for much of the year (R. E. Smith, 1968; Fitch, 1970, 1973a). Goldberg (2009a) reported one or two oviductal eggs in females from Colombia, Costa Rica, and Nicaragua, with evidence of females producing multiple clutches per year. Vitt and Zani (1996) provided a detailed ecological study on this species in southeastern Nicaragua. Vitt and Zani (1996) found the number of vitellogenic follicles varies from two to five, thus providing an estimated clutch size. Vitt and Zani (1996) also found diet consisted primarily of crickets, spiders, roaches, and katydids. Toral (2004) also observed an *H. festivus* feeding on a small frog in Costa Rica. An adult *H. festivus* was regurgitated overnight in a collecting bag (McCranie, 2016) holding one snake, *Clelia clelia* (Daudin, 1803b: 330).

Etymology.—The name *festivus* is Latin for gay, joyous, or merry, and was apparently used in reference to the bright color pattern of this species.

Specimens Examined (163, 1 skin, 2 skeletons [54]; Map 63).—**ATLÁNTIDA**: Carmelina, USNM 62972; Colorado District, UMMZ 58376–77; mountains S of Corozal, LACM 48072–76, 48083–85, LSUMZ 21622–27; Estación Forestal CURLA, USNM 563529–31, 578850; Guaymas District, UMMZ 58378, 58382, 149450 (skull from UMMZ 58378); about 12 km SSE of La Ceiba, KU 101195–96; Lancetilla, AMNH 70448, 70456, 102568, ANSP 33140–46, UMMZ 70324 (3), USNM 578851–52; Quebrada de Oro, USNM 563532–34; about 8 km ESE of Tela, FMNH 13008; Tela, MCZ R27576. **COLÓN**: Cerro Calentura, LSUMZ

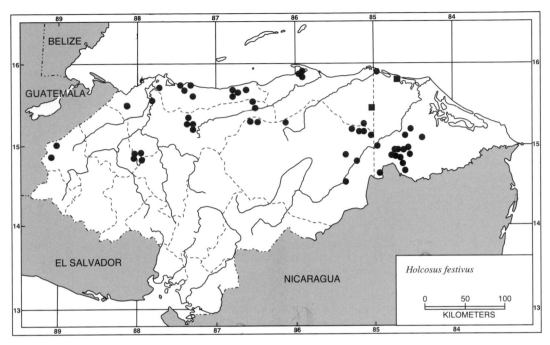

Map 63. Localities for *Holcosus festivus*. Solid circles denote specimens examined and solid squares represent accepted records.

22465; Quebrada Machín, USNM 536493 (skin, skeleton), 563535–45; about 3.2 km E of Trujillo, LSUMZ 22499; Trujillo, CM 65666; between Trujillo and Santa Fé, CM 65673–74. **COPÁN**: 19.3 km ENE of Copán, LACM 72084–85; below Quebrada Grande, SMF 79142. **CORTÉS**: Agua Azul, AMNH 70336; El Jaral, FMNH 5073; Lago de Yojoa, MSUM 4695–703; Laguna Tica-maya, FMNH 5072; Los Pinos, UF 166177, USNM 578849, 579528; W of San Pedro Sula, FMNH 5067–71, 5074. **GRACIAS A DIOS**: Bachi Kiamp, FMNH 282567, USNM 573943, FN 257006 (still in Hon-duras because of permit problems); Bodega de Río Tapalwás, USNM 563552–56; Caño Awalwás, UF 150294; Cerros de Sabaní, USNM 570387–88; Hiltara Kiamp, USNM 563546; Kalila Plapan Tingni, USNM 563547; Kipla Tingni Kiamp, USNM 563548, 573191–95; Mocorón, UTA R-46168; Palacios, BMNH 1985.1284; Río Coco, USNM 24527–28; Río Sutawala,

USNM 563551; Rus Rus, USNM 563549; Sachin Tingni Kiamp, USNM 563550; Sadyk Kiamp, USNM 573942; San San Hil Kiamp, USNM 570389–90; Sisinbila, USNM 579527; Urus Tingni Kiamp, USNM 563557–58, 565498; Warunta Tingni Kiamp, USNM 563559–60, 565499–500, 570391–92. **OLANCHO**: Caobita, SMF 80832, USNM 563563; about 40 km E of Cata-camas, TCWC 23638; Cuaca, UTA R-53193–95, 53267; between La Llorona and confluence of ríos Lagarto and Wampú, USNM 563566; Las Delicias, USNM 563564; Quebrada Las Cantinas, USNM 342390–94; Quebrada El Guásimo, SMF 80831, USNM 563565; confluence of Que-brada Siksatara and Río Wampú, USNM 563568; confluence of ríos Aner and Wampú, USNM 563561–62; confluence of ríos Sausa and Wampú, USNM 563567; confluence of ríos Yanguay and Wampú, USNM 563569–70; confluence of ríos Ya-puwás and Patuca, LSUMZ 28469; Terrero

Blanco, USNM 342395. **SANTA BÁRBARA**: 8.8 km SW of El Jaral, LACM 48077–82. **YORO**: La Libertad, USNM 570393; Montañas de Mataderos, FMNH 21781, MCZ R38924; 8.0 km W of Olanchito, UTA R-50245–47; Portillo Grande, MCZ R38925; San Francisco, MVZ 52417; Subirana Valley, MCZ R38926.

Other Records (Map 63).—**ATLÁNTIDA**: about 12 km SSE of La Ceiba, KU 101197 (Echternacht, 1971, specimen now lost). **COLÓN**: Empalme Río Chilmeca, UNAH 5459 (Cruz Díaz, 1978, specimen now lost). **GRACIAS A DIOS**: Barra Río Plátano, UNAH 5451 (Cruz Díaz, 1978, specimen now lost).

Holcosus undulatus (Wiegmann, 1834b)

Cnemidophorus undulatus Wiegmann, 1834b: 27 (three syntypes, ZMB 867–69 [see Bauer and Günther, 1994: 272]; type locality not stated, but "Mexico" inferred from title of publication containing description).

Ameiva undulata: Cope, 1862b: 63; Dunn and Emlen, 1932: 30; Meyer, 1969: 268; Wilson and Meyer, 1969: 146; Echternacht, 1970: 6; Echternacht, 1971: 40; Hahn, 1971: 111; Wilson et al., 1979a: 25; O'Shea, 1986: 43; Wilson and McCranie, 1998: 15; Köhler, 1999a: 214; Wilson et al., 2001: 134; Castañeda, 2002: 15; McCranie et al., 2002: 27; McCranie and Castañeda, 2005: 15; Castañeda, 2006: 27; Lovich et al., 2006: 14; McCranie et al., 2006: 132; Townsend, 2006a: 35; Wilson and Townsend, 2006: 105; Townsend et al., 2007: 10; Wilson and Townsend, 2007: 145; Lovich et al., 2010: 113; Townsend and Wilson, 2010b: 692.

Ameiva undulata pulchra: P. W. Smith, 1950: 55.

Ameiva undulata hartwegi: Echternacht, 1968: 152.

Ameiva festiva: Cruz [Díaz] et al., 1993: 28.

Holcosus undulatus: Harvey et al., 2012: 118; McCranie and Solís, 2013: 242; McCranie, Valdés Orellana, and Gutsche, 2013: 288; McCranie, 2014: 292; Solís et al., 2014: 133; McCranie, 2015a: 371; McCranie and Gutsche, 2016: 869.

Geographic Distribution.—*Holcosus undulatus* occurs at low, moderate, and lower portions of intermediate elevations on the Atlantic versant from southern Tamaulipas, Mexico, to southern Nicaragua and from southern Nayarit, Mexico, to central and west-central Costa Rica on the Pacific versant. It is also found on the Islas del Maíz, Nicaragua, and a few islands and cays off the coast of Belize and Quintana Roo, Mexico. In Honduras, this species occurs in open habitats at low and moderate elevations throughout most of the mainland and on a couple of islands in the Golfo de Fonseca.

Description.—The following is based on 12 males (FMNH 283706; KU 67331, 101219, 101236, 101245, 101248–49, 101263; UF 150295; USNM 24371, 563572, 580927) and ten females (KU 101216, 101218, 101221–25; SMF 77787; USNM 563576, 570396). *Holcosus undulatus* is a moderate-sized lizard (maximum recorded SVL 129 mm [Echternacht, 1971]; 127 mm in largest Honduran specimen measured [USNM 573945, a male; not used in this description]); dorsal head scales enlarged, smooth, platelike, with dorsally expanded anterior nasal scales contacting each other medially by short sutures, single frontonasal, paired prefrontals, single frontal, paired frontoparietals, 2 parietal scales plus 1 narrower interparietal scale, incomplete longitudinal furrow frequently present in interparietal scale; rostral not contacting frontonasal; postnasal contacting prefrontal; prefrontal and first superciliary separated; frontal-frontoparietal suture aligned near posterior edge of second supraocular; scales in frontoparietal region smooth, flat; 3 supraoculars (occasionally a small fourth supraocular present posteriorly); first supraocular contacting second supraocular; median pair of occipital scales not greatly enlarged, sometimes moderately enlarged, always larger than first dorsal scale row; 9–19 (14.1 ± 2.8) occipitals in 20; supratemporals moderately enlarged, separated from parietals; rostral groove absent; posterior edge of nostril passing across nasal suture, nostril opening oval; 1.0 or 1.5 rows of lateral supraocular granules, 11–32 (17.7 ± 5.2) granules on 39 sides; anterior extent of circumorbital semicircles reaches between posterior edge of, to posterior third of supraocular 3; 2–6 (3.8 ± 1.1) circumorbitals on 40 sides; 4–6 (5.3 ± 0.6) superciliaries on 31 sides, second almost always greatly elongated; 4 (rarely 5) suboculars, first divided, lower part usually contacting supralabials 3–4 (rarely contacting only third supralabial); subocular keel present; slightly enlarged scales located in front of auditory meatus; no auricular or preauricular flap or fold; 1 loreal; 5–7 (usually 6) supralabials; 5–6 infralabials; first supralabial slightly curved ventrally, shorter than second; lingual sheath present; moveable eyelid present; pupil reniform; paired chinshields, contacting infralabials, chinshields usually separated only at posterior edge by granular scales; intergular sulcus absent; 22–34 (27.8 ± 3.4) anterior gulars to point of first enlarged scale in midgular patch in 20; 7–12 (9.2 ± 1.6) posterior gulars from ultimate enlarged scale in midgular patch in 20; midgular patch of distinctively enlarged scales present, usually arranged in a central, straight row, no midgular scale much larger than others; intertympanic sulcus absent; sharp transition from anterior gulars to smaller posterior gulars; mesoptychial scales moderately enlarged, differentiated, bordered anteriorly by sharp transition to small scales; edge of gular fold not serrated; dorsal scales conical, not keeled, 220–289 (252.6 ± 18.4) granular middorsals between occipitals and first enlarged caudal scale in 20; 135–179 (155.3 ± 11.7) granules around midbody between enlarged ventrals; middorsal gran-

ules subequal in size to lateral scales; chest scales large, flat; pectoral sulcus absent; ventral body scales large, platelike, squarish, juxtaposed, smooth, in 27–32 (29.5 ± 1.5) rows between gular fold and level of hind limb, in 8 transverse rows at midbody, 1–2 progressively smaller scales present immediately lateral to outside ventral plate; paired, slightly enlarged terminal scales forming precloacal plate, those enlarged scales flanked by single small subtriangular scale, that scale flanked by granular scales; paired precloacal plates usually slightly smaller than scale anterior to them; 4–10 (5.9 ± 1.3) precloacal scales in 19; precloacal spur and postcloacal button absent; pair of slightly enlarged postcloacal scales present in males, those enlarged scales absent in females; caudal annuli complete, tail lacking dorsolateral crests; dorsolateral row of serrated caudal scales present (laterally projecting mucrons and heavily keeled); proximal subcaudals smooth; large dorsal brachial and antebrachial scales continuous; enlarged dorsal scales on brachial and antebrachial surfaces 1.5–2.0 times as wide as long, both sets of enlarged scales smooth, large antebrachial scales extending well beyond center of forearm, large brachial scales extending only slightly beyond center of upper arm; postaxial antebrachial scales noticeably enlarged ventrolaterally; postaxial patch of enlarged scales present on brachial surface; 13–18 (15.5 ± 1.0) heterogeneous, but projecting or swollen, subdigital scales on 40 sides of Finger IV; 24–36 (30.2 ± 2.8) combined subdigital scales on 40 sides of Toe IV, distal ones smooth; no row of distinctively enlarged scales between toes IV–V; small scales separating supradigital scales from subdigital scales restricted to phalangeal articulations; denticulate fringe present along postaxial edge of outer toe; Toe V reduced, claw reaching only to point of articulation of toes III–IV; 5–10 (7.4 ± 1.4) prefemoral scales in 17; heel with moderately enlarged scales; tibiotarsal shield and spur absent; 34–45 (39.8 ± 3.5) total femoral-abdominal pores in males, 30–44 (35.7 ± 4.4) in females; no gap between femoral and abdominal pores; 7 or 8 medial scales separating each set of femoral and abdominal pores from each other; SVL 83.0–117.0 (97.6 ± 10.4) mm in males, 83.0–100.0 (89.8 ± 4.6) mm in females; TAL/SVL 1.64–2.31 in eight males, 1.69–2.24 in eight females; HL/SVL 0.22–0.28 in males, 0.23–0.27 in females; HW/SVL 0.13–0.17 in 11 males, 0.13–0.16 in nine females; HD/SVL 0.13–0.16 in 11 males, 0.12–0.15 in nine females; HW/HD 0.98–1.11 in 11 males, 1.00–1.14 in nine females; SL/SVL 0.11–0.12 in 11 males, 0.10–0.11 in nine females; SHL/SVL 0.22–0.27 in males, 0.22–0.25 in females; foot L/SVL 0.31–0.42 in 11 males, 0.30–0.40 in nine females; hand L/SVL 0.11–0.16 in 11 males, 0.13–0.17 in nine females.

Color in life of an adult male (USNM 563572): middorsal swath Cinnamon-Brown (33) with scattered irregular Sepia (219) spots, except for Sepia (119) neck; lateral area of body with complex pattern of spots on variable background, with dorsalmost portion of body with Lime Green (59) large spots in line on Sepia (119) background, dorsalmost portion of body with row of Lime Green spots followed below by row of smaller Straw Yellow (56) spots on Sepia (119) background, attached to bottom of those spots are vertically elongated Lime Green bars above, grading to Sky Blue (66) below; top of head Grayish Horn Color (91), grading laterally to Trogon Yellow (153) on supralabials, chin, and throat; underside of body pale turquoise blue; dorsal surfaces of fore- and hind limb Sepia (219) with extensive Amber (36) mottling; underside of forelimb dirty cream; venter of hind limb pale turquoise blue; dorsal surface of tail Hazel (35) with scattered Sepia (119) mottling, grading to Turquoise Blue (65) ventrolaterally and pale turquoise blue midventrally.

Color in alcohol: dorsal surface of body with indistinct, broad yellowish brown to brown middorsal swath confluent with color of top of head, swath usually continuing posteriorly to at least base of tail; middorsal swath bordered laterally by narrow, white to gray dorsolateral stripe, stripe originating above eye and extending onto tail, with dorsolateral stripe usually disappearing in larger specimens; dorsolateral pale stripe bordered below by alternating, irregular, dark brown to black and yellowish cream vertical bars in males, females have dark brown to black swath bordered below by narrow, bluish gray lateral stripe instead of vertical bars, with that pattern fading in larger specimens; ventral surface of body bluish white to bluish gray in males, bluish white to nearly black in females. Juveniles have the dorsolateral pale stripe and upper lateral pale stripe interrupted, without a lateral pale stripe.

Diagnosis/Similar Species.—The combination of large platelike scales on top of the head and ventral surfaces, moveable eyelids, femoral pores in both sexes, and granular dorsal body scales distinguishes *Holcosus undulatus* from all remaining Honduran lizards, except the other Teiidae. The presence of an enlarged patch of midgular scales distinguishes *H. undulatus* from all other Teiidae, except *H. festivus*. *Holcosus festivus* has those enlarged midgular scales irregularly arranged and never in distinct longitudinal rows, has 1 greatly enlarged scale in the midgular patch, has a HW/HD of 1.07–1.46, and has a pale vertebral stripe in all but the largest specimens (versus central gular scales in more-or-less regular longitudinal rows, no greatly enlarged midgular scale, a HW/HD 0.98–1.14, and pale vertebral stripe absent in *H. undulatus*).

Illustrations (Figs. 125, 132; Plate 80; all but three as *Ameiva*).—Acevedo, 2006 (adult); Álvarez del Toro, 1983 (adult); Álvarez Solórzano and González Escamilla, 1987 (adult); Calderón-Mandujano et al., 2008 (adult); Campbell, 1998 (adult, subadult; head scales); A. H. A. Duméril et al., 1870–1909b (adult, head scales, forelimb

Plate 80.   *Holcosus undulatus.* USNM 579690. Valle: Amapala, del Tigre Island, at summit.

scales, anal plate region scales); Echternacht, 1971 (adult, subadult, head scales, precloacal and thigh region scales); Harvey et al., 2012 (adult, dorsal head scales, postaxial and antebrachial scales; as *Holcosus*); Köhler, 1999b (head scales), 2000 (adult, head scales), 2001b (adult, head scales), 2003a (adult, head scales), 2008 (adult, head scales); Köhler et al., 2005 (adult, head scales); Lee, 1996 (adult, head scales, precloacal and thigh scales), 2000 (adult, head scales, precloacal and thigh scales); Lemos-Espinal, 2015 (adult); Lemos-Espinal and Dixon, 2013 (adult; as *Holcosus*); McCranie et al., 2006 (adult, head scales); Mertens, 1952b (adult); Savage, 2002 (adult, head scales); Stafford and Meyer, 1999 (adult, subadult); Sunyer, Nicholson et al., 2013 (adult; as *Holcosus*).

Remarks.—Echternacht (1971) provided a thorough morphological review of *Holcosus undulatus* (as *Ameiva*) and concluded that most meristic differences among populations are clinal. Thus, Echternacht (1971) did not discuss subspecies, even though as many as 11 subspecies have been proposed for this wide-ranging species (Smith and Taylor, 1950b: 171, listed ten alone for Mexico). Harvey et al. (2012) provided an analysis of the Teiidae based on morphology and resurrected the generic name *Holcosus* for ten species previously placed in *Ameiva* (p. 118). Those authors also placed *H. undulatus* in the *H. undulatus* species group (p. 124). Goicoechea et al. (2016), using molecular data only, also recovered a monophyletic *Holcosus* but did not recognize any species groups within this genus.

Gray (1838a: 277) transferred *Cnemidophorus undulatus* Wiegmann (1834b) to *Ameiva*, where it remained until Harvey et al. (2012) included it in the resurrected genus *Holcosus*.

Natural History Comments.—*Holcosus undulatus* is known from near sea level to 1,240 m elevation in the Lowland Moist Forest, Lowland Dry Forest, Lowland Arid Forest, Premontane Moist Forest, and Premontane Dry Forest formations and peripherally in the Premontane Wet Forest formation. The species occurs in open areas in the mesic forest formations. It is diurnal and terrestrial and is active only during sunny periods. It is usually seen in grassy, sandy, or rocky areas; however, one was

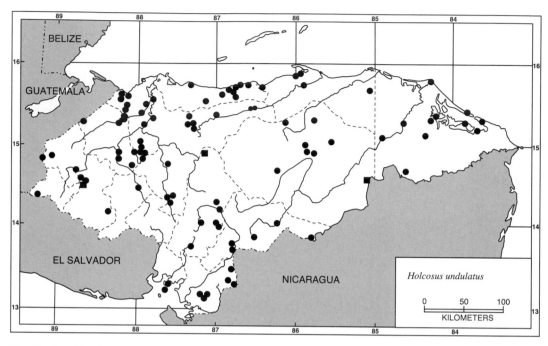

Map 64.   Localities for *Holcosus undulatus*. Solid circles denote specimens examined and solid squares represent accepted records.

basking on top of a rock wall. Retreats include under logs, rocks, and leaves, in holes in the ground, and in rock walls. It is probably active on sunny days throughout the year because it has been collected in every month of the year except January and March. Females from Costa Rica, Nicaragua, and the Yucatán Peninsula of Mexico are reported to deposit several clutches of one to seven eggs per year (Fitch, 1970, 1973a, 1985). Breeding season corresponds to wetter parts of the year (Fitch, 1985). Goldberg (2009a) found four oviductal eggs in one Costa Rican female collected in July. Lee (1996: 259) said *H. undulatus* feeds "predominately on invertebrates, especially leaf litter insects," and Savage (2002: 514), based on a literature review, said "food consists of small arthropods."

Etymology.—The name *undulatus* is derived from the Latin *undul* (wavy) and – *anus* (belonging to), and probably refers to

the "wavy" dorsolateral pattern of the species.

Specimens Examined (411, 3 skeletons, 1 hemipenis [235]; Map 64).—**ATLÁNTIDA**: mountains S of Corozal, LACM 48104–07, LSUMZ 21713; Corozal, LACM 72086; 1 km W of La Ceiba, KU 101227; 2 km SE of La Ceiba, KU 101228–38; 8 km SE of La Ceiba, KU 101239–42; 12 km SSE of La Ceiba, KU 101252–53; La Ceiba, INHS 4487; Los Planes, UTA R-41266; near Pico Bonito Lodge, USNM 579535–36; Piedra Pintada, LACM 48125, LSUMZ 43219 (hemipenis only); along Río Viejo, USNM 563571; San Marcos, USNM 570394–96. **CHOLUTECA**: La Caguasca, FN 256962 (still in Honduras because of permit problems); Cerro Guanacaure, SDSNH 72693, 72696; Finca Guayabal, UNAH 5180; Finca Monterrey, USNM 579539; La Fortuna, SDSNH 72692, 72697, UNAH 5168, 5171, 5179, 5183, 5189; La Isnaya, SDSNH

72695, UNAH 5160, 5177; La Pacaya, SDSNH 72694. **COLÓN**: Balfate, AMNH 58607–09; Río Tulito, BMNH 1985.1285; Salamá, USNM 242606–09, 242611, 242614–26, 242638–41; 1–3 km W of Trujillo, KU 101243–47; 2 km W of Trujillo, KU 101248; 0.5 km W of Trujillo, LSUMZ 21714; 2 km E of Trujillo, LACM 48108, 72087, LSUMZ 21715–16; 1 km SSW of Trujillo, KU 101249–51; about 3.2 km E of Trujillo, LSUMZ 22497–98; Trujillo, LSUMZ 22427; between Trujillo and Santa Fé, CM 65671–72. **COMAYAGUA**: 8.7 km S of Comayagua, LSUMZ 24199; La Libertad, MCZ R38923; Las Mesas, UTA R-41244; 6 km ESE of Villa San Antonio, LACM 72093. **COPÁN**: 1 km S of Copán, USNM 563572; 2 km NE of Copán, LSUMZ 24622; Copán, FMNH 28525–30, 282561, UMMZ 83035 (16); Río Amarillo, USNM 579529; Río Higuito, ANSP 22195–98. **CORTÉS**: Agua Azul, MCZ R49964, USNM 243336–39; Amapa, AMNH 70458; 3 km WSW of Cofradía, KU 67330–31; Cofradía, CM 65668; between Cofradía and Buenos Aires, SMF 77787; 1.6 km NW of El Jaral, LSUMZ 11648; 1.6 km W of El Jaral, LACM 48113–15, 48117–18, USNM 578853; 3.2 km SE of El Jaral, LSUMZ 11649; 1.6 km SE of El Jaral, LACM 48116, 48119, LSUMZ 22514, 52617; El Jaral, FMNH 5089–90; El Paraíso, UF 144749–50; Hacienda Santa Ana, FMNH 5075–87, 5148–51; 7 km SW of La Lima, KU 67333; 6.4 km NE of La Lima, LACM 48121–24; E side of Lago de Yojoa, KU 67332; Laguna Ticamaya, FMNH 5088 (16), TNHC 32085–87; Los Pinos, UF 166178, USNM 563573, 573197; Quebrada Agua Buena, FMNH 283768; San Luís, LSUMZ 72860; 3.2–4.8 km W of San Pedro Sula, LACM 48126–28, MCZ R29387–88; San Pedro Sula, USNM 24371–73; near Santa Elena, LSUMZ 38836; Tegucigalpita, USNM 563574; 6 km ENE of Villanueva, CM 65669, LSUMZ 52618; 7.2 km ENE of Villanueva, LACM 48120. **EL PARAÍSO**: Arenales, LACM 39777 (formerly UCLA 14759); El Rodeo, USNM 580924; Mapachín, USNM 579532, 579538; 1 km S of Orealí, USNM 580927; Orealí, USNM 580925–26; Valle de Jamastrán, AMNH 70333–35. **FRANCISCO MORAZÁN**: near Agua Amarillo, AMNH 70368; between Cantarranas and Talanga, UF 90053; Cantarranas, ANSP 22199–212; near El Zamorano, UF 124795; El Zamorano, AMNH 70466, 70470–73, KU 101260–67, MCZ R49766–67; 51.5 km NNE of Jícaro Galán, TNHC 32089; Río Yeguare, AMNH 70369–79; near Tegucigalpa, LSUMZ 24138; Tegucigalpa, BYU 18178, 18180, 18191, 18227–30, 18279, 18282, FMNH 5091–93, KU 192325, MCZ R100001–03, MSUM 4690. **GRACIAS A DIOS**: Barra Patuca, USNM 20310–13; Cauquira, USNM 565502–03; Krausirpe, LSUMZ 52512; Mavita, USNM 579530; Mocorón, UTA R-46174; Puerto Lempira, LACM 48093; Rus Rus, UF 150295, 150300, USNM 563575–76; Samil, USNM 573944, 579533–34; Tánsin, LACM 48094, LSUMZ 21717 (listed as from 2 km E of Trujillo, Colón, by Meyer and Wilson, 1973); Wampusirpe, LSUMZ 52511; Warunta, USNM 565501; Yahurabila, USNM 573196. **INTIBUCÁ**: 3.0 km N of Jesús de Otoro, LSUMZ 33696–97. **LEMPIRA**: Erandique, LSUMZ 52614–16; 11.3 km NNW of Gracias, LACM 48109–12; Gracias, FMNH 40865; near Gracias, UF 124748. **OCOTEPEQUE**: Río Lempa near Antigua, FMNH 283706. **OLANCHO**: Babilonia, USNM 580923; 0.5–1.0 km WNW of Catacamas, LACM 48086, LSUMZ 21696–99; 2–3 km NW of Catacamas, LACM 48087–89, LSUMZ 21700–03; 4.5 km SE of Catacamas, LACM 48090–92, LSUMZ 21704–06; 6.5 km SE of Catacamas, LSUMZ 21707–08; Catacamas, LSUMZ 52513; Cuaca, USNM 579537, UTA R-53190; El Carbón, USNM 579531; 12.2 km SW of Juticalpa, LACM 45235–37; Pataste, MSUM 4691–94. **SANTA BÁRBARA**: Cerro Negro, USNM 573945–46; La Canadá, UTA R-41245; 6.4 km SW of Quimistán, KU 67326–29; near

Quimistán, USNM 128091–92; about 9 km N of Santa Bárbara, LACM 45323; 8.0 km S of Santa Bárbara, LACM 45308–09; San Pedro Zacapa, TNHC 32611–12. **VALLE**: Coyolitos, FN 256884–85 (still in Honduras because of permit problems); Isla del Tigre, near summit, USNM 579690. **YORO**: 2 km S of Coyoles, KU 101216–26, 101254–59, 107911–14, 109973–74, 119299–301 (skeletons); 0.5 km N of Coyoles, LACM 48095–97, LSUMZ 21709; 5 km E of Coyoles, LACM 48098–103, LSUMZ 21710–11; Coyoles, CM 29354; El Progreso, UMMZ 58375; Rancho San Lorenzo, LSUMZ 21712. "HONDURAS": ANSP 9066–67, UF 51244 (skeleton), 91817–18, USNM 16429.

Other Records (Map 64).—**FRANCISCO MORAZÁN**: Marale (Townsend et al., 2007). **LEMPIRA**: Villa Verde, UNAH 2506, 2522 (Cruz [Díaz] et al., 1993). **OLANCHO**: Quebrada Kuilma (personal sight record).

ORDER CROCODYLIA WAGLER, 1830

The Order Crocodylia contains the alligators and caimans (Superfamily Alligatoroidea Gray, 1844:56; as Alligatoridae; Savage, 2017: 112 inacurately listed p. 195 as where Gray introduced that name), crocodiles (Superfamily Crocodyloidea Oppel, 1811b: 16; as Crocodilini; Savage, 2017: 112 listed p. 21 as where Oppel introduced that name, but it was first occupied in Oppel's diagnostic key on p. 16), and gavials (Gavialoidea Adams, 1854: 70; as Gavialidae; Savage, 2017: 112 listed p. 20 as where Gray introduced that name). All Crocodylia have "a robust skull, a long snout and strongly toothed jaws, a short neck, a robust cylindrical trunk extending without constriction into a thick laterally compressed tail, and short but strongly developed limbs" (Vitt and Caldwell, 2014: 545). Thus, they are easily recognizable in Honduras. Only the Alligatoroidea and Crocodyloidea occur in Honduras, each with one family, genus, and species in the country. All living members of this order are oviparous.

Remarks.—Owen (1842) has frequently been credited with the order name Crocodylia, but Savage (2017) concluded that Wagler (1830a) was the first author to validate an order name for the crocodilians (p. 130; as Crocodili).

According to Vitt and Caldwell (2014: 545), "crocodyles" is the "technically correct" popular spelling and not the usually spelled "crocodiles."

KEY TO HONDURAN SUPERFAMILIES AND FAMILIES OF THE ORDER CROCODYLIA

1A. Fourth mandibular tooth fitting into pit in upper jaw, tooth not visible when mouth closed (Fig. 139); elevated crescent-shaped, transverse preorbital ridge present on top of head (Fig. 140); snout broadly rounded in dorsal aspect (Fig. 140); ventral scales without apical pits (Fig. 141) . . . . Alligatoroidea-Alligatoridae (p. 424)

1B. Fourth mandibular tooth fitting into open groove in upper jaw, tooth visible when mouth closed (Fig. 142); no transverse preorbital ridge (Fig. 143); snout somewhat elongated in dorsal aspect (Fig. 143); ventral scutes with a single apical pit per scute (Fig. 144). . . . .. Crocodyloidea-Crocodylidae (p. 430)

CLAVE PARA LAS SUPERFAMILIAS Y FAMILIAS HONDUREÑAS DE ORDEN CROCODYLIA

1A. Cuatro diente de la mandíbula se aloja en una cavidad en la maxila, diente no es visible cuando el hocico está cerrado (Fig. 139); una cresta preocular en la parte superior de la cabeza presente (Fig. 140); hocico ampliamente redondeado en perfil dorsal (Fig. 140); escamas ventrales sin fosetas

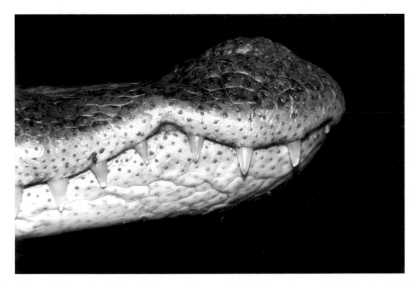

Figure 139.   Fourth mandibular tooth fitting into pit (not notch) on each side of upper jaw, each fourth tooth not visible when mouth closed. *Caiman crocodilus*. UF 120738 from Río Seco, Gracias a Dios.

apicales (Fig. 141) . . . .
. . . . Alligatoidea-Alligatoridae (p. 424)

1B. Cuatro diente de la mandíbula se adpata a una escotadura en la maxila, diente visible cuando el hocico está cerrado (Fig. 142); sin una cresta preocular (Fig. 143); hocico algo elongado en perfil dorsal (Fig. 143); placas ventrales con una foseta apical en cada placa

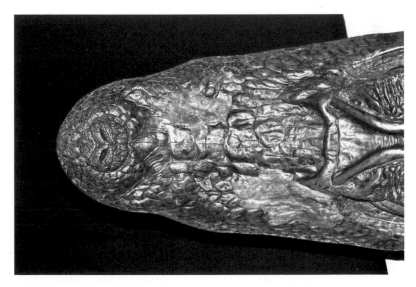

Figure 140.   Transverse ridge present just anterior to eyes on top of head and snout broadly rounded in dorsal aspect. *Caiman crocodilus*. UF 120738 from Río Seco, Gracias a Dios.

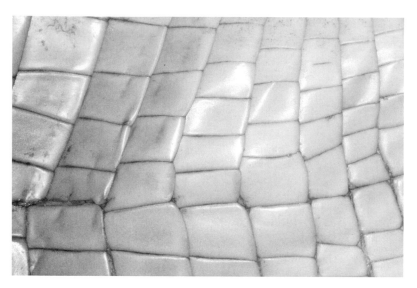

Figure 141. Ventral scales without apical pits. *Caiman crocodilus.* UF 120738 from Río Seco, Gracias a Dios.

(Fig. 144)......................
.. Crocodyloidea-Crocodylidae (p. 430)

Superfamily Alligatoroidea, Gray, 1844

Family Alligatoridae Gray, 1844

The superfamily Alligatoroidea and family Alligatoridae include the alligators and caimans and their living relatives, with only the caimans occurring in Honduras. The single Honduran member of the Alligatoridae is most easily distinguished from the only remaining crocodilian species in Honduras (*Crocodylus acutus*), by having a preorbital bony transverse ridge on top of the head and the fourth tooth of the lower

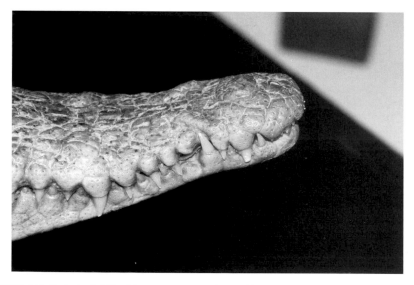

Figure 142. Fourth mandibular tooth fitting into an open groove (not a pit) on each side of upper jaw, fourth tooth visible when mouth closed. *Crocodylus acutus.* UF 20561 from Florida, USA.

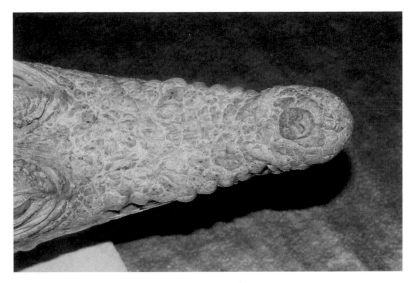

Figure 143.   No transverse ridge on top of head in front of eyes and snout relatively elongated in dorsal aspect. *Crocodylus acutus*. UF 20561 from Florida, USA.

jaw fitting into a pit on the upper jaw and not visible when the mouth is closed.

The living members of the Alligatoroidea and Alligatoridae range in the Western Hemisphere throughout most of the southeastern U.S., and from extreme southern Mexico to northern Argentina. In the Eastern Hemisphere, Alligatoridae is found only in the Yangtze River drainage of eastern China. Four genera and nine named species are included in this family, with one species occurring in Honduras.

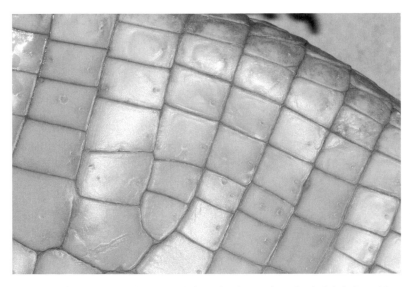

Figure 144.   Ventral scutes with a single apical pit per scale (occasional scutes have 2 apical pits). *Crocodylus acutus*. UF 20561 from Florida, USA.

Remarks.—The Alligatoroidea was placed on the Official List of Family Group Names in Zoology by the ICZN (Hemming, 1958: 91; also see Melville and Smith, 1987: 4).

Genus *Caiman* Spix, 1825

Caiman Spix, 1825: 3 (type species: *Caiman fissipes* Spix, 1825: 4 [= *Crocodilus latirostris* Daudin, 1801: 417], by subsequent designation of Schmidt, 1928b: 207).

Geographic Distribution and Content.— This genus disjunctly ranges from southern Mexico (southwestern Chiapas east of the Isthmus of Tehuantepec) to western Ecuador west of the Andes on the Pacific versant and northern Honduras to northeastern Argentina east of the Andes on the Atlantic versant. It also occurs on Trinidad and Tobago, and one species has been introduced on Isla de Juventud, Cuba, on Puerto Rico, and in southern Florida, USA. Four named species are recognized, one of which occurs in Honduras.

Etymology.—A. M. C. Duméril and Bibron (1836) applied the vernacular name Caiman to crocodilians from the southeastern U.S., the Dominican Republic, etc.; thus, the vernacular name caiman has been applied to Western Hemisphere alligator-like relatives for many years.

Caiman crocodilus (Linnaeus, 1758)

Lacerta crocodilus Linnaeus, 1758: 200 (lectotype, ZMUU specimen indicated by Lönnberg, 1896: 9, as the "type," as designated by Hoogmoed and Gruber, 1983: 379; type locality unknown, but likely from the vicinity of Paramaribo, Surinam [see Duellman, 2012: 88]).

Caiman crocodilus: Andersson, 1900: 5; Meyer, 1969: 195; Meyer and Wilson, 1973: 7; Cruz Díaz, 1978: 36; O'Shea, 1986: 31; Thorbjarnarson, 1992: 67; Busack and Pandya, 2001: 298; Wilson et al., 2001: 134; Castañeda, 2002: 36; McCranie et al., 2002: 25; Wilson and McCranie, 2003: 59; McCranie et al., 2006: 92; Wilson and Townsend, 2006: 104; Solís et al., 2014: 129; McCranie, 2015a: 364; McCranie and Gutsche, 2016: 887.

Caiman crocodilus fuscus: Klein, 1979: 5.
Caiman crocodilus chiapasius: King et al., 1990: 314.
Crocodilus acutus: McCranie, 2007a: 181.

Geographic Distribution.—*Caiman crocodilus* occurs at low elevations from southwestern Chiapas, Mexico, to extreme northwestern El Salvador and from northwestern Costa Rica to western Ecuador on the Pacific versant and from northern Honduras to eastern Peru and central and eastern Brazil on the Atlantic versant. It also occurs on the islands of Trinidad and Tobago and is introduced and established on Cuba and Isla de Juventud, Cuba, on the Puerto Rican Bank, and in southern Florida, USA. In Honduras, this species is known from the vicinity of the north coast and inland along some Caribbean river systems.

Description.—The following is based on five subadults (UF 120732–33, 120738, 120745, 120747; except where otherwise noted). *Caiman crocodilus* is a small to medium-sized crocodilian that can reach a maximum length of about 2.7 m (Medem, 1983), but adults usually reach only about 125–175 cm TL, with large males weighing up to 65 kg (Savage, 2002 in a literature review); hatchlings 21–26 cm TL (Brazaitis, 1974); snout relatively short, broad, with gradually converging sides, broadly rounded in dorsal aspect; fourth mandibular tooth fits into pit in upper jaw, not visible when mouth closed; prominent transverse, bony preorbital ridge present; upper eyelid raised into high point or large tubercles; 2 transverse rows of 3–4 scutes in each row with a total of 6–7 (6.4 ± 0.5) irregularly arranged postoccipital scutes; 4–5 (4.2 ± 0.4) transverse rows of 2–3 nuchal scales in each row with a total of 6–10 (7.8 ± 1.5); nuchal scales continuous with dorsal scutes;

Plate 81.   *Caiman crocodilus*. Adult. Gracias a Dios: Bachi Kiamp (released).

ventral collar consists of single transverse series of enlarged scales; 17–18 (17.2 ± 0.4) longitudinal rows of dorsal scutes; 6–10 (8.0 ± 1.6) dorsal transverse rows at midbody; 21–24 (22.8 ± 1.0; counted on both left and right sides) longitudinal rows of imbricate ventral scales, 10–12 (11.2 ± 0.8) scales in each row at midbody; ventral scales lack apical pits, but have double osteoderm buttons; 13–14 double rows of dorsal crested caudal whorls and 23–24 single rows of dorsal crested caudal whorls; subcaudals uniform, uninterrupted; webbing absent on forelimb, extensive on hind limb; dental formula 5+14–15/18–20 (Brazaitis, 1974).

Color in life of an adult (Plate 81; Bachi Kiamp, Gracias a Dios): dorsal surface of body Cinnamon (123A) with Burnt Umber (22) markings and several suggestions of Burnt Umber crossbands; top of head Raw Umber (23) with Burnt Umber markings; tail yellowish brown with pale Burnt Umber markings and crossbands; dorsal surfaces of fore- and hind limb Raw Umber with Burnt Umber markings; ventral surfaces of head, body, and tail yellowish brown; iris greenish brown with dark brown mottling. Color in

life of an adult (Río Warunta, near Hiltara Kiamp, Gracias a Dios): dorsal surfaces Olive-Brown (28) with few black scutes on body and black crossbands on tail; lateral surface of body reticulated with pale and dark brown, some scutes also black; subcaudal surface creamy brown; iris brown with paler brown reticulations. In general, adult dorsal coloration is pale brown or olive-brown with ventral surfaces uniformly yellow or creamy white. Dark dorsal and caudal bands usually present, with those markings fading with age. Color in life of a juvenile (4.7 km ESE of San Lorenzo Arriba, Yoro): dorsal surfaces of head, body, fore- and hind limb, and tail Grayish Horn Color (91) with Jet Black (89) irregular small streaks; venter Pale Horn Color (92); iris Sulfur Yellow (157) with black reticulations.

Color in alcohol: dorsal surfaces of head, body, and tail dark brown with darker brown to black spotting and mottling; lower lateral surface of upper jaw pale brown with dark brown spotting; lower jaw and chin pale brown with medium brown spotting; neck and ventral surface of body nearly uniform pale brown; subcaudal surface slightly darker brown than belly; dorsal surfaces of all digits dark brown with brownish black spots; webbing on hind foot with uniform dark brown scales and slightly paler brown skin.

Diagnosis/Similar Species.—*Crocodylus acutus* has the fourth mandibular tooth fitting into a notch on the upper jaw, and that tooth is visible when the mouth is closed and lacks a bony ridge between the anterior borders of the eyes (versus fourth mandibular tooth fitting into socket on upper jaw and not visible when mouth closed, and bony ridge present between anterior borders of eyes in *Caiman crocodilus*).

Illustrations (Figs. 139–141; Plate 81).— Álvarez del Toro, 1974 (adult, subadult, juvenile; as *C. sclerops* [Schneider, 1801: 162]), 1983 (adult, subadult); Brazaitis, 1974 (head); Busack and Pandya, 2001 (head, lateral body scale arrangement); Conant and Collins, 1998 (subadult); Guyer and Donnelly, 2005 (juvenile, head); Hödl, 1996 (subadult); King and Brazaitis, 1971 (dorsal and ventral views of hides); Köhler, 2000 (adult, head), 2001b (juvenile, head), 2003a (adult, juvenile, head and forebody scales), 2008 (adult, subadult, juvenile, head and forebody scales); Köhler et al., 2005 (juvenile, head); McCranie et al., 2006 (juvenile); Medem, 1962 (adult; as *C. sclerops*), 1981 (adult, juvenile, penis; as *C. sclerops*, *C. sclerops apaporiensis*, *Lacerta crocodilus* [in part]), 1983 (adult, skull; as *C. sclerops*); Mertens, 1952b (adult); Meshaka et al., 2004 (subadult head); F. D. Ross and Mayer, 1983 (drawing of dorsal scutes); J. P. Ross, 1998a (adult); Savage, 2002 (adult, bony ridge on head, anterior body scales); Trutnau and Sommerlad, 2006 (adult, head); Villa et al., 1988 (subadult); Vitt and Caldwell, 2014 (subadult).

Remarks.—Busack and Pandya (2001: 306) concluded that subspecies of *Caiman crocodilus* should not be recognized, but Venegas-Anaya et al. (2008: 619) gave an opposing view based on molecular data and

on conservation issues. Busack and Pandya (2001: 307) also elevated *C. yacare* Daudin (1801: 407), which previously was considered a subspecies of *C. crocodilus* by most workers, to full species status.

Ross, J. P. (1998b: 245) discussed discrepancies between the number of *Caiman* skins reported to CITES exported from Honduras and the published information on *Caiman* population sizes in the country. These figures are also at variance with the stated government policy on exports from Honduras.

Escobedo-Galván et al. (2015) discussed sources for inaccurate statements regarding southeastern Oaxaca, Mexico, as part of the natural range of *Caiman crocodilus* on the Pacific versant. Those authors also performed nocturnal crocodilian searches in southeastern Oaxaca and concluded that *Caiman* is not known from Oaxaca, although there are valid locality records for *Caiman* in adjacent southwestern Chiapas, Mexico.

Natural History Comments.—*Caiman crocodilus* is known from sea level to 280 m elevation in the Lowland Moist Forest and Lowland Arid Forest formations. It is nocturnal and occurs in freshwater and brackish water conditions in rivers, swamps, and lagoons. I have never seen it basking on shores of those habitats, but I have seen several adults walking on the bottoms of small rivers in the day during the dry season when that water was clear. The species usually occurs near the coast but is occasionally seen well inland in such places as the headwaters of the Río Rus Rus system, Gracias a Dios; upper and middle reaches of the Río Warunta, Gracias a Dios, including several seasonal lagoons away from the river; and middle portions of the Río Aguán Valley, Yoro. Several juveniles were seen at night in freshwater savannas and lagoons in most of those inland populations. Those sightings occurred in the months of May, July, August, and November. I also saw and photographed a subadult in a swamp near San Lorenzo Abajo, Yoro, in June, and another, about 1 m in length, was in the nearby Río San Lorenzo in March. An adult female and about 20 young were seen on the night of 6 October 2010 in a small lagoon near the confluence of the Río San Lorenzo with the Río Aguán in Yoro. That adult female was very timid and did not respond to the distress calls of her young. Another 8–10 subadults to adults were seen under a stilt house over the Laguna Warunta at night in July 2008 at Samil, Gracias a Dios. Nobody was in that house at the time, but the owner was said to feed fish to the caiman when tourists were at the locality. One subadult was seen in May 2003 in the Río Tapalwás, a Río Rus Rus tributary, in turn a Río Coco tributary. An adult female (appeared to be about 2.5 m TOL) and eight of her young were seen well upstream in the Río Warunta in May 2005. That adult female was also timid and did not respond to the distress calls of her young.

Klein (1979) and King et al. (1990) reported on their survey work on *Caiman crocodilus* in various parts of Honduras (also see Thorbjarnarson, 1992). The species apparently retains healthy populations at several Caribbean coastal localities (Klein, 1979; King et al., 1990), although survey work in 2000 by F. E. Castañeda (personal communication) indicated that some of those populations had declined drastically in recent years. The recent, uncontested destruction of previously pristine rainforests of the Río Rus Rus headwaters has had a drastic effect on *Caiman* populations in that region. Although illegal, destruction of previously pristine forests has recently occurred along several rivers and further inland in eastern Olancho and adjacent Gracias a Dios. Remains of butchered *Caiman* were commonly found at Laguna de Guaimoreto, Colón, in the 1980s and 1990s.

There are no records or published reliable sight records for *Caiman* along the Pacific versant of Honduras, despite statements in the literature of its occurrence there (Smith and Smith, 1977; Luxmoore et al., 1988). Lovich et al. (2006) reported seeing the eye shine of a caiman in the freshwater lagoon, Laguna de Agua, on the property of the Honduran naval base on Isla del Tigre in the Golfo de Fonseca in January 2006; however, that sighting certainly represents *Crocodylus acutus*, which is known to occur in that lagoon (personal communication of Naval Officer at Base). Additionally, King and Cerrato (1990) did not find *Caiman* on their crocodilian surveys on the Pacific versant in the south of Honduras.

Escobedo-Galván et al. (2010) provided a brief literature review concerning the changing population dynamics of this species in Honduras. Ferguson (1985) provided a literature summary of various parameters of *Caiman crocodilus* eggs both in captivity and in natural conditions. Savage (2002), in a literature review, wrote females deposit about 15–40 eggs from mid-June to August in a mound nest she constructs of plant material and soil. The female tends to and repairs her nest and opens it when the young vocalize upon hatching. She might also assist her young to the nearest water, with her young remaining in her vicinity for up to at least 4 months. Males are also sometimes involved in nest tending and opening of nests. Savage (2002), in his literature summation, did not cite the Thorbjarnarson (1994) extensive reproductive study on *Caiman* in the Venezuelan llanos, although Savage cited other, less extensive works on South American populations. Thorbjarnarson found a mean clutch size of 22.2 eggs and wrote an enormous amount of data on reproductive ecology in those populations. Ellis (1980) found a *Caiman* nest with 37 eggs in southern Florida, USA (month not given). Nothing has been reported on feeding behavior of this species in Honduras and little in detail has been published for all of Central America, so the following feeding data are largely based on studies of South American populations. Magnusson et al. (1987) found

Amazonian *C. crocodilus* to eat freshwater crabs, shrimp, mollusks, fish, terrestrial invertebrates, and vertebrates, with small caimans eating mostly invertebrates and large individuals mostly eating terrestrial or nonfish vertebrates and fish. Thorbjarnarson (1993) studied the diet of *Caiman* in Venezuela and found fish, mammals, freshwater snails, and freshwater crabs were the most important prey. He also noted that diet shifted both ontogenetically and seasonally, with snails and crabs consumed principally during the rainy season and mostly fish during the dry season. Smaller *Caiman* ate more insects than did larger ones. Da Silveira and Magnusson (1999) found similar prey items, as did Magnusson et al. (1987) and Thorbjarnarson (1993), but Da Silveira and Magnusson (1999) also found snakes in a few stomachs in central Amazonian caimans, with fish being the largest items consumed by mass. Grant et al. (2009) noted opportunistic feeding on suffocating fish in drying pools of water in Costa Rica after seasonal flooding of rivers during the heaviest rains of the season. Platt, Elsey, et al. (2013) reviewed literature on frugivory and seed dispersal by crocodilians, including two reports on unidentified seeds in *C. crocodilus* stomachs from Venezuela. Ellis (1980) found stomach contents of 19 adults of an introduced population in southern Florida, USA, to contain mostly fish, but a bird was in one stomach and mammal hair in another, whereas stomachs of small caimans contained frogs and tadpoles, small fish, and insects (also see Krysko et al., 2010). Friers and Flaherty (2016) reported a Blue Land Crab in the stomach of an individual from the introduced population in southern Florida. Campbell, H. W. (1973) reported on vocalization of this species in both captivity and in a natural population in Panama.

Etymology.—The name *crocodilus* is apparently classical Latin for crocodile or lizard and is derived from the Greek word *krokodeilos*. See the Etymology for this name in the genus *Crocodilus*.

Specimens Examined (29 [7]; Map 65).—**COLÓN**: Puente Río Aguán (near mouth), UF 120733, 120739; Río Chappaqua, UF 120742, 120747–48; about 12.9 km NE of Trujillo, LSUMZ 22500–01. **GRACIAS A DIOS**: Laguna Bacalar, UF 120729–32, 120735; Laguna de Brus, UF 120740, 120744: Laguna de Tánsin, LACM 48459–61, LSUMZ 21718–19; Río La Criba, UF 120737, 120746; Río Palacios, UF 120734; near mouth of Río Patuca, UF 120745; near mouth of Río Plátano, UF 120741, 120743; Río Sico, UF 120738; about 2 km S of Rus Rus, ICF (now destroyed); "no other data," UF 120736. **YORO**: near confluence of Río San Lorenzo with Río Aguán, UNAH 5673.

Other Records (Map 65).—**ATLÁNTIDA**: Río Cuero, Río Salado (King et al., 1990). **COLÓN**: Anzuelo Bridge (O'Shea, 1986); Laguna de Guaimoreto, Laguna El Lirio (King et al., 1990). **CORTÉS**: Río Chamelecón, Río Ulúa (King et al., 1990). **GRACIAS A DIOS**: Bachi Tingni (USNM HI 2804); near Barra Río Plátano, UNAH 5385 (Cruz Díaz, 1978, specimen now lost); Crique Gabú Dende, Crique La Culebra, Crique Las Flores, Criques Plaplaya, Laguna de Ébano, Laguna Jolamaya, Laguna Paptatingni, Laguna Tampatingni, Laguna Tinguitara (King et al., 1990); Krahkra (personal sight records); Laguna Biltamaira, Laguna Siksa, vicinity of Laguna de Tánsin, Laguna Tilbalacán, Laguna de Warunta, Laguntara (Klein, 1979); Leimus (Río Warunta; USNM HI 2805); Mocorón (McCranie et al., 2006); Palacios (O'Shea, 1986); Rawa Kiamp (USNM HI 2724); Sadyk Kiamp (USNM HI 2724); Samil (personal sight records); Warunta (USNM HI 2802). **YORO**: near mouth of Río San Lorenzo (USNM HI 2806); 4.7 km ESE of San Lorenzo Arriba (USNM HI 2725).

Superfamily Crocodyloidea Oppel, 1811b

Family Crocodylidae Oppel, 1811b

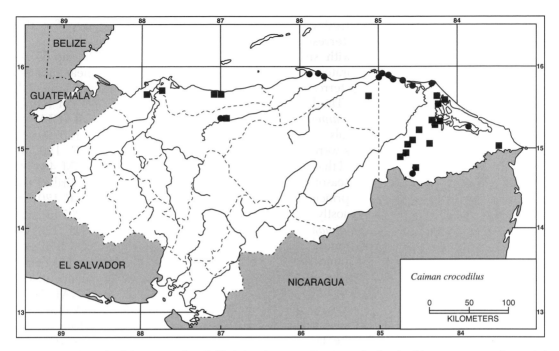

Map 65.   Localities for *Caiman crocodilus*. Solid circles denote specimens examined and solid squares represent accepted records.

The superfamily Crocodyloidea and family Crocodylidae include the crocodiles and dwarf crocodiles. Only a single species of crocodile occurs in Honduras and is most easily distinguished from the remaining crocodilian species (Alligatoridae) occurring in Honduras (*Caiman crocodilus*) by having the fourth tooth of the lower jaw fitting into an open pit in the upper jaw with the tooth visible when the mouth is closed, and in lacking a transverse bony preorbital ridge. The living members of the Crocodyloidea historically ranged in the Western Hemisphere in coastal southern Florida and the Florida Keys, USA, coastal Cuba (and adjacent islands), Hispaniola, Jamaica, and the Cayman Islands, both coasts of Mexico from northern Sinaloa and central Tamaulipas southward through Central America to northern South America from Colombia to the mouth of the Orinoco River in Venezuela east of the Andes to extreme northwestern Peru west of the Andes. In the Eastern

Hemisphere, it is distributed in tropical Africa, India, southern China, Indochina, and northern Australia. Three extant genera containing about 21 named species are included in this family.

Genus *Crocodylus* Laurenti, 1768

Crocodylus Laurenti, 1768: 53 (type species: *Crocodylus niloticus* Laurenti, 1768: 53, by subsequent designation of Brown, 1908: 113, although Brown used the spelling *Crocodilus*; also see Fitzinger 1843: 35 who credited the genus name to Cuvier, 1807: 55).

Geographic Distribution and Content.— In the Western Hemisphere, this genus occurs (or historically occurred) on Cuba, Hispaniola, Jamaica, the Cayman Islands, extreme southern Florida, USA, and from northern Sinaloa and Tamaulipas, Mexico, to Venezuela and northwestern Peru. In the Eastern Hemisphere, it occurs (or historically occurred) in Africa south of the Sahara Desert, on Madagascar, in southern and southeastern Asia, from Iran through the Philippines, Indonesia, to Palau and Fiji, and in northern Australia. Fourteen named species are recognized, one of which occurs in Honduras.

Remarks.—Boulenger (1889: 287) reported *Crocodylus moreletii* A. H. A. Duméril and Bibron (1851: 29, *In* A. M. C. Duméril and Duméril, 1851) from "Guatemala and Honduras." However, Boulenger's reference to "Honduras" is actually Belize. Neill (1971: 358) and Smith and Smith (1977: 101) speculated that the crocodile formerly inhabiting Lago de Yojoa in northwestern Honduras, might be *C. moreletii.* Wilson, McCranie, and Williams (1986: 87), however, demonstrated those crocodiles were *C. acutus.* The Lago de Yojoa population has long been extirpated, and one large adult was purposely introduced into Lago de Yojoa in 2010 but was killed shortly thereafter (J. R. Collart, personal communication).

Etymology.—A. M. C. Duméril and Bibron (1836: 49) wrote the name Crocodile is an older name than "l'erpétologie."

Crocodylus acutus Cuvier, 1807

Crocodilus acutus Cuvier, 1807: 55 (presumably six syntypes formerly in MNHN, but all now missing [see Ernst et al., 1999: 700.1]; type locality: "la grande île de *Saint-Dominque*" [= Hispaniola]).

Crocodylus americanus: Barbour, 1914: 346.

Crocodylus acutus: Stejneger, 1917: 289; Schmidt, 1924: 85; Meyer, 1969: 195; Meyer and Wilson, 1973: 7; Wilson and Hahn, 1973: 103; Klein, 1979: 10; Wilson, McCranie, and Williams, 1986: 87; Thorbjarnarson, 1989: 251; King and Cerrato, 1990: 2; King et al., 1990: 314; Thorbjarnarson, 1992: 66; Monzel, 1998: 167; Wilson and McCranie, 1998: 15; Ernst et al., 1999: 700.2; Köhler, 2000: 17; Köhler, McCranie, and Nicholson, 2000: 425; Nicholson et al., 2000:2 8; Kaiser and Grismer, 2001: 84; Kaiser et al., 2001b: 164; Lundberg, 2001: 27; Wilson et al., 2001: 134; Köhler, 2003a: 29; McCranie et al., 2005: 53; McCranie et al., 2006: 93; Wilson and Townsend, 2006: 104; Köhler, 2008: 32; Espinal and Escobedo-Galván, 2010: 210; Espinal et al., 2010: 737; Townsend and Wilson, 2010b: 692; Espinal and Escobedo-Galván, 2011: 212; McCranie and Valdés Orellana, 2014: 44; Solís et al., 2014: 129; McCranie, 2015a: 364; McCranie and Gutsche, 2016: 887.

Caiman crocodilus: Lovich et al., 2006: 15.

Geographic Distribution.—*Crocodylus acutus* occurs at low and lower limits of moderate elevations from Sinaloa, Mexico, to extreme northwestern Peru on the Pacific versant and from Tabasco, Mexico, to northeastern Venezuela on the Atlantic versant. The species is also known from southern Florida, USA, and the islands of the Cuba Bank, the Cayman Islands, Jamaica, Hispaniola, and Margarita (see Remarks). In Honduras, this species remains sporadically widespread along both coasts, including the Bay Islands, islands in the Golfo de Fonseca, and well inland along several large rivers.

Description.—*Crocodylus acutus* is a huge crocodile that historically reached a length of about 7.0 m (King and Brazaitis, 1971; Brazaitis, 1974; but that huge size not verified), but currently, largest adults usually reach about 3–4 m TL, with estimated body mass of large individuals ranging from about 900 to 1,283 kg (Rainwater et al., 2010); hatchings measure about 20–35 cm TL (Brazaitis, 1974); snout long, moderately slender, somewhat elongated; median humplike swelling present on snout; fourth tooth of lower jaw fitting into open pit or groove in upper jaw, tooth visible with mouth closed; transverse preorbital bony ridge absent; 2–6 (most often 4) postoccipi-

Plate 82. *Crocodylus acutus*. Adult. Cortés: Lago de Yojoa. Photograph by Juan R. Collart.

tal scutes arranged in single transverse row; 1–7 (usually 6) nuchal scales in 2 juxtaposed pairs flanked by a smaller scale; nuchal scales separated from postoccipitals and first row of dorsal scutes by skin; ventral collar consisting of single enlarged transverse series of scales; 14–17 (most often 16 or 17) longitudinal rows of dorsal scutes; dorsal scutes in 1–6 transverse rows at midbody; dorsal scutes flanked laterally by individual scales, those scales separated by skin; skin between enlarged scales smooth, without many raised scales; ventral scales in 25–35 longitudinal rows; single apical pit present on ventral scales, osteoderm buttons absent; 16–21 dorsal, double crested caudal whorls; 14–20 dorsal, single crested caudal whorls; subcaudals uniform, uninterrupted; slight webbing present on forelimb, extensive webbing present on hind limb; dental formula 5+13–14/15, all teeth in individual sockets (much of above taken from King and Brazaitis, 1971; Brazaitis, 1974; Platt et al., 2012).

Diagnosis/Similar Species.—*Caiman crocodilus* has the fourth mandibular tooth fitting into a socket in the upper jaw, thus that tooth is not visible when the mouth is closed, and has a transverse bony ridge between the anterior border of the eyes (versus fourth mandibular tooth fitting into notch in upper jaw and tooth visible when mouth closed, and transverse bony head ridge absent in *Crocodylus acutus*).

Illustrations (Figs. 142–144; Plate 82).— Álvarez del Toro, 1974 (adult, subadult, juvenile), 1983 (adult); Calderón-Manduja-no et al., 2008 (adult); Campbell, 1998 (adult, juvenile); Conant and Collins, 1998 (adult, subadult); Cope, 1900 (head scales, midbody scales, foot and leg scales, anal region scales); Ernst et al., 1999 (adult, dorsal scutes, skull); Escobedo-Galván et al., 2010 (adult); Espinal et al., 2010 (adult); Gutsche, 2007 (juvenile); Guyer and Don-nelly, 2005 (adult); Kaiser et al., 2001b (adult, hatching); Köhler, 1999b (subadult), 2000 (juvenile, head, tail scales), 2001b (juvenile, head), 2003a (adult, juvenile, head and forebody scales), 2008 (adult, juvenile, head and forebody scales); Köhler and Seipp, 1998 (subadult); Köhler, McCranie, and Nicholson, 2000 (juvenile); Köhler, Vesely, and Greenbaum, 2005 (adult, head); Lee, 1996 (adult, tail scales), 2000 (adult, tail scales); McCranie et al., 2005 (juvenile, head and forebody), 2006 (juvenile); Medem, 1962 (adult); Murphy, 1997 (adult); C. A. Ross and Ross, 1974 (tail scales); F. D. Ross and Mayer, 1983 (drawing of anterior portion of dorsal body scales); J. P. Ross, 1998a (adult); Savage, 2002 (adult, subadult, anterior body scales); Schmidt, 1924 (adults sunning, eggs), 1944 (adult, head), 1952 (adult); Stafford and Meyer, 1999 (adult, subadult, head, tail scales); Stejneger, 1917 (head, anterior dorsal scales); Thorbjarnarson, 1992 (adult, hatchling); Trutnau, 1986 (adult); Trutnau and Sommerlad, 2006 (adult, head, juvenile, hatchling).

Remarks.—Ernst et al. (1999) provided an overview of the morphology of *Crocodylus acutus* and included a brief literature review. Scales counts used herein for *C. acutus* are from Brazaitis (1974: 70) and Platt et al. (2012: 335). Milián-García et al. (2011: 371–372) recovered molecular evidence indicating more than one species is involved among the current concept of *C. acutus*. Should the Mexican to South American populations be shown to represent a distinct evolutionary species, the name *C. biscutatus* Cuvier (1807: 53) might be available, although the type locality of that nominal form is unknown (said to have been collected on the "Voyage au Sénégal" by Cuvier, 1807: 53). Should that name not be available, Wermuth and Mertens (1961: 359, 1977: 141), Smith and Smith (1977: 87), and Trutnau and Sommerlad (2006: 443) provided lists of other names they considered synonyms of *C. acutus*. Balaguera-Reina, Venegas-Anaya, and Densmore (2015) updated the knowledge on the current population status of *C. acutus* in part of Panama.

Natural History Comments.—*Crocodylus acutus* is known from sea level to 650 m elevation in the Lowland Moist Forest and Lowland Dry Forest formations and peripherally in the Premontane Wet Forest formation. It is a nocturnal species that also frequently basks during sunny periods on shores of rivers and lagoons. The species occurs in both fresh- and brackish water situations in rivers, swamps, lagoons, and lakes. Honduran crocodiles are notable for sometimes occurring well inland along several large rivers. King and Cerrato (1990), King et al. (1990), and Klein (1979) reported on survey work on *C. acutus* in various parts of Honduras (also see Thorbjarnarson, 1992). Schmidt (1924, 1944, 1952) discussed hunting this species at Laguna Ticamaya, Cortés, in 1923, when the species was extremely abundant at that locality. Schmidt (1924, 1952) mentioned counting 75 in view at one time with the use of binoculars at Laguna Ticamaya, at least some of which "exceeded ten feet in length" (Schmidt, 1924: 87). Schmidt (1944: 72) reported the largest crocodile collected at that locality was "eleven feet two inches" (=340 cm). Kaiser and Grismer (2001) and Kaiser et al. (2001b) reported on a breeding population on Roatán in the Bay Islands.

All remaining Honduran populations are severely depleted, and *Crocodylus* no longer occurs in many areas where it was known to be common in earlier years, even in relatively recent years (about most recent 10–15 years). A few crocodiles could still be seen during the early 2000s along the ríos Coco and Patuca, although they had been under severe hunting pressure since the early 1990s. In September 2012, I was told of the presence of an adult *Crocodylus* in the saltwater marsh and lagoon just south of the airport on Guanaja Island. That animal had been seen several times in recent months leading up to that September and its presence had become well known to locals. Also, in August 2012, a small adult *C. acutus* was captured by divers from an underwater cave in the marine waters off shore of Utila Island. That crocodile was transported to the large brackish water marsh in the middle of the island and released.

Escobedo-Galván et al. (2010) provided a brief literature review of the changing population dynamics of this species in

Honduras. Schmidt (1924: 91–92) noted a female deposited 22 eggs in a "shelving gravel beach" at Laguna Ticamaya in late April (also see Schmidt, 1952). Thorbjarnarson (1989) provided a study and literature review of reproduction in *C. acutus* based on his own fieldwork outside of Honduras and a literature review, with information from much of the species range. Females deposit about 22–60 eggs (higher numbers reported in the literature are likely a result of more than one female depositing eggs in the same nest) in a hole she usually digs in the sand or soil and typically does not cover that hole with vegetation, only soil. Females, in some cases, tend the nest until her eggs hatch, whereupon she opens the nest when hatchlings vocalize. She then carries her young in her mouth to the nearest water. Nesting takes place during the dry season. Females are known to sometimes stay in the vicinity of her young for an extended period of time, but the amount of maternal care in the species is usually minimal. Males are not known to tend nests or give much assistance to the young. Espinal and Escobedo-Galván (2010) reported female *C. acutus* were able to nest on steeply elevated slopes above the El Cajón Reservoir (Honduras) in altered habitat. They found 30 nests with 10–33 eggs or hatchlings. Espinal and Escobedo-Galván (2011) concluded that the El Cajón population is stable, although Espinal et al. (2010) found no neonates in those same populations during a survey performed subsequent to the Espinal and Escobedo-Galván (2011) survey. Platt and Thorbjarnarson (2000) studied nesting ecology of this species in coastal Belize and found egg deposition occurs in the latter half of the dry season with a mean clutch size of 22.3 eggs. Hatching in that population occurred from June to mid-July, corresponding with the onset of the wet season. Ferguson (1985) provided a literature review of various parameters of *C. acutus* eggs both in captivity and in natural populations. Cedillo

Leal et al. (2013) reported mean numbers of 34.7 and 39.3 eggs for three and six egg clutches in Oaxaca, Mexico. Budd et al. (2015) reported results of mating analysis of *C. acutus* at two sites in Costa Rica, and Balaguera-Reina, Venegas-Anaya, Sanjur, et al. (2015) studied reproductive ecology and hatchling growth rates in a population on Coiba Island, Panama.

Schmidt (1924: 91) reported "larger crocodiles" at Laguna Ticamaya, Cortés were cannibalistic, eating young crocodiles, as well as turtles. Schmidt (1924: 91) also found "a hair ball and the horny hoofs of (probably) a peccary" and small fish in stomachs of adults at the same locality. Schmidt (1924: 91) reported a stomach of a "smaller" crocodile at that locality contained "the remains of eight top-minnows." Stones were also found in stomachs of all crocodiles Schmidt examined. Schmidt (1952) also reported deer remains in stomachs of crocodiles from the same locality. Platt et al. (2002) studied hatchling diet in Belize and found they feed mostly on insects and crustaceans, with one stomach containing a partially digested fish. Thorbjarnarson (1989: 248) provided a literature review on diet in this species and stated, "diet of *C. acutus* follows the typical ontogenetic shift described for other species of crocodilians .... Hatchlings and juvenile crocodiles feed primarily on aquatic and terrestrial invertebrates ... or small fish .... With increasing size, crocodiles feed increasingly on larger vertebrate prey with fish being the dominant food item." Thorbjarnarson (1989) also listed frogs, small turtles, birds, and small mammals in the diet of subadults. Other vertebrates listed for adults by Thorbjarnarson (1989) include turtles, other *C. acutus*, peccary, snakes, domestic livestock and animals, and birds. Dogs are frequently taken by crocodiles living along several rivers in northeastern Honduras (personal observation; personal communication of several residents). Beaty and Beaty (2012) reported a juvenile *C.*

acutus in Panama feeding on an adult toad with poisonous secretions (*Rhinella horribilis*), apparently without fatal results. Platt and Rainwater (2007) reported the same feeding event with *C. moreletii* in Belize, also without fatal results. Platt, Thorbjarnarson et al. (2013) studied diet in populations of all class sizes in coastal habitats in Belize of *C. acutus* and found stomach contents contained insects, mollusks, crustaceans, fish, amphibians, reptiles, birds, and mammals. Platt, Thorbjarnarson et al. (2013: 1) concluded "hatchlings and small juveniles feed largely on insects and crustaceans, larger juveniles broaden their diet to include fish and nonfish vertebrates, subadults consume increasing amounts of crustaceans with lesser amounts of insects and nonfish vertebrates, and adults subsist primarily on marine crustaceans." Platt, Elsey et al. (2013) reviewed literature on frugivory and seed dispersal by crocodilians, including *C. acutus*. Platt et al. (2014) also reported suspected frugivory in an adult male from Belize. Acosta-Chaves et al. (2016) reported a *C. acutus* in the seawater along the shoreline in Costa Rica attempting to feed on a Brown Pelican (*Pelecanus occidentalis* [Linnaeus]).

García-Grajales and Silva (2014) studied a population of *Crocodylus* in Oaxaca, Mexico, and found a significant association with mangrove areas in juveniles, whereas subadults and adults were associated more with open waters. Platt et al. (2011) reported results of studies on size estimation, morphometrics, sex ratio, size dimorphism, and biomass of a Belizean population. Campbell, H. W. (1973) reported captive vocalization of a young crocodile from Jamaica and also provided a literature review of vocalization in this species. Savage (2002) reported three fatal attacks on humans by *C. acutus* in Costa Rica in the late 1990s. Thorbjarnarson et al. (2006) discussed a need for protection of *C. acutus* throughout its range, including six sites in Honduras. Thorbjarnarson et al. (2006) also discussed the conservation parameters of those populations. Thorbjarnarson (1988) provided ecological data on the largest remaining and endangered population of this crocodile in Haiti, part of Hispaniola, the type locality of the species.

Etymology.—The name *acutus* is a Latin word meaning "make pointed, sharpen," and alludes to the rather narrow snout of this species.

Specimens Examined (11, +7 skulls, 1 set of bones, 1 penis, 1 egg lot [9]; Map 66).— **ATLÁNTIDA**: La Ceiba, CM 29012. **COLÓN**: Balfate, AMNH 58605. **CORTÉS**: Agua Azul, AMNH 70570–71; Laguna Ticamaya, FMNH 5328 (skin), 5329 (15 eggs), 5775 (skull), 5776 (skull), 5780 (skin and skull), 11010 (skull), 11038 (skull), 13220 (skull), 22934, 31293 (penis), 35539. **ISLAS DE LA BAHÍA**: Isla de Roatán, near Roatán, FMNH 34563 (skull); "Isla de Utila," USNM 24488, 29420, 211278 (bones). "HONDURAS": "east [sic] coast," MCZ R17721.

Other Records (Map 66).—**ATLÁNTIDA**: Laguna de Los Micos, Río Cuero, Río Salado (King et al., 1990). **CHOLUTECA**: Estero de la Berbería, Río Choluteca (King et al., 1990); Estero San Bernardo (King and Cerrato, 1990; King et al., 1990). **COLÓN**: Laguna El Lirio, near mouth of Río Aguán, Río Chappaqua (King et al., 1990). **COMAYAGUA**: El Cajón Reservoir (Espinal et al., 2010). **CORTÉS**: El Cajón Reservoir, (King and Cerrato, 1990; King et al., 1990; Espinal and Escobedo-Galván, 2010); Río Chamelecón, Río Motagua, Río Ulúa (King et al., 1990). **GRACIAS A DIOS**: near mouth of Caño Awawás (UF photograph); Crique Las Flores, Laguna Bacalar, Laguna de Brus, Laguna de Ébano, Laguna Paptatingni, near mouth of Río Negro (King et al., 1990); Laguna Biltamaira, Laguna Siksa, vicinity of Laguna de Tánsin, Laguna de Warunta, Laguntara (Klein, 1979); Río Coco near Caño Awalwás (UF photograph); Río Coco near Kyras (personal sight record). **ISLAS DE LA BAHÍA**: Isla Barbareta

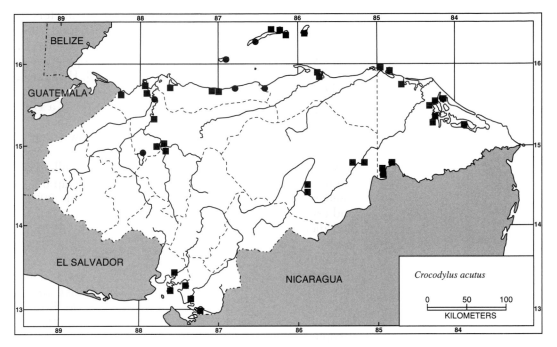

Map 66.   Localities for *Crocodilus acutus*. Solid circles denote specimens examined and solid squares represent accepted records.

(personal sight record); Isla Guanaja, lagoon near airport (purported sight records); Isla Morat (McCranie et al., 2005); Isla de Roatán, Crawfish Rock, mangroves E of New Port Royal Harbor (Kaiser et al., 2001b). **OLANCHO**: Caobita, Quebrada El Mono (Nicholson et al., 2000); Río Guayape near its confluence with Río Patuca, Río Patuca near its confluence with Río Guayape (King and Cerrato, 1990). **VALLE**: near mouth of Río Nacaome (King et al., 1990); Isla del Tigre, La Laguna (Naval Base Officer, personal communication to author; also see Lovich et al., 2006; as *Caiman crocodilus*). **YORO**: El Cajón Reservoir (Espinal et al., 2010).

ORDER TESTUDINATA BEHN, 1760

The Order Testudinata contains the turtles. Turtles are the only tetrapods that have a bony or leathery shell composed of dermal modifications that incorporate the ribs, trunk vertebrae, and portions of the pectoral girdle. The pectoral girdle is also within the ribcage under the vertebrae (Pough et al., 2015: 185) rather than external to the ribcage as in all other living reptiles. The carapace "is formed from fusion of the eight trunk vertebrae and ribs to an overlying set of dermal bones" and the plastron "arises from the fusion of parts of the sternum and pectoral girdle with external dermal bones" (Vitt and Caldwell, 2014: 523).

Remarks.—Behn (1760, Tabula Generalis) raised Klein's pre-Linnaean Testudinata to an "Ordo" level, as did Oppel (1811b: 4). Thus, Behn and Oppel both occupied Testudinata as the turtle order name. I recommend using Testudinata (Behn, 1760) for the order name over the frequently used Testudines, because the former was the spelling used by Behn (1760) and specified as an "Ordo." Also, Testudines is the family-group name for a series of turtles and thus cannot also be used at a higher level above the family-group series.

Infraorder Cryptodira Lichtenstein and von Martens, 1856

Extant turtles are divided into two infraorders, the Pleurodira Lichtenstein and von Martens (1856:2; as Pleurodera) and Cryptodira Lichtenstein and von Martens (1856: 1; as Cryptodera). The Pleurodira "retract the head and neck by laying it to the side" and the Cryptodira "retract the neck posteriorly into a medial slot within the body cavity" (Zug et al., 2001:435–436). All Honduran turtles belong to the Cryptodira and are placed in four superfamilies and seven families (taxonomy following Rhodin et al., 2008; TTWG, 2014).

Remarks.—Rhodin et al. (2008: 000.23) wrote that Lichtenstein was the sole author of the 1856 work. However, Rhodin et al. were in error with that opinion, because von Martens, at least, was also involved in authorship of the work. Lichtenstein and von Martens (1856) had used Cryptodera and Pleurodera at the infraorder level. Rhodin et al. (2008) also expressed the erroneous opinion that Lichtenstein and von Martens (1856) "was a printed catalogue distributed to zoological colleagues and museums, and not apparently sold in bookstores, so therefore not nomenclaturally available." Being sold in bookstores is not a criterion to make a 19th century publication available as a printed source for a valid publication. The TTWG (2017) has continued that opinion regarding the Lichtenstein and von Martens (1856) work. Surprisingly, the extensive Joyce et al. (2004) nomenclatural work did not cite the Lichtenstein and von Martens (1856) publication, thus ignoring the turtle names proposed or used in that work, unlike workers on lizard taxonomy who use the names proposed in Lichtenstein and von Martens (1856). Lichtenstein and von Martens (1856) is a valid publication and has to be dealt with, even though it is a tradition of turtle workers worldwide to ignore their proposed names and the entire Lichtenstein and von Mar-

tens (1856) work. Lichtenstein and von Martens (1856) also provided correct lists of the genera they considered to belong to each turtle infraorder. See Gutsche (2016) who provided a brief discussion regarding the availability of the Lichtenstein and von Martens (1856) publication and the availability of new names they proposed.

Numerous errors occur in the various TTWG lists (the three most recent were published in 2012, 2014, and 2017 and the vast majority of that inaccurate information has remained consistent in those works) relative to the names Cryptodira and Pleurodira and the correct publications where those names were proposed. In summary, and to correct those errors, Lichtenstein and von Martens (1856) coined the infraorder names Cryptodera and Pleurodera. The prevailing names currently in use are Cryptodira and Pleurodira, both spellings originated from Cope (1868: 119) and Cope (1866a: 186), respectively.

KEY TO HONDURAN SUPERFAMILIES OF THE INFRAORDER CRYPTODIRA

1A. Limbs paddlelike (Fig. 145)......
   .... Chelonioidea (p. 442)
1B. Limbs not paddlelike (Fig. 146) 2
2A. Tail long, half or more length of carapace, with a dorsal median row of large triangular scutes, with those tips dorsally directed (Fig. 147); plastron reduced and cruciform, with 11 or 12 scutes present, including widely separated abdominals of bridge elements (Fig. 148)
   .... Chelydroidea (p. 469)
2B. Tail short, less than half length of carapace, lacking dorsal median row of large, triangular-shaped scales; plastron neither reduced nor cruciform or, if reduced and slightly cruciform, only 7 or 8 scutes present 3
3A. Twelve marginal scutes present (Fig. 149) on each side; 12 plastral

Figure 145.   Paddlelike limbs present in a marine turtle. *Eretmochelys imbricata* from offshore of Isla Grande, Swan Islands, Gracias a Dios (released).

scutes present (Figs. 150, 151), with pattern of symmetrical lines (Fig.150; faded or worn in some), or isolated circles, ocelli, or mottling present (Fig. 151) ....
.... Testudinoidea (p. 501)

3B. Eleven marginal scutes present (Fig. 152) on each side; 11 plastral scutes present (Figs. 153, 154), rarely 12 in one genus, or 7 or 8 plastral scutes present in one genus (Fig. 155).... Kinosternoidea (p. 479)

Figure 146.   A turtle forefoot showing limbs non-paddlelike. *Rhinoclemmys funerea*. USNM 570468 from near Crique Wahatingni, Gracias a Dios.

Figure 147.   Long tail with large, triangular dorsal scutes. *Chelydra acutirostris*. USNM 559582 from confluence of ríos Cuyamel and Patuca, Olancho.

CLAVE PARA LAS SUPERFAMILIAS HONDUREÑAS
DE LA INFRAORDEN CRYPTODIRA

1A. Extremidades en forma de remo
(Fig. 145)...... Chelonioidea (p. 442)
1B. Extremidades no en forma de
remo (Fig. 146).................... 2

2A. Cola larga, al menos la mitad de la
longitud del carapacho, con una
hilera mediodorsal de escudos
grandes triangulares, con puntas
dorsales (Fig. 147); plastrón redu-
cido y cruciforme, con 11 o 12
placas presentes, incluyendo los

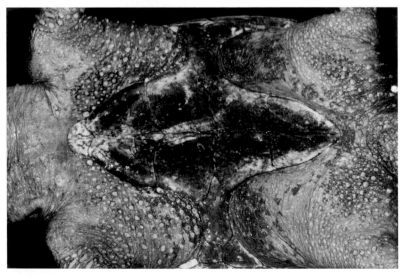

Figure 148.   Reduced and cruciform plastron present with widely separated bridge elements. *Chelydra acutirostris*. USNM 559582 from confluence of ríos Cuyamel and Patuca, Olancho.

Figure 149.   Twelve marginal scute pairs present on each side. *Trachemys scripta*. FMNH 283584 from Guanaja Island, Islas de la Bahía (introduced population).

abdominales separados de los elementos del puente (Fig. 148) . . . .
. . . . Chelydroidea (p. 469)

2B. Cola corta, menos de la mitad de la longitud del carapacho, cola sin una hilera mediodorsal de escamas

grandes en forma de triangular; plastrón no cruciforme, o si es poco cruciforme, con solamente siete u ocho escudos presentes 3

3A. Doce pares de escudos marginales presentes (Fig. 149); 12 placas

Figure 150.   Plastron with 12 scutes and dark central figure of faded symmetrical lines present. *Trachemys venusta*. USNM 559590 from Awasbila, Gracias a Dios.

Figure 151.   Plastron with 12 scutes and isolated black circles, ocelli, or mottling. *Trachemys scripta.* FMNH 283584 from near Savannah Bight, Guanaja Island, Islas de la Bahía (introduced population).

presentes en el plastrón (Figs. 150, 151), con un patrón de líneas simétricas presentes (Fig. 150); el patrón puede aparecer desvanecido o muy tenue, o patrón de círculos aislados con ocelos, o moteado (Fig. 151) . . . .
. . . . Testudinoidea (p. 501)

3B. Once pares de escudos marginales presentes (Fig. 152); 11 escudos presente en el plastrón, raramente

12 en un género (Figs. 153, 154), o 7–8 escudos presentes en un género (Fig. 155). . . . Kinosternoidea (p. 479)

Superfamily Chelonioidea Schmid, 1819

Chelonioidea contains the marine turtle families Cheloniidae Schmid (1819: 14) and Dermochelyidae Baur (1888: 422). Both families have representatives occurring in Honduran waters. Externally, the Chelonioidea are the only turtles in Honduran waters

Figure 152.   Eleven marginal scutes per side present. *Kinosternon leucostomum.* FMNH 283589 from Guanaja Island, Islas de la Bahía.

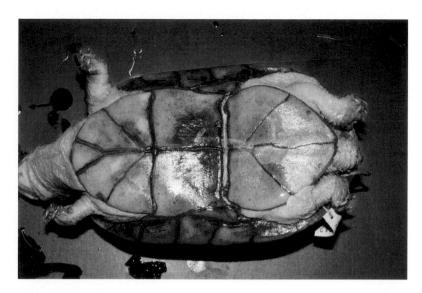

Figure 153.   Plastron with 11 scutes, 2 kinetic hinges, 0–2 inframarginal bridge scutes contacting fixed lobe on each side (between two kinetic hinges) of plastron, and fore- and hind limbs with weak webbing. *Kinosternon leucostomum.* USNM 589145 from Sisinbila, Gracias a Dios.

Figure 154.   Plastron with 11 (rarely 12) scutes. *Dermatemys mawii.** USNM 46304 from Belize. Photograph by James A. Poindexter.

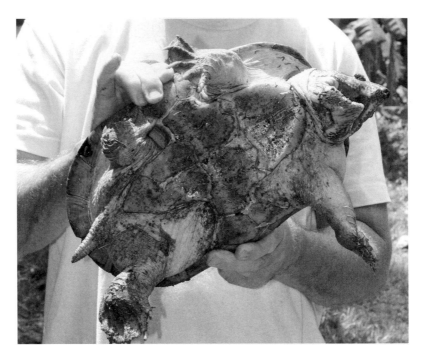

Figure 155.   Plastron reduced, with 7–8 plastral scutes present, no kinetic hinges present, and well-developed webbing on both fore- and hind limbs. *Staurotypus triporcatus*. Captive from Choloma. Cortés. Photograph by Leonardo Valdés Orellano.

that have the limbs modified into paddlelike structures. Schmid (1819) is the author of the superfamily name Chelonioidea and family name Cheloniidae by Principle of Priority (Article 23) and Principle of Coordination (Articles 36, 43, 46; ICZN, 1999).

KEY TO HONDURAN FAMILIES OF THE SUPERFAMILY CHELONIOIDEA

1A. Carapace (Fig. 156) and plastron covered with leathery skin; carapace with 5+ ridges (Fig. 156) in adults, with exception of older individuals .. Dermochelyidae (p. 465)

1B. Carapace (Fig. 157) and plastron covered with horny scutes, carapace without ridges in adults (Fig. 157) .... Cheloniidae (p. 444)

CLAVE PARA LAS FAMILIAS HONDUREÑAS DE LA SUPERFAMILIA CHELONIOIDEA

1A. Carapacho (Fig. 156) y plastrón cubierto con piel coriácea; carapa-

cho con cinco o más quillas (Fig. 156) en adultos, con excepción de los individuos muy viejas......... .... Dermochelyidae (p. 465)

1B. Carapacho (Fig. 157) y plastrón cubierto con placas queratinizadas; carapacho sin quillas en adultos (Fig. 157)...... Cheloniidae (p. 444)

Family Cheloniidae Schmid, 1819

This family of sea turtles ranges circum-globally in temperate, subtropical, and tropical oceans. The Cheloniidae can be distinguished externally from all remaining Honduran turtles, except the family Dermochelyidae, in having the limbs modified into paddlelike structures. The Cheloniidae differ from the Dermochelyidae in having a carapace and plastron covered with epidermal scutes (a shell). Five genera containing six named species are included in this family (but see Remarks in generic account for *Chelonia*). Four named species contained in four genera occur in Honduran marine waters.

Figure 156.   Carapace covered with leathery skin with five + ridges. *Dermochelys coriacea*. USNM 337799 from Brevard County, Florida, USA.

Figure 157.   Carapace covered with horny scutes without ridges, 4 pairs of costal scutes present (numbered), and nuchal scute (outlined) not contacting first pair of costal scutes. *Chelonia mydas* from Roatán Island, Islas de la Bahía (released).

Figure 158.   Five or more costal scutes present per side (numbered in image; 6 in this case). *Lepidochelys olivacea.* USNM 279321 (without locality data).

KEY TO HONDURAN GENERA OF THE FAMILY
CHELONIIDAE

1A.   Four pairs of costal scutes present (Fig. 157); nuchal scute not in contact with first costal scute (Fig. 157) 2

1B.   Five or more pairs of costal scutes present (Fig. 158); nuchal scute contacts first pair of costal scutes (Fig. 159) 3

Figure 159.   Nuchal scute contacting first pair of costal scutes (outlined). *Lepidochelys olivacea.* USNM 279321 (without locality data).

Figure 160.   Two pairs of prefrontal scales present. *Eretmochelys imbricata* from offshore of Isla Grande, Swan Islands, Gracias a Dios (released).

2A. Two pairs of prefrontal scales (Fig. 160); beak hawklike and serrations present on outside cutting edge of upper jaw (Fig. 161)..............
.... *Eretmochelys* (p. 458)

2B. A single pair of prefrontal scales (Fig. 162); beak not hawklike and no serrations present on outside cutting edge of upper jaw (Fig. 163) .... *Chelonia* (p. 454)

Figure 161.   Hawklike beak present and serrations present on each outside cutting edge of upper jaw. *Eretmochelys imbricata*. USNM 279317 (without locality data).

Figure 162.   One pair of prefrontal scales present. *Chelonia mydas*. USNM 29973 from Timor, Indonesia.

3A. Bridge most often with 3 (sometimes 4) inframarginal scutes (Fig. 164); 3–7 inframandibular scales posterior to beak on either side of lower jaw (Fig. 165) . . *Caretta* (p. 451)

3B. Bridge usually with 4 (rarely 5) inframarginal scutes (Fig. 166); single, large inframandibular scale

posterior to beak on either side of lower jaw (Fig. 167) . . . .
. . . . *Lepidochelys* (p. 463)

CLAVE PARA LOS GÉNEROS HONDUREÑOS DE LA
FAMILIA CHELONIIDAE

1A. Cuatro escudos costales presentes sobre cada lado del carapacho (Fig.

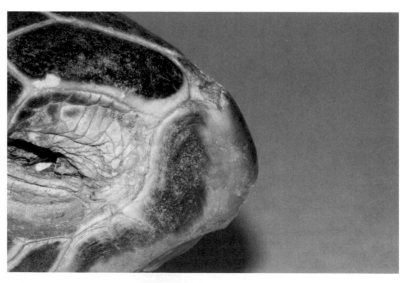

Figure 163.   Beak not hawklike and serrations absent on each outside cutting edge of upper jaw. *Chelonia mydas*. USNM 29973 from Timor, Indonesia.

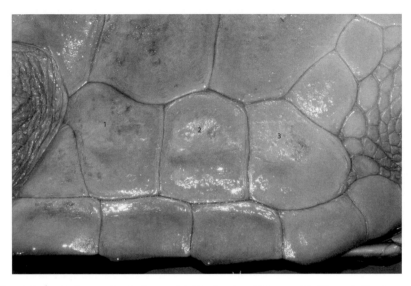

Figure 164.   Three inframarginal scutes (rarely 4) on each bridge (numbered in image; 3 in this case). *Caretta caretta*. UF 216471 (without locality data).

157); el escudo nucal no está en contacto con el primer par de los escudos costales (Fig. 157)......... 2

1B. Cinco o más escudos costales presentes sobre cada lado del carapacho (Fig. 158); el escudo nucal está en contacto con el

primer par de escudos costales (Fig. 159)........................ 3

2A. Dos pares de escamas prefrontales presentes (Fig. 160); pico conspicuo en forma de pico de halcón, aserrado a cada lado en el borde

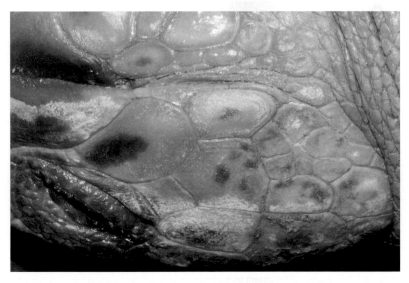

Figure 165.   Three to 7 inframandibular scales per side. *Caretta caretta*. UF 216471 (without locality data).

Figure 166.   Four inframarginal scutes on each bridge (numbered). *Lepidochelys olivacea*. USNM 14721 (without locality data).

cortante de la maxila (Fig. 161) . .
. . . . *Eretmochelys* (p. 458)
2B. Solo un par de escamas prefron-
tales presentes (Fig. 162); pico
normal, no en forma de pico de
halcón y no aserrado en el borde
cortante de la maxila (Fig. 163) . .
. . . . *Chelonia* (p. 454)
3A. Puente usualmente con 3 escudos
(raramente 4) inframarginales (Fig.

164); 3–7 escamas inframandibu-
lares presentes posteriores al pico a
cada lado de la mandíbula (Fig.
165) . . . . *Caretta* (p. 451)
3B. Puente usualmente con 4 escudos
(raramente 5) inframarginales (Fig.
166); solo una placa grande inframandibular presente posterior al
pico a cada lado de mandíbula
(Fig. 167) . . . . . . *Lepidochelys* (p. 463)

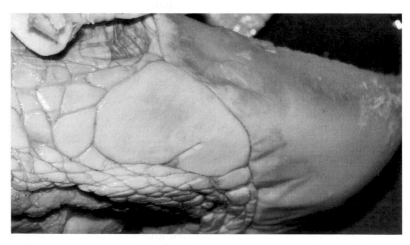

Figure 167.   Single large inframandibular scale per side. *Lepidochelys olivacea*. USNM 293033 from Oaxaca, Mexico.

Genus *Caretta* Rafinesque, 1814

Caretta Rafinesque, 1814: 66 (type species: *Caretta nasuta* Rafinesque, 1814: 66 [= *Testudo caretta* Linnaeus, 1758: 197], by monotypy).

Geographic Distribution and Content.— See the Geographic Distribution of *Caretta caretta* for this monotypic genus.

Etymology.—*Caretta* is a latinization of the French word caret (carey in Spanish and English), meaning turtle, tortoise, or sea turtle.

Caretta caretta (Linnaeus, 1758 [in part])

Testudo caretta Linnaeus, 1758: 197 (in part) (holotype not designated [see King and Burke, 1989: 18; Dodd, 1990b: 483.1; also see Remarks]; type locality: "*insulas* Americanas").
Caretta caretta: Stejneger, 1904: 715; Marin, 1984: 221; Cruz [Díaz] and Espinal, 1987: 18; Peskin, 1996: 1; Lundberg, 2000: 3; Lundberg, 2001: 27; McCranie et al., 2005: 56; McCranie et al., 2006: 95; Köhler, 2008: 55; McCranie and Valdés Orellana, 2014: 45; Solís et al., 2014: 138; McCranie, 2015a: 379; McCranie et al., 2017: 276.
Thallassochelys caretta: Lowe, 1911: 32.

Geographic Distribution.—*Caretta caretta* occurs in tropical, subtropical, and temperate waters of the Atlantic, Pacific, and Indian oceans, although it usually inhabits continental shelves, lagoons, bays, and estuaries. In Honduras, this species is known only from Caribbean waters, including the Swan Islands, and coastal regions.

Description.—*Caretta caretta* is a large sea turtle, usually with a CL of 700–950 mm and a weight of 80–200 kg (Ernst and Lovich, 2009); carapace shield-shaped, widest toward anterior end, with serrate margins and non-overlapping scutes, except scutes imbricate in some young specimens; 5 vertebrals; 5 pairs of costal scutes, with first pair in contact with nuchal scute; usually 12 pairs of marginal scutes; juvenile carapace tricarinate, with median, longitudinal knobbed ridge; adults lack ridges on carapace; plastron unhinged; hatchlings usually with 2 longitudinal plastral ridges; 12 plastral scutes, arranged in 6 pairs, with small intergular scute also present in some; bridge with 3–4 (most often 3) inframarginal scutes on each side (see Remarks) that usually lack pores; 2 pairs of prefrontal scales, sometimes with median suture, but usually with azygous scale or scales present; head large, robust, with smooth-margined lower jaw; lower jaw with 3–7 inframandibular scales posterior to beak on either side; limbs developed into paddlelike flippers, 2 small claws present on each flipper; tail of males extending beyond carapacial margin, tail of females barely reaching carapacial margin (much of above taken from Dodd, 1990a, 1990b; Ernst and Lovich, 2009).

Diagnosis/Similar Species.—*Caretta caretta* is distinguished from all remaining Honduran turtles, except the other marine turtles, by having its limbs modified into paddlelike structures. *Dermochelys coriacea* has leathery skin over a layer of connective tissue for the shell (versus shell covered with epidermal scutes or plates in *C. caretta*). *Chelonia mydas* and *Eretmochelys imbricata* have 4 pairs of costal scutes, the first pair separated from the nuchal scute (versus 5 pairs of costal scutes, with first pair in contact with nuchal scute in *C. caretta*). *Lepidochelys olivacea* has the bridge with 4 (rarely 5) inframarginal scutes on each side, each usually with a pore at its posterior border (versus bridge most often with 3 inframarginal scutes on each side, usually poreless in *C. caretta*).

Illustrations (Figs. 164, 165; Plate 83).— Bonin et al., 2006 (adult); Calderón-Mandujano et al., 2008 (adult); Carr, 1952 (adult, subadult); Conant and Collins, 1998 (adult); Dodd, 1988 (carapace, plastron); Duvernoy, 1836–1849 (adult; as *Testudo*); Ernst and Barbour, 1989 (adult); Ernst and Lovich, 2009 (adult, juvenile, hatchling);

Plate 83.   *Caretta caretta.* Juvenile. Miami Seaquarium, Miami, Florida.

Ernst et al., 1994 (adult, plastron, head); Freiberg, 1972 (adult); Fritz, 2012b (adult); Halstead, 1970 (adult, juvenile; as *C. c. gigas* [Deraniyagala, 1933: 66]); Köhler, 2003a (adult, carapace), 2008 (adult, carapace); LeBuff, 1990 (adult, juvenile, carapace, tail); Lee, 1996 (adult, hatchling, carapace, plastron), 2000 (adult, hatchling, carapace, plastron); Lemos-Espinal, 2015 (adult); Lemos-Espinal and Smith, 2009 (adult); Lutz and Musick, 1997 (head); Márquez M., 1990 (adult, head scales, carapace, plastron); McCranie et al., 2005 (adult, hatchling), 2006 (adult, carapace); Meylan et al., 2013 (adult); Murphy, 1997 (adult); Nietschmann, 1977 (adult); Pough et al., 2015 (adult); Powell, Collins, and Hooper, 1998 (carapace, plastron), 2012 (carapace, plastron); Pritchard, 1979 (adult, hatchling), 1997 (head); Pritchard and Trebbau, 1984 (adult, subadult, hatchling); Rebel, 1974 (inframarginal bridge scutes, dorsal head scales, adult); Ruckdeschel and Shoop, 2006 (adult, juvenile, hatchling, head, flipper, carapace); Savage, 2002 (adult, carapace, ventral view of bridge, ventral view of jaws, lateral view of lower jaw); Seminoff and Wallace, 2012 (juvenile, adult head); Shi, 2013 (adult, head, carapace, plastron); Smith and Smith, 1980 (carapace, plastron); Spotila, 2004 (adult, juvenile, hatchling), 2011 (hatchling); Stafford and Meyer, 1999 (adult, head scales); Vetter, 2004 (adult), 2005 (adult); Wermuth and Mertens, 1961 (adult, carapace, plastron; as *C. c. caretta*).

Remarks.—Almost all turtle books with identification keys (i.e., Ernst and Lovich, 2009) distinguish *Caretta caretta* from *Lepidochelys olivacea* by number of inframarginal scutes (3 in *Caretta* versus 4 in *Lepidochelys*). However, that character in *Caretta* has more than the expected variation, with a surprising number having 4 inframarginal bridge scutes, with 4 present on 17 of 130 sides and 5 on both sides of 1 (S. Gotte, personal communication, 14 December 2015). Gotte (personal communication, 7 August 2015) concluded that the number of inframarginal scutes is "a reasonable, though not flawless character" to distinguish *Caretta* from *Lepidochelys*.

Dodd (1990a, 1990b) provided short summarizations of the systematics and the literature on *Caretta caretta*, whereas Ernst and Lovich (2009) provided a thorough review of the biology of this species. Two subspecies of *C. caretta* are sometimes recognized, but no subspecies were recognized in the most recent turtle species list (TTWG, 2017; also see Ernst and Lovich, 2009: 38). Lundberg (2001) listed this species from Roatán, but I am unaware of any confirmed records for that island, although it certainly occurs in the waters near that island. Carr (1952: 393) made mention of little boys on the Honduran Caribbean coast using the entire carapace of large *Caretta* as boats.

Wallin (1985: 128) suggested that because Linnaeus (1758: 197) included specimens of *Caretta caretta* among his concept of *Chelonia mydas*, the name *Testudo caretta* should not be available in Linnaeus (1758). Wallin suggested Wallbaum (1782: 95) was the first to use *Testudo caretta* in its current concept, but that suggestion has not been followed in the recent literature (Fritz and Havaš, 2007; TTWG, 2017). Wallin (1985: 129) stated that *T. caretta* "has no surviving type material."

Natural History Comments.—Very little has been reported about *Caretta caretta* on Honduran shores and in Honduran waters. Marin (1984) reported sight records of this species from several coastal localities off northeastern Honduras, and Groombridge (1982) reported nesting on beaches between Puerto Cortés and La Ceiba, although the turtles' numbers are considerably diminished (also see Carr et al., 1982). Peskin (1996) reported on a conservation project at Plaplaya, Gracias a Dios, in which *C. caretta* eggs are removed from nests and artificially incubated, with the hatchlings released in the sea. Weigel (1973: 23) wrote that "loggerhead" turtles nest on the beaches that occur on Isla Grande on the Swan Islands. In general, *C. caretta* are free swimmers capable of making long migrations. Numerous records report females returning several times to nest at the same beach. Females throughout the species' range deposit about 7–220 ($x =$

99.7 in 590 clutches reported in the literature; Ernst and Lovich, 2009) eggs per clutch that are placed in a hole she digs in the beach sand. Females can nest one to seven times a season but also nest at 1–9-year intervals (Ernst and Lovich, 2009). *Caretta caretta* is "omnivorous, with a preference for invertebrates, especially gastropods, bivalves, and decapods" (Ernst and Lovich 2009: 50; those authors also presented an extensive list of food items reported for this species). Riosmena-Rodriguez and Lara-Uc (2015) reported a suspected diet shift in a *Caretta caretta* population in Baja California, Mexico. That turtle population was found to be targeting squat lobster, squid, octopus, and apparently sardines instead of predominantly feeding at the ocean surface. Meylan et al. (2013) reported that five *Caretta* tagged in Bocas del Toro, Panama, were recaptured in Nicaraguan waters, and another was taken near Cuba. Thus, *Caretta* individuals from that population also likely enter Honduran waters (also see Meylan et al., 2011). Ernst and Lovich (2009) also included much more information on reproduction, habitat, and food items of this species.

Etymology.—See the Etymology for the genus *Caretta*.

Specimens Examined (2, 1 skull, 1 egg [0]; Maps 5, 67).—**GRACIAS A DIOS**: Barra Patuca, USNM 20286 (1 egg); Islas Grande, Swan Islands, USNM 220768 (neurocranium only). **ISLAS DE LA BAHÍA**: Isla de Guanaja, near Savannah Bight, CM 62225; "Isla de Guanaja," LSUMZ 22418.

Other Records (Map 67).—**COLÓN**: Iriona (Marin, 1984). **GRACIAS A DIOS**: Laguna de Brus (Cruz [Díaz] and Espinal, 1987); Laguna de Brus, Laguna Tata (= Laguntara), near mouth of Río Kruta, near mouth of Río Negro (Marin, 1984); Plaplaya (Peskin, 1996; UF photographs). **ISLAS DE LA BAHÍA**: Isla de Guanaja, near Savannah Bight, USNM HI 2807; Isla de Roatán (Lundberg, 2001); Isla de Utila

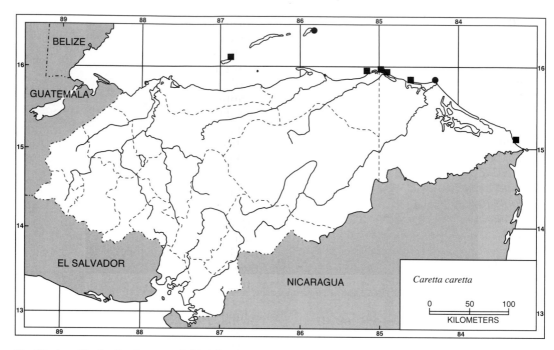

Map 67.   Localities for *Caretta caretta*. Solid circles denote specimens examined and solid squares represent accepted records. The species is also known from the Swan Islands, in the Atlantic Ocean (also see Map 5).

(Lundberg, 2000); Isla de Utila, Pumpkin Hill Beach (McCranie et al., 2005).

Genus *Chelonia* Brongniart, 1800

Chelonia Brongniart, 1800: 89 (type species: *Testudo mydas* Linnaeus, 1758: 197, by subsequent designation of Bell, 1828: 516).

Geographic Distribution and Content.—See Geographic Distribution for *Chelonia mydas* for this monotypic genus.

Remarks.—Some workers (e.g., King and Burke, 1989: 19; Iverson, 1992: 83; Pritchard, 1997: 18, 1999: 1003; Savage, 2002: 756) recognize *C. agassizii* Bocourt (1868: 122) as a distinct species for the eastern Pacific populations of *Chelonia*. However, DNA results to date have not shown that the eastern Pacific populations are distinct from other populations of *C. mydas* at the species level (see Naro-Maciel et al., 2008: 661, and references cited therein). Ernst and Lovich

(2009) presented a thorough review of the known geographic variation of *C. mydas*. Sterling et al. (2013: 9) reported that three of the 211 *Chelonia* captured in a central Pacific foraging area had the coloration and carapace shape considered consistent with eastern Pacific *Chelonia*. Those waters are well to the west of previously reported waters having the *Chelonia agassizii* morph.

Etymology.—The name *Chelonia* is formed from the Greek word *chelone* (tortoise or turtle).

Chelonia mydas (Linnaeus, 1758 [in part])

Testudo mydas Linnaeus, 1758: 197 (in part) (three or four syntypes, NHRM 19, 26, 231 [see King and Burke, 1989: 21; also see Wallin, 1985: 128], who mentioned only two NHRM specimens, plus one unnumbered carapace from Aldrovandi collection in Mus. Palazzo Poggi, Bologna, Italy [see Bauer et al.,

2013: 312]; type locality: "*insulas* Pelagi: insulam Adscensionis").

Chelonia mydas: Schweigger, 1812: 291; Carr et al., 1978: 19; Cruz [Díaz] and Espinal, 1987: 20; Peskin, 1996: 1; McCranie et al., 2005: 58; McCranie et al., 2006: 96; Köhler, 2008: 55; McCranie and Valdés Orellana, 2014: 42; Solís et al., 2014: 138; McCranie, 2015a: 379; McCranie and Gutsche, 2016: 888; McCranie et al., 2017: 277.

Chelone midas [sic]: Lowe, 1911: 32.

Chelonia mydas agassizii: Carr, 1952: 357.

Chelonia agassizi: Cruz [Díaz] et al., 1987: 341.

Geographic Distribution.—*Chelonia mydas* occurs in tropical and subtropical seas, also occasionally in temperate seas, normally between the 20° C isotherms in the Atlantic, Pacific, and Indian oceans. In Honduras, this species has been reported in both Pacific and Caribbean waters.

Description.—*Chelonia mydas* is a medium-sized to large sea turtle, usually reaching a CL of less than 1,050 mm and weighing less than 200 kg, but has been reported to reach 1,530 mm and a weight of 380 kg (Ernst and Lovich, 2009); carapace somewhat rounded, heart-shaped, broad, low, with smooth margins; dorsal scutes juxtaposed; 5 vertebrals; 4 pairs of costal scutes, with first pair separated from nuchal scute; usually 12 pairs of marginal scutes; hatchlings usually with middorsal longitudinal ridge on carapace; adults without ridges on carapace; plastron unhinged; hatchlings usually with 2 longitudinal plastral ridges; 6 pairs of plastral scutes, large intergular scute also present; bridge usually with 4 inframarginal scutes on each side, none of which have pores; 1 pair of prefrontal scales; outside cutting edge of upper jaw not serrated; horny inner surface of lower jaw serrate; 1 large inframandibular scale followed by 1 small scale posterior to beak on either side; limbs developed into paddlelike flippers, each usually with single, small claw;

tail of males with flattened, keratinized tip, tip extending well beyond margin of carapace; tail of females without keratinized tip, with projection not extending beyond margin of carapace (much of the above description taken from Hirth, 1980, and especially Ernst and Lovich, 2009).

Color in life of a subadult from Camp Bay, Roatán, Bay Islands (Plate 84): carapace Drab (27) with lineate figures consisting of Hazel (35), Prout's Brown (121A), Cinnamon (39), and other varying shades of brown on vertebrals, costals, and posterior marginals, other marginals mottled with similar colors; scales on top of head Dark Brownish Olive (129) with Hazel, Prout's Brown, and Cinnamon mottling, each scale on top of head outlined with Pale Horn Color (92); scales on dorsal surfaces of fore- and hind limb black, some mottled with gray and grayish brown, all scales outlined with Pale Horn Color; plastron yellowish white with some indistinct brown mottling; ventral surface of head and neck yellow with brown lineate pattern; ventral surfaces of fore- and hind limb yellow with brown mottling and lines, except distal portions dark brown.

Diagnosis/Similar Species.—*Chelonia mydas* is distinguished from all remaining Honduran turtles, except the other sea turtles, by having the limbs modified as paddlelike structures. *Dermochelys coriacea* has the shell covered by leathery skin over a layer of connective tissue (versus shell covered with epidermal scutes or plates in *C. mydas*). *Caretta caretta* and *Lepidochelys olivacea* have 5 or more pairs of costal scutes, the first pair contacting the nuchal scute (versus 4 pairs of costal scutes with first pair separated from nuchal scute in *C. mydas*). *Eretmochelys imbricata* has 2 pairs of prefrontal scales, the inside surface of the lower jaw smooth to only weakly serrated, and the outside cutting surface of the upper jaw serrated (versus 1 pair of prefrontal scutes, inside surface of lower jaw strongly serrated, and outside cutting surface of upper jaw not serrated in *C. mydas*).

Plate 84.   *Chelonia mydas.* Islas de la Bahía: Roatán Island, near Camp Bay (released).

Illustrations (Figs. 157, 162, 163; Plate 84).—Álvarez del Toro, 1983 (adult); Bonin et al., 2006 (adult; as *C. agassizii* and *C. mydas*); Calderón-Mandujano et al., 2008 (adult); Carr, 1952 (adult), 1968 (hatchlings, adult), 1986 (adult); Conant and Collins, 1998 (adult); Ernst and Barbour, 1989 (adult); Ernst and Lovich, 2009 (adult, plastron, head); Ernst et al., 1994 (adult, plastron, head); Freiberg, 1972 (adult); Fritz, 2012b (adult); Halstead, 1970 (adult): Köhler, 2000 (adult), 2001b (adult), 2003a (adult, head, carapace), 2008 (adult, head, carapace); Köhler et al., 2005 (adult); LeBuff, 1990 (adult); Lee, 1996 (adult, carapace, head scales), 2000 (adult, carapace, head scales); Lemos-Espinal, 2015 (adult); Lemos-Espinal and Smith, 2009 (adult); Lutz and Musick, 1997 (head); Márquez M., 1990 (adult, head scales, carapace, plastron); McCranie and Valdés Orellana, 2014 (adult, subadult); McCranie et al., 2005 (adult, hatchling, head scales), 2006 (adult, carapace, head scales); A. B. Meylan et al., 2013 (adult); P. A. Meylan et al., 2011 (head); Murphy, 1997 (adult); Nietschmann, 1977 (adult); Pough et al., 2003 (adult), 2015 (adult); Powell, Collins, and Hooper, 1998 (head scales), 2012 (head scales); Pritchard, 1979 (adult), 1997 (adult, head; as *C. agassizii* and *C. mydas*); Pritchard and Trebbau, 1984 (adult, sub-adult); Rebel, 1974 (inframarginal bridge scutes, dorsal head scales, adult); Ruckde-schel and Shoop, 2006 (adult, juvenile, hatchling, head); Sagra, 1840 (adult, cara-pace, plastron, head; as *C. [mydas] virgata* Schweigger, 1812: 291); Savage, 2002 (adult, dorsal head scales, carapace, flipper scales, lateral view of lower jaw; as *C. agassizii* and *C. mydas*); Seminoff and Wallace, 2012 (adult, juvenile, hatchling); Shi, 2013 (adult, plastron); Smith and Smith, 1980 (adult, subadult); Sowerby and Lear, 1872 (adult); Spotila, 2004 (adult, hatchling, head), 2011 (head); Stafford and Meyer, 1999 (adult, head scales); Vetter, 2004 (adult, subadult; as *C. m. mydas* and *C. m. agassizii*), 2005 (adult, subadult, juvenile; as *C. m. mydas* and *C. m. agassizii*); Vitt and Caldwell, 2014 (adult); Wagler, 1833 (adult, carapace, plastron; as *C. virgata*); Wermuth and Mertens, 1961 (adult, carapace, plastron; as *C. m. mydas*).

Remarks.—Hirth (1980) and Ernst and Lovich (2009) presented an overview of the morphology of *Chelonia mydas* and reviewed the literature on this species, with that of Ernst and Lovich (2009) being especially thorough. See Remarks for the genus for information on the systematic status of Atlantic and eastern Pacific populations of *Chelonia*. Wallin (1985: 128) reported that part of the Linnaeus (1758: 197) concept of *Testudo mydas* consisted of *Caretta caretta*.

Natural History Comments.—Very little has been reported on the habitat and behavior of this species on Honduran beaches and in Honduran waters. Cruz [Díaz] et al. (1987) reported a *Chelonia mydas* (as *C. agassizi*) nesting on 29 August 1985 at Punta Ratón, Choluteca. I saw a hatchling of this species in captivity at Rocky Point, Isla de Roatán, in November 1989. Carr et al. (1978) reported eight recoveries from 1956 to 1976 in Honduran waters of turtles tagged while nesting at Tortuguero, Costa Rica, whereas Groombridge (1982) reported 27 recovered tags in Honduran waters from turtles tagged at the same nesting ground. Carr et al. (1982) reported that *C. mydas* nests on the extensive beaches between Puerto Cortés and La Ceiba, but their numbers were greatly diminished from earlier times. Carr et al. (1982) also reported that the species nests regularly, although not abundantly, on the Cayos Vivorillos and Caratasca in the Miskito Cays off the coast of Gracias a Dios. Lewis (1941) gave a brief historical account of vessels from the Cayman Islands capturing *C. mydas* in Honduran and Nicaraguan waters. Lowe (1911) gave a general discussion on nesting habits of this turtle on Isla Grande in the Swan Islands. In general, *C. mydas* are free swimmers that are known to be capable of making long migrations. This species is the only marine turtle known to habitually bask out of water (Ernst and Lovich, 2009; also see Maxwell et al., 2014, for basking data). Free-ranging females deposit 1–18 clutches in a season ($x = 3.5$, $n = 535$) in a hole she digs in the beach sand, with reported clutch size varying from 3–238 eggs ($x = 115.3$, $n = 660$), with most females nesting every third year (Ernst and Lovich, 2009). Juvenile *C. mydas* eat much animal prey and adults are decidedly herbivorous, eating principally sea grasses and other plants (Ernst and Lovich, 2009). Lara-Uc and Riosmena-Rodriguez (2015) reported an apparent dietary shift of a Baja California, Mexico, population to octopus, squid, and apparently sardines. Meylan, A. B. et al. (2013) reported small numbers of *C. mydas* nesting in Bocas del Toro, Panama, but also gave much data on life stages of green turtles in the region and reported 16 Bocas del Toro turtles having been reported foraging in the waters off Nicaragua. Since the sea grass beds of Nicaragua extend into the waters off northeastern Honduras, it is likely that the *Chelonia* turtles tagged in Bocas del Toro also forage in Honduran waters (also see P. A. Meylan et al., 2011). Ernst and Lovich (2009) also included much additional information on reproduction, habitat, and food items of this species.

Etymology.—The name *mydas* is formed from the Greek word *mydos* (dampness) and refers to the aquatic habitat of this species.

Specimens Examined (8 heads [0]; Map 68).—**VALLE**: Golfo de Fonseca near San Lorenzo, MCZ R49401–05 (all heads only), 145747–49 (all heads only).

Other Records (Maps 5, 68).—**CHOLUTECA**: Punta Ratón (Cruz [Díaz] et al., 1987; one neonate deposited in UCR collection). **GRACIAS A DIOS**: Cayo Becerro, Cayo Caratasca, Cayo Vivorillo Grande (Carr et al., 1982); Isla Grande, Swan Islands (Lowe, 1911); Río Kruta Biological Reserve, Río Plátano Biosphere Reserve (McCranie et al., 2006). **ISLAS DE LA BAHÍA**: Isla de Guanaja, near Savannah Bight, USNM HI 2809; Isla de Roatán, Anthony's Cay (Cruz [Díaz] and Espinal, 1987); Isla de Roatán, Camp Bay, USNM HI 2808; Isla de Roatán, Rocky Point

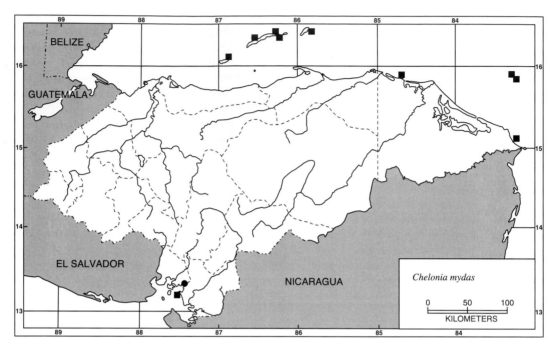

Map 68.   Localities for *Chelonia mydas*. The solid circle denotes a specimen examined and solid squares represent accepted records. The species is also known from the Swan Islands, in the Atlantic Ocean (also see Map 5).

(personal sight record); Isla de Roatán, (photographs in McCranie et al., 2005); Isla de Utila (Carr et al., 1978).

Genus *Eretmochelys* Fitzinger, 1843

Eretmochelys Fitzinger, 1843: 30 (type species: *Chelonia imbricata* Cuvier, 1816: 13 [= *Testudo imbricata* Linnaeus, 1766: 350], by original designation).

Geographic Distribution and Content.— See Geographic Distribution for *Eretmochelys imbricata* for this monotypic genus.

Remarks.—Okayama et al. (1999: 366) compared mitochondrial DNA between Indo-Pacific and Caribbean populations of *Eretmochelys* and found some haplotype differences between those two populations. As such, Iverson et al., *In* Crother (2012: 76), and Ernst and Lovich (2009: 83) used *E. i. imbricata* for the Caribbean popula-

tions and *E. i. bissa* (Rüppell, 1835: 4) for the Pacific populations, but Ernst and Lovich (2009: 83) also said "the characters used to separate these supposed races may be of little value."

Etymology.—The name *Eretmochelys* is formed from the Greek words *eretmon* (oar) and *chelys* (tortoise, turtle), and refers to the oarlike limbs of this marine turtle.

Eretmochelys imbricata (Linnaeus, 1766 [in part])

Testudo imbricata Linnaeus, 1766: 350 (in part) (apparently no surviving type specimens [see Wallin, 1985: 128; also see Remarks]; type locality: "*in Mari Americano, Asiatico*").

Eretmochelys imbricata: Fitzinger, 1843: 30; Lowe, 1911: 32; Marin, 1984: 222; Cruz [Díaz] and Espinal, 1987: 12; Peskin, 1996: 1; Lundberg, 2000: 3; Lundberg, 2001: 27; Lundberg, 2002a: 9; McCranie et al., 2005: 60; McCranie et al.,

2006: 97; Köhler, 2008: 55; Berube et al., 2012: 34; McCranie and Valdés Orellana, 2014: 42; Solís et al., 2014: 139; McCranie, 2015a: 380; McCranie and Gutsche, 2016: 889; McCranie et al., 2017: 227.
Eretmochelys imbricata squamata: Carr, 1952: 373.

Geographic Distribution.—This marine species is primarily tropical in distribution but is also known from subtropical and occasionally temperate waters in the Atlantic, Pacific, and Indian oceans. In Honduras, it is known from the Pacific waters and shoreline of the Golfo de Fonseca, as well as the Bay Islands, Swan Islands, the Cayos Vivorillos, and Cayo Gorda in Caribbean waters and the Caribbean coastline near the mainland.

Description.—*Eretmochelys imbricata* is a medium-sized sea turtle, typically reaching a CL of less than 800 mm, but has been reported to reach 1,140 mm and weigh up to 127 kg (Ernst and Lovich, 2009); carapace relatively long, narrow in adults, more heart-shaped in juveniles; margins of carapace strongly serrated posteriorly, usually with imbricate carapacial scutes (juveniles and very old animals with non-overlapping carapacial scutes); 5 vertebrals; 4 pairs of costal scutes, with first pair not in contact with nuchal scute; usually 12 pairs of marginals; hatchlings usually with mid-dorsal longitudinal ridge on carapace, ridge lost in adults; plastron unhinged; hatchlings usually with 2 longitudinal plastral ridges, ridges lost in adults; 6 pairs of plastral scutes, large intergular scute also present; bridge usually with 4 inframarginal scutes on each side, all without pores; head relatively narrow, with 2 pairs of prefrontal scales; upper jaw somewhat hooked, hawk-like in shape; outer cutting surface of upper jaw strongly serrated; inside surface of lower jaw with smooth or weakly serrated margin; 1 large inframandibular scale present posterior to beak; limbs paddlelike, each usually with 2 small claws; males with long tail extending well beyond margin of carapace, females with shorter tail extending to edge of, or just slightly beyond margin of carapace (much of the above was taken from Smith and Smith, 1980; Ernst and Lovich, 2009; supplemented by my notes and photographs).

Color in life of a subadult from Camp Bay, Roatán Island, in the Bay Islands: carapace Buff (124) or Buff with rust tinge, with elongated Sepia (119), Hazel (35), or Chestnut (32) marks, or greenish yellow and several variations of those colors; scales on top of head Sepia, some with pale brown mottling, all outlined with rusty brown seams or pale yellow seams with rust mottling; scales on dorsal surfaces of fore- and hind limb Sepia, some with reddish brown mottling, all scales outlined with yellow or yellowish brown seams; beak and scales on side of head with Sepia or reddish brown centers and pale yellow to rusty yellow margins; plastron yellowish brown with indistinct brown lineate mottling; ventral surfaces of head and neck pale yellow to rusty yellow, with brown cross-lines; ventral surfaces of fore- and hind limb yellowish brown with brown mottling, except outer edges and feet of fore- and hind limb dark brown, hind foot darker brown than forefoot. Color in life of a juvenile (from Yahurabila, Gracias a Dios): carapace Burnt Umber (22) with Cinnamon (123A) mottling; dorsal surface of head Cinnamon with grayish black along scale seams; neck gray with pale brown tubercles; all limbs grayish black with narrow pale brown line along each edge; plastron grayish black with pale brown on ridges.

Diagnosis/Similar Species.—*Eretmochelys imbricata* is distinguished from all remaining Honduran turtles, except the other sea turtles, by having the limbs modified into paddlelike structures. *Dermochelys coriacea* has the carapace covered by leathery skin over a layer of connective tissue (versus shell covered with epidermal

Plate 85.   *Eretmochelys imbricata.* Gracias a Dios: Isla Grande, Swan Islands (released).

scutes in *E. imbricata*). *Caretta caretta* and *Lepidochelys olivacea* have 5 or more pairs of costal scutes, with the first pair in contact with the nuchal scute (versus 4 pairs of costals, first pair separated from nuchal in *E. imbricata*). *Chelonia mydas* has a single pair of prefrontal scales, the inside surface of the lower jaw strongly serrated, and the outside cutting surface of the upper jaw not serrated (versus 2 pairs of prefrontal scales, inside surface of lower jaw smooth to weakly serrated, and outside cutting surface of upper jaw serrated in *E. imbricata*).

Illustrations (Figs. 145, 160, 161; Plate 85).—Álvarez del Toro, 1983 (adult); Blumenthal et al., 2009 (subadult); Bonin et al., 2006 (adult); Calderón-Mandujano et al., 2008 (adult); Carr, 1952 (adult, subadult; as *Eretmochelys imbricata squamata* Agassiz, 1857: 382), 1968 (subadult, adult), 1986 (adult); Conant and Collins, 1998 (adult); Duvernoy, 1836–1849 (adult, carapace, plastron; as *Chelonia*); Ernst and Barbour, 1989 (adult); Ernst and Lovich, 2009 (adult, juvenile, plastron); Ernst et al., 1994 (adult, plastron, head); Freiberg, 1972 (adult); Fritz, 2012b (adult); Halstead, 1970 (adult); Köhler, 2000 (subadult), 2001b (subadult),

2003a (adult, head, carapace), 2008 (adult, head, carapace); Köhler et al., 2005 (adult); LeBuff, 1990 (adult); Lee, 1996 (adult, head scales), 2000 (adult, head scales); Lemos-Espinal, 2015 (adult); Lemos-Espinal and Smith, 2009 (adult); Lutz and Musick, 1997 (adult); Márquez M., 1990 (adult, head scales, carapace, plastron); McCranie and Valdés Orellana, 2014 (subadult); McCranie et al., 2005 (adult, hatchling, head scales), 2006 (adult, head scales), 2017 (adult); Meylan et al., 2013 (adult); Murphy, 1997 (adult); Nietschmann, 1977 (adult); Pough et al., 2015 (adult); Powell, Collins, and Hooper, 1998 (head scales, carapace), 2012 (head scales, carapace); Pritchard, 1979 (adult, subadult), 1997 (adult); Pritchard and Trebbau, 1984 (adult, subadult); Rebel, 1974 (inframarginal bridge scutes, dorsal head scales, adult); Ruckdeschel and Shoop, 2006 (adult, hatchling, head); Savage, 2002 (adult, dorsal head scales, carapace, lateral view of lower jaw); Seminoff and Wallace, 2012 (adult, juvenile, hatchling, head); Shi, 2013 (adult, head, carapace, plastron); Smith and Smith, 1980 (adult); Sowerby and Lear, 1872 (subadult; as *Chelonia*); Spotila, 2004 (adult, head), 2011 (head);

Stafford and Meyer, 1999 (adult, head scales); Vetter, 2004 (adult, hatchling; as *E. i. imbricata* and *E. i. bissa*), 2005 (adult; as *E. i. imbricata* and *E. i. bissa*); Wermuth and Mertens, 1961 (adult, plastron; as *E. i. imbricata*); Witzell, 1983 (juvenile); Zug, 1966 (schematic drawing of penis).

Remarks.—Carr (1952: 376) provided measurements of an adult female from the Golfo de Fonseca, Honduras, as well as other general data of this species from that locality.

Smith and Smith (1980: 280) assumed ZMUU 130 was the holotype of *Testudo imbricata* Linnaeus. However, Wallin (1985: 128) stated that the single ZMUU specimen apparently seen by Linnaeus for his 1766 work is actually a specimen of *Chelonia mydas*. There are no surviving type specimens of *T. imbricata* (see Wallin, 1985: 128).

Natural History Comments.—The Honduran Caribbean populations of *Eretmochelys imbricata* have greatly declined in recent years (as have all sea turtle populations along both coasts), and surveys between 1982 and 1987 revealed only sparse nesting (Meylan, 1999a). I found and released a juvenile just offshore of Yahurabila, Gracias a Dios, in February 2008. Marin (1984) reported two sightings of this species from near Utila Island in the Bay Islands, and near Tela, Atlántida. Two hawksbills tagged at Tortuguero, Costa Rica, were recaptured at Isla de Guanaja and the Cayo Gorda Banks, Honduras (Meylan, 1999b). Another *Eretmochelys* that was tagged near the Dominican Republic on 20 April 1998, and recaptured and tagged again on 21 November 2000 in the same area, was killed in February 2003 off Cayo Gordo, Honduras (information gathered from tags that were purchased from a resident of Tikiraya, Gracias a Dios). Blumenthal et al. (2009) reported one that was tagged at Little Cayman Island was recaptured 6.7 yrs later in the waters off northeastern Honduras. Groombridge

(1982) reported nesting on beaches between Puerto Cortés, Cortés, and La Ceiba, Atlántida, but with greatly reduced numbers compared with those of earlier years (also see Carr et al., 1982). Berube et al. (2012) studied juvenile hawksbills in the area of Port Royal on Roatán and found the six they studied had small home ranges. Berry, F. H. (1989) reported an estimated 5,000 *E. imbricata* were captured in the marine waters of Nicaragua and Honduras during 1986–1987 and shipped to Japan, and Donnelly (1989) estimated that 10,287 hawksbill carapaces were shipped from Honduras to Japan during 1972–1986. Shells of this species can be seen in several bars and restaurants on Roatán. Weigel (1973: 23) wrote that "hawksbill" turtles nest on the beaches that occur on Isla Grande in the Swan Islands. In general, *E. imbricata* is a tropical inhabitant of relatively shallow waters (<20 m) with underwater cliffs, coral reefs, and sponge areas but sometimes enters shallower coastal waters (Ernst and Lovich, 2009). Horrocks et al. (2011) concluded that Caribbean *Eretmochelys* used the shallow continental shelf off the coasts of Nicaragua and Honduras as an important foraging ground. Hill and King (2015) reported what was apparently the first record of *E. imbricata* (a juvenile turtle) feeding on a toxic Bearded Fireworm in the sea near the US Virgin Islands. Meylan et al. (2013) mentioned monitoring satellite tracks of female *Eretmochelys* that nested on beaches in the Bocas del Toro, Panama, having traveled to Honduran waters and as far as Jamaica. Eastern Pacific populations of *Eretmochelys* are more poorly known but appear to forage in mangrove estuaries or rocky reefs (Seminoff et al., 2012). Females range wide deposit about 26–250 ($x = 140$, $n = 330$) eggs per clutch in a hole she digs in the beach sand (Ernst and Lovich, 2009). Females can nest two to six times per season, but nesting usually only occurs every 2 or 3 years (Ernst and Lovich, 2009). Langueux et al. (2003)

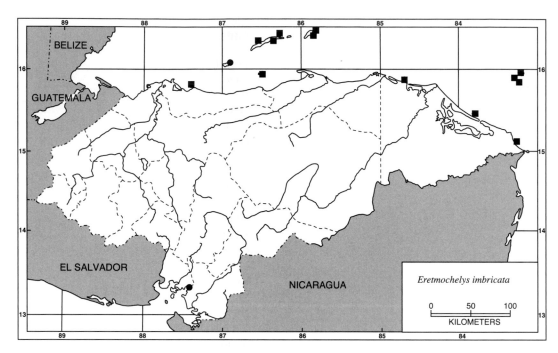

Map 69.   Localities for *Eretmochelys imbricata*. Solid circles denote specimens examined and solid squares represent accepted records. The species is also known from the Swan Islands, in the Atlantic Ocean (also see Map 5).

provided a wealth of data on nesting of *E. imbricata* on the Pearl Cays of Nicaragua and concluded those cays were one of the largest nesting sites of this species remaining in the central-western Caribbean. Cornelius (1981) briefly touched on nest poaching of hawksbills along the Pacific coast of Honduras. Gaos and Yañez (2012) also mentioned having documented *Eretmochelys* nesting on the Pacific coast of Honduras. Carrión-Cortez et al. (2013) studied juvenile habitat use and diet of a Pacific population in Costa Rica and found they feed on sponges and a tunicate and commonly used rocky reefs for foraging. Adults prefer invertebrates, particularly sponges, whereas juveniles are largely herbivorous (Ernst and Lovich, 2009). Berube et al. (2012) found sponges and octocorals were the most prevalent food items of juvenile hawksbills in the Port Royal region on Roatán. Ernst and Lovich (2009) also included much additional information on reproduction, habitat, and food items of this species.

Etymology.—The name *imbricata* is derived from the Latin *imbricatus* (overlapping, like roofing tiles and shingles) and refers to the overlapping carapacial scutes of juveniles and young adults of this species.

Specimens Examined (1, +1 hatchling, 5 heads and flippers, 4 heads only [0]; Maps 5, 69).—**CHOLUTECA** or **VALLE**: Golfo de Fonseca, AMNH 70561–65 (all heads and flippers only). **GRACIAS A DIOS**: Islas Grande, Swan Islands, MCZ R7050. **ISLAS DE LA BAHÍA**: Isla de Utila, Pumpkin Hill Beach, SMF 82640 (hatchling). **VALLE**: Golfo de Fonseca near San Lorenzo, MCZ R49406–09 (all heads only).

Other Records (Map 69).—**ATLÁNTIDA**: near Tela (Marin, 1984; estimated from coordinates given by her). **GRACIAS A DIOS**: Cayo Gorda (recovered tags); Cayos Vivorillos (shell from subadult seen in May 2011, but not taken); Isla Grande, Swan

Islands, USNM HI 2813; Río Kruta Biological Reserve, Río Plátano Biosphere Reserve (McCranie et al., 2006); Yahurabila, USNM HI 2727. **ISLAS DE LA BAHÍA**: Cayo Cochino Menor (Lundberg, 2002a); Isla de Guanaja, near Savannah Bight, USNM HI 2811–12; Isla de Guanaja (Meylan, 1999b); Isla de Roatán, Anthony's Cay (Cruz [Díaz] and Espinal, 1987); Isla de Roatán, Port Royal region (Berube et al., 2012); Isla de Roatán, Camp Bay, USNM HI 2810; "Isla de Roatán" (photographs in Lundberg, 2001; McCranie et al., 2005); Isla de Utila (Cruz [Díaz] and Espinal, 1987); near Isla de Utila (Marin, 1984; estimated from coordinates given by her).

Genus *Lepidochelys* Fitzinger, 1843

Lepidochelys Fitzinger, 1843: 30 (type species: *Thallassochelys olivacea* Fitzinger, 1843: 30 [= *Chelonia olivacea* Eschscholtz, 1829a: 15], by original designation).

Geographic Distribution and Content.— This genus occurs in waters of tropical and subtropical seas and occasionally enters temperate seas. Two named species are recognized, one of which is known to occur in Honduran marine waters.

*Etymology.—*The name *Lepidochelys* is formed from the Greek word *lepidos* (scale) and *chelys* (tortoise, turtle), and probably refers to the scutes covering the shell.

Lepidochelys olivacea (Eschscholtz, 1829a)

Chelonia olivacea Eschscholtz, 1829a: 15 (two syntypes examined by Eschscholtz, but their location not stated, probably lost [see Flores-Villela et al., 2016: 161]; type locality: "chinesischen Meere" on p. 18, but said to be "der Bai von Manilla" in Eschscholtz, 1829b: 4).
Lepidochelys olivacea: Girard, 1858b: 435 (see Remarks); Carr, 1948: 51; Carr, 1952: 403; Cruz [Díaz] et al., 1987: 341; McCranie, Valdés Orellana, and Gutsche, 2013: 289; Solís et al., 2014: 139; McCranie, 2015a: 380; McCranie and Gutsche, 2016: 889.

Geographic Distribution.—Lepidochelys olivacea occurs in tropical and subtropical waters of the Indo-West Pacific region and the eastern Pacific. It also occasionally ranges to the coasts of Chile, Alaska, South Africa (at meeting place of Pacific and Atlantic oceans), and New Zealand in the Pacific Ocean and in the Atlantic Ocean between Mauritania and Guinea in West Africa and from northeast Cuba to the Guianas in the western Atlantic. In Honduras, this species is known only from the Golfo de Fonseca waters and shorelines.

Description.—Lepidochelys olivacea is a relatively small sea turtle with a CL up to 735 mm (Ernst and Lovich, 2009) and a weight up to 60 kg (Savage, 2002; but no source of data given); carapace broad (often as broad as long), widest toward midpoint, with smooth margins and non-overlapping scutes, except posterior margins slightly serrate in hatchlings and juveniles; 5 vertebrals; about 5–6 pairs of costal scutes (one side of carapace can have more costals than other side), with first pair in contact with nuchal scute; 12–14 pairs of marginals; juvenile carapace with narrow, medial projection or partial longitudinal ridge on each vertebral scute, adult carapace without ridges; plastron unhinged; hatchlings usually with 2 longitudinal plastral ridges, those ridges lost in adults; 12 plastral scutes, arranged in 6 pairs, with small intergular scute present or absent; bridge usually with 4 inframarginal scutes on each side that usually have a posterior pore on each scute; 2 pairs of prefrontal scales; head large, relatively triangular-shaped, with smooth-margined lower jaw; lower jaw with 1 large inframandibular scale followed by 1 small scale posterior to beak on either side; limbs developed into paddlelike flippers, 2 small claws present on each flipper in young individuals, often only 1 claw on each flipper in adults; tail of males extending

Plate 86.   *Lepidochelys olivacea.* Adult female. Costa Rica: Guanacaste, Playa Naranjo. Photograph by Louis Porras.

beyond carapacial margin, tail of females extending to carapacial margin (above characters based largely on Zug et al., 1998; Ernst and Lovich, 2009).

Color in life: carapace uniformly grayish green to olive brown; dorsal surfaces of head, limbs, and tail olive to grayish brown; bridge and plastron greenish white to greenish yellow; ventral surfaces of head, limbs, and tail creamy white (brief color notes taken from Zug et al., 1998; Ernst and Lovich, 2009).

Diagnosis/Similar Species.—*Lepidochelys olivacea* is distinguished from all remaining Honduran turtles, except the other marine turtles, by having the limbs modified as paddlelike structures. *Dermochelys coriacea* has the shell covered by leathery skin over a layer of connective tissue (versus shell covered with epidermal scutes or plates in *L. olivacea*). *Chelonia mydas* and *Eretmochelys imbricata* have 4 pairs of costal scutes and the first pair separated from the nuchal scute (versus 5–6 pairs of costal scutes, first pair contacting nuchal scute in *L. olivacea*). *Caretta caretta* has the bridge with usually 3 poreless inframarginal scutes and has 3–7 inframandibular scales posterior to the beak on the lower jaw (versus bridge usually with 4 inframarginal scutes, each usually with a posterior pore, and single inframandibular scale in *L. olivacea*).

Illustrations (Figs. 158, 159, 166, 167; Plate 86).—Bonin et al., 2006 (adult); Carr, 1952 (adult); Ernst and Barbour, 1989 (adult); Ernst and Lovich, 2009 (adult); Ernst et al., 1994 (adult); Hödl, 1996 (adult); Köhler, 2001b (adult), 2003a (adult, carapace), 2008 (adult, carapace); Köhler et al., 2005 (adult); Lemos-Espinal, 2015 (adult); Lutz and Musick, 1997 (adult); Márquez M., 1990 (adult, head scales, carapace, plastron); Murphy, 1997 (adult); Pritchard, 1979 (adult), 1997 (adult); Pritchard and Trebbau, 1984 (adult, subadult, hatchling); Ruckdeschel and Shoop, 2006 (adult, hatchling, head); Savage, 2002 (adult, carapace, ventral view of bridges, ventral view of jaws, lateral view of lower jaw); Seminoff and Wallace, 2012 (adult, hatchling); Shi, 2013 (adult, head, plastron); Smith and Smith, 1980 (carapace, plastron); Spotila, 2004 (adult, hatchling); Vetter, 2004 (adult), 2005 (adult, juvenile); Wermuth and Mertens, 1961 (adult, carapace, plastron; as *L. o. olivacea*).

Remarks.—Flores-Villela et al. (2016) discovered that the original type description of *Lepidochelys olivacea* was in Eschscholtz (1829a) rather than in Eschscholtz (1829b), as previously thought.

See Remarks for *Caretta caretta* regarding variation in number of inframarginal bridge scutes in *Caretta* and *Lepidochelys*.

Zug et al. (1998) and Ernst and Lovich (2009) summarized the morphology of *Lepidochelys olivacea* and provided a literature summary, with that of Ernst and Lovich (2009) being especially thorough. Ernst and Lovich (2009: 104) said, regarding *L. olivacea*, "no subspecies are recognized, but it is possible that the eastern Pacific and southern Atlantic populations are distinct." Carr (1952: 404–405) gave some shell data for four adult females from the "Pacific coast of Honduras."

Natural History Comments.—Cruz [Díaz] et al. (1987) noted finding 279 nests of *Lepidochelys olivacea* between 15 and 30 August 1985 at Punta Ratón, Choluteca. Carr (1948) discussed egg deposition by this species at Punta Ratón (as Isla de Ratones), with much of that information repeated by Carr (1952; also see Carr, 1986). Cornelius (1981) wrote that these turtles were common in the Golfo de Fonseca from July through December and that most nesting activity was concentrated in the areas of Punta Ratón and Cedeño, Choluteca. Cornelius (1981) discussed egg poaching of *L. olivacea* at Punta Ratón and Cedeño. Langueux (1991) studied egg deposition in this species at Punta Ratón in December 1987 and the resulting economic benefit for the residents of the village. Eggs of this species were available for consumption at several restaurants along the beach at Cedeño in July 2004. No turtles could be found on the Cedeño beach during one complete night of searching in July 2004, although scattered groups of poachers were

camped along the beach. Females range-wide deposit about 30–182 ($x = 111$, $n = 428$) eggs per clutch in a hole she digs in the beach sand, and females nest one to three (usually two) times per season and are apparently capable of nesting every year (Ernst and Lovich, 2009). *Lepidochelys olivacea* apparently prefers warm and shallow marine waters, often between reefs and the shore or in large bays and lagoons, but it also migrates in the sea (Ernst and Lovich, 2009). This species is primarily carnivorous, feeding mostly on jellyfish or protochordates (Ernst and Lovich, 2009). Ernst and Lovich (2009) also included much additional information on reproduction, habitat, and food items of this species.

Several residents on small islands in the Golfo de Fonseca complained to us during 2010–2012 that rising seawaters were destroying the small beaches on those islands where these turtles used to nest. It was also extremely obvious, while standing on one of those beaches, that rising sea levels are in the process of destroying those small beaches (also see McCranie and Gutsche, 2016).

Etymology.—The name *olivacea* is formed from the Latin word *oliva* (olive tree) and refers to the characteristic olive color of this species.

Specimens Examined (46 hatchlings, 1 egg shell [0]; Map 70).—**CHOLUTECA**: Punta Ratón, UF 34483–517 (hatchlings), 69895–905 (hatchlings). **VALLE**: Isla Garrobo, UNAH 5654 (egg shell).

Other records (Map 70).—**CHOLUTECA**: Cedeño (Cornelius, 1981).

Family Dermochelyidae Baur, 1888

This family of living sea turtles is monogeneric and monospecific and occurs circumglobally in tropical, subtropical, and temperate oceans, including Honduran Caribbean coastal waters. Members of the Dermochelyidae can be distinguished externally from all remaining Honduran turtles by having the limbs modified into

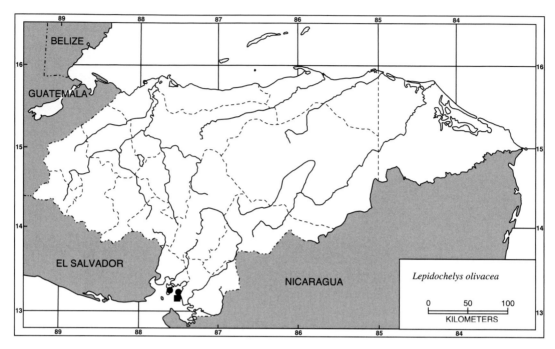

Map 70.   Localities for *Lepidochelys olivacea*. Solid circles denote specimens examined and the solid square an accepted record.

paddlelike structures and the shell composed of leathery skin over a layer of connective tissue, with adults lacking epidermal scutes.

Genus *Dermochelys* Blainville, 1816

Dermochelys Blainville, 1816: 111 (type species: *Testudo coriacea* Linnaeus, 1766: 350 [= *Testudo coriacea* Vandelli, 1761: 2], by monotypy, through Cuvier, 1829: 14 [see ICZN, 1926: 339, although erroneously including Linnaeus as author; also see Hemming, 1956a: 351–352, 1956b: 374]).

Geographic Distribution and Content.— See Geographic Distribution of *Dermochelys coriacea* for this monotypic genus.

Etymology.—The name *Dermochelys* is formed from the Greek words *derma* (skin, hide) and *chelys* (tortoise, turtle), and refers to the coarse skin covering the carapace.

Dermochelys coriacea (Vandelli, 1761)

Testudo coriacea Vandelli, 1761: 2 (see Remarks) (holotype, unnumbered specimen in collections of Mus. Zool. Univ. Padova, Italy [see Fretey and Bour, 1980: 198]; type locality: "*maris Tyrrheni oram in agro Laurentiano*").

Dermochelys coriacea: Boulenger, 1889: 10; Marin, 1984: 221; Cruz [Díaz] and Espinal, 1987: 22; Peskin, 1996: 1; McCranie et al., 2006: 99; Solís et al., 2014: 139; McCranie, 2015a: 380; McCranie and Gutsche, 2016: 893.

Geographic Distribution.—*Dermochelys coriacea* occurs in tropical to temperate oceans. In Honduras, this species has been reported from the Caribbean coastal waters near La Ceiba, Atlántida, eastward.

Description.—*Dermochelys coriacea* is a huge sea turtle (largest of all extant turtles) that typically reaches a CL between 1,320 and 1,780 mm (Spotila, 2004) but is

Plate 87.   *Dermochelys coriacea*. Adult female. Costa Rica: Guanacaste, Playa Naranjo. Photograph by Louis Porras.

reported to reach 2,438 mm (Ernst and Lovich, 2009); carapace and plastron covered by leathery skin over layer of connective tissue, except some dermal scutes present in recent hatchlings; carapace with 7 knobby, longitudinal ridges; 5 plastral ridges present in most subadults, but ridges almost or completely lost in large adults; juveniles covered with small epidermal scales, with carapacial and plastral ridges formed by yellow or white scales; upper jaw of large adults with large triangular cusp below nostril; lower jaw with median hook fitting into notch at upper jaw symphysis; 7 or more small inframandibular scales on lower jaw; limbs paddlelike, lacking claws. The above description was taken largely from Pritchard (1980) and, especially, Ernst and Lovich (2009).

Color in life: carapace brown to black, uniform or with white, cream, yellow, or pink spots laterally; dorsal surfaces of head, limbs, and tail brown to black, with scattered white, yellow, or pink spots; margins of limbs cream or white; plastron and ventral surfaces of head, limbs, and tail pale brown to creamy white, pink to cream spots present on head, limbs, and tail (brief color notes from Ernst and Lovich, 2009).

Diagnosis/Similar Species.—*Dermochelys coriacea* can be distinguished from all remaining Honduran turtles, except the other sea turtles (Cheloniidae), by having the limbs modified into paddlelike structures. It differs from the Cheloniidae by having its carapace formed by leathery skin over a layer of connective tissue (versus carapace covered with epidermal scutes or plates in Cheloniidae).

Illustrations (Fig. 156; Plate 87).— Álvarez del Toro, 1983 (adult); Bonin et al., 2006 (adult, head); Calderón-Mandujano et al., 2008 (adult); Caldwell et al., 1956 (hatchling); Carr, 1952 (adult, subadult), 1968 (hatchling, adult), 1986 (hatchling, adult); Conant and Collins, 1998 (adult); Duvernoy, 1836–1849 (subadult; as *Sphargis*); Ernst and Barbour, 1989 (adult); Ernst and Lovich, 2009 (adult, hatchling); Ernst et al., 1994 (adult); Fritz, 2012b (adult); Halstead, 1970 (adult); Köhler, 2000 (adult), 2001b (adult), 2003a (adult, carapace), 2008 (adult, carapace); Köhler et al., 2005 (adult); LeBuff, 1990 (adult); Lee, 1996 (adult, head), 2000 (adult, head); Lemos-Espinal and Smith, 2009 (adult); Lutz and Musick, 1997 (adult); Márquez M., 1990 (adult, head); McCranie et al., 2006 (juvenile, carapace); Meylan et al., 2013 (adult); Murphy, 1997 (adult, hatchling); Nietschmann, 1977 (adult); Pough et al., 2015 (adult); Pritchard, 1979 (adult), 1997 (adult); Pritchard and Trebbau, 1984 (adult, subadult, hatchling); Rebel, 1974 (adult); Ruckdeschel and Shoop, 2006 (adult, juvenile, hatchling, head); Savage, 2002 (adult, hatchling, adult and hatchling carapaces); Seminoff and Wallace, 2012 (hatchling, head); Shi, 2013 (adult, head, plastron); Spotila, 2004 (adult, hatchling), 2011 (adult, head); Stafford and Meyer, 1999 (adult); Vetter, 2004 (adult, hatchling), 2005 (adult, juvenile); Vitt and Caldwell, 2014 (adult); Wermuth and Mertens, 1961 (adult, carapace, plastron, head, limbs).

Remarks.—Bour (1979: 149, 151), Fretey and Bour (1980: 193), and Bour and Dubois (1983: 356–359) demonstrated that the author of the name *Testudo coriacea* was Vandelli (1761: 7), not Linnaeus (1766: 350), to whom most previous authors had attributed the binomial, including the ICZN (ICZN, 1926: 3; Hemming, 1956a: 351, 1956b: 374). Fretey and Bour (1980) republished part of the Vandelli (1761) *Dermochelys coriacea* description.

Comment on the name *Dermatochelys porcata*, a junior synonym of *Testudo coriacea* Vandelli (= *Dermochelys coriacea*), is also needed. The recent lists of turtles of the world have cited Wagler (1833, plate 1; see Fritz and Havaš, 2007: 175; Rhodin et al., 2008: 000.4; TTWG 2012: 000.254) as the source of the name *Dermatochelys porcata*. The TTWG (2014: 345) tried to correct those errors by saying that the "explicatio tabularum" in Wagler (1830a) was the source for the name *Dermatochelys porcata*, without realizing that the "explicatio tabularum" was published in a different book later in the same year (Wagler, 1830b). Wagler (1830b), on the first (but unnumbered) page of his explicatio tabularum, introduced the name *Dermatochelys porcata* in the legend for his table I. All references that do not specifically credit the name *Dermatochelys porcata* to the Wagler (1830b), explicatio tabularum, are in error. Also, Part 3 of the Wagler Desc. Icon. with the plate of *D. porcata* was published in 1833. The TTWG (2014) erroneously listed Part 3 as published in 1830. Parts 1 and 2 of the Wagler Desc. Icon. were published in 1828 and 1830, respectively. Gutsche and McCranie (2016) discussed the confused status regarding the years the Wagler publications appeared (1830a, 1830b, 1833), especially the extremely rare 1830b work.

Ernst and Lovich (2009) provided a thorough review of the biology and a literature review for this species. Pritchard (1980) had provided a short literature review. Ernst and Lovich (2009: 154) said the known molecular variation of *Dermochelys* "is not sufficient to recognize separate taxa at this time." Also, see Brongersma (1996: 261), who rejected the use and availability of the proposed name *Dermatochelys schlegeli* Garman (1884a: 6), for the eastern Pacific populations, should they prove distinct.

Natural History Comments.—Carr (1952) reported that the breeding season of *Dermochelys* along the north coast of Honduras extends from May to August, although it appeared not to nest anywhere in significant numbers. A leatherback turtle project run by a Peace Corps volunteer with help from MOPAWI (Mosquitia Pawisa Desarrollo de la Mosquitia) was started at Plaplaya, Gracias a Dios, in 1995 (Peskin, 1996). Nesting areas were being protected and eggs gathered for artificial incubation, with hatchlings being released into the sea. However, I do not know the current status of that protection. Females of this species deposit about 23–166 ($x = 77.4$, $n = 510$) eggs per clutch in a hole she digs in the beach sand, with females depositing 1–11

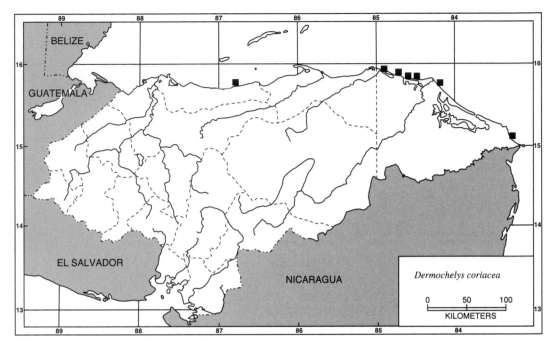

Map 71. Localities for *Dermochelys coriacea*. Solid squares denote accepted records. No voucher specimens of *Dermochelys* from Honduras are available.

clutches per season, typically with 2–4-year intervals between nesting seasons (Ernst and Lovich, 2009; those authors also presented much more information on reproduction in this species). *Dermochelys coriacea* is a pelagic species but will enter shallow coastal waters of bays and estuaries to forage (Ernst and Lovich, 2009). This species feeds primarily on planktonic invertebrates, especially oceanic jellyfish (Ernst and Lovich, 2009; those authors presented a comprehensive list of the known food items of this species).

Etymology.—The name *coriacea* is derived from the Latin *coriaceus* (of leather) and refers to the species' leatherlike carapace.

Specimens Examined (0 [0]).

Other Records (Map 71).—**ATLÁNTIDA**: La Ceiba (Marin, 1984). **GRACIAS A DIOS**: Barra Patuca, Laguna de Brus, Laguna de Ébano, Mokobila (Cruz [Díaz] and Espinal, 1987); Laguna de Ébano (UF

photograph); Plaplaya (Peskin, 1996; UF photographs); Río Kruta Biological Reserve (McCranie et al., 2006).

Superfamily Chelydroidea Gray, 1831

The turtles of this superfamily include the genera *Chelydra* Schweigger (1812: 292) and *Macrochelys* Gray (1856b: 200) of the Chelydridae (see Legler and Vogt, 2013: 355; TTWG, 2014: 339). The Chelydridae (two species of *Chelydra* in Honduras) are most easily distinguished from all remaining Honduran turtles by having a long tail (about a half or more of CL) with large, projecting triangular dorsal scutes.

Remarks.—The Principle of Coordination (Articles 36, 43, 46; ICZN, 1999) dictates that Gray (1831: 4) is the author of the superfamily name Chelydroidea.

Family Chelydridae Gray, 1831

This family of turtles occurs from southern Canada through the eastern and central

U.S., thence from southern Mexico through Central America to western Ecuador. The family Chelydridae can be distinguished externally from all remaining Honduran turtles by the combination of having strongly clawed limbs, a long tail about 50% of CL, the tail with a dorsal median row of large triangular scales, and a small and cruciform plastron with 11 or 12 scutes. Two genera comprised of six named species are included in this family, with two species in one genus occurring in Honduras.

Genus *Chelydra* Schweigger, 1812

Chelydra Schweigger, 1812: 292 (type species: *Testudo serpentina* Linnaeus, 1758: 199, by subsequent designation of Fitzinger, 1843: 29; also see ICZN, 1926: 339; Hemming, 1956b: 374).

Geographic Distribution and Content.—This genus occurs east of the Rocky Mountains from southern Canada southward through much of the U.S. to southern Florida and southern Texas, and from southern Veracruz, Mexico, southward through Central America to western Ecuador. Three named species are recognized, two of which occur in Honduras.

Remarks.—Until recently, only one species of *Chelydra*, *C. serpentina* (Linnaeus, 1758: 199), was recognized (see reviews in Ernst et al., 1988; Gibbons et al., 1988). Gibbons et al. (1988) recognized four subspecies, two of which (*C. serpentina rossignonii*, *C. s. acutirostris*) were reported to occur in western (ríos Chamelecón and Ulúa systems; Feuer, 1966: 6, speculated "at least to the Río Aguán drainage of Honduras") and eastern (Río Coco) Honduras, respectively. Those two subspecies were suggested by Feuer (1966: 190) to intergrade with one another in the intervening part of Honduras, even though specimens from much of the intervening area were (and still are) lacking. However, mitochondrial DNA data presented by Phillips et al. (1996) supported species-level distinctive-

ness for *C. rossignonii* and *C. acutirostris* in Mexico and South America, respectively. The Phillips et al. (1996) molecular data were taken from two specimens from Veracruz, Mexico (*C. rossignonii*), and two specimens from Esmeraldas, Ecuador (*C. acutirostris*). Shaffer et al. (2008) sequenced mitochondrial DNA from two additional Mexican specimens (from "Veracruz"): one from "Honduras," and one from Panama. Their recovered phylogeny also supported *C. acutirostris* and *C. rossignonii* as distinct species. External morphological data from the available Honduran *Chelydra* also demonstrates that two species are present in the country. One species has long, flat, raised projections on the neck and a third vertebral width/CL ratio of 0.29–0.48 and agrees with the description of *C. rossignonii* provided by Feuer (1966) and Ernst (2008a). The second species has low, rounded, wartlike tubercles on the neck and a third vertebral width/CL ratio of 0.24–0.28 and agrees with the description of *C. acutirostris* provided by Feuer (1966) and Ernst (2008a). *Chelydra rossignonii* is known to occur in Honduras from about the Tela, Atlántida, region westward. No specimens are known from east of Tela to validate Feuer's (1966) speculation that *C. rossignonii* ranges eastward to the Río Aguán drainage. *Chelydra acutirostris* is known in Honduras from the Río Patuca drainage system eastward to the Río Coco, the border river separating Honduras from Nicaragua.

I follow the plastral scute terminology used by Medem (1977) in my description for each of the species of *Chelydra*. Thus, in my usage there is always 5 paired plastral scutes, the humerals, pectorals, femorals, anals, and widely separated abdominals of the bridge elements, in addition to the single or sometimes paired gulars. This terminology also agrees with the revised plastral scute terminology proposed by Hutchison and Bramble (1981) for the

Figure 168.   Long, flat dermal appendages present on skin of neck. *Chelydra rossignonii*. USNM 149137 from Veracruz, Mexico.

kinosternid turtles and their closest relatives.

Etymology.—The name *Chelydra* is derived from the Greek *chelydros* (tortoise, turtle, or water serpent). The name might allude to the head and neck of this genus that is snakelike, in being long and quick striking.

KEY TO HONDURAN SPECIES OF THE GENUS
CHELYDRA

1A. Long, flat dermal appendages present on skin of neck (Fig. 168); third vertebral scute width/carapace width 0.29–0.48................
.... *rossignonii* (p. 475)
1B. Rounded, wartlike tubercles present on skin of neck (Fig. 169); third vertebral scute width/carapace width 0.24–0.28.. *acutirostris* (p. 471)

CLAVE PARA LAS ESPECIES HONDUREÑAS DEL
GÉNERO *CHELYDRA*

1A. Tubérculos largos y aplanados en el cuello (Fig. 168); ancho del tercera escudo vertebral/ancho del carapacho 0.29–0.48.............
.... *rossignonii* (p. 475)
1B. Verrugas redondeadas presentes en el cuello (Fig. 169); ancho del tercera escudo vertebral/ancho del carapacho 0.24–0.28.............
.... *acutirostris* (p. 471)

Chelydra acutirostris Peters, 1862

Chelydra serpentina var. *acutirostris* Peters, 1862: 627 (holotype, ZMB 4500 [see Bauer et al., 1995: 51]; type locality: "Guayaquil [Ecuador]."
Chelydra acutirostris: Babcock, 1932: 874; H. W. Campbell and Howell, 1965: 131; Solís et al., 2014: 139; McCranie, 2015a: 380.
Chelydra serpentina acutirostris: Feuer, 1966: 248 (in part); Gibbons et al., 1988: 420.1.
Chelydra serpentina: Meyer, 1969: 182 (in part); Meyer and Wilson, 1973: 3 (in part); Iverson, 1986: 13 (in part); Iverson, 1992: 93 (in part); Wilson and McCranie, 1998: 15 (in part); Wilson et al., 2001: 134 (in part); McCranie et al., 2006: 98 (in part); Wilson and Townsend, 2006: 104 (in part).

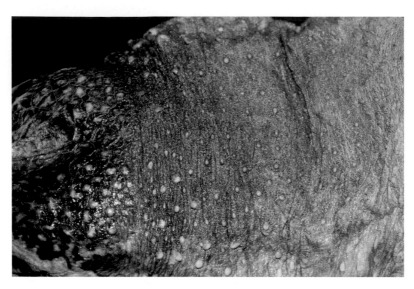

Figure 169.   Wartlike, rounded tubercles present on skin of neck. *Chelydra acutirostris.* USNM 559582 from near confluence of ríos Cuyamel and Patuca, Olancho.

Chelydra serpentina rossignonii: Gibbons et al., 1988: 420.1 (in part).
Chelydra acutirostris and/or *rossignonii*: Townsend and Wilson, 2010b: 691 (in part).

Geographic Distribution.—Chelydra acutirostris occurs at low elevations from near Catacamas, Olancho, in east-central Honduras to northwestern Colombia on the Atlantic versant and on the Pacific versant in south-central Panama and from northwestern Colombia to northwestern Ecuador. In Honduras, this species is known to occur in the ríos Patuca, Warunta, Kruta, and Coco, and their basins in the eastern portion of the mainland.

*Description.—*The following is based on three males (LACM 48449–50 [head and shell of same animal]; USNM 559582–83) and one unsexed juvenile (LACM 73821). *Chelydra acutirostris* is a huge (maximum recorded CL about 389 mm [see Medem, 1977: 58, 60]; CL 315 mm in largest Honduran specimen [LACM 48450]) freshwater turtle with a massive carapace that is broadly oval in outline and widest posterior to bridge, and provided with 3 low, longitudinal ridges composed of knobs located near to, or well posterior to midpoint of scutes, ridges becoming less prominent with age; carapace with 24 marginals, 1 nuchal, 5 vertebrals (fourth vertebral longitudinally paired in USNM 559583), and 4 pairs of costal scutes; marginals strongly serrated along posterior edge of carapace; plastron unhinged, reduced in extent, cruciform in shape, exposing neck, limbs, and tail; plastron with 5 pairs of scutes, humerals, pectorals, femorals, anals, and widely separated abdominals of bridge elements, and a single unpaired gular or sometimes with paired gulars; abdominals long, narrow, posteriorly angled, widely separated from each other, connecting plastron to inframarginals; 3 inframarginal scutes, 2–3 in contact with abdominals and marginals; head large, with dorsolaterally placed orbits; snout slightly protruding; upper jaw hooked; pair of long barbels present anteriorly on undersurface of lower jaw, 1–2 shorter pairs of barbels sometimes present at about midlength of lower jaw; posterior portion of head, including temporal region, covered with flat juxtaposed scales; neck extremely long; dorsal and lateral surfaces of skin on neck with prominent rounded, wartlike tubercles, although a few lateral neck tubercles longer and pointed; limbs large, with fully webbed feet and heavy claws; 5 claws on forelimb, 4 claws on hind limb; tail longer than plastron, longer than half of CL; dorsal surface of tail with 3 longitudinal rows of knobs, median row most prominent, median knobs formed by large, triangular scales; subcaudal surface with flattened plates in more-or-less 2 rows; cloaca located posterior to carapacial margin in males, at or anterior to carapacial margin in females; CL 162–315 (228.3 ± 78.5) mm in males, 35 mm in juvenile; PL/CL 0.73–0.78 in males, 0.72 in juvenile; width of third vertebral (along anterior margin)/CW 0.24–0.28 in males, 0.27 in juvenile; plastron forelobe L/CL 0.40–0.42 in males, 0.33 in juvenile; abdominal (bridge width of Feuer, 1966) width/CL 0.06–0.10 in males, 0.09 in juvenile.

Color in life of a newly hatched individual (USNM 579660): plastron, top of head, and dorsal surfaces of fore- and hind limb Smoke Gray (45) with paler gray and dark brown mottling; side of head pale brown with Smoke Gray stripes posterior to eye and dark brown elongated spots below eye; dirty white subocular stripe also present; marginals cream with dark brown edges; tail same color as that of body, except crest scales pale brown; plastron dark brown with large white spots; skin of bottom of head and limbs white with much dark and pale brown mottling; iris Smoke Gray (44) with dark brown lines.

Color in alcohol of adults: carapace brown to black, without pattern; plastron yellow, mottled with brown to tan, with yellow margins on some scutes; skin of dorsal surfaces brown, dorsal tubercles brown to pale brown; skin of ventral surfaces pale tan to pale brown, with cream tubercles.

Diagnosis/Similar Species.—*Chelydra acutirostris* is distinguished from all other Honduran turtles, except *C. rossignonii*, in having a tail with a dorsal median row of large triangular scales, having the tail half or more the length of the carapace, and having a cruciform plastron with 11 or occasionally 12 scutes. *Chelydra rossignonii* has long, flat dermal appendages on the skin on the neck and has a third vertebral width/CL ratio of 0.29–0.48 (versus rounded wartlike tubercles on neck and third vertebral width/CW ratio 0.24–0.28 in *C. acutirostris*).

Illustrations (Figs. 147, 148, 169; Plates 88, 89).—Guyer and Donnelly, 2005 (adult; as *C. serpentina*); Köhler, 2000 (adult; as *C. serpentina*), 2001b (adult; as *C. serpentina*), 2003a (adult, plastron; as *C. serpentina*), 2008 (adult, plastron; as *C. serpentina*); Legler and Vogt, 2013 (head, carapace, plastron); McCranie et al., 2006 (adult, plastron; as *C. serpentina*); Medem, 1962 (adult; as *C. s. acutirostris*), 1977 (adult, juvenile, hatchling, head, foot, tail, carapace, plastron; as *C. s. acutirostris*); Merchán Fornelino, 2003 (adult, head, plastron; *C. s. acutirostris*); D. Moll, 2010 (adult); Páez et al., 2012 (adult, head, plastron); Pritchard, 1979 (head; as *C. s. acutirostris*); Savage, 2002 (adult, plastron; as *C. serpentina*); Steyermark et al., 2008 (adult); Vetter, 2005 (adult).

Natural History Comments.—*Chelydra acutirostris* is known from 5 to 400 m elevation in the Lowland Moist Forest and Lowland Dry Forest formations. One was taken at night in a temporary rain-filled pond near the Río Patuca in August. An adult (subsequently lost) was collected in the headwaters of the Río Warunta in December at 2:00 a.m. as it was feeding on the carcass of an adult collard peccary (*Tayassu tajacu* [Linnaeus]). A series of plastrons were retrieved from a village; the turtles were said to have been collected from a muddy-bottomed slow-moving tributary of the Río Kruta. A recent hatchling with a distinct umbilical scar and a soft shell

Plate 88.   *Chelydra acutirostris*. Adult from Cali Zoo, Cali, Colombia. Photograph by Carlos Andrés Galvis R., courtesy of John L. Carr.

(CL 37.7 mm) was crawling on the ground in the morning about 3 hours after sunrise near a river in December. Many people in eastern Honduras eat the meat and eggs of this species, whereas other people in that region indicated they do not. Flausín et al. (1997) reported monitoring three nests of this species in Costa Rica from September to October. The three nests contained 25, 20, and 38 eggs. Medem (1977) reported a

Plate 89.   *Chelydra acutirostris*. USNM 579660, juvenile, carapace length 37.7 mm. Gracias a Dios: Rus Rus.

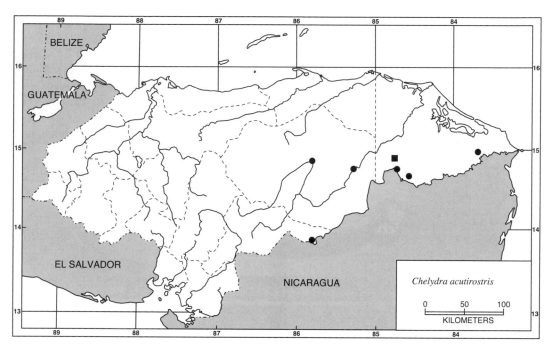

Map 72. Localities for *Chelydra acutirostris*. Solid circles denote specimens examined and the solid square represents an accepted record.

Colombian female collected in February contained 27 ovarian eggs. Moll, D. 1997, *In* Iverson et al. reported six nests in Costa Rica contained a mean of 32.5 eggs. Moll, D. (2010) reported female *C. acutirostris* at Tortuguero, Costa Rica, nested in cleared fields or in man-made clearings in otherwise dense forest. Medem (1977: 68) reported two Colombian captive individuals ate meat, fish, freshwater mollusks, aquatic plants, algae, and *plátanos*. Fallen leaves from riparian vegetation are also an important food item in this species during the dry season in Costa Rica (D. Moll and Moll, 2004). Acuña [Mesén] et al. (1983) and Acuña-Mesén (1999) said fish constituted 35% of the diet of this species in Costa Rica.

Etymology.—The name *acutirostris* is formed from the Latin *acutus* (make pointed, sharpen) and *rostrum* (snout) and refers to the somewhat pointed snout of this species compared with that of the North American *Chelydra serpentina*.

Specimens Examined (4, +1 head, 1 shell, 4 plastrons [2]; Map 72).—**EL PARAÍSO**: near Arenales, LACM 73821 (formerly UCLA 14733). **GRACIAS A DIOS**: Awasbila, USNM 559583; Rus Rus, USNM 579660; near Tikiraya, UF 137634–37 (all plastrons only); **OLANCHO**: 4.5 km SE Catacamas, LACM 48449–50 (head and shell of same individual); near confluence of ríos Cuyamel and Patuca, USNM 559582.

Other Records (Map 72).—**GRACIAS A DIOS**: Warunta Tingni Kiamp (McCranie et al., 2006, specimen lost).

Chelydra rossignonii (Bocourt, 1868)

Emysaurus rossignonii Bocourt, 1868: 121 (two syntypes, MNHN 1501, 1501a [see Gibbons et al., 1988: 420.2; also see Remarks]; type locality: "proviennent des marais de Pansos, près le Rio Polochic [Guatémala]).")

Chelydra rossignonii: Cope, 1872: 23; Schmidt, 1924: 86; Babcock, 1932:

874; Schmidt, 1946: 4; Solís et al., 2014: 139; McCranie, 2015a: 380.

Chelydra rossignoni: Werner, 1896: 344; Legler and Vogt, 2013: 358.

Chelydra serpentina rossignonii: Feuer, 1966: 246; Medem, 1977: 52; Gibbons et al., 1988: 420.1.

Chelydra serpentina acutirostris: Feuer, 1966: 248 (in part).

Chelydra serpentina: Meyer, 1969: 182 (in part); Meyer and Wilson, 1973: 3 (in part); Iverson, 1986: 13 (in part); Iverson, 1992: 93 (in part); Wilson and McCranie, 1998: 15 (in part); Wilson et al., 2001: 134 (in part); Castañeda and Marineros, 2006: 3.8; McCranie et al., 2006: 98 (in part); Wilson and Townsend, 2006: 104 (in part).

Chelydra acutirostris and/or *rossignonii*: Townsend and Wilson, 2010b: 692 (in part).

Geographic Distribution.—*Chelydra rossignonii* occurs at low and lower extremes of moderate elevations from central Veracruz, Mexico, to northwestern Honduras on the Atlantic versant. In Honduras, this species occurs in rivers and freshwater lagoons in the northwestern portion of the mainland. It is known from as far east as Tela, but I believe I got a glimpse of an adult *Chelydra* in a muddy bottom section of a badly polluted, foul smelling estuary in La Ceiba, Atlántida, as it surfaced to breathe before seeing me and rapidly going to the muddy bottom.

Description.—The following is based on nine males (AMNH 70566; FMNH 5324–26; LACM 48451; MCZ R27916, 32181; UF 123956; USNM 62980) and five females (MCZ R29078, 32182, 34331; USNM 71481, 573980). *Chelydra rossignonii* is a huge (maximum recorded CL 389 mm [MCZ R32182, a female]) freshwater turtle with a massive carapace that is broadly oval in outline, widest posterior to bridge, and provided with 3 low, longitudinal ridges composed of knobs located near to, or well posterior to midpoint of scutes, ridges becoming less prominent with age; carapace with 24 marginals, 1 nuchal, 5 vertebrals, and 4 pairs of costal scutes; marginals strongly serrated along posterior edge of carapace; plastron unhinged, reduced in extent, cruciform in shape, exposing neck, limbs, and tail; plastron with 5 pairs of scutes, consisting of humerals, pectorals, femorals, anals, and widely separated abdominals of bridge elements, plus single unpaired gular or occasionally paired gulars; abdominals long, narrow, posteriorly angled, connecting plastron to inframarginals; 3 inframarginal scutes, 2–3 in contact with abdominals and marginals; head large, with dorsolaterally placed orbits; snout slightly protruding; upper jaw hooked; pair of long barbels present anteriorly on undersurface of lower jaw, 1–2 shorter pairs of barbels sometimes present at about midlength of lower jaw; posterior portion of head, including temporal region, covered with flat juxtaposed scales; neck extremely long; dorsal and lateral surfaces on skin of neck with long, flat dermal appendages (villose of Feuer, 1966); limbs large, with fully webbed feet and heavy claws; 5 claws on forelimb, 4 claws on hind limb; tail longer than plastron, longer than half of CL; dorsal surface of tail with 3 longitudinal rows of conical knobs, median row most prominent, median knobs formed by large, triangular scales; subcaudal surface with flattened plates in more-or-less 2 rows; cloaca located posterior to carapacial margin in males, at or anterior to carapacial margin in females; CL 86–355 (228.5 ± 100.0) mm in males, 310–389 (354.0 ± 30.3) mm in females; PL/CL 0.74–0.82 in males, 0.74–0.78 in four females; width of anterior margin of third vertebral/CW 0.32–0.48 in males, 0.29–0.37 in four females; plastron forelobe L/CL 0.39–0.45 in males, 0.39–0.42 in three females; abdominal (bridge of Feuer, 1966) width/PL 0.07–0.09 in males, 0.06–0.07 in three females.

Color in alcohol: carapace brown to black, without pattern; plastron yellow, mottled with brown to tan, with yellow

Plate 90. *Chelydra rossignonii*. Cortés: near San Pedro Sula (not collected).

margins on some scutes; skin of dorsal surfaces brown, dorsal tubercles brown to pale brown; skin of ventral surfaces pale tan to pale brown, with cream tubercles.

Diagnosis/Similar Species.—*Chelydra rossignonii* is distinguished from all other Honduran turtles, except *C. acutirostris*, in having the tail with a dorsal median row of large triangular scales, having the tail about half or more length of the carapace, and having a cruciform plastron with 11 or sometimes 12 scutes. *Chelydra acutirostris* has rounded, wartlike tubercles on the skin on the neck and has a third vertebral width/CL ratio of 0.24–0.28 (versus long, flat dermal appendages on skin on neck, and third vertebral width/CL 0.29–0.48 *C. rossignonii*).

Illustrations (Fig. 168; Plate 90).— Álvarez del Toro, 1983 (adult; as *C. serpentina*); Bonin et al., 2006 (adult); Campbell, 1998 (adult, carapace; as *C. serpentina*); A. H. A. Duméril et al., 1870–1909b (juvenile; as *Emysaurus* A. M. C. Duméril and Bibron, 1835: 348); Lee, 1996 (adult, juvenile, plastron, tail scales; as *C. serpentina*), 2000 (adult, juvenile, plastron; as *C. serpentina*); Legler and Vogt, 2013

(adult, juvenile, carapace, plastron); Medem, 1977 (ventral view of juvenile; as *C. s. rossignonii*); Pritchard, 1979 (adult, as *C. s. rossignoni*); Smith and Smith, 1980 (ventral view of adult; as *C. s. rossignoni*); Steyermark et al. 2008 (adult); Vetter, 2005 (adult).

Remarks.—Bocourt (1868: 122) said two juvenile specimens of *Emysaurus rossignonii* were available for his description. Gibbons et al. (1988: 420.2) listed those as MNHN 1501 and 1051A, one number of which is an obvious typographical error. Stuart (1963: 47) had listed MNHN 1230 as the "type" of *Emysaurus rossignonii*. This confusion is likely why Legler and Vogt (2013: 359) stated, "there are two or three syntypes in the National Museum of Natural History, Paris."

Natural History Comments.—*Chelydra rossignonii* is known from about sea level to 730 m elevation in the Lowland Moist Forest and Lowland Dry Forest formations and peripherally in the Premontane Wet Forest formation. One was crossing a road during the day in March. Several people living in the vicinity of Lago de Yojoa told me that they do not eat the meat of this

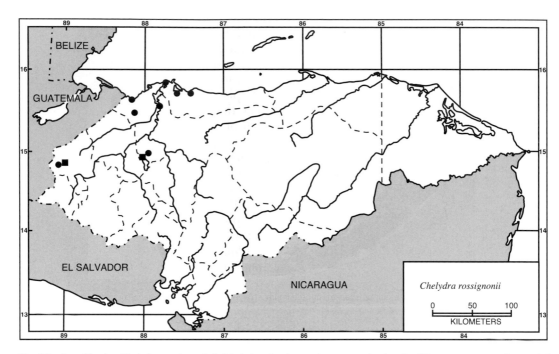

Map 73. Localities for *Chelydra rossignonii*. Solid circles denote specimens examined and solid squares represent accepted records.

turtle because they believe it is poisonous. However, those same people said they kill this turtle when they can because of its poisonous condition. Legler and Vogt (2013: 360) said one Honduran specimen was "from a freshwater estuary, partially covered with hyacinth, its mouth blocked by a sandbar." Medem (1977) reported a Honduran specimen (MCZ R29078) deposited four eggs (MCZ R31626) in captivity in March. Álvarez del Toro (1983) reported females from Chiapas, Mexico, deposit 20–30 eggs in April to June. Legler and Vogt (2013) reported six clutches from Veracruz, Mexico, varied from 26–55 eggs, with a mean of 38 (months not given). Little has been published on diet of this species, but like other *Chelydra* it probably eats a wide variety of animal and plant matter.

Etymology.—The name *rossignonii* is a patronym for Jules Rossignon, a Frenchman who lived for some time in Guatemala in the mid-1800s.

Specimens Examined (21, + 1 carapace, 1 body part and skeleton, 1 egg lot [3]; Map 73).—**ATLÁNTIDA**: road between Barra del Ulúa and Laguna de Río Tinto, USNM 573980 (carapace); Puerto Arturo, USNM 62980, 71481; Tela, AMNH 70566, MCZ R27916, 29078, 29098, 31626 (4 eggs deposited by MCZ R29098), 32181–82, 34331, 34355. **COPÁN**: 12.9 km ENE Copán, LACM 48451. **CORTÉS**: Amapá, AMNH 70541; Laguna Ticamaya, FMNH 5327; Masca, UU 3971 (body parts and skeleton only); San Pedro Sula, FMNH 5322–26. "HONDURAS": AMNH 46951, UF 123956, UTA R-23558.

Other Records (Map 73).—**COPÁN**: Río Amarillo (Castañeda and Marineros, 2006). **CORTÉS**: Lago de Yojoa near El Jaral (carapace not taken). "HONDURAS": (Werner, 1896).

Superfamily Kinosternoidea Agassiz, 1857

Kinosternoidea contains the families Dermatemydidae Gray (1870b: 711) and Kinosternidae Agassiz (1857: 249; also see Spinks et al., 2014), one of which (Kinosternidae) is represented in Honduras. Externally, the known Honduran members of this superfamily are distinguished from all remaining Honduran turtles by the combination of having non-paddlelike limbs, having short tails lacking a dorsal median row of large triangular scales, and in having fewer than 12 plastral and 11 (per side) marginal scutes.

Remarks.—As dictated by the Principle of Coordination (Articles 36, 43, 46; ICZN, 1999: 113), Agassiz (1857: 249) is the author of both the superfamily name Kinosternoidea and the family name Kinosternidae.

I use the plastral scute terminology proposed by Hutchison and Bramble (1981) for Kinosternoidea turtles in species descriptions of *Kinosternon* and *Staurotypus*.

Although *Dermatemys mawii* Gray (1847: 56), of the family Dermatemydidae Gray (1870b: 711; as Dermatemydae), is no longer considered to be part of the Honduran turtle fauna, Legler and Vogt (2013: 70) continued to say that *Dermatemys* Gray (1847: 55) occurs in the Atlantic drainages of "northern Guatemala and adjacent Honduras." Wilson and McCranie (2002) had deleted *D. mawii* from the Honduran fauna. Additionally, *D. mawii* was not included in the Honduran turtle fauna in several other listings of the Honduran reptile species or species descriptions (Köhler, 2003a, 2008; Wilson and McCranie, 2004a; McCranie, 2009, 2015a; Townsend and Wilson, 2010a; Vogt et al., 2011). In June 2011, I took color photographs of *Chelydra rossignonii*, *D. mawii*, *Staurotypus triporcatus*, *Trachemys venusta*, and *Rhinoclemmys areolata* to the environs of the Río Motagua, which forms the international border between Guatemala and Honduras in extreme northwestern Honduras. I showed those photographs to about a dozen men who were known to hunt turtles for meat and eggs. Those interviews took place in Tegucigalpita, Cortés, and along the length of a dirt road traveling north of Tegucigalpita to where it enters a swamp associated with the Río Motagua. All of those men positively identified the *Trachemys*, *Staurotypus*, and *Chelydra* as occurring in the area, but not one recognized the *Dermatemys* and *Rhinoclemmys*, thus providing some weak second-hand, negative evidence that *Dermatemys* and the *Rhinoclemmys* do not occur in the Río Motagua or in that area of Honduras. I also did not locate *Dermatemys* or *Rhinoclemmys* in my fieldwork in the Río Motagua area. Additionally, there are no records of *Dermatemys* along the length of the Río Motagua in Guatemala (Vogt et al., 2011). The nearest known locality for *D. mawii* is Lago de Izabal and associated rivers in eastern Guatemala (Vogt et al., 2011). Lago de Izabal lies only about 20 km W of the Río Motagua border with Honduras. Although the family Dermatemydidae is not considered part of the Honduran fauna, it is included in pertinent keys on the off chance a straggler might sometimes be collected by a biologist in Honduran waters. Wermuth and Mertens (1961), Meyer and Wilson (1973), Iverson and Mittermeier (1980), Smith and Smith (1980), Groombridge (1982), Iverson (1986, 1992), and Ernst and Barbour (1989) have all said that either *D. mawii* occurs in northwestern Honduras or that it is strongly suspected to occur there.

KEY TO HONDURAN FAMILIES OF THE SUPERFAMILY KINOSTERNOIDEA

1A. Zero to 2 inframarginal bridge scutes contacting fixed lobe of plastron (Fig. 153); fore- and hind limb not extensively webbed (Fig.

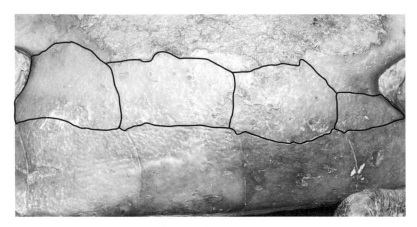

Figure 170.   Four to 6 inframarginal bridge scutes present (outlined; 4 in this case) per side. *Dermatemys mawii.** UF 65521 from Belize.

153) or, if extensively webbed (Fig. 155), fewer than 10 plastral scutes present (Fig. 155). . . . .
. . . . Kinosternidae (p. 481)
1B. At least 4 inframarginal scales contact plastron (Fig. 170); forefeet (Fig. 171) and hind feet extensively webbed . . . .
. . . . Dermatemydidae*

CLAVE PARA LAS FAMILIAS HONDUREÑAS DE LA SUPERFAMILIA KINOSTERNOIDEA

1A. A. 0–2 escudos inframarginales del puente en contacto con el lóbulo del plastrón (Fig. 153); extremidades anteriores y posteriores sin membranas interdigitales fuertemente desarrolladas (Fig. 153), o si hay membranas interdigitales

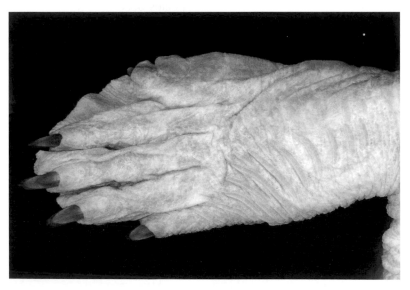

Figure 171.   Forefeet extensively webbed. *Dermatemys mawii.** USNM 46304 from Belize. Photograph by James A. Poindexter.

Figure 172.   Three high ridges present on adult carapace. *Staurotypus triporcatus*. Captive from Choloma, Cortes.

bien desarrolladas (Fig. 155), menos de diez escudos en el plastrón presentes (Fig. 155) ....
.... Kinosternidae (p. 481)

1B. Al menos 4 escudos inframarginales del puente en contacto con el plastrón (Fig. 170); extremidades anteriores (Fig. 171) y posteriores con membrana interdigital extensa .... Dermatemydidae°

Family Kinosternidae Agassiz, 1857

The distribution of this turtle family is limited to the Western Hemisphere (see Geographical Distribution for the genus *Kinosternon*). The Kinosternidae can be distinguished externally from all other Honduran turtles by having strongly clawed limbs, a short tail without a dorsal medial row of large triangular scales, only 11 pairs of marginal scutes (resulting in only 36–37 total scutes on the carapace), and either a hinged plastron with 11 scutes or a reduced

plastron with only 7–8 scutes. Four genera and 26 named species are included in this family, with two genera and two named species occurring in Honduras.

Remarks.—The ICZN (1989: 81) placed Kinosternidae Agassiz (1857: 249) on the Official List of Family-Group Names in Zoology.

KEY TO HONDURAN GENERA OF THE FAMILY
KINOSTERNIDAE

1A. Eleven plastral scutes present (Fig. 153); plastron with 2 kinetic hinges (Fig. 153); 1–3 low ridges on adult plastron; maximum CL < 200 mm .... *Kinosternon* (p. 482)

1B. Seven or 8 plastral scutes present (Fig. 155); plastron lacking 2 kinetic hinges (Fig. 155); 3 high ridges on adult carapace (Fig. 172); maximum adult CL > 200 mm .... *Staurotypus* (p. 498)

CLAVE PARA LOS GÉNEROS HONDUREÑOS DE LA
FAMILIA KINOSTERNIDAE

1A. Once escudos presentes en el
plastrón (Fig. 153); plastrón con
dos bisagras (Fig. 153); 1–3 quillas
bajas presentes en el carapacho
adulto; longitud máxima del carapacho menos que 200 mm . . . .
. . . . *Kinosternon* (p. 482)

1B. Siete o ocho escudos presentes en
el plastrón (Fig. 155); plastrón sin
dos bisagras (Fig. 155); 3 quillas
muy altas en el carapacho adulto
(Fig. 172); longitud máxima de
carapacho más que 200 mm . . . . .
. . . . *Staurotypus* (p. 498)

Genus *Kinosternon* Spix, 1824

Kinosternon **Spix, 1824: 17 (type species:
Kinosternon longicaudatum Spix, 1824:
17 [= *Testudo scorpioides* Linnaeus,
1766: 352], by subsequent designation
of Bell, 1828: 515; also see ICZN, 1989:
81).**

Geographic Distribution and Content.—
This genus ranges from southeastern Con-
necticut and Long Island westward to
southern Arizona and the lower Colorado
River, USA, southward through Mexico (but
not Baja California) and Central America
and to western Paraguay and northern
Argentina in South America. About 18
named species are included in this genus,
two of which occur in Honduras.

Remarks.—Iverson et al. (2013) provided
a phylogenetic analysis based on molecular
studies of turtles usually placed in the genus
Kinosternon. Those workers found *Kinosternon* to be paraphyletic, and to rectify
their findings, they described the new genus
Cryptochelys for a clade of turtles formerly
included in *Kinosternon*. Subsequently,
Spinks et al. (2014) produced a multilocus
phylogeny of these turtles that suggested a
return to the "traditional classification" of
the "mud" turtles (one family and four

genera: *Claudius* Cope, 1866a: 187, *Kinosternon*, *Sternotherus* Gray, 1825: 211, and
Staurotypus).

Legler (1965: 623) wrote about "satisfactory microhabitats" for *Kinosternon angustipons* Legler (1965: 617) probably existing
almost continuously from northeastern
Honduras to Colón, Panama. Whereas
seemingly true about the satisfactory habitat
in northeastern Honduras, considerable
recent fieldwork in that region of Honduras
has failed to locate a population of this turtle
there. However, *K. angustipons* has subsequently been collected in Nicaragua and
Panama.

Stejneger (1941: 458) recorded *Kinosternon acutum* Gray (1831: 34) from "Honduras," but that record is actually from Belize
(Schmidt, 1941: 488; Iverson, 1976: 258).

Etymology.—The name *Kinosternon* is
derived from the Greek words *kineo* (move)
and *sternon* (breast, chest, sternum). The
name alludes to the hinged plastron.

KEY TO HONDURAN SPECIES OF THE GENUS
KINOSTERNON

1A. Intergular scute of plastron curves
outward on anterior edge on dorsal
surface and is broader than completely exposed ventral portion
(Fig. 173); opposable patches of
ridges present on posterior thigh
and calf in adult males (Fig. 174);
carapace never with more than 1
low, weak median ridge (Fig. 175)
. . . . *leucostomum* (p. 492)

1B. Intergular scute of plastron not
curving outward on anterior edge
of dorsal surface and not broader
than completely exposed ventral
portion (Fig. 176); no opposable
patches of ridges on thigh and calf
in adult males (Fig. 177); 3 rather
weak, low ridges on carapace,
usually indicated only posteriorly
in old individuals (Fig. 178). . . . . . .
. . . . *albogulare* (p. 484)

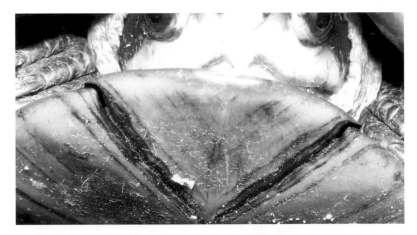

Figure 173.   Intergular scute of plastron curved outward (outlined) on dorsal surface, thus broader than completely exposed ventral portion of plastron at anterior edge. *Kinosternon leucostomum.* FMNH 283589 from Guanaja Island, Islas de la Bahía.

CLAVE PARA LAS ESPECIES HONDUREÑAS DEL
GÉNERO *KINOSTERNON*

1A. Escudo intergular del plastrón más ancho en la parte dorsal anteriormente y más ancha que la parte ventral (Fig. 173); machos adultos con parches de crestas (órganos estriduladores) en las superficies de los muslos y piernas (Fig. 174); carapacho nunca con más de una quilla medial, la loma muy baja (Fig. 175)...... *leucostomum* (p. 492)

1B. Escudo intergular anterior del plastrón no está más ancha en la parte dorsal que la parte ventral (Fig. 176); machos adultos sin parches de crestas en las superfi-

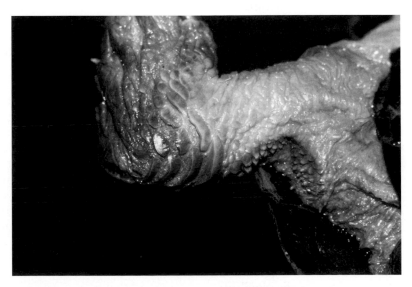

Figure 174.   Male opposable clasping organs, formed by ridges, present on each thigh and calf. *Kinosternon leucostomum.* USNM 589142 from Cuaca, Olancho.

Figure 175.　Single median, low ridge on carapace (arrow), on at least anterior half. *Kinosternon leucostomum.* FMNH 212485 from El Ocotal, Olancho.

cies de los muslos y piernas (Fig. 177); carapacho usualmente con tres quillas bajas, en adultos viejos usualmente son evidentes solamente en la parte posterior del

carapacho (Fig. 178) . . . . *albogulare* (p. 484)

Kinosternon albogulare (Bocourt, 1870)

Cinosternon albogulare Bocourt, 1870: 24, *In* A. H. A. Duméril et al., 1870–1909a

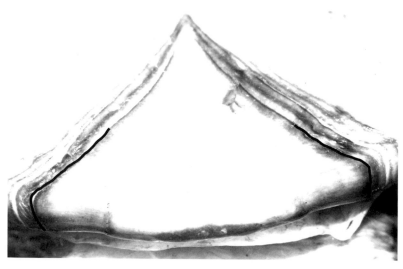

Figure 176.　Intergular scute not curved outward (outlined) on dorsal surface, thus not broader than completely exposed ventral portion of plastron at anterior edge. *Kinosternon albogulare.* USNM 580765 from El Rodeo, El Paraíso.

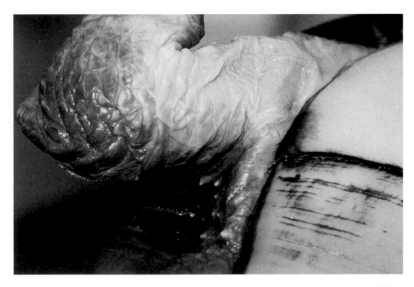

Figure 177.   Male opposable clasping organs absent on each thigh and calf. *Kinosternon albogulare*. FMNH 283748 from Río Lempa at Antigua, Ocotepeque.

(holotype MNHN 1760 [Berry and Iverson, 2001b: 725.5]; type locality: "S. Jose [Costa Rica]").

Cinosternum albogulare: Cope, 1875: 153.

Kinosternon cruentatum: Dunn and Emlen, 1932: 25.

Kinosternon albogulare: Wettstein, 1934: 14; McCranie, 2015a: 380.

Kinosternon scorpioides albogulare: Dunn and Saxe, 1950: 146; Berry, 1978: 315, 317; Berry and Iverson, 2001b: 725.2; Berry and Iverson, 2011: 063.3; Forero-

Figure 178.   Three, low, weakly raised ridges on carapace (arrows). *Kinosternon albogulare*. USNM 589139 from Sisinbila, Gracias a Dios.

Medina and Castaño-Mora, 2011: 064.3; Iverson et al., 2013: 932.
Kinosternon scorpioides subsp.: Meyer, 1966: 173.
Kinosternon scorpioides: Meyer, 1969: 186; Hahn, 1971: 111; Meyer and Wilson, 1973: 4; Wilson et al., 1979a: 25; Iverson, 1986: 124; Iverson, 1992: 233; Wilson and McCranie, 1998: 15; Köhler, 1999a: 214; Wilson et al., 2001: 134; Castañeda, 2002: 38; McCranie et al., 2002: 25; Lovich et al., 2006: 15; McCranie et al., 2006: 107; Wilson and Townsend, 2006: 104; Townsend et al., 2007: 10; Wilson and Townsend, 2007: 145; Townsend and Wilson, 2010b: 692; McCranie, 2014: 293; Solís et al., 2014: 139; Spinks et al., 2014: 258.
Kinosternon cruentatum albogulare: Vetter, 2005: 59.

Geographic Distribution.—*Kinosternon albogulare* occurs at low and moderate elevations approximately from El Salvador to western Panama on the Pacific versant and on the Atlantic versant apparently from the Yucatán Peninsula into Belize, in several interior valleys in central Honduras, and from northeastern Honduras to central Costa Rica (but see Remarks). It also occurs on Isla San Andrés, Colombia (likely a human introduction). In Honduras, this species occurs on the Pacific versant in the southern portion of the country and at several open forest localities (including lowland pine savanna) on the Atlantic versant in the central and northeastern portions of the country.

Description.—The following is based on ten males (SDSNH 72806; UF 137152, 137155, 137158, 137160; USNM 559580, 561998, 570542, 573081, 573320) and 11 females (UF 137153–54, 137156–57, 137159, 137161, 150885; USNM 559579, 559581, 561996–97). *Kinosternon albogulare* is a small (maximum recorded CL 190 mm [Vetter, 2005]; largest CL in Honduran specimen 170 mm [UF 137155, a male]) turtle with 3 low ridges on carapace, those ridges present only posteriorly in old individuals and only very indistinct; carapace oval, with 1 nuchal, 5 vertebrals, 8 costals, and 22 marginals; carapacial sides straight; costal 4 narrowly or not contacting marginal 11; anterior and posterior marginals slightly flared; marginal 10 higher than marginal 9, usually higher than marginal 11; vertebral 1 usually wider than long, rarely longer than wide, not in contact with, or narrowly in contact with marginal 2; vertebrals 1–4 with distinct posterior notches at midline; plastron with 2 kinetic hinges, 1 each anterior and posterior to posterior humeral; kinetic hinges completely close shell; plastron with 11 scutes, consisting of paired gulars, anterior humerals, posterior humerals, femorals, and anals, and a single intergular; intergular scute not curving outward dorsally (thus, not away from entire exposed ventral portion) at anterior edge; anals with little or no medial notch; plastron concave to slightly concave in males, slightly convex or flat in females; 1–2 axillary scutes and 1 inguinal scute present, axillary and inguinal scutes in contact (72.5%) or narrowly separated; 0–2 inframarginal scutes of bridge contacting fixed lobe of plastron; head relatively small, snout slightly projecting; upper jaw weakly to strongly hooked; 1–3 pairs of gular barbels present, anterior pair largest; feet webbed, but amount of webbing variable by population, apparently with those living in rivers having slightly more webbing than those living in ponds and swamps; 5 claws on forelimb, 4 claws on hind limb; elevated patches of ridges absent on posterior thigh and calf in males and females; tail much longer in males than in females, cloacal opening well posterior to plastral margin in males, but not posterior to carapacial margin, at or just posterior to plastral margin in females; tails of both sexes with terminal spines, spines largest in males; CL 127–170 (139.7 ± 17.5) mm in males, 103–165 (127.5 ± 17.7) mm in females; CH/CL 0.39–0.53 in males, 0.38–0.50 in females; PL/CL 0.91–0.94 in males, 0.90–0.98 in females; plastral fore-lobe L/CL 0.28–0.33 in males, 0.30–0.33 in females; PW at anterior hinge/CW 0.71–0.86 in males, 0.69–0.85 in females; PW at midfemoral/CW at midfemoral 0.72–0.94 in males, 0.68–0.81 in females; intergular scute L/CL 0.27–0.33 in males, 0.26–0.34 in females; first vertebral W/second vertebral W 0.83–1.04 in males, 0.77–1.07 in females; interposterior humeral seam L/PL 0.26–0.40 in males, 0.30–0.35 in females; inter-posterior humeral seam L/plastral forelobe L 0.74–1.30 in males, 0.84–1.04 in females; intergular scute L/plastral forelobe L 0.49–0.58 in males, 0.42–0.60 in females; BL/CL 0.25–0.38 in males, 0.26–0.33 in females; anterior width of posterior plastral lobe/CL 0.37–0.47 in males, 0.41–0.62 in females; interanal scute seam L/CL 0.27–0.33 in males, 0.27–0.34 in females.

Color in life of an adult female (USNM 559581): carapace Grayish Horn Color (91); plastron and bridge Sulphur Yellow (57), anterior and posterior margins of underside of carapace slightly darker yellow; dorsal surface of head Sepia (119); side of head Sepia with orange to yellow punctuations; chin yellow with brown lines; neck Sepia above, pale yellow below; limbs and tail dark gray-brown above and below.

Color in alcohol: carapace dark brown with darker brown seams; plastron yellow to tan, with dark brown seams; axillary, ingui-nal, and marginals of bridge yellow to tan, with dark brown seams, marginals in some also with brown upper edges; dorsal and lateral surfaces of head and neck dark brown, cream to pale brown spots and/or reticulations present laterally and usually dorsally; jaw sheaths cream to yellowish brown, with or without brown vertical streaks; chin and ventral surface of neck cream to pale brown, with some brown mottling; limbs dark brown on inside and outside surfaces, pale gray to grayish brown on ventral surfaces; tail dark brown dorsally, dark brown to grayish brown ventrally.

Diagnosis/Similar Species.—*Kinosternon albogulare* is distinguished from all remain-ing Honduran turtles, except *K. leucosto-mum*, in having a hinged, oblong plastron with 11 scutes and the carapace with 11 marginal scutes per side. *Kinosternon leu-costomum* has a single medial low ridge on the carapace, the plastral intergular curving outward dorsally at anterior edge and away from the entire exposed ventral portion, and males have elevated patches of horny ridges on the posterior thigh and calf (versus 3 low to poorly developed ridges on carapace in all but oldest individuals, intergular not curving outward dorsally at anterior edge, and males lacking elevated patches of horny scales on posterior thigh and calf in *K. albogulare*).

Illustrations (Figs. 176–178; Plate 91).—Berry and Iverson, 2001b (adult, head; as *K. scorpioides* [fig. 2 and possibly fig. 1; see Remarks]), 2011 (adult, head, plastron; as *K. s. albogulare*); Castañeda and Mora, 2010 (adult; as *K. scorpioides*); Forero-Medina and Castaño-Mora, 2011 (adult, plastron; as *K. s. albogulare*); Köhler, 2003a (adult, plastron; as *K. scorpioides*, fig. 50), 2008 (adult, plastron; as *K. scorpioides*, figs. 55, 59–60); Köhler et al., 2005 (adult; as *K. scorpioides*); McCranie et al., 2006 (adult; as *K. scorpioides*); Mertens, 1952b (adult; as *K. c. cruentatum*); Páez et al., 2012 (adult, plastron; as *K. s. albogulare*); Savage, 2002 (adult; as *K. scorpioides*); Vetter, 2005 (adult; as *K. c. albogulare* and *K. c. cruentatum* "Honduras–Nicaragua form").

Remarks.—*Kinosternon albogulare* is usually placed as one of four subspecies of *K. scorpioides* (Linnaeus, 1766: 352; i.e., Iverson, 1991: 2; Berry and Iverson, 2001b: 725.5, 2011: 63.2; TTWG, 2014: 350). Iverson et al. (2013: 935) elevated one of those four subspecies (*K. s. abaxillare* Baur, 1925: 462, *In* Stejneger, 1925) to a species (but see Spinks et al., 2014). Additionally, the three remaining subspecies of *K. scorpioides* were recovered as paraphyletic with respect to *K. integrum* (Le Conte,

Plate 91.   *Kinosternon albogulare.* USNM 561997, adult female, carapace length = 135 mm. Choluteca: El Madreal.

1854: 183) and *K. oaxacae* Berry and Iverson (1980: 314), thus suggesting "that *K. scorpioides* likely represents a multispecies complex" (Iverson et al., 2013: 935). Spinks et al. (2014), using multilocus phylogenetic analysis, also recovered *K. scorpioides*, including one specimen from northeastern Honduras, as paraphyletic with regard to *K. s. cruentatum*. Thus, those three names proposed as subspecies of *K. scorpioides* are likely candidates for species status. Those names are: *Testudo scorpioides, Cinosternon cruentatum* A. M. C. Duméril and Bibron (1851: 16, *In* A. M. C. Duméril and Duméril, 1851), and *Cinosternon albogulare*. I am including all Honduran populations as *K. albogulare* because of the Spinks et al. study (see above) and a molecular study underway that also indicates that *K. scorpioides* is paraphyletic (F. Köhler et al., unpublished data). According to Berry and Iverson (2011), *K. cruentatum* of Mexico, Guatemala, and Belize rarely (but see Iverson, 1976; Lee, 1996; and below) have the axillary scute in contact with the inguinal scute, whereas Honduran *K. albogulare* have those scutes in contact in 72.5% of the 21 specimens checked for that

character. Berry and Iverson (2011) stated *K. albogulare* usually has those scutes in contact. South American *K. scorpioides* cannot completely close the plastron, whereas Honduran populations can completely close their shells. Thus, in this work I recognize the species *K. scorpioides* as restricted to South America and probably southeastern Panama, *K. albogulare* from El Salvador and the Yucatán Peninsula to eastern Panama, and *K. cruentatum* from Mexico, Belize, and Guatemala, in contradiction to the ranges given in Berry and Iverson (2011; see below).

As stated above, according to Berry and Iverson (2011: 063.5), *Kinosternon cruentatum* rarely has the axillary and inguinal scutes in contact, whereas those scales are usually in contact with each other in *K. albogulare*. Berry and Iverson (2011, fig. 4) gave the geographic distribution of the nominal form *K. albogulare* as from Honduras and El Salvador to the Canal Zone, Panama, and that of *K. cruentatum* as from Mexico, Guatemala, and Belize. However, there are obvious problems with those statements, along with the data in the literature for the diagnostic axillary and inguinal contact character. The exact region where the geographic distribution of *K. cruentatum* ends and where that of *K. albogulare* begins and ends is in need of investigation. Despite Belize, eastern Guatemala, and the Yucatán Peninsula lying within the geographic distribution of *K. cruentatum* according to Berry and Iverson (2011), *K. cruentatum* is not supposed to have contact between the axillary and inguinal scutes. Iverson (1976) had reported that specimens from Belize and adjacent Mexico and eastern Guatemala have the axillary and inguinal scutes in contact in 91% of specimens, and Lee (1996: 162) said Yucatán specimens have those 2 scutes "generally in firm contact" (curiously, Iverson, 1976: 260 was not cited in Berry and Iverson, 2011). Campbell (1998), in discussing eastern Guatemalan specimens, did not mention that diagnostic character. Thus, there is an unusual amount of contradiction in the literature regarding the axillary and inguinal scute character in areas purportedly inhabited by *K. albogulare* or *K. cruentatum*.

In summary, Berry (1978: 147) and Berry and Iverson (2001b: 725.6, 2011: 63.5) say the axillary and inguinal scutes rarely contact each other in *K. cruentatum* (Berry, 1978, gave 7% of specimens), whereas the 21 Honduran specimens examined for this character have those 2 scales in contact 72.5% of the time. Berry (1978: 160) and Berry and Iverson (2001b: 725.5, 2011: 63.5) say the axillary and inguinal scutes are usually in contact (in 86% of individuals) in *K. albogulare*; thus, the Honduran specimens are more similar to *K. albogulare* in that character. Also, according to the data in Iverson (1976) and Lee (1996), *K. albogulare* also occurs in Belize, eastern Guatemala, and the outer Yucatán Peninsula of Mexico. Legler and Vogt (2013: 136) said the axillary scute "is rarely in contact with inguinal" scute in Mexican *K. cruentatum*, but that data appears not to be original and to having been taken from Berry and

Iverson (2001b). Most Honduran specimens have orange or red dots on the head, another character used by Berry and Iverson (2011: 63.5) to help define *K. albogulare*.

Berry and Iverson (2001b, 2011) extensively reviewed the literature on the *Kinosternon scorpioides* species complex, but those authors offered a contradictive distribution of *K. albogulare* and *K. cruentatum* according to data offered in other publications.

Acuña-Mesén (1994) found shell shape differences in *Kinosternon albogulare* between subhumid and mesic populations in Costa Rica and attributed it to adaptation to their habitats. Turtles in mesic habitats had lower shells, an adaptation to their more stream occurring habitats, whereas those from the subhumid forest had more rounded shells for their pond habitats. My impression of Honduran populations seems to support the lower shell for stream-dwelling *K. albogulare*, as opposed to higher shells in pond-dwelling *K. albogulare* and according to the published molecular data, apparently of the same species. Unpublished molecular data for Honduran populations indicate only one species is involved in the country. Additionally, Honduran stream-dwelling *Kinosternon* seem to have slightly more interdigital webbing than do those from pond habitats.

Natural History Comments.—*Kinosternon albogulare* is known from near sea level to 1,240 m elevation in the Lowland Moist Forest, Lowland Dry Forest, Lowland Arid Forest, Premontane Moist Forest, and Premontane Dry Forest formations. The species occurs in seasonal swampy and low-lying wet areas, seasonal ponds and streams, and permanent freshwater lagoons, swamps, streams, and in some rivers. It is both nocturnal and diurnal and is most active during periods of rain. However, it has been found active in February, April, and May during the dry season and will crawl across land when their habitat dries to search for wetter habitats (also see Teska, 1976 and Castañeda and Mora, 2010 for Costa Rican populations). Months of collection of active individuals during the rainy season are from May to November. One was under a log in a dry streambed at the height of the dry season in April. One adult that was crossing a dirt road during the dry season in May had damp mud on the posterior three-quarters of the carapace, indicating it might have left a drying habitat looking for a more mesic place. Adults used to be commonly seen in large road puddles on the dirt road between Rus Rus and Awasbila, Gracias a Dios, but a sharp increase in vehicular traffic around 2003, mostly associated with the cocaine business, appears to have decimated those populations. I have never seen individuals of this species basking above the waterline, as occasionally seen in *K. leucostomum.* Berry and Iverson (2011) also provided a summary of habitat data of other Central American populations of this turtle. Two freshly deposited eggs and two eggs containing embryos nearly ready to hatch were found in two shallow excavations in rotten logs in a swampy area in northeastern Honduras in June. Recently hatched specimens were found in shallow water–containing depressions and small, muddy, and temporary ponds in June to August. Iverson (2010) studied reproduction in the *K. albogulare–K. cruentatum* complex of turtles from southeastern Mexico and Belize and found that females deposit multiple clutches of one to four eggs per clutch, with reproduction apparently continuous between August and June, thus spanning some of the late rainy season and most of the dry season. Eggs hatch during the wet season from June to August. Iverson (2010) also provided a wealth of other reproductive data on the *K. scorpioides* species complex from throughout its northern geographical distribution, including a female from "northwestern Honduras" (p. 255; also see Berry and Iverson, 2011). Goode (1994) recorded multiple clutches (one to three per female)

of one to eight eggs in captivity in *K. albogulare–K. cruentatum,* some animals of which were from "Honduras." Dunn and Saxe (1950) reported finding one egg of this species on Isla San Andrés, Colombia, in April or May (where the egg was found was not given). Forero-Medina and Castaño-Mora (2011) reported two nests with one and two eggs on Isla San Andrés. Based on examination of stomach contents of specimens in the UU museum collection, Legler (1966) considered the *K. scorpioides* complex members to be opportunistic feeders that are primarily herbivorous in the wild. Moll, D. (1990) reported that a population from Belize was primarily insectivorous but also ate small snails, fish, and various plant matter. Acuña [Mesén] et al. (1983) stated this species ate tadpoles, freshwater crustaceans, insects, and freshwater gastropods in Costa Rica. Berry and Iverson (2011) reviewed literature on diet in this species, but some of those populations represent the related *K. cruentatum* and *K. scorpioides* and noted that members of this complex are primarily carnivorous predators and scavengers but will also eat plant matter. Forero-Medina and Castaño-Mora (2011: 064.3; as *K. s. albogulare*) summed up its diet as "omnivorous and occasionally a scavenger." Recorded items in its diet include fruits, terrestrial and aquatic invertebrates (mainly mollusks and arthropods), and dipteran larvae. Legler and Vogt (2013: 137) reported "adults from the lower Yucatán Peninsula, which were not eating mollusks, seemed to have narrower heads than those from Belize that were eating mollusks." Those authors went on to say that James F. Berry "demonstrated statistically that this was true."

Kinosternon albogulare occurs sympatrically with *K. leucostomum* in several freshwater lagoons in northeastern Honduras. Legler and Vogt (2013) did not report sympatric occurrence between the *K. scorpioides* group member and *K. leucostomum* in Mexico. Some people living in the

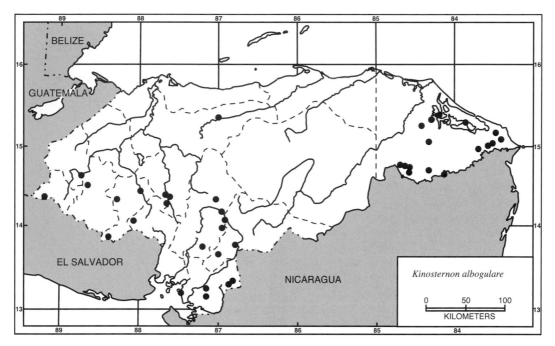

Map 74.   Localities for *Kinosternon albogulare*. Solid circles denote specimens examined.

Mosquitia of northeastern Honduras use this species as a food source. People that I have been with put the entire live animal on its back in coals of a fire and then extract the meat from the shell when cooked. However, some Miskitos have told me that eating that meat had made them sick to the stomach.

Etymology.—The name *albogulare* is derived from the Latin *albus* (white), *gula* (throat), and suffix *-arius* (pertaining to). The name alludes to the white throat as described in the holotype by Bocourt (1870).

Specimens Examined (83, + 2 shell fragments, 1 skeleton, 1 egg, 2 embryos [24]; Map 74).—**CHOLUTECA**: 1.6 km N of Cedeño, KU 209309; 1.0 km N of Cedeño, LSUMZ 34042; El Madreal, USNM 561997, 579659; Finca La Libertad, SDSNH 72804; Quebrada del Horno, SDSNH 72805–06, UNAH 5274; near Quebrada La Florida, UNAH 5277, 5284. **COMAYAGUA**: 4.8 km SSE of Comayagua,

LACM 48448; 3 km W of Comayagua, TCWC 23825; near Comayagua, AMNH 70549. **EL PARAÍSO**: El Rodeo, USNM 580765; 11.3 km S of Güinope, TCWC 19230. **FRANCISCO MORAZÁN**: Cantarranas, ANSP 24776; Caserío Los Encinitos, UNAH (1); El Zamorano, AMNH 70548, MCZ R48770–72, 49749–51, UF 150885, UU 7634–37, 7638 (skeleton), 7639–40; 11 km NE of San Antonio de Oriente, LSUMZ 24633; 5 km N of Talanga, TCWC 30128. **GRACIAS A DIOS**: 42 km NE of Awasbila, USNM 559581; between Awasbila and Rus Rus, UF 137152–53; Calpo, UF 137158–61; Canco, UF 137664–65 (both shell fragments); Crique Sikiatingni, UF 137157; Leimus (Río Coco), LACM 73825; Leimus (Río Warunta), USNM 573979; 1.2 km SW of road to Pranza on Rus Rus-Puerto Lempira road, USNM 559579; Krahkra, USNM 573321; Mocorón, UTA R-42646, 53526; Rus Rus, USNM 559580, 570542; Samil, USNM 573320, 573978; Sisinbila,

USNM 589139; Tánsin, LACM 48440–46, LSUMZ 21720–27; Tikiraya, UF 137154–56, 137174 (egg and 2 embryos); Warunta, USNM 573081, 589140. **INTIBUCÁ**: 1.4 km N of Jesús de Otoro, LSUMZ 34043; near San Juan, KU 209308; 5 km E of Santa Lucía, USNM 570474. **LA PAZ**: La Paz, FN 256880 (still in Honduras because of permit problems); Marcala, CM 62034–35. **LEM-PIRA**: 21.7 km NNW of Gracias, LACM 48447; Gracias, FMNH 283767. **OCOTE-PEQUE**: Río Lempa at Antigua, FMNH 283748. **YORO**: 4.7 km ESE of San Lorenzo Arriba, USNM 561996, 561998.

Kinosternon leucostomum (A. M. C. Duméril and Bibron, 1851)

Cinosternon leucostomum A. M. C. Duméril and Bibron, 1851: 17, *In* A. M. C. Duméril and Duméril, 1851 (holotype not stated; type locality: "N.-Orléans; Mexique; Rio-Sumasinta [Amér. centr.]; ... Amér. septentr.? ... Valleé de la Madeleine [N. Grenada], ... Santa-Fé de Bogota [N. Grenade]"; Schmidt [1941: 488] restricted the type locality to "Rio Usumacinta, Peten, Guatemala" based on A. M. C. Duméril and Bibron's misspelling of that river as "Rio-Sumasinta [Amér. centr.]"; lectotype, MNHN 9087 [formerly MNHN 8311], designated by Stuart, 1963: 49 "by fiat of restriction of type locality" of Schmidt, 1941: 488 [also see Berry and Iverson, 2001a: 724.1]).
Kinosternum [sic] *leucostomum*: Le Conte, 1854: 183; Cruz Díaz, 1978: 23.
Kinosternon leucostomum: Gray, 1856a: 46 (see Berry and Iverson, 2001a: 724.2); Stuart, 1934: 5; Meyer, 1966: 174; Meyer, 1969: 183; Meyer and Wilson, 1973: 4; Iverson, 1992: 228; Wilson and Cruz Díaz, 1993: 15; Grillitsch et al., 1996: 93; Cruz D[íaz], 1998: 29, *In* Bermingham et al., 1998; Monzel, 1998: 167; Wilson and McCranie, 1998: 15; Köhler, 2000: 27; Köhler, McCranie, and Nicholson, 2000: 425; Nicholson et al., 2000: 29; Wilson et al., 2001:

134; Castañeda, 2002: 37; Lundberg, 2002a: 12; McCranie, Castañeda, and Nicholson, 2002: 25; McCranie, 2005: 20; McCranie et al., 2005: 64; Boback et al., 2006: 239; Castañeda, 2006: 26; Castañeda and Marineros, 2006: 3.8; McCranie et al., 2006: 105; Townsend, 2006a: 34; Wilson and Townsend, 2006: 104; Wilson and Townsend, 2007: 145; Townsend and Wilson, 2010b: 692; McCranie et al., 2011: 565; Valdés Orellana and McCranie, 2011a: 565; Solís et al., 2014: 139; McCranie, 2015a: 380.
Cinosternum [sic] *leucostomum*: Werner, 1896: 345; Siebenrock, 1907: 581.
Cinosternon brevigulare: Atkinson, 1907: 152.
Cinosternon cobanum: Atkinson, 1907: 152.
Kinosternon leucostomum leucostomum: Berry, 1978: 320; Berry and Iverson, 2001a: 724.2.
Cryptochelys leucostoma: Iverson et al., 2013: 933; McCranie and Valdés Orellana, 2014: 44.

Geographic Distribution.—*Kinosternon leucostomum* occurs at low and moderate elevations from central Veracruz, Mexico, to north-central Colombia on the Atlantic versant and from west-central Costa Rica to southwestern Ecuador on the Pacific versant. It also occurs on the Islas del Maíz (Corn Islands), Nicaragua, and on Guanaja Island and the Cayos Cochinos, Honduras. In Honduras, this species is widely distributed across the northern and eastern portions of the mainland as well as the islands just mentioned.

Description.—The following is based on ten males (UF 137145, 137148, 137150; USNM 559561, 559564–65, 559568, 559570–71, 559575) and ten females (UF 137146; USNM 559563, 559566, 559569, 559572–73, 559576–78, 570473). *Kinosternon leucostomum* is a small (maximum recorded CL 214 mm [Legler and Vogt, 2013]; largest CL in Honduran specimen 150 mm [UF 137145, a male]) turtle with a low ridge on the carapace, ridge can be absent or barely visible in old individuals; carapace oval, with 1 nuchal, 5 vertebrals, 8 costals, and 22 marginals; carapacial sides straight; costal 4 contacting marginal 11; marginals anterior and posterior to bridge moderately to distinctively flared; marginal 10 higher than marginal 9, usually lower than marginal 11; vertebral 1 usually wider than long, rarely longer than wide, contacting marginal 2; vertebrals 1–4 lack distinct posterior notches at midline, but vertebral 1 and/or 2 sometimes with weak posterior notches; plastron with 2 kinetic hinges, 1 each anterior and posterior to posterior humeral; plastron completely closes shell, or only partially closes shell in adults of some populations (see Remarks); plastron with 11 scutes, consisting of paired gulars, anterior humerals, posterior humerals, femorals, and anals, and single intergular; plastral intergular curving outwardly dorsally, away from entire exposed ventral portion at anterior edge; anals with slight or no medial notch; plastron concave to slightly concave in males, slightly convex or flat in females; 1–2 axillary scutes and 1 inguinal scute present, axillary and inguinal scutes usually not in contact (in contact in 15%); 0–2 inframarginal bridge scutes contacting fixed lobe of plastron; head relatively small, snout slightly projecting; upper jaw weakly to strongly hooked; 1–4 pairs of gular barbels present, anterior pair largest; feet weakly to distinctively webbed (see Remarks); 5 claws on forelimb, 4 claws on hind limb; elevated patches of scales (clasping organs) present on posterior thigh and calf in males, although usually weakly developed; tail much longer in males than in females, cloacal opening well posterior to plastral margin in males (but not posterior to carapacial margin), at or just posterior to plastral margin in females; tails of both sexes with terminal spines, spines larger in males; CL 114–150 (129.4 ± 10.4) mm in males, 110–145 (126.9 ± 11.6) mm in females; maximum carapace depth/CL 0.32–0.38 in

males, 0.35–0.40 in females; PL/CL 0.86–0.92 in males, 0.86–0.98 in females; plastral forelobe L/CL 0.28–0.32 in males, 0.28–0.33 in females; PW at anterior hinge/CW 0.67–0.74 in males, 0.67–0.79 in females; PW at midfemoral/CW at midfemoral 0.60–0.72 in males, 0.63–0.76 in females; intergular scute L/CL 0.12–0.22 in males, 0.11–0.16 in females; first vertebral W/second vertebral W 0.96–1.17 in males, 0.89–1.25 in females; interposterior humeral (fixed lobe) seam L/PL 0.23–0.25 in males, 0.21–0.26 in females; interposterior humeral (fixed lobe) seam L/plastral forelobe L 0.63–0.74 in males, 0.66–0.79 in females; intergular scute L/plastral forelobe L 0.38–0.71 in males, 0.38–0.53 in females; BL/CL 0.23–0.27 in males, 0.24–0.31 in females; anterior width of posterior plastral lobe/CL 0.33–0.40 in males, 0.34–0.43 in females; interanal scute seam L/CL 0.25–0.30 in males, 0.27–0.32 in females.

Color in life of an adult male (USNM 559575): carapacial scutes Grayish Horn Color (91) centrally, Sepia (119) on marginals; bridge and plastron Yellow Ocher (123C), anterior and posterior margins of underside of carapace slightly darker yellow; head speckled yellow and dark brown; upper rhamphotheca brown with yellow flecking; lower rhamphotheca brown with yellow lines; neck gray-brown above, pale yellow below; limbs gray-brown above, pale yellow below; tail gray-brown above and below. Color in life of another adult male (USNM 559561): carapace dark brown; bridge and plastron pale yellow with overlay of rust brown; head dark brown above; temporal region yellow-tan, reticulated with dark brown; side of head brown; jaws yellow-tan; chin pale yellow; limbs gray-brown above, pale yellow below; tail dark brown above, paler brown below.

Color in alcohol: carapace dark brown with darker brown seams, except marginals sometimes paler brown with dark brown seams; plastron yellow to tan, with dark brown seams, brown mottling sometimes

Plate 92.   *Kinosternon leucostomum*. FMNH 283589. Islas de la Bahía: Guanaja Island, Mitch.

present on some to most scutes; axillary, inguinal, and marginals of bridge yellow to tan, with dark brown seams, marginals in some specimens also mottled with brown; dorsal and lateral surfaces of head and neck dark brown, cream to pale brown spots and/or reticulations present laterally and usually dorsally; pale postorbital stripe can be distinct, obscure, or absent; jaw sheaths cream to pale yellow, with or without brown vertical streaks; chin and ventral surface of neck cream to pale brown, sometimes with brown mottling; limbs dark brown on inside and outside surfaces, pale gray to grayish brown on ventral surfaces; tail brown dorsally and ventrally, tail tubercles cream in some.

Diagnosis/Similar Species.—*Kinosternon leucostomum* is distinguished from all other Honduran turtles, except *K. albogulare*, in having a hinged, oblong plastron with 11 scutes and 11 marginal scutes per side on the carapace. *Kinosternon albogulare* has 3, low carapacial ridges, but those ridges usually only barely visible posteriorly in old individuals, the intergular not curving outward dorsally at its anterior edge, and males lack elevated patches of horny ridges

on the posterior thigh and calf (versus only 1 low medial ridge on carapace, intergular curving outward dorsally at anterior edge, and males with elevated patches of horny ridges on posterior thigh and calf in *K. leucostomum*).

Illustrations (Figs. 152, 153, 173–175; Plate 92).—Álvarez del Toro, 1983 (adult); Berry and Iverson, 2001a (adult); Bonin et al., 2006 (head); Calderón-Mandujano, 2008 (adult); Campbell, 1998 (adult, carapace, plastron); Castañeda and Mora, 2015 (sexual dichromatic head); A. H. A. Duméril, 1852a (adult, carapace, plastron: as *Emys* A. M. C. Duméril, 1805: 76); Ernst and Barbour, 1989 (adult); Günther, 1885, *In* Günther, 1885–1902 (head, carapace, plastron; as *Cinosternon* Wagler, 1830a: 137); Guyer and Donnelly, 2005 (adult, head, plastron); Hödl, 1996 (adult); Hutchison and Bramble, 1981 (plastron); Köhler, 2000 (adult, head, plastron), 2001b (adult), 2003a (adult, plastron), 2008 (adult, plastron); Lee, 1996 (adult, plastron), 2000 (adult, plastron); Legler, 1965 (plastron, relationships of vertebral 5 with marginals 10–11); Legler and Vogt, 2013 (adult, juvenile, head, carapace, plastron); McCranie and Valdés

Orellana, 2014 (adult; as *Cryptochelys*); McCranie et al., 2005 (adult, plastron), 2006 (adult, plastron); D. Moll, 2010 (adult); E. O. Moll, 1979 (plastron and comparative egg size; fig. 8, not fig. 9 as indicated); Neill, 1965 (adult; as *K. leucostomum* and *K. mopanum* Neill, 1965: 117); Pough et al., 2015 (adult); Pritchard, 1967 (adult), 1979 (adult); Savage, 2002 (adult, male thigh and calf spines); Shi, 2013 (adult, head, plastron); Smith and Smith, 1980 (adult, subadult); Stafford and Meyer, 1999 (adult, juvenile, shell); Vetter, 2005 (adult; as *K. l. leucostomum*); Wermuth and Mertens, 1961 (adult, carapace, plastron).

Remarks.—Two subspecies of *Kinosternon leucostomum*, *K. l. leucostomum* and *K. l. postinguinale* (Cope, 1887: 23), were recognized in the latest revision of the species (Berry, 1978: 179; also see Berry and Iverson, 2001a: 724.2). However, Savage (2002: 748), in discussing Berry's (1978) work, stated "his [Berry's] data clearly document the existence of a number of north-south character clines that make any attempt at delimiting geographic races within this species ambiguous." Savage did not present any original data on Costa Rican *K. leucostomum* to support his claim. Specimens from Honduras are very variable, with specimens showing characteristics of both subspecies as defined by Berry and Iverson (2001a: 724.2; see below). Iverson et al. (2013: 935) stated "more thorough geographic sampling [molecular] is needed for the wide-ranging species *leucostomum*, since preliminary morphological data indicate the existence of undescribed variation."

Berry (1978), Iverson (1991), and Berry and Iverson (2001a) diagnosed those two subspecies on the basis of morphology, with *Kinosternon l. leucostomum* the form purported to occur in Honduras. Most of the proposed differences between those two subspecies are based on minor mean ratios. My measurements and ratios for 20 Honduran specimens show much overlap be-

tween the diagnostic characters of each subspecies. Also, one of the diagnostic characters, the carapace height, "carapace relative high" (Berry, 1978: 179 for *K. l. leucostomum*) and "carapace relatively depressed" (Berry, 1978: 185 for *K. l. postinguinale*), was shown to be influenced by adaptation to different habitats in Costa Rican populations of a different species (*K. albogulare*; see Acuña-Mesén, 1994). Honduran populations of *K. albogulare* occurring in streams have lower shells than populations occurring in pond habitats, which have higher shells. *Kinosternon leucostomum* occurs in streams and ponds in Honduras and also appear to have slight differences in carapace height between those two habitat types. Thus, some of that supposed diagnostic characteristic might be influenced by different habitats available to the turtles. Stream and small river-dwelling *K. leucostomum* in Honduras also seem to have slightly more interdigital webbing than do those from pond habitats.

Some Honduran populations have distinct interdigital webbing (i.e., El Paraíso and Cuaca, Olancho, populations), whereas interdigital webbing is absent in some other populations (i.e., Sisinbila, Gracias a Dios). Also, adults can't completely close the shell in the El Paraíso population, but adults from most other Honduran populations are capable of completely closing their shells. However, molecular study of several Honduran populations, including the El Paraíso population that can not completely close its shell, indicates only one molecular species is involved (F. Köhler, unpublished data).

Berry and Iverson (2001a) extensively reviewed the literature on this species. Iverson (1976) presented a morphological comparison of this species to *Kinosternon albogulare* (as *K. cruentatum*) in Belize, adjacent Mexico, and eastern Guatemala. Hernández-Guzmán et al. (2014) reported on the chromosomes of *K. leucostomum* from Tabasco, Mexico. Castañeda and Mora (2015) reported ventral head color differences between males and females in a Costa Rican population.

The photographs of *Kinosternon leucostomum* from Colombia in Páez et al. (2012) do not resemble Honduran specimens of that species.

Natural History Comments.—*Kinosternon leucostomum* is known from near sea level to 1,120 m elevation in the Lowland Moist Forest, Lowland Dry Forest, Lowland Arid Forest, Premontane Wet Forest, and Premontane Moist Forest formations. It is commonly seen walking on bottoms of ponds and lakes and normally slow-moving streams and rivers (when not rain swollen). It can also be found crawling on the forest floor and crossing roads, especially during rainy weather, but also during the dry season when it is apparently searching for a wet site, to replace a site that had dried, in which to spend the remainder of the dry season. One was found on the Cayos Cochinos near an estuary (Wilson and Cruz Díaz, 1993). Retreats include beneath logs, in piles of debris in the forest, and in piles of discarded rice plants in fields. The species is both nocturnal and diurnal and is active throughout the year in areas of permanent water. I have also occasionally seen individuals of this species basking on objects above the waterline. Hatchlings have been found in June and July. *Kinosternon leucostomum* occurs sympatrically with *K. albogulare* in several freshwater lagoons in northeastern Honduras. Legler and Vogt (2013) did not report sympatric occurrence between these two species or species complexes in Mexico. Some, but not all, people living in the Mosquitia of northeastern Honduras use this species as a food source (see *K. albogulare* for notes on the methods for cooking these small turtles). Morales-Verdeja and Vogt (1997) studied the reproductive cycle of *K. leucostomum* in Veracruz, Mexico, and found that nesting occurred from late August to March, thus spanning much of the dry season. Moll, E. O., and Legler (1971) reported Panamanian and Costa Rican females contained one or two eggs per clutch and are capable of multiple clutches per season and suggested reproductive activity throughout the year in those populations. Legler and Vogt (2013) reported one to five eggs per clutch, with the potential for females in a Mexican population to deposit one to seven clutches per year. Clutches are placed in shallow holes dug by the female or are simply covered with leaves (E. O. Moll and Legler, 1971; D. Moll, 2010; Legler and Vogt, 2013). Moll, E. O., and Legler (1971: 89) reported "approximately equal amounts of plants (chiefly *Elodea* and grass) and animal material (chiefly snails with a few insects)" were consumed in a Panamanian population. Legler (1966), based on examination of stomach contents of UU museum specimens, regarded *K. leucostomum* as an opportunistic feeder, but primarily herbivorous. Moll, D. (1990) reported a population in Belize was primarily insectivorous but also ate small snails and plant matter. Vogt and Guzman (1988) found two populations of *K. leucostomum* in Veracruz, Mexico, that were omnivorous generalists with a wide range of food items taken, depending on availability (also see Legler and Vogt, 2013; Ceballos et al., 2016). Legler and Vogt (2013) also discussed variation in natural history parameters in three Mexican populations of *K. leucostomum*.

Etymology.—The name *leucostomum* is derived from the Greek words *leukos* (white) and *stoma* (mouth), and alludes to the pale-colored jaw sheaths in many specimens of this species (especially juveniles).

Specimens Examined (144, 3 shell fragments, 4 skeletons, 2 embryos [28]; Map 75).—**ATLÁNTIDA**: Corozal, LSUMZ 21728–29, 21860–61; 18.3 km E of La Ceiba, AMNH 71386; La Ceiba, USNM 62981–90; Lancetilla, AMNH 71387; Punta Sal, USNM 579657; Tela, MCZ R27912, 29125–31, USNM 76871. **COLÓN**: Balfate,

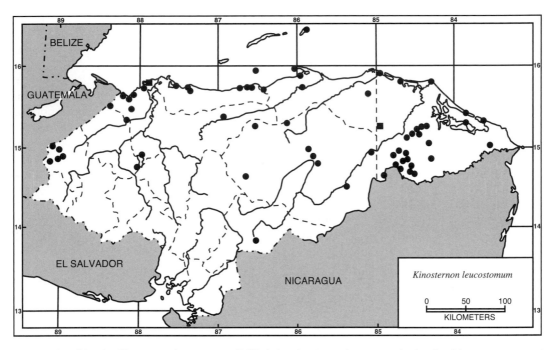

Map 75. Localities for *Kinosternon leucostomum*. Solid circles denote specimens examined and solid squares represent accepted records.

AMNH 58606, 58671–72; Finca Sonora, MCZ R20476; Puerto Castillo, LSUMZ 27733; Trujillo, CM 62026–33, LSUMZ 33659, 34045. **COPÁN**: 12.9 km ENE of Copán, LACM 48437–38; 19.3 km ENE of Copán, LACM 48439; 30.6 km SW of La Florida, LACM 48436; La Playona, USNM 559561, 561995; Laguna del Cerro, USNM 559563. **CORTÉS**: 3.2 km N of Agua Azul, LSUMZ 28513–14; Agua Azul, AMNH 70545–47, 124036 (2 embryos), TCWC 19232; near Cofradía, USNM 102886–87; El Paraíso, UF 144665; N end of Lago de Yojoa, AMNH 70544; Masca, UU 8591–94, 8595–96 (both skeletons), 8597–602, 8603–04 (both skeletons), 8605–08; San Pedro Sula, FMNH 5315, UF 123957; Santa Teresa, USNM 579656; 2.9 km E of Tegucigalpita, USNM 559562; Tulián, UU 8589–90. **EL PARAÍSO**: Mapachín, USNM 578923–24. **GRACIAS A DIOS**: Awasbila, USNM 559578; Bachi Kiamp, FMNH 282718; Bodega de Río Tapalwás, USNM

559572–74; Cabeceras de Río Rus Rus, USNM 573080; Caño Awalwás, UF 137145, USNM 559570–71, 559575–77; Cauquira, UF 137666–67 (both shell fragments); near Coco, USNM 589143; Crique Sikiatingni, UF 137148–51; Crique Wahatingni, UF 137146; Crique Yulpruan, USNM 570472; near Cueva de Leimus, USNM 589144; Hiltara Kiamp, USNM 562916; Kakamuklaya, USNM 573077; Kaska Tingni, USNM 559567–69; Laguna Baraya, ROM 19268; Leimus (Río Warunta), FMNH 282717; about 30 km S of Mocorón, UTA R-42648; Mocorón, UTA R-42647, 52342, 53525, 55512; Palacios, BMNH 1986:41; Río Coco, USNM 24540–41; Rus Rus, ICS (now destroyed); Sadyk Kiamp, USNM 565395, 573977, 579654; Sisinbila, USNM 589145; Tánsin, LACM 48432–35, LSUMZ 21859; Tikiraya, UF 137147; Usus Paman, USNM 573319; Warunta Tingni Kiamp, USNM 561994, 570473; Yahurabila, USNM 573078–79. **ISLAS DE LA BAHÍA**: Cayo

Cochino Mayor, La Ensenada, KU 220134; Isla de Guanaja, Mitch, FMNH 283589. **OLANCHO**: 7 km E of Azacualpa, UF 90019; Callejón, LACM 73823; 1 km SE of Catacamas, LACM 48431; Cuaca, USNM 589142; El Ocotal, USNM 589141; 3.2 km W of Galeras, TCWC 19231; 6.2 km N of La Unión, USNM 559564; Quebrada El Guásimo, USNM 559565–66; Río Kosmako, USNM 559592 (shell fragments); **SANTA BÁRBARA**: SW corner of Lago de Yojoa, USNM 578922. **YORO**: 5 km S of San Patricio, USNM 578921.

Other Records (Map 75).—**COPÁN**: Río Amarillo (Castañeda and Marineros, 2006). **CORTÉS**: Puerto Cortés (Atkinson, 1907). **GRACIAS A DIOS**: Raudal Pomokir, UNAH 5509 (Cruz Díaz, 1978, specimen now lost). **ISLAS DE LA BAHÍA**: Cayo Cochino Menor (Boback et al., 2006). "HONDURAS": NMW 1703 (Grillitsch et al., 1996); Werner, 1896; Siebenrock, 1907.

Genus *Staurotypus* Wagler, 1830a

Staurotypus Wagler, 1830a: 137 (type species: *Terrapene triporcata* Wiegmann, 1828, col. 364, by monotypy).

Geographic Distribution and Content.— This genus ranges from central Veracruz, Mexico, to northern Honduras and the Isthmus of Tehuantepec region, Oaxaca, Mexico, to south-central El Salvador. Two named species are recognized, one of which occurs in fresh- and brackish water near the coast of northwestern Honduras.

Remarks.—Gutsche and McCranie (2016) cleared up the publication dates and content of three Wagler works (1830a, 1830b, 1833) involving *Staurotypus*, all of which have been confused in the various TTWG lists of the turtles of the world.

Etymology.—The name *Staurotypus* is derived from the Greek words *stauros* (cross) and *typos* (figure, impression, model, shape). The name alludes to the slightly cruciform and relatively small plastron (Wagler, 1830a).

Staurotypus triporcatus (Wiegmann, 1828)

Terrapene triporcata Wiegmann, 1828, col. 364 (holotype, ZMB 127 [see Iverson, 1983: 328.1; Smith and Smith, 1980: 47; Gutsche, 2016: 158]; type locality: "Río Alvarado" [Veracruz, Mexico]).

Staurotypus triporcatus: Wagler, 1830b, Explicatio Tabularum; Meyer, 1969: 189; Meyer and Wilson, 1973: 5; Iverson, 1983: 328.1; Iverson, 1986: 134; Iverson, 1992: 239; Wilson and McCranie, 1998: 15; Wilson et al., 2001: 134; Wilson and McCranie, 2003: 59; McCranie et al., 2006: 216; Wilson and Townsend, 2006: 104; Townsend and Wilson, 2010b: 692; Castañeda et al., 2013: 309; Legler and Vogt, 2013: 91; Solís et al., 2014: 139; McCranie, 2015a: 380.

Geographic Distribution.—*Staurotypus triporcatus* occurs at low elevations from central Veracruz, Mexico, to northwestern Honduras on the Atlantic versant. In Honduras, this species is known from a few localities near the north coast of the northwestern portion of the country.

Description.—The following is based on three males (USNM 102884, 524353, 579661) and one female (UU 6417). *Staurotypus triporcatus* is a large (maximum recorded CL 400 mm [Iverson, 1983]; largest CL in Honduran specimen 285 mm [USNM 102884, a male]) turtle with a strongly tricarinate carapace, especially in adults; carapace oval, with 2 nuchals, 5 vertebrals, 8 costals, and 22 marginals; carapacial sides straight; costal 4 broadly contacting marginal 11; anterior and posterior marginals not flared; marginal 10 slightly higher than marginals 9 and 11; vertebral 1 longer than wide, not in contact with marginal 2; vertebrals 1–4 with distinct posterior notches at midline; plastron with weak hinge anterior to paired humerals, hinge absent posterior to paired humerals; plastron with 7–8 scutes, consisting of paired gulars, paired humerals, paired femorals, 1 anal, plus tiny intergular some-times present (see Remarks for family); anal without medial notch; plastron slightly concave in males, flat in female; 1 large axillary scute broadly contacting 1 large inguinal scute, axillary and inguinal scutes spanning and forming bridge from margin-als; head large, snout distinctively project-ing; nasal scale slightly emarginated posteriorly; upper jaw not hooked, lower jaw slightly hooked; 1 pair of large gular barbels present; feet strongly webbed; 5 claws on forelimb, 4 claws on hind limb; elevated widened, ridged scales present (clasping organs) on posterior thigh and calf in males; tail much longer in males than in females; cloacal opening posterior to car-apacial margin in males, anterior to carapa-cial margin in females; tails of both sexes with terminal spines, spines larger in males; CL 93–285 (212.3 ± 104.2) mm in males, 111 mm in female; PL/CL 0.61–0.68; interhumeral seam L/PL 0.21–0.36; BL/CL 0.11–0.28.

Color in life of a juvenile male (USNM 579661): carapace Glaucous (79) with Sepia (119) blotches, also with Light Drab (119C) mottling on each scute, except marginals slightly paler than Glaucous; top of head Sepia with dusty white spotting and blotch-ing; side of head similar in color to top of head, except white more extensive than Sepia; ventral surface of head cream with dark brown lines and stippling; skin of nape mottled Sepia and Pale Horn Color (92) dorsally, paler brown and cream laterally, cream with brown flecking ventrally; web-bing of all digits Pale Horn Color with Sepia flecking, remainder of dorsal surface of feet mottled Sepia and grayish brown; plastron Pale Horn Color with large, indistinct brown blotch on each scute; exposed skin of venter with grayish tinge and flecked with brown.

Color in alcohol of a juvenile male (USNM 579661): carapace pale brownish gray with large dark brown blotches on each scute; dark brown carapacial blotches with paler brown interior mottling; plastron, axillary, and inguinal scutes pale yellow with brownish gray blotches on each scute, much smaller dark brown blotches also present on each scute; dorsal and lateral surfaces of head black with cream lines and blotches; ventral surface of head cream with black mottling; dorsal surface of neck black with cream spotting or mottling; lateral surface of neck cream with black mottling; ventral surface of neck pale brown with brown flecking; dorsal surfaces of limbs brownish gray with some dark brown spots, especially on digits; webbing pale brown with dark brown mottling; ventral surfaces of limbs brownish gray with indistinct pale brown mottling; exposed ventral skin brownish gray with pale brown mottling anterior to plastron; dorsal surface of tail dark brown with double row of pale brown conical tubercles; ventral surface of tail brownish gray. Adults are darker overall than are juveniles, and have pale brown or cream markings on the head that are less distinct and have the carapace nearly uniformly brown.

Diagnosis/Similar Species.—*Staurotypus triporcatus* is distinguished from all remain-ing Honduran turtles by the combination of having a reduced plastron with only 7 or 8 scutes, a short tail, and 3 strong ridges on the adult carapace.

Illustrations (Figs. 155, 172; Plates 93, 94).—Álvarez del Toro, 1983 (adult); Álvarez Solórzano and González Escamilla, 1987 (adult); Bonin et al., 2006 (adult, plastron); Bramble et al., 1984 (head); Calderón-Mandujano et al., 2008 (adult); Campbell, 1998 (adult, carapace, plastron); Ernst and Barbour, 1989 (adult); Gray, 1856a (adult); Gutsche, 2016 (holotype); Holman, 1964 (adult); Hutchison and Bramble, 1981 (plastron); Köhler, 2000 (adult), 2003a (adult, carapace), 2008 (adult, carapace); Lee, 1996 (adult), 2000 (adult); Legler and Vogt, 2013 (adult, juvenile, head, carapace, plastron); D. Moll and Moll, 2004 (adult); Neill and Allen, 1959 (carapace; as

Plate 93. *Staurotypus triporcatus.* Adult (in captivity). Cortés: Choloma.

S. salvini [sic] Gray, 1864: 127); Pritchard, 1967 (adult), 1979 (adult, hatchling); Shi, 2013 (adult, head, carapace, plastron); Smith and Smith, 1980 (adult, carapace, plastron); Stafford and Meyer, 1999 (adult); Vetter, 2005 (adult, juvenile); Vogt, 1997 (adult); Wagler, 1830b (adult, carapace, plastron), 1833 (adult, carapace, plastron); Wermuth and Mertens, 1961 (adult, carapace, plastron); Zug, 1966 (schematic drawing of penis).

Remarks.—Iverson (1983) provided an overview of the morphology of *Staurotypus triporcatus*, along with a brief review of the

Plate 94. *Staurotypus triporcatus.* USNM 579661, juvenile. Cortés: Choloma.

literature. Hernández-Guzmán et al. (2014) reported on chromosomes of *S. triporcatus* from Tabasco, Mexico. Terán-Juárez et al. (2015) reported a specimen from extreme southwestern Tamaulipas, Mexico, that, in their opinion, was likely introduced.

Natural History Comments.—*Staurotypus triporcatus* is known from near sea level to 100 m elevation in the Lowland Moist Forest and Lowland Dry Forest formations. The Atlántida specimen is a shell found near a village. Some people living within the range of this species consider it to be a delicacy, whereas others say they do not eat its meat. One (USNM 579661) is a captive-bred juvenile from parents collected at Choloma, Cortés. Moll, D., and Moll (2004) said this turtle is crepuscular and/or nocturnal. Vogt (1997) stated that in Veracruz, Mexico, this species occurs in a variety of permanent water habitats, including lakes, ponds, lagoons, slow-moving rivers, and swamps, including mangrove swamps (also see Legler and Vogt, 2013). Legler and Vogt (2013) reported that aerial basking is not known in this species. Legler and Vogt (2013) said the overall number of eggs per clutch is 4–17. Those authors also gave a mean of 9.8 eggs in 18 clutches from Veracruz (shelled eggs found from late August to mid-March with peak in October) and a mean number of 8.1 in 59 clutches from Chiapas, Mexico. Females produced one to six clutches per reproductive season, with a mean number of 8.4 eggs (October to February) per clutch in 77 natural populations combined from Veracruz and Chiapas, and a mean number of 3.7 clutches per year in 26 females (Legler and Vogt, 2013). Vogt (1997) reported that *S. triporcatus* in Veracruz, Mexico, deposit 6–17 hard-shelled eggs, with a peak time occurring between October and November. Goode (1994) reported clutches of 1–18 eggs in this species in captivity, with females depositing one to five clutches per nesting season. Vogt and Guzman (1988: 37) reported this species in

Veracruz, Mexico, to be a "mollusk specialist under some conditions," but that it also consumed a significant proportion of seeds. Vogt (1997) also reported that *S. triporcatus* in Veracruz, Mexico, eat *Kinosternon*, hard seeds, mollusks, and fruit. Moll, D. (1990) studied a Belezean population of *S. triporcatus* and found that it feeds on large gastropods and other turtles, primarily *K. leucostomum* and *K. cruentatum* (but see Remarks for *K. albogulare*). Legler and Vogt (2013) reported this species to be largely carnivorous in all but one of the sites studied (apparently a summary of the data presented by Vogt and mentioned above). Castañeda et al. (2013) reported predation on a *S. triporcatus* by a jaguar (*Panthera onca* [Linnaeus]) in Parque Nacional Jeannette Kawas in northwestern Honduras.

Etymology.—The name *triporcatus* is formed from the Latin *tri* (thrice), *porca* (ridge between two furrows), and *-atus* (provided with, having the nature of, pertaining to). The name alludes to 3 high ridges on the carapace of this species.

Specimens Examined (3, 1 shell [1]; Map 76).—**ATLÁNTIDA**: Canál Martinez, USNM 524353 (shell). **CORTÉS**: Choloma, USNM 579661; Estero Prieto, UU 6417; San Pedro Sula, USNM 102884.

Other Records (Map 76).—**ATLÁNTIDA:** Cerro Agua Caliente, Parque Nacional Jeannette Kawas (Castañeda et al., 2013).

Superfamily Testudinoidea Batsch, 1788

Testudinoidea contains the families Emydidae Rafinesque (1815: 75; as Emidania), Geoemydidae Theobald (1868: 9), Platysternidae Gray (1869: 208), and Testudinidae Batsch (1788: 437; as Testudines; also see TTWG, 2014). Only the Emydidae and Geoemydidae have representatives in Honduras. Externally, the Honduran members can be distinguished from all other Honduran turtles by the combination of having non paddlelike limbs, short tails that lack a median dorsal row of large triangular scales, and by having 12 plastral scutes.

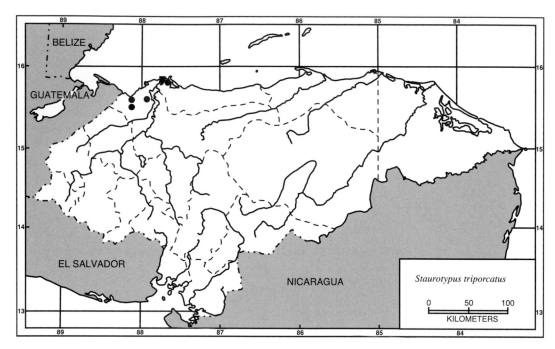

Map 76.   Localities for *Staurotypus triporcatus*. Solid circles denote specimens examined and the solid square represents an accepted record.

Remarks.—Batsch (1788: 437) coined Testudines as an equivalent of a family-series name. Thus, The Principle of Coordination (Articles 36, 43, 46) in IUCZ (1999) dictates that Batsch (1788) is also the author of the superfamily name Testudinoidea. Therefore, Testudines Batsch, or a derivative, cannot apply to an order or infraorder name, as is frequently seen in the literature.

KEY TO HONDURAN FAMILIES OF THE SUPERFAMILY TESTUDINOIDEA

1A. Triturating surface of upper jaw with a ridge (Fig. 179); plastron figure frequently composed of dark, symmetrical lines (Fig. 150) or isolated black circles, ocelli, or mottling (Fig. 151); adults lack distinct median longitudinal ridge on carapace (Fig. 180) . . . .
. . . . Emydidae (p. 503)

1B. Triturating surface of upper jaw without a ridge (Fig. 181); plastron either black or dark brown with a yellow border or yellow with a brown to black central figure; adults usually with distinct median longitudinal ridge on carapace (Fig. 182) . . . . . . Geoemydidae (p. 528)

CLAVE PARA LAS FAMILIAS HONDUREÑAS DE LA SUPERFAMILIA TESTUDINOIDEA

1A. Superficie trituradora de la maxila con un borde (Fig. 179); la figura del plastrón frecuentemente formada por líneas oscuras simétricas (Fig. 150), o círculos, u ocelos o moteado (Fig. 151); adultos sin cresta media en el carapacho (Fig. 180) . . . . Emydidae (p. 503)

1B. Superficie trituradora de la maxila sin un borde (Fig. 181); plastrón negro o color pardo oscuro con una

Figure 179.   Triturating surface of upper jaw with a ridge (arrow). *Trachemys venusta*. USNM 564160 from Sachin Tingni Kiamp, Gracias a Dios.

franja amarilla, o amarillo con una figura central de parda a negra; adultos usualmente con una cresta media en el carapacho (Fig. 182) .... Geoemydidae (p. 528)

Family Emydidae Rafinesque, 1815

This family of turtles ranges in the Western Hemisphere from southern Canada, through much of the U.S., much of Mexico and Central America, and into

Figure 180.   Carapace without distinct median longitudinal ridge in adults. *Trachemys venusta*. USNM 559590 from Awasbila, Gracias a Dios.

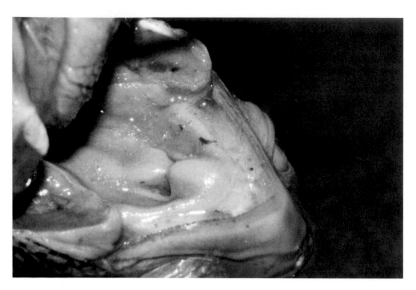

Figure 181.   Triturating surface of upper jaw without a ridge. *Rhinoclemmys pulcherrima.* USNM 580761 from near Amapala, del Tigre Island, Valle.

South America as far as southern Brazil, Uruguay, and northeastern Argentina (the South American distribution is highly disjunct). It is also distributed in the Bahamas and the Greater Antilles. In the Eastern Hemisphere, the family occurs in Europe, western Asia, and northern Africa. The Emydidae is distinguished externally from all other Honduran turtles by the combination of having clawed limbs with strongly webbed feet, a short tail without a medial dorsal row of large triangular scales, an

Figure 182.   Carapace with distinct median longitudinal ridge in adults. *Rhinoclemmys pulcherrima.* FMNH 283550 from between Nuevo Ocotepeque and Antigua, Ocotepeque.

oblong carapace normally with 38 scutes, a large oblong and unhinged plastron with 12 scutes, and a ridge on the alveolar surface of the upper jaw. Twelve genera containing about 46 named species (but see Remarks) are included in this family, with three named species placed in one genus occurring in Honduras, one of which is introduced through the pet trade.

Genus *Trachemys* Agassiz, 1857

Trachemys Agassiz, 1857: 434 (type species: *Testuda scripta* Thundberg, in Schoepff, 1792: 16; see Rhodin and Carr, 2009: 12; also see Remarks).

Geographic Distribution and Content.— The genus *Trachemys* occurs in the Bahama Islands and Greater Antilles, the U.S. as far north as southern Michigan and as far west as New Mexico, southward through Mexico and Central America to Brazil and Argentina. Many populations on the mainland of Mexico and Central and South America appear to be disjunct, with an especially wide disjunction in eastern Brazil. About 13 named species are included in this genus, three of which occur in Honduras, with one of those introduced.

Remarks.—Brown (1908: 114) designated *Testudo scabra* Linnaeus (1758: 198) as type species of the genus *Trachemys* and Lindholm (1929: 28) designated *Emys troostii* *Holbrook (1836: 55) as the type species of *Trachemys*. However, Rhodin and Carr (2009: 8) noted that the type specimen of *T. scabra* Linnaeus is "most probably" a specimen of *Rhinoclemmys punctularia* (Daudin, 1801: 249 [p. 349 erroneously given on page of introduction of this name]). Rhodin and Carr (2009: 12) also rediscovered and redescribed the holotype of *Testudo scripta*, which for a long time was thought to be lost, thus resolving the long-term confusion surrounding the correct type species of the genus *Trachemys*.

The nomenclature of the genera and species of Emydidae occurring in Central America has an extremely convoluted history that remains controversial to this day. The majority of the 20th century literature variously places these turtles in the genera *Pseudemys* Gray (1856b: 197) or *Chrysemys* Gray (1844: 27). However, during the late 1980s through the 2000s, most authors have used *Trachemys* as the genus for these turtles (Savage, 2002: 769, being one exception). Likewise, the specific names used for these turtles occurring in the region have also suffered an unstable history. Thus, species of this complex of turtles occurring in Central America and Honduras were variously called *T. ornata* (Gray, 1830: 12, *In* Gray, 1830–1831), *T. scripta* (Thundberg, 1792: 16, *In* Schoepff, 1792), *T. venusta* (Gray, 1856a: 24), or *T. emolli* (Legler, 1990: 93). Seidel (2002: 289) concluded that *Trachemys venusta* (Gray) and *T. emolli* (Legler) were the correct names for these turtles in the region of Honduras, Nicaragua, and Costa Rica. Most workers generally followed that decision until Fritz et al. (2012: 129; also see Fritz, 2012a, 2012b) recovered a phylogeny based on DNA analyses that suggested *T. venusta* was nested within *T. ornata*. Unfortunately, the five tissue samples of *T. ornata* they used were all from near Acapulco, Guerrero, Mexico, a locality where the turtles are thought to be introduced; obviously, natural populations from as near as the type locality of *T. ornata* as possible would be preferable (see below), but were not available to Fritz et al. (2012). Fritz et al. (2012) did not morphologically examine any supposedly *T. ornata* specimens from that introduced population and only assumed them to be *T. ornata*. Thus, using that assumption for the basis of their taxonomy, the Atlantic versant Central American turtles would become *T. ornata*, whereas those from the Pacific versant from just west of the Isthmus of Tehuantepec, Mexico, to northwestern Costa Rica would become *T. grayi*. As a result, *T. emolli* nested within *T. grayi*. McCranie, Köhler et al. (2013: 23) added two sequences from the first Honduran *Trachemys* definitely found on the Pacific versant to the Fritz et al. (2012) database. Those workers recovered a phylogeny that generally agreed with that of Fritz et al. (2012) for the Central American turtles, with the Honduran Pacific versant turtles nested with the Fritz et al. *T. grayi emolli*, now also known to be in error. Parham et al. (2015) sequenced tissues of *T. ornata* from the region of its type locality in Sinaloa, Mexico. Those new tissue sequence data showed *T. ornata* to not nest with *T. venusta*, as the Acapulco, Guerrero, Mexico, tissue sequences did in Fritz et al. (2012). Instead, the Acapulco specimens nested within *T. venusta* in the Parham et al. (2015) analysis. Parham et al. (2015) also examined photographs of living specimens of the Acapulco population and reidentified them as *T. venusta*, thereby identifying a fatal point with the Fritz et. (2012) results. Thus, the Atlantic versant populations of *Trachemys* in Mexico and Central America should be called *T. venusta* as suggested by Seidel (2002). Also, Parham et al. (2013, 2015) demonstrated that the Pacific versant populations in Nicaragua and Costa Rica should be called *T. emolli*. Molecular analysis of two Honduran Pacific versant *Trachemys* (McCranie, Köhler et al., 2013) demonstrated they clustered with *T. emolli*. Thus, Honduran Pacific versant *Trachemys* included in this work should be called *T. emolli* Legler.

Legler (*In* Legler and Vogt, 2013: 247) stated that the recognition of all Mesoamerican *Trachemys* as subspecies of *T. scripta* was a natural classification. However, the limited molecular studies and other recent studies on the *Trachemys* group have shown that classification to be anything but natural (J. T. Jackson et al., 2008: 132–133; Fritz et al., 2012: 129; McCranie, Köhler, et al., 2013: 26–27; Parham et al., 2013: 179; TTWG, 2014: 445).

More than one species of *Trachemys* apparently do not naturally occur sympatrically in natural populations (Seidel and Ernst, 2006; Kraus, 2009), but when human introduction of one species into the range of another *Trachemys* species occurs, hybridization is known to occur (Seidel and Ernst, 2006: 831.15; Kraus, 2009: 88; Parham et al., 2013: 184). Thus, when more than one species of *Trachemys* are placed in mixed pools in captivity, hybridization would be expected, although one notable exception is the report in Álvarez del Toro (1983: 46) that males and females of *T. grayi* and *T. venusta* captured in Chiapas, Mexico, ignored each other in mixed captive pools and that no interbreeding had been observed. Previous phylogenetic studies of species of *Trachemys* based on molecular data from pet trade, turtle farms, or unknown origin, and without voucher specimens, have complicated the resulting evidence in the phylogenetic relationships of these turtles (see Fong et al., 2007: 457; McCranie, Köhler et al., 2013: 28, for discussions). Hopefully, in the future, phylogenetic studies will use only animals with known collecting localities associated with voucher specimens placed in an institutional collection (i.e., like that of Parham et al., 2013; also see Parham et al., 2015). Also, it is hoped that those vouchers are examined for their morphological characters.

Intentional releases of formerly captive *Trachemys* into natural habitats containing a native *Trachemys* population are known to have occurred in Honduras (see Remarks for *T. scripta*). Also, Acuña-Mesén (1992: 157) reported collecting *Trachemys* from various localities in the "north and south" of Costa Rica and placing them on a farming ranch. One of the objectives of that project was "restocking of natural areas, especially when the species shows critically low density levels." At least two biological species of *Trachemys* (*T. emolli* and *T. venusta*) are known to occur naturally in

Figure 183.   Adult male with elongated foreclaws. *Trachemys scripta.* UF 30105 from Georgia, USA.

Costa Rica (Fritz et al., 2012; McCranie, Köhler et al., 2013; Parham et al., 2013, 2015). Mixing captives from various localities apparently of more than one species in a single captive breeding pool for the possible release into the wild of resulting offspring should be condemned. It is also likely that some of the undocumented molecular data used in various phylogenetic studies mentioned above might have come from those mixed captive populations (see Fong et al., 2007: 457, for a discussion of some of the consequences that can be expected from such releases).

Female *Trachemys* are usually much larger than males, but some Honduran males of *T. venusta* collected in the early to mid-1900s are unusually large and even approach the sizes of adult females.

Etymology.—The name *Trachemys* is formed from the Greek words *trachys* (rough, uneven) and *emys* (freshwater tortoise, turtle), as stated by Agassiz (1857: 434), in discussing the carapace: "At first smooth, they afterwards assume radiating ridges, up to the seventh or eighth year; and, finally, longitudinal ridges and rugosities prevail upon the scales."

KEY TO HONDURAN SPECIES OF THE GENUS *TRACHEMYS*

1A. Plastron with isolated black circles, ocelli, or mottling (Fig. 151), especially obvious in juveniles and subadults; adult males with elongated claws on forelimbs (Fig. 183) .... *scripta* (p. 513)

1B. Juveniles, subadults, and some adults with plastron pattern of brown symmetrical lines (Fig. 150); adult males without elongated claws on forelimbs (Fig. 184) 2

2A. Symphyseal stripe can be connected to both neck stripes (Fig. 185); postorbital stripe can be strongly constricted medially above tympanum (Fig. 186) .... *emolli* (p. 508)

2B. Symphyseal stripe usually not connected to both neck stripes (Fig. 187); postorbital stripe not strongly constricted medially above tympanum (Fig. 188) ...... *venusta* (p. 518)

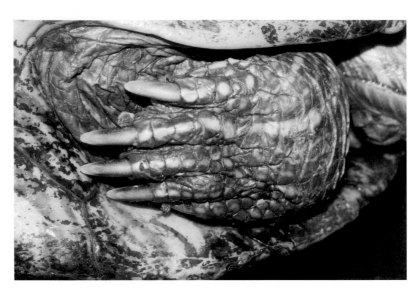

Figure 184.   Adult male foreclaws not elongated. *Trachemys venusta.* USNM 562003 from Quebrada San Lorenzo, Yoro.

CLAVE PARA LAS ESPECIES HONDUREÑAS DEL
GÉNERO *TRACHEMYS*

1A. Figura del plastrón compuesta de
círculos aislados, ocelos o moteado
de color negro (Fig. 151), patrón
especialmente evidente en jóvenes
y subadultos; machos adultos con
garras alargadas en las extremidades anteriores (Fig. 183) ......
.... *scripta* (p. 513)
1B. Jóvenes, subadultos y algunos adultos con la figura del plastrón
frecuentemente compuesta de
líneas oscuras simétricas (Fig.
150); machos adultos sin garras
alargadas en las extremidades anteriores (Fig. 184) 2
2A. Raya (symphyseal) que sale de la
comisura de la boca puede estar
conectada con las dos rayas de
cuello (Fig. 185); la raya postorbital
puede estar fuertemente constreñida medialmente por arriba del
tímpano (Fig. 186).... *emolli* (p. 508)
2B. Raya (symphyseal) que sale de la
comisura de la boca usualmente no
está conectada con las dos rayas de

cuello (Fig. 187); raya postorbital
usualmente no está fuertemente
constreñida medialmente por arriba el tímpano (Fig. 188) ....
.... *venusta* (p. 518)

Trachemys emolli (Legler, 1990)

Pseudemys scripta grayi: Pritchard, 1979:
115.
Pseudemys scripta emolli Legler, 1990: 91
(holotype; UU 6728; type locality: "Río
Tepetate, 2.5 km northeast of Granada,
Granada Province, Nicaragua").
Trachemys emolli: Seidel, 2002: 289;
McCranie and Gutsche, 2016: 891.
Trachemys grayi emolli: McCranie, Köhler,
et al., 2013: 21.
Trachemys grayi: Solís et al., 2014: 139;
McCranie, 2015a: 381.

Geographic Distribution.—*Trachemys
emolli* occurs at low elevations on the Pacific
versant (coastal plain) from the Golfo de
Fonseca in extreme southeastern El Salvador to Panama (see Remarks). The species
also occurs on the Atlantic versant in
southwestern Nicaragua in the western
portion of the Río San Juan flowing into

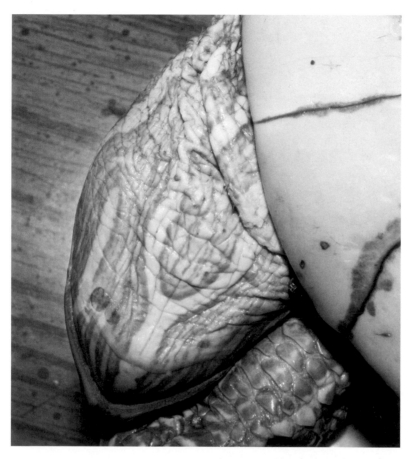

Figure 185. Symphyseal stripe connected with both neck stripes in this image. Symphyseal stripe connected to neck stripe condition said to be diagnostic of *T. emolli* by its describer, but condition occurs in only one of two available adults. *Trachemys emolli*. USNM 578925 from El Faro, Choluteca.

Lago de Nicaragua, and possibly in the coastal plain of the Río San Juan in northeastern Costa Rica (see Remarks). In Honduras, this species is known from a freshwater river, a freshwater lagoon, and a brackish water estuary in the extreme southern portion of the country, including Isla del Tigre in the Golfo de Fonseca.

Description.—The following is based on two females (USNM 578925–26). *Trachemys emolli* is a large (maximum recorded CL 540 mm [Ernst, 2008b]; 330 mm CL in largest Honduran specimen [USNM 578925]) turtle with a broad, oval carapace; carapace with 1 nuchal, 5 vertebral, 8 costal,

and 24 marginal scutes; carapace widest at level of seam between marginals 8–9, highest at about midlength of vertebral 3; carapacial sides slightly bowed; posterior marginals slightly serrated; each vertebral wider than long; carapace texture somewhat roughened due to growth rings; nuchal narrow; carapace very shallowly notched posteriorly; plastron unhinged, with 12 scutes, including paired gulars, humerals, pectorals, abdominals, femorals, and anals; gulars without anterior projections; anals with wide medial notch; plastron connected to carapace by bridge; a small axillary and inguinal scute present; head moderately

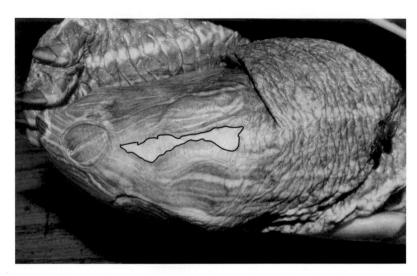

Figure 186.   Each postorbital stripe constricted above tympanum (outlined) in this image. *Trachemys emolli.* Postorbital stripe strongly constricted above tympanum condition said to be diagnostic of *T. emolli* by its describer, but condition occurs in only one of two available adults. USNM 578925 from El Faro, Choluteca.

sized, snout projecting, rather pointed; upper jaw serrated laterally, shallowly notched; triturating surface of upper jaw with ridge; feet strongly webbed; 5 claws on forelimb, 4 claws on hind limb; foreleg covered with about 7 rows of large scales on outer surface of lower segment; cloacal opening anterior to carapacial margin; CL 330 mm in one; CH/CL 0.39 in one; CH/CW 0.43–0.51; CW/CL 0.76 in one; first

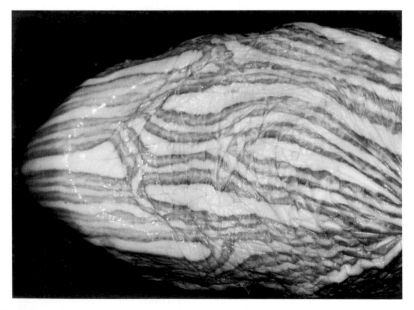

Figure 187.   Symphyseal stripe not connected with both neck stripes. *Trachemys venusta.* USNM 564160 from near Sachin Tingni Kiamp, Gracias a Dios.

Figure 188.   Each postorbital stripe not constricted medially above tympanum. *Trachemys venusta*. USNM 580392 from Camp Bay, Roatán, Islas de la Bahía.

vertebral width/first vertebral L 0.78–0.79; PL/CL 0.92 in one; BL/CL 0.36 in one; plastron width at humeropectoral seam/PL 0.48–0.50; plastron width at midfemoral scute/PL 0.51–0.52; abdominal scute L/PL 0.25 in both.

Color in life of an adult female (USNM 578925; Plate 95): carapace Sayal Brown (223C) with pale orange lines outlining vertebral scutes and dark brown large costal spots; nuchals and marginals Sayal Brown with large dark brown blotches and Orange Rufous (132C) central vertical lines; top of head Grayish Olive (43) with short orange lines; side of head Grayish Olive with Orange Yellow (18) stripes; Orange Yellow postorbital stripe strongly constricted above level of tympanum; dorsal surfaces of fore- and hind limb dark brown; plastron yellowish cream with large pattern of indistinct brown, mostly symmetrical lines; ventral surface of head brown with pale orange symphyseal stripe connected to pale orange neck stripes; ventral surfaces of fore- and hind limb brown with pale orange stripes: iris Yellowish Olive-Green (50) with yellow around upper and lower edges, yellow line

Plate 95.   *Trachemys emolli*. USNM 578925, adult female, carapace length = 330 mm. Choluteca: El Faro.

extending anteriorly and posteriorly from yellow rims.

Color in alcohol: carapace dark brown with pale, dark brown bordered ocelli on costals and vertebrals; plastron yellowish brown with obscure brown, irregular central figure consisting of irregular lines; bridge yellowish brown with dark brown, paler brown centered ocelli on underside of marginals and elongated alternating yellowish brown and dark brown lines along lateral edges of plastron; head brownish olive with numerous, dark bordered, yellow or yellowish brown stripes on all surfaces, stripes continuing onto neck; postorbital and primary orbitocervical stripes separated from eye, former slightly broader than latter, both continuing onto neck; yellowish brown postorbital stripe strongly constricted medially above tympanum in one, not constricted in one; yellowish brown symphyseal stripe connected to both sides of neck stripes in one, not connected in one; forelimb brownish olive with yellowish brown stripes; hind limb same color as that of forelimb on outer surface, yellowish brown with brownish olive stripes on inside surface; tail brownish olive with yellowish brown stripes.

Diagnosis/Similar Species.—*Trachemys emolli* is distinguished from all remaining Honduran turtles, except the other *Trachemys* and one species of *Rhinoclemmys*, by the combination of having strongly clawed limbs with webbed feet and an oblong and large plastron with 12 scutes. The members of the genus *Rhinoclemmys* lack a ridge on triturating surface of the upper jaw (versus triturating ridge present in *T. emolli*). Additionally, *R. funerea* is the only Honduran *Rhinoclemmys* with strongly webbed toes but has a mostly black plastron with a pale midseam (versus plastron yellow or pale brown, usually with darker brown symmetrical lines in *T. emolli*). *Trachemys scripta* has black blotches or ocelli on the plastron and males have elongated claws on the forelimb, at least during the breeding season (versus dark brown figure of symmetrical lines on plastron, except in some old individuals in which figure fades, and males lack unusually long claws on forelimb in *T. emolli*). *Trachemys venusta* usually has the symphyseal stripe separated from both neck stripes and has the postorbital stripe usually not strongly constricted above the tympanum (versus the morphologically poorly defined *T. emolli* with the symphyseal stripe connected or not to both neck stripes, and postorbital stripe strongly constricted or not above tympanum in *T. emolli*).

Illustrations (Figs. 185, 186; Plate 95).— Bonin et al., 2006 (adult); Bour, 2003 (head, juvenile), 2004 (adult); Ernst, 2008b (adult, plastron, head); Legler, 1990 (head pattern, carapace, plastron; as *Pseudemys scripta emolli*); McCord et al., 2010 (adult and plastron, juvenile and plastron); McCranie, Köhler et al., 2013 (carapace, head; as *T. grayi emolli*); Rogner, 2008 (juvenile); Seidel and Ernst, 2017 (head; as *T. grayi emolli*); Vetter, 2005 (adult, juvenile, head, plastron).

Remarks.—Substantially more fieldwork is needed in Honduras to settle taxonomic problems regarding the identity of the species of *Trachemys* occurring in the country, as well as to better understand their geographic distribution.

Honduran *Trachemys emolli* from the southern part of the country differ significantly from southern coastal plain Mexican *T. grayi* (data for *T. grayi* from Legler and Vogt, 2013: 272) in the pattern of the postorbital stripes and location of the ocelli on the marginal scutes, thus supporting the molecular data recovered by Parham et al. (2013) that two species exist among those two nominal forms (*T. emolli, T. grayi*; also see fig. 1 in Rodrigues and Diniz-Filho, 2016). *Trachemys emolli* in the southern Pacific area of Honduras have or do not have the postorbital stripe strongly constricted or divided above the tympanum (as opposed to narrow and sometimes vague and without constriction above the tympanum in many Honduran *T. venusta*). The nature of the symphyseal stripe in the two Honduran *T. emolli* is also too variable to use as a diagnostic character, as is usually stated. Thus, there seems to be no reliable morphological character to distinguish *T. emolli* from the Atlantic versant *T. venusta* in Honduras; the stated nature of the

symphyseal and postorbital stripes are too variable, although they are readily separated from each other by molecular characters. In addition to those two characters showing much variation, considerable confusion exists in the literature concerning those character states in *T. emolli* from Nicaragua and Costa Rica (see below). The population described by McCord et al. (2010) from Panama as *T. venusta panamensis* is more closely related to *T. emolli* than it is to *T. venusta* (see Fritz et al. 2012).

Contradictory information exists in the literature about the nature of the symphyseal stripe in *Trachemys emolli*. Legler (1990: 93), in the type description of *T. scripta emolli*, stated "symphyseal stripe usually (90.5%) connected to neck stripes," whereas Ernst (2008b: 846.2), in a review of *T. emolli*, stated "symphyseal stripe forks posteriorly, and usually (in 90% of specimens) is interrupted anterior to the split," which seems to mean that the symphyseal stripe is usually not connected to the neck stripe (fig. 7.6D in Legler, 1990, shows both conditions in same specimen). Clearly, much work needs to be done on both morphological and molecular data on the populations of *Trachemys* in Central America, most notably along the headwaters of the Río Choluteca, Honduras, along the lower reaches of the rivers and marshes in Valle and Choluteca, Honduras, and along the length of the Río San Juan in southern Nicaragua and northeastern Costa Rica.

Ibarra Portilla et al. (2009: 111) reported *Trachemys emolli* from the Golfo de Fonseca in extreme southeastern El Salvador. McCranie, Köhler et al. (2013: 23) also reported *T. emolli* in southern Honduras. *Trachemys emolli* also occurs in western Nicaragua and northwestern Costa Rica. Savage recognized only one species of *Trachemys* (curiously as *Chrysemys ornata*) in Costa Rica, even though several phylogenetic analyses had demonstrated that *T. emolli* was a valid taxon. Savage (2002) used the genus name *Chrysemys* for these Costa

Rican turtles, even though results of several molecular analyses had previously demonstrated contrariwise. Ernst (2008b) provided an overview of the morphology and a literature review of *T. emolli*.

Pritchard (1979: 115) stated he was given a carapace of *Pseudemys scripta grayi* from Tegucigalpa, Honduras. Apparently, based on that report, McCord et al. (2010: 47) mapped *T. emolli* (as *T. venusta grayi*) as occurring through southern Honduras and northward to near Tegucigalpa. However, the few specimens from the vicinity of Tegucigalpa I have seen most closely resemble the more northerly Honduran specimens of *T. venusta* (also see Remarks for *T. venusta* for information on molecular data on a specimen from the vicinity of Tegucigalpa), in having the symphyseal stripe not connected with the neck stripes (a supposedly, but extremely variable, diagnostic character of *T. emolli*). Unfortunately, there is also much variation within the symphyseal stripe–neck stripe condition in northern Honduran *T. venusta*, with those stripes connected in some (i.e., UU 6160, UNAH [SMR] 456, USNM 580392). On the other hand, one specimen from "Tegucigalpa" (UU 6157) has a thin postorbital stripe that is greatly interrupted above the tympanum, a supposedly diagnostic *T. emolli* character known to occur in the Pacific coastal plain of southern Honduras. A thorough DNA study of as many Honduran populations of *Trachemys* as possible is needed to resolve this issue, including from the bones of the carapace of Pritchard's specimen that remains in the Pritchard collection (P. Pritchard, personal communication). One potential problem for that study, however, would be the releasing of captive pets that are no longer wanted in various places in Honduras. Those former pets could be *T. emolli*, *T. scripta*, or *T. venusta*, thus helping create problems associated with crossbreeding between those species.

A population of this species in northwestern Costa Rica was shown to have a CL of about 35 mm at hatching (Cabrera Peña et al., 1997; also see Merchán Fornelino, 2002).

Natural History Comments.—*Trachemys emolli* is known only from near sea level to 5 m elevation in the Lowland Dry Forest formation. Two specimens of this diurnal and heavily aquatic species were collected in a brackish water estuary in July. Numerous individuals were also concentrated in a relatively deep and isolated pool in the freshwater Río Negro flowing into that estuary during low water levels in April, the height of the dry season. One decomposed juvenile was also seen in a lagoon on Isla del Tigre in July, and two shells were collected on the shore of that same lagoon in March 2 years later. Legler (1990) reported that the type series of *Pseudemys scripta emolli* (= *T. emolli*) was collected in shallow, fresh water near a lakeshore and buried in moist humus in lakeside vegetation. Ibarra Portillo et al. (2009) reported a dead adult *T. emolli* lodged between rocks at about 2 m depth in seawater of the Golfo de Fonseca near an island in extreme southeastern El Salvador. Those authors also reported another *T. emolli* from an estuary in that same area. Pritchard (1993) found a range of 12–33 eggs per clutch, with an average of "about 20" eggs in a population from northwestern Costa Rica, and D. Moll (1994: 110) reported 15–25 ($x =$ 20) eggs in a population from Lago de Nicaragua, with months of egg deposition not given. Mora and Ugalde (1991) noted finding predation on some of those nests in March and April at the same site Pritchard (1993) visited. Moll, E. O., and Legler (1971) reported gut contents of six *T. emolli* from the Pacific drainage of Nicaragua contained vegetation (93%) and insects (6.6%).

Etymology.—The name *emolli* is a patronym honoring Edward O. Moll, who helped collect the holotype of this species and also

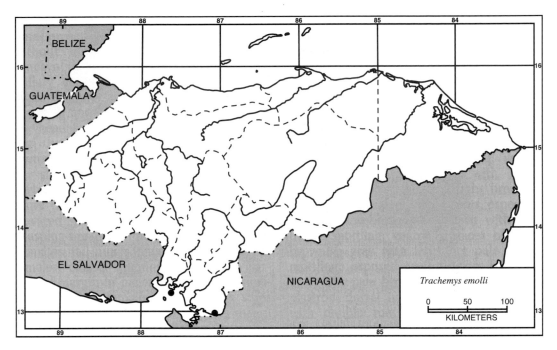

Map 77.   Localities for *Trachemys emolli*. Solid circles denote specimens examined.

did extensive fieldwork with Legler on *Trachemys* populations in Panama.

Specimens Examined (2, +1 group of shell fragments [0]; Map 77)—**CHOLUTECA**: El Faro, USNM 578925–26. **VALLE**: La Laguna, USNM 581903 (shell fragments).

Trachemys scripta (Thundberg, *In* Schoepff, 1792)

Testudo scripta Thundberg, 1792: 16, *In* Schoepff, 1792 (holotype not designated, but a juvenile was figured in plate III, figs. 4, 5; Rhodin and Carr [2009: 12] rediscovered the juvenile illustrated in Schoepff [1792] and confirmed it to be the *T. scripta* holotype [UUZM 7455]; type locality: not stated, but restriction by Schmidt, 1953: 102, to "Charleston, South Carolina" generally accepted; see Seidel and Ernst, 2006).

Trachemys scripta: Gray, 1863: 181; McCranie and Valdés Orellana, 2014:

44; Solís et al., 2014: 139; McCranie, 2015a: 381.

Geographic Distribution.—*Trachemys scripta* occurs naturally from southeastern Virginia and Illinois southward to northern Florida and westward to New Mexico, U.S., and southward from Texas, USA, through eastern Coahuila to central Nuevo León and northeastern Tamaulipas, Mexico. The species has also been introduced through the pet trade to many places in the world, including Honduras.

Description.—The following is based on one female (FMNH 283584). *Trachemys scripta* is a large (maximum recorded CL 302 mm [Tucker et al., 2006]), 220 mm CL in Honduran specimen) turtle with a broad, oblong carapace; carapace with 1 nuchal, 5 vertebral, 8 costal, and 24 marginal scutes; carapace widest at level of seam between marginals 4–5, highest at level of seam between marginals 3–4; carapacial sides slightly bowed; posterior marginals slightly

Plate 96.   *Trachemys scripta*. FMNH 283584, adult female, carapace length = 220 mm. Islas de la Bahía: Guanaja Island, near Savannah Bight (introduced).

serrated; vertebrals 2–5 wider than long, vertebral 1 longer than wide; carapace texture somewhat roughened due to growth rings; nuchal narrow; carapace very shallowly notched posteriorly; plastron unhinged, with 12 scutes, consisting of paired gulars, humerals, pectorals, abdominals, femorals, and anals; gulars without anterior projections; anals with wide medial notch; plastron flat, connected to carapace; small axillary and inguinal scute present; head large, snout projecting, rather pointed; upper jaw finely serrated laterally, not notched; triturating surface of upper jaw with ridge; feet strongly webbed; 5 claws on forelimb, 4 on hind limb; foreleg covered with 7 rows of large scales on outer surface of lower segment; cloacal opening anterior to carapacial margin; CL 220 mm; CH/CL 0.41; CW/CL 0.64; CH/CW 0.64; first vertebral width/first vertebral L 0.61; PL/CL 0.94; BL/CL 0.36; plastron width at humeropectoral seam/PL 0.46; plastron width at midfemoral scute/PL 0.44; abdominal scute L/PL 0.24.

Color in life of that adult female (FMNH 283584; Plate 96): carapace Grayish Horn Color (91) with Sepia (219) blotches on vertebrals and vertical lines on marginals and costals; postorbital stripe Spectrum Red (11); ground color of head greenish brown with pale yellow subocular stripe; skin of groin and ventral surfaces of fore- and hind limb dark brown with pale yellow–dirty white mottling and scales; subcaudal surface dark brown with yellow longitudinal lines and crosslines; dorsal and lateral surfaces of neck Olive Brown (28) with yellow lines; chin and venter of neck dark brown with pale yellow lines; top of head Olive Brown; anterior surface of forelimb Olive Brown with 3 pale yellow lines; iris greenish olive; plastron and bridge pale orange with dark brown rounded and longitudinal spots.

Color in alcohol (FMNH 283584): carapace brown with dark brown posteriorly directed V-shaped blotches on vertebrals; dark brown vertical narrow bands on costals; marginals with dark brown borders along each seam; plastron yellowish brown with dark brown blotches on fore- and hind lobe and dark brown longitudinal blotches on about lateral two-thirds of abdominals; dark brown elongated blotches also present near lateral edges of pectorals and abdominals; bridge pale orange with black oblong blotches and gray lineate markings; head dark brown with numerous, dark bordered, yellowish brown stripes on all surfaces, stripes continuing onto neck; brown postorbital stripe and pale brown primary orbito-cervical stripe contacting eye border, former much broader than latter, both continuing onto neck; postorbital stripe broader posteriorly, narrowing anteriorly until pointed at point of contact with eye border, not medially constricted above tympanum; pale yellow symphyseal stripe connected to both sides of neck stripes; forelimb dark brown with pale yellowish brown longitudinal stripes dorsally, mostly dark brown with a few yellowish brown scales ventrally; hind limb dark brown dorsally and ventrally, with yellowish brown stripe along outer edge, some yellowish brown scales also present ventrally, suggesting incomplete stripes; tail dark brown with yellow to yellowish brown stripes.

Diagnosis/Similar Species.—*Trachemys scripta* is distinguished from all remaining Honduran turtles, except the other *Trachemys* and one species of *Rhinoclemmys*, by the combination of having strongly clawed limbs, strongly webbed hind feet, and an oblong and large plastron with 12 scutes. All *Rhinoclemmys* lack a ridge on the triturating surface of the upper jaw (versus triturating ridge present in *T. scripta*). Additionally, *R. funerea* is the only Honduran *Rhinoclemmys* that has strongly webbed toes, but that species has a mostly black plastron with a yellow midseam (versus plastron yellow or pale brown with black blotches or mottling in *T. scripta*). *Trachemys emolli* and *T. venusta* have a brown figure of symmetrical lines on the plastron, except in some old individuals in which that plastral figure has faded or worn, have some shade of yellow postorbital stripes, and males lack elongated claws on the forelimb, thus lacking Titillation courtship behavior (versus plastron with black blotches or mottling, red postorbital stripes present, and males with elongated claws on forelimb, thus having Titillation courtship behavior in *T. scripta*).

Illustrations (Figs. 149, 151, 183; Plate 96).—Bonin et al., 2006 (adult, head, juvenile); Bour, 2003 (head, juvenile, head pattern, carapace, plastron; as *T. s. scripta, T. s. elegans* [Wied, 1839: 213]); Carr, 1952 (adult, juvenile, carapace, plastron; as *Pseudemys s. scripta, P. s. elegans,* and *P. s. troostii*); Ernst, 1990 (adult; as *T. s. scripta, T. s. elegans,* and *T. s. troostii* [°Holbrook, 1836: 55]); Ernst and Barbour, 1972 (adult, head, plastron; as *Chrysemys s. scripta, C. s. elegans,* and *C. s. troostii*); Ernst and Lovich, 2009 (adult, juvenile, plastron); Ernst et al., 1994 (adult, head pattern, hatchling, plastron; as *T. s. scripta, T. s. elegans*); Fritz, 2012b (adult; as *T. s. scripta, T. s. elegans*); Legler and Vogt, 2013 (juvenile, head, carapace, plastron, elongated male claws; as *T. s. elegans*); McCranie and Valdés Orellana, 2014 (adult); Rogner, 2008 (adult; as *T. s. elegans*); Rhodin and Carr, 2009 (juvenile holotype); Schoepff, 1792 (juvenile holotype); Seidel and Ernst, 2006 (adult, juvenile, head; as *T. s. scripta, T. s. elegans,* and *T. s. troostii*), 2017 (carapace, plastron); Vetter, 2004 (adult, juvenile, carapace, plastron; as *T. s. scripta, T. s. elegans, T. s. troostii,* and "southern Texas form"); Wermuth and Mertens, 1961 (adult, carapace, plastron; as *P. s. elegans*).

Remarks.—Populations of *Trachemys scripta* are introduced into Honduras via the pet trade and are apparently from the southeastern U.S. (see TTWG, 2014: 364; as *T. s. elegans*). Hatchlings were frequently shipped to Honduras, as well as to many other countries in the world, to be sold in pet shops. Adults of *T. scripta* can now be seen in several small Honduran roadside zoos, where they were donated by former owners or by people who captured them nearby. It is highly likely that many of those pet *T. scripta* have been released at various localities in Honduras. Those releases have the potential to contaminate the native *Trachemys* gene pool throughout Honduras through crossbreeding. I witnessed, and strongly complained in vain, about the release of one adult *T. scripta* and several adult *T. venusta* into the waters of the Cuero y Salado Wildlife Reserve near La Ceiba by a La Ceiba resident who no longer wanted those turtles as pets.

Ernst and Lovich (2009) and Seidel and Ernst (2006) provided overviews of the morphology and literature reviews of *Trachemys scripta*.

The IUCN (1985c) official decision regarding the name *Testudo scripta* is no longer valid.

Natural History Comments.—*Trachemys scripta* is known from near sea level to 1,480 m elevation in the Lowland Moist Forest and Premontane Moist Forest formations. One specimen of this introduced turtle was crossing a road in August, and another was on the ground during the morning near Savannah Bight, Isla de Guanaja, in September after a heavy rain the previous night. It is not known if there are any breeding and established populations of *T. scripta* in Honduras, but some of those released individuals are likely crossbreeding with natural *Trachemys* in some places. This is a diurnal turtle that occurs in many freshwater habitats but prefers quiet waters. It also sometimes occurs in salt marshes in the U.S. The species frequently basks in the sun on many objects above the water surface. It also sleeps at night either on the bottom or floating on the surface (those data from Ernst and Lovich, 2009, for natural populations in the U.S.). Ernst and Lovich (2009) also provided a thorough survey on reproduction in this species, as briefly discussed below. Females excavate a nest with their hind limbs in an open, unshaded area in which she deposits her eggs. Clutch sizes are known to range from 1 to 30 eggs (mean 10.5 eggs for 234 reported clutches). Nesting generally occurs between April and July, with May and June

the peak months. Ernst and Lovich (2009: 464) also provided an excellent review of the diet of this species and called it "an opportunistic omnivore subsisting on a wide ranging diet of various plants and animal foods." A captive individual found as an adult in a small Honduran stream feeds on just about anything offered to it, including bananas, fruits, and various cooked meats (Fernando Marquez, personal communication).

Etymology.—The name *scripta* is from the Latin *scriptus* (write) "presumably in reference to the letter-like markings on the carapace" Seidel and Ernst (2006: 831.14).

Specimens Examined (2 [0]; Map 78)— **CORTÉS**: 11.9 km N of Cofradía, UNAH (SMR) 454. **ISLAS DE LA BAHÍA**: Isla de Guanaja, near Savannah Bight, FMNH 283584.

Other Records (Map 78).—**ATLÁNTIDA**: La Ceiba (juvenile offered for sale said to be from a badly polluted estuary in town). **FRANCISCO MORAZÁN**: Santa Lucía (individual captured in nearby river [F. Marquez, personal communication], but locality too high in elevation to establish a breeding population).

Trachemys venusta (Gray, 1856a)

Emys venusta Gray, 1856a: 24 (lectotype, BMNH 1947.3.4.80 [designated by Smith and Smith, 1980: 494–495; as 1845.8.5.26]; type locality: "Honduras" [see Remarks]).

Emys valida Le Conte, 1859: 7 (holotype, ANSP 216 [see Malnate, 1971: 354]; type locality: "Honduras" [see Remarks]); Troschel, 1860: 270; Malnate, 1971: 354.

Pseudemys [sp.]: Schmidt, 1924: 86.

Pseudemys ornata: Dunn and Emlen, 1932: 25.

Pseudemys scripta ornata: Carr, 1938: 135.

Pseudemys scripta subsp.: Meyer, 1966: 174.

Chrysemys ornata: Meyer, 1969: 189; Meyer and Wilson, 1973: 5; Wilson

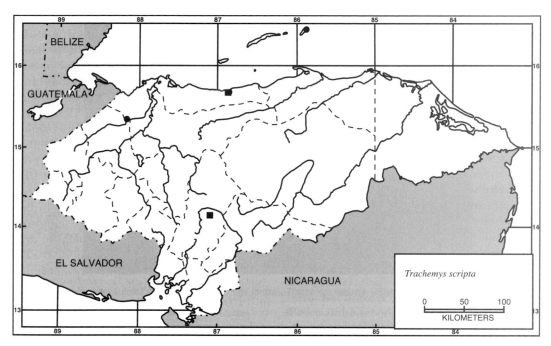

Map 78. Localities for *Trachemys scripta*. Solid circles denote specimens examined and solid squares represent accepted records (introduced species).

and Hahn, 1973: 103; Cruz Díaz, 1978: 24; O'Shea, 1986: 29.

Pseudemys ornata ornata: Wermuth and Mertens, 1977: 56.

Trachemys scripta: Iverson, 1986: 106; McCranie, 1990: 95; Wilson and Cruz Díaz, 1993: 16; Cruz D[íaz], 1998: 29, *In* Bermingham et al., 1998; Monzel, 1998: 167; Wilson and McCranie, 1998: 15; Köhler, 2000: 24; Köhler, McCranie, and Nicholson, 2000: 425; Nicholson et al., 2000: 29; Wilson et al., 2001: 134; Lundberg, 2002a: 12; Köhler, 2003a: 40; Köhler, 2008: 42.

Trachemys scripta ssp.: Seidel, 1988: 37.

Trachemys scripta venusta: Iverson, 1992: 209.

Trachemys venusta: Seidel, 2002: 289; McCranie et al., 2005: 62; Ernst and Seidel, 2006: 832.1:1 (in part); McCranie et al., 2006: 100; Townsend, 2006a: 34; Wilson and Townsend, 2006: 104; Townsend et al., 2007: 10; Wilson and Townsend, 2007: 145; Frazier et al.,

2010: 510; Townsend and Wilson, 2010b: 691; Parham et al., 2013: 178; Parham et al., 2015: 360.

Trachemys venusta venusta: Ernst and Seidel, 2006: 832.3.

Trachemys venusta uhrigi McCord, Joseph-Ouni, Hagen, and Blanck, 2010: 43 (holotype, UF 157800; type locality: "in the Río Chamelecón, 3.0 km south of San Pedro Sula, Honduras"); Pasachnik and Chavarria, 2011: 429.

Trachemys ornata: McCranie, Köhler, et al., 2013: 27; McCranie and Valdés Orellana, 2014: 43; Solís et al., 2014: 139; McCranie, 2015a: 381.

Geographic Distribution.—Trachemys venusta occurs at low and moderate elevations on the Atlantic versant from northern Tamaulipas, Mexico, to at least Panama, including Cozumel and other islands. It also occurs on the Pacific versant in the Acapulco, Guerrero, Mexico region (introduced). The taxonomy of the northern South

American populations previously assigned to *T. venusta* is poorly understood and at least some represent other species (U. Fritz, personal communication). In Honduras, this species is widely distributed along the Atlantic versant of the northern and eastern portions of the mainland and on Islas de Guanaja, Roatán, and the Cayo Cochinos and possibly Utila in the Bay Islands. The species also occurs in the upper portion of the Pacific versant Río Choluteca and tributaries in south-central Honduras, but these populations might represent recent introductions by man.

Description.—The following is based on 12 males (FMNH 5316, 5318; MCZ R31484; TCWC 19234–35; UF 137164, 137167; USNM 102885, 129587, 129608, 562002, 580392) and 11 females (UF 137163, 137165–66, 137633; USNM 559590, 562003, 570541, 573338–40, 573358). *Trachemys venusta* is a large (maximum recorded CL 440 mm [Legler and Vogt, 2013]; 342 mm CL in largest Honduran male [USNM 102885] and 335 mm CL in largest Honduran female [UF 137633]) turtle with a broad, oval carapace; carapace with 1 nuchal, 5 vertebral, 8 costal, and 24 marginal scutes; carapace widest at level of seam between marginals 8–9, highest at about midlength of vertebral 3; carapacial sides bowed in some, straight in others, and angled outward in others; posterior marginals not serrated, but sometimes flared; single, flattened vertebral ridge present in juveniles to small adults; each vertebral wider than long; carapace texture somewhat roughened due to growth rings or mostly smooth; nuchal narrow; carapace shallowly notched posteriorly; plastron unhinged, with 12 scutes, consisting of paired gulars, humerals, pectorals, abdominals, femorals, and anals; gulars without anterior projections; anals with wide medial notch; plastron slightly concave posteriorly in males; plastron connected to carapace; a small axillary and inguinal scute present; head moderately sized, snout projecting,

rather pointed in both sexes; upper jaw not, or usually only finely serrated laterally (strongly serrated in occasional individuals), shallowly notched; triturating surface of upper jaw with a ridge; feet strongly webbed; 5 claws on forelimb, 4 on hind limb; foreleg covered with about 7 rows of large scales on outer surface of lower segment; cloacal opening at or posterior to carapacial margin in males, anterior to carapacial margin in females; CL 138–342 (229.7 ± 86.7) mm in males, 96–335 (202.9 ± 92.6) mm in females; CH/CL 0.30–0.45 in males, 0.31–0.43 in females; CH/CW 0.42–0.59 in males, 0.45–0.62 in females; CW/CL 0.68–0.79 in males, 0.65–0.85 in females; first vertebral width/first vertebral L 0.77–1.16 in males, 0.81–1.26 in females; PL/CL 0.85–0.99 in males, 0.73–0.96 in females; BL/CL 0.33–0.41 in males, 0.32–0.40 in females; plastron width at humeropectoral seam/PL 0.42–0.50 in males, 0.40–0.52 in females; plastron width at midfemoral scute/PL 0.29–0.53 in males, 0.42–0.51 in females; abdominal scute L/PL 0.18–0.34 in males, 0.21–0.27 in females.

Color in life of an adult male (USNM 562002): carapace Brownish Olive (29) with Blackish Neutral Gray (82) spot in center of each costal scute, each spot encircled by narrow Robin Rufous (340) ring; marginals similarly colored as costals, except dark spot in posteroventral corner of each marginal and rust colored line just inside edge of each marginal; plastron horn colored; bridge with several longitudinal pale brown enclosed pale yellow stripes; ventral portion of each marginal pale yellow with pale brown lines; head Brownish Olive above, with several paler longitudinal lines; postorbital stripe Robin Rufous with dark gray edging, stripe extending onto neck; dark gray edged Sulphur Yellow (57) primary orbitocervical (mandibular) stripe extending from ventral edge of eye posterior to angle of mouth onto neck; area between those 2 stripes with dark gray-edged pale olive lines; another dark gray-edged Sulphur Yellow line extending from in front of eye to edge of mouth and another extending from nostril to edge of mouth, area between those lines with several similarly colored, more slender lines; neck Grayish Olive (43) with Sulphur Yellow lines; dorsal surfaces of forelimb, hind limb, and tail Olive-Green (Basic) (46) with Sulphur Yellow transverse stripes; posterior of thigh with transverse Sulphur Yellow lines; iris pale yellowish green with horizontal dark gray stripe across pupil. Color in life of an adult female (FMNH 283585): carapace greenish brown with pale orange ocelli, with central Sepia (219) blotches on costals and marginals; head and neck pale green with yellow lines, including supratemporal stripe; skin of remaining surfaces dark brown with yellow lines; plastron yellowish orange with medium brown symmetrical lined pattern; bridge and marginals yellowish orange with greenish brown ocelli and lines. Color in life of another adult female (KU 220133) was described by Wilson and Cruz Díaz (1993: 16): "carapace yellowish tan; plastron pale yellow; bridge yellow with gray spots; head with olive green, pale yellow, and black striping, postorbital blotch pale olive brown; neck with pale yellow and gray stripes; front limbs with yellow and dark gray striping; hind limbs dark gray with pale yellow stripes; posterior thigh pale yellow with longitudinal gray striping."

Color in alcohol: carapace dark brown to black, with pale bordered ocelli on costals and vertebrals; plastron yellow to pale brown, usually with brown to dark brown figure of symmetrical lines, figure fading or disappearing with age or wear, older specimens can have a pale brown plastron with medium brown mottling or large medium brown spots; bridge yellow with dark brown, paler brown centered ocelli on underside of marginals and along lateral edge of plastron, although ocelli fade and sometimes disappear with aging and ocelli are sometimes indistinct in adults of some populations; plastron yellow-brown with brown symme-

trical lines on much of plastron, those lines becoming faded or worn in older individuals; head greenish olive to black, with numerous, dark bordered, yellow or yellowish brown stripes on all surfaces, primary lateral stripes continuing onto neck; yellow or yellowish brown postorbital and primary orbitocervical stripes contact orbit, latter slightly broader than, or about same width as former; postorbital stripe usually not constricted medially above tympanum; yellow or yellowish brown symphyseal stripe usually not connected to yellow or yellowish brown neck stripe (those stripes connected in occasional specimens); forelimb greenish olive to dark brown, with yellow stripes; hind limb same color as forelimb on outer surface, yellow with dark brown stripes on inside surface; tail greenish olive to dark brown, with yellow stripes.

Diagnosis/Similar Species.—*Trachemys venusta* is distinguished from all remaining Honduran turtles, except the other *Trachemys* and one species of *Rhinoclemmys*, by the combination of having strongly clawed limbs, webbed hind feet, and an oblong and large plastron with 12 scutes. *Rhinoclemmys* lack a ridge on the triturating surface of the upper jaw (versus triturating ridge present in *T. venusta*). Also, only *R. funerea* among Honduran species of *Rhinoclemmys* has strongly webbed toes, but that species has a mostly black plastron with a yellow midseam (versus plastron yellow or pale brown with darker brown symmetrical lines in *T. venusta*). *Trachemys emolli* has the symphyseal stripe connected with both neck stripes in one of two specimens and the postorbital stripe strongly constricted above the tympanum in one of two (versus symphyseal stripe usually not connected to both neck stripes and postorbital stripe usually not strongly constricted above tympanum in *T. venusta*; see Remarks). *Trachemys scripta* has black blotches or ocelli on the plastron and has a red postorbital stripe, and males have elongated claws on each forelimb (versus dark brown figure of

Plate 97.   *Trachemys venusta*. USNM 580392, adult male, carapace length = 160 mm. Islas de la Bahía: Roatán Island, near Camp Bay.

symmetrical lines on plastron, except in some old individuals where figure fades or is worn, yellow to yellowish brown postorbital stripe, and males lacking unusually long claws on forelimb in *T. venusta*).

Illustrations (Figs. 150, 179, 180, 184, 187, 188; Plate 97).—Bonin et al., 2006 (adult); Bour, 2003 (adult, plastron, juvenile; as *T. v. venusta*), 2004 (adult; as *T. v. venusta*); Calderón-Mandujano et al., 2008 (head); Campbell, 1998 (adult, carapace, plastron; as *T. scripta*); Ernst, 1990 (adult; as *T. s. venusta*); Ernst and Barbour, 1989 (adult; as *T. s. venusta*); Ernst and Seidel, 2006 (adult; as *T. v. venusta*); Gray, 1856a (adult; as *Emys venusta*); Günther, 1885, *In* Günther, 1885–1902 (head, carapace, plastron; as *Emys salvini*); Köhler, 1999b (juvenile; as *T. scripta*), 2000 (adult, juvenile; as *T. scripta*), 2001b (juvenile; as *T. scripta*), 2003a (adult, juvenile; as *T. scripta*), 2008 (adult, juvenile; as *T. scripta*); Lee, 1996 (adult, plastron; as *T. scripta*), 2000 (adult, plastron; as *T. scripta*); Legler, 1990 (head pattern; as *Pseudemys s. venusta*); Legler and Vogt, 2013 (adult, juvenile, head, carapace, plastron; as *T. s. venusta*);

McCord et al., 2010 (carapace, plastron, head pattern; as *T. v. uhrigi*); McCranie and Valdés Orellana, 2014 (adult; as *T. ornata*); McCranie et al., 2005 (adult, shell), 2006 (adult, carapace, plastron); D. Moll, 2010 (adult, hatchling); D. Moll and Moll, 2004 (adult, hatchling; as *T. scripta venusta*); E. O. Moll, 1979 (plastron and comparative egg size; as *P. scripta* but fig. 9, not fig. 8 as indicated); E. O. Moll and Legler, 1971 (head, hatchling; as *P. scripta*); Parham et al., 2015 (adult, juvenile); Pasachnik and Chavarria, 2011 (hatchling being consumed by a *Ctenosaura oedirhina*; as *T. v. uhrigi*); Seidel and Ernst, 2017 (carapace, plastron); Smith and Smith, 1980 (neck pattern, shell; as *Pseudemys s. venusta*); Stafford and Meyer, 1999 (adult; as *T. scripta*); Townsend et al., 2007 (adult); Vetter, 2005 (adult, juvenile; as *T. venusta, T. v. venusta*).

Remarks.—Smith and Smith (1980: 494–495) designated BMNH 1845.8.5.26 (= 1947.3.4.80) as the lectotype of *Emys venusta* Gray, because Gray (1873b: 48) "appeared to regard no. [BMNH] 1845.8.5.26 as the type" since Gray listed "Honduras. Dyson. 45.8.5.26" for his *Emys venusta*. Smith and Taylor (1950a: 319) had earlier erroneously restricted the type locality of *Emys venusta* to "Honduras." The Dyson specimens in the BMNH collection from "Honduras" actually were collected in Belize (including some old BMNH specimens with locality data simply as Honduras), which formerly was known as British Honduras (see Schmidt, 1941: 503; Fugler, 1968: 97; McCranie, *In* McCranie and Wilson, 2002: 27; McCord et al., 2010: 40; McCranie, 2011a: 243). McCord et al. (2010) further clarified that the lectotype of *Emys venusta* is from Belize, as did Legler and Vogt (2013: 266). That latter group of authors also restricted the type locality of *E. venusta* Gray to "Belize City, Belize." Despite that information, Parham et al. (2015: 363), in an excellent and valuable publication, continued to use "Honduras" as the type locality for *T. venusta*. Peculiarly, Ernst and Seidel (2006), in their detailed review of *T. venusta*, did not mention this inaccurate type locality of "Honduras."

Fritz et al. (2012: 133) mentioned the misleading "coloration and pattern characters" from Atlantic versant *Trachemys venusta* in Honduras, as so evident in the McCord et al. (2010) effort. The publication by McCord et al. (2010) contains many errors and would have greatly benefited from scientifically sound peer review of the manuscript before it was published in a popular magazine. One serious error in McCord et al. (2010) is the discussion of the lectotype and syntypes of *Emys venusta* that adds to the confusion already existing in the literature concerning this series. McCord et al. (2010: 39), in discussing BMNH 1845.8.5.26, state "However in 1873 he [= Gray, 1873b] referred to only one syntype as '*Emys venusta*:' stuffed specimen 'e' (1845.8.5.26) in the British Museum of Natural History, labeled 'Charming Emys' for its beautiful pattern." First of all, Gray (1873b: 48; as *Callichelys venusta*) did not refer to the holotype of *E.*

venusta as "stuffed specimen e" and according to Ernst and Seidel (2006: 832.1), who examined BMNH 1845.8.5.26, it is "alcoholic adult parts and shell." Gray (1873b: 48; also see Gray, 1856a: 25) did refer to a specimen "e. Half-grown" from "Honduras. Mr. Dyson Collection," but apparently indicated it is an alcohol specimen because he stated "stuffed" among his data for stuffed specimens. Adding to the already existing confusion in the literature, Legler and Vogt (2013: 266) said BMNH 1845.8.5.26 is a stuffed specimen (thus, confusion still exists in the literature, when in reality the case should be clear concerning the actual preserved state of the lectotype of *Emys venusta*). Legler and Vogt (2013: 266) also said the specimen of *E. venusta* figured by Gray (1856a: table XIIA) is "a mirror image" of the lectotype.

Seidel (2002: 289) resurrected *Trachemys venusta* from the synonymy of *T. scripta* and applied *T. v. venusta* to the Honduran specimens. See Remarks for *Trachemys* for a discussion of the Fritz et al. (2012) results that changed the landscape in Central America *Trachemys* taxonomy. Parham et al. (2015), a more thoroughly accomplished study based on additional molecular gene characters and morphological data sets, reliably demonstrated that Atlantic versant turtles should be *T. venusta*. Ernst and Seidel (2006) provided an overview of the morphology of *T. venusta* and a thorough review of the literature on *T. venusta*. Hernández-Guzmán et al. (2014) reported on the chromosomes of "*T. scripta*" from Tabasco, Mexico, but that study likely involved *T. venusta*.

The distribution given by Ernst (1990) in his text for *Trachemys v. venusta* (p. 64 and several other of his recognized subspecies occurring outside of Honduras) does not match that shown on his distribution map (p. 62). Wilson and Hahn (1973: 103) reported *T. venusta* from Utila Island in the Bay Islands on the basis of a "shell in the possession of a private individual" (also see Cruz Díaz, 2008: 41). A cautionary note should be added here. In September 2012, I was told a young man living in Utila Town had two jicoteas (*Trachemys*) thought to be from Utila Island he was keeping as pets. I tracked down the person and was told by him that both of the turtles were captured on the mainland of northern Honduras and, furthermore, one had recently escaped. However, I have heard reports of a *Trachemys* population living in the brackish water channels and swamps in the interior of the island of Utila. An effort to trap turtles in those waters should be undertaken. The presence of a natural population of *Trachemys* on Isla de Utila still needs to be investigated and confirmed if one does or does not occur.

McCord et al. (2010: 43) described the new subspecies *Trachemys venusta uhrigi* for the *Trachemys venusta* population on the Atlantic versant of Honduras. However, the diagnostic characters given for that nominal form by those authors conceal the fact that much more variation occurs in Honduran specimens than admitted by those authors, including the symphyseal stripe connected to, versus not connected to the neck stripes for the Pacific versant *T. emolli* and Atlantic versant *Trachemys* of Honduras. Thus, *T. v. uhrigi* was placed in the synonymy of *T. venusta* by McCranie, Köhler et al. (2013: 26), not withstanding the untenable assignment of some Colombian populations of *Trachemys* to *T. v. uhrigi* by Bock et al., *in* Páez et al. (2012). McCord et al. (2010) also described the new subspecies, *T. v. panamensis* from the Pacific versant of Panama, but that nominal form is more closely related to the *T. emolli* clade than it is to the *T. venusta* clade (see Fritz et al., 2012). McCord et al. (2010: 45, 46) also described *T. v. iversoni* from the Yucatán, Mexico. All of the *Trachemys* subspecies described by McCord et al. (2010) need to be reevaluated using molecular techniques and morphological data, as does *T. v. cataspila* Günther (1885: 4, *In*

Günther, 1885–1902) from Mexico. The Parham et al. (2013, 2015) studies recovered a paraphyletic *T. v. cataspila* in regard to some *T. v. venusta* populations, as did the study by Rodrigues and Diniz-Filho (2016). McCord et al. (2010: 40) also inaccurately placed *Emys valida* Le Conte (1859: 7; type locality "Honduras") in the synonymy of *T. v. grayi* (Bocourt, 1868: 121), but *T. grayi* appears to not occur in Honduras. Because *E. valida* is an older name than *E. grayi*, McCord et al. (2010: 40) designated *E. valida* "as a nomen oblitum to avoid taxonomic instability." The principal reason McCord et al. (2010) considered *E. valida* to represent *T. grayi* (= *T. emolli*; see above) is because of the strongly serrated lower jaw in the *E. valida* holotype, which was purported to resemble *T. grayi* (= *T. emolli*) more than *T. venusta*. However, the degree of lower jaw serration is variable, with some Honduran Atlantic versant specimens (*T. venusta*) having strongly serrated lower jaws similar to those of the Pacific population (*T. emolli*). *Trachemys valida* is a much older name than *T. emolli* and thus would be available over the rather recently described *T. emolli* should they be found to be conspecific. The McCord et al. (2010) *nomen oblitum* act regarding *T. grayi* is premature. It also remains possible the *E. valida* holotype is from the Atlantic versant and thus would then be a junior synonym of *T. venusta*. Since the holotype of *E. valida* (ANSP 216) was apparently preserved in alcohol (Le Conte, 1859: 7), an analysis of its DNA seems possible and would likely resolve which nominal form it is allied with. A molecular and more thorough study of the *E. valida* holotype (ANSP 216) than that of McCord et al. (2010) would be desirable. Another problem is, now that we know *T. grayi* does not occur in Honduras, Pacific populations from Guatemala to Costa Rica need to be reinvestigated both morphologically and genetically. Also, the placement of the Honduran type specimen of *E. valida* in the synonymy of *T. grayi* is not tenable.

Returning to the Colombian *Trachemys* population that was considered *T. v. uhrigi* in Páez et al. (2012): The *T. v. uhrigi* photographs in that work do not resemble any *Trachemys* in Honduras assigned to *T. v. uhrigi* by McCord et al. (2010). Ceballos and Brand (2014) studied morphological characters of 89 Colombian *Trachemys* and concluded that those turtles did not resemble the description of *T. v. uhrigi* in McCord et al. (2010). Ceballos and Brand (2014) also included those two photographs from Páez et al. (2012). Apparently, Ceballos and Brand (2014) were not aware that McCranie, Köhler et al. (2013) had already synonymized *T. v. uhrigi* with *T. venusta*. More recently, Vargas-Ramírez et al. (2017) described the north-western Colombian Río Atrato population as the new species *T. medemi*.

The large population of *Trachemys venusta* at West Bay (Gumbalimba Park and Reserve), Isla de Roatán in the Bay Islands, is a mixture of native Roatán turtles and mainland Honduran turtles brought to West Bay for that tourist attraction. The population at Camp Bay on eastern Roatán is said by locals to be natural. Three specimens of *T. venusta* from Guanaja Island, in the Honduran Bay Islands (FMNH 283585–86; UNAH), are nearly identical to a specimen of *T. venusta* from the Atlantic coastal plain from northwestern Honduras in the mitochondrial gene ND4 (Frank Köhler, in litteris). Recent tissues and voucher specimens of *T. venusta* are now available from all of the major Bay Islands and Cayos Cochinos, except Utila Island.

In December 2012, I saw approximately 15 *Trachemys* basking while floating at the surface of the badly polluted, foul-smelling estuary at La Ceiba. All turtles I was able to see well enough had the head and carapace patterns of the native *T. venusta*. Although I offered 500 Lempiras to several passing men for one of those turtles (a significant sum of money to the poor of La Ceiba), none were willing to enter the badly polluted water in an effort to collect one.

Molecular data in Parham et al. (2013: 178) allied a *Trachemys* specimen from a tributary of the upper Río Choluteca on the Pacific versant of south-central Honduras to *Trachemys venusta* of the Atlantic versant and not *T. emolli* of the Pacific versant, as would be expected. As discussed elsewhere in this paper, the popularity of *Trachemys* as pets, especially when babies, and then their subsequent release has muddied the current chances of ever completely knowing the natural distributions of the native *Trachemys* of Honduras. Also, as humans consume the meat and eggs of *Trachemys* in Honduras, natural *Trachemys* populations in some places have been decimated, or possibly even extirpated. One likely example might be the population occurring in the Río Choluteca and its tributaries in south-central Honduras (the Río Choluteca passes through the center of the capital city of Tegucigalpa).

Natural History Comments.—*Trachemys venusta* is known from near sea level to 1,000 m elevation in the Lowland Moist Forest, Lowland Dry Forest, Lowland Arid Forest, and Premontane Dry Forest formations and peripherally in the Premontane Wet Forest and Premontane Moist Forest formations. This largely diurnal turtle was collected along and in rivers, freshwater lagoons, in streams, in swampy areas, next to and in estuaries, and crossing roads. Another was on the floor of a closed-canopy broadleaf rainforest locality during the day during a period of heavy rains. One adult was captured from a fairly deep pool in the otherwise shallow Río Rus Rus near its headwaters during the latter part of the dry season in May. That same pool was also occupied by about ten adult *Rhinoclemmys funerea*. *Trachemys venusta* is commonly seen sunning itself along the ríos Coco, Kruta, Patuca, and Warunta, and in the Laguna de Warunta in the Mosquitia of eastern Honduras. *Trachemys venusta,* along the lower reaches near the mouth of the Río Warunta, is frequently seen basking out of water on logs and other accumulated debris above the river's waterline. However, fewer and fewer *T. venusta* are seen basking the further one gets upriver from the mouth of that river where vegetation becomes close canopy broadleaf forest. Rather, those turtles shift their habitats to lagoons and swamps (temporarily and/or permanently) away from the river, where they remain common all the way to that river's headwaters near Awasbila, Gracias a Dios. Those diminishing populations upriver and along a river have apparently been interpreted in some literature to signify that those turtles do not occur upriver in some regions (i.e., E. O. Moll and Legler, 1971). The turtle also remains in similar habitats away from the Río Rus Rus all the way to its headwaters. *Trachemys venusta* has been collected in every month of the year, except March and October; thus, the species is likely active throughout the year on sunny days. Individuals of *T. venusta* bite aggressively when handled. Moll, E. O., and Legler (1971: 14; as *Pseudemys scripta*) reported that this species was collected and observed in Honduras "in estuaries which were temporarily closed to the sea but periodically inundated by salt water." Specimens (KU 214779; USNM 580392) were collected in similar situations on Roatán. The *Trachemys* specimens found on the Cayos Cochinos are thought to be waifs from the Río Cangrejal on the northern Honduran mainland near La Ceiba that reach the Cayos Cochinos during times of heavy rainfall (Chad Montgomery, personal communication). Montgomery also believes those turtles cannot survive on the Cayos Cochinos because of the lack of permanent fresh and brackish water on those small islands. On the other hand, the *T. venusta* population on Guanaja Island seems to be breeding, as one adult and one hatchling were found in a permanent freshwater swamp near Savannah Bight. Moll, D.

(2010) reported females of the *T. venusta* population at Tortuguero, Costa Rica, would migrate northward with river currents and enter the sea and travel south before emerging to deposit their eggs in cocoplum scrub on the upper beach. The spent females then generally reentered the sea and again traveled southward before emerging again to travel overland to reach the freshwater Tortuguero canal where they spend the remainder of the year. Females deposit 5–32 eggs per clutch in nest cavities they dig in the sand (freshwater river, lake beaches, and even sea beaches) from December to May during the dry season, with the hatchlings emerging in the rainy season (E. O. Moll and Legler, 1971, Panama; D. Moll and Moll, 1990, southern Mexico, Belize, Panama; Vogt, 1990, Veracruz, Mexico; D. Moll, 1994, southern Mexico and Central America; Cabrera Peña et al., 1996, Costa Rica). Females are capable of producing one to six clutches per year (E. O. Moll and Legler, 1971; Legler, 1973; D. Moll and Moll, 1990; Vogt, 1990). Moll, D. (1990) studied a population of *T. venusta* in Belize and found the species to be primarily herbivorous, but it also ate insects, crustaceans, and fish. Moll, E. O., and Legler (1971: 2, Panama) stated that turtles "of all ages are opportunistic and tend to feed on the most readily available food. *Elodea* [Waterweed] and aquatic grasses constitute the bulk of the adult diet on the Río Charges whereas *Najas* [naiads], grasses, and some animal food comprise the diet of juveniles" (also see Legler, 1973). Acuña-Mesén et al. (1983) noted that *Trachemys* in eastern Costa Rica eat the introduced water hyacinth *Eichhornia crassipes*. Legler and Vogt (2013: 267) called *T. venusta* in Mexico "opportunistic omnivores, consuming most of the plants and animals that occur in or fall into natural aquatic habitats. Notable are aquatic insects, crustaceans, and fish (live or as carrion). However, 80% of the diet in

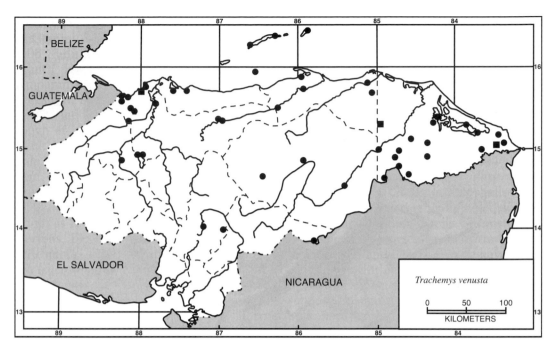

Map 79. Localities for *Trachemys venusta*. Solid circles denote specimens examined and solid squares represent accepted records.

Veracruz and Chiapas was found to be plants: leaves, flowers, fruits, and seeds."

Etymology.—The name *venusta* is derived from the Latin *Venus* (charming) and the Latin suffix *–tus* (pertaining to) and most likely refers to the numerous colorful markings, especially on the head, in this turtle (called the Charming Emys by Gray, 1856a: 24).

Specimens Examined (106, + 24 shells or shell parts, 3 heads only, 7 skulls, 2 soft parts, 3 egg lots [13]; Map 79).—**ATLÁNTIDA**: between Barra de Ulúa and Laguna del Río Tinto, USNM 580760 (shell); Puerto Arturo, USNM 62991; Tela, MCZ R31484, 34356; lagoon along W side of mouth of Río Lean, UNAH (SMR 0456; tissues MVZ 263395 given as voucher in Parham et al., 2013). **COLÓN**: Finca Sonora, MCZ R20477; Los Andes, ANSP 26663; Río Blanquito, USNM 129586 (shell), 129587 (shell and skull), 129588 (shell), 129608–09, 129624, 134440; Río Paulaya, BMNH

1986.42–44; near Sabá, USNM 562002; Trujillo, LSUMZ 33658. **CORTÉS**: Agua Azul, AMNH 70551–59, 70567, USNM 243359–67, 243368 (19 eggs), 243369, 243370 (2 hatchlings, 2 eggs or egg shells); Cofradía, CM 62091; 3.2 km SSE of El Jaral, LACM 47205; 10 km SE of El Jaral, UF 90020; El Jaral, FMNH 5320 (shell); Lago de Yojoa, AMNH 70560; Laguna San Idelfonso, FMNH 283808; Laguna Ticamaya, CM 62092–93, FMNH 5316–17, 5321 (soft parts only), 5330 (6 eggs), 284696; 8 km S of San Pedro Sula, UF 105425; near San Pedro Sula, USNM 102885; San Pedro Sula, FMNH 5318, 5319 (skull and shell), UF 105421; 2.0 km W of Tegucigalpita, USNM 330190; Tulián, UU 6159–60. **EL PARAÍSO:** near Arenales, LACM 73822. **FRANCISCO MORAZÁN**: El Zamorano, UF 150886; Tegucigalpa, UU 6158. **GRACIAS A DIOS**: Awasbila, USNM 559590 (shell); Bachi Kiamp, USNM 573341; Cabeceras de Río Rus

Rus, USNM 570541; Calpo, UF 137165–67; Canco, UF 137648–53 (all shell fragments); Caño Awalwás, USNM 559591 (head and shell); Laguna Kohunta, UF 137164; Mavita, USNM 589138; about 30 km S of Mocorón, UTA R-52340–41; Mocorón, UTA R-42661; Puerto Lempira, LACM 48454–55, LSUMZ 21479; Río Coco, USNM 24536–38; confluence of ríos Wampú and Patuca, USNM 319987–88; Rus Rus, USNM 562004 (shell); near Sachin Tingni Kiamp, USNM 564160 (head and shell); Tánsin, LSUMZ 21480–81, USNM 573338–40, 573360 (shell); Tikiraya, UF 137633 (shell), 137163, 137429; Warunta, USNM 573076, 573358, FN 256943–44 (still in Honduras because of permit problems). **ISLAS DE LA BAHÍA**: Cayo Cochino Mayor, La Ensenada, KU 220133; Cayo Cochino Menor, FMNH 282715 (shell fragments), 282720, USNM 570530 (shell fragments); Isla de la Guanaja, near Savannah Bight, FMNH 283585–86; Isla de Roatán, Camp Bay, USNM 580392; Isla de Roatán, West Bay, KU 214779 (shell). **OLANCHO**: 3.2 km W of Galeras, TCWC 19234–35; 13.1 km NE of Juticalpa, KU 200508; Río Patuca across from Quebrada EL Guásimo, SMF 80837. **SANTA BÁRBARA**: 8.0 km S of Santa Bárbara, LACM 45310. **YORO**: Quebrada San Lorenzo, USNM 562003; 3.5 km S San Lorenzo Arriba, USNM 580363. "HONDURAS": ANSP 216 (head and limbs), UF 66716, 151692–93 (shells), 150740, 154050–52 (skeletons), USNM 327980–81, 328088, 328464–66 (skulls); UU 6157 (purchased at market in Tegucigalpa).

Other Records (Map 79).—**CORTÉS**: El Paraíso (Townsend, 2006a); Estero Prieto, UU 6161 (specimen listed as a shell in UU catalogue, but could not be found in November 2015); Tulián, UU 7928 (specimen listed as a shell in UU catalogue, but could not be found in November 2015). **GRACIAS A DIOS**: Krahkra (UF slide of burnt shell of a turtle used as a food source by locals); Raudal Pomokir, UNAH 5511 (Cruz Díaz, 1978, specimen now lost). **ISLAS DE LA BAHÍA**: Isla de Roatán, West Bay (personal sight records; also see Pasachnik and Chavarria, 2011).

Family Geoemydidae Theobald, 1868

This turtle family ranges in the Western Hemisphere from southern Sonora and southern Veracruz, Mexico, through Central America to western Ecuador and northeastern Brazil. In the Eastern Hemisphere, it occurs in northwestern Africa, from Europe to western Asia and the Middle East, from southern Asia to China, Japan, the Philippines, and on islands of the Sunda Shelf. The Geoemydidae can be distinguished externally from all other Honduran turtles by the combination of having strongly clawed limbs, a short tail without a dorsal medial row of large triangular scales, a large and oblong plastron with 12 scutes, the carapace normally with 28 scutes, and by lacking a ridge on the alveolar surface of upper jaw. Nineteen genera comprised of 68 named species are included in this family. Four species placed in one genus occur in Honduras.

Genus *Rhinoclemmys* Fitzinger, 1835

Rhinoclemmys Fitzinger, 1835: 108, 115 (type species: *Emys dorsata* Schweigger, 1812: 297 [= *Testudo dorsata* Schoepff, 1801: 136, a subjective synonym of *Testudo punctularia* Daudin, 1801: 249; see ICZN, 1963: 187, 1985b: 152; Rhodin and Carr, 2009: 7], by subsequent designation of Lindholm, 1929: 283).

Geographic Distribution and Content.— This genus ranges from southern Sonora and southern Veracruz, Mexico, to western Ecuador and northern Brazil, including Trinidad and Tobago. Nine named species are recognized, four (one only questionably) occur in Honduras.

Etymology.—The name *Rhinoclemmys* is derived from the Greek words *rhinos* (nose,

Figure 189.   Tip of upper jaw hooked. *Rhinoclemmys annulata.* USNM 570466 from near San San Hil, Gracias a Dios.

snout, beak, bill) and *klemmys* (tortoise, turtle). The name alludes to the protuberant snout exhibited by some individuals of the type species.

KEY TO HONDURAN SPECIES OF THE GENUS *RHINOCLEMMYS*

1A. Tip of upper jaw hooked (Fig. 189) or nearly level . . . . . . *annulata* (p. 533)

1B. Tip of upper jaw notched (Fig. 190) 2

2A. Adult hind feet heavily webbed (Fig. 191); adult plastron largely black with yellow line along mid-seam and bridge largely black (Fig. 192) . . . . *funerea* (p. 540)

2B. Adult hind feet with little or no webbing (Fig. 193); adult plastron yellow to pale brown, at least on outer edges, and bridge with significant yellow to yellowish brown pigment present 3

3A. Head pattern of a single pair of supratemporal stripes posterior to eye (Fig. 194); bridge usually yellow to pale brown without extensive dark coloration (Fig. 195) . . . . *areolata* (p. 536)

3B. Head pattern of 2 or 3 red stripes, with at least 1 stripe crossing tip of snout in Honduran specimens (Fig. 196), although those red stripes sometimes reduced to series of dashes extralimital to Honduras; bridge with extensive dark coloration only centrally (Fig. 197) . . . . *pulcherrima* (p. 544)

CLAVE PARA LAS ESPECIES HONDUREÑAS DEL GÉNERO *RHINOCLEMMYS*

1A. Punta del pico superior en forma de gancho (Fig. 189) o poco curvada . . . . *annulata* (p. 533)

1B. Punta del pico superior con una muesca (Fig. 190). 2

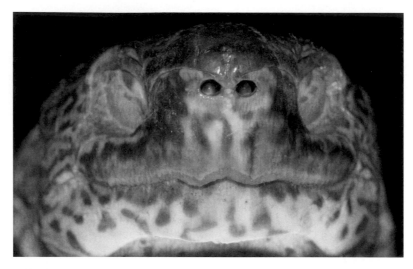

Figure 190.   Tip of upper jaw notched. *Rhinoclemmys funerea*. USNM 579654 from Sadyk Kiamp, Gracias a Dios.

2A. Membranas interdigitales amplias en las extremidades posteriores (Fig. 191); plastrón de los adultos casi completamente negro con una línea longitudinal medial de color amarillo y el puente casi completamente negro (Fig. 192) ....

.... *funerea* (p. 540)

2B. Membrana interdigitales débilmente desarrolladas o ausente en las extremidades posteriores (Fig. 193); plastrón de adultos amarillo a amarillo-pardo claro, al menos en los bordes exteriores; el puente de color amarillo hasta amarillo-pardo.. 3

3A. Patrón de la cabeza con un par de rayas supratemporales por detras

Figure 191.   Hind feet with well-developed webbing. *Rhinoclemmys funerea*. USNM 573075 from Kakamuklaya, Gracias a Dios.

Figure 192.   Plastron and each bridge largely black, with a yellow line along midseam of plastron. *Rhinoclemmys funerea*. USNM 559588 from Caño Awalwás, Gracias a Dios.

del ojo (Fig. 194); el puente usualmente amarillo a pardo claro, sin
color oscuro (Fig. 195). . . . .
. . . . *areolata* (p. 536)

3B. Patrón de la cabeza con 2 o 3 rayas
supratemporales rojas, al menos
una cruzando la punta del pico en
especímenes Hondureñas (Fig.

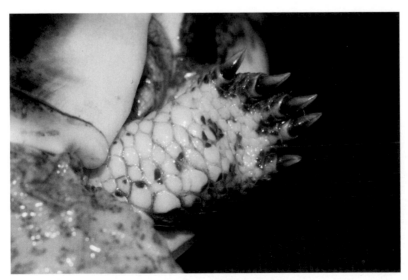

Figure 193.   Hind feet without, or with only slight webbing, even in adults. *Rhinoclemmys annulata*. USNM 570465 from Auka Kiamp, Gracias a Dios.

Figure 194.   Head pattern of a single pair of supratemporal stripes (1 stripe on each side of head) posterior to eye. *Rhinoclemmys areolata*. UF 109842 from Belize.

Figure 195.   Each bridge without extensive dark coloration. *Rhinoclemmys areolata*. UF 67524 from "Honduras."

Figure 196. Head pattern of 2 or 3 red stripes, at least one of which crosses tip of snout. *Rhinoclemmys pulcherrima*. USNM 580763 from El Faro, Choluteca.

196), pero ocasionalmente las rayas rojas reducidas a series de guiones en populaciones fuera de Honduras; el puente con color oscuro intenso solamente en el centro (Fig. 197)....... *pulcherrima* (p. 544)

Rhinoclemmys annulata (Gray, 1860)

Geoclemmys annulata Gray, 1860: 231 (three syntypes, BMNH 1946.1.22.56, 1947.3.5.58–59 [see Ernst, 1978: 116]; type locality: "Esmeraldas, Ecuador").

Figure 197. Each bridge with rather extensive dark central coloration and much of plastron unpatterned. *Rhinoclemmys pulcherrima*. USNM 580762 from Isla Inglasera, Valle.

Rhinoclemys [sic] *annulata*: Gray, 1863: 183 (first use of present combination, but with generic name misspelled).
Rhinoclemmys annulata: Gray, 1869: 189 (first use of present combination correctly spelled); Meyer, 1969: 192; Meyer and Wilson, 1973: 6; Ernst, 1980a: 250.1; Iverson, 1986: 63; Iverson, 1992: 155; Köhler, McCranie, and Nicholson, 2000: 425; Nicholson et al., 2000: 29; Wilson et al., 2001: 134; Castañeda, 2002: 15; McCranie, Castañeda, and Nicholson, 2002: 25; McCranie et al., 2006: 102; Wilson and Townsend, 2006: 104; Solís et al., 2014: 139; McCranie, 2015a: 381.
Geoemyda annulata: Dunn and Emlen, 1932: 25.
Callopsis annulata: Ernst, 1978: 116.

Geographic Distribution.—*Rhinoclemmys annulata* occurs at low and moderate elevations from possibly north-central Honduras to northwestern Colombia on the Atlantic versant and from central Panama to western Ecuador on the Pacific versant. It is known in Honduras from low elevations in the northeastern and questionably in the north-central portion of the country.

Description.—The following is based on four males (USNM 559584, 570466–67, 579651) and ten females (FMNH 282716, 282719; USNM 559585–86, 562001, 565394, 570465, 570539, 573074, 573342). *Rhinoclemmys annulata* is a moderately large (maximum recorded CL 204 mm [Ernst, 1978]; 178 mm CL in largest Honduran specimen [USNM 570539, a female]) turtle with a broad, oval, and not or somewhat flattened carapace; carapace with 1 nuchal, 5 vertebrals, 8 costals, and 24 marginal scutes; carapace widest at level of marginals 6–8, highest at seam between vertebrals 2–3 or at about midlength of vertebral 2; carapacial sides straight or slightly bowed; posterior marginals slightly serrated and sometimes flared in adults, flared in juveniles; single, flattened vertebral ridge present; each vertebral wider than long; carapace texture rough due to growth rings; nuchal narrow, bifurcated posteriorly; carapace notched posteriorly; plastron unhinged, with 12 scutes, consisting of paired gulars, humerals, pectorals, abdominals, femorals, and anals; gulars with 1–2 slight anterior projections in adults; anals with wide medial notch; plastron slightly concave in males, flat in females, anteriorly upturned in both sexes, but more so in females; a small axillary and inguinal scute present; head relatively small, snout slightly projecting; upper jaw serrated laterally, slightly hooked; triturating surface of upper jaw without a ridge; hind foot slightly webbed to not webbed; 5 claws on forelimb, 4 claws on hind limb; foreleg covered with about 7–10 rows of large scales; cloacal opening posterior to carapacial margin in males, anterior to carapacial margin in females; CL 104–165 (145.8 ± 28.1) mm in males, 85–178 (122.6 ± 37.4) mm in females; MSH/CL 0.30–0.38 in males, 0.40–0.48 in females; CW/CL 0.70–0.80 in males, 0.69–0.84 in females; second vertebral W/second vertebral L 1.18–1.45 in males, 1.20–1.58 in females; PL/CL 0.86–1.14 in males, 0.87–0.97 in females; humeral suture L/gular suture L 0.90–1.51 in males, 0.95–1.66 in females.

Color in life of an adult female (USNM 559586): carapace mottled Grayish Horn Color (91) and Dark Brownish Olive (129); plastron Dusky Brown (19) medially and on bridge, Buff-Yellow (53) around central Dusky Brown figure; top of head Dusky Brown, sides of head Dusky Brown with pale yellow stripes; front limb yellow with dark gray stripes; hind limb dark gray with pale yellow stripes on inside surface.

Color in alcohol: carapace tan to dark brown, with pale brown blotches on lateral edges of costals and sometimes vertebrals; vertebral ridge tan to brown, with yellow blotches; plastron black with yellow border; bridge dark brown to black; head brown to dark brown; supratemporal pale brown stripe present, poorly to well defined, not

Plate 98.   *Rhinoclemmys annulata.* FMNH 282716, adult female, carapace length = 93 mm. Gracias a Dios: Sadyk Kiamp.

reaching eye, slightly downward angled to nape where it continues as neck stripe; well-defined pale brown stripe present from posterior edge of eye to tympanum where stripe meets similarly colored stripe from lower jaw, these 2 stripes fuse and continue onto side of neck as single stripe; poorly to well-defined pale brown stripe extending from anterior edge of eye to tip of snout; lower jaw cream, usually mottled with brown; chin cream to yellow, mottled with tiny brown spots; neck brown with pale brown dorsal and lateral stripes; forelimb yellow with rows of dark brown spots forming interrupted stripes; hind limb brown on outside, yellow on inside, both sides with brown spots or mottling; tail brown with yellow dorsal stripes.

Diagnosis/Similar Species.—*Rhinoclemmys annulata* is distinguished from all remaining Honduran turtles, except the other *Rhinoclemmys*, by the combination of having strongly clawed toes, no or only slight webbing between the toes, a short tail without a dorsal medial row of large triangular scutes, and a large and oblong plastron with 12 scutes. *Rhinoclemmys annulata* is the only Honduran species of

Rhinoclemmys with a hooked upper jaw. Also, *R. areolata* has a yellow to brown bridge, *R. funerea* has strongly webbed toes in adults, and *R. pulcherrima* has 2 or 3 red stripes on the head, with at least 1 crossing the tip of the snout in all Honduran populations (versus significant dark brown to black bridge, toe webbing absent or only slightly developed in adults, and red lines, if present, short and no red lines crossing snout in *R. annulata*).

Illustrations (Figs. 189, 193; Plate 98).— Ernst and Barbour, 1989 (adult); Gray, 1870a (head); Guyer and Donnelly, 2005 (adult, plastron); Köhler, 2000 (adult), 2001b (adult), 2003a (adult), 2008 (adult); McCranie et al., 2006 (adult, plastron); Medem, 1956 (adult; as *Geoemyda* Gray 1834: 100), 1962 (adult, subadult, carapace, plastron; as *Geoemyda*); Mittermeier, 1971 (adult); D. Moll, 2010 (adult); Pritchard, 1979 (hatchling); Savage, 2002 (adult, plastron); Schmidt, 1946 (anterior carapace and head); Vetter, 2005 (adult); Wermuth and Mertens, 1961 (adult, carapace, plastron; as *Geoemyda*).

Remarks.—Ernst (1978: 116–117) reviewed the systematics of *Rhinoclemmys*

annulata. Ernst (1980a) provided an overview of the species' morphology and reviewed its literature. Although *R. annulata* has been reported at least 14 times in the literature (beginning in 1932) as occurring in Honduras, Legler and Vogt (2013: 338) still stated the geographical distribution of this species extended "from Ecuador to Nicaragua." Wettstein (1934: 18) reported this species from "Honduras ?," perhaps questioning the locality data given by Dunn and Emlen (1932; for AMNH 136563). That AMNH record is based on a group of bones carrying a locality from north-central Honduras, so that identification remains uncertain.

The photographs of *Rhinoclemmys annulata* from Colombia in Páez et al. (2012) do not resemble Honduran specimens in color and pattern.

Natural History Comments.—*Rhinoclemmys annulata* is known from 20 to 540 m elevation in the Lowland Moist Forest formation. This diurnal turtle was usually found active on the forest floor in rainy weather in May, July, August, October, and December. Several others were active on the forest floor during periods of extremely dry weather in February and May, and another was basking in the sun on a small log completely surrounded by water about 1 m deep in a temporary pond in July. Another was apparently resting or sleeping at night in November, half buried in a muddy backwater area of a small stream. Ticks were present on the carapace of each adult I collected (also see Ernst and Ernst, 1977, and references cited therein). This turtle is rather timid and rarely attempts to bite, even when provoked. Females have been reported to deposit one or two relatively large eggs (most often one) per clutch, with egg deposition possibly occurring throughout the year (Ernst and Barbour, 1989). Females deposit their egg(s) on the forest floor and characteristically only scrape a little litter over them before abandoning the nest (D. Moll, 2010).

Mittermeier (1971: 487) studied the diet of a Panamanian population and found the species to be "predominantly or wholly herbivorous" and that it "feed[s] on shrubs, ferns, *Selaginella*, and seedlings of various kinds." Moll, D., and Jansen (1995: 124) reported *R. annulata* in Costa Rica to forage in wet forest habitats and treefall areas and reported the most important foods by volume were "various pteridophytes … , vine seedlings, fruits and seeds, tree seedlings, and leaves from forest trees." Mittermeier (1971: 487) also said "Captive turtles refused to eat their natural foods but ate papaya, apples, bread, cantaloupe, and particularly bananas. Adults refused meat of all kinds, but a small juvenile occasionally ate scraps of cooked chicken."

Etymology.—The name *annulata* is derived from the Latin *annulatus* (ringed, circular) and probably refers to the growth rings on the carapace.

Specimens Examined (17, + 1 group of bones [1]); Map 80).—**COLÓN**: Los Andes, ANSP 26657; Quebrada Machín, USNM 559584; about 4 km SE of Trujillo, AMNH 136563 (bones). **GRACIAS A DIOS**: Auka Kiamp, USNM 570465; Bachi Kiamp, FMNH 282719, USNM 573342, 579651, FN 257021 (still in Honduras because of permit problems); Bodega de Río Tapalwás, USNM 559586; Cabeceras de Río Rus Rus, USNM 573074; near confluence of Crique Ibantara with Río Rus Rus, USNM 570539; near Crique Wahatingni, USNM 570467; Rus Rus, USNM 559585; Sadyk Kiamp, FMNH 282716; near San San Hil Kiamp, USNM 570466; Warunta Tingni Kiamp, USNM 562001, 565394. **OLANCHO**: Quebrada El Guásimo, SMF 80835.

Rhinoclemmys areolata (A. M. C. Duméril and Bibron, 1851)

Emys areolata A. M. C. Duméril and Bibron, 1851: 10, *In* A. M. C. Duméril and Duméril, 1851 (holotype, MNHN 9424 [see Ernst, 1978: 117; Bour, 2007: 30];

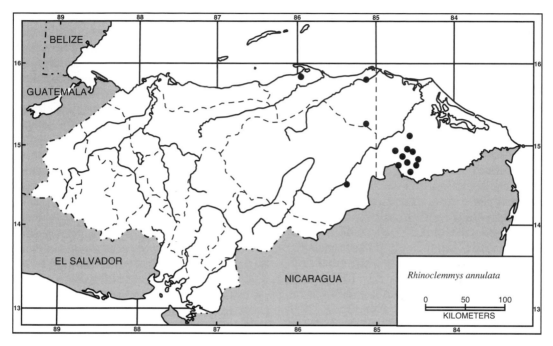

Map 80.   Localities for *Rhinoclemmys annulata*. Solid circles denote specimens examined.

type locality: "Province du Peten (Amér. centr.)" [Guatemala]).

Rhinoclemys [sic] *areolata*: McDowell, 1964: 267 (first use of present combination, but with generic name misspelled).

Rhinoclemmys areolata: Smith and Taylor, 1966: 12 (first use of present combination correctly spelled); Iverson, 1986: 64; Pérez-Higareda and Smith, 1987: 118; Iverson, 1992: 156; Wilson and McCranie, 1998: 15; Wilson et al., 2001: 134 (in part); McCranie et al., 2006: 103 (in part); Vogt et al., 2009: 022.3; Townsend and Wilson, 2010b: 691; Solís et al., 2014: 139; McCranie, 2015a: 381.

Geographic Distribution.—*Rhinoclemmys areolata* occurs at low elevations from southern Veracruz, Mexico, possibly to about the San Pedro Sula region of northwestern Honduras on the Atlantic versant. The species also occurs on Isla Cozumel,

Mexico, and the Turneffe Atoll, Belize. It is known in Honduras based only on animal trade specimens purported to be from the region of San Pedro Sula, Cortés.

Description.—The following is based on one shell (UF 67524), except that of the soft parts, which is based on UF 10984, an adult female from Belize. *Rhinoclemmys areolata* is a moderately large (maximum recorded CL 240 mm [Charruau et al., 2014]; 88 mm CL in only possibly Honduran specimen [from an animal dealer]) turtle with a broad, oval, and somewhat flattened carapace; carapace with 1 nuchal, 5 vertebrals, 8 costals, and 24 marginal scutes; carapace widest at level of marginal 6, highest at seam between vertebrals 2–3; carapacial sides straight or slightly bowed; posterior marginals slightly serrated and flared; single, flattened vertebral ridge present; each vertebral wider than long; carapace texture rough due to growth rings; nuchal narrow, bifurcated posteriorly; carapace notched posteriorly; plastron unhinged, with 12 scutes, consisting of paired gulars, humerals, pectorals, abdominals, femorals, and anals; gulars with 2 slight anterior projections; anals with wide medial notch; plastron slightly concave, anteriorly upturned; small axillary and inguinal scute present; head relatively small, snout projecting; upper jaw weakly serrated laterally, notched medially; lower jaw more serrated than upper jaw; triturating surface of upper jaw without a ridge; hind foot slightly webbed; 5 claws on forelimb, 4 claws on hind limb; foreleg covered with about 7 rows of large scales; cloacal opening anterior to carapacial margin; CL 93 mm; MSH/CL 0.42; CW/CL 0.89; second vertebral W/second vertebral L 1.73; PL/CL 0.95; humeral suture L/gular suture L 1.06.

Color of UF 67524 (dry shell): carapace dark brown with significant medium brown mottling, especially around scute edges; vertebral ridge dark brown with some medium brown mottling; plastron with large centered dark brown pattern with medium brown mottling; lateral edges of plastron pale brown to yellow; bridge medium brown with paler brown mottling.

Diagnosis/Similar Species.—*Rhinoclemmys areolata* is distinguished from all remaining Honduran turtles, except the other *Rhinoclemmys*, by the combination of having strongly clawed toes, only slight adult toe webbing, a short tail without a dorsal medial row of large triangular scales, and a large and oblong plastron with 12 scutes. *Rhinoclemmys annulata* has the tip of the upper jaw hooked and has significant dark brown to black pigment on the bridge (versus upper jaw notched and yellow to pale brown bridge without significant dark pigment in *R. areolata*). *Rhinoclemmys funerea* has strongly webbed toes in adults, has a dark brown to black plastron with a yellow medial seam, and has a dark brown bridge (versus no or only weak adult toe webbing, yellow to pale brown plastron with central dark area, and bridge yellow to pale

brown without significant dark pigment in *R. areolata*). *Rhinoclemmys pulcherrima* has a head pattern of 2 or 3 red stripes, with at least 1 crossing the tip of the snout in all Honduran specimens, and the bridge with extensive dark pigment (versus red lines, if present, short and not crossing snout and bridge yellow to pale brown without significant dark pigment in *R. areolata*).

Illustrations (Figs. 194, 195; Plate 99).— Álvarez del Toro, 1983 (adult); Álvarez Solórzano and González Escamilla, 1987 (adult); Bonin et al., 2006 (adult); Bour, 2007 (adult, carapace, plastron; as *Emys*); Calderón-Mandujano et al., 2008 (adult); Campbell, 1998 (adult, hatchling); A. H. A. Duméril, 1852a (adult, carapace, plastron; as *Emys*); Ernst and Barbour, 1989 (adult); Günther, 1885, *In* Günther, 1885–1902 (head; as *Emys*); Köhler, 2003a (adult; turtle in his fig. 36 identified as *R. areolata* actually a juvenile *R. pulcherrima*), 2008 (adult); Lee, 1996 (adult), 2000 (adult); Legler and Vogt, 2013 (adult, carapace, plastron, head); McCranie et al., 2006 (adult); Pérez-Higareda and Smith, 1987 (adult); Pritchard, 1967 (adult; as *Geoemyda*), 1979 (adult); Smith and Smith, 1980 (adult, carapace, plastron); Stafford and Meyer, 1999 (adult); Vetter, 2005 (adult); Vogt et al., 2009 (adult, juvenile, hatchling); Wermuth and Mertens, 1961 (adult, carapace, plastron; as *Geoemyda*); Zug, 1966 (schematic drawing of penis).

Remarks.—Vogt et al. (2009) provided a thorough review of *Rhinoclemmys areolata* (Ernst, 1980b, had earlier provided a brief literature review of this species). Several authors have plotted a record of this species from San Pedro Sula, Honduras, on their maps or have stated that it is known from that locality (i.e., Iverson, 1986: 64, 1992: 156; Pérez-Higareda and Smith, 1987: 118) despite those records being based only on animals sent to the U.S. by an animal dealer located in San Pedro Sula. I tentatively include this species herein in the Honduran herpetofauna based on those pet trade

Plate 99.   *Rhinoclemmys areolata*. Guatemala: Izabal, Motagua Valley. Photograph by Jonathan A. Campbell.

animals. I once visited the compound of the San Pedro Sula animal dealer and saw about a hundred *R. pulcherrima* in a pit but was unable to find a single *R. areolata*. Also, I showed photographs of *R. areolata* to several people living in northwestern Honduras near the Guatemalan border. None of those people appeared to know this turtle as occurring in that area, thus providing some second-hand, negative information that *R. areolata* does not occur there. Also, my recent fieldwork in that area of Honduras did not discover a population of *R. areolata*. Legler and Vogt (2013: 339) also stated this species occurs in "extreme northwestern Honduras (near Tela)." There are no specimens of any species of *Rhinoclemmys* known from the Tela region, nor is Tela located in extreme northwestern Honduras. See Remarks for *R. funerea* for a discussion of a misidentified specimen of *R. funerea* (misidentified as *R. areolata*) from northeastern Honduras.

Natural History Comments.—*Rhinoclemmys areolata* is purported to occur at about 100 m elevation in the Lowland Dry Forest formation. Nothing is known about

the natural history of *R. areolata* in Honduras other than two eggs were deposited by a pet trade animal purported to be from the San Pedro Sula region (month not recorded). Vogt et al. (2009: 022.3) stated that *R. areolata* "inhabits savanna, thorn scrub woodland, broadleaf forest, fallow agricultural land, and marshes throughout its range. In Belize, *R. areolata* is especially abundant in lowland pine communities that are characterized by a mosaic of pine forest and savanna vegetation. This habitat is associated with nutrient poor, acidic soils and dominated by *Pinus caribaea*" (also see Legler and Vogt, 2013). Female clutch size usually consists of one egg, but two eggs are occasionally produced, with up to four clutches per season (Vogt et al., 2009, and references cited therein; also see Legler and Vogt, 2013). Egg deposition occurs in May to October (Vogt et al., 2009). The species "feeds primarily on herbaceous plants, but also consumes fruits, insects, and occasionally carrion" (Vogt et al., 2009: 022.1). Other known foods include eggshells, crayfish, and mammal scat (Vogt et al., 2009,

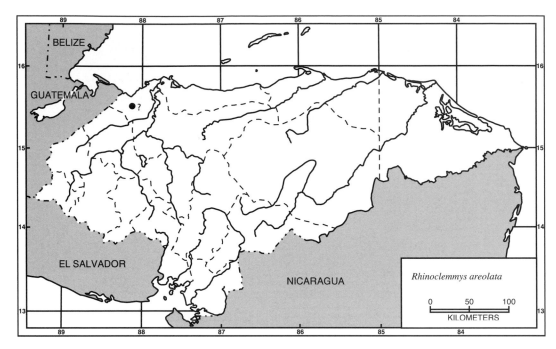

Map 81. Locality for *Rhinoclemmys areolata*. The solid circle denotes a questionable locality record for a shell examined that originated from an animal dealer.

and references cited therein; also see Legler and Vogt, 2013).

Etymology.—The name *areolata* is derived from the Latin *areolatus* (with small spaces) and apparently alludes to the pattern of the growth rings on the carapace. Legler and Vogt (2013: 338) stated the name alluded "to a conspicuous pale spot on each lateral scute in juveniles," but A. M. C. Duméril and Bibron, *In* A. M. C. Duméril and Duméril (1851), mentioned having only one specimen available of their *Emys areolata*, a "dry adult female" (Ernst, 1978: 117); thus, the authors of the name *R. areolata* could not have based their choice of the name on any aspect of juvenile coloration.

Specimens Examined (1 shell, 2 egg shells [0]; Map 81).—**CORTÉS**: purported to be from San Pedro Sula (see Remarks), UF unnumbered (two open eggs deposited by captive female). "HONDURAS": UF 67524 (shell only).

Rhinoclemmys funerea (Cope, 1875)

Chelopus funereus Cope, 1875: 154 (four syntypes, USNM 45900–01, 46134–35 [see Reynolds et al., 2007: 21]; type locality: "Limon" [Costa Rica]).

Rhinoclemys [sic] *funerea*: McDowell, 1964: 267 (first use of present combination, but with generic name misspelled).

Geomyda funerea: Campbell and Howell, 1965: 132.

Rhinoclemmys funerea: Meyer, 1969: 192 (first use of present combination correctly spelled, but in an unpublished dissertation); Meyer and Wilson, 1973: 6 (first published use of present combination correctly spelled); Ernst, 1981a: 263.1; Iverson, 1986: 66; Iverson, 1992: 158; Wilson et al., 2001: 134 (in part); Wilson and McCranie, 2003: 59; Wilson et al., 2003: 17; Wilson and Townsend, 2006: 104 (in part); McCranie 2007a: 181; Vargas-Ramírez et al., 2013: 242;

Solís et al., 2014: 139; McCranie, 2015a: 381.
Callopsis areolata: Ernst, 1978: 118 (in part).
Callopsis funerea: Ernst, 1978: 119 (in part).
Rhinoclemmys areolata: Ernst, 1980b: 251.1 (in part); Wilson et al., 2001: 134 (in part); McCranie et al., 2006: 103 (in part); Wilson and Townsend, 2006: 104 (in part).

Geographic Distribution.—*Rhinoclemmys funerea* occurs at low elevations from northeastern Honduras to central Panama on the Atlantic versant. The species is known in Honduras from the Río Plátano and its tributaries eastward to the Río Coco, which forms the border with Nicaragua in northeastern Honduras.

Description.—The following is based on five males (UF 137638; USNM 559588, 562918, 570540, 573359) and nine females (UF 137162; USNM 24539, 559587, 570468, 570471, 579652–54, 580393). *Rhinoclemmys funerea* is a large (maximum recorded CL 325 mm [Ernst, 1978]; 304 mm CL in largest Honduran specimen [USNM 570540, a male]) turtle with a broad, oval, and somewhat dome-shaped carapace; carapace with 1 nuchal, 5 (occasionally 6) vertebral, 8 costals, and 24 marginal scutes; carapace widest at level of marginal 6, highest at seam between vertebrals 2–3 or at about midlength of vertebral 2; carapacial sides straight; posterior marginals slightly serrated, flared; single, flattened vertebral ridge present, somewhat worn in largest adults, distinct in juveniles and subadults; each vertebral wider than long; carapace texture somewhat roughened due to growth rings or mostly smooth in adults, strongly rugose in juveniles; nuchal narrow, bifurcated posteriorly; carapace notched posteriorly; plastron unhinged, with 12 scutes, consisting of paired gulars, humerals, pectorals, abdominals, femorals, and anals; gulars with or without anterior projection in adults, 3 anterior projections present on each gular in juveniles; anals with wide medial notch; plastron concave in males, nearly flat in females, upturned anteriorly in both sexes; a small axillary and inguinal scute present; head moderately sized, snout slightly projecting; upper jaw serrated laterally, notched medially; triturating surface of upper jaw without a ridge; hind foot strongly webbed in adults, weakly webbed in juveniles and small subadults; 5 claws on forelimb, 4 claws on hind limb; foreleg covered with about 9 rows of large scales; cloacal opening at or posterior to carapacial margin in males, anterior to carapacial margin in females; CL 273–304 (290.8 ± 12.2) mm in males, 68–241 (126.1 ± 57.5) mm in females; MSH/CL 0.31–0.42 in males, 0.32–0.42 in females; CW/CL 0.56–0.67 in males, 0.67–0.86 in females; second vertebral W/second vertebral L 1.09–1.13 in four males, 1.14–1.43 in females; PL/CL 0.87–0.96 in males, 0.91–0.98 in females; humeral suture L/gular suture L 0.50–0.80 in four males, 0.32–0.68 in females.

Color in life of an adult male (USNM 559588): carapace Blackish Neutral Gray (82); plastron Blackish Neutral Gray with pale yellow line along midline; bridge Blackish Neutral Gray; top of head Sepia (119), postocular region Drab-Gray (119D), remainder of side of head Yellow Ocher (123C), both dorsal and lateral surfaces of head with Sepia lines; chin Yellow Ocher with Sepia spotting; forelimb Trogon Yellow (153) with scattered black spots and short lines on inside surface and Spectrum Orange (17) on outside surface; hind limb largely Sepia on outside surface, with pale orange edging along posterior edge, inside surface same color as that of inside of front limb; tail brown above, pale orange with Sepia spotting below.

Color in alcohol: carapace uniformly dark brown to black; plastron black with yellow midseam dividing paired scutes, outer edges of some scutes also yellow; bridge uniformly dark brown, with or without yellow bar on

Plate 100.   *Rhinoclemmys funerea.* USNM 579654, adult female, carapace length = 135 mm. Gracias a Dios: between Sadyk Kiamp and Bachi Kiamp.

lateral edge of each marginal; head brown with elongated yellow bar between eye and tympanum; yellow line on upper jaw extending from level below eye to tympanum; tympanum and temporal region mottled yellow and black; neck brown with thin black stripe laterally; lower jaw and chin yellow with large, small, or both sizes of black irregular spots; forelimb yellow with black spots on inside surface, dark brown on outside surface; hind limb dark brown on outside surface, yellow with dark brown spots on inside surface; tail yellow with dark brown stripes.

Diagnosis/Similar Species.—*Rhinoclemmys funerea* is distinguished from all remaining Honduran turtles, except *Trachemys* and other species of *Rhinoclemmys*, by the combination of having strongly clawed toes, a short tail without a dorsal medial row of large triangular scales, and having a large and oblong plastron with 12 scutes. The three species of *Trachemys* have a ridge on the triturating surface of the upper jaw and have symmetrically lined, mottled, or with ocelli or spotting on the plastron (versus no triturating ridge and

plastron black with yellow midseam in *R. funerea*). Adult *R. funerea* is the only Honduran *Rhinoclemmys* with strongly webbed toes and also has a mostly black plastron with a yellow midseam. Also, *R. annulata* has the upper jaw hooked (versus upper jaw notched in *R. funerea*). Also, *R. areolata* usually has a yellow bridge (versus bridge largely black in *R. funerea*). Also, *R. pulcherrima* has a head pattern of 2 or 3 red stripes, with at least 1 stripe crossing the tip of the snout in Honduran specimens (versus no red lines crossing snout in *R. funerea*).

Illustrations (Figs. 146, 190–192; Plate 100).—Bonin et al., 2006 (juvenile, head); Gutsche, 2007 (adult); Guyer and Donnelly, 2005 (adult, subadult, plastron); Köhler, 1999b (adult), 2000 (adult), 2001b (juvenile), 2003a (juvenile), 2008 (juvenile); McCranie et al., 2006 (adult, plastron); Merchán [Fornelino] and Mora, 2000 (adult, plastron); D. Moll, 2010 (adult); D. Moll and Moll, 2004 (adult); Pritchard, 1967 (adult; as *Geoemyda*), 1979 (adult); Savage, 2002 (adult, plastron; see Remarks); Sletto, 1999 (adult, plastron); Vetter, 2005 (adult); Wermuth and Mertens, 1961 (adult, carapace, plastron; as *Geoemyda*); Wilson et al., 2003 (adult).

Remarks.—*Rhinoclemmys funerea* was included in the Honduran herpetofauna by Meyer and Wilson (1973: 6) based on a single specimen collected in the Río Coco (LACM 73824). The Río Coco forms the border between much of eastern Honduras and northern Nicaragua. Although the Río Coco specimen was supposedly collected at Krasa, Nicaragua, Campbell and Howell (1965: 131) stated, "records from localities along the Río Coco may apply to Nicaragua or Honduras." Subsequently, specimens have been collected in Honduras in tributaries of the ríos Coco and Warunta, and Sletto (1999: 27–28) documented its presence in the Río Plátano with a photograph. People living along the Río Kruta indicated to me that this turtle also occurs in tributaries of that river.

Ernst (1978: 119–120) reviewed the systematics of *Rhinoclemmys funerea*. Ernst (1981a) provided an overview of the species' morphology and reviewed the literature. Honduran specimens have extensive black pigmentation on the bridges, unlike the schematic drawing showing color pattern in Costa Rican specimens of *R. funerea* in Savage (2002: 765). Savage (2002: 766) also stated that *R. funerea* has a "free fleshy flap along the outer margin of the foot." However, a flap is present along the outer margin of the foot in only a few of the Honduran specimens examined (i.e., USNM 559587).

Ernst (1978: 118, 1980b: 251.1) questioned the record of *Rhinoclemmys areolata* from northeastern Honduras. McCranie et al. (2006: 103) accepted that record without examining the actual specimen. Examination of the specimen in question (USNM 24539) in August 2012 for this work revealed it to be a subadult *R. funerea*.

Natural History Comments.—*Rhinoclemmys funerea* is known from about sea level to 190 m elevation in the Lowland Moist Forest formation. One was underwa-

ter beneath a large log during early afternoon in August in the Caño Awalwás near its confluence with the Río Coco. Two were found by snorkeling during the afternoon below logs on the bottom of the Río Rus Rus in January and May (dry season when water is relatively clear), and others were in the open in shallow rivers during the day and at night in January, February, May, and from October to December. One was collected during the afternoon from a rain-swollen river (about 2 m deep) in July after it came to the surface to breathe; that turtle quickly dove when it noted my presence but was captured about 0.5 m below the surface by quickly diving into the water. Another was collected at night in August as it was feeding on leaves of a shrub growing about 0.5 m from a small river that flows into the Caño Awalwás. When startled, this specimen bolted toward the river but was captured as it was about to enter the water. About 10 adults (only two collected) were seen in a fairly deep pool in an otherwise shallow Río Rus Rus along its headwaters in the day in the latter part of the dry season in May. An adult *Trachemys venusta* was also collected from the same pool. A juvenile was feeding at night on grass alongside a river at 10:50 a.m. in February, another juvenile was active on the ground at night in the forest about 20 m from a river in November, another juvenile was partially buried by day in mud in October in the forest near a river, and another juvenile was swimming in a river during the day in October. My observations on *R. funerea* in Honduras indicate that it is primarily nocturnal and occurs only in broadleaf forest but does make daytime movements in shallow rivers. It also surfaces in deeper rivers to breathe during the day. It is occasionally seen sunning on debris in small rivers and streams in open places in otherwise closed canopy broadleaf rainforest. This aquatic turtle is apparently capable of terrestrial movement along sizeable inclines, as USNM 559588 was collected about 1 km upstream from a sizeable waterfall about 10 m high in a small river with steep banks in most places. This turtle is timid and, in my experience, never attempts to bite, even when provoked in the face with a finger. Sletto (1999) documented that people living in the Río Plátano Biosphere Reserve eat this species and its eggs. People living in the vicinities of the ríos Coco, Kruta, Rus Rus, and Warunta also eat this turtle and its eggs and say it is the best tasting freshwater turtle in the area. Moll, E. O., and Legler (1971) reported that Panamanian females deposit one to six eggs per clutch with about one to four clutches per season from at least April (dry season) through July (rainy season). Eggs are deposited on the ground in closed canopy broadleaf forest, and females cover their eggs with leaves (D. Moll and Moll, 2004, and references cited therein) or only scrape a little leaf litter over them before abandoning the nest (Moll, 2010). Merchán [Fornelino] and Fournier (2007) reported one to seven eggs deposited between May and October by Costa Rican captive females. Moll, E. O., and Legler (1971: 89) reported that stomachs of individuals from Nicaragua and Panama "contained fruits, grasses, and parts of broad-leaved plants." Those authors also reported that captives would "eat a wide variety of meats, fruit, and vegetables." Acuña [Mesén] et al. (1983) said the diet of this species in Costa Rica includes various fruits (guavas, papayas, oranges) and grasses. Moll, D., and Jansen (1995: 123) reported *R. funerea* in Costa Rica regularly leaves the water at night to forage on riverbanks. Food items recorded at that site were "aquatic plants, and leaves and fruits from terrestrial riparian vegetation which fall into the water" and "inundated terrestrial vegetation during flood periods, and riparian vegetation and fruits" while foraging on land.

Etymology.—The name *funerea* is derived from the Latin *funereus* (pertaining to burial, funeral) and alludes to the black coloration on the head, neck, carapace, and plastron mentioned by Cope (1875) for the syntypes of this species.

Specimens Examined (16, + 5 heads, 1 head and shell, 1 shell, 1 carapace [1]; Map 82).—**GRACIAS A DIOS**: Cabeceras de Río Rus Rus, USNM 570540, 573359 (shell); Caño Awalwás, USNM 559587–88; Caño Sucio, USNM 562917 (carapace); Concho Kiamp, FN 257045 (still in Honduras because of permit problems); Kakamuklaya, USNM 573075; Kipla Tingni Kiamp, USNM 570469 (head); Rawa Kiamp, USNM 573976 (head); Río Coco, USNM 24539; Río Coco near Krasa, LACM 73824; Río Rus Rus near Rus Rus, UF 137162, 137638, USNM 579652; Río Tapalwás near Crique Wahatingni, USNM 570468; Río Warunta between Sadyk Kiamp and Bachi Kiamp, USNM 579654; Río Warunta near Hiltara Kiamp, USNM 562918 (head and shell); Sachin Tingni Kiamp, USNM 564158 (head), 570470–41; Sisinbila, USNM 579653; Warunta, USNM 580393; Warunta Tingni Kiamp, USNM 564159 (head).

Other Records (Map 82).—**GRACIAS A DIOS**: Río Plátano (Sletto, 1999).

Rhinoclemmys pulcherrima (Gray, 1856a)

Emys pulcherrimus Gray, 1856a: 25 (holotype, BMNH 1947.3.5.52 [see Ernst, 1978: 125]; type locality: "Mexico," restricted to "vicinity of San Marcos, Guerrero, Mexico" by Ernst, 1978: 125, in his revision of the genus).

Rhinoclemmys pulcherrima: Gray, 1873a: 145; Meyer, 1969: 192; Meyer and Wilson, 1973: 6; Iverson, 1986: 69; Wilson et al., 1991: 69; Iverson, 1992: 161; Wilson and McCranie, 1998: 15; Köhler, 1999a: 214; Köhler, 2000: 22; Köhler, McCranie, and Nicholson, 2000: 425; Nicholson et al., 2000: 29; Wilson et al., 2001: 134; Köhler, 2003a: 37; Lovich et al., 2006: 13; McCranie et al., 2006: 104; Wilson and Townsend, 2006: 104; Wilson and Townsend,

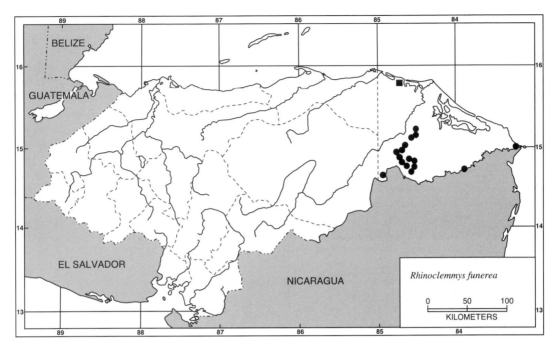

Map 82.   Localities for *Rhinoclemmys funerea*. Solid circles denote specimens examined and the solid square represents an accepted record.

2007: 145; Köhler, 2008: 39; Townsend and Wilson, 2010b: 692; McCranie and Valdés Orellana, 2011c: 566; McCranie, 2014: 292; Solís et al., 2014: 139; McCranie, 2015a: 381; McCranie and Gutsche, 2016: 892.

Geoemyda pulcherrima subsp.: Meyer, 1966: 174.

Callopsis pulcherrima: Ernst and Ernst, 1977: 140; Wilson et al., 1979a: 25.

Callopsis pulcherrima incisa: Ernst, 1978: 129.

Rhinoclemmys pulcherrima incisa: Ernst, 1981b: 275.1; Sites et al., 1981: 263.

Geographic Distribution.—*Rhinoclemmys pulcherrima* occurs at low and moderate elevations from southern Sonora, Mexico, to central Costa Rica on the Pacific versant (but see Remarks). It also occurs on the Atlantic versant in disjunct populations from eastern Guatemala to western Nicaragua. In Honduras, this species occurs widely on the Pacific versant and in several isolated populations in naturally open areas on the Atlantic versant.

Description.—The following is based on ten males (FMNH 283551; LSUMZ 36599; MCZ R49743; TCWC 19233, 23650; USNM 102888, 102892–93, 579655, 580361) and 12 females (AMNH 70569; CM 62172; LSUMZ 24596–98; USNM 102889–91, 559589, 561999–200, 580763). *Rhinoclemmys pulcherrima* is a moderately large (maximum recorded CL 235 mm [USNM 580763, a female]) turtle with a broad, oval, and somewhat flattened carapace; carapace with 1 nuchal, 5 vertebrals, 8 costals, and 24 marginal scutes; carapace widest at level of marginals 6–7, highest at seam between vertebrals 2–3; carapacial sides straight; posterior marginals slightly serrated, flared; single, flattened vertebral ridge present; each vertebral wider than long or about equal in width and length; carapace texture rough due to growth rings; nuchal narrow, bifurcated posteriorly; carapace notched posteriorly; plastron un-hinged, with 12 scutes, consisting of paired gulars, humerals, pectorals, abdominals, femorals, and anals; gulars without anterior projection in adults; anals with wide medial notch; plastron slightly concave in adult males, flat in females, and slightly upturned anteriorly in females; small axillary and inguinal scutes present; head relatively small, snout slightly projecting; upper jaw serrated laterally, notched; triturating sur-face of upper jaw without a ridge; feet not webbed, or with basal webbing at most; 5 claws on forelimb, 4 claws on hind limb; foreleg covered with about 10–11 rows of large scales; cloacal opening at, or posterior to carapacial margin in males, anterior to carapacial margin in females; CL 130–164 (144.9 ± 11.2) mm in males, 125–235 (175.3 ± 31.2) mm in females; CH/CL 0.35–0.43 in males, 0.35–0.55 in females; CW/CL 0.71–0.86 in males, 0.64–0.98 in females; second vertebral W/second vertebral L 1.06–1.57 in males, 0.97–1.37 in females; PL/CL 0.84–0.96 in males, 0.86–1.03 in females; humeral suture L/gular suture L 0.37–1.14 in males, 0.38–1.03 in females.

Color in life of an adult female (USNM 561999): costal scutes of carapace Buff (24) with Sepia (119)-edged C-shaped Spectrum Orange (17) figure; vertebral scutes similar in color to costal scutes, but with less defined central figure; marginal scutes also similarly colored to costal scutes, but orange on upper edges; bridge scutes with similar vivid coloration; plastron Buff with elongate Brownish Olive (29) central figure; top of head Olive (30) with Sepia-edged Flame Scarlet (15) line beginning on temporal region and continuing inside of upper eyelid and across tip of snout; similarly colored line begins on temporal area and continuing along outer edge of upper eyelid across nostrils; another line of similar color begins at angle of jaw, continuing along upper edge of rhamphotheca; lower jaw Spectrum Orange with Sepia blotching; dorsal surfac-es of fore- and hind limb Sepia with

Spectrum Orange stripes; iris gray-white horizontally, with Spectrum Orange upper edge.

Color in alcohol: carapace varies from pale brown with dark brown "circular" lines and "straight" dashes to uniformly brown (except for short, laterally directed pale brown stripe on lateral edge of some costals); plastron yellow with brown mottled central area; bridge brown with yellow mottling or 1 yellow bar on each marginal; head olive brown with pale pink middorsal stripe extending from between orbits nearly to level above nostrils, not meeting pale pink stripe extending from each eye and crossing snout, latter stripe extends across eye onto nape; pale pink stripe extending from each nostril to each eye; pale pink stripe extending posteriorly from level below nostril along upper jaw to tympanum area; 2 pale pink stripes extending posteri-orly from eye to tympanum area; lower jaw and chin cream with dark brown lines, spots, and ocelli on chin; forelimb yellow with rows of dark brown spots; hind limb brown on outside, yellow on inside surface with dark brown spots; tail yellow with dark brown dorsal stripes.

Diagnosis/Similar Species.—*Rhino-clemmys pulcherrima* is distinguished from all remaining Honduran turtles, except the other *Rhinoclemmys*, by a combination of having strongly clawed toes, no or only slight toe webbing in adults, a short tail without a dorsal median row of large triangular scales, and a large, oblong plas-tron with 12 scutes. *Rhinoclemmys pulcher-rima* is the only Honduran *Rhinoclemmys* with several long red lines on the head with at least 1 crossing the snout.

Illustrations (Figs. 181, 182, 196, 197; Plate 101).—Álvarez del Toro, 1983 (adult); Bonin et al., 2006 (adult, plastron); Bour, 2007 (adult, carapace, plastron; as *Emys incisa* Bocourt, 1868: 121); A. H. A. Duméril et al., 1870–1909b (adult, head, carapace, plastron; as *Emys incisa*); Ernst, 1978 (carapace, plastron; as *Callopsis p.*

Plate 101. *Rhinoclemmys pulcherrima*. FMNH 283550. Ocotepeque: between Antigua and Nuevo Ocotepeque.

incisa); Ernst and Barbour, 1989 (adult); Gray, 1856a (young; as *Emys*); Günther, 1885, *In* Günther, 1885–1902 (adult, carapace, plastron; as *Emys*); Köhler, 2000 (adult), 2003a (adult; Honduran specimen only), 2008 (adult; Honduran specimen only); Köhler et al., 2005 (adult); Legler and Vogt, 2013 (head; as *R. p. incisa*); McCranie et al., 2006 (adult, plastron); Pritchard, 1967 (adult; as *Geoemyda pulcherrima* only), 1979 (adult; as *R. pulcherrima* only); Smith and Smith, 1980 (adult, juvenile); Vetter, 2005 (adult, juvenile, carapace, plastron; as *R. p. incisa, R. p. pulcherrima*); Wermuth and Mertens, 1961 (adult, carapace, plastron; as *Geoemyda p. incisa*).

Remarks.—Ernst (1978) reviewed the systematics of *Rhinoclemmys pulcherrima* and (1981b) provided an overview of the species' morphology and reviewed its literature. Four subspecies are recognized by some workers (Ernst, 1978: 125–130, 1981b: 275.1; Iverson, 1992: 161; David, 1994: 63), with *R. p. incisa* purported to be the one occurring in Honduras. Sites et al. (1981: 263) used a Honduran specimen of *R. pulcherrima* (TCWC 56993) in their study

of the biochemical systematics of the genus. The molecular analysis of Le and McCord (2008: 761) suggested the current concept of *R. pulcherrima* might represent more than one evolutionary species, with candidates for species status being the southernmost populations assigned to *R. p. manni* (Dunn, 1930: 33) and *R. p. rogerbarbouri* (Ernst, 1978: 127, the northernmost populations); however, Le and McCord (2008) did not study the nominal form *R. p. pulcherrima*. The possibility of more than one evolutionary species being involved in this complex might explain the wide variation in the red stripes on the head in some extralimital populations. However, I have never seen a Honduran specimen of *R. pulcherrima* without a complete red stripe crossing the snout.

Natural History Comments.—*Rhinoclemmys pulcherrima* is known from near sea level to 1,480 m elevation in the Lowland Dry Forest, Lowland Arid Forest, and Premontane Dry Forest formations and peripherally in the Lowland Moist Forest, Premontane Wet Forest, and Premontane Moist Forest formations. This predominantly diurnal species was collected in swamps, rain-filled depressions, shallow man-made ditches, and terrestrial situations, including crossing roads, from March to August and in November. One turtle that was not collected was active at night in a small, muddy, temporary pond filled with cow droppings in June in southern Ocotepeque. Two others from the same locality that also were not collected were feeding on human feces during midafternoon. Two others were partially buried in mud under a log in May and another was active in a tomato field in October. One (USNM 559589) collected in a stream in 1999 in primary rainforest along the Río Patuca was likely washed downstream from the subhumid Guayape-Guayambre Valley by floodwaters resulting from Hurricane Mitch in October 1998. That turtle escaped from a bag about 100 m from where it was collected, but 2 days later it was in the same deep pool in the stream where it was first found. Individuals of this turtle are timid, and I have never seen one attempt to bite, even when provoked with a finger to the face. Legler and Vogt (2013: 347–348) reported one to three eggs, mean 1.5, in 102 clutches from a captive population in Oaxaca, Mexico. Monge-Nájera et al. (1988) reported two Costa Rican captives deposited one to three eggs per clutch in shallow holes in the ground dug by her (but see my Remarks concerning *R. p. manni*). Ernst (1983: 419) wrote that this species "is probably an omnivore, but with stronger preferences toward plant foods. The wild foods have not been recorded, but captives readily eat a variety of domestic fruits and vegetables, earthworms, fish, beef strips, and canned dog food. When given a choice, they usually chose plant food over meats" (also see Legler and Vogt, 2013). Hainz (2008) reported that captives from Nicaragua and Costa Rica (but see Remarks concerning *R. p. manni*) ate dandelion leaves, Chinese cabbage, lettuce, tubifex, bloodworms, aquatic snails, and chopped liver with powdered eggshells.

Etymology.—The name *pulcherrima* is derived from the Latin word *pulcherrimus* (prettiest) and refers to the colorful head and shell markings of the juvenile holotype of this species (given the non-Latinized name "The Dotted Emys" by Gray, 1856a).

Specimens Examined (61, + 4 shell fragments, 3 skeletons [11]; Map 83).— **CHOLUTECA**: 1.6 km N of Cedeño, KU 209310; 1.6 km N of Choluteca, LACM 140053; El Faro, USNM 580763; near San Marcos de Colón, UNAH 5276; 0.3 km N of Nicaragua border on CA Hwy 2, UF 65634 (shell fragments). **COMAYAGUA**: 4.8 km W of Comayagua, TCWC 23826. **COPAN**: Copán, UNAH (SMR) 425. **CORTÉS**: 2.0 km N of Agua Azul, LSUMZ 28508; near Cofradía, USNM 102888–93; Cofradía, CM 57186, 62172. **EL PARAÍSO**: El Rodeo, UNAH (1); between El Rodeo and Orealí, UNAH (1); Mapachín, USNM 578920. **FRANCISCO MORAZÁN**: El Zamorano, AMNH 70569; Lepaterique, CAS 152995; 16.1 km NE of Talanga, TCWC 19233; Talanga, UF 33122; 27 km NW of Tegucigalpa, AMNH 150101, LSUMZ 24595; near Tegucigalpa, FMNH 72586; Villa San Francisco, USNM 580764 (shell fragments). **OCOTEPEQUE**: Río Lempa at Antigua, FMNH 283551; between Nuevo Ocotepeque and Antigua, FMNH 283550. **OLANCHO**: 2 km E of Campamento, TCWC 23650; 6.5 km SE of Catacamas, LACM 48352–53; Quebrada El Guásimo, USNM 559589. **SANTA BÁRBARA**: 15 km SW of Cofradía, LSUMZ 24596; near Quimistán, USNM 128098. **VALLE**: near Amapala, USNM 580761; 6 km E of El Amatillo, TCWC 22312; Isla Comandante, USNM 580358; Isla Exposición, Playona Exposición, USNM 579655; Isla Exposición, W side, USNM 580359; Isla Garrobo, USNM 580360; Isla Inglasera, USNM 580762; Isla Zacate Grande, LSUMZ 36599; 1.6 km S of Jícaro Galán, LACM 48350; near Jícaro Galán, TCWC 56993; 2 km E of Nacaome, LSUMZ 24597; Punta El Molino, USNM 580362; Punta Novillo,

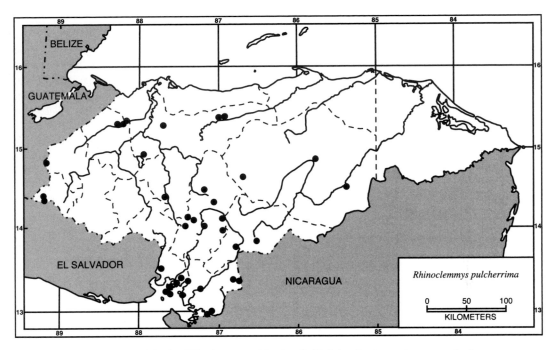

Map 83.   Localities for *Rhinoclemmys pulcherrima*. Solid circles denote specimens examined.

USNM 580361; 4.8 km E of San Lorenzo, LSUMZ 24598; 11.1 km SSW of San Lorenzo, TCWC 61762; "Pacific Coast," MCZ R49742–48. **YORO**: Morazán, MCZ R38792; near Río Aguán S of San Lorenzo Arriba, USNM 562000; near San Lorenzo Abajo, USNM 561999. "HONDURAS": BYU 42864, UF 54909 (bones and shell), 54979 (skeleton), 55442 (skeleton; shell fragments), 55574 (shell), 65634 (shell fragments).

Other Records.—"HONDURAS" Peter Pritchard private collection 145, 390 (Ernst, 1978).

SPECIES OF PROBABLE OCCURRENCE IN HONDURAS

Future collecting in Honduras should reveal the presence of several species of reptiles not presently reported from the country. The following list is conservative;

each of these species almost certainly occurs or, at least in one case, used to occur within the country. Additionally, three of the seven taxa in this list are based on confident personal sight records.

Lepidophyma smithii Bocourt, 1876: 401

Köhler et al. (2005) recorded this species from Cerro Montecristo, El Salvador. That mountain range extends across the Honduran border into southwestern Ocotepeque, Honduras. Köhler et al. (2005) gave an elevational range of 200 to 1,240 m for this species on Cerro Montecristo in El Salvador. Thus, appropriate elevations should be searched for this species in extreme southwestern Ocotepeque.

Gehyra mutilata (Wiegmann, 1834a: 238)

A single preserved adult specimen of this species (LACM 47300) from La Lima, Cortés, Honduras (15°26′N, 87°55′W; 40

m elevation) was collected sometime between 1964 and 1968 by the late H. E. Ostmark, a former resident of La Lima, who is also known to have purposely released live specimens of this species in the La Lima area. One specimen of that lizard was given to J. R. Meyer. This species complex was recently divided into two species based largely on molecular data (see Zug, 2013: 84): *G. mutilata* and *G. insularis* (Girard, 1858a: 195). I am not certain which one of those two species the Honduran specimen represents. The species complex occurs in India, Sri Lanka through Indochina to Japan, the Philippines, Indonesia, Mascarine, Madagascar, New Guinea, Melanesia, Micronesia, and Polynesia (Bauer, 1994: 92). The species complex has also been introduced into numerous localities, including Mexico and the Hawaiian Islands, USA (Hawaiian Island population likely *G. insularis*). McKeown (1996: 70) reported that before the arrival of *Hemidactylus frenatus* on the Hawaiian Islands, *G. "mutilata"* lived equally well in edificarian and uninhabited areas but has now been replaced in the former habitat by *H. frenatus*. *Hemidactylus frenatus* is now common in edificarian situations throughout much of Honduras, including the La Lima area. Given the situation reported for the Hawaiian Islands, it is possible that a *G. mutilata* complex species persists today in places away from buildings in the La Lima area. Smith, M. A. (1935) provided illustrations of important head and hind foot characters, and Zug (1991) presented drawings of hind foot characters of *G. mutilata–G. insularis* that will help distinguish it from the otherwise similar species of *Hemidactylus*. Figures 25 and 26 in this paper demonstrate a lack of enlarged tubercles on the dorsal and lateral surfaces, lack of a claw on Digit I of the hind limbs, and oblong toe pads in *G. mutilata*. Figures 23 and 24 in this paper demonstrate numerous tubercles on those surfaces in two of the three species of *Hemidactylus* occurring in Honduras. A clawed terminal lamella on each Digit I on hind limbs and elongated toe pads occur in Honduran species of *Hemidactylus*, all of which are also introduced. Those are some of the diagnostic characters to distinguish *G. mutilata–G. insularis* from the species of *Hemidactylus*.

Aristelliger georgeensis (Bocourt, 1873: 41, *In* A. H. A. Duméril et al., 1870–1909a)

The Honduran population (Cayos Vivorillos) formerly referred to this species is being described as a new species by workers in the S. Blair Hedges laboratory. However, *A. georgeensis* likely occurs on the Cayos Zapotillas northwest of Puerto Cortés, Cortés (various ones of those islands are claimed by Belize, Guatemala, and Honduras). The Honduran islands in that island chain have never been collected herpetologically.

Diploglossus bilobatus (O'Shaughnessy, 1874: 257)

Two animals I am certain were this species were seen at Bodega de Río Tapalwás (14°56'N, 84°32'W) north of Rus Rus, Gracias a Dios. The lizards were in primary broadleaf rain forest, but the area has now been cleared for some distance away from the river close to where those lizards were seen. The locality lies at 180 m elevation, and the lizards were uncovered by tearing apart small rotten, standing tree stumps. Upon being uncovered, both lizards immediately ran out of site. Strangely, the two animals were seen less than 1 hour apart on the midmorning of 28 May 2003. Despite spending about 1 year of real time working in rainforests in the Mosquitia of northeastern Honduras, including seven trips to the Bodega region, I never saw that lizard species again. Myers (1973) provided a detailed description, a drawing of the diagnostic claw sheath character, and a photograph of an adult that should prove helpful in verifying the identification of that lizard should one happen to turn up again.

Diploglossus rozellae (Smith, 1942: 372)

Campbell and Camarillo R. (1994) recorded this species from a Guatemalan locality in the Sierra Espíritu Santo near the Honduran border. *Diploglossus rozellae* should also occur on the Honduran side of the border in the Sierra Espíritu Santo in the departments of Copán and Santa Bárbara in northwestern Honduras. Campbell and Camarillo R. (1994) gave an elevational range of from near sea level to 1,350 m for this species. Compare Figs. 52 and 53 herein for illustrations of diagnostic characters to distinguish *D. rozellae* from other *Diploglossus* species in Honduras (see Figs. 49–51 herein to compare with other species of *Diploglossus*).

Diploglossus sp.

A large, smooth-scaled diploglossid resembling *Diploglossus monotropis* (Kuhl, 1820: 128) in size was seen at about 900 m elevation at about 2:00 p.m. on a sunny day on the forest floor of a steep slope of a hillside above Quebrada de Las Marías, Olancho (15°18′N, 85°21′W). *Diploglossus monotropis* is known to occur from southern Nicaragua to western Ecuador. The date of that observation was 3 August 1998. The lizard was crawling downslope in a serpentine fashion on top of thick leaf litter and keeping a safe distance between itself and me as I was pursing it. After about 30 seconds of pursuit, the lizard jumped onto a tree and quickly climbed the side opposite me. After reaching the tree, I jumped and blindly hit the lizard hard enough to dislodge it from the tree. Upon hitting the ground, the lizard quickly disappeared in the thick leaf litter. I sat down and waited for about 0.5 hour, but never saw the lizard again. The following day two field companions and I returned to the same tree and walked in the same vicinity for about 3 hours without seeing the lizard again. In May 2010, a field companion and I returned to the locality and spent much of a sunny afternoon walking the same slope without seeing such a lizard. The lizard I saw lacked red markings on the lateral surfaces that *D. monotropis* has (see photographs in Savage, 2002; A. A. Schmidt, 2011), instead appearing an overall brown color. It would be extremely rewarding if this unknown lizard species could be rediscovered and captured.

Leposoma Spix (1825: 24) sp.

A lizard strongly resembling the gymnophthalmid genus *Leposoma* was seen at 950 m elevation on 9 August 1994 at Quebrada Las Cantinas, Olancho (15°09′N, 86°43′W), in the northwestern portion of Parque Nacional La Muralla. The lizard was under a small log on top of leaf litter next to a small, but rain-swollen river. Immediately on being uncovered, the lizard jumped into the adjacent river and disappeared under the brown water. A search with my hands below debris on the river bottom failed to find the lizard. The animal was small (about 30–40 mm SVL), rather stout, brown, and had strongly keeled lateral scales. I can think of no other genus occurring in Central America that the lizard resembled more. Another gymnophthalmid genus, *Neusticurus* A. M. C. Duméril and Bibron (1839: 61), is known to be semiaquatic but is a much longer lizard (to about 100 mm SVL) than the short, stout lizard I saw.

DISTRIBUTION OF THE LIZARDS, CROCODILES, AND TURTLES IN HONDURAS

DISTRIBUTION WITHIN DEPARTMENTS

The distribution of the 126 species of lizards, crocodiles, and turtles of Honduras by departments is shown in Table 2 (Map 84 shows distributions of these departments). Five of those lizard species and one of those turtle species are human-aided introductions

TABLE 2. DISTRIBUTION BY DEPARTMENTS OF THE 126 SPECIES OF THE LIZARDS, CROCODILES, AND TURTLES KNOWN FROM HONDURAS. DEPARTMENT ABBREVIATIONS ARE: ATL = ATLÁNTIDA; CHO = CHOLUTECA; COL = COLÓN; COM = COMAYAGUA; COP = COPÁN; COR = CORTÉS; EP = EL PARAÍSO; FM = FRANCISCO MORAZÁN; GAD = GRACIAS A DIOS; INT = INTIBUCÁ; IDB = ISLAS DE LA BAHÍA; LAP = LA PAZ; LEM = LEMPIRA; OCO = OCOTEPEQUE; OLA = OLANCHO; SB = SANTA BÁRBARA; VAL = VALLE; AND YOR = YORO. SYMBOLS AND A ABBREVIATION IN THE SPECIES LIST ARE: X = WITH VOUCHER SPECIMEN(S); O = SIGHT (BY AUTHOR) OR ACCEPTABLE LITERATURE RECORD(S); P = PHOTOGRAPH(S) AVAILABLE; AND I = INTRODUCED POPULATION, OTHER THAN SPECIES INTRODUCED TO HONDURAN TERRITORY.

Species (126)	ATL	CHO	COL	COM	COP	COR	EP	FM	GAD	INT	IDB	LAP	LEM	OCO	OLA	SB	VAL	YOR	Total
Lizards (107)																			
Lepidophyma flavimaculatum	X	—	X	X	X	X	—	—	X	—	—	—	—	—	X	X	—	X	9
Lepidophyma mayae	—	X	—	—	X	—	—	—	—	—	—	—	—	—	—	—	—	—	2
Coleonyx mitratus	X	X	X	—	X	X	X	X	—	—	—	—	—	X	X	—	X	X	11
Hemidactylus frenatus (I)	X	X	X	—	X	X	X	X	X	—	X	X	X	X	X	X	X	X	16
Hemidactylus haitianus (I)	—	—	—	—	—	—	—	—	—	—	—	—	X	—	—	—	—	—	1
Hemidactylus mabouia (I)	—	—	X	—	—	—	—	—	X	—	X	—	—	—	X	—	—	—	3
Phyllodactylus palmeus	—	—	—	—	—	—	—	—	—	—	X	—	—	—	—	—	—	—	1
Phyllodactylus paralepis	—	X	—	—	—	—	—	—	—	—	—	—	—	—	—	—	—	—	1
Phyllodactylus tuberculosus	X	—	X	—	—	X	X	X	X	—	—	—	—	—	X	X	X	X	4
Thecadactylus rapicauda	X	—	X	—	—	X	—	—	X	—	X	—	—	—	X	X	—	X	8
Aristelliger sp. A	—	—	—	—	X	—	—	—	—	—	—	—	—	—	—	—	—	—	1
Aristelliger nelsoni	—	X	X	—	—	—	X	X	X	—	—	—	—	—	X	—	X	—	1
Gonatodes albogularis	—	X	X	—	—	—	—	—	—	—	X	—	—	—	X	X	X	—	7
Sphaerodactylus alphus	—	—	—	—	—	—	—	X	—	—	—	—	—	—	—	—	—	—	1
Sphaerodactylus continentalis	X	X	X	—	X	X	—	X	X	—	—	—	—	—	X	X	—	X	9
Sphaerodactylus dunni	X	X	X	—	—	X	—	—	X	—	—	—	—	—	—	X	—	X	5
Sphaerodactylus exsul	—	—	—	X	—	—	—	—	—	—	—	—	—	—	—	—	—	—	1
Sphaerodactylus glaucus	—	—	—	—	—	—	—	—	—	—	X	—	—	—	—	—	—	—	1
Sphaerodactylus guanajae	—	—	—	—	—	—	—	—	—	—	X	—	—	—	—	—	—	—	1
Sphaerodactylus leonardovaldesi	—	X	X	—	—	—	—	—	—	—	X	—	—	—	—	X	—	—	1
Sphaerodactylus millepunctatus	—	—	—	—	—	—	—	—	X	—	X	—	—	—	X	—	—	—	3
Sphaerodactylus poindexteri	—	—	—	—	—	—	—	—	—	—	X	X	—	—	—	—	—	—	1
Sphaerodactylus rosaurae	—	—	—	—	—	—	—	—	—	—	X	—	—	—	—	—	—	—	1
Abronia montecristoi	—	—	—	X	—	—	—	—	—	—	—	—	—	—	—	—	—	—	1
Abronia salvadorensis	P	—	—	—	—	—	—	—	—	X	—	X	X	X	X	—	—	—	2
Mesaspis moreletii	P	—	—	—	—	X	X	X	X	X	—	X	X	X	X	—	—	—	8+1
Diploglossus bivittatus	—	—	—	—	—	—	—	—	—	X	—	X	X	X	X	—	—	—	4
Diploglossus montanus	—	—	—	—	X	X	—	—	—	X	—	—	—	—	—	—	—	—	1
Diploglossus scansorius	—	—	—	—	—	—	—	—	—	—	—	—	—	—	—	—	X	—	1
Basiliscus plumifrons	—	O	—	—	—	—	—	—	X	—	—	—	—	—	X	—	—	—	2+1
Basiliscus vittatus	X	X	X	X	X	X	X	X	X	X	—	X	X	X	X	X	X	X	17
Corytophanes cristatus	—	X	X	—	—	X	—	—	—	—	—	—	—	—	X	X	—	X	7
Corytophanes hernandesii	—	—	—	—	—	—	—	—	X	—	—	—	—	—	—	X	—	X	2
Corytophanes percarinatus	—	—	—	—	—	—	—	—	—	—	—	—	—	X	—	—	—	—	1
Laemanctus julioi	—	—	—	—	—	—	O	X	—	—	—	—	—	—	X	—	—	—	1+1
Laemanctus longipes	—	—	—	X	—	—	—	X	—	—	—	—	—	—	X	—	—	—	2

TABLE 2. CONTINUED.

Species (126)	ATL	CHO	COL	COM	COP	COR	EP	FM	GAD	INT	IDB	LAP	LEM	OCO	OLA	SB	VAL	YOR	Total
Laemanctus serratus	X	—	—	—	—	—	—	—	—	—	—	—	—	—	—	—	—	—	0
Laemanctus waltersi	X(1)	—	—	—	—	X	—	—	—	—	—	—	—	—	—	—	—	—	2
Anolis allisoni	—	—	—	—	—	X	—	—	—	—	X	—	—	—	—	—	—	—	2
Norops amplisquamosus	X	—	X	—	—	X	—	—	X	—	—	—	—	—	X	X	—	X	1
Norops beckeri	—	—	—	—	—	X	—	—	—	—	—	—	—	—	—	—	—	—	7
Norops bicaorum	X	—	X	—	X	X	X	X	X	—	X	X	—	—	X	P	—	X	1
Norops biporcatus	—	—	X	—	X	X	X	—	X	—	—	—	—	—	X	X	—	—	9+1
Norops capito	—	—	—	—	—	—	—	—	X	—	—	X	X	—	X	—	—	—	6
Norops carpenteri	—	X	X	X	X	X	X	X	X	X	—	X	X	—	X	X	—	—	1
Norops crassulus	—	X	X	—	X	X	X	X	X	X	—	—	—	X	—	X	—	—	5
Norops cupreus	—	—	—	X	X	X	—	—	—	X	—	X	—	—	—	—	—	—	6
Norops cusuco	X	—	X	X	X	X	X	X	—	X	—	X	X	X	X	X	—	X	3
Norops heteropholidotus	X	—	X	X	X	X	X	—	X	X	—	X	—	X	X	X	—	X	2
Norops johnmeyeri	—	—	—	—	—	—	—	—	—	—	—	—	—	—	X	—	—	—	2
Norops kreutzi	—	—	—	—	—	—	—	—	—	—	X	X	X	—	X	—	—	—	2
Norops laeviventris	X	—	—	X	X	X	X	—	—	X	—	X	—	—	—	X	—	—	10
Norops lemurinus	—	—	X	—	X	X	X	—	—	—	—	X	X	—	X	X	—	X	11
Norops limifrons	—	—	—	—	—	—	—	—	—	—	—	X	—	—	—	—	—	—	3
Norops loveridgei	X	—	—	X	X	X	X	X	—	X	—	X	X	—	X	X	—	X	2
Norops mccraniei	—	—	—	X	X	X	X	X	—	X	—	X	—	—	X	X	—	X	12
Norops morazani	—	—	—	—	—	—	—	—	—	—	—	—	—	—	X	—	—	—	1
Norops muralla	—	—	—	—	—	—	—	—	—	—	—	—	—	X	—	—	—	—	1
Norops nelsoni	—	—	—	—	—	—	—	—	X	—	—	—	—	—	X	X	—	—	3
Norops ocelloscapularis	—	—	—	X	X	—	—	—	—	—	X	—	—	—	—	—	—	—	2
Norops oxylophus	—	—	—	X	X	—	—	—	X	X	—	X	X	—	X	X	—	—	3
Norops petersii	—	—	—	X	X	—	—	—	—	—	—	—	—	—	X	X	—	—	1
Norops pijolense	X	—	—	—	—	—	—	—	—	—	—	—	—	—	—	—	—	X	2
Norops purpurgularis	—	—	—	—	—	—	—	—	—	—	—	—	—	—	X	X	—	X	3
Norops quaggulus	—	—	—	—	X	X	X	—	X	—	—	—	—	—	—	—	—	—	1
Norops roatanensis	—	—	—	—	—	—	—	—	—	—	X	—	—	—	—	X	—	—	3
Norops rodriguezii	X	—	—	X	X	X	X	—	X	X	—	X	—	—	X	X	—	—	1
Norops rubribarbaris	—	—	—	X	X	X	X	—	—	X	—	X	—	X	X	X	—	X	4
Norops sagrei (1)	X	—	—	—	—	X	X	X	X	—	—	—	—	—	—	—	—	—	2
Norops sminthus	X	—	—	X	—	—	X	—	—	—	—	—	—	—	—	—	—	—	4
Norops uniformis	X	—	X	X	X	X	X	X	X	X	X	X	—	—	X	X	—	—	14
Norops unilobatus	—	—	—	—	—	—	—	—	—	—	X	X	—	—	X	X	X	—	1
Norops utilensis	—	—	—	—	—	—	X	X	X	X	—	X	—	—	X	X	—	—	1
Norops wampuensis	—	—	—	—	—	—	—	—	—	—	—	—	—	—	X	—	—	X	6
Norops wellbornae	—	X	—	—	—	—	—	X	—	—	—	—	X	X	X	X	X	—	1
Norops wermuthi	—	—	—	—	—	—	—	—	—	—	—	—	—	—	—	—	—	—	2
Norops wilsoni	X	—	X	—	—	—	—	—	—	—	—	—	—	—	—	—	—	—	2

TABLE 2. CONTINUED.

Species (126)	ATL	CHO	COL	COM	COP	COR	EP	FM	GAD	INT	IDB	LAP	LEM	OCO	OLA	SB	VAL	YOR	Total
Norops yoroensis	X	—	—	X	—	X	—	X	—	—	—	—	—	—	X	X	—	X	7
Norops zeus	X	—	X	—	—	X	—	—	—	—	—	—	—	—	—	—	—	X	4
Ctenosaura bakeri	—	—	—	—	—	—	—	—	—	—	X	—	—	—	—	—	—	—	1
Ctenosaura flavidorsalis	—	—	—	—	—	—	—	—	—	X	—	X	—	—	—	—	—	X	2
Ctenosaura melanosterna	—	—	—	—	—	—	—	—	—	X	X	—	—	—	—	—	—	—	2
Ctenosaura oedirhina	—	—	—	—	—	—	—	—	—	—	X	—	—	—	—	—	—	—	1
Ctenosaura quinquecarinata	—	X	—	—	X	X	X	X	X	X	X	—	X	—	X	X	—	X	3
Ctenosaura similis	X	X	X	X	—	X	X	X	X	X	X	—	—	—	X	X	X	X	15
Iguana iguana	X	X	X	O	X	X	—	X	X	X	X	—	X	—	X	X	X	—	12+1
Leiocephalus varius (1)	—	—	—	—	—	—	—	—	—	—	—	—	—	—	—	—	—	—	1
Sceloporus esperanzae	—	—	—	—	—	—	—	—	—	—	—	X	—	X	—	—	—	X	2
Sceloporus hondurensis	X	X	X	X	—	X	X	X	X	X	X	X	X	—	X	X	—	—	14
Sceloporus schmidti	—	—	—	—	—	X	—	—	—	—	—	—	—	—	—	—	—	—	3
Sceloporus squamosus	X	X	—	X	X	X	X	X	X	X	—	X	X	X	X	X	X	X	11
Sceloporus variabilis	X	X	—	X	X	X	X	X	X	X	—	X	X	X	X	X	X	X	15
Polychrus gutturosus	X	—	X	X	X	X	X	—	X	X	—	X	—	—	—	—	—	—	2
Marisora brachypoda	X	X	X	X	X	X	X	X	X	X	X	X	X	—	X	X	X	X	17
Marisora roatanae	—	—	—	—	—	—	—	—	—	—	—	—	—	—	—	—	—	—	1
Scincella assata	—	—	—	—	—	—	—	—	—	X	O	—	—	—	—	—	—	—	1
Scincella cherriei	X	X	X	X	X	X	X	X	X	—	—	X	X	—	X	X	—	X	14+1
Scincella incerta	—	—	—	—	—	X	—	—	—	—	—	—	—	X	X	X	X	X	4
Mesoscincus managuae	X	X	—	—	—	—	X	—	—	—	—	—	—	—	—	—	—	—	3
Plestiodon sumichrasti	—	X	X	—	—	X	X	X	—	X	X	—	—	—	X	X	X	X	11
Gymnophthalmus speciosus	—	X	X	—	—	X	X	X	X	X	X	—	X	—	X	X	X	X	1
Ameiva fuliginosa	—	—	—	—	—	—	—	—	—	—	—	—	—	—	—	—	—	—	1
Aspidoscelis deppii	X	X	X	X	X	X	X	X	X	X	X	X	X	—	X	X	X	X	16
Aspidoscelis motaguae	—	X	—	—	X	—	X	X	—	X	—	—	—	—	X	X	—	X	6
Cnemidophorus ruatanus	X	—	X	—	X	X	—	—	X	X	X	—	—	—	—	—	—	X	7
Holcosus festivus	X	X	X	—	X	X	—	X	X	X	—	—	—	—	X	X	X	X	8
Holcosus undulatus	X	X	X	X	X	X	X	X	X	X	—	X	X	X	X	X	X	X	16
Lizard subtotals	33	21	30	18	31	42	26	30	37	22	28	15	15	12	37	32	14	35	478
Literature or sight records	0	0	1	1	0	0	1	0	0	0	1	0	0	0	0	1	0	0	4
Photographs	1	0	0	0	0	0	0	0	0	0	0	0	0	0	0	1	0	0	2
Total	34	21	31	19	31	42	27	30	37	22	29	15	15	12	37	33	14	35	484
Crocodiles (2)																			
Caiman crocodilus	O	—	X	—	—	O	—	X	X	—	—	—	—	—	—	—	—	X	3+2
Crocodylus acutus	X	O	X	O	—	X	—	—	P	—	X	—	—	—	O	—	O	O	4+5+1
Crocodile subtotals	1	0	2	0	0	1	0	0	1	0	1	0	0	0	1	0	0	1	7
Literature or sight records	1	1	0	1	0	0	0	1	0	0	0	0	0	0	0	0	1	1	7
Photographs	0	0	0	0	0	0	0	0	1	0	0	0	0	0	0	0	0	0	1
Total	2	1	2	1	0	2	0	2	2	0	1	0	0	0	1	0	1	2	15

TABLE 2. Continued.

Species (126)	ATL	CHO	COL	COM	COP	COR	EP	FM	GAD	INT	IDB	LAP	LEM	OCO	OLA	SB	VAL	YOR	Total
Turtles (17)																			
Caretta caretta	—	—	O	—	—	—	—	—	X	—	X	—	—	—	—	—	—	—	2+1
Chelonia mydas	—	O	—	—	—	—	—	—	O	—	O	—	—	—	—	—	X	—	1+3
Eretmochelys imbricata	O	X	—	—	—	—	—	—	X	—	X	—	—	—	—	—	X	—	4+1
Lepidochelys olivacea	—	X	—	—	—	—	—	—	—	—	—	—	—	—	—	—	X	—	2
Dermochelys coriacea	O	—	—	—	—	—	—	—	P	—	—	—	—	—	—	—	—	—	0+2
Chelydra acutirostris	—	—	—	—	—	X	X	—	X	—	—	—	—	—	—	—	—	—	3
Chelydra rossignonii	X	—	—	—	X	—	—	—	X	—	—	—	—	—	—	—	—	—	3
Kinosternon albogulare	—	X	—	X	—	X	X	X	—	X	—	X	X	—	X	—	—	X	10
Kinosternon leucostomum	X	—	X	—	X	X	X	—	X	—	X	—	—	—	X	X	—	X	10
Staurotypus triporcatus	—	—	—	—	—	—	—	—	—	—	—	—	—	—	X	—	X	—	2
Trachemys emolli	—	X	—	—	—	X	—	—	—	—	—	—	—	—	—	—	—	—	2
Trachemys scripta (1)	O	—	X	—	—	—	—	O	—	—	—	—	—	—	—	—	—	—	1+2
Trachemys venusta	X	—	X	—	—	X	X	X	X	—	X	—	—	—	X	X	—	X	10
Rhinoclemmys annulata	X	—	—	—	—	—	—	—	X	—	—	—	—	X	—	—	—	—	3
Rhinoclemmys areolata	—	—	—	—	—	X	—	—	—	—	—	—	—	—	—	—	—	—	1
Rhinoclemmys funerea	—	—	—	—	—	—	—	—	X	—	—	—	—	—	—	—	—	—	1
Rhinoclemmys pulcherrima	—	X	—	X	X	X	X	X	—	—	—	—	—	X	X	X	X	X	11
Turtle subtotals	4	5	3	2	3	7	5	3	8	1	4	1	1	2	5	3	5	4	66
Literature or sight records	3	1	1	0	0	0	0	1	1	0	1	0	0	0	0	0	0	0	8
Photographs	0	0	0	0	0	0	0	0	1	0	0	0	0	0	0	0	0	0	1
Total	7	6	4	2	3	7	5	4	10	1	5	1	1	2	5	3	5	4	75
All subtotals	38	26	35	20	34	50	31	33	46	23	33	16	16	14	42	35	19	40	551
Literature or sight records	4	2	2	2	0	1	1	1	1	0	2	0	0	0	1	1	1	1	20
Photographs	1	0	0	0	0	0	0	0	2	0	0	0	0	0	0	1	0	0	4
Total	43	28	37	22	34	51	32	34	49	23	35	16	16	14	43	37	20	41	575

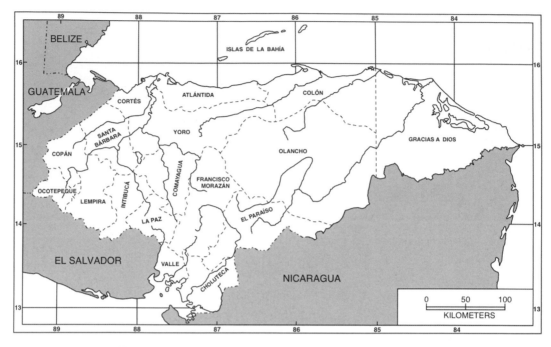

Map 84. Map of Honduras showing the boundaries of the 18 departments.

to Honduran territory. Additionally, only one species of Dactyloidae was treated in a species account in the taxonomy section above, but all 41 species are treated in all distribution and conservation sections, despite 39 having been treated in similar sections by McCranie and Köhler (2015). The total number of species by departments ranges from a high of 51 species in Cortés to a low of 14 in Ocotepeque. The second highest total for a department is Gracias a Dios with 49 species. Those two most speciose departments have all been relatively well studied but have largely different faunas because of different ecological regimes. Cortés has terrain that extends from sea level to 2,242 m elevation above sea level. Its ecological regimes are also varied and include Lowland Moist Forest along the north coast, Lowland Dry Forest in the Chamelecón and Ulúa plains, Premontane Wet Forest in the vicinity of Lago de Yojoa and on the north-facing slopes in the Sierra de Omoa, Premontane Moist Forest on

many south-facing slopes in the Sierra de Omoa, and Lower Montane Wet Forest in higher elevations of the Sierra de Omoa. The second highest species total is in the department of Gracias a Dios (49 species), which, with the exception of a couple of low mountain peaks in its southwestern portion, lies in lowlands, much of which are below 200 m elevation. Additionally, with the exception of the Swan Islands and several small islands off the northeastern coast, Gracias a Dios lies in Lowland Moist Forest. That department has several types of habitats in the Lowland Moist Forest (i.e., closed canopy rainforest, pine savanna, etc.; see discussion of this forest formation in McCranie, 2011a), several of which have their own distinctive faunas. Also, a major contributor to the high species number for Gracias a Dios are the five species known only from the Swan Islands or Cayos Miskitos.

Of the group of departments with the lowest total number of species, Ocotepeque (with 14 species), La Paz and Lempira (each with 16), Valle (20), Comayagua (22), and Intibucá (23) all appear to be a combination of being understudied and having somewhat depauperate faunas. Five of those six departments, Valle the exception, contain extensive areas of somewhat depauperate Premontane Moist Forest. Valle, on the other hand, lies entirely in the somewhat depauperate subhumid region of the Pacific versant. However, Choluteca, Valle's neighbor to the east, contains eight species more than are known from Valle. Valle, unlike Choluteca, lacks a relatively large area of low mountains. An even more obvious indication of the understudied status of Valle is the lack of any records for the widespread and common *Basiliscus vittatus* from there. Valle is the only Honduran department where *B. vittatus* has not been recorded. Comayagua, in addition to its somewhat depauperate Premontane Moist Forest, also appears understudied when compared with its well-studied neighbors to the north (Cortés and Yoro with 51 and 41 species, respectively) and south (Francisco Morazán with 34 species).

Many of these same departmental patterns with the reptilian fauna under consideration herein were demonstrated by similar analyses of amphibian (McCranie, 2007b) and snake (McCranie, 2011a) faunas. McCranie (2011a) also found the Cortés and Gracias a Dios high total pattern for snakes, as well as the low numbers for the groups of southwestern departments. However, the Islas de la Bahía have the second lowest total number of snakes. Additionally, Valle also seems to be significantly understudied with concern to snakes and amphibians. An updated table of departmental distribution of the amphibians (McCranie, personal data) over that of McCranie (2007b) demonstrates that Olancho has the highest species total, followed by Cortés, Atlántida, and Gracias a Dios, in that order (all are in the mesic Atlantic versant). The lowest total of amphibians occurs on the Swan Islands and the Bay Islands, followed by the subhumid southern departments of Valle and Choluteca. The same group of southwestern departments, as found in the other two analyses, are also among those with the lowest total species of amphibians.

DISTRIBUTION WITHIN ECOLOGICAL (FOREST) FORMATIONS AND BY ELEVATION

The distribution of the 121 species of lizards, crocodiles, and nonmarine turtles of Honduras in nine ecological (forest) formations (Holdridge, 1967) is indicated in Table 3 (Map 85 shows general distributions of those ecological formations). Montane Rainforest formation is excluded from consideration because its herpetofauna has been poorly studied and there are no reptiles recorded from it. Montane Rainforest regions are difficult to access and are limited to the highest peaks of the Cordillera de Celaque and cerros El Pital and Santa Bárbara. The following distributional categories are used: WIDESPREAD (occurs widely in a particular forest formation in Honduras, as well as in at least one other forest formation); RESTRICTED (restricted to a single forest formation in Honduras); and PERIPHERAL (barely enters a particular forest formation in Honduras). Forest formations, the number of reptilian species under consideration herein known from each formation, and their distribution categories (Tables 3 and 4) are as follows (from highest to lowest number of species in each formation; formation abbreviations explained in Table 3): LMF 73 (48 widespread; 22 restricted; 3 peripheral); PWF 49 (36 widespread; 13 peripheral); LDF 45 (37 widespread; 3 restricted; 5 peripheral); PMF 35 (24 widespread; 1 restricted; 10 peripheral); LAF 30 (30 widespread); LMWF 26 (17 widespread; 4 restricted; 5 peripheral); PDF 26 (24 widespread; 2 restricted); LMMF 14 (7 widespread; 3 restricted; 4 peripheral); and LDF(WI) 9 (3

TABLE 3. DISTRIBUTION OF THE HONDURAN LIZARD, CROCODILE, AND TURTLE SPECIES WITHIN NINE ECOLOGICAL (FOREST) FORMATIONS (I = INTRODUCED). THE FIVE MARINE TURTLE SPECIES ARE NOT INCLUDED, ALTHOUGH THEY DO OCCASIONALLY NEST ON A FEW BEACHES IN THE COUNTRY. THE MONTANE RAINFOREST FORMATION IS ALSO NOT INCLUDED IN THIS TABLE. ABBREVIATIONS USED IN THIS AND SOME SUBSEQUENT TABLES ARE AS FOLLOWS: LMF = LOWLAND MOIST FOREST; LDF = LOWLAND DRY FOREST; LAF = LOWLAND ARID FOREST; PWF = PREMONTANE WET FOREST; PMF = PREMONTANE MOIST FOREST; PDF = PREMONTANE DRY FOREST; LMWF = LOWER MONTANE WET FOREST; LMMF = LOWER MONTANE MOIST FOREST; LDF (WI) = LOWLAND DRY FOREST, WEST INDIAN SUBREGION; M = METERS; W = WIDESPREAD IN THAT PARTICULAR FORMATION; P = PERIPHERAL IN THAT FORMATION; R = RESTRICTED TO THAT FORMATION. WHEN AN ELEVATIONAL RANGE BEGINS WITH 0, IT IS MEANT TO CONVEY ABOUT SEA LEVEL.

Species (121)	LMF	LDF	LAF	PWF	PMF	PDF	LMWF	LMMF	LDF (WI)	Total	Elevational Range (m)
Lizards (107)											
Lepidophyma flavimaculatum	W	P	—	W	—	—	—	—	—	3	0–ca. 1,400
Lepidophyma mayae	W	—	—	W	—	—	—	—	—	2	435–1,040
Coleonyx mitratus	W	W	W	W	—	W	—	—	—	5	0–ca. 1,400
Hemidactylus frenatus (I)	W	W	W	—	W	W	—	—	W	6	0–1,340
Hemidactylus haitianus (I)	—	—	—	—	—	R	—	—	—	1	950
Hemidactylus mabouia (I)	W	—	—	—	—	—	—	—	W	2	0
Phyllodactylus palmeus	R	—	—	—	—	—	—	—	—	1	0–30
Phyllodactylus paralepis	R	—	—	—	—	—	—	—	—	1	0–30
Phyllodactylus tuberculosus	—	W	W	—	W	W	—	—	—	4	0–1,200
Thecadactylus rapicauda	W	W	W	P	—	—	—	—	—	4	0–750
Aristelliger sp. A	—	—	—	—	—	—	—	—	R	1	0
Aristelliger nelsoni	—	—	—	—	—	—	—	—	R	1	0–10
Gonatodes albogularis	W	W	—	—	—	W	—	—	—	3	0–1,000
Sphaerodactylus alphus	R	—	—	—	—	—	—	—	—	1	0–15
Sphaerodactylus continentalis	W	W	W	P	W	W	—	—	—	6	0–1,100
Sphaerodactylus dunni	W	W	W	—	—	—	—	—	—	3	60–280
Sphaerodactylus exsul	—	—	—	—	—	—	—	—	R	1	0
Sphaerodactylus glaucus	—	—	—	—	R	—	—	—	—	1	600
Sphaerodactylus guanajae	R	—	—	—	—	—	—	—	—	1	0–30
Sphaerodactylus leonardovaldesi	R	—	—	—	—	—	—	—	—	1	0–30
Sphaerodactylus millepunctatus	R	—	—	—	—	—	—	—	—	1	0–190
Sphaerodactylus poindexteri	R	—	—	—	—	—	—	—	—	1	0–10
Sphaerodactylus rosaurae	R	—	—	—	—	—	—	—	—	1	0–20
Abronia montecristoi	—	—	—	—	—	—	R	—	—	1	1,370
Abronia salvadorensis	—	—	—	—	—	—	—	R	—	1	2,020–2,125
Mesaspis moreletii	—	—	—	—	—	—	W	W	—	2	1,450–2,530
Diploglossus bivittatus	—	—	—	W	—	—	—	W	—	2	1,330–2,100
Diploglossus montanus	—	—	—	W	—	—	W	—	—	2	915–1,780
Diploglossus scansorius	—	—	—	—	P	—	W	—	—	2	1,550–1,590
Basiliscus plumifrons	R	—	—	—	—	—	—	—	—	1	40–225
Basiliscus vittatus	W	W	W	W	W	W	—	—	—	6	0–1,400
Corytophanes cristatus	W	—	—	W	—	—	—	—	—	2	0–1,300
Corytophanes hernandesii	W	—	—	W	—	—	—	—	—	2	ca. 150–1,000
Corytophanes percarinatus	—	—	—	—	W	—	—	P	—	2	1,350–1,700
Laemanctus julioi	—	—	—	—	—	R	—	—	—	1	650–1,000
Laemanctus longipes	W	—	—	W	P	—	—	—	—	3	600–1,200
Laemanctus serratus	—	—	—	—	—	—	—	—	—	0	—
Laemanctus waltersi	W	P	—	W	—	—	—	—	—	3	0–700
Anolis allisoni	R	—	—	—	—	—	—	—	—	1	0–30
Norops amplisquamosus	—	—	—	—	—	—	R	—	—	1	1,530–1,990
Norops beckeri	W	P	—	W	—	—	—	—	—	3	0–ca. 1,400
Norops bicaorum	R	—	—	—	—	—	—	—	—	1	0–20
Norops biporcatus	W	—	—	W	P	—	—	—	—	3	0–1,050
Norops capito	W	—	—	W	—	—	—	—	—	2	0–1,300
Norops carpenteri	R	—	—	—	—	—	—	—	—	1	30–40
Norops crassulus	—	—	—	—	W	—	W	W	—	3	1,200–2,285
Norops cupreus	W	W	W	W	W	—	—	—	—	5	0–1,300
Norops cusuco	—	—	—	P	—	—	W	—	—	2	1,350–1,990

TABLE 3.   CONTINUED.

Species (121)	LMF	LDF	LAF	PWF	PMF	PDF	LMWF	LMMF	LDF (WI)	Total	Elevational Range (m)
Norops heteropholidotus	—	—	—	—	—	—	—	R	—	1	1,860–2,200
Norops johnmeyeri	—	—	—	P	—	—	W	—	—	2	1,300–2,000
Norops kreutzi	—	—	—	W	—	—	W	—	—	2	980–1,690
Norops laeviventris	—	—	—	W	W	—	W	W	—	4	1,000–2,000
Norops lemurinus	W	W	W	W	—	—	—	—	—	4	0–960
Norops limifrons	W	—	—	W	—	—	—	—	—	2	0–900
Norops loveridgei	P	—	—	W	—	—	W	—	—	3	ca. 550–1,600
Norops morazani	—	—	—	P	—	—	W	—	—	2	1,275–2,150
Norops mccraniei	—	W	W	W	W	W	—	P	—	6	200–1,900
Norops muralla	—	—	—	P	—	—	W	—	—	2	1,440–1,740
Norops nelsoni	—	—	—	—	—	—	—	—	R	1	0–10
Norops ocelloscapularis	—	—	—	W	—	—	P	—	—	2	1,040–1,550
Norops oxylophus	R	—	—	—	—	—	—	—	—	1	60–225
Norops petersii	—	—	—	W	—	—	W	—	—	2	1,300–1,550
Norops pijolense	—	—	—	W	—	—	W	—	—	2	1,180–2,050
Norops purpurgularis	—	—	—	—	—	—	R	—	—	1	1,550–2,040
Norops quaggulus	W	—	—	W	—	—	—	—	—	2	60–840
Norops roatanensis	R	—	—	—	—	—	—	—	—	1	0–30
Norops rodriguezii	W	W	—	W	P	—	—	—	—	4	0–1,200
Norops rubribarbaris	—	—	—	—	—	—	R	—	—	1	1,600–1,800
Norops sagrei (I)	W	W	—	—	—	—	—	—	—	2	0–100
Norops sminthus	—	—	—	—	P	—	—	W	—	2	1,450–1,900
Norops uniformis	W	—	—	W	—	—	P	—	—	3	30–1,370
Norops unilobatus	W	W	W	P	W	W	—	—	—	6	0–1,320
Norops utilensis	R	—	—	—	—	—	—	—	—	1	0–8
Norops wampuensis	R	—	—	—	—	—	—	—	—	1	95–110
Norops wellbornae	—	W	W	—	—	W	—	—	—	3	0–1,000
Norops wermuthi	—	—	—	—	—	—	—	R	—	1	1,800
Norops wilsoni	W	—	—	W	—	—	P	—	—	3	0–980
Norops yoroensis	—	—	—	W	P	—	W	—	—	3	650–1,600
Norops zeus	W	P	—	W	—	—	—	—	—	3	0–900
Ctenosaura bakeri	R	—	—	—	—	—	—	—	—	1	0
Ctenosaura flavidorsalis	—	W	—	—	—	W	—	—	—	2	350–920
Ctenosaura melanosterna	W	—	W	—	—	—	—	—	—	2	0–300
Ctenosaura oedirhina	R	—	—	—	—	—	—	—	—	1	0–20
Ctenosaura quinquecarinata	—	W	W	—	—	W	—	—	—	3	95–1,000
Ctenosaura similis	W	W	W	—	P	W	—	—	—	5	0–1,300
Iguana iguana	W	W	W	—	W	W	—	—	W	6	0–800
Leiocephalus varius (I)	—	—	—	—	—	—	—	—	R	1	0
Sceloporus esperanzae	—	—	—	—	P	—	—	W	—	2	1,530–1,900
Sceloporus hondurensis	P	—	—	W	P	—	W	W	—	5	650–2,530
Sceloporus schmidti	—	—	—	W	—	—	W	—	—	2	600–2,240
Sceloporus squamosus	—	W	W	—	W	W	—	—	—	4	0–1,470
Sceloporus variabilis	P	W	W	P	W	W	P	P	—	8	0–1,760
Polychrus gutturosus	R	—	—	—	—	—	—	—	—	1	10–190
Marisora brachypoda	W	W	W	W	W	W	—	—	—	6	0–1,510
Marisora roatanae	R	—	—	—	—	—	—	—	—	1	0–20
Scincella assata	—	R	—	—	—	—	—	—	—	1	370
Scincella cherriei	W	W	W	W	W	W	P	P	—	8	0–1,860
Scincella incerta	—	—	—	W	—	—	W	—	—	2	1,100–1,670
Mesoscincus managuae	—	W	W	—	W	—	—	—	—	3	0–920
Plestiodon sumichrasti	W	—	—	W	—	—	—	—	—	2	30–880
Gymnophthalmus speciosus	W	W	W	—	W	W	—	—	—	5	0–1,320
Ameiva fuliginosa	—	—	—	—	—	—	—	—	R	1	0
Aspidoscelis deppii	W	W	W	—	W	W	—	—	—	5	0–900
Aspidoscelis motaguae	—	W	W	—	W	W	—	—	—	4	ca. 50–950

TABLE 3. CONTINUED.

Species (121)	LMF	LDF	LAF	PWF	PMF	PDF	LMWF	LMMF	LDF (WI)	Total	Elevational Range (m)
Cnemidophorus ruatanus	W	W	—	—	—	—	—	—	—	2	0–400
Holcosus festivus	W	P	—	W	—	—	—	—	—	3	0–1,400
Holcosus undulatus	W	W	W	P	W	W	—	—	—	6	0–1,240
Subtotals	61	35	25	44	30	23	26	14	9	267	0–2,530
Crocodiles (2)											
Caiman crocodilus	W	—	W	—	—	—	—	—	—	2	0–280
Crocodylus acutus	W	W	—	P	—	—	—	—	—	3	0–650
Subtotals	2	1	1	1	0	0	0	0	0	5	0–650
Turtles (12)											
Chelydra acutirostris	W	W	—	—	—	—	—	—	—	2	5–400
Chelydra rossignonii	W	W	—	P	—	—	—	—	—	3	0–730
Kinosternon albogulare	W	W	W	—	W	W	—	—	—	5	0–1,240
Kinosternon leucostomum	W	W	W	W	W	—	—	—	—	5	0–1,120
Staurotypus triporcatus	W	W	—	—	—	—	—	—	—	2	0–100
Trachemys emolli	—	R	—	—	—	—	—	—	—	1	5
Trachemys scripta (I)	W	—	—	—	W	—	—	—	—	2	0–1,480
Trachemys venusta	W	W	W	P	P	W	—	—	—	6	0–1,000
Rhinoclemmys annulata	R	—	—	—	—	—	—	—	—	1	20–540
Rhinoclemmys areolata	—	R	—	—	—	—	—	—	—	1	100
Rhinoclemmys funerea	R	—	—	—	P	—	—	—	—	1	0–190
Rhinoclemmys pulcherrima	P	W	W	P	P	W	—	—	—	6	0–1,480
Subtotals	10	9	4	4	5	3	0	0	0	35	0–1,480
Totals	73	45	30	49	35	26	26	14	9	307	0–2,530

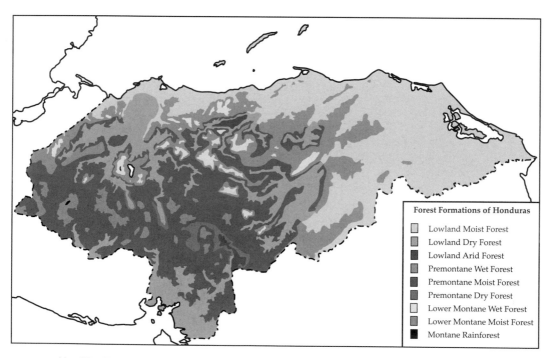

Map 85. Forest formations (modified from Holdridge, 1967) of Honduras (in color in McCranie, 2011a).

TABLE 4. THE 121 LIZARD, CROCODILE, AND NON-MARINE TURTLE SPECIES NUMBERS FOR THE THREE DISTRIBUTIONAL CATEGORIES IN EACH OF THE NINE ECOLOGICAL FORMATIONS. SEE TABLE 3.

| Formations | Distributional Categories | | | | | |
| | Widespread | | Restricted | | Peripheral | |
	N	%	N	%	N	%
LMF	48	65.8	22	30.1	3	4.1
LDF	37	82.2	3	6.7	5	11.1
LAF	30	100.0	0	0.0	0	0.0
PWF	36	73.5	0	0.0	13	26.5
PMF	24	66.7	1	2.8	10	30.5
PDF	24	92.3	2	7.7	0	0.0
LMWF	17	65.4	4	15.4	5	19.2
LMMF	7	50.0	3	21.4	4	28.6
LDF(WI)	3	33.3	6	66.7	0	0.0
Totals	226	73.4	41	13.3	41	13.3

widespread; 6 restricted). The two formations with the highest number of species are mesic (LMF, PWF). Of the two other mesic formations, one (LMWF) is only the sixth highest, and the other (LMMF) is only the eighth highest. The LMWF and LMMF formations are cloud forest localities, and their total number of species is negatively influenced by their relatively high elevations (see below). The mean number of formations inhabited by the 121 nonmarine species is 2.5.

With respect to the three reptilian orders, the 107 species of lizards (Squamata) collectively occur in all nine formations, as follows (highest to lowest): LMF 61 (39 widespread; 20 restricted; 2 peripheral); PWF 44 (35 widespread; 9 peripheral); LDF 35 (29 widespread; 1 restricted; 5 peripheral); PMF 30 (21 widespread; 1 restricted; 8 peripheral); LMWF 26 (17 widespread; 4 restricted; 5 peripheral); LAF 25 (25 widespread); PDF 23 (21 widespread; 2 restricted); LMMF 14 (7 widespread; 3 restricted; 4 peripheral); and LDF(WI) 9 (3 widespread; 6 restricted). The two crocodiles (Crocodylia) collectively occur in only four of nine formations, as follows: LMF 2 (2 widespread); LDF 1 (1 widespread); LAF 1 (1 widespread); and PWF 1 (1 peripheral) formations. The 12

species of nonmarine turtles (Testudinata) collectively occur in only six of nine formations, as follows (highest to lowest): LMF 10 (7 widespread; 2 restricted; 1 peripheral); LDF 9 (7 widespread; 2 restricted); PMF 5 (3 widespread; 2 peripheral); LAF 4 (4 widespread); PWF 4 (1 widespread; 3 peripheral); and PDF 3 (3 widespread). The greatest total number of formations inhabited by a member of each order under study herein is: Squamata 8 (*Sceloporus variabilis* and *Scincella cherriei*); Crocodylia 3 (*Crocodylus acutus*); and Testudinata 6 (*Trachemys venusta* and *Rhinoclemmys pulcherrima*). The mean number of formations inhabited by the species of each order is: lizards 2.5; crocodiles 2.5; and turtles (nonmarine) 2.9.

Table 4 summarizes absolute and relative numbers for each of the three distributional categories relative to the nine ecological formations under study. The following conclusions can be made based on those data:

1. WIDESPREAD species are most numerous in eight of nine ecological formations, the exception being LDF(WI), in which the percentage of widespread species is only 33.3%. That figure is the result of that ecological formation being very small and occurring only on isolated islands in the Caribbean. The percent representations of the remaining eight formations range from 50.0% in the LMMF formation to 100.0% in the LAF formation, with a mean value of 73.6% for all nine formations. The relatively low percentage for the LMMF formation is the result of a sizable representation of both restricted (21.4%) and peripheral (28.6%) species in that formation. The relatively low figure for the LMWF (65.4%) formation is because of a relatively sizable representation of both restricted (15.4%) and peripheral (19.2%) species in that formation. Widespread species make up the entire fauna in LAF and almost the entire

fauna in the PDF (92.3%) formation, with the latter having only two restricted and no peripheral species.

2. The largest percentage of RESTRICTED species is found on the isolated islands in the Caribbean of the LDF(WI) formation (66.7%). The next largest percentage is in the LMF formation (30.1%), with relatively sizable representations being found in the LMMF and LMWF formations (21.4% and 15.9%, respectively). The percent representations for restricted species range from 0.0% in the LAF and PWF formations to 66.7% in the LDF(WI) formation, with a mean value of 13.3%. With the exception of the isolated island LDF(WI), restricted species are generally relatively increasingly less common on the mainland the drier the formation.

3. PERIPHERAL species vie with restricted species for being least commonly represented in a given ecological formation. They are absent in LAF, PDF, and LDF(WI) formations. Peripheral species are most common in PMF (30.5%), LMMF (28.6%), PWF (26.5%), and LMWF (19.2%) formations. Peripheral species average 13.3% per formation.

As a group, lizards, crocodiles, and nonmarine turtles in Honduras are known to range from sea level to 2,530 m elevation (Table 3). Lizards are known from sea level to 2,530 m (*Mesaspis moreletii* and cf. *Sceloporus hondurensis*), crocodiles from sea level to 650 m (*Crocodylus acutus*), and nonmarine turtles from sea level to 1,480 m (*Trachemys scripta* and *Rhinoclemmys pulcherrima*; the 1,480 m elevation record for *Trachemys scripta* [an introduced species in Honduras] is a released animal that would probably not survive long at that elevation). Placing each lizard, crocodile, and turtle species into the elevational categories of Stuart (1963) illustrates the following, broken down by order: 1. Low elevations (0–600 m)—67 lizards, 2 croco-

diles, 12 nonmarine turtles; 2. Moderate elevations (601–1,500 m)—66 lizards, 1 crocodile, 6 nonmarine turtles; 3. Intermediate elevations (1,501–2,700 m)—34 lizards, 0 crocodiles, 0 turtles. Mean elevational spans are as follows: lizards 638.1 m, crocodiles 465.0 m, and nonmarine turtles 687.9 m. Pincheira-Donoso et al. (2013) pointed out a similar worldwide failure of crocodilians and turtles to adapt to colder climates, with most species dropping out by 650 m elevation.

The number of species of lizard (including four human-aided introductions of species), crocodile, and nonmarine turtle species (including one human-aided species introduction) found at various elevations declines more or less gradually and consistently with an increase in elevation (Table 5). There are, however, two slight increases of two species between elevations of 401 and 600 m and four species between 501 and 700. As can be seen in Table 5, marked decreases in species numbers occur between elevations of 0 and 200 m (30 species, of which 22 are lizards and 8 are turtles), 901 and 1,100 m (10 species, of which 9 are lizards and 1 is a turtle), 1,301 and 1,500 m (7 lizards), 1,501 and 1,700 (6 lizards), 1,701 and 1,900 (4 lizards), 1,801 and 2,000 (5 lizards), 1,901 and 2,100 (4 lizards), and 2,101 and 2,300 (3 lizards).

The decrease between 0 and 200 m is primarily due to a relatively large number of species (30) that are only known to occur at very low elevations along the northern coast of Honduras, in the Mosquitia, or in insular environments. The decrease from 901 to 1,100 m (10 species) is primarily due to loss of several species of lizards whose elevational ranges begin near sea level and extend no farther than a third or quarter of the way into the moderate elevations. Another significant drop in species numbers occurs at the upper limits of premontane vegetation in the country between 1,301 and 1,500 m (7 species), before cloud forest

TABLE 5. NUMBERS OF LIZARD, CROCODILE, AND TURTLE SPECIES (INCLUDING MARINE SPECIES) FOUND AT VARIOUS ELEVATIONS IN HONDURAS.

Elevational Segments (m)	Lizards	Crocodiles	Turtles	Totals
0–100	69	2	17	88
101–200	47	2	9	58
201–300	43	2	8	53
301–400	42	1	8	51
401–500	41	1	7	49
501–600	43	1	7	51
601–700	48	1	6	55
701–800	47	—	6	53
801–900	45	—	5	50
901–1,000	43	—	5	48
1,001–1,100	34	—	4	38
1,101–1,200	34	—	4	38
1,201–1,300	33	—	3	36
1,301–1,400	34	—	2	36
1,401–1,500	27	—	2	29
1,501–1,600	28	—	—	28
1,601–1,700	22	—	—	22
1,701–1,800	21	—	—	21
1,801–1,900	17	—	—	17
1,901–2,000	12	—	—	12
2,001–2,100	8	—	—	8
2,101–2,200	6	—	—	6
2,201–2,300	3	—	—	3
2,301–2,400	2	—	—	2
2,401–2,500	2	—	—	2
2,501–2,600	2	—	—	2

vegetation is reached. Beyond this point, species begin to reach their upper elevational range limitations relatively rapidly, as mean annual temperatures drop below the lower limit of thermal tolerance for most reptiles. It is not known whether the two lizards reported from 2,530 m, or any other lizard species, occur above that elevation. Elevations above 2,530 m do occur on cerros Celaque and El Pital and on Montaña Santa Bárbara, so it would be interesting to determine whether any lizards are part of the unknown Montane Rainforest herpetofauna.

DISTRIBUTION WITHIN PHYSIOGRAPHIC REGIONS

The distribution of the Honduran species of lizards (including four human-aided species introductions), crocodiles, and nonmarine turtles (including one human-aided species introduction) in 12 physiographic regions (Maps 5, 86) is indicated in Table 6. McCranie (2011a) included 11 physiographic regions in his study of the Honduran snake fauna. A Cayos Miskitos physiographic region is added herein to accommodate the lizard *Aristelliger* sp. A. The physiographic regions and number of reptilian species known for each region are as follows (from highest number of species to lowest number; physiographic region abbreviations as in Table 6): NC 78; SC 47; MC 40; UCP 35; NDP 33; BI 31; ANP 25; PLR 24; MP 19; CC 12; SI 8; and CM 1. The two regions with the highest number of species are also, by far, the two largest regions. Likewise, the four smallest regions also harbor the lowest numbers of species. The Northern Cordillera is the largest and most complex area in topography and vegetation, containing six of the nine ecological formations; thus, it is not surprising that it contains the largest number of species.

With reference to the reptilian orders under study herein, the 107 lizard species (Squamata) collectively inhabit all 12 regions, as follows (highest to lowest): NC 67; SC 41; MC 32; UCP 28; BI 27; NDP 26; PLR 20; ANP 19; MP 14; CC 10; SI 8; and CM 1. The two crocodile species (Crocodylia) occur collectively in 8 of 12 physiographic regions, as follows (highest to lowest): UCP 2; NDP 2; ANP 2; MC 2; NC 2; PLR 1; MP 1; and BI 1. The 12 nonmarine turtles (Testudinata) collectively occupy 10 of 12 regions, as follows (highest to lowest): NC 9; MC 6; SC 6; UCP 5; NDP 5; MP 4; PLR 3; ANP 3; BI 3; and CC 2. The highest total number of physiographic regions inhabited by a lizard is 11 (*Iguana iguana*), a crocodilian is 8 (*Crocodylus acutus*), and a nonmarine turtle is 9 (*Kinosternon leucostomum* and *Trachemys venusta*).

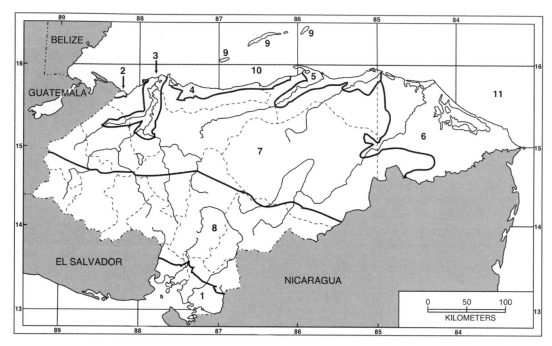

Map 86.   Physiographic regions of Honduras. (1) Pacific lowland region; (2) Motagua Plain of Caribbean lowland region; (3) Ulúa-Chamelecón Plain of Caribbean lowland region; (4) Nombre de Dios Piedmont of Caribbean lowland region; (5) Aguán-Negro Plain of Caribbean lowland region; (6) Mosquito Coast of Caribbean lowland region; (7) Northern Cordillera of Serranía region; (8) Southern Cordillera of Serranía region; (9) Bay Islands; (10) Cayos Cochinos; (11) Cayos Miskitos.

DISTRIBUTION WITHIN ECOPHYSIOGRAPHIC AREAS

McCranie (2011a) analyzed the Honduran snake fauna in 33 ecophysiographic areas. Two areas were combined (areas 2 and 3) because of their continuous nature along the same river, the middle and upper Río Choluteca valleys. Those two areas are also combined into one area for this study. In this work, I add three ecophysiographic areas: the Yoro Highlands, San Esteban Valley, and Cayos Miskitos (Table 7; Maps 5, 87) that indicate the distribution of the lizard, crocodile, and nonmarine turtle species in 36 areas (including the combined Middle and Upper Choluteca Valley; thus, only 35 areas are shown in Table 8) of the 40 ecophysiographic areas recognized in Honduras (ecophysiographic areas briefly defined in Table 8). Four ecophysiographic areas are not included in Table 7 (those with

an asterisk in Table 8), because their reptile faunas are poorly known. The ecophysiographic areas and their known number of lizard, crocodile, and nonmarine turtle species are as follows: (Table 7; from highest number to lowest; area names for each number are shown in Table 8):

Area 21—41 species (32 lizards; 2 crocodiles; 7 turtles);

Area 26—36 species (30 lizards; 2 crocodiles; 4 turtles);

Area 27—35 species (29 lizards; 2 crocodiles; 4 turtles);

Area 22—32 species (27 lizards; 2 crocodiles; 3 turtles);

Area 29—29 species (22 lizards; 1 crocodile; 6 turtles);

Area 31—29 species (24 lizards; 1 crocodile; 4 turtles);

Area 9—27 species (23 lizards; 4 turtles);

Area 16—27 species (27 lizards);

Area 30—27 species (26 lizards; 1 turtle);
Area 15—26 species (25 lizards; 1 turtle);
Area 3—25 species (22 lizards; 3 turtles);
Area 1—24 species (20 lizards; 1 crocodile; 3 turtles);
Area 24—24 species (19 lizards; 1 crocodile; 4 turtles);
Area 14—22 species (18 lizards; 4 turtles);
Area 8—20 species (19 lizards; 1 turtle);
Area 28—19 species (14 lizards; 1 crocodile; 4 turtles);
Area 12—18 species (13 lizards; 1 crocodile; 4 turtles);
Area 35—18 species (16 lizards; 1 crocodile; 1 turtle);
Area 13—16 species (13 lizards; 3 turtles);
Area 36—15 species (13 lizards; 1 crocodile; 1 turtle);
Area 37—14 species (10 lizards; 1 crocodile; 3 turtles);
Area 5—12 species (10 lizards; 2 turtles);
Area 38—12 species (10 lizards; 2 turtles);
Area 11—11 species (11 lizards);
Area 32—11 species (11 lizards);
Area 7—10 species (9 lizards; 1 turtle);
Area 17—10 species (10 lizards);
Area 23—10 species (9 lizards; 1 turtle);
Area 39—9 species (9 lizards);
Area 10—7 species (7 lizards);
Area 19—5 species (5 lizards);
Area 20—4 species (4 lizards);
Area 18—3 species (3 lizards);
Area 33—3 species (3 lizards); and
Area 40—1 species (1 lizard).

The four areas with the greatest number of species are all low-elevation Atlantic versant mesic areas (areas 21, 26, 27, and 22) with 32–41 species. Two areas (29 and 31) are tied for fifth greatest number of species, with 29 each. Area 29 is a lowland subhumid Atlantic versant locality with several mesic gallery forest corridors, and area 31 is a moderate-elevation humid Atlantic versant locality. Thus, it appears that elevation and amount of precipitation, along with size of a given area, are all affecting the total number of species

present in a given area. Among the orders studied herein, *Basiliscus vittatus* (Squamata) is known to occur in 25 of 36 areas, *Crocodylus acutus* (Crocodylia) in 12 areas, and *Kinosternon leucostomum* (Testudinata) in 17 areas.

BROAD PATTERNS OF GEOGRAPHIC DISTRIBUTION

The 11 broad patterns of geographical distribution used herein for all 126 species are as follows:

A. Northern terminus of range in US and southern terminus in Central America south of Nicaraguan Depression;
B. Northern terminus of range in Mexico north of Isthmus of Tehuantepec and southern terminus in South America;
C. Northern terminus of range in Mexico north of Isthmus of Tehuantepec and southern terminus in Central America south of Nicaraguan Depression;
D. Northern terminus of range in Mexico north of Isthmus of Tehuantepec and southern terminus in Nuclear Middle America;
E. Northern terminus of range in Nuclear Middle America and southern terminus in South America;
F. Northern terminus of range in Nuclear Middle America and southern terminus in Central America south of Nicaraguan Depression;
G. Restricted to Nuclear Middle America (exclusive of Honduran endemics);
H. Endemic to Honduras (including insular endemics);
I. Marine species;
J. Insular species (not including Honduran endemic species);
K. Introduced species.

The allocation of species to these categories is as follows:

A—*Sceloporus variabilis* (more than one species most likely involved under that nominal form);

TABLE 6. DISTRIBUTION OF THE HONDURAN LIZARD, CROCODILE, AND NON-MARINE TURTLE SPECIES BY 12 PHYSIOGRAPHIC REGIONS (I = INTRODUCED). ABBREVIATIONS USED IN THIS AND SOME SUBSEQUENT TABLES ARE AS FOLLOWS: PLR = PACIFIC LOWLAND REGION; MP = MOTAGUA PLAIN; UCP = ULÚA-CHAMELECÓN PLAIN; NDP = NOMBRE DE DIOS PIEDMONT; ANP = AGUÁN-NEGRO PLAIN; MC = MOSQUITO COAST; NC = NORTHERN CORDILLERA; SC = SOUTHERN CORDILLERA; BI = BAY ISLANDS; CC = CAYOS COCHINOS; SI = SWAN ISLANDS; CM = CAYOS MISKITOS.

Species (121)	PLR	MP	UCP	NDP	ANP	MC	NC	SC	BI	CC	SI	CM	Total
Lizards (107)													
Lepidophyma flavimaculatum	—	X	X	X	—	X	X	—	—	—	—	—	5
Lepidophyma mayae	—	—	—	—	—	—	X	—	—	—	—	—	1
Coleonyx mitratus	X	—	X	X	—	—	X	X	X	—	—	—	6
Hemidactylus frenatus (I)	X	X	X	X	—	X	X	X	X	X	X	—	10
Hemidactylus haitianus (I)	—	—	—	—	—	—	—	X	—	—	—	—	1
Hemidactylus mabouia (I)	—	—	—	—	—	—	X	—	X	—	X	—	3
Phyllodactylus palmeus	—	—	—	—	—	—	—	—	X	X	—	—	2
Phyllodactylus paralepis	—	—	—	—	—	—	—	—	X	—	—	—	1
Phyllodactylus tuberculosus	X	—	—	—	—	—	—	X	—	—	—	—	2
Thecadactylus rapicauda	—	—	X	X	X	X	X	—	X	—	—	—	6
Aristelliger sp. A	—	—	—	—	—	—	—	—	—	—	—	X	1
Aristelliger nelsoni	—	—	—	—	—	—	—	—	—	—	X	—	1
Gonatodes albogularis	X	—	—	—	—	X	X	X	—	—	—	—	4
Sphaerodactylus alphus	—	—	—	—	—	—	—	—	X	—	—	—	1
Sphaerodactylus continentalis	—	X	X	X	X	—	X	X	X	—	—	—	7
Sphaerodactylus dunni	—	—	—	X	X	X	X	—	—	—	—	—	4
Sphaerodactylus exsul	—	—	—	—	—	—	—	—	—	—	X	—	1
Sphaerodactylus glaucus	—	—	—	—	—	—	—	X	—	—	—	—	1
Sphaerodactylus guanajae	—	—	—	—	—	—	—	—	X	—	—	—	1
Sphaerodactylus leonardovaldesi	—	—	—	—	—	—	—	—	X	—	—	—	1
Sphaerodactylus millepunctatus	—	—	—	—	—	X	X	—	—	—	—	—	2
Sphaerodactylus poindexteri	—	—	—	—	—	—	—	—	X	—	—	—	1
Sphaerodactylus rosaurae	—	—	—	—	—	—	—	—	X	X	—	—	2
Abronia montecristoi	—	—	—	—	—	—	X	—	—	—	—	—	1
Abronia salvadorensis	—	—	—	—	—	—	—	X	—	—	—	—	1
Mesaspis moreletii	—	—	—	—	—	—	X	X	—	—	—	—	2
Diploglossus bivittatus	—	—	—	—	—	—	—	X	—	—	—	—	1
Diploglossus montanus	—	—	—	—	—	—	X	—	—	—	—	—	1
Diploglossus scansorius	—	—	—	—	—	—	X	—	—	—	—	—	1
Basiliscus plumifrons	—	—	—	—	—	X	X	—	—	—	—	—	2
Basiliscus vittatus	X	X	X	X	X	X	X	X	X	X	—	—	10
Corytophanes cristatus	—	—	X	X	X	X	X	—	—	—	—	—	5
Corytophanes hernandesii	—	—	—	—	—	—	X	—	—	—	—	—	1
Corytophanes percarinatus	—	—	—	—	—	—	—	X	—	—	—	—	1
Laemanctus julioi	—	—	—	—	—	—	—	X	—	—	—	—	1
Laemanctus longipes	—	—	—	—	—	—	X	X	—	—	—	—	2
Laemanctus serratus	—	—	—	—	—	—	—	—	—	—	—	—	0
Laemanctus waltersi	—	X	X	X	—	—	—	—	—	—	—	—	3
Anolis allisoni	—	—	X	—	—	—	—	—	X	X	—	—	3
Norops amplisquamosus	—	—	—	—	—	—	X	—	—	—	—	—	1
Norops beckeri	—	—	X	X	X	X	X	—	—	—	—	—	5
Norops bicaorum	—	—	—	—	—	—	—	—	—	X	—	—	1
Norops biporcatus	—	—	—	X	X	X	X	X	—	—	—	—	5
Norops capito	—	—	—	—	X	X	X	X	—	—	—	—	4
Norops carpenteri	—	—	—	—	—	X	—	—	—	—	—	—	1
Norops crassulus	—	—	—	—	—	X	X	—	—	—	—	—	2
Norops cupreus	X	—	—	—	—	X	X	X	—	—	—	—	4
Norops cusuco	—	—	—	—	—	—	X	—	—	—	—	—	1
Norops heteropholidotus	—	—	—	—	—	—	—	X	—	—	—	—	1
Norops johnmeyeri	—	—	—	—	—	—	X	—	—	—	—	—	1
Norops kreutzi	—	—	—	—	—	—	X	—	—	—	—	—	1
Norops johnmeyeri	—	—	—	—	—	—	X	—	—	—	—	—	1
Norops laeviventris	—	—	—	—	—	—	X	X	—	—	—	—	2

TABLE 6. CONTINUED.

Species (121)	PLR	MP	UCP	NDP	ANP	MC	NC	SC	BI	CC	SI	CM	Total
Norops lemurinus	—	X	X	X	X	X	X	X	—	X	—	—	8
Norops limifrons	—	—	—	—	X	X	X	—	—	—	—	—	3
Norops loveridgei	—	—	—	—	—	—	X	—	—	—	—	—	1
Norops mccraniei	—	—	X	—	X	X	X	X	—	—	—	—	5
Norops morazani	—	—	—	—	—	—	X	—	—	—	—	—	1
Norops muralla	—	—	—	—	—	—	X	—	—	—	—	—	1
Norops nelsoni	—	—	—	—	—	—	—	—	—	—	X	—	1
Norops ocelloscapularis	—	—	—	—	—	—	X	—	—	—	—	—	1
Norops oxylophus	—	—	—	—	—	X	X	—	—	—	—	—	2
Norops petersii	—	—	—	—	—	—	X	—	—	—	—	—	1
Norops pijolense	—	—	—	—	—	—	X	—	—	—	—	—	1
Norops purpurgularis	—	—	—	—	—	—	X	—	—	—	—	—	1
Norops quaggulus	—	—	—	—	—	X	X	—	—	—	—	—	2
Norops roatanensis	—	—	—	—	—	—	—	—	X	—	—	—	1
Norops rodriguezii	—	X	X	—	—	—	X	—	—	—	—	—	3
Norops rubribarbaris	—	—	—	—	—	—	X	—	—	—	—	—	1
Norops sagrei (I)	—	—	X	X	—	—	—	—	X	—	—	—	3
Norops sminthus	—	—	—	—	—	—	—	X	—	—	—	—	1
Norops uniformis	—	X	—	—	—	—	X	—	—	—	—	—	2
Norops unilobatus	—	X	X	X	X	X	X	X	X	—	—	—	8
Norops utilensis	—	—	—	—	—	—	—	—	X	—	—	—	1
Norops wampuensis	—	—	—	—	—	—	X	—	—	—	—	—	1
Norops wellbornae	X	—	—	—	—	—	X	—	—	—	—	—	2
Norops wermuthi	—	—	—	—	—	—	X	—	—	—	—	—	1
Norops wilsoni	—	—	X	X	X	—	X	—	—	—	—	—	4
Norops yoroensis	—	—	—	—	—	—	X	—	—	—	—	—	1
Norops zeus	—	—	X	X	—	—	X	—	—	—	—	—	3
Ctenosaura bakeri	—	—	—	—	—	—	—	—	X	—	—	—	1
Ctenosaura flavidorsalis	X	—	—	—	—	—	—	X	—	—	—	—	2
Ctenosaura melanosterna	—	—	—	—	—	—	X	—	—	X	—	—	2
Ctenosaura oedirhina	—	—	—	—	—	—	—	—	X	—	—	—	1
Ctenosaura quinquecarinata	X	—	—	—	—	—	—	X	—	—	—	—	2
Ctenosaura similis	X	X	X	X	X	X	X	X	X	—	—	—	9
Iguana iguana	X	X	X	X	X	X	X	X	X	X	X	—	11
Leiocephalus varius (I)	—	—	—	—	—	—	—	—	—	—	X	—	1
Sceloporus esperanzae	—	—	—	—	—	—	—	X	—	—	—	—	1
Sceloporus hondurensis	—	—	—	—	—	—	X	X	—	—	—	—	2
Sceloporus schmidti	—	—	—	—	—	—	X	X	—	—	—	—	2
Sceloporus squamosus	X	—	X	—	—	—	—	X	—	—	—	—	3
Sceloporus variabilis	X	—	X	—	—	X	X	X	—	—	—	—	5
Polychrus gutturosus	—	—	X	—	—	X	X	—	—	—	—	—	3
Marisora brachypoda	X	—	X	X	X	X	X	X	X	—	—	—	8
Marisora roatanae	—	—	—	—	—	—	—	—	X	—	—	—	1
Scincella assata	X	—	—	—	—	—	—	—	—	—	—	—	1
Scincella cherriei	—	X	X	X	—	X	X	X	—	X	—	—	7
Scincella incerta	—	—	—	—	—	—	X	—	—	—	—	—	1
Mesoscincus managuae	X	—	—	—	—	—	—	X	—	—	—	—	2
Plestiodon sumichrasti	—	—	—	X	—	—	X	—	—	—	—	—	2
Gymnophthalmus speciosus	X	—	—	—	—	X	X	X	X	—	—	—	5
Ameiva fuliginosa	—	—	—	—	—	—	—	—	—	—	X	—	1°
Aspidoscelis deppii	X	—	X	X	X	X	X	X	—	—	—	—	7
Aspidoscelis motaguae	X	—	—	—	—	—	X	X	—	—	—	—	3
Cnemidophorus ruatanus	—	X	X	X	X	X	X	—	X	X	—	—	8
Holcosus festivus	—	—	X	X	X	X	X	—	—	—	—	—	5
Holcosus undulatus	X	X	X	X	X	X	X	X	—	—	—	—	8
Subtotals	20	14	28	26	20	32	67	41	27	10	8	1	294

TABLE 6.   CONTINUED.

Species (121)	PLR	MP	UCP	NDP	ANP	MC	NC	SC	BI	CC	SI	CM	Total
Crocodiles (2)													
Caiman crocodilus	—	—	X	X	X	X	X	—	—	—	—	—	5
Crocodylus acutus	X	X	X	X	X	X	X	—	X	—	—	—	8
Subtotals	1	1	2	2	2	2	2	0	1	0	0	0	13
Turtles (12)													
Chelydra acutirostris	—	—	—	—	—	X	X	X	—	—	—	—	3
Chelydra rossignonii	—	X	X	X	—	—	X	—	—	—	—	—	4
Kinosternon albogulare	X	X	—	—	—	X	X	X	—	—	—	—	5
Kinosternon leucostomum	—	X	X	X	X	X	X	X	X	X	—	—	9
Staurotypus triporcatus	—	—	X	X	—	—	—	—	—	—	—	—	2
Trachemys emolli	X	—	—	—	—	—	—	—	—	—	—	—	1
Trachemys scripta (I)	—	—	—	X	—	—	—	X	X	—	—	—	3
Trachemys venusta	—	X	X	X	X	X	X	X	X	X	—	—	9
Rhinoclemmys annulata	—	—	—	—	X	X	X	—	—	—	—	—	3
Rhinoclemmys areolata	—	—	—	—	—	—	X	—	—	—	—	—	1
Rhinoclemmys funerea	—	—	—	—	—	X	X	—	—	—	—	—	2
Rhinoclemmys pulcherrima	X	—	X	—	—	—	X	X	—	—	—	—	4
Subtotals	3	4	5	5	3	6	9	6	3	2	0	0	46
Totals	24	19	35	33	25	40	78	47	31	12	8	1	355

B—*Basiliscus vittatus, Norops biporcatus, Iguana iguana, Crocodylus acutus* (isolated populations also occur in extreme southern Florida and the West Indies; however, more than one species is likely involved, see Milián-García et al., 2011), *Kinosternon leucostomum, Trachemys venusta*;

C—*Lepidophyma flavimaculatum, Phyllodactylus tuberculosus, Norops lemurinus, Marisora brachypoda* (more than one species involved in this complex); *Scincella cherriei, Aspidoscelis deppii, Holcosus undulatus, Kinosternon albogulare, Rhinoclemmys pulcherrima* (more than one species probably involved);

D—*Sphaerodactylus glaucus, Corytophanes hernandesii, Laemanctus longipes, L. serratus, Norops laeviventris, N. petersii, N. uniformis, Scincella assata, Plestiodon sumichrasti, Aspidoscelis motaguae, Chelydra rossignonii, Staurotypus triporcatus, Rhinoclemmys areolata*;

E—*Thecadactylus rapicauda, Gonatodes albogularis, Corytophanes cristatus, Polychrus gutturosus, Gymnophthal-*

mus speciosus, Holcosus festivus, Caiman crocodilus, Chelydra acutirostris, Rhinoclemmys annulata;

F—*Coleonyx mitratus, Sphaerodactylus millepunctatus, Basiliscus plumifrons, Norops capito, N. carpenteri, N. cupreus, N. limifrons, N. oxylophus, N. quaggulus, N. unilobatus, Ctenosaura quinquecarinata, C. similis, Sceloporus squamosus, Mesoscincus managuae, Trachemys emolli, Rhinoclemmys funerea*;

G—*Lepidophyma mayae, Sphaerodactylus continentalis, Abronia montecristoi, Mesaspis moreletii, Diploglossus bivittatus, D. montanus, Corytophanes percarinatus, Norops beckeri, N. crassulus, N. heteropholidotus, N. mccraniei, N. ocelloscapularis, N. rodriguezii, N. wellbornae, N. wermuthi, Ctenosaura flavidorsalis, Sceloporus hondurensis, S. schmidti, Scincella incerta, Cnemidophorus ruatanus*;

H—*Phyllodactylus palmeus, P. paralepis, Aristelliger* sp. A, *A. nelsoni, Sphaerodactylus alphus, S. dunni, S. exsul, S. guanajae, S. leonardovaldesi, S. poindexteri, S. rosaurae, Abronia salvador-*

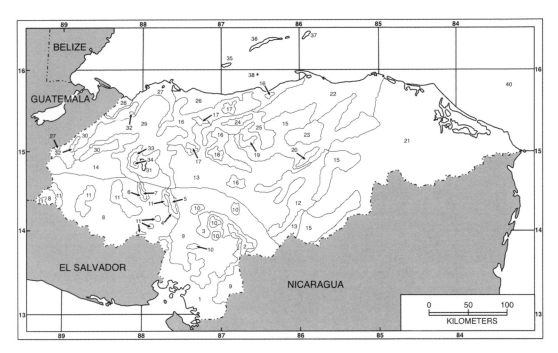

Map 87.   Ecophysiographic areas of Honduras: 1 = Pacific Lowlands; 2 = Middle Choluteca Valley; 3 = Upper Choluteca Valley; 4 = Comayagua Valley Rim; 5 = Comayagua Valley; 6 = Otoro Valley Rim; 7 = Otoro Valley; 8 = Southwestern Uplands; 9 = Southeastern Uplands; 10 = Southeastern Highlands; 11 = Southwestern Highlands; 12 = Guayape-Guayambre Valley; 13 = Northeastern Uplands; 14 = Northwestern Uplands; 15 = Eastern Caribbean Slope; 16 = Central Caribbean Slope; 17 = North-central Highlands; 18 = Yoro Highlands; 19 = Ocote Highlands; 20 = Agalta Highlands; 21 = Eastern Caribbean Lowlands; 22 = East-central Caribbean Lowlands; 23 = San Esteban Valley; 24 = Aguán Valley; 25 = Aguán Valley Rim; 26 = West-central Caribbean Lowlands; 27 = Western Caribbean Lowlands; 28 = Lower Motagua Valley; 29 = Sula Valley; 30 = Western Caribbean Slope; 31 = Yojoa Uplands; 32 = Northwestern Highlands; 33 = Santa Bárbara Highlands; 34 = Santa Bárbara Peak; 35 = Utila Island; 36 = Roatán Island; 37 = Guanaja Island; 38 = Cayos Cochinos; 40 = Cayos Vivorillos. See Map 5 for the location of area 39, the Swan Islands (Islas del Cisne).

ensis, *Diploglossus scansorius, Laemanctus julioi, L. waltersi, Norops amplisquamosus, N. bicaorum, N. cusuco, N. johnmeyeri, N. kreutzi, N. loveridgei, N. morazani, N. muralla, N. nelsoni, N. pijolense, N. purpurgularis, N. roatanensis, N. rubribarbaris, N. sminthus, N. utilensis, N. wampuensis, N. wilsoni, N. yoroensis, N. zeus, Ctenosaura bakeri, C. melanosterna, C. oedirhina, Sceloporus esperanzae, Marisora roatanae;*

I—*Caretta caretta, Chelonia mydas, Eretmochelys imbricata, Lepidochelys olivacea, Dermochelys coriacea;*

J—*Anolis allisoni, Ameiva fuliginosa;*

K—*Hemidactylus frenatus, H. haitianus, H. mabouia, Norops sagrei, Leiocephalus varius, Trachemys scripta.*

The number of species in each distribution category is as follows (highest to lowest):

H—39 species (31.0% of total of 126 species of reptilian fauna under study herein);

G—20 species (15.9%);

F—16 species (12.7%);

D—13 species (10.3%);

C—9 species (7.1%);

E—9 species (7.1%);

B—6 species (4.7%);

K—6 species (4.7%);

I—5 species (4.0%);

TABLE 7. THE LIZARDS, CROCODILES, AND NON-MARINE TURTLES DISTRIBUTIONAL RECORDS FOR 36 ECOPHYSIOGRAPHIC AREAS IN HONDURAS. AREAS 2 AND 3 ARE COMBINED FOR THIS ANALYSIS (THUS ONLY 35 LISTED). SEE TABLE 8 FOR THE CORRESPONDING AREA NAME FOR EACH AREA NUMBER. I = INTRODUCED SPECIES TO HONDURAN TERRITORY.

Species (126)	1	3	5	7	8	9	10	11	12	13	14	15	16	17	18	19	20	21	22	23	24	26	27	28	29	30	31	32	33	35	36	37	38	39	40	Total
Lizards (107)																																				
Lepidophyma flavimaculatum									X			X	X					X	X			X	X	X	X	X	X									11
Lepidophyma mayae																							X			X										2
Coleonyx mitratus	X	X	X		X				X	X		X								X	X	X	X	X	X		X			X	X		X			13
Hemidactylus frenatus (I)	X	X			X	X			X	X	X	X	X					X		X	X	X	X	X	X	X	X			X	X	X	X	X		19
Hemidactylus haitianus (I)		X																																		1
Hemidactylus mabouia (I)																						X									X	X		X		3
Phyllodactylus palmeus																															X	X	X			3
Phyllodactylus paralepis																																	X			1
Phyllodactylus tuberculosus	X	X					X												X																	3
Thecadactylus rapicauda												X						X	X	X	X	X	X	X	X		X			X					9
Aristelliger sp. A																																		X		1
Aristelliger nelsoni																																		X		1
Gonatodes albogularis	X	X				X			X									X			X	X	X	X	X	X	X									6
Sphaerodactylus alphus																																X				1
Sphaerodactylus continentalis		X							X	X	X	X	X					X	X		X	X	X	X	X	X	X									14
Sphaerodactylus dunni																			X		X	X	X	X	X							X		X		6
Sphaerodactylus exsul																									X											1
Sphaerodactylus glaucus											X																									1
Sphaerodactylus guanajae																																X				1
Sphaerodactylus leonardovaldesi																																	X			1
Sphaerodactylus millepunctatus																		X	X																	2
Sphaerodactylus poindexteri																		X												X	X					1
Sphaerodactylus rosaurae																		X										X		X	X	X				3
Abronia montecristoi																												X								1
Abronia salvadorensis								X	X																											1
Mesaspis moreletii								X	X						X	X	X											X	X							7
Diploglossus bivittatus					X														X		X	X	X													2
Diploglossus montanus																			X		X	X	X		X	X	X									3
Diploglossus scansorius											X			X					X		X															2
Basiliscus plumifrons																		X	X		X	X	X													2
Basiliscus vittatus	X	X	X	X	X	X			X	X	X	X	X					X	X	X	X	X	X	X	X	X	X			X	X	X	X	X		25
Corytophanes cristatus												X	X					X	X		X	X	X	X	X	X	X									9
Corytophanes hernandesii			X															X								X										2
Corytophanes percarinatus									X																											2
Laemanctus julioi		X			X																															1
Laemanctus longipes											X	X																								2

TABLE 7. CONTINUED.

Species (126)	1	3	5	7	8	9	10	11	12	13	14	15	16	17	18	19	20	21	22	23	24	26	27	28	29	30	31	32	33	35	36	37	38	39	40	Total
Laemanctus serratus																																				0
Laemanctus waltersi																								X	X	X	X									5
Anolis allisoni																															X	X	X	X		5
Norops amplisquamosus																						X					X									1
Norops beckeri											X	X	X					X	X		X	X	X		X			X								8
Norops bicaorum																														X						1
Norops biporcatus					X			X	X	X	X	X	X					X	X		X	X	X	X	X	X	X									11
Norops capito																		X	X			X	X	X	X	X	X									7
Norops carpenteri																		X																		1
Norops crassulus	X	X			X	X		X		X							X	X				X														3
Norops cupreus						X				X		X	X					X	X							X	X									8
Norops cusuco																										X										2
Norops heteropholidotus								X																												1
Norops johnmeyeri														X											X	X	X									2
Norops kreutzi														X																						2
Norops laeviventris					X	X	X	X				X	X		X				X		X	X	X	X	X	X	X									9
Norops lemurinus					X	X	X	X	X	X	X	X	X				X	X	X	X	X	X	X	X	X	X	X		X		X					12
Norops limifrons																			X			X														3
Norops loveridgei			X	X	X	X		X	X	X	X	X	X	X						X		X			X		X									3
Norops mccraniei			X	X	X	X		X	X	X	X	X	X			X	X	X	X	X	X				X		X									18
Norops morazani															X	X					X															2
Norops muralla												X				X																				2
Norops nelsoni																																		X		1
Norops ocelloscapularis																								X		X		X								2
Norops oxylophus																	X																			1
Norops petersii																								X		X		X								2
Norops pijolense													X	X																						2
Norops purpurgularis													X	X																						1
Norops quaggulus												X						X																		3
Norops roatanensis											X																				X					5
Norops rodriguezii																								X	X	X			X		X					1
Norops rubribarbaris												X	X																		X	X				5
Norops sagrei (1)																												X								5
Norops sminthus	X	X				X				X	X	X	X					X	X	X	X	X	X	X	X	X	X									2
Norops uniformis	X	X																																		5
Norops unilobatus						X				X	X	X	X					X	X	X	X	X	X	X	X	X	X									2
Norops utilensis																														X	X					5
Norops wampuensis																		X	X								X									16
Norops wellbornae	X	X			X	X												X	X																	1
Norops wermuthi						X																														4

TABLE 7.   CONTINUED.

Species (126)	1	3	5	7	8	9	10	11	12	13	14	15	16	17	18	19	20	21	22	23	24	26	27	28	29	30	31	32	33	35	36	37	38	39	40	Total
Norops wilsoni	—	—	—	—	—	—	—	—	—	—	—	—	X	—	—	—	—	—	—	—	—	X	—	—	—	X	—	—	—	—	—	—	—	—	—	3
Norops yoroensis	—	—	—	—	—	—	—	—	—	—	—	—	X	X	—	—	—	—	—	—	—	—	—	—	—	X	—	—	—	—	—	—	—	—	—	3
Norops zeus	—	—	—	—	—	—	—	—	—	—	—	—	X	—	—	—	—	—	—	—	—	—	X	—	X	—	—	—	—	—	—	—	—	—	—	4
Ctenosaura bakeri	—	—	—	—	X	—	—	—	—	—	—	—	—	—	—	—	—	—	—	—	—	—	—	—	—	—	—	—	—	—	—	—	—	—	—	1
Ctenosaura flavidorsalis	X	X	X	—	X	—	—	—	—	—	—	—	—	—	—	—	—	—	—	—	—	—	—	—	—	—	—	—	—	—	—	—	—	—	—	3
Ctenosaura melanosterna	—	—	—	—	—	—	—	—	—	—	—	—	—	—	—	—	—	—	—	—	X	—	—	—	—	—	—	—	—	—	—	—	X	—	—	2
Ctenosaura oedirhina	—	—	—	—	—	—	—	—	—	—	—	—	—	—	—	—	—	—	—	—	—	—	—	—	—	—	—	—	—	—	X	—	—	—	—	1
Ctenosaura quinquecarinata	X	X	—	—	—	—	—	—	—	—	—	—	—	—	—	—	—	—	—	—	—	—	—	—	—	—	—	—	—	X	—	—	—	—	—	3
Ctenosaura similis	X	X	X	X	X	X	—	—	X	X	X	X	X	X	X	X	—	X	X	X	X	X	X	X	X	X	—	—	—	X	X	X	X	—	—	18
Iguana iguana	X	X	—	—	—	—	—	—	—	—	X	X	X	X	—	—	—	X	X	—	X	X	X	X	X	—	X	—	—	X	X	X	X	X	—	16
Leiocephalus varius (1)	—	—	—	—	—	—	—	—	—	—	—	—	—	—	—	—	—	—	—	—	—	—	—	—	—	—	—	—	—	—	—	—	X	—	—	1
Sceloporus esperanzae	—	—	—	—	—	—	X	X	—	—	—	—	—	—	—	—	—	—	—	—	—	—	—	—	—	—	—	—	—	—	—	—	—	—	—	2
Sceloporus hondurensis	—	—	—	—	X	X	X	X	—	X	X	X	X	X	X	X	X	X	—	—	—	—	—	X	—	X	X	X	X	—	—	X	—	—	—	15
Sceloporus schmidti	—	—	—	—	X	—	—	X	—	—	—	—	—	—	—	—	—	X	—	—	—	—	—	—	—	X	X	X	—	—	—	—	—	—	—	2
Sceloporus squamosus	X	X	X	—	X	X	—	—	—	X	X	X	X	X	X	—	—	X	X	X	X	X	X	X	X	X	X	—	—	—	—	—	—	—	—	7
Sceloporus variabilis	X	X	X	—	X	X	—	—	—	X	X	X	X	X	X	X	—	X	X	X	X	X	X	X	X	X	X	X	—	—	—	—	—	—	—	16
Polychrus gutturosus	—	—	—	X	X	—	—	—	—	—	—	—	X	—	—	—	—	—	—	—	—	X	X	—	—	—	—	—	—	—	—	—	—	—	—	2
Marisora brachypoda	X	X	—	X	X	—	—	X	X	X	X	X	X	X	—	—	—	X	X	X	X	X	X	X	X	X	X	—	—	X	X	X	—	—	—	21
Marisora roatanae	—	—	—	—	—	—	—	—	—	—	—	—	—	—	—	—	—	—	—	—	—	—	—	—	—	—	—	—	—	—	X	—	—	—	—	1
Scincella assata	X	—	—	—	—	—	—	—	—	—	—	—	—	—	—	—	—	—	—	—	—	—	—	—	—	—	—	—	—	—	—	—	—	—	—	1
Scincella cherriei	—	X	—	—	X	X	—	—	X	X	X	X	X	X	—	—	—	X	X	X	X	X	X	X	X	X	X	—	—	—	—	X	—	—	—	19
Scincella incerta	—	—	—	—	X	—	X	—	—	—	—	—	X	X	—	—	—	—	—	—	—	—	—	—	—	X	X	—	—	—	—	—	—	—	—	5
Mesoscincus managuae	X	X	—	—	—	X	—	—	—	—	—	—	—	—	—	—	—	X	—	—	—	X	—	—	—	—	—	—	—	—	—	—	—	—	—	3
Plestiodon sumichrasti	—	—	X	—	—	X	—	—	X	—	—	—	X	—	—	—	—	—	—	—	X	X	—	—	—	—	—	—	—	—	—	—	—	—	—	2
Gymnophthalmus speciosus	X	X	—	—	—	—	—	—	—	—	—	—	—	—	—	—	—	X	X	X	X	X	X	X	—	X	—	—	—	X	X	X	—	X	—	11
Ameiva fuliginosa	—	—	—	—	—	—	—	—	—	—	—	—	—	—	—	—	—	—	—	—	—	—	—	—	X	—	—	—	—	—	—	—	—	—	—	1
Aspidoscelis deppii	X	X	X	X	X	X	—	X	X	X	X	X	—	—	—	—	—	X	X	X	X	X	X	X	X	—	—	—	—	—	—	—	—	—	—	14
Aspidoscelis motaguae	X	X	—	X	X	X	—	—	—	X	—	—	—	—	—	—	—	X	X	X	X	X	X	X	—	—	—	—	—	—	—	—	—	—	—	6
Cnemidophorus ruatanus	—	—	—	—	—	—	—	—	—	—	—	—	X	X	—	—	—	X	X	X	X	X	X	X	X	X	X	X	—	X	X	X	X	—	—	9
Holcosus festivus	—	—	—	—	—	—	—	—	X	X	X	X	X	X	—	—	—	X	X	X	X	X	X	X	X	X	X	—	—	—	—	—	—	—	—	9
Holcosus undulatus	X	X	X	X	X	X	—	X	X	X	X	X	X	X	X	—	—	X	X	X	X	X	X	X	X	X	X	X	X	X	X	X	X	X	—	21
Subtotals	20	22	10	9	19	23	7	11	13	13	18	25	27	10	3	5	4	32	27	9	19	30	29	14	22	26	24	11	3	16	13	10	10	9	1	544
Crocodiles (2)																																				
Caiman crocodilus	X	—	—	—	—	—	—	X	—	—	—	—	—	—	—	—	—	X	X	—	X	X	X	X	—	—	—	—	—	X	X	—	—	—	—	5
Crocodylus acutus	X	—	—	—	—	—	—	X	—	—	—	—	—	—	—	—	—	X	X	—	X	X	X	X	X	X	X	—	—	X	X	X	—	—	—	12
Subtotals	1	0	0	0	0	0	0	1	0	0	0	0	0	0	0	0	0	2	0	0	1	2	2	1	1	1	1	0	0	1	1	1	0	0	0	17

TABLE 7. CONTINUED.

Species (126)	1	3	5	7	8	9	10	11	12	13	14	15	16	17	18	19	20	21	22	23	24	26	27	28	29	30	31	32	33	35	36	37	38	39	40	Total
Turtles (12)																																				
Chelydra acutirostris	—	—	—	—	—	—	—	—	X	—	—	—	—	—	—	—	—	X	—	—	—	—	—	—	—	—	—	—	—	—	—	—	—	—	—	2
Chelydra rossignonii	—	—	—	—	—	—	—	—	—	—	—	—	—	—	—	—	—	X	X	—	X	X	X	—	X	—	X	—	—	—	—	—	—	—	—	6
Kinosternon albogulare	X	X	X	X	X	X	—	—	X	—	—	—	—	—	—	—	—	—	—	—	X	—	—	—	—	—	—	—	—	—	—	—	—	—	—	8
Kinosternon leucostomum	—	—	—	X	X	X	—	—	X	X	X	—	X	—	—	—	—	X	X	X	X	X	X	X	X	X	X	—	—	—	—	X	—	—	—	17
Staurotypus triporcatus	—	—	—	—	—	—	—	—	X	X	X	—	—	—	—	—	—	—	—	—	—	—	—	—	—	—	—	—	—	—	—	—	—	—	—	3
Trachemys emolli	X	—	—	—	—	—	—	—	—	—	—	—	—	—	—	—	—	—	—	—	—	—	—	—	—	—	—	—	—	—	—	—	—	—	—	1
Trachemys scripta (1)	—	X	—	—	—	X	—	—	—	X	—	—	—	—	—	—	—	—	—	—	—	—	—	—	—	—	—	—	—	—	—	—	—	—	—	3
Trachemys venusta	—	X	—	—	—	—	—	—	X	X	X	—	—	—	—	—	—	X	—	—	X	X	X	X	X	—	X	—	—	X	X	X	X	—	—	15
Rhinoclemmys annulata	—	—	—	—	—	—	—	—	—	—	—	—	—	—	—	—	—	—	—	—	—	X	X	—	—	—	—	—	—	—	—	—	—	—	—	2
Rhinoclemmys areolata	—	—	—	—	—	—	—	—	—	—	—	—	—	—	—	—	—	—	—	—	—	—	—	—	X	—	—	—	—	—	—	—	—	—	—	1
Rhinoclemmys funerea	—	—	—	—	—	—	—	—	—	—	—	—	—	—	—	—	—	X	—	—	—	—	—	—	—	—	—	—	—	—	—	—	—	—	—	1
Rhinoclemmys pulcherrima	X	X	X	X	—	X	—	—	X	X	X	—	—	—	—	—	—	X	—	—	—	—	—	—	X	—	X	—	—	X	—	X	—	—	—	11
Subtotals	3	3	2	1	1	4	0	0	4	3	4	0	0	0	0	0	0	7	3	1	4	4	4	4	6	1	4	0	0	1	1	3	2	0	0	71
Totals	24	25	12	10	20	27	7	11	18	16	22	26	27	10	3	5	4	41	32	10	24	36	35	19	29	27	29	11	3	18	15	14	12	9	1	632

TABLE 8. CHARACTERISTICS OF 40 ECOPHYSIOGRAPHIC AREAS OF HONDURAS WITH DOMINANT FOREST FORMATIONS AND VERSANTS INDICATED. AREAS INDICATED WITH AN ASTERISK (°) ARE NOT INCLUDED IN THE ANALYSIS HEREIN. SEE TABLE 3 FOR EXPLANATION OF FOREST FORMATION ABBREVIATIONS.

Area No.	Area Name	Forest Formation	Versant
1	Pacific Lowlands	LDF	Pacific
2	Middle Choluteca Valley	PDF	Pacific
3	Upper Choluteca Valley	PDF	Pacific
4°	Comayagua Valley Rim	PDF	Atlantic
5	Comayagua Valley	LDF	Atlantic
6°	Otoro Valley Rim	PDF	Atlantic
7	Otoro Valley	LDF	Atlantic
8	Southwestern Uplands	PMF	Pacific[1]
9	Southeastern Uplands	PMF	Pacific[1]
10	Southeastern Highlands	LMMF	Pacific[1]
11	Southwestern Highlands	LMMF	Pacific[1]
12	Guayape-Guayambre Valley	LDF	Atlantic
13	Northeastern Uplands	PMF	Atlantic
14	Northwestern Uplands	PMF	Atlantic
15	Eastern Caribbean Slope	PWF	Atlantic
16	Central Caribbean Slope	PWF	Atlantic
17	North-central Highlands	LMWF	Atlantic
18	Yoro Highlands	LMWF	Atlantic
19	Ocote Highlands	LMWF	Atlantic
20	Agalta Highlands	LMWF	Atlantic
21	Eastern Caribbean Lowlands	LMF	Atlantic
22	East-central Caribbean Lowlands	LMF	Atlantic
23	San Esteban Valley	LDF	Atlantic
24	Aguán Valley	LAF	Atlantic
25°	Aguán Valley Rim	LDF	Atlantic
26	West-central Caribbean Lowlands	LMF	Atlantic
27	Western Caribbean Lowlands	LMF	Atlantic
28	Lower Motagua Valley	LMF	Atlantic
29	Sula Valley	LDF	Atlantic
30	Western Caribbean Slope	PWF	Atlantic
31	Yojoa Uplands	PWF	Atlantic
32	Northwestern Highlands	LMWF	Atlantic
33	Santa Bárbara Highlands	LMWF	Atlantic
34°	Santa Bárbara Peak	MR	Atlantic
35	Utila Island	LMF	Atlantic
36	Roatán Island	LMF	Atlantic
37	Guanaja Island	LMF	Atlantic
38	Cayos Cochinos	LMF	Atlantic
39	Swan Islands	LDF (WI)	Atlantic
40	Cayos Miskitos	LDF (WI)	Atlantic

[1] The northernmost portions of these areas are on the Atlantic versant.

J—2 species (1.6%);
A—1 species (0.8%).

This summary indicates that the largest category (H) contains those species endemic to Honduras (39), all of which are lizards. The next largest category (G) contains Nuclear Middle American endemics (20 species, exclusive of Honduran endemics), all of which are also lizards. The next highest category is F (16 species, with 14 lizards and 2 turtles), which contains species with northern terminus in Nuclear Middle America and southern terminus in Central America south of the Nicaraguan Depression. Thus, the two largest categories contain only species that are either Honduran endemics or are endemic to Nuclear Middle America (59 species, 46.8% of total). The third largest category (F) contains only species that do not occur north of Nuclear Middle America and southward no further than Central America (16 species, 12.7% of total).

HONDURAS AS A DISTRIBUTIONAL ENDPOINT

In addition to 39 Honduran endemic species under study (all lizards), analysis of overall geographic distribution of the 81 remaining native species (six introduced species not included) occurring in the country reveals that an additional 37 lizard and turtle species have their known distributional ranges terminating somewhere in Honduras (Table 9). Twenty-three of those species extend into Honduras from the west and/or north, and 14 extend into Honduras from the east and/or south. Further analysis of the 37 nonendemic and nonintroduced species with their ranges terminating in Honduras reveals the following:

Category 1. From west and/or north. Twelve of these 23 species are Nuclear Middle American Endemics, all but one of which have their known distributional ranges terminating somewhere in the western half of Honduras. That single remaining species of these Nuclear Middle American Endemics extends across the northern half of the country to east-central Honduras. Of the remaining 11 of these

TABLE 9. THE LIZARD AND TURTLE SPECIES HAVING THEIR KNOWN GEOGRAPHICAL DISTRIBUTION ENDING ON THE MAINLAND OF HONDURAS (EXCEPT FOR THE 39 HONDURAN ENDEMICS AND THE SIX INTRODUCED SPECIES). SEE TEXT.

From West and North	From East and South
Lepidophyma mayae	*Sphaerodactylus millepunctatus*
Sphaerodactylus continentalis	*Basiliscus plumifrons*
Sphaerodactylus glaucus	*Ctenosaura quinquecarinata*
Abronia montecristoi	*Norops carpenteri*
Diploglossus montanus	*Norops cupreus*
Corytophanes hernandesii	*Norops limifrons*
Corytophanes percarinatus	*Norops oxylophus*
Laemanctus serratus	*Norops quaggulus*
Norops crassulus	*Norops wermuthi*
Norops heteropholidotus	*Sceloporus hondurensis*
Norops laeviventris	*Polychrus gutturosus*
Norops ocelloscapularis	*Chelydra acutirostris*
Norops petersii	*Rhinoclemmys annulata*
Norops rodriguezii	*Rhinoclemmys funerea*
Norops uniformis	
Ctenosaura flavidorsalis	
Sceloporus schmidti	
Scincella assata	
Scincella incerta	
Plestiodon sumichrasti	
Chelydra rossignonii	
Staurotypus triporcatus	
Rhinoclemmys areolata	
Total 23	Total 14

23 species, all but one have their known distributional ranges extending from Mexico north of the Isthmus of Tehuantepec to terminate somewhere in the western half of Honduras. Fifteen of the species in Category 1 are confined to the Atlantic versant, three are confined to the Pacific Versant, four occur on both versants, and the distribution of one species is not known.

Category 2. From east and/or south. Three of these 14 species have their known distributional ranges extending from Honduras to terminate in South America, eight have their southern distribution terminating in Central America south of the Nicaraguan Depression, and three are Middle American Endemics. Four of these 14 species are restricted in Honduras to the Mosquitia region of northeastern Honduras, whereas five others barely occur outside the Mosquitia to southeastern, east-central, or north-central Honduras. Eleven of these 14 species occur on the Atlantic versant, one on the Pacific versant, and two on both versants.

In summary, Honduras has a relatively high percentage of lizard and turtle species whose known geographical distributions end somewhere in the country. Including the 39 Honduran Endemics, there are 76 such species (63.3%) of the total lizard, crocodile, and turtle fauna known from the country (exclusive of six introduced lizard and turtle species). That figure is greater than the 51.9% of the snake fauna found in a similar analysis of snakes by McCranie (2011a). The updated figure for the amphibians is 55.6% (my unpublished data), a figure in between those of the two reptile groups.

CONSERVATION STATUS OF THE LIZARDS, CROCODILES, AND TURTLES OF HONDURAS

VULNERABILITY GAUGES

McCranie (2011a) used a gauge for calculating environmental vulnerability of each snake species then known to occur in Honduras (also see Townsend and Wilson, 2010a). That gauge was modified from an earlier series of gauges developed in recent years (see references listed in Townsend and Wilson, 2010a; McCranie, 2011a; McCranie and Köhler, 2015). Each gauge has three components, with the gauge for lizards, crocodiles, and turtles discussed as follows.

The first component deals with extent of geographic distribution of each species as follows (one Honduran species, *Ameiva fuliginosa*, does not fit into any of these categories) but is included herein in the high vulnerability component (that *Ameiva* is extirpated from Honduran territory):

1—widespread in and outside of Honduras;

2—distribution peripheral to Honduras, but widespread elsewhere;

3—distribution restricted to Nuclear Middle America (exclusive of Honduran endemics);

4—distribution restricted to Honduras; and

5—known only from vicinity of type locality.

The second component deals with extent of ecological distribution of each species on the basis of a slightly modified version of forest formations of Holdridge (1967) by the following scale (omitting the Montane Rainforest formation, from which no reptile species are known):

1—occurs in eight formations;

2—occurs in seven formations;

3—occurs in six formations;

4—occurs in five formations;

5—occurs in four formations;

6—occurs in three formations;

7—occurs in two formations; and

8—occurs in one formation.

The third component considers degree of human persecution as follows:

1—fossorial and usually escapes human notice, or normally ignored by humans;

2—semifossorial or nocturnal arboreal or aquatic, nonvenomous and usually non-mimicking, frequently escapes human notice;

3—terrestrial and/or arboreal or aquatic, generally ignored or not seen by humans;

4—terrestrial and/or arboreal or aquatic, thought to be harmful, usually killed on sight;

5—species or mimics thereof, killed on sight;

6—exploited for hides and/or meat and/or eggs.

The composite environmental vulnerability scores (EVS; used either in singular or plural form, as determined by context) for Honduran lizards, crocodiles, and turtles (Table 10) range from a low of 4 (*Scincella*

TABLE 10. ENVIRONMENTAL VULNERABILITY SCORES (EVS) FOR THE 126 SPECIES OF LIZARDS, CROCODILES, AND TURTLES KNOWN FROM HONDURAS. NUMBERS FOR EACH GAUGE ARE EXPLAINED IN THE TEXT. THE TABLE IS BROKEN INTO THREE PARTS: LOW VULNERABILITY SPECIES (EVS OF 4–9); MEDIUM VULNERABILITY SPECIES (EVS OF 10–13); AND HIGH VULNERABILITY SPECIES (EVS OF 14–19).

Species	Geographic Distribution	Ecological Distribution	Human Persecution	Total Score
Low (34 species)				
Lepidophyma flavimaculatum	1	6	2	9
Coleonyx mitratus	1	4	2	7
Hemidactylus frenatus (introduced)	1	3	3	7
Thecadactylus rapicauda	1	5	2	8
Phyllodactylus tuberculosus	1	5	2	8
Sphaerodactylus continentalis	1	3	2	6
Basiliscus vittatus	1	3	3	7
Norops beckeri	1	6	1	8
Norops biporcatus	1	6	1	8
Norops capito	1	7	1	9
Norops cupreus	1	4	1	6
Norops laeviventris	1	5	1	7
Norops lemurinus	1	5	1	7
Norops limifrons	1	7	1	9
Norops mccraniei	3	3	1	7
Norops rodriguezii	2	5	1	8
Norops sagrei (introduced)	1	7	1	9
Norops uniformis	2	6	1	9
Norops unilobatus	1	3	1	5
Norops wellbornae	1	6	1	8
Ctenosaura similis	1	4	3	8
Sceloporus squamosus	1	5	3	9
Sceloporus variabilis	1	1	3	5
Marisora brachypoda	1	3	3	7
Scincella cherriei	1	1	2	4
Mesoscincus managuae	1	6	2	9
Gymnophthalmus speciosus	1	4	2	7
Aspidoscelis deppii	1	4	3	8
Aspidoscelis motaguae	1	5	3	9
Holcosus undulatus	1	3	3	7
Chelydra rossignonii	1	6	2	9
Kinosternon albogulare	1	4	3	8
Kinosternon leucostomum	1	4	3	8
Rhinoclemmys pulcherrima	1	3	3	7
Medium (54 species)				
Lepidophyma mayae	2	7	2	11
Hemidactylus haitianus (introduced)	2	8	3	13
Hemidactylus mabouia (introduced)	2	7	3	12
Gonatodes albogularis	1	6	3	10
Sphaerodactylus dunni	4	6	2	12
Sphaerodactylus glaucus	2	8	2	12
Sphaerodactylus millepunctatus	1	8	2	11
Mesaspis moreletii	3	7	3	13
Diploglossus bivittatus	3	7	3	13
Diploglossus montanus	3	7	3	13
Basiliscus plumifrons	1	8	3	12
Corytophanes cristatus	1	7	3	11
Corytophanes hernandesii	2	7	3	12
Corytophanes percarinatus	3	7	3	13
Laemanctus longipes	1	6	3	10
Laemanctus serratus	1	8	3	12
Laemanctus waltersi	4	6	3	13
Anolis allisoni	1	8	1	10

Table 10.   Continued.

Species	Geographic Distribution	Ecological Distribution	Human Persecution	Total Score
Norops carpenteri	2	8	1	11
Norops crassulus	3	6	1	10
Norops cusuco	4	7	1	12
Norops heteropholidotus	3	8	1	12
Norops johnmeyeri	4	7	1	12
Norops kreutzi	4	7	1	12
Norops loveridgei	4	6	1	11
Norops morazani	4	7	1	12
Norops muralla	5	7	1	13
Norops ocelloscapularis	3	7	1	11
Norops oxylophus	2	8	1	11
Norops petersii	2	7	1	10
Norops pijolense	5	7	1	13
Norops purpurgularis	4	8	1	13
Norops quaggulus	2	7	1	10
Norops sminthus	4	7	1	12
Norops wermuthi	4	8	1	13
Norops wilsoni	4	6	1	11
Norops yoroensis	4	6	1	11
Norops zeus	4	6	1	11
Ctenosaura flavidorsalis	3	7	3	13
Ctenosaura quinquecarinata	1	6	3	10
Iguana iguana	1	3	6	10
Leiocephalus varius (introduced)	2	8	3	13
Sceloporus hondurensis	3	7	3	13
Sceloporus schmidti	3	5	3	11
Polychrus gutturosus	1	8	3	12
Scincella assata	2	8	2	12
Scincella incerta	3	7	2	12
Plestiodon sumichrasti	1	7	2	10
Cnemidophorus ruatanus	3	7	3	13
Holcosus festivus	1	6	3	10
Crocodylus acutus	1	6	6	13
Trachemys venusta	1	3	6	10
Rhinoclemmys annulata	1	8	3	12
Rhinoclemmys areolata	1	8	3	12
High (38 species)				
Phyllodactylus palmeus	4	8	2	14
Phyllodactylus paralepis	5	8	2	15
Aristelliger sp. A	5	8	2	15
Aristelliger nelsoni	5	8	2	15
Sphaerodactylus alphus	5	8	2	15
Sphaerodactylus exsul	5	8	2	15
Sphaerodactylus guanajae	5	8	2	15
Sphaerodactylus leonardovaldesi	4	8	2	14
Sphaerodactylus poindexteri	5	8	2	15
Sphaerodactylus rosaurae	4	8	2	14
Abronia montecristoi	3	8	3	14
Abronia salvadorensis	4	8	3	15
Diploglossus scansorius	4	7	3	14
Laemanctus julioi	5	8	3	16
Norops amplisquamosus	5	8	1	14
Norops bicaorum	5	8	1	14
Norops nelsoni	5	8	1	14
Norops roatanensis	5	8	1	14
Norops rubribarbaris	5	8	1	14

TABLE 10.   CONTINUED.

Species	Geographic Distribution	Ecological Distribution	Human Persecution	Total Score
Norops utilensis	5	8	1	14
Norops wampuensis	5	8	1	14
Ctenosaura bakeri	5	8	6	19
Ctenosaura melanosterna	4	7	3	14
Ctenosaura oedirhina	4	8	6	18
Sceloporus esperanzae	4	7	3	14
Marisora roatanae	5	8	3	16
Ameiva fuliginosa[1]	3	8	3	14
Caiman crocodilus	1	7	6	14
Caretta caretta[2]	2	8	6	16
Chelonia mydas[2]	2	8	6	16
Eretmochelys imbricata[2]	2	8	6	16
Lepidochelys olivacea[2]	2	8	6	16
Dermochelys coriacea[2]	2	8	6	16
Chelydra acutirostris	1	7	6	14
Staurotypus triporcatus	1	7	6	14
Trachemys emolli	2	8	6	16
Trachemys scripta (introduced)	1	7	6	14
Rhinoclemmys funerea	2	8	6	16

[1] Extirpated from Honduran territory.
[2] Marine species.

cherriei) to a high of 19 (*Ctenosaura bakeri*). The number of species attaining various EVS scores are as follows:

EVS 4 – 1 species;
EVS 5 – 2 species;
EVS 6 – 2 species;
EVS 7 – 10 species;
EVS 8 – 10 species;
EVS 9 – 9 species;
EVS 10 – 11 species;
EVS 11 – 11 species;
EVS 12 – 18 species;
EVS 13 – 14 species;
EVS 14 – 19 species;
EVS 15 – 8 species;
EVS 16 – 9 species;
EVS 17 – 0 species;
EVS 18 – 1 species;
EVS 19 – 1 species.

As with previous gauges for reptilian and/or amphibian species of Honduras, the present gauge is divided into three categories of environmental vulnerability, as indicated in Table 11. However, the present categories differ from that of McCranie and

Köhler (2015) as a result of rethinking and making slight adjustments, in that scores of 13 herein are considered a medium vulnerability score as opposed to that of a high vulnerability score. There are 34 species of low vulnerability (27.0%), 54 species of medium vulnerability (42.9%), and 38 species of high vulnerability (30.2%).

Of the 39 Honduran endemic species, 14 (*Sphaerodactylus dunni, Laemanctus waltersi, Norops cusuco, N. johnmeyeri, N. kreutzi, N. loveridgei, N. morazani, N. muralla, N. pijolense, N. purpurgularis, N. smithus, N. wilsoni, N. yoroensis,* and *N. zeus*) are classified as having medium vulnerability. All but one (*N. kreutzi*) of these 14 species are also thought to have stable populations (Table 11) remaining somewhere in their respective ranges. The remaining 25 Honduran endemics fall into the high vulnerability category. However, all but 6 of those 25 species are thought to have stable populations (Table 11), having adapted to somewhat degraded habitat, and are still common at one or more localities (*Phyllodactylus palmeus, P. paralepis, Aris-*

TABLE 11. CURRENT STATUS OF POPULATIONS OF HONDURAN LIZARD ENDEMICS AND SPECIES OTHERWISE RESTRICTED TO NUCLEAR MIDDLE AMERICA. STABLE = AT LEAST SOME POPULATIONS STABLE; DECLINING = ALL POPULATIONS BELIEVED TO BE DECLINING.

Species	Stable	Declining
Honduran endemics (39 species)		
Phyllodactylus palmeus	X	
Phyllodactylus paralepis	X	
Aristelliger sp. A	X	
Aristelliger nelsoni	X	
Sphaerodactylus alphus	X	
Sphaerodactylus dunni	X	
Sphaerodactylus exsul	X	
Sphaerodactylus guanajae	X	
Sphaerodactylus leonardovaldesi	X	
Sphaerodactylus poindexteri	X	
Sphaerodactylus rosaurae	X	
Abronia salvadorensis		X
Diploglossus scansorius		X
Laemanctus julioi		X
Laemanctus waltersi	X	
Norops amplisquamosus		X
Norops bicaorum	X	
Norops cusuco	X	
Norops johnmeyeri	X	
Norops kreutzi		X
Norops loveridgei	X	
Norops morazani	X	
Norops muralla	X	
Norops nelsoni	X	
Norops pijolense	X	
Norops purpurgularis	X	
Norops roatanensis	X	
Norops rubribarbaris	X	
Norops sminthus	X	
Norops utilensis	X	
Norops wampuensis		X
Norops wilsoni	X	
Norops yoroensis	X	
Norops zeus	X	
Ctenosaura bakeri		X
Ctenosaura melanosterna	X	
Ctenosaura oedirhina	X	
Sceloporus esperanzae	X	
Marisora roatanae	X	
Honduran species otherwise restricted to Nuclear Middle America (20 species)		
Lepidophyma mayae		X
Sphaerodactylus continentalis	X	
Abronia montecristoi		X
Mesaspis moreletii	X	
Diploglossus bivittatus		X
Diploglossus montanus		X
Corytophanes percarinatus		X
Norops beckeri	X	
Norops crassulus	X	
Norops heteropholidotus	X	
Norops mccraniei	X	

TABLE 11. CONTINUED.

Species	Stable	Declining
Norops ocelloscapularis	X	
Norops rodriguezii	X	
Norops wellbornae	X	
Norops wermuthi	X	
Ctenosaura flavidorsalis	X	
Sceloporus hondurensis	X	
Sceloporus schmidti	X	
Scincella incerta	X	
Cnemidophorus ruatanus	X	

telliger sp. A, *A. nelsoni* (but see that species account), *Sphaerodactylus alphus*, *S. exsul*, *S. guanajae*, *S. leonardovaldesi*, *S. poindexteri*, *S. rosaurae*, *Norops bicaorum*, *N. nelsoni*, *N. roatanensis*, *N. rubribarbaris*, *N. utilensis*, *Ctenosaura melanosterna*, *C. oedirhina*, *Sceloporus esperanzae*, and *Marisora roatanae*). The six remaining Honduran endemics are all thought to have declining populations (Table 11) because of habitat degradation (*Abronia salvadorensis*, *Diploglossus scansorius*, *Laemanctus julioi*, *Norops amplisquamosus*, *N. wampuensis*, and *Ctenosaura bakeri*). *Ctenosaura bakeri* is also persecuted for its meat and eggs.

Of 20 Honduran species restricted to Nuclear Middle America (non-Honduran endemics), five (*Sphaerodactylus continentalis*, *Norops beckeri*, *N. mccraniei*, *N. rodriguezii*, and *N. wellbornae*) are classified as having low vulnerability scores. Those species are also thought to have stable populations remaining in Honduras because of suitable forest remaining in its habitat and by also having adapted well to somewhat degraded habitat. Fourteen of these 20 species are classified as having medium vulnerability, 10 of which appear to have some stable populations remaining in Honduras either because of suitable remaining habitat somewhere (*Mesaspis moreletii*, *Norops crassulus*, *N. heteropholidotus*, *N. ocelloscapularis*, *Sceloporus hondurensis*, *Scincella incerta*, and *Cnemidophorus ruatanus*) or adaptation to somewhat degraded habitat

(*Norops wermuthi, Ctenosaura flavidorsalis,* and *Sceloporus schmidti*). Both cases seem to occur for *Sceloporus hondurensis* and *S. schmidti* and the *Cnemidophorus*. The remaining four Nuclear Middle American Endemics classified as having medium vulnerability all appear to have declining populations because of habitat degradation (*Lepidophyma mayae, Diploglossus bivittatus, D. montanus,* and *Corytophanes percarinatus*). The final Nuclear Middle American Endemic (*Abronia montecristoi*) is classified as a species of high vulnerability. The only known Honduran locality of *Abronia montecristoi* has been severely degraded in recent years, but seemingly suitable habitat does remain on several isolated mountain peaks between the Honduran *A. montecristoi* locality and its type locality in El Salvador, where the possibility remains that the species remains on at least one peak.

Poe (2016) criticized these EVS methods as used by McCranie and Köhler (2015) for the anoles but did not provide a better method for dealing with multiple species.

CITES CATEGORIES AND SPECIES OCCURRING IN HONDURAS

Eleven species of lizards, crocodiles, and turtles occurring in Honduras are listed on one of the three CITES appendices (version valid from 5 February 2015). Four lizards (*Ctenosaura bakeri, C. melanosterna, C. oedirhina,* and *Iguana iguana*) are placed on Appendix II. Those three species of *Ctenosaura* are in the High Vulnerability category in the EVS study recovered in this work, but another 38 species occur in Honduras with High Vulnerability scores. Those three *Ctenosaura* species (see discussion for the genus *Ctenosaura*) have a high EVS because of the small territorial area in which they occur. One (*C. oedirhina*) is restricted to a single island and some satellite islands, one (*C. bakeri*) is restricted to a single island, and one (*C. melanosterna*) has populations on one island and in a single area on the mainland. *Ctenosaura oedirhina* is the most common and most easily seen lizard on Roatán Island and some satellite islands and appears to be expanding its territory and population numbers on Roatán Island (personal observation, 1983–2013), despite inaccurate IUCN reports saying otherwise. Also, *C. oedirhina* has never been documented to have international trade and is also not endangered in its habitat. *Ctenosaura bakeri* is endangered on Utila Island because of the human destruction of its habitat, development of the beaches where egg deposition occurs, and consumption of its meat and eggs by humans. *Ctenosaura melanosterna* remains common on Cayo Cochino Pequeña and also occurs on at least one satellite island. The *C. melanosterna* population in the Río Aguán Valley on the mainland is endangered only because of the uncontested habitat destruction taking place there over the years, but at an accelerated rate during the most recent 10–12 years. Because no evidence exists that international trade took place in any of these species of *Ctenosaura*, they are misplaced (see Discussion of the genus *Ctenosaura*) on a list requiring international trade (Convention on International Trade in Endangered Species of Wild Fauna and Flora). Thus, their CITES listing provides no role in protecting those species and their habitats. The only way to save those species for the future is probably to undertake a breeding and head start program on Roatán for *C. oedirhina* and in the Aguán Valley for *C. melanosterna*. The fourth Honduran CITES lizard, *Iguana iguana*, has received benefit from its CITES listing but needs efforts to stop the uncontested out-of-hand habitat destruction taking place. Thus, the major problem facing the three *Ctenosaura* species and the *Iguana* species on the CITES appendices is not international trade of any of those species, but the lack of any attempt to stop the ongoing, accelerating to the point of being out-of-hand, and rampart illegal forest devastation of the few remaining forests in Honduras.

Both crocodilian species occurring in Honduras used to be involved in the international trade, but the current government needs to deal with the growing illegal problem of direct human-caused forest devastation. *Crocodylus acutus* is of Medium Vulnerability and *Caiman crocodilus* of High Vulnerability on the vulnerability scores.

The five marine turtles occurring in Honduran waters were involved in international trade. But they are in the same position as the two crocodilian species in having no protection inside Honduras. The egg poaching on Honduran beaches also needs to be stopped by governmental action. Additionally, eggs of *Lepidochelys olivacea* are openly sold in restaurants along the south coast. All five marine turtles are of High Vulnerability in my vulnerability studies.

IUCN RED LIST CATEGORIES

Each of the 126 species of lizards, crocodiles, and turtles known to occur in Honduran territory was placed in one of five categories (Table 12) using criteria developed by the IUCN. Examination of Table 12 demonstrates that 29 of the 39 Honduran endemic species are classified either as Critically Endangered (14 species), Endangered (6 species), or Vulnerable (9 species). Two Honduran Endemic species (*Laemanctus waltersi, N. zeus*) are included in Least Concern category because they occur on private property used for ecotourism that still has healthy secondary forest. Those two species also occur in the still forested foothills of Pico Bonito National Park. One Honduran Endemic species (*N. sminthus*) is included in the Least Concern category because it occurs in the well-forested Parque Nacional La Tigra, a major water source for the capital city of Tegucigalpa. Of the 20 Nuclear Middle American endemics,

one is classified as Critically Endangered, two as Endangered, four as Vulnerable, three as Near Threatened, and nine as species of Least Concern. Of the remaining 67 species with a more widespread geographical distribution, one lizard (*Ameiva fuliginosa*) is extirpated from Honduran territory and the status of its only remaining population is poorly known, but it still occurs on at least one Colombian island (see McCranie and Gotte, 2014); both crocodiles are Endangered, and all five marine turtles are classified as Critically Endangered. Additionally, one lizard is listed as Vulnerable, two lizards and two turtles as Near Threatened, and 56 species as Least Concern. Thus, in total there are 21 Critically Endangered species (one of which is extirpated from Honduran territory), 10 Endangered species, 14 Vulnerable species, 13 Near Threatened species, and 68 species of Least Concern.

The IUCN (2012) has provided assessments of only 53 of the 126 species (42.1%) of lizards, crocodiles, and turtles occurring in Honduras (see Table 12 herein). Although, presumably, the same criteria were used in those IUCN assessments as were used in construction of Table 12, different categories were found in this paper for 25 of those 53 species (47.2%) between the two efforts. Actual and sufficient fieldwork with the species in question is critical to evaluating the IUCN criteria. For example, I have made field trips targeting five of the six species of *Ctenosaura* occurring in Honduras; my findings based on those trips disagree with those IUCN assessments for three species. All *Ctenosaura* species, under targeted fieldwork, proved to retain robust populations in their favored habitats. With recent targeted efforts to find the smaller *Ctenosaura* species of the *C. quinquecarinata* and *C. palearis* species groups, especially those that occur on the mainland, all are now known to have much broader distributions and much more robust and apparently healthy populations than indicat-

TABLE 12.   IUCN RED LIST CATEGORIES FOR THE 126 HONDURAN SPECIES OF LIZARDS, CROCODYILES, AND TURTLES. I = INTRODUCED SPECIES.

Species	Critically Endangered	Endangered	Vulnerable	Near Threatened	Least Concern
Lizards (107 species)					
Lepidophyma flavimaculatum	—	—	—	—	X
Lepidophyma mayae	—	B2ab(iii)	—	—	—
Coleonyx mitratus	—	—	—	—	X
Hemidactylus frenatus (I)	—	—	—	—	X
Hemidactylus haitianus (I)	—	—	—	—	X
Hemidactylus mabouia (I)	—	—	—	—	X
Phyllodactylus palmeus	—	—	B2ab(iii)	—	—
Phyllodactylus paralepis	B2ab(iii)	—	—	—	—
Phyllodactylus tuberculosus	—	—	—	—	X
Thecadactylus rapicauda	—	—	—	—	X
Aristelliger sp. A	B2ab(iii)	—	—	—	—
Aristelliger nelsoni	B2ab(ii)	—	—	—	—
Gonatodes albogularis	—	—	—	—	X
Sphaerodactylus alphus	B2ab(iii)	—	—	—	—
Sphaerodactylus continentalis	—	—	—	—	X
Sphaerodactylus dunni	—	—	B2ab(iii)	—	—
Sphaerodactylus exsul	B2ab(ii)	—	—	—	—
Sphaerodactylus glaucus	—	—	—	—	X
Sphaerodactylus guanajae	—	B2ab(iii)	—	—	—
Sphaerodactylus leonardovaldesi	—	B2ab(iii)	—	—	—
Sphaerodactylus millepunctatus	—	—	—	—	X
Sphaerodactylus poindexteri	B2ab(iii)	—	—	—	—
Sphaerodactylus rosaurae	—	—	B2ab(iii)	—	—
Abronia montecristoi	B2ab(iii)	—	—	—	—
Abronia salvadorensis	B2ab(iii)	—	—	—	—
Mesaspis moreletii	—	—	—	—	X
Diploglossus bivittatus	—	—	—	X	—
Diploglossus montanus	—	B2ab(iii)	—	—	—
Diploglossus scansorius	B2ab(iii)	—	—	—	—
Basiliscus plumifrons	—	—	—	—	X
Basiliscus vittatus	—	—	—	—	X
Corytophanes cristatus	—	—	—	—	X
Corytophanes hernandesii	—	—	—	—	X
Corytophanes percarinatus	—	—	—	X	—
Laemanctus julioi	—	B2ab(iii)	—	—	—
Laemanctus longipes	—	—	—	—	X
Laemanctus serratus	—	—	—	X	—
Laemanctus waltersi	—	—	—	—	X
Anolis allisoni	—	—	—	—	X
Norops amplisquamosus	B1ab(v)	—	—	—	—
Norops beckeri	—	—	—	—	X
Norops bicaorum	—	—	—	X	—
Norops biporcatus	—	—	—	—	X
Norops capito	—	—	—	—	X
Norops carpenteri	—	—	—	—	X
Norops crassulus	—	—	—	—	X
Norops cupreus	—	—	—	—	X
Norops cusuco	—	B2ab(iii)	—	—	—
Norops heteropholidotus	—	—	—	—	X
Norops johnmeyeri	—	—	—	X	—
Norops kreutzi	—	B2ab(iii)	—	—	—
Norops laeviventris	—	—	—	—	X
Norops lemurinus	—	—	—	—	X
Norops limifrons	—	—	—	—	X
Norops loveridgei	—	—	—	X	—
Norops mccraniei	—	—	—	—	X

Table 12. Continued.

Species	Critically Endangered	Endangered	Vulnerable	Near Threatened	Least Concern
Norops morazani	—	—	B2ab(iv)	—	—
Norops muralla	—	—	B2ab(iv)	—	—
Norops nelsoni	—	—	—	X	—
Norops ocelloscapularis	—	—	—	X	—
Norops oxylophus	—	—	—	—	X
Norops petersii	—	—	—	—	X
Norops pijolense	—	—	B2ab(iii)	—	—
Norops purpurgularis	—	—	B2ab(iii)	—	—
Norops quaggulus	—	—	—	—	X
Norops roatanensis	—	—	—	X	—
Norops rodriguezii	—	—	—	—	X
Norops rubribarbaris	—	—	B2ab(iii)	—	—
Norops sagrei (I)	—	—	—	—	X
Norops sminthus	—	—	—	—	X
Norops uniformis	—	—	—	—	X
Norops unilobatus	—	—	—	—	X
Norops utilensis	B2ab(iii)	—	—	—	—
Norops wampuensis	B2ab(iii)	—	—	—	—
Norops wellbornae	—	—	—	—	X
Norops wermuthi	—	—	B2ab(iii)	—	—
Norops wilsoni	—	—	—	—	X
Norops yoroensis	—	—	—	X	—
Norops zeus	—	—	—	—	X
Ctenosaura bakeri	B2ab(ii)	—	—	—	—
Ctenosaura flavidorsalis	—	—	B2ab(iii)	—	—
Ctenosaura melanosterna	—	B2ab(iii)	—	—	—
Ctenosaura oedirhina	B2ab(iii)	—	—	—	—
Ctenosaura quinquecarinata	—	—	—	—	X
Ctenosaura similis	—	—	—	—	X
Iguana iguana	—	—	—	—	X
Leiocephalus varius (I)	—	—	—	X	—
Sceloporus esperanzae	—	—	B2ab(iii)	—	—
Sceloporus hondurensis	—	—	—	—	X
Sceloporus schmidti	—	—	B2ab(iii)	—	—
Sceloporus squamosus	—	—	—	—	X
Sceloporus variabilis	—	—	—	—	X
Polychrus gutturosus	—	—	—	—	X
Marisora brachypoda	—	—	—	—	X
Marisora roatanae	B2ab(iii)	—	—	—	—
Scincella assata	—	—	B2ab(iii)	—	—
Scincella cherriei	—	—	—	—	X
Scincella incerta	—	—	B2ab(iii)	—	—
Mesoscincus managuae	—	—	—	—	X
Plestiodon sumichrasti	—	—	—	—	X
Gymnophthalmus speciosus	—	—	—	—	X
Ameiva fuliginosa[1]	B2ab(ii)	—	—	—	—
Aspidoscelis deppii	—	—	—	—	X
Aspidoscelis motaguae	—	—	—	—	X
Cnemidophorus ruatanus	—	—	—	—	X
Holcosus festivus	—	—	—	—	X
Holcosus undulatus	—	—	—	—	X
Subtotals	16	8	14	11	58
Crocodiles (2)					
Caiman crocodilus	—	A2acd	—	—	—
Crocodylus acutus	—	A2acd	—	—	—
Subtotals	0	2	0	0	0

TABLE 12. CONTINUED.

Species	Critically Endangered	Endangered	Vulnerable	Near Threatened	Least Concern
Turtles (17)					
Caretta caretta	A2acd	—	—	—	—
Chelonia mydas	A2acd	—	—	—	—
Eretmochelys imbricata	A2acd	—	—	—	—
Lepidochelys olivacea	A2acd	—	—	—	—
Dermochelys coriacea	A2acd	—	—	—	—
Chelydra acutirostris	—	—	—	—	X
Chelydra rossignonii	—	—	—	—	X
Kinosternon albogulare	—	—	—	—	X
Kinosternon leucostomum	—	—	—	—	X
Staurotypus triporcatus	—	—	—	X	—
Trachemys emolli	—	—	—	X	—
Trachemys scripta (I)	—	—	—	—	X
Trachemys venusta	—	—	—	—	X
Rhinoclemmys annulata	—	—	—	—	X
Rhinoclemmys areolata	—	—	—	—	X
Rhinoclemmys funerea	—	—	—	—	X
Rhinoclemmys pulcherrima	—	—	—	—	X
Subtotals	5	0	0	2	10
Totals	21	10	14	13	68

[1] Extirpated from Honduran territory.

ed in the IUCN assessments. Another example of the importance of fieldwork for assessing conservation status is that my results for the five marine turtles occurring in Honduras disagree with the conclusions of the IUCN for four of the five marine turtles. I have almost no fieldwork experience with marine turtles, but the IUCN assessments were made by biologists presumably with much more fieldwork knowledge than I have and on a more global basis and, thus, are preferable to my assessments. Also important is fieldwork to gain knowledge of how effective a given country is in combating ongoing habitat destruction. Currently, the Honduran government has absolutely no interest in trying to stop, or even slow down, the accelerated and out-of-hand forest and habitat destruction.

SPECIES OCCURRING IN PROTECTED AREAS

McCranie (2011a: 572, 576; also see McCranie and Köhler, 2015: 272) discussed the window dressing of the protected areas of Honduras, which, in reality, generally offers no form of protection for habitat, including the plants and animals, in those paper parks. A few exceptions are those parks (i.e., Cerro La Tigra, Pico Bonito) that actually receive some protection, especially from several NGOs, because those areas are major water sources for the cities Tegucigalpa and La Ceiba, respectively. McCranie and Townsend (2011) discussed the point of the windward side of Pico Bonito, which is visible to the authorities from La Ceiba, and the major water source for La Ceiba and numerous nearby coastal villages, which have pristine forest remaining. On the other hand, the leeward side of those mountains is not visible to the authorities from the La Ceiba side. Thus, those forests are almost completely denuded, resulting in human populations in the villages away from the coast in that area of Honduras being left with unprotected, degraded water sources, along with the animal inhabitants suffering from loss of habitat—that, despite those leeward slopes being part of that national park. At least four of the amphibian species endemic to the now degraded interior not visible from the La Ceiba side are extinct (McCranie, personal observation).

Another problem is the generally poor state of knowledge concerning these parks. Of the four recent unpublished governmental reports discussing these parks (Sánchez et al., 2002; Anonymous, 2006; Cruz [Díaz], 2008; ICF, 2010), no two agree with each other on what parks exist, their sizes, and the extent of their boundaries. These problems regarding the general lack of knowledge about the parks seriously hamper the biologist interested in knowing the fauna and flora of any given protected area, regardless of the amount of field work the biologist has undertaken.

Despite these problems, I offer some information on the reptile species covered in this paper with regard to their occurrence in protected areas. McCranie (2011a) listed 30 protected areas that still had some suitable habitat and with reptile collections made within them. McCranie and Köhler (2015) deleted one of those areas (Area 23, Rus Rus) because it was no longer being considered a protected area and was suffering from the rapid and critical recent deforestation taking place. Additionally, McCranie and Köhler (2015) added 11 other protected areas not included in McCranie (2011a). Two marine parks (Areas 39 and 41) are also included herein to cover the marine turtles. Table 13 demonstrates the distributions of the lizard, crocodile, and turtle species in those 41 areas, and Table 14 includes a list of those areas with main forest type(s), estimated sizes, and brief comments. Map 88 shows the general location of 40 of those 41 areas (see Map 5 for the location of the remaining area, the Swan Islands). Map 89 shows the location of many collecting sites.

Examination of Table 13 shows that the number of species within an area ranges from 1 (Area 28) or 2 (Areas 18, 22, 23, 26, 27, 29, 33, 35) up to 32 (Area 20), with an average of 3.3 for the 126 species covered herein. I have had experience in all but

three of these areas (17, 24, and 37). On the basis of anoles, which are usually commonly seen, McCranie and Köhler (2015) concluded there were 15 areas that needed additional study (Areas 11, 15–17, 22–25, 27–30, 32, 33, and 35). That conclusion likely remains the case for all species treated herein. A surprisingly high number (25; 23.4%) of lizard species are not known from any protected area in Honduras and 3 (17.6%) turtle species are not known from a protected area. Thus, of the total lizard, crocodile, and turtle species known to occur in Honduras, 28 (22.2%) of 126 species are not known from any of these 41 protected areas, at least on paper.

Fourteen of 25 (56.0%) lizard species not known from a protected area are Honduran endemics and thus are not offered protection anywhere within their known geographical distributions. However, 12 of those 14 species (*Phyllodactylus paralepis*, *Aristelliger* sp. A, *Sphaerodactylus alphus*, *S. guanajae*, *S. leonardovaldesi*, *S. poindexteri*, *Norops bicaorum*, *N. roatanensis*, *N. utilensis*, *Ctenosaura oedirhina*, *Sceloporus esperanzae*, and *Marisora roatanae*) are thought to have stable populations (the *Ctenosaura* appears to have stable, if not increasing, populations remaining, despite the significantly inaccurate declining status published in Pasachnik, Ariani-Sánchez et al., 2010). The remaining two Honduran endemics not occurring in a protected area (*Laemanctus julioi* and *Ctenosaura bakeri*) are thought to have declining populations because of habitat destruction, with the *Ctenosaura* also being hunted for its meat and eggs. Another 2 of the 25 species of lizards not occurring in a protected area are Middle American Endemics (*Norops wermuthi* and *Ctenosaura flavidorsalis*), and both are thought to have stable populations. The latter is an extremely common lizard in degraded and cattle-grazed areas within its range in Honduras. The remaining nine species not known from a protected area all have wide distributions outside of Hondu-

TABLE 13. OCCURRENCE OF THE 126 LIZARD, CROCODILE, AND TURTLE SPECIES IN THE VARIOUS PROTECTED AREAS IN HONDURAS THAT STILL HAVE SOME FOREST OR SUITABLE HABITAT REMAINING, INCLUDING MARINE HABITATS FOR THE MARINE TURTLES. REFER TO TABLE 14 FOR THE CORRESPONDING AREA FOR EACH NUMBER. I = HUMAN-AIDED INTRODUCTION.

Species (126)	1	2	3	4	5	6	7	8	9	10	11	12	13	14	15	16	17	18	19	20	21	22	23	24	25	26	27	28	29	30	31	32	33	34	35	36	37	38	39	40	41	Totals
Lizards (107)																																										
Lepidophyma flavimaculatum	X		X	X		X	X		X			X	X		X	X		X		X																X				X		13
Lepidophyma mayae			X																																							1
Coleonyx mitratus						X				X																												X	X			3
Hemidactylus frenatus (I)				X										X																				X				X	X			5
Hemidactylus haitianus (I)																																					X					0
Hemidactylus mabouia (I)																																					X					1
Phyllodactylus palmeus																																			X							1
Phyllodactylus paralepis																																										0
Phyllodactylus tuberculosus																																					X				1
Thecadactylus rapicauda				X		X						X	X					X	X														X							X		8
Aristelliger sp. A																																										0
Aristelliger nelsoni																																					X					1
Gonatodes albogularis																				X																		X				2
Sphaerodactylus alphus																																										0
Sphaerodactylus continentalis	X			X		X							X	X				X															X					X				7
Sphaerodactylus dunni	X			X		X							X					X																					X			4
Sphaerodactylus exsul																																					X					1
Sphaerodactylus glaucus																																										0
Sphaerodactylus guanajae																																										0
Sphaerodactylus leonardovaldesi																																										0
Sphaerodactylus millepunctatus												X						X	X															X								4
Sphaerodactylus poindexteri																																										0
Sphaerodactylus rosaurae																																				X						1
Abronia montecristoi			X																																							1
Abronia salvadorensis																									X																	1
Mesaspis moreletii	X			X		X	X							X				X							X	X	X	X	X													11
Diploglossus bivittatus																																	X									1
Diploglossus montanus				X	X	X																				X																3
Diploglossus scansorius																																				X						1
Basiliscus plumifrons												X						X												X												4
Basiliscus vittatus	X			X		X	X	X				X	X		X	X		X								X	X	X		X		X	X	X		X		X	X			16
Corytophanes cristatus	X			X		X	X	X				X	X		X	X		X								X	X			X		X						X	X			12
Corytophanes hernandesii						X																																				1
Corytophanes percarinatus											X														X																	2
Laemanctus julioi																																										0
Laemanctus longipes			X				X																																			2

TABLE 13. CONTINUED.

Species (126)	1	2	3	4	5	6	7	8	9	10	11	12	13	14	15	16	17	18	19	20	21	22	23	24	25	26	27	28	29	30	31	32	33	34	35	36	37	38	39	40	41	Totals
Laemanctus serratus																																										0
Laemanctus waltersi														X																							X			X		3
Anolis allisoni																																		X		X						2
Norops amplisquamosus	X																																									1
Norops beckeri				X			X									X			X	X																		X				7
Norops bicaorum																																										0
Norops biporcatus			X	X	X	X	X		X			X		X				X	X	X		X	X	X					X													12
Norops capito			X	X	X	X	X		X			X						X	X	X			X							X												11
Norops carpenteri		X																																								0
Norops crassulus	X						X		X			X		X				X					X			X								X								5
Norops cupreus			X		X		X		X			X						X					X	X						X												8
Norops cusuco			X		X																						X				X											2
Norops heteropholidotus												X												X	X		X						X									4
Norops johnmeyeri			X		X																																					2
Norops kreutzi																																			X							1
Norops laeviventris	X		X	X		X	X	X	X									X	X		X					X																10
Norops lemurinus	X		X	X		X	X	X	X			X		X				X	X	X											X					X						10
Norops limifrons									X			X						X	X											X												5
Norops loveridgei														X																					X							2
Norops mccraniei			X	X		X	X	X	X	X	X							X	X	X	X	X	X	X				X							X	X						17
Norops morazani										X																																1
Norops muralla							X																																			1
Norops nelsoni																																					X					1
Norops ocelloscapularis			X		X																																					2
Norops oxylophus												X										X																				2
Norops petersii			X		X																																					2
Norops pijolense															X																											1
Norops purpurgularis														X																						X						2
Norops quaggulus																			X	X	X																					3
Norops roatanensis																																										0
Norops rodriguezii			X			X																																				2
Norops rubribarbaris																		X																								1
Norops sagrei (1)																																										0
Norops sminthus								X	X																			X														3
Norops uniformis			X		X	X								X																												3
Norops unilobatus					X	X								X		X		X		X				X						X	X		X	X						X		10
Norops utilensis																																										0
Norops wampuensis																					X																					1
Norops wellbornae																																							X			1
Norops wermuthi																																										0

TABLE 13. CONTINUED.

Species (126)	1	2	3	4	5	6	7	8	9	10	11	12	13	14	15	16	17	18	19	20	21	22	23	24	25	26	27	28	29	30	31	32	33	34	35	36	37	38	39	40	41	Totals
Norops wilsoni	X													X		X				X																				X		5
Norops yoroensis		X			X									X	X																				X							5
Norops zeus	X		X											X			X													X									X			6
Ctenosaura bakeri																																										0
Ctenosaura flavidorsalis																																					X					1
Ctenosaura melanosterna																																										0
Ctenosaura oedirhina																																										0
Ctenosaura quinquecarinata																																										0
Ctenosaura similis	X							X																									X					X				4
Iguana iguana					X		X						X							X	X								X				X		X		X	X	X			9
Leiocephalus varius (1)																																					X					1
Sceloporus esperanzae																																										0
Sceloporus hondurensis						X	X		X	X	X					X				X	X								X					X		X						8
Sceloporus schmidti				X	X	X												X																								3
Sceloporus squamosus																																						X				1
Sceloporus variabilis			X		X	X															X												X					X				6
Polychrus gutturosus																																										0
Marisora brachypoda			X										X	X						X	X								X		X		X		X				X	X		10
Marisora roatanae																																										0
Scincella assata																																										0
Scincella cherriei	X		X	X	X	X		X				X			X	X			X	X	X									X			X		X				X			13
Scincella incerta				X	X							X		X	X																		X									4
Mesoscincus managuae																																						X				1
Plestiodon sumichrasti													X																				X		X				X			3
Gymnophthalmus speciosus	X																			X													X					X				4
Ameiva fuliginosa																																				X						1
Aspidoscelis deppii	X											X																		X	X		X		X		X					5
Aspidoscelis motaguae																																										0
Cnemidophorus ruatanus													X		X	X				X											X		X		X		X					6
Holcosus festivus	X		X	X		X	X					X	X		X	X			X	X	X								X	X			X		X		X		X			13
Holcosus undulatus			X		X							X				X			X	X	X									X			X		X		X		X			10
Subtotals	15	3	16	19	11	24	16	6	12	4	5	4	19	23	3	8	4	2	13	23	21	2	2	4	3	2	1	2	3	10	5	2	14	2	11	9	8	14	16	0		363
Crocodiles (2)																																										
Caiman crocodilus																														X	X		X									4
Crocodylus acutus												X								X											X							X	X		X	5
Subtotals	0	0	0	0	0	0	0	0	0	0	0	1	0	0	0	0	0	0	0	2	0	0	0	0	0	0	0	0	0	1	2	0	1	0	0	0	0	1	1	0	1	9

TABLE 13. CONTINUED.

Species (126)	1	2	3	4	5	6	7	8	9	10	11	12	13	14	15	16	17	18	19	20	21	22	23	24	25	26	27	28	29	30	31	32	33	34	35	36	37	38	39	40	41	Totals
Turtles (17)																																										
Caretta caretta																			X	X																		X		X	X	4
Chelonia mydas																			X																		X	X	X		X	5
Eretmochelys imbricata																																	X	X	X		X	X				5
Lepidochelys olivacea																																							X			1
Dermochelys coriacea													X						X																							1
Chelydra acutirostris																	X												X													2
Chelydra rossignonii																	X																									1
Kinosternon albogulare			X																										X		X		X									2
Kinosternon leucostomum						X					X							X	X								X		X		X		X									7
Staurotypus triporcatus																X																										1
Trachemys emolli																																										0
Trachemys scripta (I)																																										0
Trachemys venusta						X						X							X	X							X		X		X		X		X							7
Rhinoclemmys annulata												X							X								X															2
Rhinoclemmys areolata																																										0
Rhinoclemmys funerea												X							X										X		X											3
Rhinoclemmys pulcherrima																																							X			1
Subtotals	0	0	1	0	0	2	0	0	0	0	0	0	5	0	0	2	0	0	7	2	0	0	0	0	0	0	0	0	5	0	4	0	4	0	5	3	4	0	2	3	4	0	3	40
Totals	15	3	17	19	11	26	16	6	12	4	5	25	23	3	8	6	2	13	32	23	2	2	4	3	2	1	3	16	7	2	3	16	7	2	19	2	11	11	19	16	4	412

TABLE 14. Honduran Protected areas with some suitable or old second growth forest or habitat remaining, including marine waters. Area abbreviations: PN = Parque Nacional (National Park); R = Reserva (Reserve); RA = Reserva Antopológica (Anthropological Reserve); RB = Reserva Biológica (Biological Reserve); RVS = Reserva de Vida Silvestre (Wildlife Refuge); PNM = Parque Nacional Marino (National Marine Park); AUM = Área de Uso Múltiple (Multiple Use Area); JB = Jardín Botánica (Botanical Garden). Forest type abbreviations: LDF(WI) = Lowland Dry Forest, West Indian subregion; LMF = Lowland Moist Forest; LMMF = Lower Montane Moist Forest; LMWF = Lower Montane Wet Forest; MR = Montane Rainforest; PMF = Premontane Moist Forest; and PWF = Premontane Wet Forest. An asterisk (°) preceding a number indicates that I have not collected in that area. Area sizes in hectares (ha) are from Sánchez et al. (2002), unless otherwise noted.

Area (ha)	Main Forest Type(s)	Comments
1. PN Capiro y Calentura (5,566)	LMF, PWF	Locked gate across only road accessing this park offers protection to existing forest.
2. PN Celaque (26,639)	PMF, LMMF	Large tract of forest S of visitor's center. Western section in Ocotepeque badly deforested.
3. PN Cerro Azul (15,574)	PWF, LMMF	Best tract of forest remaining is along western side above San Isidro. Much of eastern portion badly deforested, except highest peak above about 1,600 m elev.ation near Quebrada Grande
4. PN Cerro Azul Meámbar (20,789)	PWF, LMWF	Large tracts of both forest types present.
5. PN Cusuco (17,908)	PWF, PMF, LMWF	Core zone around El Cusuco well protected, much of lower elevations around flanks of core zone under heavy human pressure.
6. PN El Merendón (35,182)	LMF, PWF	Some good forest remaining in more interior areas.
7. PN La Muralla (14,941)	PWF, LMWF	LMWF above visitor's center well protected, some of harder to reach PWF still preserved
8. PN La Tigra (8,768)	PMF, LMMF	LMMF well protected, but much of PMF damaged and/or burned annually.
9. PN Montaña Botaderos (38,214)	PWF, LMWF	Largely deforested with good forest remaining only in highest reaches.
10. PN Montaña de Comayagua (18,273)	PWF, PMF, LMMF	Much of park highly disturbed, some pine forest remaining.
11. PN Montaña de Yoro (15,468)	LMWF	Tracts of forest remaining in harder to reach areas, but largely deforested.
12. PN Montecristo-Trifinio (1,534)	LMMF	Some of higher reaches still forested.
13. PN Patuca (376,448)	LMF	Areas along rivers and trails heavily deforested. Forest remaining in more interior areas
14. PN Pico Bonito (56,473)	LMF, PWF, LMWF	Northern and western slopes and upper reaches retain much forest, but interior along Río Viejo and tributaries much impacted by humans and Hurricane Mitch.
15. PN Pico Pijol (11,453)	PWF, LMWF	Forest still remaining in some upper reaches and in a few pockets in lower elevations.
16. PN Punta Izopo (6,405)	LMF	Little forest remaining. Lagoons and swamps make up much of park.
°17. PN Punta Sal (also called PN Jeannette Kawas) (37,996)	LMF	Lagoons and swamps make up much of park. Apparently some pockets of forest remain.
18. PN Santa Bárbara (13,236)	LMWF, MR	Tracts of forest remain only above about 2,100 m elevation.
19. PN Sierra de Agalta (51,837)	PWF, LMWF	Much forest remains in core zone; area around flanks of core zone much disturbed.
20. R Biósfera Río Plátano (829,779)	LMF	Some core zone broadleaf forest remains, but all of buffer zone badly impacted, including broadleaf forest, pine savanna, and freshwater lagoons and swamps.
21. RA Tawahka (252,079)	LMF	Forest remains in places, but forests along all rivers and trails badly impacted.
22. RB Cerro El Uyuca (817)	PMF, LMMF	Much pine forest remains.
23. RB Cordillera de Montecillos (13,191)	PMF, LMMF	A little forest remaining in highest reaches.

TABLE 14. CONTINUED.

Area (ha)	Main Forest Type(s)	Comments
°24. RB El Chile (6,280)	PMF, LMMF	Apparently much of reserve still forested.
25. RB El Pital (1,799)	LMMF	Almost entirely deforested for agriculture. Unusually heavy pesticide used.
26. RB Guajiquiro (7,368)	LMMF	Limited patches of forest remaining.
27. RB Guisayote (ha area not given in Sánchez et al.)	LMMF	Limited patches of forest remaining.
28. RB Hierba Buena (3,522)	PMF, LMMF	Heavily impacted for agriculture; good forest remaining on higher elevations of Cerro Cantagallo.
29. RB Monserrat (2,241)	PMF, LMMF	Limited patches of forest remaining, except on top at communications tower.
30. RB Opalaca (14,953)	LMMF	Almost entirely deforested for agriculture.
31. RB Río Kruta (115,107)	LMF	Area along Río Kruta and Río Coco denuded, but more inland broadleaf swamp forest and freshwater marshes of little use to humans.
32. RVS Cuero y Salado (7,948)	LMF	Lagoons and swamps make up much of park.
33. RVS Erapuca (7,317)	LMMF	Tracts of little-disturbed forest remain above about 2,000 m elevation.
34. RVS Laguna de Caratasca (133,749)	LMF	Reserve made up largely of pine savanna and cocotales, though largely disturbed, still offers habitat for several species otherwise occurring only in denuded subhumid forest in south and interior valleys of country.
35. RVS Mixcure (7,766)	LMMF	Higher portions show little human impact.
36. RVS Texíguat (15,810)	LMWF	Little primary forest remaining. Much of reserve now crop fields. Heavy logging also present.
°37. PNM Cayos Cochinos (ha area not given in Sánchez et al.)	LWF	Marine park also protects two largest islands. Primary forest remaining in parts of both main islands.
38. PNM Swan Islands [Islas del Cisne] (ha area not given in Sánchez et al.)	LDF(WI)	Marine park also includes these difficult-to-reach islands. Isla Pequeña lacks good landing beaches and mostly covered by karsted limestone of little use to humans. Much primary shrub forest remaining.
39. AUM Isla del Tigre; PNM Archipiélago del Golfo de Fonseca (49.95 km² according to ICF, 2010)	LDF	AUM—Much of mountain slopes of Isla del Tigre with second growth forest; PNM—Also said to protect land areas on islands. Terrestrial areas of all remaining islands badly degraded.
40. JB Lancetilla (1,010)	LMF	Botanical gardens with many species of introduced trees and some good secondary forest preserved.
41. Islas de la Bahía	Marine	Sánchez et al. list a PNM for the Islas de la Bahía, and Reserva Marina (Marine Reserve) for each of Guanaja and Utila, as well as for several areas on Roatán. These areas are all combined here as PNM Islas de la Bahía. Terrestrial habitats all degraded.

ras; two of those species are introduced (*Hemidactylus haitianus*, *Norops sagrei*) and thus are of no concern for conservation. Of the seven remaining species not known from a Honduran protected area (*Sphaerodactylus glaucus, Laemanctus serratus, Norops carpenteri, Ctenosaura quinquecarinata, Polychrus gutturosus, Scincella assata*, and *Aspidoscelis motaguae*), all but two (*L. serratus, P. gutturosus*) appear to have stable populations remaining in degraded areas in Honduras. *Laemanctus serratus* is not known from any precise locality in Honduras, and the *Polychrus* species needs closed-canopy forests to maintain its existence, which are in extreme danger of being eliminated.

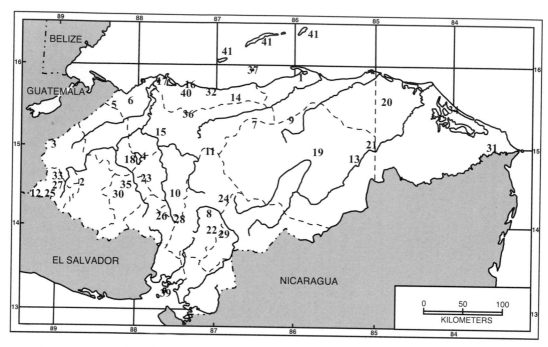

Map 88. The general locations of some protected areas of Honduras with forest or other habitats remaining: 1 = Capiro and Calentura National Park; 2= Celaque National Park; 3 = Cerro Azul National Park; 4 = Cerro Azul Meámbar National Park; 5 = Cusuco National Park; 6 = El Merendón National Park; 7 = La Muralla National Park; 8 = La Tigra National Park; 9 = Montaña Botaderos National Park; 10 = Montaña de Comayagua National Park; 11 = Montaña de Yoro National Park; 12 = Montecristo-Trifinio National Park; 13 = Patuca National Park; 14 = Pico Bonito National Park; 15 = Pico Pijol National Park; 16 = Punta Izopo National Park; 17 = Punta Sal National Park; 18 = Santa Bárbara National Park; 19 = Sierra de Agalta National Park; 20 = Río Plátano Biosphere Reserve; 21 = Tawahka Anthropological Reserve; 22 = Cerro El Uyuca Biological Reserve; 23 = Cordillera de Montecillos Biological Reserve; 24 = El Chile Biological Reserve; 25 = El Pital Biological Reserve; 26 = Guajiquiro Biological Reserve; 27 = Guisayote Biological Reserve; 28 = Hierba Buena Biological Reserve; 29 = Monserrat Biological Reserve; 30 = Opalaca Biological Reserve; 31 = Río Kruta Biological Reserve; 32 = Cuero and Salado Wildlife Refuge; 33 = Erapuca Wildlife Refuge; 34 = Laguna de Caratasca Wildlife Refuge; 35 = Mixcure Wildlife Refuge; 36 = Texíguat Wildlife Refuge; 37 = Cayos Cochinos Marine National Park; 39 = Islas del Tigre Multiple Use Area; 40 = Lancetilla Botanical Gardens; 41 = marine areas around the Islas de la Bahía which are part of the Parque Nacional Marino of the Islas de la Bahía. Also see Map 5 for the location of area 38 (the Swan Islands or Islas del Cisne).

Both crocodile species are known from protected areas, as are 14 of the 17 turtle species. Of the three turtle species not known from a protected area, one (*Trachemys scripta*) is an introduced species and one (*Rhinoclemmys areolata*) has no definitely known localities in Honduras. The third of these three turtle species is the freshwater turtle *Trachemys emolli*, which appears to remain common at some degraded localities.

As noted in the introduction to this section, almost all protected areas in Honduras exist on paper only, and those few exceptions have only select parts within their boundaries protected for the time being. Until the Honduran government takes these protected areas seriously and implements some action for preservation of their forests and beaches, all conservation-related activities in the country are acts of futility. Unfortunately, the most likely scenario is the continued lack of action toward stopping or even slowing down the ongoing forest and beach devastation. Continued lack of much-needed action on the part of the Honduran governmental officials is almost certainly the future of Honduras.

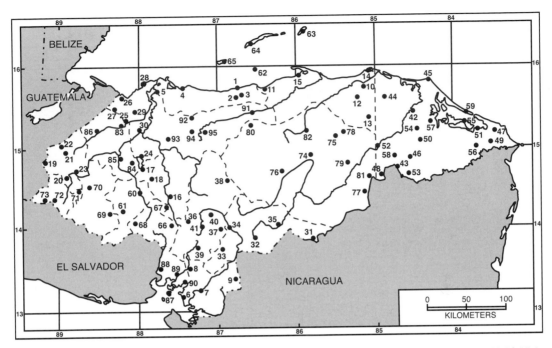

Map 89.   Selected localities in Honduras. Localities mapped are: ATLÁNTIDA—(1) La Ceiba, (2) Quebrada de Oro, (3) Río Viejo, (4) Tela, (5) Toloa Creek; CHOLUTECA—(6) Cedeño, (7) Choluteca, (8) Pespire, (9) San Marcos de Colón; COLÓN—(10) Amarillo, (11) Balfate, (12) Barranco, (13) Empalme Río Chilmeca, (14) Sangrelaya, (15) Trujillo; COMAYAGUA—(16) Comayagua, (17) Pito Solo, (18) Siguatepeque; COPÁN—(19) Copán, (20) Cucuyagua, (21) La Florida, (22) Quebrada Grande, (23) Santa Rosa de Copán; CORTÉS—(24) Agua Azul, (25) Cofradía, (26) Cuyamel, (27) El Cusuco, (28) Puerto Cortés, (29) San Pedro Sula, (30) Villanueva; EL PARAÍSO—(31) Arenales, (32) Danlí, (33) Güinope, (34) Ojo de Agua, (35) Valle de Jamastrán; FRANCISCO MORAZÁN—(36) Cerro Cantagallo, (37) El Zamorano, (38) Guaimaca, (39) Sabana Grande, (40) San Juancito, (41) Tegucigalpa; GRACIAS A DIOS—(42) Ahuás, (43) Awasbila, (44) Baltiltuk, (45) Barra Patuca, (46) Bodega del Río Tapalwás, (47) Canco, (48) Caño Awalwás, (49) Krahkra, (50) Mocorón, (51) Puerto Lempira, (52) Quebrada Waskista, (53) Rus Rus, (54) Sadyk Kiamp, (55) Tánsin, (56) Tikiraya, (57) Warunta, (58) Warunta Tingni Kiamp, (59) Yahurabila; INTIBUCÁ—(60) Jesús de Otoro, (61) La Esperanza; ISLAS DE LA BAHÍA—(62) Cayos Cochinos, (63) Guanaja, (64) Roatán, (65) Utila; LA PAZ—(66) Guajiquiro, (67) La Paz, (68) Marcala; LEMPIRA—(69) Erandique, (70) Gracias; OCOTEPEQUE—(71) Belén Gualcho, (72) El Volcán, (73) Nueva Ocotepeque; OLANCHO—(74) Catacamas, (75) Dulce Nombre de Culmí, (76) Juticalpa, (77) Kauroahuika, (78) La Colonia, (79) Matamoros, (80) Parque Nacional La Muralla Centro de Visitantes, (81) Quebrada Kuilma, (82) San Esteban; SANTA BÁRBARA—(83) Quimistán, (84) San José de Los Andes, (85) Santa Bárbara, (86) Sula; VALLE—(87) Amapala, (88) Goascorán, (89) Nacaome, (90) San Lorenzo; YORO—(91) Coyoles, (92) La Fortuna, (93) Pico Pijol, (94) Portillo Grande, (95) Yoro. All collecting localities are included in a Glossary in the Supplemental Data.

ACKNOWLEDGMENTS

Collecting and exportation permits have been provided over the years by personnel of the Dirección General de Recursos Naturales Renovables, then the Departamento de Áreas Protegidas y Vida Silvestre (DAPVS)/Administración Forestal del Estado, Corporación Hondureña de Desarrollo Forestal (AFE-COHDEFOR), and starting in 2013 by the Instituto Nacional de Conservación y Desarrollo Forestal, Áreas Protegidas y Vida Silvestre (ICF), all in Tegucigalpa. Licenciado (Lic.) Iris Acosta, Lic. Carla Cárcamo de Martínez, Lic. Franklin E. Castañeda, Lic. Saíd Laínez, and Leonardo Valdés Orellana were helpful from 2005 to 2013.

Field assistance in pursuit of Honduran reptiles has been provided over the years by the following individuals: Damian Almendarez, Mardo Bancequel, Breck Bartholomew, Franklin E. Castañeda, Juan R. Collart, Gustavo A. Cruz Díaz, Gary Dodge, Mario R. Espinal, Gerardo A. Flores, Steve

W. Gotte, Dalmacia Green, the late Emiliano Green, Emiliano Green, Jr., the late Mario Guiffaro, Alexander Gutsche, Eric Hedl, John Himes, Gunther Köhler, Mario Francisco Lacoth, Melissa Lacoth Montoya, Tomás Manzanares, Emiliano Meráz, Kirsten E. Nicholson, Louis Porras, Javier Rodriguez, John Rindfleish, José M. Solís, Josiah H. Townsend, Leonardo Valdés Orellana, Rony Valle Ocho, Kenneth L. Williams, and Larry D. Wilson.

One or more copies of difficult-to-obtain literature were provided by: Aaron Bauer, Wolfgang Böhme, John L. Carr, Lee Fitzgerald, Steve W. Gotte, Alexander Gutsche, S. Blair Hedges, Gunther Köhler, Monika Laudahn, Angela B. Marion, James A. Poindexter, and Jay M. Savage.

The following colleagues provided one or more figures, photographs of museum specimens for verifying identification, or photographs in life used, or considered for use for this work: Christopher C. Austin, John L. Carr, Juan Ramon Collart, Carl Franklin, Carlos Andrés Galvis R., Rachel Grill, Alexander Gutsche, Toby Hibbits, Gunther Köhler, Jeffrey Lovich, José Martínez, Chris Phillips, James A. Poindexter, Louis Porras, Alan Resetar, Stephen P. Rogers, José Rosado, Jeffrey Seminoff, Jack Sites, Javier Sunyer, Greg Taylor, Gabriel N. Ugueto, Leonardo Valdés Orellana, and Dan Wylie. Shannen L. Robson searched the UU collection in vain for two *Trachemys venusta* shells purported to be in that collection. Patrick Campbell (BMNH) supplied several images for identification, for which I am especially grateful.

The following curators and other museum personnel facilitated loans or responded with lists of their Honduran reptile holdings: Margaret Arnold, David Dickey, Linda S. Ford, Darrel Frost, David Kizirian, Lauren Vonnahme (AMNH); Ted Daeschler, Ned S. Gilmore (ANSP); Jack W. Sites, William W. Tanner (BYU); Jens Vindum (CAS); the late Clarence J. McCoy, Stephen P. Rogers (CM); Kathleen M. Kelly, Alan Resetar (FMNH); Chris Mayer, John E. Petzing, Chris A. Phillips (INHS); Rafe Brown, Andrew Campbell, William E. Duellman, Jamie Oaks, John Simmons, Linda Trueb (KU); Kent Beaman, Rick Feeney, Jeff Seigel (LACM); Christopher Austin, James A. McGuire, Diana R. Reynolds, Eric N. Rittmeyer (LSUMZ); James Hanken, Joe Martinez, José P. Rosado (MCZ); H. F. Faraji, Carol Spencer, David B. Wake (MVZ); Laura Abraczinskas (MSUM); Amy Lathrop, Ross MacCulloch, Robert W. Murphy (ROM); Brad Hollingsworth, Robert E. Lovich, Dustin Wood (SDSNH); David Lintz (SMBU); Linda Acker, Gunther Köhler, Sebastian Lotzkat (SMF); Toby Hibbits, R. Kathryn Vaughan (TCWC); David C. Cannatella, Travis La Duc (TNHC); Harold Dundee (TU); Rosanne Humphrey, Mariko Kageyma, Christy McCain, the late Hobart M. Smith (UCM); Kenneth L. Krysko (UF); Chris Phillips, Steven D. Sroka (UIMNH); Arnold G. Kluge, Ronald Nussbaum, Gregory Schneider (UMMZ); Gustavo A. Cruz Díaz, Sofía Nuñez (UNAH); Steve W. Gotte, Roy W. McDiarmid, James A. Poindexter (USNM); Jonathan A. Campbell, Carl J. Franklin, Eric N. Smith (UTA); John M. Legler, Eric Rickart (UU).

Very early parts of this manuscript were read and improved upon as follows: Xantusiidae (Robert Bezy); Gymnophthalmidae (Charles J. Cole, Tiffany M. Doan); and Scincomorpha (Aurélien Miralles). Leonardo Valdés Orellana corrected errors in my first attempt at writing the Spanish version of the dichotomous keys (claves). Oscar Flores-Villela used his professional biological background and his native Spanish language to much improve those claves. More recent versions of various sections (September–December 2015) were read and improved on by colleagues as follows: Xantusioidea (R. Alexander Pyron, Jack Sites); Gekkomorpha (Aaron M. Bauer); Anguioidea (Edmund D. Brodie III, Jonathan A. Campbell); Corytophanidae (R.

Alexander Pyron); Dactyloidae (Gunther Köhler); Iguanidae (Alexander Gutsche, Gunther Köhler); Leiocephalidae (Robert Powell); Phrynosomatidae (Jay M. Savage, Jack Sites, Eric N. Smith); Polychrotidae (R. Alexander Pyron); Scincomorpha (S. Blair Hedges); Teiformata (Michael B. Harvey); Crocodylia (Steve W. Gotte, Steven G. Platt); and Testudinata (James R. Buskirk, John L. Carr, James F. Parham). Steve W. Gotte and Jay M. Savage read the sections before the beginning of the species accounts and those following the end of those species accounts. I am especially grateful to Eric N. Smith for providing his newly generated tree showing the results of various Central American populations of *Sceloporus* "*malachiticus*," including many Honduran tissues collected by me for a project he and I have planned. I especially appreciate the time and effort of all of those colleagues listed above that went into reviewing and improving the quality and accuracy of this work. However, I am completely responsible for the contents of the final product.

Colleagues who answered various inquiries for information include: Aaron Bauer, Juan R. Collart, Carl J. Franklin, Thierry Frétey, Steve W. Gotte, Alexander Gutsche, S. Blair Hedges, Toby J. Hibbits, John Iverson, David Kizirian, Gunther Köhler, Amy Lathrop, Jeffrey Lovich, Robert Lovich, Ross MacCulloch, Julio E. Mérida, Stesha Pasachnik, James A. Poindexter, R. Alexander Pyron, Alan Resetar, Jay M. Savage, Jeffrey Seminoff, Eric N. Smith, Hobart M. Smith, Lauren Vonnahme, John J. Wiens, and George R. Zug. Alan Resetar (FMNH) and the Barbour Fund of the Museum of Comparative Zoology helped finance two recent trips. I am especially grateful to Steve W. Gotte for the work he did voluntarily on the variation in number of inframarginal bridge scutes in series of *Caretta caretta* and *Lepidochelys olivacea* housed in the USNM collection. Alexander Gutsche provided pertinent data regarding his knowledge about various aspects on the species of *Ctenosaura*.

Jay M. Savage (San Diego State University) assembled the Honduran SDSNH specimens at the laboratory in his house awaiting my visit. Ron Heyer, Steve W. Gotte, and James A. Poindexter went out of their way to obtain several museum loans for me to examine while I was on a visit to the USNM in October 2009. Leonardo Valdés Orellana helped with data gathering on visits to several museums in October 2009. Alexander Gutsche provided an English translation on the diet habitats of *Ctenosaura bakeri* included in his dissertation.

Bayardo Aleman, Franklin E. Castañeda, Mario R. Espinal, Leonnel Marineros, and Leonardo Valdés Orellana donated to me various reptile specimens they collected in Honduras. Chad Montgomery collected tissues of *Cnemidophorus ruatanus* and *Trachemys venusta* on the Cayos Cochinos that helped me on two projects. Leonardo Valdés Orellana collected the holotype of the new species of *Laemanctus*.

I am sorry if I left out any names of persons that should have been acknowledged in one of the above paragraphs. I sincerely apologize to those persons and stress any omissions were unintentional.

I am using the same base map used in previous works in this series to plot the localities of the Honduran reptiles. I am grateful to Kraig Adler, Tim Perry, and the Society for the Study of Amphibians and Reptiles (SSAR) for permission to use that base map.

This paper would not have been possible without the help of the many colleagues listed above. The paper also benefitted from the previous work of many authors acknowledged in the Literature Cited and credited throughout this work. However, all specimen data used in the species accounts were taken by me, unless otherwise stated. Despite that help, it remains certain some errors on my part remain.

LITERATURE CITED

Included is a list of all references in the text. Plates are listed only if they are not included among the numbered pages, and the titles are copied from the title page, and not the covers. An asterisk (*) at the beginning of an entry means that I have not seen that publication in any form, and the information in the citation and the citation itself, was taken from the literature.

Acevedo, A. A., M. Lampo, and R. Cipriani. 2016. The cane or marine toad, *Rhinella marina* (Anura: Bufonidae): two genetically morphologically distinct species. *Zootaxa* 4103: 574–586.

Acevedo, M. 2006. Anfibios y reptiles de Guatemala: una breve síntesis con bibliografía, PP. 487–524 IN: E. B. Cano, editor. *Biodiversidad de Guatemala*. Vol. I. Ciudad de Guatemala: Universidad del Valle de Guatemala.

Acosta-Chaves, V. J., K. R. Russell, and C. Sartini. 2016. Nature notes. *Crocodylus acutus*. Predation. *Mesoamerican Herpetology* 3: 140–142.

Acosta-Chaves, V. J., N. Solís-Miranda, and C. L. Barrio-Amorós. 2015. Nature notes. *Thecadactylus rapicauda*. Preying on large insects. *Mesoamerican Herpetology* 2: 197–199.

Acuña-Mesén, R. A. 1992. Potential exploitation of captive slider turtles (*Trachemys scripta*) in Costa Rica: a preliminary study. *Brenesia* 38: 157–158.

Acuña-Mesén, R. A. 1994. Variación morfometrica y caracteristicas ecologicas del habitat de la Tortuga Candado *Kinosternon scorpioides* en Costa Rica (Chelonia, Kinosternidae). *Revista Brasileira de Biologia* 54: 537–547.

Acuña-Mesén, R. A. 1999. *Conservación y ecología de las Tortugas terrestres, semiacuáticas y acuáticas (de agua dulce y marinas) de Costa Rica*. Primera reimpresión. San José, Costa Rica: Editorial Universidad Estatal a Distancia.

Acuña [Mesén], R., A. Castaing, and F. Flores. 1983. Aspectos ecológicos de la distribución de las tortugas terrestres y semiacuáticas en el Valle Central de Costa Rica. *Revista de Biologia Tropical* 31: 181–192.

Adams, A. 1854. *A Manual of Natural History, for the Use of Travellers; Being a Description of the Families of the Animal and Vegetable Kingdoms: With Remarks on the Practical Study of Geology and Meteorology. To which are Appended Directions for Collecting and Dissecting*. London: John van Voorst.

Agassiz, L. 1857. *Contributions to the Natural History of the United States of America. First Monograph. In Three parts.—I. Essay on Classification.—II. North American Testudinata*. Vol. I. Boston, Massachusetts: Little, Brown, and Company.

Ahl, E. 1939. Ueber eine Sammlung von Reptilien aus El Salvador. *Sitzungsberichte der Gesellschaft Naturforschender Freunde zu Berlin* 1939: 245–249.

Alemán, B. M., and J. Sunyer. 2014. Nature notes. *Aspidoscelis deppii*. Diet. *Mesoamerican Herpetology* 1: 155–156.

Alemán, B. M., and J. Sunyer. 2015. Nature notes. *Hemidactylus frenatus* Schlegel, 1836 In A. M. C. Duméril & Bibron, 1836. Predation attempt. *Mesoamerican Herpetology* 2: 518–519.

Álvarez del Toro, M. 1974. *Los Crocodylia de Mexico (Estudio Comparativo)*. Mexico City, Mexico: Instituto Mexicano de Recursos Naturales Renovables.

Álvarez del Toro, M. 1983 (dated 1982). *Los Reptiles de Chiapas. Tercera Edición, corregida y aumentada*. Tuxtla Gutierrez, Mexico: Publcación del Instituto de Histotia Natural.

Álvarez Solórzano, T., and M. González Escamilla. 1987. *Atlas Cultural de México. Fauna*. San Mateo Tecoloapán, México: Instituto Nacional de Anthropología e Historia.

Anderson, C., and K. M. Enge. 2012. Natural history notes. *Ctenosaura similis* (Gray's Spiny-tailed Iguana) and *Iguana iguana* (Green Iguana). Carrion Feeding. *Herpetological Review* 43: 131.

Andersson, L. G. 1900. Catalogue of Linnean type-specimens of Linnaeus's Reptilia in the Royal Museum in Stockholm. *Bihang till Konigl Svenska Vetenskapsakademiens Handlingar* 26: 1–29.

Anonymous. 1994. *Evaluación Ecológica Rápida (EER) Parque Nacional "El Cusuco" y Cordillera del Merendón*. San Pedro Sula, Honduras: Fundación Ecologista "Hector Rodrigo Pastor Fasquelle."

Anonymous. 2002 Sep 11. World Court asked to consider boundary. Miami Herald; Sect. A:16.

Anonymous. 2006. *Atlas Geográfica de Honduras*. Edición 2006–2007. Tegucigalpa, Honduras: Ediciones Ramsés.

Anonymous. 2008. Análisis del Potencial de Desarrollo en Islas del Cisne. Tegucigalpa, Honduras: Unpublished Report to Secretaria de Turismo e Instituto Hondureño de Turismo y Secretaria de Recursos Naturales y Ambiente.

Ashton, K. G. 2003. Sexing *Cnemidophorus* lizards using a potential scale character. *Herpetological Review* 34: 109–111.

Atkinson, D. A. 1907. Notes on a collection of batrachians and reptiles from Central America. *The Ohio Naturalist* 7: 151–157.

Auth, D. L. 1994. Checklist and bibliography of the amphibians and reptiles of Panama. *Smithsonian Herpetological Information Service* 98: 1–59.

Avila-Pires, T. C. S. 1995. Lizards of Brazilian Amazonia (Reptilia: Squamata). *Zoologische Verhandelingen* 299: 1–706.

Babcock, H. L. 1932. The American snapping turtles of the genus *Chelydra* in the collection of the Museum of Comparative Zoology, Cambridge, Mass., U.S.A. *Proceedings of the Zoological Society of London* 44: 873–874.

Bailey, J. W. 1928. A revision of the lizards of the genus *Ctenosaura*. *Proceedings of the United States National Museum* 73: 1–58, plates 1–30.

Baird, S. F. 1859a (dated 1858). Description of new genera and species of North American lizards in the Museum of the Smithsonian Institution. *Proceedings of the Academy of Natural Sciences of Philadelphia* 10: 253–256.

Baird, S. F. 1859b. Reptiles of the boundary, with notes by the naturalists of the survey. IN: *United States and Mexican Boundary Survey, Part II, Zoology of the Boundary*. Under the order of Lieut. Col. W. H. Emory. Washington: Department of the Interior.

Baird, S. F., and C. Girard. 1852. Descriptions of new species of reptiles, collected by the U. S. Exploring Expedition under the command of Capt. Charles Wilkes, U. S. N. *Proceedings of the Academy of Natural Sciences of Philadelphia* 6: 174–177.

Balaguera-Reina, S. A., M. Venegas-Anaya, and L. D. Densmore, III. 2015. Miscellaneous notes. *Crocodilus acutus* in Panama: a status report. *Mesoamerican Herpetology* 2: 566–571.

Balaguera-Reina, S. A., M. Venegas-Anaya, O. I. Sanjur, H. A. Lessios, and L. D. Densmore, III. 2015. Reproductive ecology and hatchling growth rates of the American Crocodile (*Crocodylus acutus*) on Coiba Island, Panama. *South American Journal of Herpetology* 10: 10–22.

Barbour, T. 1914. A contribution to the zoögeography of the West Indies, with especial reference to amphibians and reptiles. *Memoirs of the Museum of Comparative Zoölogy* 44: 209–359, 1 plate.

Barbour, T. 1921a. Some reptiles from Old Providence Island. *Proceedings of the New England Zoölogical Club* 7: 81–85.

Barbour, T. 1921b. *Sphaerodactylus*. *Memoirs of the Museum of Comparative Zoölogy* 47: 217–278, plates 1–26.

Barbour, T. 1926. *Reptiles and Amphibians. Their Habits and Adaptations*. Revised Edition. Boston, Massachusetts: Houghton Mifflin Company.

Barbour, T. 1928. Reptiles from the Bay Islands. *Proceedings of the New England Zoölogical Club* 10: 55–61.

Barbour, T. 1930a. Some faunistic changes in the Lesser Antilles. *Proceedings of the New England Zoölogical Club* 11: 73–85.

Barbour, T. 1930b. A list of Antillean reptiles and amphibians. *Zoologica, Scientific Contributions of the New York Zoological Society* 11: 61–116.

Barbour, T. 1935. A second list of Antillean reptiles and amphibians. *Zoologica, Scientific Contributions of the New York Zoological Society* 19: 77–141.

Barbour, T. 1937. Third list of Antillean reptiles and amphibians. *Bulletin of the Museum of Comparative Zoölogy* 82: 77–166.

Barbour, T., and A. Loveridge. 1929a. Typical reptiles and amphibians in the Museum of Comparative Zoölogy. *Bulletin of the Museum of Comparative Zoölogy* 69: 205–360.

Barbour, T., and A. Loveridge. 1929b. On some Honduranian and Guatemalan snakes with the description of a new arboreal pit viper of the genus *Bothrops*. *Bulletin of the Antivenin Institute of America* 3: 1–3.

Barbour, T., and A. Loveridge. 1946. First supplement to typical reptiles and amphibians. *Bulletin of the Museum of Comparative Zoölogy* 96: 59–214.

Barbour, T., and G. K. Noble. 1915. A revision of the lizards of the genus *Ameiva*. *Bulletin of the Museum of Comparative Zoölogy* 59: 417–479.

Barbour, T., and B. Shreve. 1934. A new race of rock iguana. *Occasional Papers of the Boston Society of Natural History* 8: 197–198.

Barrio-Amorós, C., and R. A. Ojeda. 2015. Nature notes. *Iguana iguana*. Predation by Tayras (*Eira barbara*). *Mesoamerican Herpetology* 2: 112–114.

Bartelt, U., and S. Prassel. 2009. Bemerkungen zur Haltung und Nachzucht von *Sceloporus olloporus*—Lilabauch Stachelleguan (Smith 1937). *Iguana Rundschreiben* 22: 15–20.

Batsch, A. J. G. C. 1788. *Versuch einer Anleitung, zur Kenntniß und Geschichte der Thiere und Mineralien, für akademische Boriesungen entworsen, und mit den nöthigsten abbisdungen dersehen. Erster Theil. Allgemeine Geschichte der Natur; besondre der Säugthiere, Vögel, Amphibien und Fische*. Jena, Germany: Akademischen Buchandlung.

Bauer, A. M. 1994. Familia Gekkonidae (Reptilia, Sauria). Part I Australia and Oceania. *Das Tierreich* 109: i–xiii, 1–306.

Bauer, A. M. 2013. Geckos: *The Animal Answer Guide*. Baltimore, Maryland: The Johns Hopkins University Press.

Bauer, A. M., and K. Adler. 2001. The dating and correct citation of A. F. A. Wiegmann's "Amphibien" section of Meyen's *Reise um die Erde*, with a bibliography of Wiegmann's herpetological publications. *Archives of Natural History* 28: 313–326.

Bauer, A. [M.] and R. Günther. 1991. An annotated type catalogue of the geckos (Reptilia: Gekkonidae) in the Zoological Museum, Berlin. *Mittei-*

lungen aus dem Zoologisches Museum in Berlin 67: 279–310.

Bauer, A. M., and R. Günther. 1994. An annotated type catalogue of the teiid and microteiid lizards in the Zoological Museum, Berlin (Reptilia: Squamata: Teiidae and Gymnophthalmidae). *Mitteilungen aus dem Zoologisches Museum in Berlin* 70: 267–280.

Bauer, A. M., and A. P. Russell. 1993a. *Aristelliger. Catalogue of American Amphibians and Reptiles* 565.1–565.4.

Bauer, A. M., and A. P. Russell. 1993b. *Aristelliger georgeensis. Catalogue of American Amphibians and Reptiles* 568.1–568.2.

Bauer, A. M., and A. P. Russell. 1993c. *Aristelliger praesignis. Catalogue of American Amphibians and Reptiles* 571.1–571.4.

Bauer, A. M., and R. A. Sadlier. 2000. The herpetofauna of New Caledonia. *Society for the Study of Amphibians and Reptiles, Contributions to Herpetology* 17: i–xii, 1–310, plates 1–24.

Bauer, A. M., A. Ceregato, and M. Delfino. 2013. The oldest herpetological collection in the world: the surviving amphibian and reptile specimens of the Museum of Ulisse Aldrovandi. *Amphibia-Reptilia* 34: 305–321.

Bauer, A. M., D. A. Good, and W. R. Branch. 1997. The taxonomy of the southern African leaf-toed geckos (Squamata: Gekkonidae), with a review of Old World *"Phyllodactylus"* and the description of five new genera. *Proceedings of the California Academy of Sciences* 49: 447–497.

Bauer, A. M., R. Günther, and M. Klipfel. 1995. Synopsis of the herpetological taxa described by Wilhelm Peters, PP. 39–81. IN: *The Herpetological Contributions of Wilhelm C. H. Peters (1815–1883). Society for the Study of Amphibians and Reptiles, Facsimile Reprints in Herpetology* (no number given for Facsimile).

Bauer, A. M., T. R. Jackman, E. Greenbaum, V. B. Giri, and A. de Silva. 2010. South Asia supports a major endemic radiation of *Hemidactylus* geckos. *Molecular Phylogenetics and Evolution* 57: 343–352.

Baur, G. 1888. Osteologische Notizen über Reptilien. (Fortsetzung III). *Zoologischer Anzeiger* 11: 417–424.

Beaty, L., and S. Beaty. 2012. Natural history notes. *Rhinella marina* (Cane Toad). Crocodile predation. *Herpetological Review* 43: 471.

Behn, F. D. 1760. *Jacob Theodor Kleins, Klassifikation und kurze Geschichte der vierfüßigen Thiere, aus dem lateinischen Übersetzt, und mit zusätzen Vermehret, nebst einer Vorrede.* Lübeck: Jonas Schmidt.

Bell, E. L., H. M. Smith, and D. Chiszar. 2003. An annotated list of the species-group names applied to the lizard genus *Sceloporus. Acta Zoologica Mexicana (n.s.)* 90: 103–174.

Bell, T. 1825. On a new genus of Iguanidae. *Zoological Journal* 2: 204–208.

Bell, T. 1828. Characters of the order, families, and genera of the Testudinata. *Zoological Journal* 3: 513–516.

Bender, C. 1995. Auf der Suche nach dem Utila-Schwarzleguan *Ctenosaura bakeri. Iguana Rundschreiben* 8: 10–13.

Bergmann, P. J., and A. P. Russell. 2003. Lamella and scansor numbers in *Thecadactylus rapicauda* (Gekkonidae): patterns revealed through correlational analysis and implications for systematic and functional studies. *Amphibia-Reptilia* 24: 379–385.

Bergmann, P. J., and A. P. Russell. 2007. Systematics and biogeography of the widespread neotropical geckkonid genus *Thecadactylus* (Squamata), with the description of a new cryptic species. *Zoological Journal of the Linnean Society* 149: 339–370.

Bermingham, E., A. Coates, G. Cruz Díaz, L. Emmons, R. B. Foster, R. Leschen, G. Seutin, S. Thorn, W. Wcislo, and B. Werfel. 1998. Geology and terrestrial flora and fauna of Cayos Cochinos, Honduras. *Revista de Biologia Tropical* 46(Suppl. 4): 15–37.

Berry, F. H. 1989. Socioeconomic importance of sea turtles. Exploitation, PP. 33–37. IN: L. Ogren et al., editors. *Proceedings of the Second Western Atlantic Turtle Symposium.* Panama City, Florida: NOAA Technical Memorandum NMFS-SEFC-226.

Berry, J. F. 1978. Variation and systematics in the *Kinosternon scorpioides* and *K. leucostomum* complexes (Reptilia: Testudines: Kinosternidae) of Mexico and Central America [dissertation]. Salt Lake City: University of Utah.

Berry, J. F., and J. B. Iverson. 1980. A new species of mud turtle, genus *Kinosternon*, from Oaxaca, Mexico. *Journal of Herpetology* 14: 313–320.

Berry, J. F., and J. B. Iverson. 2001a. *Kinosternon leucostomum. Catalogue of American Amphibians and Reptiles* 724.1–724.8.

Berry, J. F., and J. B. Iverson. 2001b. *Kinosternon scorpioides. Catalogue of American Amphibians and Reptiles* 725.1–725.11.

Berry, J. F., and J. B. Iverson. 2011. *Kinosternon scorpioides* (Linnaeus 1766)—Scorpion Mud Turtle, PP. 063.1–0.63.15. IN: A. G. J. Rhodin, P. C. H. Pritchard, P. P. van Dijk, R. A. Saumure, K. A. Buhlmann, J. B. Iverson, and R. A. Mittermeier, editors. *Conservation Biology of Freshwater Turtles and Tortoises: A Compilation Project of the IUCN/SSC Tortoise and Freshwater Turtle Specialist Group.* Lunenburg, Massachusetts: Chelonian Research Foundation. *Chelonian Research Monographs* 5.

Berthold, A. A. 1845. Ueber verschiedene neue oder seltene Reptilien aus Neu-Granada und Crustaceen aus China. *Nachricten von der Georg-Augusts Universität un der Königlchen Gesellschaft der Wissenschaften, Göttingen* 3: 37–48.

Berthold, A. A. 1846. *Über verschiedene neue oder seltene Reptilien aus Neu-Granada und Crustaceen aus China*. Göttingen, Germany: Dieterichsehen Buchhandlung.

Berube, M. D., S. G. Dunbar, K. Rützler, and W. K. Hayes. 2012. Home range and foraging ecology of juvenile Hawksbill Sea Turtles (*Eretmochelys imbricata*) on inshore reefs of Honduras. *Chelonian Conservation and Biology* 11: 33–43.

Bezy, R. L. 1973. A new species of the genus *Lepidophyma* (Reptilia: Xantusiidae) from Guatemala. *Contributions in Science, Natural History Museum, Los Angeles County* 239: 1–7.

Bezy, R. L. 1989a. Morphological differentiation in unisexual and bisexual xantusiid lizards of the genus *Lepidophyma* in Central America. *Herpetological Monographs* 3: 61–80.

Bezy, R. L. 1989b. Night lizards. The evolution of habitat specialists. *Terra* 28: 29–34.

Bezy, R. L., and J. L. Camarillo R. 2002. Systematics of xantusiid lizards of the genus *Lepidophyma*. *Contributions in Science, Natural History Museum of Los Angeles County* 493: 1–41.

Binn, E. 2014. Utila—neues vom Swamper bzw. einir Swampine. *Iguana Rundschreiben* 27: 29–32.

Binns, J. 2003. Utila: home to three native iguanas. *Iguana* 10: 27, 30.

Binns, J. 2007. [Photographs]. *Iguana* 14: 249 and inside front cover.

Blainville, H. de. 1816. Prodrome d'une nouvelle distribution systématique du règne animal. *Bulletin des Sciences par La Société Philomathique de Paris* 3: 113–120 (erroneously numbered PP. 105–112).

Blumenthal, J. M., T. J. Austin, C. D. L. Bell, J. B. Bothwell, A. C. Broderick, G. Ebanks-Petrie, J. A. Gibb, K. E. Luke, J. R. Olynik, M. F. Orr, J. L. Soloman, and B. J. Godley. 2009. Ecology of Hawksbill turtles, *Eretmochelys imbricata*, on a western Caribbean foraging ground. *Chelonian Conservation and Biology* 8: 1–10.

Boback, S. M., C. E. Montgomery, R. N. Reed, and S. Green. 2006. Geographic distribution. *Kinosternon leucostomum* (White-lipped Mud Turtle, Tortuga Amarilla). *Herpetological Review* 37: 239.

Bocourt, M. [F.] 1868. Description de quelques chéloniens nouveaux appartenant a la faune Mexicaine. *Annales des Sciences Naturelles, Zoologie et Paléontologie* 10: 121–122.

Bocourt, M. F. 1871. Description de quelques gerrhonotes nouveaux provenant du Mexique et de l'Amérique Centrale. *Nouvelles Archives du Muséum d'Histoire Naturelle de Paris* 7: 101–108.

Bocourt, M. F. 1873a. Caractères d'une espèce nouvelle d'iguaniens le *Sceleporus* [sic] *acathhinus* [sic]. *Annales des Sciences Naturelles, Zoologie et Paléontologie* 17: 1.

Bocourt, M. F. 1873b. Note sur quelques espèces nouvelles d'iguaniens du genre *Sceleporus* [sic]. *Annales des Sciences Naturelles, Zoologie et Paléontologie* 17: 1–2.

Bocourt, M. F. 1876. Sur quelques reptiles de l'Isthme de Tehuantepec (Mexique) donnés par M. Sumichrast au Muséum. *Journal de Zoologie, Paris* 5: 386–411.

Bocourt, M. F. 1881. Description d'un Ophidien Opotérodonte, appartenant au genre *Catodon* (2). *Bulletin des Sciences par La Société Philomathique de Paris* 5: 81–82.

Bocourt, M. F. 1882. Note sur les espèces appartenant au genre *Ctenosaura*. *Le Naturaliste* 4: 47.

Boettger, O. 1893. Bericht über die Leistungen in der Herpetologie während des Jahres 1892. *Archiv für Naturgeschichte* 59: 65–172.

Bogert, C. M. 1949. Thermoregulation and eccritic body temperatures in Mexican lizards of the genus *Sceloporus*. *Anales del Instituto de Biología, Mexico* 20: 415–426.

Bogert, C. M. 1959. How reptiles regulate their body temperature. *Scientific American* 200: 105–120.

Böhme, W. 1988. Zur Genitalmorphologie der Sauria: funktionelle und stammesgeschichtliche Aspekte. *Bonner Zoologische Monographien* 27: 1–176.

Böhme, W. 2010. A list of the herpetological type specimens in the Zoologisches Forschungsmuseum Alexander Koenig, Bonn. *Bonn Zoological Bulletin* 59: 79–108.

Böhme, W., and W. Bischoff. 1984. Die Wirbeltiersammlungen des Museums Alexander Koenig. III. Amphibien und Reptilien. *Bonner Zoologische Monographien* 19: 151–213.

Bonaparte, C. L. 1838a. Synopsis Vertebratorum Systematis. *Nouvi Annali delle Scienze Naturali, Bologna* 1: 105–133.

Bonaparte, C. L. 1838b. Amphibiorum. Tabula analytica. *Nouvi Annali delle Scienze Naturali, Bologna* 1: 391–393.

Bonaparte, C. L. 1840. A new systematic arrangement of vertebrated animals. *Transactions of the Linnean Society of London* 18: 247–304.

Bonfiglio, F., R. L. Balestrin, and L. H. Cappellari. 2006. Diet of *Hemidactylus mabouia* (Sauria, Gekkonidae) in urban areas of southern Brazil. *Biociências* 14: 107–111.

Bonin, F., B. Devaux, and A. Dupré. 2006. *Turtles of the World*. London: A & C Black.

Boonman, J. 2000. *Ctenosaura melanosterna* in captivity. *Pod@cris* 1: 100–104.

Boulenger, G. A. 1877. Étude monographique du genre *Laemanctus* et description d'une espèce nouvelle. *Bulletin de la Société Zoologique de France* 2: 460–466, plate VII.

Boulenger, G. A. 1882 (dated 1881). Description of a new species of *Anolis* from Yucatan. *Proceedings of the Zoological Society of London* 1881: 921–922.

Boulenger, G. A. 1883. Remarks on the Nyctisaura. *Annals and Magazine of Natural History* 12: 308.

Boulenger, G. A. 1884. Synopsis of the families of existing Lacertilia. *Annals and Magazine of Natural History* 14: 117–122.

Boulenger, G. A. 1885a. *Catalogue of the Lizards in the British Museum (Natural History)*. 2nd ed. Vol. I. *Geckonidae, Eublepharidae, Uroplatidae, Pygopodidae, Agamidae*. London: Printed by Order of Trustees of British Museum (Natural History).

Boulenger, G. A. 1885b. *Catalogue of the Lizards in the British Museum (Natural History)*. 2nd ed. Vol. II. *Iguanidae, Xenosauridae, Zonuridae, Anguidae, Anniellidae, Helodermatidae, Varanidae, Xantusiidae, Teiidae, Amphisbaenidae*. London: Printed by Order of Trustees of British Museum (Natural History).

Boulenger, G. A. 1889. *Catalogue of the Chelonians, Rhynchocephalians, and Crocodiles in the British Museum (Natural History)*. New Edition. London: Printed by Order of Trustees of British Museum (Natural History).

Boulenger, G. A. 1893. Reptilia and Batrachia. *Zoological Record* 29: 1–41.

Boulenger, G. A. 1895 (dated 1894). Second report on additions to the lizard collection in the Natural-History Museum. *Proceedings of the Zoological Society of London* 31: 722–736, plates XLVII–XLIX.

Boulenger, G. A. 1897. Reptilia and Batrachia. *Zoological Record* 33: 1–38.

Boulenger, G. A. 1902. Reptilia and Batrachia. *Zoological Record* 38: 1–35.

Boulenger, G. A. 1911. Descriptions of new reptiles from the Andes of South America, preserved in the British Museum. *Annals and Magazine of Natural History* 7: 19–25.

Bour, R. 1979 (dated 1978). Les tortues actuelles de Madagascar (République malgache): liste systématique et description de deux sous-espècies nouvelles (Reptilia–Testudines). *Bulletin de la Société d'études scientifiques de l'Anjou* 1978: 141–154.

Bour, R. 2003. Le genre *Trachemys*: systématique et répartition. *Manouria* 6: 2–9.

Bour, R. 2004. Die Schmuckscildkröten der Gattung *Trachemys*: Systematik und Verbreitung. *Sacalia* 4: 5–23.

Bour, R. 2007. The type specimens of *Rhinoclemmys areolata* (Duméril & Bibron, 1851), *R. pulcherrima incisa* (Bocourt, 1868), and *R. punctularia* (Daudin, 1801). *Emys* 14: 28–34.

Bour, R., and A. Dubois. 1983. Nomenclatural availability of *Testudo coriacea* Vandelli, 1761: a case against a rigid application of the rules to old, well-known zoological works. *Journal of Herpetology* 17: 356–361.

Bramble, D. M., J. H. Hutchison, and J. M. Legler. 1984. Kinosternid shell kinesis: structure, function and evolution. *Copeia* 1984: 456–475.

Brandley, M. C., H. Ota Fls T. Hikida, A. Nieto Montes de Oca, M. Fería-Ortíz, X. Guo, and Y. Wang. 2012. The phylogenetic systematics of blue-tailed skinks (*Plestiodon*) and the family Scincidae. *Zoological Journal of the Linnean Society* 165: 163–189.

Brandley, M. C., A. Schmitz, and T. W. Reeder. 2005. Partitioned Bayesian analyses, partition choice, and the phylogenetic relationships of scincid lizards. *Systematic Biology* 54: 373–390.

Brandley, M. C., Y. Wang, X. Guo, A. Nieto Montes de Oca, M. Fería-Ortíz, T. Hikida, and H. Ota. 2010. Bermuda as an evolutionary life raft for an ancient lineage of endangered lizards. *PLOS ONE* 5: e1135 (pp. 1–4).

Brattstrom, B. H., and T. R. Howell. 1954. Notes on some collections of reptiles and amphibians from Nicaragua. *Herpetologica* 10: 114–123.

Braun, D. 1993. Erfahrungen bei der Pflege und Zucht von *Ctenosaura palearis*. *Iguana Rundschreiben* 12: 13–18.

Brazaitis, P. 1974. The identification of living crocodilians. *Zoologica, Scientific Contributions of the New York Zoological Society* 58: 59–101.

Brongersma, L. D. 1996. On the availability of the name *Dermochelys coriacea schlegelii* [sic] (Garman, 1884) as a species or subspecies of Leatherback Turtle. *Chelonian Conservation and Biology* 2: 261–265.

Brongniart, A. 1800. Essai d'une classification naturelle des reptiles, par le citoyen. IIᵉ. Partie. Formation et disposition des genres. *Bulletin des Sciences par La Société Philomathique de Paris* 2: 89–91, plate VI.

Brown, A. E. 1908. Generic types of Neartic Reptilia and Amphibia. *Proceedings of the Academy of Natural Sciences of Philadelphia* 60: 112–127.

Brygoo, E. R. 1989. Les types d'Iguanidés (Reptiles, Sauriens) du Muséum national d'Histoire naturelle. Catalogue critique. *Bulletin du Muséum National d'Histoire Naturelle, Paris* 11: 1–112.

Buckley, L. J., and R. W. Axtell. 1990. *Ctenosaura palearis*. *Catalogue of American Amphibians and Reptiles* 491.1–491.3.

Buckley, L. J., and R. W. Axtell. 1997. Evidence for the specific status of the Honduran lizards formerly referred to *Ctenosaura palearis* (Reptilia: Squamata: Iguanidae). *Copeia* 1997: 138–150.

Budd, K. M., J. R. Spotila, and L. A. Mauger. 2015. Preliminary mating analysis of American Crocodiles, *Crocodylus acutus*, in Las Baulas, Santa Rosa, and Palo Verde National Parks, Guanacaste, Costa Rica. *South American Journal of Herpetology* 10: 4–9.

Burger, W. L. 1952. Notes on the Latin American skink, *Mabuya mabouya*. *Copeia* 1952: 185–187.

Burghardt, G. M., and A. S. Rand, editors. 1982. *Iguanas of the World. Their Behavior, Ecology, and Conservation*. Park Ridge, New Jersey: Noyes Publications.

Burt, C. E. 1931. A study of the teiid lizards of the genus *Cnemidophorus* with special reference to their phylogenetic relationships. *Smithsonian Institution, United States National Museum Bulletin* 154: i–viii, 1–286.

Burt, C. E., and M. D. Burt. 1931. South American lizards in the collection of the American Museum of Natural History. *Bulletin of the American Museum of Natural History* 61: 227–395.

Burt, C. E., and G. S. Myers. 1942. Neotropical lizards in the collection of the Natural History Museum of Stanford University. *Stanford University Publications, University Series, Biological Sciences* 8: 277–324.

Busack, S. D., and S. Pandya. 2001. Geographic variation in *Caiman crocodilus* and *Caiman yacare* (Crocodylia: Alligatoridae): systematic and legal implications. *Herpetologica* 57: 294–312.

Butterfield, B. P., J. B. Hauge, A. Flanagan, and J. M. Walker. 2009. Identity, reproduction, variation, ecology, and geographic origin of a Florida adventive: *Cnemidophorus lemniscatus* (Rainbow Whiptail Lizard, Sauria: Teiidae). *The Southeastern Naturalist* 8: 45–54.

Buurt, G. van. 2006. Conservation of amphibians and reptiles in Aruba, Curaçao and Bonaire. *Applied Herpetology* 3: 307–321.

Cabrera Peña, J., J. R. Rojas M., M. G. Galeano M., and V. Meza G. 1996. Mortalidad embrionaria y éxito de eclosión en huevos de *Trachemys scripta* (Testudines: Emydidae) incubados en un área natural protegida. *Revista de Biologia Tropical* 44: 841–846.

Cabrera Peña, J., J. R. Rojas Morales, and M. Muñoz Rodriguez. 1997. Tamaño post-eclosión de los neonates de *Trachemys scripta* (Testudines: Emydidae). *Revista de Biologia Tropical* 44/45: 667–669.

Caceres, D. A. 1993. Representantes más Comunes de la Herpetofauna del "Parque Nacional La Tigra" [thesis]. Tegucigalpa, Honduras: Universidad Pedagogica Nacional "Francisco Morazán."

Cadle, J. E. 2012. Cryptic species within the *Dendrophidion vinitor* complex in Middle America (Serpentes: Colubridae). *Bulletin of the Museum of Comparative Zoology* 160: 183–240.

Cadle, J. E., and J. M. Savage. 2012. Systematics of the *Dendrophidion nuchale* complex (Serpentes: Colubridae) with the description of a new species from Central America. *Zootaxa* 3513: 1–50.

Calderón-Mandujano, R. R., H. Bahena Basave, and S. Calmé. 2008. *Guía de Los Anfibios y Reptiles de la Reserva de la Biósfera de Sian Ka'an y zonas aledañas = Amphibians and Reptiles of Sian Ka'an Biosphere Reserve and surrounding areas*. 2nd ed. Quintana Roo, Mexico: COMPACT, ECOSUR, CONABIO y SHM A.C.

Caldwell, D. K., A. Carr, and T. R. Hellier, Jr. 1956 (dated 1955). A nest of the Atlantic Leatherback Turtle, *Dermochelys coriacea coriacea* (Linnaeus), on the Atlantic coast of Florida, with a summary of American nesting records. *Quarterly Journal of the Florida Academy of Sciences* 18: 279–284.

Camp, C. L. 1923. Classification of the lizards. *Bulletin of the American Museum of Natural History* 48: 289–480.

Campbell, H. W. 1973. Observations on the acoustic behavior of crocodilians. *Zoologica, Scientific Contributions of the New York Zoological Society* 58: 1–11.

Campbell, H. W., and T. R. Howell. 1965. Herpetological records from Nicaragua. *Herpetologica* 21: 130–140.

Campbell, J. A. 1998. *Amphibians and Reptiles of Northern Guatemala, the Yucatán, and Belize*. Norman: University of Oklahoma Press.

Campbell, J. A., and J. L. Camarillo R. 1994. A new lizard of the genus *Diploglossus* (Anguidae: Diploglossinae) from Mexico, with a review of the Mexican and northern Central American species. *Herpetologica* 50: 193–209.

Campbell, J. A., and D. R. Frost. 1993. Anguid lizards of the genus *Abronia*: revisionary notes, descriptions of four new species, a phylogenetic analysis, and key. *Bulletin of the American Museum of Natural History* 216: 1–121.

Campbell, J. A., and J. P. Vannini. 1989. Distribution of amphibians and reptiles in Guatemala and Belize. *Proceedings of the Western Foundation of Vertebrate Zoology* 4: 1–21.

Campbell, J. A., M. Sasa, M. Acevedo, and J. R. Mendelson, III. 1998. A new species of *Abronia* (Squamata: Anguidae) from the high Cuchumatanes of Guatemala. *Herpetologica* 54: 221–234.

Canseco-Márquez, L., G. Gutiérrez-Mayen, and A. A. Mendoza-Hernández. 2008. A new species of night lizard of the genus *Lepidophyma* (Squamata: Xantusiidae) from the Cuicatlán Valley, Oaxaca, México. *Zootaxa* 1750: 59–67.

Carr, A. F., Jr. 1938. Notes on the *Pseudemys scripta* complex. *Herpetologica* 1: 131–135.

Carr, A. F., Jr. 1939. A geckonid lizard new to the fauna of the United States. *Copeia* 1939: 232.

Carr, A. F., Jr. 1948. Sea turtles on a tropical island. *Fauna* 10: 50–54.

Carr, A. F., Jr. 1952. *Handbook of Turtles. The Turtles of the United States, Canada, and Baja California.* Ithaca, New York: Comstock Publishing Associates.

Carr, A. F., Jr. 1968. *The Turtle. A Natural History of Sea Turtles.* London: Cassell & Company.

Carr, A. F., Jr. 1986. *The Sea Turtle. So Excellent a Fishe.* Austin: University of Texas Press.

Carr, A. F., Jr., M. H. Carr, and A. B. Meylan. 1978. The ecology and migrations of sea turtles, 7. The west Caribbean Green Turtle colony. *Bulletin of the American Museum of Natural History* 162: 1–46.

Carr, A. F., Jr., A. Meylan, J. Mortimer, K. Bjorndal, and T. Carr. 1982. Surveys of sea turtle populations and habitats in the western Atlantic. Panama City, Florida: NOAA Technical Memorandum NMFS-SEFC 91.

Carranza, S., and E. N. Arnold. 2006. Systematics, biogeography, and evolution of *Hemidactylus* geckos (Reptilia: Gekkonidae) elucidated using mitochondrial DNA sequences. *Molecular Phylogenetics and Evolution* 38: 531–545.

Carrión-Cortés, J., C. Canales-Cerro, R. Arauz, and R. Riosmena-Rodríguez. 2013. Habitat use and diet of juvenile Eastern Pacific Hawksbill turtles (*Eretmochelys imbricata*) in the north Pacific coast of Costa Rica. *Chelonian Conservation and Biology* 12: 235–245.

de Carvalho, C. M. 1997. Uma nova espécie de microteiídeo de gênero *Gymnophthalmus* do estado do Roraima, Brasil (Sauria, Gymnophthalmidae). *Papéis Avulsos de Zoologia* 40: 161–174.

Casas-Andreau, G., and H. M. Smith. 1990. Historia nomenclatorial y status taxonómico de *Abronia ochoterenai* y *Abronia lythrochila* (Lacertilia: Anguidae), con una clave de identificación para el grupo *aurita*. *Anales Instituto de Biología, Universidad Nacional Autónoma de México, Serie Zoología* 61: 317–326.

Castañeda, F. E. 2002. Anfibios y Reptiles del Área Protegida Propuesta Rus-Rus, La Mosquitia. Informe Técnico. Tegucigalpa, Honduras. Unpublished Report Submitted to Administración Forestal del Estado, Corporación Hondureña de Desarrollo Forestal.

Castañeda, F. E. 2006. *Herpetofauna del Parque Nacional Sierra de Agalta, Honduras.* Available from: International Resources Group, Washington, D.C.

Castañeda, F. E., and L. Marineros. 2006. La herpetofauna de la Zona de Río Amarillo, Copán, Honduras, PP. 3.1–3.13 + 1 map. IN: O. Komar, J. P. Arce, C. Begley, F. E. Castañeda, K. Eisermann, R. J. Gallardo, and L. Marineros. *Evaluación de la biodiversidad del Parque*

Arqueológico y Reserva Forestal Río Amarillo (Copán, Honduras). San Salvador, El Salvador: SalvaNATURA.

Castañeda, F. E., and J. M. Mora. 2010. Impact of fire on a wetland population of the Scorpion Mud Turtle (*Kinosternon scorpioides*) in northwestern Costa Rica, PP. 705–715. IN: L. D. Wilson, J. H. Townsend, and J. D. Johnson, editors. *Conservation of Mesoamerican Amphibians and Reptiles.* Eagle Mountain, Utah: Eagle Mountain Publishing, LC.

Castañeda, F. E., and J. M. Mora. 2015. Nature notes. *Kinosternon leucostomum.* Sexual dimorphism. *Mesoamerican Herpetology* 2: 204–205.

Castañeda, F. E., J. R. McCranie, and L. A. Herrera. 2013. Natural history notes. *Staurotypus triporcatus* (Giant Musk Turtle: Guao de Tres Filas). Predation. *Herpetological Review* 44: 309.

Castiglia, R., A. M. R. Bezerra, O. Flores-Villela, F. Annesi, A. Muñoz, and E. Gornung. 2013. Comparative cytogenetics of two species of ground skinks: *Scincella assata* and *S. cherriei* (Squamata: Scincidae: Lygosominae) from Chiapas, Mexico. *Acta Herpetologica* 8: 69–73.

Ceballos, C. P., and W. A. Brand. 2014. Morphology and conservation of the Mesoamerican Slider (*Trachemys venusta*, Emydidae) from the Atrato River Basin, Colombia = Morfología y conservación de la tortuga hicotea Mesoamericana (*Trachemys venusta*, Emydidae) del río Atrato, Colombia. *Acta Biológica Colombiana* 19: 483–488.

Ceballos, C. P., D. Zapata, C. Alvarado, and E. Rincón. 2016. Morphology, diet, and population structure of the Southern White-lipped Mud Turtle *Kinosternon leucostomum postinguinale* (Testudines: Kinosternidae) in the Nus River drainage, Colombia. *Journal of Herpetology* 50: 374–380.

Cedillo Leal, C., J. García Grajales, J. C. Martinéz González, F. Briones Encinia, and E. Cienfuegas Rivas. 2013. Aspectos ecológicas de la anidación de *Crocodylus acutus* (Reptilia: Crocodylidae) en dos localidades de la costa de Oaxaca, México. *Acta Zoológica Mexicana (n.s.)* 29: 164–177.

Charruau, P., A. H. Escobedo-Galván, and M. A. Morales-Garduzza. 2014. Natural history notes. *Rhinoclemmys areolata* (Furrowed Wood Turtle). Maximum size and mass. *Herpetological Review* 45: 487.

Chen, X., S. Huang, P. Guo, G. R. Colli, A. Nieto Montes de Oca, L. J. Vitt, R. A. Pyron, and F. T. Burbrink. 2013. Understanding the formation of ancient intertropical disjunct distributions using Asian and neotropical hinged-teeth snakes (*Sibynophis* and *Scaphiodontophis*: Serpentes: Colubridae). *Molecular Phylogenetics and Evolution* 66: 254–261.

Cochran, D. M. 1938. Reptiles and amphibians from the Lesser Antilles collected by Dr. S. T. Danforth. *Proceedings of the Biological Society of Washington* 51: 147–156.

Cochran, D. M. 1941. The herpetology of Hispaniola. *Smithsonian Institution, United States National Museum, Bulletin* 177: i–vii, 1–398.

Cochran, D. M. 1961. Type specimens of reptiles and amphibians in the United States National Museum. *Smithsonian Institution, United States National Museum, Bulletin* 220: i–xv, 1–291.

Cole, C. J., H. C. Dessauer, C. R. Townsend, and M. G. Arnold. 1990. Unisexual lizards of the genus *Gymnophthalmus* (Reptilia: Tciidae) in the neotropics: genetics, origin, and systematics. *American Museum Novitates* 2994: 1–29.

Cole, C. J., C. R. Townsend, R. P. Reynolds, R. D. MacCulloch, and A. Lathrop. 2013. Amphibians and reptiles of Guyana, South America: illustrated keys, annotated species accounts, and a biogeographic synopsis. *Proceedings of the Biological Society of Washington* 125: 317–620.

Cole, N. C., C. G. Jones, and S. Harris. 2005. The need for enemy-free space: the impact of an invasive gecko on island endemics. *Biological Conservation* 125: 467–474.

Conant, R. 1984. A new subspecies of the pit viper, *Agkistrodon bilineatus* (Reptilia: Viperidae) from Central America. *Proceedings of the Biological Society of Washington* 97: 135–141.

Conant, R., and J. T. Collins. 1998. *A Field Guide to Reptiles & Amphibians. Eastern and Central North America.* 3rd ed., expanded. Boston, Massachusetts: Houghton Mifflin Company.

Conrad, J. L. 2008. Phylogeny and systematics of Squamata (Reptilia) based on morphology. *Bulletin of the American Museum of Natural History* 310: 1–182.

Cooper, W. E., Jr., and J. J. Habegger. 2001. Lingually mediated discrimination of prey, but not plant chemicals, by the Central American anguid lizard, *Mesaspis moreletii. Amphibia-Reptilia* 22: 81–90.

Cope, E. D. 1861a. Notes and descriptions of anoles. *Proceedings of the Academy of Natural Sciences of Philadelphia* 13: 208–215.

Cope, E. D. 1861b. On the Reptilia of Sombrero and Bermuda. *Proceedings of the Academy of Natural Sciences of Philadelphia* 13: 312–314.

Cope, E. D. 1862a (dated 1861). On the genera *Panolopus, Centropyx, Aristelliger* and *Sphaerodactylus. Proceedings of the Academy of Natural Sciences of Philadelphia* 13: 494–500.

Cope, E. D. 1862b. Synopsis of the species of *Holcosus* and *Ameiva*, with diagnoses of new West Indian and South American Colubridae. *Proceedings of the Academy of Natural Sciences of Philadelphia* 14: 60–82.

Cope, E. D. 1862c. Contributions to neotropical saurology. *Proceedings of the Academy of Natural Sciences of Philadelphia* 14: 176–188.

Cope, E. D. 1862d. Catalogue of the reptiles obtained during the explorations of the Parana, Paraguay, Vermejo, and Uraguay rivers, by Capt. Thos. J. Page, U. S. N.; and of those procured by Lieut. N. Michler, U. S. Top. Eng., Commander of the expedition conducting the survey of the Atrato River. *Proceedings of the Academy of Natural Sciences of Philadelphia* 14: 346–359.

Cope, E. D. 1863. Descriptions of new American Squamata, in the Museum of the Smithsonian Institution, Washington. *Proceedings of the Academy of Natural Sciences of Philadelphia* 15: 100–106.

Cope, E. D. 1865a (dated 1864). Contributions to the herpetology of tropical America. *Proceedings of the Academy of Natural Sciences of Philadelphia* 16: 166–181.

Cope, E. D. 1865b (dated 1864). On the limits and relations of the Raniformes. *Proceedings of the Academy of Natural Sciences of Philadelphia* 16: 181–183.

Cope, E. D. 1865c (dated 1864). On the characters of the higher groups of Reptilia Squamata—and especially of the *Diploglossa. Proceedings of the Academy of Natural Sciences of Philadelphia* 16: 224–231.

Cope, E. D. 1866a (dated 1865). Third contribution to the herpetology of tropical America. *Proceedings of the Academy of Natural Sciences of Philadelphia* 17: 185–198.

Cope, E. D. 1866b. Fourth contribution to the herpetology of tropical America. *Proceedings of the Academy of Natural Sciences of Philadelphia* 18: 123–132.

Cope, E. D. 1867 (dated 1866). Fifth contribution to the herpetology of tropical America. *Proceedings of the Academy of Natural Sciences of Philadelphia* 18: 317–323.

Cope, E. D. 1868. An examination of the Reptilia and Batrachia obtained by the Orton Expedition to Equador [sic] and the upper Amazon, with notes on other species. *Proceedings of the Academy of Natural Sciences of Philadelphia* 20: 96–140.

Cope, E. D. 1869 (dated 1868). Sixth contribution to the herpetology of tropical America. *Proceedings of the Academy of Natural Sciences of Philadelphia* 20: 305–313.

Cope, E. D. 1870 (dated 1869). Seventh contribution to the herpetology of tropical America. *Proceedings of the American Philosophical Society* 11: 147–169, plates IX–XI.

Cope, E. D. 1871 (dated 1870). Eighth contribution to the herpetology of tropical America. *Proceedings of the American Philosophical Society* 11: 553–559.

Cope. E. D. 1872. Synopsis of the species of the Chelydrinae. *Proceedings of the Academy of Natural Sciences of Philadelphia* 24: 22–29.

Cope, E. D. 1875. *On the Batrachia and Reptilia of Costa Rica*. Philadelphia: Published by the author. PP. 93–154, plates 23–28 (also published in 1877 [but dated 1876] in *Journal of the Academy of Natural Sciences of Philadelphia* 8: 93–154, plates 23–28; Murphy et al. [2007] discussed the various issues of this important publication, giving information on when those issues were released).

Cope, E. D. 1877. Tenth contribution to the herpetology of tropical America. *Proceedings of the American Philosophical Society* 17: 85–98.

Cope, E. D. 1884 (dated 1883). Notes on the geographical distribution of Batrachia and Reptilia in western North America. *Proceedings of the Academy of Natural Sciences of Philadelphia* 35: 10–35.

Cope, E. D. 1885. A contribution to the herpetology of Mexico. *Proceedings of the American Philosophical Society* 22: 379–404.

Cope, E. D. 1887. Catalogue of batrachians and reptiles of Central America and Mexico. *Bulletin of the United States National Museum* 32: 1–98.

Cope, E. D. 1892a. On *Tiaporus*, a new genus of Teiidae. *Proceedings of the American Philosophical Society* 30: 132–133, 1 plate.

Cope, E. D. 1892b. A synopsis of the species of the teïd genus *Cnemidophorus*. *Transactions of the American Philosophical* 17: 27–52, plates VI–XIII.

Cope, E. D. 1893. Second addition to the knowledge of the Batrachia and Reptilia of Costa Rica. *Proceedings of the American Philosophical Society* 31: 333–347 (separately published in 1893, in advance of volume published in 1894).

Cope, E. D. 1900. *The Crocodilians, Lizards, and Snakes of North America*. Washington: Government Printing Office. From the Annual Report of the Board of Regents of the Smithsonian Institution, Report of the U. S. National Museum for the Year ending June 30, 1898, PP. 153–1294, plates 1–36.

Cornelius, S. E. 1981. Status of sea turtles along the Pacific coast of Middle America, PP. 211–219. IN: K. A. Bjorndal, editor. *Biology and Conservation of Sea Turtles. Proceedings of the World Conference on Sea Turtle Conservation*; 1979 Nov 26–30; Washington: Smithsonian Institution Press.

Costa-Campos, C. E., and M. F. M. Furtado. 2013. Natural history notes. *Hemidactylus mabouia* (Tropical House Gecko). Cannibalism. *Herpetological Review* 44: 673–674.

Cover, J. F., Jr. 1986. Life history notes. *Basiliscus plumifrons* (Crested Green Basilisk Lizard). Food. *Herpetological Review* 17: 19.

Crother, B. I., editor. 2012. Scientific and standard English names of amphibians and reptiles of North America north of Mexico, with comments regarding confidence in our understanding. 7th ed. *Society for the Study of Amphibians and Reptiles, Herpetological Circular* 39: 1–92.

Cruz Díaz , G. A. 1978. Herpetofauna del Río Plátano [dissertation]. Tegucigalpa: Universidad Nacional Autónoma de Honduras.

Cruz [Díaz], G. A. 2008. Áreas protegidas de Honduras. Tegucigalpa, Honduras: Unpublished Report to Secretaría de Turismo.

Cruz [Díaz], G. A., and M. Espinal. 1987. Situación de las Tortugas Marinas en el Mar Atlántico de Honduras. Tegucigalpa, Honduras: Unpublished Report to WATS II.

Cruz [Díaz], G. A. M. Espinal, and O. Meléndez. 1987. Primer registro de anidamiento de la tortuga marina *Chelonia agassizi* en Punta Ratón, Honduras. *Revista de Biologia Tropical* 35: 341–343.

Cruz [Díaz], G. A. L. Girón, S. Flores, and V. Henríquez 2006. Evaluación de la herpetofauna en las partes que formarán el Área Protegida Trinacional Montecristo en territorio guatemalteco y hondureño, PP. 3.1–3.15. IN: *Consultoría para Ejecutar una Evaluación Ecológica Rápida (EER) en las Partes que Formarán el Área Protegida Trinacional de Montecristo en Territorio Guatemalteco y Hondureño*. San Salvador, El Salvador: SalvaNATURA.

Cruz [Díaz], G. A. V. J. Lopez, and S. Rodriguez. 1993. Primer inventario de mamíferos, reptiles y anfibios del Parque Nacional de Celaque. Siguatepeque, Honduras: Conservación y Silvicultura de Especies Forestales de Honduras, Serie Miscelanea de CONSEFORH 34.

Cruz [Díaz], G. A. L. D. Wilson, and J. Espinosa. 1979. Two additions to the reptile fauna of Honduras, *Eumeces managuae* Dunn and *Agkistrodon bilineatus* (Gunther [sic]), with comments on *Pelamis platurus* (Linnaeus). *Herpetological Review* 10: 26–27.

Cuvier, G. 1807. Sur les différentes espèces de crocodiles vivans et sur leurs caractères distinctifs. *Annales du Muséum d'Histoire Naturelle, Paris* 10: 8–66, 2 plates.

Cuvier, G. 1816 (dated 1817). *Le Règne Animal. Distribué D'Après son Organisation, pour Servir de Base a l'Histoire Naturelle des Animaux et D'Introduction a L'Anatomie Comparée. Tome II, Contenant les Reptiles, les Poissons, les Mollusques et les Annélides*. Paris: Deterville.

Cuvier, G. 1829. *Le Règne Animal. Distribué D'Après son Orgnisation, pour Servir de Base a L'Histoire Naturelle des Animaux et D'Introduction a L'Anatomie Comparée*. Nouvelle Édition, Revue et Augmentée. Tome II. Paris: Deterville.

Da Silveira, R., and W. E. Magnusson. 1999. Diets of Spectacled and Black Caiman in the Anavilhanas

Archipelago, Central Amazonia, Brazil. *Journal of Herpetology* 33: 181–192.

Daudin, F. M. 1801 (dated 1802). *Histoire Naturelle, Générale et Particulière des Reptiles; Ouvrage faisant suite à l'Histoire Naturelle générale et particulière, composée par Leclerc de Buffon, et rédigée par C. S. Sonnini, membre de plusieurs Sociétés savantes. Tome Second.* Paris: F. Dufart.

Daudin, F. M. 1802. *Histoire Naturelle, Générale et Particulière des Reptiles; Ouvrage faisant suite à l'Histoire Naturelle générale et particulière, composée par Leclerc de Buffon, et rédigée par C. S. Sonnini, membre de plusieurs Sociétés savantes. Tome Quatrième.* Paris: F. Dufart.

Daudin, F. M. 1803a. *Histoire Naturelle, Générale et Particulière des Reptiles; Ouvrage faisant suite aux Ouvres de Leclerc de Buffon, et partie du Cours complet d'Histoire naturelle rédigé par C. S. Sonnini, membre de plusieurs Sociétés savantes. Tome Cinquième.* Paris: F. Dufart.

Daudin, F. M. 1803b. *Histoire Naturelle, Générale et Particulière des Reptiles; Ouvrage faisant suite aux Ouvres de Leclerc de Buffon, et partie du Cours complet d'Histoire naturelle rédigé par C. S. Sonnini, membre de plusieurs Sociétés savantes. Tome Sixième.* Paris: F. Dufart.

Daudin, F. M. 1803c. *Histoire Naturelle, Générale et Particulière des Reptiles; Ouvrage faisant suite aux Ouvres de Leclerc de Buffon, et partie du Cours complet d'Histoire naturelle rédigé par C. S. Sonnini, membre de plusieurs Sociétés savantes. Tome Septième.* Paris: F. Dufart.

David, P. 1994. Liste des reptiles actuels du monde I. Chelonii. *Dumerilia* 1: 7–127.

Davis, W. B., and J. R. Dixon. 1961. Reptiles (exclusive of snakes) of the Chilpancingo region, Mexico. *Proceedings of the Biological Society of Washington* 74: 37–56.

Daza, J. D., and A. M. Bauer. 2012. A new amber-embedded sphaerodactyl gecko from Hispaniola, with comments on morphological synapomorphies of the Sphaerodactylidae. *Breviora* 529: 1–28.

Deppe, W. 1830. *Preis-Verzeichniss der Säugethiere, Vögel, Amphibien, Fische und Krebse, welche von den Herren Deppe und Schiede in Mexico gesammelt worden, und bei dem unterzeichneten Bevollmächtigten in Berlin gegen baare Zahlung in Preuss. Courant zu erhalten sind.* Berlin: Privately Printed by the Author.

de Queiroz, K. 1982. The scleral ossicles of sceloporine iguanids: a reexamination with comments on their phylogenetic significance. *Herpetologica* 38: 302–311.

de Queiroz, K. 1987a. A new spiny-tailed iguana from Honduras, with comments on relationships within *Ctenosaura* (Squamata: Iguania). *Copeia* 1987: 892–902.

de Queiroz, K. 1987b. Phylogenetic systematics of iguanine lizards. A comparative osteological study. Berkeley: University of California Press, University of California Publications. *Zoology* 118.

de Queiroz, K. 1990a. *Ctenosaura bakeri. Catalogue of American Amphibians and Reptiles* 465.1–465.2.

de Queiroz, K. 1990b. *Ctenosaura oedirhina. Catalogue of American Amphibians and Reptiles* 466.1–466.2.

de Queiroz, K. 1995. Checklist and key to the extant species of Mexican iguanas (Reptilia: Iguanidae). *Publicaciones Especiales del Museo de Zoología, Universidad Nacional Autónoma de México* 9: i, 1–48.

Deraniyagala, P. E. P. 1933. The loggerhead turtles (Carettidae) of Ceylon. *The Ceylon Journal of Science, Section B—Zoology and Geology* 18: 61–72, plate V.

Dial, B. E., and L. L. Grismer. 1992. A phylogenetic analysis of physiological-ecological character evolution in the lizard genus *Coleonyx* and its implications for historical biogeographic reconstruction. *Systematic Biology* 41: 178–195.

Díaz Pérez, J. A., J. A. Dávila Suárez, D. M. Alvarez García, and A. C. Sampedro Marin. 2012. Dieta de *Hemidactylus frenatus* (Sauria: Gekkonidae) en un área urbana de la región Caribe Colombiana. *Acta Zoológica Mexicana (n.s.)* 28: 613–616.

Dibble, C. J., A. Boyd, M. E. Ogle, G. R. Smith, and J. A. Lemos-Espinal. 2007. Natural history notes. *Phyllodactylus tuberculosus* (Yellow-bellied Gecko). Diet. *Herpetological Review* 38: 81.

Diener, E. 2007. Die Erdspitznatter *Oxybelis aeneus* und de Lianennatter *Leptophis mexicanus* als Prädatoren der Schwarzleguane *Ctenosaura similis* und *C. bakeri. Elaphe* 15: 59–62.

Dion, K., and L. Porras. 2014. Nature notes. *Ctenosaura similis.* Diet. *Mesoamerican Herpetology* 1: 157–158.

Dirksen, L. 2004. The last dozen. *Iguana* 11: 21–22.

Dirksen, L., and A. Gutsche. 2006. Beobachtungen zur Saurophagie bei *Ctenosaura bakeri* (Squamata: Iguanidae). *Elaphe* 14: 51–52.

Dixon, J. R. 1960. The discovery of *Phyllodactylus tuberculosus* (Reptilia: Sauria) in Central America, the resurrection of *P. xanti*, and description of a new gecko from British Honduras. *Herpetologica* 16: 1–11.

Dixon, J. R. 1964. The systematics and distribution of lizards of the genus *Phyllodactylus* in North and Central America. *New Mexico State University, Scientific Bulletin* 64-1: i–iv, 1–139.

Dixon, J. R. 1968. A new species of gecko (Sauria: Gekkonidae) from the Bay Islands, Honduras. *Proceedings of the Biological Society of Washington* 81: 419–426.

Dodd, C. K., Jr. 1988. Synopsis of the biological data on the Loggerhead Sea Turtle *Caretta caretta*

(Linnaeus 1758). Washington: Fish and Wildlife Service, United States Department of the Interior, *Biological Report* 88(14).

Dodd, C. K., Jr. 1990a. *Caretta. Catalogue of American Amphibians and Reptiles* 482.1–482.2.

Dodd, C. K., Jr. 1990b. *Caretta caretta. Catalogue of American Amphibians and Reptiles* 483.1–483.7.

Dodd, C. K., Jr., and R. Franz. 1996. Species richness and biogeography of the herpetofauna in the Exuma Cays Land and Sea Park, Bahamas, PP. 359–369. IN: R. Powell and R. W. Henderson, editors. *Contributions to West Indian Herpetology: A Tribute to Albert Schwartz. Society for the Study of Amphibians and Reptiles, Contributions to Herpetology* 12: 1–457, plates 1–8.

Donnelly, M. 1989. Socioeconomic importance of sea turtles. International trade in tortoiseshell, PP. 38–49. IN: L. Ogren et al., editors. *Proceedings of the Second Western Atlantic Turtle Symposium.* Panama City, Florida: NOAA Technical Memorandum NMFS-SEFC 226.

Dornburg, A., D. L. Warren, T. Iglesias, and M. C. Brandley. 2011. Natural history observations of the ichthyological and herpetological fauna on the island of Curaçao (Netherlands). *Bulletin of the Peabody Museum of Natural History* 52: 181–186.

Dowling, H. G., and W. E. Duellman. 1978. Systematic herpetology: a synopsis of families and higher categories. New York: Herpetological Information Search Systems, *Publications in Herpetology* 7.

Dowling, H. G., T. C. Majupuria, and F. W. Gibson. 1971. The hemipenis of the Onion-Tail Gecko *Thecadactylus rapicaudus* (Houttuyn). *Herpetological Review* 3: 110.

Dubois, A., and R. Bour. 2012 (dated 2011). The authorship and date of the familial nomen Ranidae (Amphibia, Anura). *Alytes* 27: 154–160.

Duellman, W. E. 2012. Linnaean names in South American herpetology. *International Society for the History and Bibliography of Herpetology* 9: 87–97.

Duellman, W. E., and B. Berg. 1962. Type specimens of amphibians and reptiles in the Museum of Natural History, The University of Kansas. *University of Kansas Publications, Museum of Natural History* 15: 183–204.

Duellman, W. E., and J. Wellman. 1960. A systematic study of the lizards of the *deppei* group (genus *Cnemidophorus*) in Mexico and Guatemala. *Miscellaneous Publications Museum of Zoology, University of Michigan* 111: 1–80, plate I.

Duellman, W. E., and R. G. Zweifel. 1962. A synopsis of the lizards of the *sexlineatus* group (genus *Cnemidophorus*). *Bulletin of the American Museum of Natural History* 123: 155–210, plates 24–31.

Dugès, A. 1891. *Eumeces altamirani,* A. Dug. *La Naturaleza* 1: 485–486, plate XXXII.

Duméril, A. H. A. 1852a. Description des reptiles nouveaux ou imparfaitement connus de la collection du Muséum d'Histoire Naturelle et remarques sur la classification et les charactères des reptiles. Premier Mémoire. Ordre des chéloniens et premières de l'ordre des sauriens (crocodiliens et cameleoniens). *Archives du Muséum d'Histoire Naturelle, Paris* 6: 209–264, plates XIV–XXII.

Duméril, A. H. A. 1852b. Note sur un nouveaux genre de Reptiles Sauriens, de la famille des Chalcidiens (le Lépidophyme), et sur le rang que les Amphisbéniens doivent occuper dans la classe des Reptiles. *Revue et Magasin de Zoologie* 4: 401–414, 1 plate.

Duméril, A. H. A. 1856. Description des reptiles nouveaux ou imparfaitement connus de la collection du Muséum d'Histoire Naturelle et remarques sur la classification et les caractères des reptiles. Deuxième Mémoire. Troisième, quatrième et cinquième familles de l'ordre des sauriens (Geckotiens, Varaniens et Iguaniens). *Archives du Muséum d'Histoire Naturelle, Paris* 8: 437–588, 1 plate.

Duméril, A. H. A., M. F. Bocourt, and F. Mocquard. 1870–1909a. *Études sur les Reptiles. Recherches Zoologiques pour servir a l'Histoire de la Faune de l'Amérique Centrale et du Mexique. Mission Scientifique au Mexique et dans l'Amérique Centrale. Troisième Partie.* 1re Section. Texte. Paris: Imprimeire Nationale.

Duméril, A. H. A., M. F. Bocourt, and F. Mocquard. 1870–1909b. *Recherches Zoologiques pour servir a L'Histoire de la Faune de L'Amérique Centrale et du Mexique. Troisième Partie.* 1re Section. Atlas. Paris: Imprimeire Impériale.

Duméril, A. M. C. 1805 (dated 1806). *Zoologie Analytique, ou Méthode Naturelle de Classification des Animaux, rendue plus facile a L'Aide de Tableaux Synoptiques.* Paris: Allais.

Duméril, A. M. C., and G. Bibron. 1835. *Erpétologie Générale ou Histoire Naturelle Complète des Reptiles. Tome Second. Contenant l'Histoire de Toutes les Espèces des Quatre Premières Familles de l'ordre des tortues ou chéloniens, et les généralités de celui des lézards ou sauriens.* Paris: Librairie Encyclopédique de Roret.

Duméril, A. M. C., and G. Bibron. 1836. *Erpétologie Générale ou Histoire Naturelle Complète des Reptiles. Tome Troisième. Contenant l'Histoire de Toutes les Espèces des Quatre Premières Familles de l'ordre des Lézards ou Sauriens, Savoir: les Crocodiles, les Caméléons, les Geckos et les Varans.* Paris: Librairie Encyclopédique de Roret.

Duméril, A. M. C., and G. Bibron. 1837. *Erpétologie Générale ou Histoire Naturelle Complète des Reptiles. Tome Quatrième. Contenant l'Histoire de Quarante-Six Genres et de Cent Quarante-Six Espèces de la Famille des Iguaniens, de l'Ordre des Sauriens.* Paris: Librairie Encyclopédique de Roret.

Duméril, A. M. C., and G. Bibron. 1839. *Erpétologie Générale ou Histoire Naturelle Complète des Reptiles. Tome Cinquième, Contenant l'Histoire de Quatre-Vingt-Trois Genres et de Deux Cent Sept Espèces des Trois Dernières Familles de l'ordre des Sauriens, Savoir: Les Lacertiens, les Chalcidiens et les Scincoïdiens.* Paris: Librairie Encyclopédique de Roret.

Duméril, A. M. C., and A. H. A. Duméril. 1851. *Catalogue Méthodique de la Collection des Reptiles du Muséum d'Histoire Naturelle de Paris.* Paris: Gide et Baudry.

Duméril, A. M. C., G. Bibron, and A. H. A. Duméril. 1854. *Erpétologie Générale ou Histoire Naturelle Complète des Reptiles. Tome Septième. Première Partie. Comprenant l'Histoire des Serpents non Venimeux.* Paris: Librairie Encyclopédique de Roret.

Dunn, E. R. 1928. A tentative key and arrangement of the American genera of Colubridae. *Bulletin of the Antivenin Institute of America* 2: 18–24.

Dunn, E. R. 1930. A new *Geomyda* from Costa Rica. *Proceedings of the New England Zoölogical Club* 12: 31–34.

Dunn, E. R. 1933a. A new lizard from Nicaragua. *Proceedings of the Biological Society of Washington* 46: 67–68.

Dunn, E. R. 1933b. Amphibians and reptiles from El Valle de Anton, Panama. *Occasional Papers of the Boston Society of Natural History* 8: 65–79.

Dunn, E. R. 1934. Notes on *Iguana. Copeia* 1934: 1–4, 1 unnumbered plate.

Dunn, E. R. 1936. Notes on American mabuyas. *Proceedings of the Academy of Natural Sciences of Philadelphia* 87: 533–557.

Dunn, E. R., and M. T. Dunn. 1940. Generic names proposed in herpetology by E. D. Cope. *Copeia* 1940: 69–76.

Dunn, E. R., and J. T. Emlen, Jr. 1932. Reptiles and amphibians from Honduras. *Proceedings of the Academy of Natural Sciences of Philadelphia* 84: 21–32.

Dunn, E. R., and L. H. Saxe, Jr. 1950. Results of the Catherwood-Chaplin West Indies Expedition, 1948. Part V. Amphibians and reptiles of San Andrés and Providencia. *Proceedings of the Academy of Natural Sciences of Philadelphia* 102: 141–165.

Duvernoy, M. 1836–1849. Les Reptiles. IN: *Le Règne Animal distribue d'après son Organisation, pour servir de base a l'Histoire Naturelle des Animaux, et d'introduction a l'Anatomie Comparée par Georges Cuvier. Un Réunion de disciples de Cuvier.* Paris: Fortin, Masson et Cⁱᵉ.

Echelle, A. A., A. F. Echelle, and H. S. Fitch. 1971. A new anole from Costa Rica. *Herpetologica* 27: 354–362.

Echternacht, A. C. 1968. Distributional and ecological notes on some reptiles from northern Honduras. *Herpetologica* 24: 151–158.

Echternacht, A. C. 1970. Taxonomic and ecological notes on some Middle and South American lizards of the genus *Ameiva* (Teiidae). *Breviora* 354: 1–9.

Echternacht, A. C. 1971. Middle American lizards of the genus *Ameiva* (Teiidae) with emphasis on geographic variation. *University of Kansas Museum of Natural History, Miscellaneous Publication* 55: 1–86.

Eisenberg, T., and G. Köhler. 1996. Lebensweise, Pflege und Nachzucht von *Sceloporus squamosus* Bocourt, 1874. *Sauria* 18: 11–14.

Ellis, T. M. 1980. *Caiman crocodilus*: an established exotic in south Florida. *Copeia* 1980: 152–154.

English, T. M. S. 1912. Some notes on the natural history of Grand Cayman. *Handbook of Jamaica* 1912: 598–600.

Ernst, C. H. 1978. A revision of the neotropical turtle genus *Callopsis* (Testudines: Emydidae: Batagurinae). *Herpetologica* 34: 113–134.

Ernst, C. H. 1980a. *Rhinoclemmys annulata. Catalogue of American Amphibians and Reptiles* 250.1–250.2.

Ernst, C. H. 1980b. *Rhinoclemmys areolata. Catalogue of American Amphibians and Reptiles* 251.1–251.2.

Ernst, C. H. 1981a. *Rhinoclemmys funerea. Catalogue of American Amphibians and Reptiles* 263.1–263.2.

Ernst, C. H. 1981b. *Rhinoclemmys pulcherrima. Catalogue of American Amphibians and Reptiles* 275.1–275.2.

Ernst, C. H. 1983. *Rhinoclemmys pulcherrima* (Tortuga Roja, Red Turtle), PP. 418–419. IN: D. H. Janzen, editor. *Costa Rican Natural History.* Chicago, Illinois: University of Chicago Press.

Ernst, C. H. 1990. Systematics, taxonomy, variation, and geographic distribution of the slider turtle, PP. 57–67. IN: J. W. Gibbons, editor. *Life History and Ecology of the Slider Turtle.* Washington: Smithsonian Institution Press.

Ernst, C. H. 2008a. Systematics, taxonomy, and geographic distribution of the snapping turtles, Family Chelydridae, PP. 5–13. IN: A. C. Steyermark, M. S. Finkler, and R. J. Brooks, editors. *Biology of the Snapping Turtle* (Chelydra serpentina). Baltimore, Maryland: The Johns Hopkins University Press.

Ernst, C. H. 2008b. *Trachemys emolli. Catalogue of American Amphibians and Reptiles* 846.1–846.3.

Ernst, C. H., and R. W. Barbour. 1972. *Turtles of the United States*. Lexington: University Press of Kentucky.

Ernst, C. H., and R. W. Barbour. 1989. *Turtles of the World*. Washington: Smithsonian Institution Press.

Ernst, C. H., and E. M. Ernst. 1977. Ectoparasites associated with neotropical turtles of the genus *Callopsis* (Testudines, Emydidae, Batagurinae). *Biotropica* 9: 139–142.

Ernst, C. H., and J. E. Lovich. 2009. *Turtles of the United States and Canada*. 2nd ed. Baltimore, Maryland: The Johns Hopkins University Press.

Ernst, C. H., and M. E. Seidel. 2006. *Trachemys venusta. Catalogue of American Amphibians and Reptiles* 832.1–832.12.

Ernst, C. H., R. W. Barbour, and J. E. Lovich. 1994. *Turtles of the United States and Canada*. Washington: Smithsonian Institution Press.

Ernst, C. H., J. W. Gibbons, and S. S. Novak. 1988. *Chelydra. Catalogue of American Amphibians and Reptiles* 419.1–419.4.

Ernst, C. H., F. D. Ross, and C. A. Ross. 1999. *Crocodylus acutus. Catalogue of American Amphibians and Reptiles* 700.1–700.17.

Eschscholtz, F. 1829a. Beschreibungen dreier neuer Meerschildkröten. *Die Quatember* 1(1): 10–18.

Eschscholtz, F. 1829b. *Zoologischer Atlas, enthaltend Abbildungen und Beschreibungen neuer Thierarten, während des Flottcapitains von Kotzebue zweiter Reise um die Welt, auf der Russisch-Kaiserlichen en Kriegsschlupp Predpriaetië in den Jahren 1823–1826*. Erst Heft. Berlin: G. Reimer.

Escobedo-Galván, A. H., G. Casas-Andreau, and G. Barrios-Quiroz. 2015. On the occurrence of *Caiman crocodilus* in Oaxaca, Mexico: a misunderstanding for over 140 years. *Mesoamerican Herpetology* 2: 220–223.

Escobedo-Galván, A. H., M. Venegas-Anaya, M. R. Espinal, S. G. Platt, and F. Buitrago. 2010. Conservation of crocodilians in Mesoamerica, PP. 746–757. IN: L. D. Wilson, J. H. Townsend, and J. D. Johnson, editors. *Conservation of Mesoamerican Amphibians and Reptiles*. Eagle Mountain, Utah: Eagle Mountain Publishing LC.

Espinal, M. R. 1993. Reptiles y anfibios, Refugio de Vida Silvestre "La Muralla." PP. 21, 171–178. IN: Anonymous. Tegucigalpa, Honduras. Unpublished Report submitted to Projecto Paseo Pantera.

Espinal, M. R., and A. H. Escobedo-Galván. 2010. Natural history notes. *Crocodylus acutus* (American Crocodile). Nesting ecology. *Herpetological Review* 41: 210–211.

Espinal, M. R., and A. H. Escobedo-Galván. 2011. Population status of the American Crocodile

(*Crocodylus acutus*) in the El Cajon Reservoir, Honduras. *The Southwestern Naturalist* 56: 212–215.

Espinal, M. R., J. R. McCranie, and L. D. Wilson. 2001. The herpetofauna of Parque Nacional La Muralla, Honduras, PP. 100–108. IN: J. D. Johnson, R. G. Webb, and O. A. Flores-Villela, editors. *Mesoamerican Herpetology: Systematics, Zoogeography, and Conservation. Centennial Museum Special Publication, University of Texas at El Paso*, 1: i–iv, 1–200.

Espinal, M. R., J. M. Mora, and F. Leiva. 2010. Abundance and distribution of the American Crocodile (*Crocodilus acutus*) at El Cajón Reservoir, Honduras, and the development of an integrated management plan for conservation, PP. 734–745. IN: L. D. Wilson, J. H. Townsend, and J. D. Johnson, editors. *Conservation of Mesoamerican Amphibians and Reptiles*. Eagle Mountain, Utah: Eagle Mountain Publishing, LC.

Espinal, M. R., J. M. Solís, C. O'Reilly, L. Marineros, and H. Vega. 2014. New distributional records for amphibians and reptiles from the department of Santa Bárbara, Honduras. *Mesoamerican Herpetology* 1: 300–303.

Espinal, M. R., J. M. Solís, C. O'Reilly, and R. Valle. 2014. New distributional records for amphibians and reptiles from the department of Choluteca, Honduras. *Mesoamerican Herpetology* 1: 298–300.

Etheridge, R. E. 1982. Checklist of the iguanine and Malagasy iguanid lizards, PP. 7–37. IN: G. M. Burghardt and A. S. Rand, editors. *Iguanas of the World. Their Behavior, Ecology, and Conservation*. Park Ridge, New Jersey: Noyes Publications.

Farr, W. L. 2011. Distribution of *Hemidactylus frenatus* in Mexico. *The Southwestern Naturalist* 56: 265–273.

Faulkner, S., N. Belal, S. A. Pasachnik, C. R. Bursey, and S. R. Goldberg. 2012. Natural history notes. *Ctenosaura bakeri* (Utila Spiny-tailed Iguana). Endoparasites. *Herpetological Review* 43: 332.

Ferguson, M. W. J. 1985. Reproductive biology and embryology of the crocodilians, PP. 329–491. IN: C. Gans, F. Billett, and P. F. A. Maderson, editors. *Biology of the Reptilia*. Vol. 14. *Development*. New York: A. John Wiley & Son.

Feuer, R. C. 1966. Variation in Snapping Turtles, *Chelydra serpentina* Linnaeus: a study in quantitative systematics [dissertation]. Salt Lake City: University of Utah.

Fischer, J. G. 1884. Herpetologische Bemerkungen. *Abhandlungen aus dem Gebiete der Naturwissenschaften herausgegben vom Naturwissenschaftlichen Verein in Hamburg* 8: 3–11, plate VII.

Fitch, H. S. 1970. Reproductive cycles in lizards and snakes. *University of Kansas Museum of Natural History, Miscellaneous Publication* 52: 1–247.

Fitch, H. S. 1973a. A field study of Costa Rican lizards. *The University of Kansas Science Bulletin* 50: 39–126.

Fitch, H. S. 1973b. Population structure and survivorship in some Costa Rican lizards. *Occasional Papers of the Museum of Natural History, The University of Kansas* 18: 1–41.

Fitch, H. S. 1983. *Sphenomorphus cherriei* (Escincela Parda, Skink), PP. 422–425. IN: D. H. Janzen, editor. *Costa Rican Natural History*. Chicago, Illinois: The University of Chicago Press.

Fitch, H. S. 1985. Variation in clutch and litter size in New World reptiles. *University of Kansas Museum of Natural History, Miscellaneous Publication* 76: 1–76.

Fitch, H. S., and R. W. Henderson. 1978. Ecology and exploitation of *Ctenosaura similis*. *The University of Kansas Science Bulletin* 51: 483–500.

Fitch, H. S., R. W. Henderson, and D. M. Hillis. 1982. Exploitation of iguanas in Central America, PP. 397–417. IN: G. M. Burghardt and A. S. Rand, editors. *Iguanas of the World. Their Behavior, Ecology, and Conservation*. Park Ridge, New Jersey: Noyes Publications.

Fitzinger, L. 1826. *Neue Classification der Reptilien nach ihren Natürlichen Verwandtschaften nebst einer Verwandtschafts-Tafel und einem Verzeichnisse der Reptilien—Sammlung des K. K. Zoologischen Museum's zu Wien*. Vienna: J. G. Heubner.

Fitzinger, L. 1835. Entwurf einer Systematischen Anordnung der Schildkröten nach den Grundsätzen der Natürlichen Methode. *Annelen des naturhistorischen Museums in Wien* 1: 103–128.

Fitzinger, L. 1843. *Systema Reptilium. Fasciculus Primus. Amblyglossae*. Vindobonae [Vienna]: Braumüller et Seidel Bibliopolas.

Flausín, L. P., R. Acuña-Mesén, and E. Araya. 1997. Natalidad de *Chelydra serpentina* (Testudines: Chelydridae) en Costa Rica. *Revista de Biologia Tropical* 44/45: 663–666.

Flores-Villela, O., K. Adler, and T. G. Eimermachor. 2016. Identity of three new sea turtles named by J. Friedrich Eschscholtz. *Chelonian Conservation and Biology* 15: 157–162.

Flower, S. S. 1928. Reptilia and Amphibia. *Zoological Record* 65: 1–72.

Fong, J. J., J. F. Parham, H. Shi, B. L. Stuart, and R. L. Carter. 2007. A genetic survey of heavily exploited, endangered turtles: caveats on the conservation value of trade animals. *Animal Conservation* 10: 452–460.

Forero-Medina, G., and O. V. Castaño-Mora. 2011. *Kinosternon scorpioides albogulare* (Duméril and Bocourt 1870)—White-Throated Mud Turtle, Swanka Turtle, PP. 064.1–0.64.5. IN: A. G. J. Rhodin, P. C. H. Pritchard, P. P. van Dijk, R. A. Saumure, K. A. Buhlmann, J. B. Iverson, and R. A. Mittermeier, editors. *Conservation Biology of Freshwater Turtles and Tortoises: A Compilation Project of the IUCN/SSC Tortoise and Freshwater Turtle Specialist Group*. Lunenburg, Massachusetts: Chelonian Research Foundation. *Chelonian Research Monographs* 5.

Franklin, C. J. 2000. Geographic distribution. *Hemidactylus frenatus*. *Herpetological Review* 31: 53.

Frazier, J. A., C. E. Montgomery, and S. M. Boback. 2010. Geographic distribution. *Trachemys venusta* (Mesoamerican Slider). *Herpetological Review* 41: 510.

Frazier, J. A., N. Pollock, M. Holding, and C. E. Montgomery. 2011. Geographic distribution. *Sphaerodactylus rosaurae* (Bay Island Least Gecko). *Herpetological Review* 42: 391.

Freiberg, M. A. 1972. Los reptiles (Reptilia), PP. 447–634. IN: L. Cendrero, editor. *Zoología Hispanoamericana. Vertebrados*. Mexico: Editorial Porrúa, S.A.

Frenkel, C. 2006. *Hemidactylus frenatus* (Squamata: Gekkonidae): call frequency, movement and condition of tail in Costa Rica. *Revista de Biologia Tropical* 54: 1125–1130.

Fretey, J., and R. Bour. 1980. Redécouverte du type de *Dermochelys coriacea* (Vandelli) (Testudinata, Dermochelyidae). *Bolletin Zoologie* 47: 193–205.

Friers, J., and J. P. Flaherty. 2016. Natural history notes. *Caiman crocodilus* (Spectacled Caiman). Diet. *Herpetological Review* 47: 131.

Fritts, T. H. 1969. The systematics of the parthenogenetic lizards of the *Cnemidophorus cozumela* complex. *Copeia* 1969: 519–535.

Fritz, U., editor. 2012a. *Die Schildkröten Europas. Ein umfassendes Handbuch zur Biologie, Verbreitung und Bestimmung (Teil 1)*. Wiebelsheim: Aula-Verlag GmbH.

Fritz, U., editor. 2012b. *Die Schildkröten Europas. Ein umfassendes Handbuch zur Biologie, Verbreitung und Bestimmung (Teil 2)*. Wiebelsheim: Aula-Verlag GmbH.

Fritz, U., and P. Havas. 2007. Checklist of Chelonians of the World. *Vertebrate Zoology, Dresden* 57: 149–368.

Fritz, U., H. Stuckas, M. Vargas-Ramírez, A. K. Hundsdörfer, J. Maran, and M. Päckert. 2012. Molecular phylogeny of Central and South American slider turtles: implications for biogeography and systematics (Testudines: Emydidae: *Trachemys*). *Journal Zoological Systematics and Evolutionary Research* 50: 125–136.

Frost, D. R., and R. Etheridge. 1989. A phylogenetic analysis and taxonomy of iguanian lizards (Reptilia: Squamata). *University of Kansas Museum of Natural History, Miscellaneous Publication* 81: 1–65.

Frost, D. R., R. Etheridge, D. Janies, and T. A. Titus. 2001. Total evidence, sequence alignment, evolution of polychrotid lizards, and a reclassification of the Iguania (Squamata: Iguania). *American Museum Novitates* 3343: 1–38.

Fugler, C. M. 1968. The distributional status of *Anolis sagrei* in Central America and northern South America. *Journal of Herpetology* 1: 96–98.

Fürbringer, M. 1900. Zur vergleichenden Anatomie des Brustschulterapparates und der Schultermuskeln. *Jenaische Zeitschrift für Naturwissenschaft* 34: 215–718, plates XIII–XVII.

Gadow, H. 1906. A contribution to the study of evolution based upon the Mexican species of *Cnemidophorus*. *Proceedings of the Zoological Society of London* 1906: 277–375, plate XX.

Gamble, T., A. M. Bauer, G. R. Colli, E. Greenbaum, T. R. Jackman, L. J. Vitt, and A. M. Simons. 2011. Coming to America: multiple origins of New World geckos. *Journal of Evolutionary Biology* 24: 231–244.

Gamble, T., A. M. Bauer, E. Greenbaum, and T. R. Jackman. 2008. Out of the blue: a novel, trans-Atlantic clade of geckos (Gekkota, Squamata). *Zoologica Scripta* 37: 355–366.

Gamble, T., E. Greenbaum, T. R. Jackman, A. P. Russell, and A. M. Bauer. 2012. Repeated origin and loss of adhesive toepads in geckos. *PLOS ONE* 7: e39429 (pp. 1–20).

Gandola, R., and C. Hendry. 2011. Natural history notes. *Ctenosaura oedirhina* (Roatán Spiny-tailed Iguana). Diet. *Herpetological Review* 42: 428–429.

Gaos, A. R., and I. L. Yañez. 2012. Saving the Eastern Pacific Hawksbill from extinction, PP. 244–262. IN: J. A. Seminoff and B. P. Wallace, editors. *Sea Turtles of the Eastern Pacific. Advances in Research and Conservation*. Tucson: The University of Arizona Press.

García, A., and G. Ceballos. 1994. *Guía de Campo de los Reptiles y Anfibios de la Costa de Jalisco, Mexico = Field Guide to the Reptiles and Amphibians of the Jalisco Coast, Mexico*. Mexico City, Mexico: Fundación Ecologica de Cuixmala, A.C. Instituto de Biologia U.N.A.M.

García-Grajales, J., and A. B. Silva. 2014. Abundancia y estructura poblacional de *Crocodylus acutus* (Reptilia: Crocodylidae) en la Laguna Palmasola, Oaxaca, México. *Revista de Biologia Tropical* 62: 165–172.

García-Padilla, V. Mata-Silva, D. DeSantis and L. D. Wilson. 2015. Reptilia: Squamata (lizards). Family Iguanidae. *Ctenosaura similis* Gray, 1830. *Mesoamerican Herpetology* 2: 540–541.

García-Vázquez, U. O., and M. Feria-Ortiz. 2006. Skinks of Mexico. *Reptilia* (Spain) 49: 74–79.

García-Vázquez, U. O., L. Canseco-Márquez, and J. L. Aguilar-López. 2010. A new species of night lizard of the genus *Lepidophyma* (Squamata: Xantusiidae) from southern Puebla, México. *Zootaxa* 2657: 47–54.

Garman, S. 1884a. The North American reptiles and batrachians. A list of the species occurring north of the Isthmus of Tehuantepec, with references. *Bulletin of the Essex Institute* 16: 1–46.

Garman, S. 1884b (dated 1883). The reptiles and batrachians of North America. Part I Ophidia-Serpentes. *Memoirs of the Museum of Comparative Zoölogy* 53(3): i–xxxi, 1–185, Plates 1–IX.

Garman, S. 1887a. On the reptiles and batrachians of Grand Cayman. *Proceedings of the American Philosophical Society* 24: 273–277.

Garman, S. 1887b. On West Indian reptiles. Iguanidae. *Bulletin of the Essex Institute* 19: 25–50.

Gauthier, J. 1982. Fossil xenosaurid and anguid lizards from the early Eocene Wasatch Formation, southeast Wyoming, and a revision of the Anguioidea. *Contributions to Geology, University of Wyoming* 21: 7–54.

Gauthier, J., M. Kearney, and R. L. Bezy. 2008. Homology of cephalic scales in xantusiid lizards, with comments on night lizard phylogeny and morphological evolution. *Journal of Herpetology* 42: 708–722.

Gees, K. 2003. Brief history of the Iguana Station. *Iguana* 10: 30–32.

Geoffroy Saint-Hilaire, I. 1829. Description des reptiles qui se trouvent en Égypte, PP. 1–96. IN: E. Geoffroy Saint-Hilaire, and I. Geoffroy Saint-Hilaire, editors. *Description de l'Égypte, ou Receuil des Observations et des Recherches qui ont été faites en Égypte pendant l'Expédition de l'armée Française. Seconde Édition. Tome Vingt-Quartriéme. Histoire Naturelle. Zoologie*. Paris: Imprimerie C. L. F. Panckoucke.

Gibbons, J. W., S. S. Novak, and C. H. Ernst. 1988. *Chelydra serpentina. Catalogue of American Amphibians and Reptiles* 420.1–420.4.

Gicca, D. F. 1983. *Enyaliosaurus quinquecarinatus. Catalogue of American Amphibians and Reptiles* 329.1–329.2.

Girard, C. 1858a (dated 1857). Descriptions of some new reptiles, collected by the United States Exploring Expedition, under the command of Capt. Charles Wilkes, U. S. N. Fourth Part.— Including the species of sauriens, exotic to North America. *Proceedings of the Academy of Natural Sciences of Philadelphia* 9: 195–199.

Girard, C. 1858b. *United States Exploring Expedition. During the Years 1838, 1839, 1840, 1841, 1842. Under the Command of Charles Wilkes, U.S.N. Herpetology*. Philadelphia: J. B. Lippincott & Co (there has been some confusion about the authorship of this work. The "official" issue listed S. F. Baird as the author, but Girard did the work and was correctly listed as the author of the "unofficial" issue).

Giugliano, L. G., C. de Campos Nogueira P. H. Valdujo, R. G. Collevatti, and G. R. Colli. 2013.

Cryptic diversity in South American Teiinae (Squamata, Teiidae) lizards. *Zoologica Scripta* 42: 473–487.

Gmelin, J. F. 1789 (dated 1788). *Caroli a Linné Systema Naturae per Regna Tria Naturae, Secundum Classes, Ordines, Genera, Species; cum Characteribus, Differeniis, Synonymis, Locis Linné, Carl von. Tomos Primus. Editio Decima tertia, reformata.* Tom. I. Pars III. *Amphibia.* Lipsiae [Leipzig]: Georg. Emanuel Beer, PP. 1033–1125.

Goicoechea, N., D. R. Frost, I. de La Riva, K. C. M. Pellegrino, J. Sites, Jr., M. T. Rodrigues, and J. M. Padial. 2016. Molecular systematics of teioid lizards (Teioidea/Gymnophthalmoidea: Squamata) based on analysis of 48 loci under tree-alignment and similarity-alignment. *Cladistics* 2016: 1–48.

Goldberg, S. R. 2008. Reproductive cycle of the Brown Forest Skink, *Sphenomorphus cherriei* (Squamata: Scincidae), from Costa Rica. *Texas Journal of Science* 60: 317–321.

Goldberg, S. R. 2009a. Reproductive cycles of three Ameiva species, *Ameiva festiva, Ameiva quadrilineata*, and *Ameiva undulata* (Squamata: Teiidae) from Central America. *Bulletin of the Maryland Herpetological Society* 45: 7–13.

Goldberg, S. R. 2009b. Reproduction in the Blackbelly Racerunner, *Aspidoscelis deppei* (Squamata: Teiidae) from Costa Rica. *Bulletin of the Maryland Herpetological Society* 45: 17–20.

Goldberg, S. R. 2009c. Reproduction in the Yellow-spotted night lizard, *Lepidophyma flavimaculatum* (Squamata, Xantusiidae), from Costa Rica. *Phyllomedusa* 8: 59–62.

Goldberg, S. R. 2009d. Reproductive cycle of the Central American Mabuya, *Mabuya unimarginata* (Squamata: Scincidae), from Costa Rica. *Texas Journal of Science* 61: 147–151.

Goldberg, S. R., and C. R. Bursey. 2003. Natural history notes. *Mabuya unimarginata* (Central American Mabuya). Endoparasites. *Herpetological Review* 34: 369.

Goldberg, S. R., C. R. Bursey, and S. A. Pasachnik. 2011. Natural history notes. *Ctenosaura oedirhina* (Roatan Spinytail Iguana). Endoparasites. *Herpetological Review* 42: 600–601.

Good, D. A. 1988. Phylogenetic relationships among gerrhonotine lizards. An analysis of external morphology. Vol. 121. Berkeley: University of California Press, UC Publications in Zoology.

Good, D. A. 1989. Allozyme variation and phylogenetic relationships among the species of *Mesaspis* (Squamata: Anguidae). *Herpetologica* 45: 227–232.

Good, D. A., A. M. Bauer, and R. Günther. 1993. An annotated type catalogue of the anguimorph lizards (Squamata: Anguidae, Helodermatidae,

Varanidae, Xenosauridae) in the Zoological Museum, Berlin. *Mitteilungen aus dem Zoologisches Museum in Berlin* 69: 45–56.

Goode, J. M. 1994. Reproduction in captive neotropical musk and mud turtles (*Staurotypus triporcatus, S. salvinii* [sic], and *Kinosternon scorpioides*), PP. 275–295. IN: J. B. Murphy, K. Adler, and J. T. Collins, editors. *Captive Management and Conservation of Amphibians and Reptiles. Society for the Study of Amphibians and Reptiles, Contributions to Herpetology* 11: 1–405, 1 plate.

Gosse, P. H. 1850. Descriptions of a new genus and six new species of saurian reptiles. *Annals and Magazine of Natural History* 6: 344–348.

Govender, Y., M. C. Muñoz, L. A. Ramírez Camejo, A. R. Puente-Rolón, E. Cuevas, and L. Sternberg. 2012. An isotopic study of diet and muscles of the Green Iguana (*Iguana iguana*) in Puerto Rico. *Journal of Herpetology* 46: 167–170.

Grant, C. 1932. The hemidactyls of the Porto Rico region. *The Journal of the Department of Agriculture of Puerto Rico* 16: 51–58, plate X.

Grant, C. 1941 (dated 1940). The herpetology of the Cayman Islands. *Bulletin of the Institute of Jamaica*, Science Series 2: i–iv, 1–56, plates I–VI.

Grant, C. 1957. The gecko *Hemidactylus frenatus* in Acapulco, Mexico. *Herpetologica* 13: 153.

Grant, P. B. C., T. R. Lewis, T. C. Laduke, and C. Ryall. 2009. Natural history notes. *Caiman crocodilus* (Spectacled Caiman). Opportunistic foraging. *Herpetological Review* 40: 80–81.

Gravendyck, M., P. Ammermann, R. E. Marschang, and E. F. Kaleta. 1998. Paramyxoviral and reoviral infections of iguanas on Honduran islands. *Journal of Wildlife Diseases* 34: 33–38.

Gravenhorst, J. L. C. 1833. Über *Phrynosoma orbicularis, Trapelus hispidus, Phrynocephalus helioscopus, Corythophanes cristatus*, und *Chamaeleopsis hernandesii. Nova Acta Academiae Caesareae Leopoldino Carolinae, Halle* 16: 911–958, plates LXIII–LXV.

Gray, J. E. 1825. A synopsis of the genera of reptiles and Amphibia, with a description of some new species. *Annals of Philosophy* 10: 193–217.

Gray, J. E. 1827a. A synopsis of the genera of saurian reptiles, in which some new genera are indicated, and the others reviewed by actual examination. *The Philosophical Magazine* 2: 54–58.

Gray, J. E. 1827b. A description of a new genus and some new species of saurian reptiles; with a revision of the species of chameleons. *The Philosophical Magazine* 2: 207–214.

Gray, J. E. 1828. *Spicilegia Zoologica; or Original Figures and Short Systematic Descriptions of New and Unfigured Animals.* Part I. London: Treüttel, Würtz, and Co.

Gray, J. E. 1830–1831. A synopsis of the species of the class Reptilia, PP. 1–110. IN: E. Griffith and E.

Pidgeon. *The Class Reptilia Arranged by the Baron Cuvier, with Specific Descriptions.* London: Whittaker, Treacher, and Co.

Gray, J. E. 1831. *Synopsis Reptilium; or Short Descriptions of the Species of Reptiles.* Part I. *Cataphracta. Tortoises, Crocodiles, and Enaliosaurians.* London: Treuttel, Wurtz and Co.

Gray, J. E. 1834. [Exhibition of several reptiles accompanied by notes]. *Proceedings of the Zoological Society of London* 1834: 99–101.

Gray, J. E. 1837. [General arrangement of the Reptilia]. *Proceedings of the Zoological Society of London* 1837: 131–132.

Gray, J. E. 1838a. Catalogue of the slender-tongued saurians, with descriptions of many new genera and species. *Annals and Magazine of Natural History* 1: 274–283.

Gray, J. E. 1838b. Catalogue of the slender-tongued saurians, with descriptions of many new genera and species. *Annals and Magazine of Natural History* 1: 388–394.

Gray, J. E. 1839a. Catalogue of the slender-tongued saurians, with descriptions of many new genera and species. *Annals and Magazine of Natural History* 2: 287–293.

Gray, J. E. 1839b. Catalogue of the slender-tongued saurians, with descriptions of many new genera and species. *Annals and Magazine of Natural History* 2: 331–337.

Gray, J. E. 1842. Description of some new species of reptiles, chiefly from the British Museum collection. *The Zoological Miscellany* 1842: 57–59.

Gray, J. E. 1844. *Catalogue of the Tortoises, Crocodiles, and Amphisbaenians, in the Collection of the British Museum.* London: Printed by Order of the Trustees, British Museum.

Gray, J. E. 1845a. Description of a new genus of night lizards from Belize. *Annals and Magazine of Natural History* 16: 162–163.

Gray, J. E. 1845b. *Catalogue of the Specimens of Lizards in the Collection of the British Museum.* London: Printed by Order of the Trustees, British Museum.

Gray, J. E. 1847. Description of a new genus of Emydae. *Proceedings of the Zoological Society of London* 15: 55–56.

Gray, J. E. 1849. *Catalogue of the Specimens of Snakes in the Collection of the British Museum.* London: Printed by Order of the Trustees, British Museum.

Gray, J. E. 1852. Descriptions of several new genera of reptiles, principally from the collection of H.M.S. Herald. *Annals and Magazine of Natural History* 10: 437–440.

Gray, J. E. 1856a (dated 1855). *Catalogue of Shield Reptiles in the Collection of the British Museum.* Part I. *Testudinata (Tortoises).* London: Printed

by Order of the Trustees, British Museum (Natural History).

Gray, J. E. 1856b (dated 1855). On some new species of freshwater tortoises from North America, Ceylon and Australia in the collection of the British Museum. *Proceedings of the Zoological Society of London* 1855: 197–202.

Gray, J. E. 1860. Description of a new species of *Geoclemmys* from Ecuador. *Proceedings of the Zoological Society of London* 28: 231–232, plate XXIX.

Gray, J. E. 1863. Notes on American Emydidae, and Professor Agassiz's observations on my catalogue of them. *Annals and Magazine of Natural History* 12: 176–183.

Gray, J. E. 1864. Description of a new species of *Staurotypus* (*S. salvini*) from Guatemala. *Proceedings of the Zoological Society of London* 1864: 127–128.

Gray, J. E. 1869. Notes on the families and genera of tortoises (Testudinata), and on the characters afforded by the study of their skulls. *Proceedings of the Zoological Society of London* 37: 165–225, plate XV.

Gray, J. E. 1870a. On the family Dermatemydae, and a description of a living species in the Gardens of the Society. *Proceedings of the Zoological Society of London* 38: 711–716, Plate XLII.

Gray, J. E. 1870b. Notes on the species of *Rhinoclemmys* in the British Museum. *Proceedings of the Zoological Society of London* 38: 722–724.

Gray, J. E. 1873a. Notes on tortoises. *Annals and Magazine of Natural History* 11: 143–149.

Gray, J. E. 1873b. *Hand-list of the Specimens of Shield Reptiles in the British Museum.* London: Printed by Order of the Trustees, British Museum (Natural History).

Grazziotin, F. G., H. Zaher, R. W. Murphy, G. Scrocchi, M. A. Benavides, Y-P. Zhang, and S. L. Bonatto. 2012. Molecular phylogeny of the New World Dipsadidae (Serpentes: Colubroidea): a reappraisal. *Cladistics* 1: 1–23.

Greene, H. W. 1969. Reproduction in a Middle American skink, *Leiolopisma cherriei* (Cope). *Herpetologica* 25: 55–56.

Greer, A. E., Jr. 1970. A subfamilial classification of scincid lizards. *Bulletin of the Museum of Comparative Zoology* 139: 151–183.

Greer, A. E., Jr. 1974. The generic relationships of the scincid lizard genus *Leiolopisma* and its relatives. *Australian Journal of Zoology, Supplementary Series* 31: 1–67.

Greer, A. E., Jr., and D. G. Broadley. 2000. Six characters of systematic importance in the scincid lizard genus *Mabuya*. *Hamadryad* 25: 1–12.

Greer, A. E., Jr., and R. A. Nussbaum. 2000. New character useful in the systematics of the scincid lizard genus *Mabuya*. *Copeia* 2000: 615–618.

Greer, A. E., Jr., and G. Shea. 2003. Secondary temporal scale overlap pattern: a character of possible broad systematics importance in sphenomorphine skinks. *Journal of Herpetology* 37: 545–549.

Griffith, H., A. Ngo, and R. W. Murphy. 2000. A cladistic evaluation of the cosmopolitan genus *Eumeces* Wiegmann (Reptilia, Squamata, Scincidae). *Russian Journal of Herpetology* 7: 1–16.

Grillitsch, H., E. Schleiffer, and F. Tiedemann. 1996. Katalog der Trochenpräparate der Herpetologischen Sammlung des Naturhistorischen Museums in Wien. Stand: 31. Dezember 1995. *Selbstverlag Naturhistorisches Museum Wien*, Kataloge Band 11, Vertebrata Heft 5.

Grismer, L. L. 1988. Phylogeny, taxonomy, classification, and biogeography of Eublepharid geckos, PP. 369–469. IN: R. Estes and G. Pregill, editors. *Phylogenetic Relationships of the Lizard Families. Essays Commemorating Charles L. Camp*. Stanford, California: Stanford University Press.

Grismer, L. L., L. L. Grismer, K. M. Marson, A. B. Matteson, E. J. R. Sihotang, K. M. Crane, J. Dayov, T. A. Mayer, A-L. Simpson, and H. Kaiser. 2001. New herpetological records for the Islas de la Bahía, Honduras. *Herpetological Review* 32: 134–135.

Groombridge, B. compiler. 1982. *The IUCN Amphibia—Reptilia Red Data Book. Part 1. Testudines, Crocodylia, Rhynchocephalia*. Gland, Switzerland: International Union for Conservation of Nature and Natural Resources (IUCN).

Guibé, J. 1954. *Catalogue des Types de Lézards du Muséum National d'Histoire Naturelle*. Bayeux, France: Imprimerie Colas.

Gundy, G. C., and G. Z. Wurst. 1976. The occurrence of parietal eyes in Recent Lacertilia (Reptilia). *Journal of Herpetology* 10: 113–121.

Günther, A. [C. L. G.]. 1858. *Catalogue of Colubrine Snakes in the Collection of the British Museum*. London: Printed by Order of the Trustees of British Museum.

Günther, A. C. L. G. 1859. On the reptiles from St. Croix, West Indies, collected by Messrs. A. and E. Newton. *Annals and Magazine of Natural History* 4: 209–217, 1 plate.

Günther, A. [C. L. G.]. 1863. Third account of new species of snakes in the collection of the British Museum. *Annals and Magazine of Natural History* 12: 348–365, plates V–VI.

Günther, A. [C. L. G.]. 1872. Seventh account of new species of snakes in the collection of the British Museum. *Annals and Magazine of Natural History* 9: 13–37, plates III–VI.

Günther, A. C. L. G. 1885–1902. Reptilia and Batrachia, IN: O. Salvin and F. D. Godman, editors. *Biologia Centrali-Americana; or, Contributions to the Knowledge of the Fauna and Flora of Mexico and Central America*. London: R. H. Porter and Dulau & Co.

Gutman, A. 2002. A new threat to the Utila Island Iguana. *Iguana Times* 9: 59–64.

Gutman, A. 2003. The Zen of swamping—adventures on Utila. *Iguana* 10: 32–35.

Gutsche, A. 2003. Utila Island swampers. *Iguana* 10: 28–29.

Gutsche, A. 2005a. Freilanduntersuchungen zur Populations- und Nahrungsökologie des Utila-Leguans (*Ctenosaura bakeri* Stejneger 1901) [dissertation]. Berlin: Humbolt-Universität.

Gutsche, A. 2005b. Beobachtungen zu natürlichen Inkubationsbedingungen von *Cnemidophorus lemniscatus* (Linnaeus, 1758) (Sauria: Teiidae) auf der Isla de Utila, Honduras. *Sauria* 27: 13–16.

Gutsche, A. 2005c. Natural history notes. *Ctenosaura bakeri* (Utila Spiny-tailed Iguana). Predation. *Herpetological Review* 36: 317.

Gutsche, A. 2005d. Distribution and habitat utilization of *Ctenosaura bakeri* on Utila. *Iguana* 12: 142–151.

Gutsche, A. 2005e. Freilanduntersuchungen als Basis für ein Arten- und Lebensraummanagement des Utila- Schwarzleguans (*Ctenosaura bakeri* Stejneger 1901). *Treffpunkt Biologische Vielfalt* 6: 239–243.

Gutsche, A. 2006. Population structure and reproduction in *Ctenosaura bakeri* on Isla de Utila. *Iguana* 13: 108–115.

Gutsche, A. 2007. Von Leguaninseln und Regenwäldern: Biodiversitätsforschung in Honduras. *Terraria (Wissenschaft)* 8: 40–43.

Gutsche, A. 2012. Bewohner des Elfenwaldes: herpetofauna am Pico La Picucha im Nationalpark Sierra de Agalta, Honduras. *Elaphe* 2012: 68–71.

Gutsche, A. 2016. Arend Friedrich August Wiegmann (1802–1841) und sein Beitrag zur Herpetologie der Neotropis am Zoologischen Museum Berlin. *Mertensiella* 23: 156–169.

Gutsche, A., and F. Köhler. 2008. Phylogeography and hybridization in *Ctenosaura* species (Sauria, Iguanidae) from Caribbean Honduras: insights from mitochondrial and nuclear DNA. *Zoosystematics and Evolution* 84: 245–253.

Gutsche, A., and G. Köhler. 2004. A fertile hybrid between *Ctenosaura similis* (Gray, 1831) and *C. bakeri* Stejneger, 1901 (Squamata: Iguanidae) on Isla de Utila, Honduras. *Salamandra* 40: 201–206.

Gutsche, A., and J. R. McCranie. 2009. Geographic distribution. *Hemidactylus mabouia* (Wood Slave). *Herpetological Review* 40: 112–113.

Gutsche, A., and J. R. McCranie. 2016. Johann Georg Wagler and the "Natürliches System der Amphibien." *Bibliotheca Herpetologica* 12: 41–49.

Gutsche, A., and M. Ohl. 2003. Ungewöhnlicher Todesfall eines Blattfingergeckos (*Phyllodactylus palmeus* Dixon, 1968). *Elaphe* 11: 48–51.

Gutsche, A., and W. J. Streich. 2009. Demography and endangerment of the Utila Island Spiny-Tailed Iguana, *Ctenosaura bakeri*. *Journal of Herpetology* 43: 105–113.

Gutsche, A., F. Mutschmann, W. J. Streich, and H. Kampen. 2012. Ectoparasites in the endangered Utila spiny-tailed iguana (*Ctenosaura bakeri*). *The Herpetological Journal* 22: 157–161.

Guyer, C. and M. A. Donnelly. 2005. *Amphibians and Reptiles of La Selva, Costa Rica, and the Caribbean Slope. A Comprehensive Guide*. Berkeley: University of California Press.

Hahn, D. E. 1971. Noteworthy herpetological records from Honduras. *Herpetological Review* 3: 111–112.

Hahn, D. E., and L. D. Wilson. 1976. New records for amphibians and reptiles for the department of Colon [sic], Honduras. *Herpetological Review* 7: 179.

Hainz, P. 2008. Haltung und Zucht von *Rhinoclemmys pulcherrima manni* (Dunn, 1930). *Sacalia* 19: 24–40.

Hallmen, M., and S. Hallmen. 2011. Notiz über eine Beobachtung zur Saurophagie beim Utila-Schwarzleguan *Ctenosaura bakeri*. *Elaphe* 2011: 11–13.

Hallmen, S. 2011. Neues aus Utila—reloaded. *Iguana Rundschreiben* 24: 5–16.

Hallowell, E. 1854a (dated 1852). Description of new species of Reptilia from western Africa. *Proceedings of the Academy of Natural Sciences of Philadelphia* 6: 62–65.

Hallowell, E. 1854b (dated 1852). Descriptions of new species of reptiles inhabiting North America. *Proceedings of the Academy of Natural Sciences of Philadelphia* 6: 177–182.

Hallowell, E. 1855. Contributions to South American herpetology. *Journal of the Academy of Natural Sciences of Philadelphia* 3: 33–36, plates III–IV.

Hallowell, E. 1857 (dated 1856). Notes on reptiles in the collection of the ANS of Philad'a. *Proceedings of the Academy of Natural Sciences of Philadelphia* 8: 221–238.

Hallowell, E. 1861 (dated 1860). Report upon the Reptilia of the North Pacific Exploring Expedition, under command of Capt. John Rogers [sic], U. S. N. *Proceedings of the Academy of Natural Sciences of Philadelphia* 12: 480–510.

Halstead, B. W. 1970. *Poisonous and Venomous Marine Animals of the World*. Vol. 3–*Vertebrates Continued*. Washington: United States Government Printing Office.

Hardy, L. M., and R. W. McDiarmid. 1969. The amphibians and reptiles of Sinaloa, México. *University of Kansas Publications Museum of Natural History* 18: 39–252, plates 1–8.

Harrington, S. M., D. H. Leavitt, and T. W. Reeder. 2016. Squamate phylogenetics, molecular branch lengths, and molecular apomorphies: a response to McMahan et al. *Copeia* 104(3): 702–707.

Harris, D. M. 1982a. The *Sphaerodactylus* (Sauria: Gekkonidae) of South America. *Occasional Papers of the Museum of Zoology, University of Michigan* 704: 1–31.

Harris, D. M. 1982b. The phenology, growth, and survival of the Green Iguana, *Iguana iguana*, in northern Colombia, PP. 150–161. IN: G. M. Burghardt and A. S. Rand, editors. *Iguanas of the World. Their Behavior, Ecology, and Conservation*. Park Ridge, New Jersey: Noyes Publications.

Harris, D. M., and A. G. Kluge. 1984. The *Sphaerodactylus* (Sauria: Gekkonidae) of Middle America. *Occasional Papers of the Museum of Zoology, University of Michigan* 706: 1–59.

Hartdegen, R. 1998. Cone-headed lizards. *Reptile & Amphibian Magazine* 56: 42–47.

Hartweg, N., and J. A. Oiver. 1937. A contribution to the herpetology of the Isthmus of Tehuantepec. I. The scelopori of the Pacific slope. *Occasional Papers of the Museum of Zoology, University of Michigan* 356: 1–9.

Harvey, M. B., G. N. Ugueto, and R. L. Gutberlet, Jr. 2012. Review of teiid morphology with a revised taxonomy and phylogeny of the Teiidae (Lepidosauria: Squamata). *Zootaxa* 3459: 1–156.

Hasbún, C. R. 2001. *Herpetofaunal biodiversity of the endangered spiny-tailed lizards of the* Ctenosaura quinquecarinata/flavidorsalis *complex: geographic variation, mtDNA phylogeography and systematics* [thesis]. Hull, U.K.: The University of Hull.

Hasbún, C. R., and G. Köhler. 2001. On the identity of the holotype of *Ctenosaura quinquecarinata* (Gray 1842) (Reptilia, Squamata, Iguanidae). *Senckenbergiana biologica* 81: 247–255.

Hasbún, C. R., and G. Köhler. 2009. New species of *Ctenosaura* (Squamata, Iguanidae) from southeastern Honduras. *Journal of Herpetology* 43: 192–204.

Hasbún, C. R., A. Gómez, G. Köhler, and D. H. Lunt. 2005. Mitochondrial DNA phylogeography of the Mesoamerican spiny-tailed lizards (*Ctenosaura quinquecarinata* complex): historical biogeography, species status and conservation. *Molecular Ecology* 14: 3095–3107.

Hasbún, C. R., G. Köhler, J. R. McCranie, and A. Lawrence. 2001. Additions to the description of *Ctenosaura flavidorsalis* Köhler & Klemmer, 1994 and its occurrence in south-western Honduras, El Salvador, and Guatemala (Squamata: Sauria: Iguanidae). *Herpetozoa* 14: 55–63.

Hecht, M. K. 1951. Fossil lizards of the West Indian genus *Aristelliger* (Gekkonidae). *American Museum Novitates* 1538: 1–33.

Hecht, M. K. 1952. Natural selection in the lizard genus *Aristelliger*. *Evolution* 6: 112–124.

Hedges, S. B. 2014. The high-level classification of skinks (Reptilia, Squamata, Scincomorpha). *Zootaxa* 3765: 317–338.

Hedges, S. B., and C. E. Conn. 2012. A new skink fauna from Caribbean islands (Squamata, Mabuyidae, Mabuyinae). *Zootaxa* 3288: 1–244.

Hedges, S. B., A. B. Marion, K. M. Lipp, J. Marin, and N. Vidal. 2014. A taxonomic framework for typhlopid snakes from the Caribbean and other regions (Reptilia, Squamata). *Caribbean Herpetology* 49: 1–61.

Helmus, M. R., D. L. Mahler, and J. B. Losos. 2014. Island biogeography of the Anthropocene. *Nature* 513: 543–548 + extended data.

Hemming, F., editor.1956a. Direction 56. Completion and in certain cases correction of entries relating to the names of genera belonging to the Classes Pisces, Amphibia and Reptilia made in the *Official List of Generic Names in Zoology* in the period up to the end of 1936. *Opinions & Declarations Rendered by the International Commission on Zoological Nomenclature* 1D(D.17): 337–364.

Hemming, F., editor.1956b. Direction 56. Addition to the *Official List of Specific Names in Zoology* (a) of the specific names of forty-seven species belonging to the Classes Cyclostomata, Pisces, Amphibia and Reptilia, each of which is the type species of a genus, the name of which was placed on the *Official List of Generic Names in Zoology* in the period up to the end of 1936 and (b) of the specific name of one species of the Class Amphibia which is currently treated as a senior subjective synonym of the name of such a species. *Opinions and Declarations Rendered by the International Commission on Zoological Nomenclature* 1D(D.18): 365–388.

Hemming, F., editor. 1958. Direction 97. Determination under the Plenary Powers of the specific name to be used for the North American Alligator and of the spelling to be used for that name (Class Reptilia) (*Opinion supplementary to Opinion 92*). *Opinions & Declarations Rendered by the International Commission on Zoological Nomenclature* 1F(F.8): 87–126.

Henderson, R. W. 1973. Ethoecological observations of *Ctenosaura similis* (Sauria: Iguanidae) in British Honduras. *Journal of Herpetology* 7: 27–33.

Henderson, R. W., and R. Powell. 2004. Thomas Barbour and the Utowana voyages (1929–1934) in the West Indies. *Bonner zoologische Beiträge* 52: 297–309.

Henderson, R. W., and R. Powell. 2009. *Natural History of West Indian Reptiles and Amphibians*. Gainesville: University Press of Florida.

Hendry, C., and R. Gandola. 2011. Natural history notes. *Ctenosaura oedirhina* (Roatán Spiny-tailed Iguana). Saurophagy. *Herpetological Review* 42: 273–274.

Henkel, F.-H., and W. Schmidt. 1991. *Geckos. Biologie, Haltung und Zucht*. Stuttgart: Verlag Eugen Ulmer.

°Hernández, F. 1849. *Rerum Medicarum Novae Hispaniae Thesaurus seu Plantarum Animalium Mineralium Mexicanorum Historia*. Rome: Mascardi.

Hernández-Guzman, J., J. R. Indy, G. S. Yasui, and L. Arias-Rodriguez. 2014. Los cromosomas de las tortugas tropicales: *Kinosternon leucostomum*, *Trachemys scripta* y *Staurotypus triporcatus* (Testudines: Kinosternidae/Emydidae). *Revista de Biologia Tropical* 62: 671–688.

Hertz, P. E., Y. Arima, A. Harrison, R. B. Huey, J. B. Losos, and R. E. Glor. 2013. Asynchronous evolution of physiology and morphology in *Anolis* lizards. *Evolution* 67: 2101–2113.

Heuvel, W. van den. 1997. De Zwarte Leguaan *Ctenosaura palearis* in zijn natuurlijke omgeving en de voortplanting in het terrarium. *Lacerta* 55: 155–159.

Heuvel, W. van den. 2003. Beobachtungen an *Ctenosaura melanosterna* Buckley & Axtell, 1997 in seinem natürlichen Verbreitungsgebiet und im Terrarium. *Elaphe* 11: 33–39.

Heuvel, W. van den and T. Leenders. 1998. Een andere kijk op de ernstig bedreigde *Ctenosaura bakeri*, de Zwarte Leguaan van Utila. *Lacerta* 56: 90–97.

Heyborne, W. H., and A. Mahan. 2017. Natural History Notes. *Hemidactylus frenatus* (Common House Gecko). Tail bifurcation. *Herpetological Review* 48: 437–438.

Hidalgo, H. 1983. Two new species of *Abronia* (Sauria: Anguidae) from the cloud forests of El Salvador. *Occasional Papers of the Museum of Natural History, The University of Kansas* 105: 1–11.

Hill, J., and C. King. 2015. Natural history notes. *Eretmochelys imbricata* (Hawksbill Turtle). Diet. *Herpetological Review* 46: 617–618.

Hirth, H. F. 1963a (dated 1962). Food of *Basiliscus plumifrons* on a tropical strand. *Herpetologica* 18: 276–277.

Hirth, H. F. 1963b. The ecology of two lizards on a tropical beach. *Ecological Monographs* 33: 83–112.

Hirth, H. F. 1980. *Chelonia mydas*. Catalogue of American Amphibians and Reptiles 249.1–249.4.

Hödl, W. 1996. Die Reptilien- und Amphibienfauna Costa Ricas, PP. 56–76. IN: P. Sehnal and H. Zettel, editors. *Esquinas-Nationalpark. Der Regenwald der Österreicher in Costa Rica*. Vienna: Naturhistorisches Museum.

°Holbrook, J. E. 1836. *North American Herpetology; or, a Description of the Reptiles Inhabiting the United States*. Vol. I. Philadelphia: J. Dobson.

Holdridge, L. R. 1967. *Life Zone Ecology.* Revised Edition. San José, Costa Rica: Tropical Science Center.

Hollingsworth, B. D. 1998. The systematics of chuckwallas (*Sauromalus*) with a phylogenetic analysis of other iguanid lizards. *Herpetological Monographs* 12: 38–191.

Hollingsworth, B. D. 2004. The evolution of iguanas. An overview of relationships and a checklist of species, PP. 19–44. IN: A. C. Alberts, R. L. Carter, W. K. Hayes, and E. P. Martins, editors. *Iguanas. Biology and Conservation.* Berkeley: University of California Press.

Holman, J. A. 1964. Observations on dermatemyid and staurotypine turtles from Veracruz, Mexico. *Herpetologica* 19: 277–279.

Honda, M., H. Ota, G. Köhler, I. Ineich, L. Chirio, S.-L. Chen, and T. Hikida. 2003. Phylogeny of the lizard subfamily Lygosominae (Reptilia: Scincidae), with special reference to the origin of the New World taxa. *Genes Genetic Systematics* 78: 71–80.

Hoogmoed, M. S. 1973. *Notes on the herpetofauna of Surinam IV. The lizards and amphisbaenians of Surinam.* The Hague: Dr. W. Junk b.v., Publishers.

Hoogmoed, M. S., and U. Gruber. 1983. Spix and Wagler type specimens of reptiles and amphibians in the Natural History Musea in Munich (Germany) and Leiden (The Netherlands). *Spixiana Supplement* 9: 319–415.

Horrocks, J. A., B. H. Krueger, M. Fastigi, E. L. Pemberton, and K. L. Ekert. 2011. International movements of adult female Hawksbill Turtles (*Eretmochelys imbricata*): first results from the Caribbean's Marine Turtle Tagging Centre. *Chelonian Conservation and Biology* 10: 18–25.

Houttuyn, M. 1782. Het onderscheid der Salamanderen van de Haagdissen in 't algemeen, en van de Gekkoos in 't byzonder, aangetoond. *Zeeuwsch genootschap der Wetenschappen Middleburg, Verhandelingen* 9: 305–336, 1 plate.

Hudson, D. M. 1981. Blood parasitism incidence among reptiles of Isla de Roatan, Honduras. *Journal of Herpetology* 15: 377–379.

Hunsaker, D., II. 1967 (dated 1966). Notes on the population expansion of the house gecko, *Hemidactylus frenatus. The Philippine Journal of Science* 95: 121–122.

Hurtado, L. A., C. A. Santamaria, and L. A. Fitzgerald. 2014. The phylogenetic position of the critically endangered Saint Croix ground lizard *Ameiva polops*: revisiting molecular systematics of West Indian *Ameiva. Zootaxa* 3794: 254–262.

Hutchison, J. H., and D. M. Bramble. 1981. Homology of the plastral scales of the Kinosternidae and related turtles. *Herpetologica* 37: 73–85.

Huy, A. 2008. Erfahrungen und Nachzucht des Fünfkiel-Schwarzleguans (*Ctenosaura quinquecarinata*). *Iguana Rundschreiben* 21: 5–9.

Hynková, I., Z. Starostová, and D. Frynta. 2009. Mitochondrial DNA variation reveals recent evolutionary history of main *Boa constrictor* clades. *Zoological Science* 26: 623–631.

Ibarra Portillo, R., V. Henríquez, and E. Greenbaum. 2009. Geographic distribution. *Trachemys emolli* (Moll's Slider). *Herpetological Review* 40: 111.

[ICF] Instituto Nacional de Conservación y Desarrollo Forestal, Áreas Protegidas y Vida Silvestre. 2010. Plan de Manejo Parque Nacional Marino Archipiélago del Golfo de Fonseca. Periodo 2010–2014. Tegucigalpa, Honduras: Unpublished report for the ICF.

[ICZN] International Commission on Zoological Nomenclature. 1926. Opinion 92. Sixteen generic names of Pisces, Amphibia, and Reptilia placed in the Official Lists of Generic Names. Washington: Smithsonian Institution, *Smithsonian Miscellaneous Collections* 73: 339–340.

ICZN. 1963. Opinion 660. Suppression under the plenary powers of seven specific names of turtles (Reptilia, Testudines). *The Bulletin of Zoological Nomenclature* 20: 187–190.

ICZN. 1985a. Opinion 1300. Teiidae Gray, 1827 given nomenclatural precedence over Ameividae Fitzinger, 1826 (Reptilia, Sauria). *The Bulletin of Zoological Nomenclature* 42: 130–133.

ICZN. 1985b. Opinion 1309. *Geoemyda* Gray, 1834, and *Rhinoclemmys* Fitzinger, 1835 (Reptilia: Testudines): conserved. *The Bulletin of Zoological Nomenclature* 42: 152–153.

ICZN. 1985c. Opinion 1313. *Testudo scripta* Schoepff, 1792 and *Emys cataspila* Günther, 1885 (Reptilia, Testudines): conserved. *The Bulletin of Zoological Nomenclature* 42: 160–161.

ICZN. 1989. Opinion 1534. *Sternotherus* Gray, 1825 and *Pelusios* Wagler, 1830 (Reptilia: Testudines): conserved. *The Bulletin of Zoological Nomenclature* 46: 81–82.

ICZN. 1999. *International Code of Zoological Nomenclature.* 4th ed. London: International Trust for Zoological Nomenclature.

Inturriaga, M., and R. Marrero. 2013. Feeding ecology of the Tropical House Gecko *Hemidactylus mabouia* (Sauria: Gekkonidae) during the dry season in Havana, Cuba. *Herpetology Notes* 6: 11–17.

[IUCN] International Union for Conservation of Nature. 2012. IUCN Red List of Threatened Species [Internet]. Gland, Switzerland: IUCN; [cited 2015 Jul 18]. Available from: http://www.webcitation.org/6Ezvq1n6z.

Iverson, J. B. 1976. The genus *Kinosternon* in Belize (Testudines: Kinosternidae). *Herpetologica* 32: 258–262.

Iverson, J. B. 1980. Colic modifications in iguanine lizards. *Journal of Morphology* 163: 79–93.

Iverson, J. B. 1983. *Staurotypus triporcatus. Catalogue of American Amphibians and Reptiles* 328.1–328.2.

Iverson, J. B. 1986. *A Checklist with Distribution Maps of the Turtles of the World.* First Edition. Richmond, Indiana: Privately Printed.

Iverson, J. B. 1991. Phylogenetic hypotheses for the evolution of modern kinosternine turtles. *Herpetological Monographs* 5: 1–27.

Iverson, J. B. 1992. *A Revised Checklist with Distribution Maps of the Turtles of the World.* Richmond, Indiana: Privately Printed.

Iverson, J. B. 2010. Reproduction in the Red-Cheeked Mud Turtle (*Kinosternon scorpioides cruentatum*) in southeastern Mexico and Belize, with comparisons across the species range. *Chelonian Conservation and Biology* 9: 250–261.

Iverson, J. B., and R. A. Mittermeier. 1980. Dermatemydidae. River Turtles. *Dermatemys* Gray Central American River Turtle. *Dermatemys mawii* Gray Central American River Turtle. *Catalogue of American Amphibians and Reptiles* 237.1–237.4.

Iverson, J. B., H. Higgins, A. Sirulnik, and C. Griffiths. 1997. Local and geographic variation in the reproductive biology of the snapping turtle (*Chelydra serpentina*). *Herpetologica* 53: 96–117.

Iverson, J. B., M. Le, and C. Ingram. 2013. Molecular phylogenetics of the mud and musk turtle family Kinosternidae. *Molecular Phylogenetics and Evolution* 69: 929–939.

Jablonski, D. 2015. Nature notes. *Gonatodes albogularis.* Communal egg laying. *Mesoamerican Herpetology* 2: 195–196.

Jackson, J. F. 1973. Notes on the population biology of *Anolis tropidonotus* in a Honduran highland pine forest. *Journal of Herpetology* 7: 309–311.

Jackson, J. T., D. E. Starkey, R. W. Guthrie, and M. R. J. Forstner. 2008. A mitochondrial DNA phylogeny of extant species of the genus *Trachemys* with resulting taxonomic implications. *Chelonian Conservation and Biology* 7: 131–135.

Jadin, R. C., F. T. Burbrink, G. A. Rivas, L. J. Vitt, C. L. Barrio-Amorós, and R. P. Guralnick. 2013. Finding arboreal snakes in an evolutionary tree: phylogenetic placement and systematic revision of the neotropical birdsnakes. *Journal of Zoological Systematics and Evolutionary Research* 52: 257–264.

Jadin, R. C., J. H. Townsend, T. A. Castoe, and J. A. Campbell. 2012. Cryptic diversity in disjunct populations of Middle American montane pitvipers: a systematic reassessment of *Cerrophidion godmani. Zoologica Scripta* 41: 455–470.

James, E. 1823 COMPILER. *Account of an Expedition from Pittsburgh to the Rocky Mountains, Performed in the Years 1819 and '20, by Order of the Hon. J. C. Calhoun, Sec'y of War: Under the Command of Major Stephen H. Long. From the Notes of Major Long, Mr. T. Say, and other Gentlemen of the Exploring Party.* Vol. II. Philadelphia: H. C. Carey and I. Lea.

Jiménez, A., and J. M. Savage. 1962. A new blind snake (genus *Typhlops*) from Costa Rica. *Revista de Biologia Tropical* 10: 199–203.

Jones, J. P. 1927. Descriptions of two new scelopori. *Occasional Papers of the Museum of Zoology, University of Michigan* 186: 1–7.

Joyce, W. G., J. F. Parham, and J. A. Gauthier. 2004. Developing a protocol for the conversion of rank-based taxon names to phylogenetically defined clade names, as exemplified by turtles. *Journal of Paleontology* 78: 989–1013.

Kaiser, H., and L. L. Grismer. 2001. Crocodiles in Islas de la Bahía. *Crocodile Specialist Group, Newsletter* 20: 83–85.

Kaiser, H., C. Cole, A. B. Matteson, T. A. Mayer, A.-L. Simpson, J. D. Wray, and L. L. Grismer. 2001a. Natural history notes. *Ctenosaura oedirhina* (Roatán Spiny-Tailed Iguana). Behavior. *Herpetological Review* 32: 253–254.

Kaiser, H., E. J. R. Sihotang, K. M. Marson, K. M. Crane, J. Dayov, and L. L. Grismer. 2001b. A breeding population of American crocodiles, *Crocodylus acutus*, on Roatán, Islas de la Bahía, Honduras. *Herpetological Review* 32: 164–165.

King, F. W. 1962. Systematics of Lesser Antillean lizards of the genus *Sphaerodactylus. Bulletin of the Florida State Museum, Biological Sciences* 7: 1–52.

King, F. W., and P. Brazaitis. 1971. Species identification of commercial crocodilian skins. *Zoologica, Scientific Contributions of the New York Zoological Society* 56: 15–70.

King, F. W., and R. L. Burke, editors. 1989. *Crocodilian, Tuatara, and Turtle Species of the World. A Taxonomic and Geographic Reference.* Washington: Association of Systematics Collections.

King, F. W., and C. Cerrato. 1990. Survey of the crocodilians of Honduras: an addendum. A re-evaluation of the results of the 1989 survey conducted under the auspices of the Convention on International Trade in Endangered Species of Wild Fauna and Flora and the Honduras Secretaria de Recursos Naturales Renovables. Tegucigalpa, Honduras: Unpublished Report to CITES and Honduran Secretaria de Recursos Naturales Renovables.

King, F. W., M. R. Espinal, and C. A. Cerrato. 1990. Distribution and status of the crocodilians of Honduras. Results of a survey conducted for the Convention on International Trade in Endangered Species of Wild Fauna and Flora and the

Honduras Secretaria de Recursos Naturales Renovables, PP. 313–354. IN: *Crocodiles: Proceedings of the 10th Working Meeting of the Crocodile Specialist Group of the Species Survival Commission of IUCN—The World Conservation Union convened at Gainesville, Florida, U.S.A., 23 to 27 April 1990.* 1: i–xvi, 1–354. Gland, Switzerland: IUCN.

Kizirian, D. A., and C. J. Cole. 1999. Origin of the unisexual lizard *Gymnophthalmus underwoodi* (Gymnophthalmidae) inferred from mitochondrial DNA nucleotide sequences. *Molecular Phylogenetics and Evolution* 11: 394–400.

Klauber, L. M. 1945. The geckos of the genus *Coleonyx* with descriptions of new subspecies. *Transactions of the San Diego Society of Natural History* 10: 133–216.

Klein, E. H. 1979. Los cocodrilidos de Honduras: su biologia [sic] y estado actual con recomendaciones para su manejo. Tegucigalpa, Honduras: Unpublished Report submitted to Dirección General de Recursos Naturales Renovables.

Klein, E. H. 1982. Reproduction of the green iguana (*Iguana iguana* L.) in the Tropical Dry Forest of southern Honduras. *Brenesia* 19/20: 301–310.

Kliment, P. 2011. Haltung und Biologie des Mittelamerikanischen Krallengeckos *Coleonyx mitratus* (Peters, 1863). *Sauria* 33: 3–18.

Kluge, A. G. 1969. The evolution and geographical origin of the New World *Hemidactylus mabouia-brookii* complex (Gekkonidae, Sauria). *Miscellaneous Publications Museum of Zoology, University of Michigan* 138: 1–78.

Kluge, A. G. 1975. Phylogenetic relationships and evolutionary trends in the Eublepharine lizard genus *Coleonyx*. *Copeia* 1975: 24–35.

Kluge, A. G. 1993. *Gekkonoid Lizard Taxonomy*. San Diego, California: International Gecko Society.

Kluge, A. G. 1995. Cladistic relationships of sphaerodactyl lizards. *American Museum Novitates* 3139: 1–23.

Kluge, A. G. 2001. Gekkotan lizard taxonomy. *Hamadryad* 26: 1–209.

Kluge, A. G., and R. A. Nussbaum. 1995. A review of African-Madagascan gekkonid lizard phylogeny and biogeography (Squamata). *Miscellaneous Publications Museum of Zoology, University of Michigan* 183: i–iv, 1–20.

Kober, I. 2012. *Basiliscus plumifrons*, der Stirnlappenbasilisk: Erfahrungen aus 20 Jahren Haltung und Vermehrung im Terrarium. *Elaphe* 2012: 36–45.

Koch, C., P. J. Venegas, A. Garcia-Bravo, and W. Böhme. 2011. A new bush anole (Iguanidae, Polychrotinae, *Polychrus*) from the upper Marañon basin, Peru, with a redescription of *Polychrus peruvianus* (Noble, 1924) and additional information on *P. gutturosus* Berthold, 1845. *ZooKeys* 141: 79–107.

Köhler, G. 1991. *Ctenosaura similis* (Gray). *Sauria* Supplement 13: 193–196.

Köhler, G. 1993a. *Schwarze Leguane. Freilandbeobachtungen, Pflege, und Zucht.* Hanau, Germany: Verlag Gunther Köhler.

Köhler, G. 1993b. *Der Grüne Leguan. Freilandbeobachtungen, Pflege, Zucht und Erkrankungen.* Offenbach, Germany: Herpeton Verlag Elke Köhler.

Köhler, G. 1993c. *Basilisken. Freilandbeobachtungen, Pflege und Zucht.* Hanau, Germany: Verlag Gunther Köhler.

Köhler, G. 1994a. Sobre la sistematica y ecologia [sic] de *Ctenosaura bakeri* y *C. oedirhina* (Sauria: Iguanidae). Estudios de campo realizados en las Islas de la Bahía, Honduras. Tegucigalpa, Honduras: Unpublished Report submitted to AFE-COHDEFOR.

Köhler, G. 1994b. Ecology, status, and conservation of the Utila spiny-tailed iguana *Ctenosaura bakeri*. *Iguana Times* 3: 12–13.

Köhler, G. 1994c. Schutzprojekt Utila-Schwarzleguan *Ctenosaura bakeri*. *Iguana Rundschreiben* 7: 51–53.

Köhler, G. 1994d. [Cover photograph]. *Iguana Rundschreiben* 7: back cover.

Köhler, G. 1994e. Vom Aussterben bedroht: Utila-Schwarzleguan. *Deutsche Aquarien- und Terrarien- Zeitschrift, Berlin (DATZ)* 47: 688–689.

Köhler, G. 1995a. Die Lokalnamen der Leguane der Gattung *Ctenosaura* Wiegmann, 1828. *Sauria* 17: 11–12.

Köhler, G. 1995b. Freilanduntersuchungen zur Morphologie und Ökologie von *Ctenosaura bakeri* und *C. oedirhina* auf den Islas de la Bahia, Honduras, mit Bemerkungen zur Schutzproblematik. *Salamandra* 31: 93–106.

Köhler, G. 1995c. *Ctenosaura palearis* Stejneger. *Sauria, Supplement* 17: 329–332.

Köhler, G. 1995d. De soorten Zwarte Leguanen (*Ctenosaura*). *Lacerta* 54: 13–28.

Köhler, G. 1995e. Schutz- und Forschungsprojekt Utila-Schwarzleguan Bericht 1995. *Mitteilungen Zoologische Gesellschaft für Arten- und Populationsschutz* 11: 9–11.

Köhler, G. 1995f. Schutzprojekt *Ctenosaura bakeri*—Utila-Schwarzleguan. *Iguana Rundschreiben* 8: 8–9.

Köhler, G. 1995g. Neue Schwarzleguane der Gattung *Ctenosaura* aus Honduras und Mexico. *Iguana Rundschreiben* 8: 21–23 (also front cover photograph).

Köhler, G. 1995h. Schutz- und Forschungsprojekt Utila-Schwarzleguan Bericht 1995. *Iguana Rundschreiben* 8: 8–13.

Köhler, G. 1995i. Zur Systematik und Ökologie der Schwarzleguane (Gattung Ctenosaura) [dissertation]. Frankfurt am Main, Germany: Johann-Wolfgang-Goethe-Universität.

Köhler, G. 1995j. Zwei neue Schwarzleguane aus Honduras und Mexico. *Deutsche Aquarien- und Terrarien-Zeitschrift, Berlin (DATZ)* 48: 618–619.

Köhler, G. 1995k. Schutz- und Forschungsprojekt Utila-Schwarzleguan. Bericht 1995. *Elaphe* 3: 73–78.

Köhler, G. 1995l. Leguáni rodu *Ctenosaura*—pozorování v prírode, ohrození, chov a rozmnozování v teráriu. *Niedeliana* 1: 5–14.

Köhler, G. 1995m. Eine neue Art der Gattung *Ctenosaura* (Sauria: Iguanidae) aus dem südlichen Campeche, Mexico. *Salamandra* 31: 1–14.

Köhler, G. 1996a. Haltungsfehler bei Grünen Leguanen. *Deutsche Aquarien- und Terrarien-Zeitschrift, Berlin (DATZ)* 49: 99–102.

Köhler, G. 1996b. Das Portrait. *Cnemidophorus lemniscatus* (Linnaeus). *Sauria* 18: 1–2.

Köhler, G. 1996c. A new species of anole of the *Norops pentaprion* group from Isla de Utila, Honduras (Reptilia: Sauria: Iguanidae). *Senckenbergiana biologica* 75: 23–31.

Köhler, G. 1996d. Additions to the known herpetofauna of Isla de Utila (Islas de la Bahia [sic], Honduras) with the description of a new species of the genus *Norops* (Reptilia: Sauria: Iguanidae). *Senckenbergiana biologica* 76: 19–28.

Köhler, G. 1996e. Notes on a collection of reptiles from El Salvador collected between 1951 and 1956. *Senckenbergiana biologica* 76: 29–38.

Köhler, G. 1996f. Schutz- und Forschungsprojekt Utila-Schwarzleguan. *Deutsche Aquarien- und Terrarien-Zeitschrift, Berlin (DATZ)* 49: 181–183.

Köhler, G. 1996g. BNA-Projektreisen Schutz- und Forschungsprojekt Utila-Schwarzleguan. *BNA/aktuell* 1996: 65–68.

Köhler, G. 1997a. Schutz-und Forschungsprojekt Utila-Schwarzleguan. *Elaphe* 5: 73–76.

Köhler, G. 1997b. Schutz- und Forschungsprojekt Utila-Leguan Bericht 1996. *Iguana Rundschreiben* 10: 10–17.

Köhler, G. 1997c. [Cover photograph]. *Iguana Rundschreiben* 10: back cover.

Köhler, G. 1997d. Conservation and research project: Utila Iguana. *Iguana Times* 6: 10–13.

Köhler, G. 1997e. Bericht des Jahres 1996 zum Schutz- und Forschungsprojekt: "Utila-Schwarzleguan." *Mitteilungen* 13: 20–21.

Köhler, G. 1997f. *Inkubation von Reptilieneiern. Grundlagen, Anleitungen, Erfahrungen.* Offenbach, Germany: Herpeton, Verlag Elke Köhler.

Köhler, G. 1998a. Das Schutz- und Forschungsprojekt Utila-Schwarzleguan. *Natur und Museum* 128: 44–49.

Köhler, G. 1998b. Further additions to the known herpetofauna of Isla de Utila (Islas de la Bahia [sic], Honduras) with notes on other species and a key to the amphibians and reptiles of the island (Amphibia, Reptilia). *Senckenbergiana biologica* 77: 139–145.

Köhler, G. 1998c. *Ctenosaura bakeri* Stejneger. *Sauria.* 20(Suppl.): 417–420.

Köhler, G. 1998d. Herpetologische Beobachtungen in Honduras I. Die Islas de la Bahía. *Natur und Museum* 128: 372–383.

Köhler, G. 1998e. *Der Grüne Leguan. Biologie, Pflege, Zucht, Erkrankungen.* Offenbach, Germany: Herpeton, Verlag Elke Köhler.

Köhler, G. 1998f. Schutz- und Forschungsprojekt Utila-Schwarzleguan: die Nachzucht von *Ctenosaura bakeri* Stejneger, 1901 im ex-situ-Zuchtprogramm. *Salamandra* 34: 227–238.

Köhler, G. 1998g. Artenschutz in Honduras. Das Schutzprojekt Utila-Leguan. *BNA/aktuell* 1998: 49–57.

Köhler, G. 1998h. Vom Aussterben bedroht-der Leguan von Utila. *Reptilia* (Germany) 3: 51–56.

Köhler, G. 1998i. Schutz- und Forschungsprojekt Utila-Schwarzleguan, PP. 121–126. IN: D. Jelden, I. Sprotte, and M. Gruschwitz, editors. *Nachhaltige Nutzung.* Bonn: Bundesamt für Naturschutz.

Köhler, G. 1999a. Herpetologische Beobachtungen in Honduras II. Das Comayagua–Becken. *Natur und Museum* 129: 212–217.

Köhler, G. 1999b. The amphibians and reptiles of Nicaragua. A distributional checklist with keys. *Courier Forschungsinstitut Senckenberg* 213: 1–121.

Köhler, G. 1999c. Endangered Utila iguana needs help. *The Vivarium* 10: 36.

Köhler, G. 1999d. *Basilisken. Helmleguane und Kronenbasilisken. Lebensweise, Pflege, Zucht.* 2., neu bearbeitete, stark erweiterte Auflage. Offenbach, Germany: Herpeton, Verlag Elke Köhler.

Köhler, G. 1999e. Amphibien und Reptilien im Hochland von Nicaragua. *Deutsche Aquarien- und Terrarien-Zeitschrift, Berlin (DATZ)* 52: 48–54.

Köhler, G. 1999f. Artenschutz in Honduras: das Schutzprojekt Utila-Leguan. *Kleine Senckenbergiana-Rheinbach* 32: 77–84.

Köhler, G. 1999g. Illegale Mangrovenzerstörung auf Utila. Aufruf zu protestbriefen. *Reptilia (Germany)* 4: 7–8.

Köhler, G. 1999h. Das Leguanportrait. Der Pazifische Helmleguane *Corytophanes percarinatus*. *Iguana Rundschreiben* 12: 13–14.

Köhler, G. 2000. *Reptilien und Amphibien Mittelamerikas. Band 1: Krokodile, Schildkröten, Echsen.* Offenbach, Germany: Herpeton, Verlag Elke Köhler.

Köhler, G. 2001a. Geographic distribution. *Hemidactylus frenatus* (House Gecko). *Herpetological Review* 32: 57.

Köhler, G. 2001b. *Anfibios y Reptiles de Nicaragua*. Offenbach, Germany: Herpeton, Verlag Elke Köhler.

Köhler, G. 2001c. Das Portrait. *Sphaerodactylus millepunctatus* Hallowell. *Sauria* 23: 1–2.

Köhler, G. 2002. *Schwarzleguane. Lebensweise, Pflege, Zucht*. Offenbach, Germany: Herpeton, Verlag Elke Köhler.

Köhler, G. 2003a. *Reptiles of Central America*. Offenbach, Germany: Herpeton, Verlag Elke Köhler.

Köhler, G. 2003b. Das Leguanportrait. *Polychrus gutturosus* Berthold 1845. *Iguana Rundschreiben* 16: 5–8.

Köhler, G. 2004a. Conservation status of spiny-tailed iguanas (genus *Ctenosaura*), with special emphasis on the Utila Iguana (*C. bakeri*). *Iguana* 11: 206–211.

Köhler, G. 2004b. *Ctenosaura flavidorsalis* [Internet]. The IUCN Red List of Threatened Species. Version 2015.2. Gland, Switzerland: IUCN; [cited 2017 August 17]. Available from: www.iucnredlist. org.

Köhler, G. 2008. *Reptiles of Central America*. 2nd ed. Offenbach, Germany: Herpeton, Verlag Elke Köhler.

Köhler, G. 2012. *Color Catalogue for Field Biologists. Bilingual Edition: English/Español*. Offenbach, Germany: Herpeton, Verlag Elke Köhler.

Köhler, G. 2016. Beiträge senckenbergischer Herpetologen in Frankfurt zur Erforschung der Amphibien und Reptilien Lateinamerikas. *Mertensiella* 23: 307–332.

Köhler, G., and E. Blinn. 2000. Natürlichen Bastardierung zwischen *Ctenosaura bakeri* und *Ctenosaura similis* auf Utila, Honduras. *Salamandra* 36: 77–79.

Köhler, G., and J. A. Ferrari. 2011a. Geographic distribution. *Sphaerodactylus dunni* (NCN). *Herpetological Review* 42: 113.

Köhler, G., and J. A. Ferrari. 2011b. Geographic distribution. *Hemidactylus brookii* (NCN). *Herpetological Review* 42: 240.

Köhler, G., and M. Fried. 2012. Natural history notes. *Sceloporus variabilis* (Rose-bellied Lizard). Prey. *Herpetological Review* 43: 651–652.

Köhler, G., and C. R. Hasbún. 2001. A new species of spiny-tailed iguana from Mexico formerly referred to *Ctenosaura quinquecarinata* (Gray 1842) (Reptilia, Squamata, Iguanidae). *Senckenbergiana biologica* 81: 257–267.

Köhler, G., and P. Heimes. 2002. *Stachelleguane. Lebensweise, Pflege, Zucht*. Offenbach, Germany: Herpeton, Verlag Elke Köhler.

Köhler, G., and K. Klemmer. 1994. Eine neue Schwarzleguanart der Gattung *Ctenosaura* aus La Paz, Honduras. *Salamandra* 30: 197–208.

Köhler, G., and J. R. McCranie. 2001. Two new species of anoles from northern Honduras (Reptilia, Squamata, Polychrotidae). *Senckenbergiana biologica* 81: 235–245.

Köhler, G., and M. Obermeier. 1998. A new species of anole of the *Norops crassulus* group from central Nicaragua (Reptilia: Sauria: Iguanidae). *Senckenbergiana biologica* 77: 127–137.

Köhler, G., and D. Rittmann. 1998. Beobachtungen bei der erstmaligen Nachzucht des Roatán-Schwarzleguans *Ctenosaura oedirhina* de Queiroz, 1987. *Herpetofauna* 20: 5–10.

Köhler, G., and R. Seipp. 1998. Eine Expedition in den Randbereich des Biosphärenreservats Bosawas, Nicaragua. *Natur und Museum* 128: 170–175.

Köhler, G., and B. Streit. 1996. Notes on the systematic status of the taxa *acanthura, pectinata*, and *similis* of the genus *Ctenosaura* (Reptilia: Sauria: Iguanidae). *Senckenbergiana biologica* 75: 33–43.

Köhler, G., and M. Vesely. 1996. Freilanduntersuchungen zur Morphologie und Lebensweise von *Ctenosaura palearis* in Honduras und Guatemala. *Herpetofauna* 18: 23–26.

Köhler, G., and M. Vesely. 2010. A revision of the *Anolis sericeus* complex with the resurrection of *A. wellbornae* and the description of a new species (Squamata: Polychrotidae). *Herpetologica* 66: 207–228.

Köhler, G., and M. Vesely. 2011. A new species of *Thecadactylus* from Sint Maarten, Lesser Antilles (Reptilia, Squamata, Gekkonidae). *ZooKeys* 118: 97–107.

Köhler, G., R. Gómez Trejo Pérez, C. B. P. Petersen, and F. R. Méndez de la Cruz. 2014. A revision of the Mexican *Anolis* (Reptilia, Squamata, Dactyloidae) from the Pacific versant of the Isthmus de Tehuantepec in the states of Oaxaca, Guerrero, and Puebla, with the description of six new species. *Zootaxa* 3862: 1–210.

Köhler, G., J. R. McCranie, and L. Marineros. 2009. Geographic distribution. *Hemidactylus brooki* (NCN). *Herpetological Review* 40: 451.

Köhler, G., J. R. McCranie, and K. E. Nicholson. 2000. Eine herpetologische Expedition in den Patuca-Nationalpark, Honduras. *Natur und Museum* 130: 421–425.

Köhler, G., J. R. McCranie, and L. D. Wilson. 1999. Two new species of anoles of the *Norops crassulus* group from Honduras (Reptilia: Sauria: Polychrotidae). *Amphibia-Reptilia* 20: 279–298.

Köhler, G., J. R. McCranie, and L. D. Wilson. 2001. A new species of anole from western Honduras

(Squamata: Polychrotidae). *Herpetologica* 57: 247–255.

Köhler, G., M. J. Rodríguez Bobadilla, and S. B. Hedges. 2016. A new dune-dwelling lizard of the genus *Leiocephalus* (Iguania, Leiocephalidae) from the Dominican Republic. *Zootaxa* 4126: 517–532.

Köhler, G., M. Salazar Saavedra, J. Martinez, G. Lopez, and J. Sunyer. 2013. First record of *Aspidoscelis motaguae* (Sackett, 1941) (Reptilia: Squamata: Teiidae) from Nicaragua. *Check List* 9: 475.

Köhler, G., F. Schmidt, R. Schröter, and R. Siemer. 1999. Untersuchungen zur Variation mehrerer Reptilienarten aus El Salvador unter Berücksichtigung der Verwendbarkeit von Pholidosemerkmalen zur individuellen Wiedererkennung. *Salamandra* 35: 227–242.

Köhler, G., W. Schroth, and B. Streit. 2000. Systematics of the *Ctenosaura* group of lizards (Reptilia: Sauria: Iguanidae). *Amphibia-Reptilia* 21: 177–191.

Köhler, G., J. H. Townsend, and C. B. P. Petersen. 2016. A taxonomic revision of the *Norops tropidonotus* complex (Squamata, Dactyloidae), with the resurrection of *N. spilorhipis* (Álvarez del Toro and Smith, 1956) and the description of two new species. *Mesoamerican Herpetology* 3: 7–41.

Köhler, G., M. Vesely, and E. Greenbaum. 2005 (dated 2006). *The Amphibians and Reptiles of El Salvador*. Malabar, Florida: Krieger Publishing Company.

Köhler, J. J., S. Poe, M. J. Ryan, and G. Köhler. 2015. *Anolis marsupialis* Taylor 1956, a valid species from southern Pacific Costa Rica (Reptilia, Squamata, Dactyloidae). *Zootaxa* 3915: 111–122.

Kraus, F. F. 2009. *Alien Reptiles and Amphibians. A Scientific Compendium and Analysis*. New York: Springer Science.

Kronauer, D. J. C., P. J. Bergmann, J. M. Mercer, and A. P. Russell. 2005. A phylogeographically distinct and deep divergence in the widespread neotropical turnip-tailed gecko, *Thecadactylus rapicauda*. *Molecular Phylogenetics and Evolution* 34: 431–437.

Krysko, K. L., M. Granatosky, Z. W. Fratto, J. L. Kline, and M. R. Rockford. 2010. Natural history notes. *Caiman crocodilus* (Spectacled Caiman). Prey. *Herpetological Review* 41: 348–349.

Krysko, K. L., K. W. Larson, D. Diep, E. Abellana, and E. R. McKercher. 2009. Diet of the nonindigenous Black Spiny-Tailed Iguana, *Ctenosaura similis* (Gray 1831) (Sauria: Iguanidae) in southern Florida. *Florida Scientist* 72: 48–58.

Krysko, K. L., C. M. Sheehy, III, and A. N. Hooper. 2003. Interspecific communal oviposition and reproduction of four species of lizards (Sauria:

Gekkonidae) in the lower Florida Keys. *Amphibia-Reptilia* 24: 390–396.

Kuhl, H. 1820. *Beiträge zur Zoologie und vergleichenden Anatomie*. Frankfurt am Main, Germany: Verlag der Hermannschen Buchhandlung.

Kundert, F. 1974. *Fasicination. Schlangen und Echsen. Serpents et Lézards. Snakes and Lizards*. Spreitenbach, Switzerland: Verlag F. Kundert.

Lacepède, B. G. E. 1788. *Histoire Naturelle des Quadrupèdes Ovipares et des Serpens*. Tome Premier. Paris: Hôtel de Thou.

Lacepède, B. G. E. 1789. *Histoire Naturelle des Serpens*. Tome Second. Paris: Hôtel de Thou.

Lang, M. 1989. Phylogenetic and biogeographic patterns of basiliscine iguanians (Reptilia: Squamata: "Iguanidae"). *Bonner Zoologische Monographien* 28: 1–172.

Langueux, C. J. 1991. Economic analysis of sea turtle eggs in a coastal community on the Pacific coast of Honduras, PP. 136–144. IN: J. G. Robinson and K. H. Redford, editors. *Neotropical Wildlife Use and Conservation*. Chicago, Illinois: University of Chicago Press.

Langueux, C. J., C. L. Campbell, and W. A. McCoy. 2003. Nesting and conservation of the Hawksbill Turtle, *Eretmochelys imbricata*, in the Pearl Cays, Nicaragua. *Chelonian Conservation and Biology* 4: 588–602.

Lara-Uc, M., and R. Riosmena-Rodriguez. 2015. Natural history notes. *Chelonia mydas agassizii* (East Pacific Green Sea Turtle). Diet. *Herpetological Review* 46: 617.

Lattanzio, M. S. 2014. Temporal and ontogenetic variation in the escape response of *Ameiva festiva* (Squamata: Teiidae). *Phyllomedusa* 13: 17–27.

Lattanzio, M. S., and T. C. LaDuke. 2012. Habitat use and activity budgets of Emerald Basilisks (*Basiliscus plumifrons*) in northeast Costa Rica. *Copeia* 2012: 465–471.

Laurenti, J. N. 1768. *Specimen Medicum, Exhibens Synopsin Reptilium Emendatam cum Experimentis circa Venena et Antidota Reptilium Austriacorum, quod Authoritate et Consensu*. Vienna: Joannes Thomae Nobilis de Trattnern.

Lawson, R., J. B. Slowinski, B. I. Crother, and F. T. Burbrink. 2005. Phylogeny of the Colubroidea (Serpentes); new evidence from mitochondrial and nuclear genes *Molecular Phylogenetics and Evolution* 37: 581–601.

Lazell, J. D., Jr. 1973. The lizard genus *Iguana* in the Lesser Antilles. *Bulletin of the Museum of Comparative Zoology* 145: 1–28.

Le, M., and W. P. McCord. 2008. Phylogenetic relationships and biogeographical history of the genus *Rhinoclemmys* Fitzinger, 1835 and the monophyly of the turtle family Geoemydidae (Testudines: Testudinoidea). *Zoological Journal of the Linnean Society* 153: 751–767.

LeBuff, C. R., Jr., 1990. *The Loggerhead Turtle in the Eastern Gulf of Mexico.* Sanibel, Florida: Caretta Research, Inc.

Le Conte, J. 1854. Description of four new species of *Kinosternum.* *Proceedings of the Academy of Natural Sciences of Philadelphia* 7: 180–190.

Le Conte, J. 1859. Description of two new species of tortoises. *Proceedings of the Academy of Natural Sciences of Philadelphia* 11: 4–7.

Lee, J. C. 1996. *The Amphibians and Reptiles of the Yucatán Peninsula.* Ithaca, New York: Comstock Publishing Associates, Cornell University Press.

Lee, J. C. 2000. *A Field Guide to the Amphibians and Reptiles of the Maya World. The Lowlands of Mexico, Northern Guatemala, and Belize.* Ithaca, New York: Comstock Publishing Associates, Cornell University Press.

Leenders, T. A. A. M., and G. J. Watkins-Colwell. 2004. Notes on a collection of amphibians and reptiles from El Salvador. *Postilla* 231: 1–31.

Legler, J. M. 1965. A new species of turtle, genus *Kinosternon,* from Central America. *University of Kansas Publications Museum of Natural History* 15: 615–625, plates 26–28.

Legler, J. M. 1966. Notes on the natural history of a rare Central American turtle, *Kinosternon angustipons* Legler. *Herpetologica* 22: 118–122.

Legler, J. M. 1973. Studies on the life history and ecology of a neotropical slider turtle, *Pseudemys scripta,* in Panamá. *National Geographic Society Research Reports 1966 Projects*: 143–146.

Legler, J. M. 1990. The genus *Pseudemys* in Mesoamerica: taxonomy, distribution, and origins, PP. 82–105. IN: J. W. Gibbons, editor. *Life History and Ecology of the Slider Turtle.* Washington: Smithsonian Institution Press.

Legler, J. M., and R. C. Vogt. 2013. *The Turtles of Mexico. Land and Freshwater Forms.* Berkeley: University of California Press.

Lemos-Espinal, J. A., editor. 2015. Amphibians and reptiles of the US—Mexico Border States = Anfibios y reptiles de los estados de la frontera México—Estados Unidos. College Station: Texas A&M University Press.

Lemos-Espinal, J. A., and J. R. Dixon. 2013. *Amphibians and Reptiles of San Luis Potosí.* Eagle Mountain, Utah: Eagle Mountain Publishing, LC.

Lemos-Espinal, J. A., and H. M. Smith. 2009. *Claves para los Anfibios y Reptiles de Sonora, Chihuahua, y Coahuila, México = Keys to the Amphibians and Reptiles of Sonora, Chihuahua and Coahuila, Mexico.* Mexico City: Universidad Nacional Autónoma de México.

Lever, C. 2003. *Naturalized Reptiles and Amphibians of the World.* Oxford: Oxford University Press.

Lewis, C. B. 1941 (dated 1940). The Cayman Islands and marine turtle. *Bulletin of the Institute of Jamaica, Science Series* 2: 56–65.

Lichtenstein, M. H. C. E., and C. E. von Martens. 1856. *Nomenclator Reptilium et Amphibiorum Musei Zoologici Berolinensis. Namenverzeichniss der in der zoologischen Sammlung der Königlichen Universität zu Berlin aufgestellten Arten von Reptilien und Amphibien nach ihren Ordnungen, Familien und Gattungen.* Berlin: Buchdruckerei Königlichen Akademie Wissenschaften.

Lindholm, W. A. 1929. Revidiertes Verzeichnis der Gattungen der rezenten Schildkröten nebst Notizen zur Nomenklatur einiger Arten. *Zoologischer Anzeiger* 81: 275–295.

Liner, E. A., and G. Casas-Andreau, editors. 2008. *Nombres Estándar en Español en Inglés y Nombres Científicos de los Anfibios y Reptiles de México. Segunda Edición = Standard Spanish, English and Scientific Names of the Amphibians and Reptiles of Mexico. Second Edition. Society for the Study of Amphibians and Reptiles, Herpetological Circulars* 38: i–iv, 1–162.

Linkem, C. W., A. C. Diesmos, and R. M. Brown. 2011. Molecular systematics of the Philippine forest skinks (Squamata: Scincidae: *Sphenomorphus*): testing morphological hypotheses of interspecific relationships. *Zoological Journal of the Linnean Society* 163: 1217–1243.

Linnaeus, C. 1758. *Systema Naturae per Regna Tria Naturae, Secundum Classes, Ordines, Genera, Species, cum Characteribus, Differentiis, Synonymis, Locis.* Tomus I. Pars I. Editio Decima, Reformata. Stockholm: Laurentii Salvii.

Linnaeus, C. 1766. *Systema Naturae per Regna Tria Naturae, Secundum Classes, Ordines, Genera, Species, cum Characteribus, Differentiis, Synonymis. Locis.* Tomus I. Pars. I. Editio Duodecima, Reformata. Stockholm: Laurentii Salvii.

Lisle, H. F., de, R. A. Nazarov, L. R. G. Raw, and J. Grathwohl. 2013. *Gekkota. A Catalog of Recent Species.* Privately printed by the authors.

Lönnberg, E. 1896. Linnean type-specimens of birds, reptiles, batrachians and fishes in the Zoological Museum of the R. University in Uppsala. *Bihang till Konigl Svenska Vetenskapsakademiens Handlingar* 22: 1–45.

Loveridge, A. 1947. Revision of the African lizards of the family Gekkonidae. *Bulletin of the Museum of Comparative Zoölogy* 98: 1–469, plates 1–7.

Lovich, R. E., T. Akre, M. Ryan, S. Nuñez, G. Cruz, G. Borjas, N. J. Scott, S. Flores, W. del Cid, A. Flores, C. Rodriguez, I. R. Luque-Montes, and R. Ford. 2010. New herpetofaunal records from southern Honduras. *Herpetological Review* 41: 112–115.

Lovich, R., T. Akre, M. Ryan, N. Scott, and R. Ford. 2006. Herpetofaunal surveys of Cerro Guanacaure, Montaña La Botija, and Isla del Tigre protected areas in southern Honduras. Washington: Unpublished Report to the International Resource Group.

Lowe, P. R. 1911. *A Naturalist on Desert Islands.* London: Witherby & Co.

Luja, V. H. 2006. Natural history notes. *Mabuya unimarginata* (Central American Mabuya). Reproduction. *Herpetological Review* 37: 469.

Lundberg, M. 2000. Herpetofaunan på Isla de Utila, Honduras. *Snoken* 30: 2–8.

Lundberg, M. 2001. Herpetofaunan på Roatán, Honduras. *Snoken* 31: 20–29.

Lundberg, M. 2002a. Herpetofaunan på Hog Islands, Honduras. *Snoken* 32: 4–13.

Lundberg, M. 2002b. Herpetofaunan på Guanaja, Honduras. *Snoken* 32: 4–12.

Lundberg, M. 2003. Besöki nationalparken La Tigra, Honduras. *Snoken* 33: 25–29.

Lutz, P. L., and J. A. Musick, editors. 1997. *The Biology of Sea Turtles.* Boca Raton, Florida: CRC Press.

Luxmoore, R., B. Groombridge, and S. Broad, editors. 1988. *Significant Trade in Wildlife: A Review of Selected Species in CITES Appendix II.* Vol. 2: *Reptiles and Invertebrates.* Cambridge, U.K.: International Union for Conservation of Nature and Natural Resources.

Lynn, W. G. 1944. Notes on some reptiles and amphibians from Ceiba, Honduras. *Copeia* 1944: 189–190.

MacLean, W. P., R. Kellner, and H. Dennis. 1977. Island lists of West Indian amphibians and reptiles. *Smithsonian Herpetological Information Service* 40: 1–47.

Magnusson, W. E., E. Vieira da Silva and A. P. Lima. 1987. Diet of Amazonian crocodilians. *Journal of Herpetology* 21: 85–95.

Mahler, D. L., and M. Kearney. 2006. The palatal dentition in squamate reptiles: morphology, development, attachment, and replacement. *Fieldiana. Zoology New Series* 108: 1–61.

Malnate, E. V. 1971. A catalog of primary types in the herpetological collections of the Academy of Natural Sciences, Philadelphia (ANSP). *Proceedings of the Academy of Natural Sciences of Philadelphia* 123: 345–375.

Malone, C. L., and S. K. Davis. 2004. Genetic contributions to Caribbean iguana conservation, PP. 45–57. IN: A. C. Alberts, R. L. Carter, W. K. Hayes, and E. P. Martins, editors. *Iguanas. Biology and Conservation.* Berkeley: University of California Press.

Marcellini, D. L. 1971. Activity patterns of the gecko *Hemidactylus frenatus. Copeia* 1971: 631–635.

Marcellini, D. L. 1974. Acoustic behavior of the gekkonid lizard, *Hemidactylus frenatus. Herpetologica* 30: 44–52.

Marin, M. 1984. The National Report for the Country of Honduras = El reporte nacional por el pais de Honduras, PP. 220–224. IN: P. Bacon, F. H. Berry, K. Bjorndal, H. Hirth, L. Ogren, and M. Weber, editors. *Proceedings of the Western Atlantic Turtle Symposium. Symposium on Sea Turtle Research of the Western Atlantic (Populations and Socioeconomics).* Vol. 3; 1983 Jul 17–22; San José, Costa Rica. Miami, Florida: University of Miami Press.

Markezich, A. L., C. J. Cole, and H. C. Dessauer. 1997. The blue and green whiptail lizards (Squamata: Teiidae: *Cnemidophorus*) of the Peninsula de Paraguana, Venezuela: systematics, ecology, descriptions of two new taxa, and relationships to whiptails of the Guianas. *American Museum Novitates* 3207: 1–60.

Márquez, M., R. 1990. FAO Species Catalogue. Vol. 11: Sea turtles of the World. An annotated and illustrated catalogue of sea turtle species known to date. Rome: Food and Agriculture Organization of the United Nations, *FAO Fisheries Synopsis* 125.

Martin, P. S. 1958. A biogeography of reptiles and amphibians in the Gomez Farias Region, Tamaulipas, Mexico. *Miscellaneous Publications Museum of Zoology, University of Michigan* 101: 1–102, plates I–VII.

Marx, H. 1958. Catalogue of type specimens of reptiles and amphibians in Chicago Natural History Museum. *Fieldiana: Zoology* 36: 409–496.

Maslin, T. P., and D. M. Secoy. 1986. A checklist of the lizard genus *Cnemidophorus* (Teiidae). *Contributions in Zoology, University of Colorado Museum* 1: 1–60.

Mata-Silva, D. DeSantis, García-PadillaE. and L. D. Wilson. 2015. Reptilia: Squamata (lizards). Family Teiidae. *Aspidoscelis motaguae* (Sackett, 1941). *Mesoamerican Herpetology* 2: 541–543.

Mather, C. M., and J. W. Sites, Jr. 1985. *Sceloporus variabilis. Catalogue of American Amphibians and Reptiles* 373.1–373.3.

Maturana, H. R. 1962. A study of the species of the genus *Basiliscus. Bulletin of the Museum of Comparative Zoology* 128: 1–33.

Mausfield, P., A. Schmitz, W. Böhme, B. Misof, D. Vrcibradic, and C. F. D. Rocha. 2002. Phylogenetic affinities of *Mabuya atlantica* Schmidt, 1945, endemic to the Atlantic Ocean Archipelago of Fernando de Noronha (Brazil): necessity of partitioning the genus *Mabuya* Fitzinger, 1826 (Scincidae: Lygosominae). *Zoologischer Anzeiger* 241: 281–293.

Maxwell, S. M., J. W. E. Jeglinski, F. Trillmich, D. P. Costa, and P. T. Raimondi. 2014. The influence of weather and tides on the land basking behavior of Green Sea Turtles (*Chelonia mydas*) in the Galapagos Islands. *Chelonian Conservation and Biology* 13: 247–251.

McCord, W. P., M. Joseph-Ouni, C. Hagen, and T. Blanck. 2010. Three new subspecies of *Trachemys venusta* (Testudines: Emydidae) from Honduras, northern Yucatán (Mexico), and Pacific coastal Panama. *Reptilia (GB)* 71: 39–49.

McCoy, C. J. 1968. A review of the genus *Laemanctus* (Reptilia, Iguanidae). *Copeia* 1968: 665–678.

McCranie, J. R. 1990. Geographic distribution. *Trachemys scripta* (Neotropical Slider). *Herpetological Review* 21: 95.

McCranie, J. R. 2005 (dated 2004). The herpetofauna of Parque Nacional Cerro Azul, Honduras (Amphibia, Reptilia). *The Herpetological Bulletin* 90: 10–21.

McCranie, J. R. 2007a. Herpetological fieldwork in the lowland rainforests of northeastern Honduras: pleasure or how quickly we forget? *Iguana* 14: 172–183.

McCranie, J. R. 2007b. Distribution of the amphibians of Honduras by departments. *Herpetological Review* 38: 35–39.

McCranie, J. R. 2009. Amphibians and Reptiles of Honduras. Listas Zoológicas Actualizadas San Pedro [Internet]. Costa Rica: Universidad de Costa Rica, Museo de Zoología UCR; [cited 2017 May 1]. Available from: http://museo.biologia.ucr.ac.cr/Listas/Anteriores/HerpHon.htm.

McCranie, J. R. 2011a. The Snakes of Honduras: Systematics, Distribution, and Conservation. *Society for the Study of Amphibians and Reptiles, Contributions to Herpetology* 26: i–x, 1–714, plates 1–20.

McCranie, J. R. 2011b. A new species of *Tantilla* of the *taeniata* species group (Reptilia, Squamata, Colubridae, Colubrinae) from northeastern Honduras. *Zootaxa* 3037: 37–44.

McCranie, J. R. 2014. First departmental records of amphibians and reptiles from Intibucá, Lempira, and Ocotepeque in southwestern Honduras. *Herpetological Review* 45: 291–293.

McCranie, J. R. 2015a. A checklist of the amphibians and reptiles of Honduras, with additions, comments on taxonomy, some recent taxonomic decisions, and areas of further studies needed. *Zootaxa* 3931: 352–386.

McCranie, J. R. 2015b. Bibliography of McCranie's 2015 checklist of amphibians & reptiles of Honduras. *Smithsonian Herpetological Information Service* 146: 1–30.

McCranie, J. R. 2016. Nature notes. *Clelia clelia* (Daudin, 1803). Predation. *Mesoamerican Herpetology* 3: 492–493.

McCranie, J. R., and F. E. Castañeda. 2005. The herpetofauna of Parque Nacional Pico Bonito, Honduras. *Phyllomedusa* 4: 3–16.

McCranie, J. R., and M. R. Espinal. 1998. Geographic distribution. *Corytophanes hernandezii* (Hernandez's Helmeted Basilisk). *Herpetological Review* 29: 174.

McCranie, J. R., and S. W. Gotte. 2014. An investigation into the Swan Island Honduras collecting event of *Tiaporus fuliginosus* Cope (Reptilia: Teiidae) and its systematic status. *Proceedings of the Biological Society of Washington* 127: 543–556 (front cover photograph also).

McCranie, J. R., and A. Gutsche. 2016. The herpetofauna of islands in the Golfo de Fonseca and adjacent waters, Honduras. *Mesoamerican Herpetology* 3: 842–899.

McCranie, J. R., and S. B. Hedges. 2012. Two new species of geckos from Honduras and resurrection of *Sphaerodactylus continentalis* Werner from the synonymy of *Sphaerodactylus millepunctatus* Hallowell (Reptilia, Squamata, Gekkonoidea, Sphaerodactylidae). *Zootaxa* 3492: 65–76.

McCranie, J. R., and S. B. Hedges. 2013a. Two additional new species of *Sphaerodactylus* (Reptilia, Squamata, Gekkonoidea, Sphaerodactylidae) from the Honduran Bay Islands. *Zootaxa* 3694: 40–50.

McCranie, J. R., and S. B. Hedges. 2013b. A new species of *Phyllodactylus* (Reptilia, Squamata, Gekkonoidea, Phyllodactylidae) from Isla de Guanaja in the Honduran Bay Islands. *Zootaxa* 3694: 51–58.

McCranie, J. R., and S. B. Hedges. 2013c. A review of the *Cnemidophorus lemniscatus* group in Central America (Squamata: Teiidae), with comments on other species in the group. *Zootaxa* 3722: 301–316.

McCranie, J. R., and S. B. Hedges. 2016. Molecular phylogeny and taxonomy of the *Epictia goudotii* species complex (Serpentes: Leptotyphlopidae: Epictinae) in Middle America and northern South America. *PeerJ* 3: e1551;DOI 7717/peerj:1551 (pp. 1–27).

McCranie, J. R., and G. Köhler. 1999. Geographic distribution. *Sphenomorphus assatus* (Red Forest Skink). *Herpetological Review* 30: 111.

McCranie, J. R., and G. Köhler. 2001. A new species of anole from eastern Honduras related to *Norops tropidonotus* (Reptilia, Squamata, Polychrotidae). *Senckenbergiana biologica* 81: 227–233.

McCranie, J. R., and G. Köhler. 2004a. *Laemanctus longipes*. Catalogue of American Amphibians and Reptiles 795.1–795.4.

McCranie, J. R., and G. Köhler. 2004b. *Laemanctus serratus*. Catalogue of American Amphibians and Reptiles 796.1–796.5.

McCranie, J. R., and G. Köhler. 2015. The anoles (Reptilia: Squamata: Dactyloidae: *Anolis*: *Norops*) of Honduras. Systematics, distribution, and conservation. *Bulletin of the Museum of Comparative Zoology, Special Publications Series* 1: 1–292.

McCranie, J. R., and S. M. Rovito. 2011. Geographic distribution. *Hemidactylus frenatus* (Common House Gecko). *Herpetological Review* 42: 241.

McCranie, J. R., and J. M. Solís. 2013. Additions to the amphibians and reptiles of Parque Nacional Pico Bonito, Honduras: with an updated nomenclatural list. *Herpetology Notes* 6: 239–243.

McCranie, J. R., and J. H. Townsend. 2011. Description of a new species of worm salamander (Caudata, Plethodontidae, *Oedipina*) in the subgenus *Oedipinola* from the central portion of the Cordillera Nombre de Dios, Honduras. *Zootaxa* 2990: 59–68.

McCranie, J. R., and L. Valdés Orellana. 2011a. Geographic distribution. *Celestus bivittatus* (NCN). *Herpetological Review* 42: 240.

McCranie, J. R., and L. Valdés Orellana. 2011b. Geographic distribution. *Lepidophyma mayae* (Maya Night Lizard). *Herpetological Review* 42: 241.

McCranie, J. R., and L. Valdés Orellana. 2011c. Geographic distribution. *Rhinoclemmys pulcherrima* (Painted Wood Turtle; Casco Rojo). *Herpetological Review* 42: 566.

McCranie, J. R., and L. Valdés Orellana. 2014. New island records and updated nomenclature of amphibians and reptiles from the Islas de la Bahía, Honduras. *Herpetology Notes* 7: 41–49.

McCranie, J. R., and J. Villa. 1993. A new genus for the snake *Enulius sclateri* (Colubridae: Xenodontinae). *Amphibia-Reptilia* 14: 261–267.

McCranie, J. R., and L. D. Wilson. 1996. A new arboreal lizard of the genus *Celestus* (Squamata: Anguidae) from northern Honduras. *Revista de Biología Tropical* 44: 259–264.

McCranie, J. R., and L. D. Wilson. 1998. Geographic distribution. *Corytophanes percarinatus* (Keeled Helmeted Basilisk). *Herpetological Review* 29: 174.

McCranie, J. R., and L. D. Wilson. 1999. Status of the anguid lizard *Abronia montecristoi* Hidalgo. *Journal of Herpetology* 33: 127–128.

McCranie, J. R., and L. D. Wilson. 2000. Geographic distribution. *Hemidactylus mabouia* (Tropical Gecko). *Herpetological Review* 31: 113.

McCranie, J. R., and L. D. Wilson. 2002. The amphibians of Honduras. *Society for the Study of Amphibians and Reptiles, Contributions to Herpetology* 19: i–x, 1–625, plates 1–20.

McCranie, J. R., F. E. Castañeda, and K. E. Nicholson. 2002. Preliminary results of herpetofaunal survey work in the Rus Rus region, Honduras: a proposed biological reserve. *The Herpetological Bulletin* 81: 22–29.

McCranie, J. R., G. A. Cruz [Díaz] and P. A. Holm. 1993. A new species of cloud forest lizard of the *Norops schiedei* group (Sauria: Polychrotidae) from northern Honduras. *Journal of Herpetology* 27: 386–392.

McCranie, J. R., R. Dionicio, J. Ramos, L. Valdés Orellana, J. E. Merida, and G. A. Cruz [Díaz]. 2014. Eight new departmental records of lizards and snakes (Reptilia: Squamata) from subhumid areas in El Paraíso, Honduras, and morphometry of the poorly known pitviper *Agkistrodon howardgloydi*. *Cuadernos de Investigación, UNED Research Journal* 6: 99–104.

McCranie, J. R., A. Harrison, and L. Valdés Orellana. 2017. Updated population and habitat comments about the reptiles of the Swan Islands, Honduras. *Bulletin of the Museum of Comparative Zoology* 161: 265–284.

McCranie, J. R., F. Köhler, A. Gutsche, and L. Valdés Orellana. 2013. *Trachemys grayi emolli* (Testudines, Emydidae) in Honduras and its systematic relationships based on mitochondrial DNA. *Zoosystema Evolution* 89: 21–29.

McCranie, J. R., G. Köhler, and L. D. Wilson. 2000. Two new species of anoles from northwestern Honduras related to *Norops laeviventris* (Wiegmann 1834) (Reptilia, Squamata, Polychrotidae). *Senckenbergiana biologica* 80: 213–223.

McCranie, J. R., K. E. Nicholson, and G. Köhler. 2002 (dated 2001). A new species of *Norops* (Squamata: Polychrotidae) from northwestern Honduras. *Amphibia-Reptilia* 22: 465–473.

McCranie, J. R., J. H. Townsend, and L. D. Wilson. 2004. *Corytophanes hernandesii*. Catalogue of the American Amphibians and Reptiles 790.1–790.6.

McCranie, J. R., J. H. Townsend, and L. D. Wilson. 2006. *The Amphibians and Reptiles of the Honduran Mosquitia*. Malabar, Florida: Krieger Publishing Company.

McCranie, J. R., L. Valdés Orellana and A. Gutsche. 2011. Geographic distribution. *Kinosternon leucostomum* (White-lipped Mud Turtle; Pochitoque). *Herpetological Review* 42: 565.

McCranie, J. R., L. Valdés Orellana and A. Gutsche. 2013. New departmental records for amphibians and reptiles in Honduras. *Herpetological Review* 44: 288–289.

McCranie, J. R., L. D. Wilson, and G. Köhler. 2005. *The Amphibians & Reptiles of the Bay Islands and Cayos Cochinos, Honduras*. Salt Lake City, Utah: Bibliomania!

McCranie, J. R., L. D. Wilson, and K. L. Williams. 1992. A new species of anole of the *Norops crassulus* group (Sauria: Polychridae) from northwestern Honduras. *Caribbean Journal of Science* 28: 208–215.

McCranie, J. R., L. D. Wilson, and K. L. Williams. 1993. Another new species of lizard of the *Norops schiedei* group (Sauria: Polychrotidae) from northern Honduras. *Journal of Herpetology* 27: 393–399.

McDowell, S. B. 1964. Partition of the genus *Clemmys* and related problems in the taxonomy of the aquatic Testudinidae. *Proceedings of the Zoological Society of London* 143: 239–279.

McDowell, S. B. 1987. Systematics, PP. 3–50. IN: R. A. Seigel, J. T. Collins, and S. S. Novak, editors. *Snakes: Ecology and Evolutionary Biology*. New York: Macmillan Publishing Company.

McKeown, S. 1996. *A Field Guide to Reptiles and Amphibians in the Hawaiian Islands*. Los Osos, California: Diamond Head Publishing, Inc.

McMahan, C. D., L. R. Freeborn, W. C. Wheeler, and B. I. Crother. 2015. Forked tongues revisited: molecular apomorphies support morphological hypotheses of Squamate evolution. *Copeia* 2015: 525–529.

Medem, F. 1956. Informe sobre reptiles Colombianos (I). Noticia sobre el primer hallazgo de la tortuga *Geomyda annulata* (Gray) en Colombia. *Caldasia* 7: 317–325.

Medem, F. 1962. La distribución geográfica y ecología de los Crocodylia y Testudinata en el Departamento del Chocó. *Revista de la Academia Colombiana de Ciencias Exactas, Fisicas y Naturales* 11: 279–304, 56 plates.

Medem, F. 1977. Contribución al conocimiento sobre la taxonómia, distribución geográfica y ecologia de la tortuga "bache" (*Chelydra serpentina acutirostris*). *Caldasia* 12: 41–101.

Medem, F. 1981. *Los Crocodylia de Sur America*. Vol. I. *Los Crocodylia de Colombia*. Bogotá, Colombia: Ministerio de Educación Nacional, COLCIENCIAS.

Medem, F. 1983. *Los Crocodylia de Sur America*. Vol. II. *Venezuela–Trinidad–Tobago–Guyana–Suriname–Guayana Francesca–Ecuador–Perú–Bolivia–Brasil–Paraguay–Argentina–Uruguay*. Bogotá, Colombia: Universidad Nacional de Colombia y COLCIENCIAS.

Meerwarth, H. 1901. Die westindischen Reptilien und Batrachier des Naturhistorischen Museums in Hamburg. *Mitteilungen aus dem Naturhistorischen Museum in Hamburg* 18: 1–41, plates I–II.

Melville, R. V., and J. D. D. Smith. 1987. *Official Lists and Indexes of Names and Works in Zoology*. London: The International Trust for Zoological Nomenclature.

Mendoza-Quijano, F., O. Flores-Villela, and J. W. Sites, Jr. 1998. Genetic variation, species status, and phylogenetic relationships in rose-bellied lizards (*variabilis* group) of the genus *Sceloporus* (Squamata: Phrynosomatidae). *Copeia* 1998: 354–366.

Merchán Fornelino, M. 2002. Estudio biométrico de juveniles de tortuga "jicotea" (*Trachemys scripta emolli*) en Costa Rica. *Revista Española de Herpetología* 16: 11–17.

Merchán Fornelino, M. 2003. *Chelydra serpentina acutirostris*. Biology and conservation of the snapping turtle in Nicaragua and Costa Rica. *Reptilia (GB)* 28: 60–65.

Merchán [Fornelino], M. and R. Fournier. 2007. Natural history notes. *Rhinoclemmys funerea* (Black Wood Turtle). Reproduction. *Herpetological Review* 38: 72.

Merchán [Fornelino], M. and J. M. Mora. 2000. *Rhinoclemmys funerea*. Biology, distribution and conservation of the Black Wood Turtle in Costa Rica. *Reptilia (GB)* 13: 31–38.

Merrem, B. 1820. *Versuch eines Systems der Amphibien. Tentamen Systematis Amphibiorum*. Marburg, Germany: Johann Christian Krieger.

Mertens, R. 1937. Reptilien und Amphibien aus dem südlichen Inner-Afrika. *Abhandlungen der Senckenbergischen Naturforschenden Gesellschaft* 435: 1–23.

Mertens, R. 1952a. Neues über die Reptilienfauna von El Salvador. *Zoologischer Anzeiger* 148: 87–93.

Mertens, R. 1952b. Die Amphibien und Reptilien von El Salvador, auf Grund der Reisen von R. Mertens und A. Zilch. Frankfurt, Germany: Senckenbergische Naturforschende Gesellschaft, *Abhandlungen* 487.

Meshaka, W. E., Jr., B. P. Butterfield, and J. B. Hauge. 2004. *The Exotic Amphibians and Reptiles of Florida*. Malabar, Florida: Krieger Publishing Company.

Meyer, F. A. A. 1795. *Synopsis Reptilium, nouam ipsorum sistens generum methodum, nec non Gottingensium huius ordinis animalium enumerationem*. Gottingae [Göttingen]: Vandenhoek et Ruprecht.

Meyer, J. R. 1966. Records and observations on some amphibians and reptiles from Honduras. *Herpetologica* 22: 172–181.

Meyer, J. R. 1969. *A biogeographic study of the amphibians and reptiles of Honduras* [dissertation]. Los Angeles: University of Southern California.

Meyer, J. R., and L. D. Wilson. 1972 (dated 1971). Taxonomic studies and notes on some Honduran amphibians and reptiles. *Bulletin of the Southern California Academy of Sciences* 70: 106–114.

Meyer, J. R, and L. D. Wilson. 1973. A distributional checklist of the turtles, crocodilians, and lizards of Honduras. *Contributions in Science, Natural History Museum, Los Angeles County* 244: 1–39.

Meylan, A. B. 1999a. Status of the Hawksbill Turtle (*Eretmochelys imbricata*) in the Caribbean region. *Chelonian Conservation and Biology* 3: 177–184.

Meylan, A. B. 1999b. International movements of immature and adult Hawksbill turtles (*Eretmochelys imbricata*) in the Caribbean region. *Chelonian Conservation and Biology* 3: 189–194.

Meylan, A. B., P. A. Meylan, and C. Ordoñez Espinosa. 2013. Sea turtles of Bocas del Toro Province and the Comarca Ngöbe-Buglé, Republic of Panamá. *Chelonian Conservation and Biology* 12: 17–33.

Meylan, P. A., A. B. Meylan, and J. A. Gray. 2011. The ecology and migration of sea turtles 8. Tests of the developmental habitat hypothesis. *Bulletin of the American Museum of Natural History* 357: 1–70.

Mijares-Urrutia, A., and A. Arends R. 2000. Herpetofauna of Estado Falcón, northwestern Venezuela: a checklist with geographical and ecological data. *Smithsonian Herpetological Information Service* 123: 1–30.

Milián-García, Y., M. Venegas-Anaya, R. Frias-Soler, A. J. Crawford, R. Ramos-Targarona, R. Rodríguez-Soberón, M. Alonso-Tabet, J. Thorbjarnarson, O. I. Sanjur, G. Espinosa-López, and E. Bermingham. 2011. Evolutionary history of Cuban crocodiles *Crocodylus rhombifer* and *Crocodylus acutus* inferred from multilocus markers. *Journal of Experimental Zoology* 315: 358–375.

Miller, C. M. 1997. Natural history notes. *Eumeces sumichrasti* (NCN). Brood. *Herpetological Review* 28: 151–152.

Miller, M. R. 1966. The cochlear duct of lizards. *Proceedings of the California Academy of Sciences* 33: 255–359.

Miralles, A. 2004. Natural history notes. *Lepidophyma flavimaculatum* (Yellow-Spotted Night Lizard). Placentophagia. *Herpetological Review* 35: 170.

Miralles, A. 2006. A new species of *Mabuya* (Reptilia, Squamata, Scincidae) from the Caribbean island of San Andrés, with a new interpretation of nuchal scales: a character of taxonomic importance. *The Herpetological Journal* 16: 1–7.

Miralles, A., and S. Carranza. 2010. Systematics and biogeography of the neotropical genus *Mabuya*, with special emphasis on the Amazonian skink *Mabuya nigropunctata* (Reptilia, Scincidae). *Molecular Phylogenetics and Evolution* 54: 857–869.

Miralles, A., J. C. Chaparro, and M. B. Harvey. 2009. Three rare and enigmatic South American skinks. *Zootaxa* 2012: 47–68.

Miralles, A., G. R. Fuenmayor, C. Bonillo, W. E. Schargel, T. Barros, J. E. García-Perez, and C. L. Barrio-Amorós. 2009. Molecular systematics of Caribbean skinks of the genus *Mabuya* (Reptilia, Scincidae), with descriptions of two new species from Venezuela. *Zoological Journal of the Linnean Society* 156: 598–616.

Mittermeier, R. A. 1971. Notes on the behavior and ecology of *Rhinoclemys* [sic] *annulata* Gray. *Herpetologica* 27: 485–488.

Mittleman, M. B. 1950. The generic status of *Scincus lateralis* Say, 1823. *Herpetologica* 6: 17–20.

Mittleman, M. B. 1952. A generic synopsis of the lizards of the subfamily Lygosominae. *Smithsonian Miscellaneous Collections* 117: 1–35.

Mocquard, M. F. 1905. Note-préliminaire sur une collection de reptiles et de batrachiens offerte au muséum par M. Maurice de Rothschild. *Bulletin du Muséum National d'Histoire Naturelle, Paris* 11: 285–290.

Moll, D. 1990. Population sizes and foraging ecology in a tropical freshwater stream turtle community. *Journal of Herpetology* 24: 48–53.

Moll, D. 1994. The ecology of sea beach nesting in slider turtles (*Trachemys scripta venusta*) from Caribbean Costa Rica. *Chelonian Conservation and Biology* 1: 107–116.

Moll, D. 2010. The backdoor turtles of Tortuguero. *IRCF, Reptiles and Amphibians* 17: 103–107.

Moll, D., and K. P. Jansen. 1995. Evidence for a role in seed dispersal by two tropical herbivorous turtles. *Biotropica* 27: 121–127.

Moll, D., and E. O. Moll. 1990. The slider turtle in the Neotropics: adaptation of a temperate species to a tropical environment, PP. 152–161. IN: J. W. Gibbons, editor. *Life History and Ecology of the Slider Turtle*. Washington: Smithsonian Institution Press.

Moll, D., and E. O. Moll. 2004. *The Ecology, Exploitation, and Conservation of River Turtles*. New York: Oxford University Press.

Moll, E. O. 1979. Reproductive cycles and adaptations, PP. 305–331. IN: M. Harless and H. Morlock, editors. *Turtles. Perspectives and Research*. New York: John Wiley & Sons.

Moll, E. O., and J. M. Legler. 1971. The life history of a neotropical slider turtle, *Pseudemys scripta* (Schoepff), in Panama. *Bulletin of the Los Angeles County Museum of Natural History* 11: 1–102.

Monge-Nájera, J., B. Morera, and M. Chávez. 1988. Nesting behaviour of *Rhinoclemmys pulcherrima* in Costa Rica (Testudines: Emydidae). *The Herpetological Journal* 1: 308.

Montgomery, C. E., S. M. Boback, S. E. W. Green, M. A. Paulissen, and J. M. Walker. 2011. *Cnemidophorus lemniscatus* (Squamata: Teiidae) on Cayo Cochino Pequeño, Honduras: extent of island occupancy, natural history, and conservation status. *Herpetological Conservation and Biology* 6: 10–24.

Montgomery, C. E., S. A. Pasachnik, L. E. Ruyle, J. A. Frazier, and S. E. W. Green. 2015. Natural history of the black-chested spiny-tailed iguanas, *Ctenosaura melanosterna* (Iguanidae), on Cayo

Cochino Menor, Honduras. *The Southwestern Naturalist* 59: 280–285.

Montgomery, C. E., R. N. Reed, H. J. Shaw, S. M. Boback, and J. M. Walker. 2007. Distribution, habitat, size, and color pattern of *Cnemidophorus lemniscatus* (Sauria: Teiidae) on Cayo Cochino Pequeño, Honduras. *The Southwestern Naturalist* 52: 38–45.

Monzel, M. 1998. Zoogeographische Untersuchungen zur Herpetofauna der Islas de la Bahía (Honduras) [thesis]. Ommersheim, Germany: Universität Saarlandes.

Mora, J. M. 2010. Natural history of the Black Spiny-tailed Iguana (*Ctenosaura similis*) at Parque Nacional Palo Verde, Costa Rica, with comments on the conservation of the genus *Ctenosaura*, PP. 716–733. IN: L. D. Wilson, J. H. Townsend, and J. D. Johnson, editors. *Conservation of Mesoamerican Amphibians and Reptiles*. Eagle Mountain, Utah: Eagle Mountain Publishing, LC.

Mora, J. M., and A. N. Ugalde. 1991. A note on the population status and exploitation of *Pseudemys scripta emolli* (Reptilia: Emydidae) in northern Costa Rica. *Bulletin of the Chicago Herpetological Society* 26: 111.

Mora, J. M., F. H. G. Rodrigues, L. I. López, and L. D. Alfaro. 2015. Nature notes. *Ctenosaura similis*. Cannibalism. *Mesoamerican Herpetology* 2: 107–109.

Morales-Verdeja, S. A., and R. C. Vogt. 1997. Terrestrial movements in relation to aestivation and the annual reproductive cycle of *Kinosternon leucostomum*. *Copeia* 1997: 123–130.

Moravec, J., L. Kratochvíl, Z. S. Amr, D. Jandzik, J. Smíd, and Gvozdík. 2011. High genetic differentiation within the *Hemidactylus turcicus* complex (Reptilia: Gekkonidae) in the Levant, with comments on the phylogeny and systematics of the genus. *Zootaxa* 2894: 21–38.

Moreau de Jonnès, M. 1818. Monographie du Mabouia des murailles, ou *Gecko Mabouia* des Antilles. *Bulletin des Sciences par La Société Philomathique de Paris* 3: 138–139.

Morgan, G. S. 1985. Taxonomic status and relationships of the Swan Island Hutia, *Geocapromys thoracatus* (Mammalia: Rodentia: Capromyidae), and the zoogeography of the Swan Islands vertebrate fauna. *Proceedings of the Biological Society of Washington* 98: 29–46.

Moyne, L. 1938. *Atlantic Circle*. London: Blackie & Son, Ltd.

Muelleman, P. J., C. E. Montgomery, E. N. Taylor, J. A. Frazier, J. C. Ahle, and S. M. Boback. 2009. Geographic distribution. *Hemidactylus frenatus* (Common House Gecko). *Herpetological Review* 40: 452.

Murphy, J. C. 1997. *Amphibians and Reptiles of Trinidad and Tobago*. Malabar, Florida: Krieger Publishing Company.

Murphy, R. W., A. Smith, and A. Ngo. 2007. The versions of Cope's *Batrachia and Reptilia of Costa Rica*. *Bibliotheca Herpetologica* 7: 12–16.

Myers, C. W. 1973. Anguid lizards of the genus *Diploglossus* in Panama, with the description of a new species. *American Museum Novitates* 2523: 1–20.

Myers, C. W. 2011. A new genus and new tribe for *Enicognathus melanauchen* Jan, 1863, a neglected South American snake (Colubridae: Xenodontinae), with taxonomic notes on some Dipsadinae. *American Museum Novitates* 3715: 1–33.

Myers, C. W., and W. Böhme. 1996. On the type specimens of two Colombian poison frogs described by A. A. Berthold (1845), and their bearing on the locality "Provinz Popayan." *American Museum Novitates* 3185: 1–20.

Myers, C. W., and M. A. Donnelly. 1991. The lizard genus *Sphenomorphus* (Scincidae) in Panama, with description of a new species. *American Museum Novitates* 3027: 1–12.

Myers, C. W., and S. B. McDowell. 2014. New taxa and cryptic species of neotropical snakes (Xenodontinae), with commentary on hemipenes as generic and specific characters. *Bulletin of the American Museum of Natural History* 385: 1–112.

Naccarato, A. M., J. B. Dejarnette, and P. Allmann. 2015. Successful establishment of a non-native species after an apparent single introduction event: investigating ND4 variability in introduced Black Spiny-Tailed Iguanas (*Ctenosaura similis*) in southwestern Florida. *Journal of Herpetology* 49: 230–236.

Naro-Maciel, E., M. Le, N. N. FitzSimmons, and G. Amato. 2008. Evolutionary relationships of marine turtles: a molecular phylogeny based on nuclear and mitochondrial genes. *Molecular Phylogenetics and Evolution* 49: 659–662.

Neill, W. T. 1965. New and noteworthy amphibians and reptiles from British Honduras. *Bulletin of the Florida State Museum, Biological Sciences* 9: 77–130.

Neill, W. T. 1971. *The Last of the Ruling Reptiles. Alligators, Crocodiles, and their Kin*. New York: Columbia University Press.

Neill, W. T., and R. Allen. 1959. Studies on the amphibians and reptiles of British Honduras. *Publications of the Research Division Ross Allen's Reptile Institute* 2: 1–76.

Neill, W. T., and R. Allen. 1962. Reptiles of the Cambridge Expedition to British Honduras, 1959–60. *Herpetologica* 18: 79–91.

Nicholson, K. E., B. I. Crother, C. Guyer, and J. M. Savage. 2012. It is time for a new classification of anoles (Squamata: Dactyloidae). *Zootaxa* 3477: 1–108.

Nicholson, K. E., B. I. Crother, C. Guyer, and J. M. Savage. 2014. Anole classification: a response to Poe. *Zootaxa* 3814: 109–120.

Nicholson, K. E., J. R. McCranie, and G. Köhler. 2000. Herpetofaunal expedition to Parque Nacional Patuca: a newly established park in Honduras. *The Herpetological Bulletin* 72: 26–31.

Nietschmann, B. 1977 (dated 1976). *Memorias de Arrecife Tortuga. Historia Natural y Económica de las Tortugas en el Caribe de América Central.* Vol. 2. San José: Serie Geografía Naturaleza.

Niewiarowski, P. H., A. Stark, B. McClung, B. Chambers, and T. Sullivan. 2012. Faster but not stickier: invasive house geckos can out-sprint resident Mournful geckos in Moorea, French Polynesia. *Journal of Herpetology* 46: 194–197.

Nogales, M., A. Martín, B. R. Tershy, C. J. Donlan, D. Veitch, N. Puerta, B. Wood. and J. Alonso. 2004. A review of feral cat eradication on islands. *Conservation Biology* 18: 310–319.

Noonan, B. P., J. B. Pramuk, R. L. Bezy, E. A. Sinclair, K. de Queiroz, and J. W. Sites, Jr. 2013. Phylogenetic relationships within the lizard clade Xantusiidae: using trees and divergence times to address evolutionary questions at multiple levels. *Molecular Phylogenetics and Evolution* 69: 109–122.

Okayama, T., R. Díaz-Fernández, Y. Baba, M. Halim, O. Abe, N. Azeno, and H. Koike. 1999. Genetic diversity of the Hawksbill Turtle in the Indo-Pacific and Caribbean regions. *Chelonian Conservation and Biology* 3: 362–367.

Oken, L. 1817. Cuviers und Okens Zoologien neben einander gestellt. *Isis* 8: cols. 1145–1186 (the intervening columns are irregularly and often inaccurately numbered).

Oldham, J. C., and H. M. Smith. 1983. Relationships among iguanine lizards (Sauria: Iguanidae) as suggested by appendicular myology. *Bulletin of the Maryland Herpetological Society* 19: 73–82.

Oliveira Nogueira, C. H. de, C. A. Figueiredo-de-Andrade, and N. Nascimento de Freitas. 2013. Death of a juvenile snake *Oxyrhopus petolarius* (Linnaeus, 1758) after eating an adult house gecko *Hemidactylus mabouia* (Moreau de Jonnès, 1818). *Herpetology Notes* 6: 39–43.

Oliver, J. A. 1937. Notes on a collection of amphibians and reptiles from the state of Colima, Mexico. *Occasional Papers of the Museum of Zoology, University of Michigan* 360: 1–28, plate I.

Oppel, M. 1811a (dated 1810). Suite du Ier. Memoire sur la classification des reptiles. *Annales du Muséum d'Histoire Naturelle, Paris* 16: 376–393.

Oppel, M. 1811b. *Die Ordnungen, Familien und Gattungen der Reptilien als Prodrom einer Naturgeschichte Derselben.* München: Joseph Lindauer.

O'Shaughnessy, A. W. E. 1874. Description of a new species of lizard of the genus *Celestus*. *Annals and Magazine of Natural History* 14: 257–258.

O'Shaughnessy, A. W. E. 1875. Descriptions of new species of Geckotidae in the British Museum collection. *Annals and Magazine of Natural History* 16: 262–266.

O'Shea, M. [T.]. 1986. Operation Raleigh. Herpetological survey of the Rio Paulaya and Laguna Bacalar regions northeastern Honduras, Central America. April-June 1985. London: Unpublished Final Report to British Museum of Natural History.

O'Shea, M. T. 1989. New departmental records for northeastern Honduran herpetofauna. *Herpetological Review* 20: 16.

Owen, R. 1842 (dated 1841). Report on British fossil reptiles. *Report Eleventh Meeting of the British Association for the Advancement of Science, London 1841*: 60–204.

Páez, V. A., M. A. Morales-Betancourt, C. A. Lasso, O. V. Castaño-Mora, and B. C. Bock, editors. 2012. *Biología y Conservación de las tortugas continentales de Colombia.* Bogotá: Instituto de Investigación de Recursos Biológicos Alexander von Humbolt.

Parham, J. F., T. J. Papenfuss, J. R. Buskirk, G. Parra-Olea, J-Y. Chen, and W. B. Simison. 2015. *Trachemys ornata* or not *ornata*: reassessment of a taxonomic revision for Mexican *Trachemys*. *Proceedings of the California Academy of Sciences* 62: 359–367.

Parham, J. F., T. J. Papenfuss, P. P. van Dijk, B. S. Wilson, C. Marte, L. Rodriguez Schettino, and W. B. Simison. 2013. Genetic introgression and hybridization in Antillean freshwater turtles (*Trachemys*) revealed by coalescent analyses of mitochondrial and cloned nuclear markers. *Molecular Phylogenetics and Evolution* 67: 176–187.

Parker, H. W. 1940. Undescribed anatomical structures and new species of reptiles and amphibians. *Annals and Magazine of Natural History* 5: 257–274.

Pasachnik, S. [A.]. 2006. Ctenosaurs of Honduras: notes from the field. *Iguana* 13: 264–271.

Pasachnik, S. A. 2011a. On the iguana trail. *IRCF Reptiles & Amphibians, Conservation and Natural History* 18: 106–109.

Pasachnik, S. A. 2011b. Geographic distribution. *Hemidactylus frenatus* (Common House Gecko). *Herpetological Review* 42: 391.

Pasachnik, S. A. 2011c. Natural history notes. *Ctenosaura oedirhina* (Roatan's Spiny-tailed Iguana). Limb regeneration. *Herpetological Review* 42: 600.

Pasachnik, S. A., and D. Ariano. 2010. CITES Appendix II listing for the *Ctenosaura palearis* clade: developing conservation policies in Central America. *IRCF Reptiles & Amphibians, Conservation and Natural History* 17: 136–139.

Pasachnik, S. A., and Z. Chavarria. 2011. Natural history notes. *Ctenosaura oedirhina* (Roatan's Spiny-Tailed Iguana). Diet. *Herpetological Review* 42: 429.

Pasachnik, S. A., and J. P. Corneil. 2011. Natural history notes. *Ctenosaura similis* (Black Spiny-tailed Iguana). Diet. *Herpetological Review* 42: 601–602.

Pasachnik, S. A., D. Ariani-Sánchez, J. Burgess, C. E. Montgomery, J. R. McCranie, and G. Köhler. 2010. *Ctenosaura oedirhina*—de Queiroz, 1987 [Internet]. IUCN Red List Assessment. Gland, Switzerland: IUCN; [cited 2017 November 1]. Available from: www.iucnredlist.org (A new assessment was released without including McCranie as an assessor in 2017, but the contents continue to be inaccurate).

Pasachnik, S. A., J. A. Dannof-Burg, E. E. Antúnez, and J. P. Corneil. 2014. Local knowledge and use of the Valle de Aguán Spiny-Tailed iguana, *Ctenosaura melanosterna*, in Honduras. *Herpetological Conservation and Biology* 9: 436–447.

Pasachnik, S. A., A. C. Echternacht, and B. M. Fitzpatrick. 2010. Gene trees, species and species trees in the *Ctenosaura palearis* clade. *Conservation Genetics* 11: 1767–1781.

Pasachnik, S. A., A. C. Echternacht, and B. M. Fitzpatrick. 2011. Population genetics of the Honduran spiny-tailed iguana *Ctenosaura melanosterna*: implications for conservation and management. *Endangered Species Research* 14: 113–126.

Pasachnik, S. A., B. M. Fitzpatrick, T. J. Near, and A. C. Echternacht. 2009. Gene flow between an endangered endemic iguana, and its wide spread relative, on the island of Utila, Honduras: when is hybridization a threat. Does hybridization threaten an endangered iguana? *Conservation Genetics* 10: 1247–1254.

Pasachnik, S. A., C. E. Montgomery, A. Martinez, N. Belal, S. Clayson, and S. Faulkner. 2012. Body size, demography, and body condition in Utila Spiny-Tailed iguanas, *Ctenosaura bakeri*. *Herpetological Conservation and Biology* 7: 391–398.

Pasachnik, S. A., C. E. Montgomery, L. E. Ruyle, J. P. Corneil, and E. E. Antúnez. 2012. Morphological and demographic analyses of the Black-Chested Spiny-Tailed iguana, *Ctenosaura melanosterna*, across their range: implications for population level management. *Herpetological Conservation and Biology* 7: 399–406.

Pelcastre Villafuerte, L., and O. A. Flores-Villela. 1992. Lista de especies y localidades de recolecta de la herpetofauna de Veracruz, México. *Publicaciones Especiales del Museo de Zoología, Universidad Nacional Autónoma de México* 4: 25–96.

Pérez-Higareda, G., and H. M. Smith. 1987. Comments on geographic variation in *Rhinoclemmys areolata* (Testudines). *Bulletin of the Maryland Herpetological Society* 23: 113–118.

Pérez-Higareda, G., and R. C. Vogt. 1985. A new subspecies of arboreal lizard, genus *Laemanctus*, from the mountainous region of Los Tuxtlas, Veracruz, Mexico (Lacertilia, Iguanidae). *Bulletin of the Maryland Herpetological Society* 21: 139–144.

Pérez-Ramos, E., and L. Saldaña-de-la Riva. 2008. Morphological revision of lizards of the *formosus* group, genus *Sceloporus* (Squamata: Sauria) of southern México, with description of a new species. *Bulletin of the Maryland Herpetological Society* 44: 77–97.

Peskin, J. 1996. Proyecto de conservación de tortugas marinas. Plaplaya, Depto. Gracias a Dios. Reserva Biósfera del Río Plátano. Informe final—1996. Resultados y comparaciones de los datos de 1995 y 1996. Tegucigalpa, Honduras: Unpublished Report to MOPAWI/Cuerpo de Paz.

Peters, J. A. 1948. The northern limit of the range of *Laemanctus serratus*. *Natural History Miscellanea, The Chicago Academy of Sciences* 27: 1–3.

Peters, J. A. 1952. Catalogue of type specimens in the herpetological collections of the University of Michigan Museum of Zoology. *Occasional Papers of the Museum of Zoology, University of Michigan* 539: 1–55.

Peters, J. A. 1967. The lizards of Ecuador, a check list and key. *Proceedings of the United States National Museum* 119: 1–49.

Peters, J. A., and R. Donoso-Barros. 1970. Catalogue of the Neotropical Squamata part II. Lizards and amphisbaenians. *United States National Museum Bulletin* 297: i–viii, 1–293.

Peters, W. 1862. Über einen neuen *Phyllodactylus* aus Guayaquil. *Monatsberichte der Preussischen Akademie der Wissenschaften zu Berlin* 1862: 626–627.

Peters, W. 1863a. Über einen neuen Gecko, *Brachydactylus mitratus* aus Costa Rica. *Monatsberichte der Preussischen Akademie der Wissenschaften zu Berlin* 1863: 41–44.

Peters, W. 1863b. Derselbe machte eine Mittheilung über einige neue Arten der Saurier-Gattung *Anolis*. *Monatsberichte der Preussischen Akademie der Wissenschaften zu Berlin* 1863: 135–149.

Peters, W. 1864. Über einige neue Säugethiere (*Mormops, Macrotus, Vesperus, Molossus, Capromys*), Amphibien (*Platydactylus, Otocryptis, Euprepes, Ungalia, Dromicus, Tropidonotus, Xenodon, Hylodes*) und Fische (*Sillago, Sebastes,*

Channa, Myctophum, Carassius, Barbus, Ca-poëta, Poecilia, Saurenchelys, Leptocephalus). *Monatsberichte der Preussischen Akademie der Wissenschaften zu Berlin* 1864: 381–399.

Phillips, C. A., W. W. Dimmick, and J. L. Carr. 1996. Conservation genetics of the common snapping turtle (*Chelydra serpentina*). *Conservation Biology* 10: 397–405.

Pianka, E. R., and L. J. Vitt 2003. *Lizards. Windows to the Evolution of Diversity*. Berkeley: University of California Press.

Pincheira-Donoso, D., A. M. Bauer, S. Meiri, and P. E. Uetz. 2013. Global taxonomic diversity of living reptiles. *PLOS ONE* 8: e59741 (pp. 1–10).

Pinto-Sánchez, N. R., M. L. Calderón-Espinosa, A. Miralles, A. J. Crawford, and M. P. Ramírez-Pinilla. 2015. Molecular phylogenetics and biogeography of the Neotropical skink genus *Mabuya* Fitzinger (Squamata: Scincidae) with emphasis on Colombian populations. *Molecular Phylogenetics and Evolution* 93: 188–211.

Platt, S. G., and T. R. Rainwater. 2007. Notes on the consumption of *Bufo marinus* (Anura: Bufonidae) by *Crocodylus moreletii* in northern Belize. *Brenesia* 67: 79–81.

Platt, S. G., and J. B. Thorbjarnarson. 2000. Nesting ecology of the American Crocodile in the coastal zone of Belize. *Copeia* 2000: 869–873.

Platt, S. G., C. Chenot-Rose, V. Rose, and T. R. Rainwater. 2014. Natural history notes. *Crocodilus acutus* (American Crocodile). Frugivory. *Herpetological Review* 45: 120–121.

Platt, S. G., R. M. Elsey, H. Liu, T. R. Rainwater, J. C. Nifong, A. E. Rosenblatt, M. R. Heithaus, and F. J. Mazzotii. 2013. Frugivory and seed dispersal by crocodilians: an overlooked form of saurochory? *Journal of Zoology* 291: 87–99.

Platt, S. G., T. R. Rainwater, and J. B. Thorbjarnarson. 2002. Natural history notes. *Crocodilus acutus* (American Crocodile). Hatchling diet. *Herpetological Review* 33: 202–203.

Platt, S. G., T. R. Rainwater, J. B. Thorbjarnarson, and E. R. Hekkala. 2010. Natural history notes. *Iguana iguana* (Green Iguana). Nesting. *Herpetological Review* 41: 493–494.

Platt, S. G., T. R. Rainwater, J. B. Thorbjarnarson, and D. Martin. 2011. Size estimation, morphometrics, sex ratio, sexual size dimorphism, and biomass of *Crocodylus acutus* in the coastal zone of Belize. *Salamandra* 47: 179–192.

Platt, S. G., J. B. Thorbjarnarson, and T. R. Rainwater. 2012 (dated 2010). Scalation of the American Crocodile, *Crocodylus acutus* (Crocodylidae, Crocodilia), from the coastal zone of northern Belize. *Caribbean Journal of Science* 46: 332–338.

Platt, S. G., J. B. Thorbjarnarson, T. R. Rainwater, and D. R. Martin. 2013. Diet of the American Crocodile (*Crocodylus acutus*) in marine envi-ronments of coastal Belize. *Journal of Herpetology* 47: 1–10.

Poe, S. 2013. 1986 Redux: new genera of anoles (Squamata: Dactyloidae) are unwarranted. *Zootaxa* 3626: 295–299.

Poe, S. 2016. Book review: James R. McCranie and Gunther Köhler; The anoles of Honduras: systematics, distribution, and conservation. *The Quarterly Review of Biology* 91: 227–228.

Pope, C. H. 1935. *The Reptiles of China. Turtles, Crocodilians, Snakes, Lizards*. New York: The American Museum of Natural History.

Porras, L. W., L. D. Wilson, G. W. Schuett, and R. S. Reiserer. 2013. A taxonomic reevaluation and conservation assessment of the common cantil, *Agkistrodon bilineatus* (Squamata: Viperidae): a race against time. *Amphibian & Reptile Conservation* 7: 48–73.

Pough, F. H., R. M. Andrews, J. E. Cadle, M. L. Crump. A. H. Savitzky, and K. D. Wells. 2003 (dated 2004). *Herpetology*. 3rd ed. Upper Saddle River, New Jersey: Pearson Prentice Hall.

Pough, F. H., R. M. Andrews, M. L. Crump. A. H. Savitzky, K. D. Wells, and M. C. Brandley. 2015 (dated 2016). *Herpetology*. 4th ed. Sunderland, Massachusetts: Sinauer Associates, Inc.

Poulin, B., G. Lefebvre, and A. S. Rand. 1995. Natural history notes. *Hemidactylus frenatus* (House Gecko). Foraging. *Herpetological Review* 26: 205.

Powell, R. 1993. Comments on the taxonomic arrangement of some Hispaniolan amphibians and reptiles. *Herpetological Review* 24: 135–137.

Powell, R. 2003. Species profile: Utila's reptiles. *Iguana* 10: 36–38.

Powell, R. 2004. Species profile: Saw-scaled Curlytail (*Leiocephalus carinatus*). *Iguana* 11: 153.

Powell, R., and R. W. Henderson. 2003. Some historical perspectives, PP. 9–20. IN: R. W. Henderson and R. Powell, editors. *Islands and the Sea: Essays on Herpetological Exploration in the West Indies. Society for the Study of Amphibians and Reptiles, Contributions to Herpetology* 20: i–viii, 1–304.

Powell, R., and R. W. Henderson. 2012. Swan Islands, PP. 91–92. IN: R. Powell, and R. W. Henderson, editors. Island lists of West Indian amphibians and reptiles. *Florida Museum of Natural History, Bulletin* 51: 85–166.

Powell, R., and S. A. Maxey. 1990. *Hemidactylus brookii. Catalogue of American Amphibians and Reptiles* 493.1–493.3.

Powell, R., and J. S. Parmerlee, Jr. 1993. *Hemidactylus haitianus* Meerwarth 1901. An endemic West Indian house gecko. *Dactylus* 2: 54–55.

Powell, R., J. T. Collins, and E. D. Hooper, Jr. 1998. *A Key to the Amphibians and Reptiles of the Continental United States and Canada*. Lawrence: University Press of Kansas.

Powell, R., J. T. Collins, and E. D. Hooper, Jr. 2012. *Key to the Herpetofauna of the Continental United States and Canada*. Second Edition, Revised and Updated. Lawrence: University Press of Kansas, Lawrence.

Powell, R., R. I. Crombie, and H. E. A. Boos. 1998. *Hemidactylus mabouia*. *Catalogue of American Amphibians and Reptiles* 674.1–674.11.

Powell, R., R. W. Henderson, K. Adler, and H. A. Dundee. 1996. An annotated checklist of West Indian amphibians and reptiles, PP. 51–94. IN: R. Powell and R. W. Henderson, editors. *Contributions to West Indian Herpetology: A Tribute to Albert Schwartz*. *Society for the Study of Amphibians and Reptiles, Contributions to Herpetology* 12: 1–457.

Powell, R., J. S. Parmerlee, Jr., M. A. Rice, and D. D. Smith. 1990. Ecological observations of *Hemidactylus brookii haitianus* Meerwarth (Sauria: Gekkonidae) from Hispaniola. *Caribbean Journal of Science* 26: 67–70.

Pregill, G. K. 1992. Systematics of the West Indian lizard genus *Leiocephalus* (Squamata: Iguania: Tropiduridae). *University of Kansas Museum of Natural History, Miscellaneous Publication* 84: 1–69.

Presch, W. 1978. Descriptions of the hemipenial morphology in eight species of microteiid lizards (Family Teiidae, Subfamily Gymnophthalminae). *Herpetologica* 34: 108–112.

Pritchard, P. C. H. 1967. *Living Turtles of the World*. Neptune, New Jersey: TFH Publications, Inc.

Pritchard, P. C. H. 1979. *Encyclopedia of Turtles*. Neptune, New Jersey: T.F.H. Publications, Inc. Ltd.

Pritchard, P. C. H. 1980. [*Dermochelys*] *Dermochelys coriacea*. *Catalogue of American Amphibians and Reptiles* 238.1–238.4.

Pritchard, P. C. H. 1993. A ranching project for freshwater turtles in Costa Rica. *Chelonian Conservation and Biology* 1: 48.

Pritchard, P. C. H. 1997. Evolution, phylogeny, and current status, PP. 1–28. IN: P. L. Lutz and J. A. Musick, editors. *The Biology of Sea Turtles*. Boca Raton, Florida: CRC Press.

Pritchard, P. C. H. 1999. Status of the Black Turtle. *Conservation Biology* 13: 1000–1003.

Pritchard, P. C. H., and P. Trebbau. 1984. The turtles of Venezuela. *Society for the Study of Amphibians and Reptiles, Contributions to Herpetology* 2: i–viii, 1–403 pp., plates 1–47, Maps 1–16.

Pyron, R. A., and V. Wallach. 2014. Systematics of the blindsnakes (Serpentes: Scolecophidia: Typhlopoidea) based on molecular and morphological evidence. *Zootaxa* 3829: 1–81.

Pyron, R. A., F. T. Burbrink, G. R. Colli, A. Nieto Montes de Oca, L. J. Vitt, C. A. Kuczynski, and J. J. Wiens. 2011. The phylogeny of advanced snakes (Colubroidea), with discovery of a new subfamily and comparison of support methods for likelihood trees. *Molecular Phylogenetics and Evolution* 58: 329–342.

Pyron, R. A., F. T. Burbrink, and J. J. Wiens. 2013. A phylogeny and revised classification of Squamata, including 4161 species of lizards and snakes. *BMC Evolutionary Biology* 13: 93 (48 pp., 28 figures).

Pyron, R. A., R. G. Reynolds, and F. T. Burbrink. 2014. A taxonomic revision of boas (Serpentes: Boidae). *Zootaxa* 3846: 249–260.

Rabb, G. B. 1957. A new race of the iguanid lizard *Leiocephalus carinatus* from Cayman Brac, B. W. I. *Herpetologica* 13: 109–110.

Raddi, G. 1822. Continuazione della descrizione dei rettili brasiliani. Indicati nella Memoria inserita nel secondo Fascicolo delle Memorie di Fisica del precedente. XVIII. *Societá Italiana delle Scienze* 18: 56–73.

Rafinesque, C. S. 1814. Prodrono di Erpetologia Siciliana. *Specchio delle Scienze o Giornale Enciclopedico di Sicilia* 2: 65–67, 102–104.

Rafinesque, C. S. 1815. *Analyse de la Nature ou Tableau de l'Univers et des Corps Organisés*. Palerme: Privately Printed by Author.

Rainwater, T. R., L. D. Barrantes, J. R. Bolaños Montero, S. G. Platt, and B. R. Barr. 2010. Natural history notes. *Crocodylus acutus* (American Crocodile). Adult mass. *Herpetological Review* 41: 489–490.

Ramírez-Bautista, A. 1994. Manual y Claves Illustradas de los Anfibios y Reptiles de la Región de Chamela, Jalisco, México. *Cuadernos Instituto de Biología, Universidad Nacional Autónoma de México, México* 23: 1–127, 1–16 plates.

Ramos Galdamez, J., L. G. Zuñiga, and M. R. Espinal. 2016. Other Contributions. A new locality for *Corytophanes hernandesii* (Wiegmann, 1831), (Squamata: Corytophanidae) in western Honduras, with comments on its distribution. *Mesoamerican Herpetology* 3: 1041–1044.

Rand, A. S. 1954. Variation and predator pressure in an island and a mainland population of lizards. *Copeia* 1954: 260–262.

Rand, A. S., and H. W. Greene. 1982. Latitude and climate in the phenology of reproduction in the Green Iguana, *Iguana iguana*, PP. 142–149. IN: G. M. Burghardt and A. S. Rand, editors. *Iguanas of the World. Their Behavior, Ecology, and Conservation*. Park Ridge, New Jersey: Noyes Publications.

Rebel, T. P., editor. 1974. *Sea Turtles and the Turtle Industry of the West Indies, Florida, and the Gulf of Mexico*. Revised Edition. Coral Gables, Florida: University of Miami Press.

Reed, R. N., S. Green, S. M. Boback, and C. E. Montgomery. 2006. Natural history notes. *Cteno-*

saura melanosterna (Black-chested Ctenosaur). Predation. *Herpetological Review* 37: 84.

Reeder, T. W. 1990. *Eumeces managuae*. *Catalogue of American Amphibians and Reptiles* 467.1–467.2.

Reeder, T. W., and J. J. Wiens. 1996. Evolution of the lizard family Phrynosomatidae as inferred from diverse types of data. *Herpetological Monographs* 10: 43–84.

Reeder, T. W., C. J. Cole, and H. C. Dessauer. 2002. Phylogenetic relationships of whiptail lizards of the genus *Cnemidophorus* (Squamata: Teiidae): a test of monophyly, reevaluation of karyotypic evolution, and review of hybrid origins. *American Museum Novitates* 3365: 1–61.

Reeder, T. W., T. M. Townsend, D. G. Mulcahy, B. P. Noonan, P. L. Wood, Jr., J. W. Sites, Jr., and J. J. Wiens. 2015. Integrated analyses resolve conflicts over squamate reptile phylogeny and reveal unexpected placements for fossil taxa. *PLOS ONE* 10: 13.71/journal.pone.0118199 (pp. 1–22).

Regan, C. T. 1916. Reptilia and Batrachia. *Zoological Record* 51: 1–20.

Reynolds, R. G., M. L. Niemiller, and L. J. Revell. 2014. Toward a tree-of-life for the boas and pythons: multilocus species-level phylogeny with unprecedented taxon sampling. *Molecular Phylogenetics and Evolution* 71: 201–213.

Reynolds, R. P., S. W. Gotte, and C. H. Ernst. 2007. Catalog of type specimens of recent Crocodilia and Testudines in the National Museum of Natural History, Smithsonian Institution. *Smithsonian Contributions to Zoology* 626: 1–49.

Rhodin, A. G. J., and J. L. Carr. 2009. A quarter millenium [sic] of uses and misuses of the turtle name *Testudo scabra*: identification of the type specimens of *T. scabra* Linnaeus 1758 (= *Rhinoclemmys punctularia*) and *T. scripta* Thunberg, *In* Schoepff 1792 (= *Trachemys scripta scripta*). *Zootaxa* 2226: 1–18.

Rhodin, A. G. J., P. P. van Dijk, and J. F. Parham. 2008. Turtles of the world: annotated checklist of taxonomy and synonymy, PP. 000.1–000.38. IN: A. G. J. Rhodin, P. C. H. Pritchard, P. P. van Dijk, R. A. Saumure, K. A. Buhlmann, and J. B. Iverson, editors. *Conservation Biology of Freshwater Turtles and Tortoises: A Compilation Project of the IUCN/SSC Tortoise and Freshwater Turtle Specialist Group*. Lunenburg, Massachusetts: Chelonian Research Foundation. *Chelonian Research Monographs* 5.

Riosmena-Rodriguez, R., and M. Lara-Uc. 2015. Natural history notes. *Caretta caretta* (Loggerhead Sea Turtle). Diet. *Herpetological Review* 46: 616–617.

Rittmann, D. 2007. Langjährige Haltung, Pflege und Nachzucht des Roatan-Schwarzleguans *Ctenosaura oedirhina* de Queiroz, 1987 sowie Beo-

bachtungen in dessen Lebensraum. *Elaphe* 15: 33–40.

Rivas Fuenmayor, G., G. N. Ugueto, A. M. Bauer, T. Barros, and J. Manzanilla. 2005. Expansion and natural history of a successful colonizing gecko in Venezuela (Reptilia: Gekkonidae: *Hemidactylus mabouia*) and the discovery of *H. frenatus* in Venezuela. *Herpetological Review* 36: 121–125.

Rivero, J. A. 1998. *Los Anfibios y Reptiles de Puerto Rico = The Amphibians and Reptiles of Puerto Rico*. Segunda Edición Revisada. San Juan: Editorial de la Universidad de Puerto Rico.

Rivero-Blanco, C. V. 1979. The Neotropical Lizard Genus *Gonatodes* Fitzinger (Sauria: Sphaerodactylinae) [dissertation]. College Station: Texas A&M University.

Rivero-Blanco, C. V., and W. E. Schargel. 2012. A strikingly polychromatic new species of *Gonatodes* (Squamata: Sphaerodactylidae) from northern Venezuela. *Zootaxa* 3518: 66–78.

Roberts, W. E. 1997. Behavioral observations of *Polychrus gutturosus*, a sister taxon of anoles. *Herpetological Review* 28: 184–185.

Rödder, D., M. Solé, and W. Böhme. 2008. Predicting the potential distributions of two alien invasive house geckos (Gekkonidae: *Hemidactylus frenatus*, *Hemidactylus mabouia*). *North-Western Journal of Zoology* 4: 236–246.

Rodrigues, J. F. M., and J. A. F. Diniz-Filho. 2016. Ecological opportunities, habitat, and past climatic fluctuations influenced the diversification of modern turtles. *Molecular Phylogenetics and Evolution* 101: 352–358.

Rogner, M. 2008. *Schildkröten. Biologie, Haltung, Vermehrung*. Stuttgart, Germany: Eugen Ulmer KG.

Rösler, H. 1998. Bemerkungen zur Fortpflanzung von *Gonatodes albogularis fuscus* (Hallowell, 1855) aus Costa Rica (Sauria: Gekkonidae). *Gekkota* 1: 176–183.

Rösler, H. 2000. Kommentierte Liste der rezent, subrezent und fossil bekannten Gecko-Taxa (Reptilia: Gekkonomorpha). *Gekkota* 2: 28–153.

Ross, C. A., and F. D. Ross. 1974. Caudal scalation of Central American *Crocodylus*. *Proceedings of the Biological Society of Washington* 87: 231–234.

Ross, F. D., and G. C. Mayer. 1983. On the dorsal armor of the Crocodilia, PP. 305–331. IN: A. G. J. Rhodin and K. Miyata, editors. *Advances in Herpetology and Evolutionary Biology. Essays in Honor of Ernest E. Williams*. Cambridge, Massachusetts: Museum of Comparative Zoology.

Ross, J. P., editor. 1998a. *Crocodiles: Status Survey and Conservation Action Plan*. 2nd ed. Gland, Switzerland: IUCN/Species Survival Commission, Crocodile Specialist Group.

Ross, J. P. 1998b. Report on crocodilian trade from Latin America, PP. 243–253. IN: *Proceedings of* the 14th Working Meeting of the Crocodile Specialist Group of the Species Survival Commission of IUCN–The World Conservation Union convened at Singapore, 13–17 July 1998. Gland, Switzerland: IUCN. i–x, 1–410 pp.

Rossman, D. A., and D. A. Good. 1993. Herpetological type specimens in the Museum of Natural Science, Louisiana State University. *Occasional Papers of the Museum of Natural Science, Louisiana State University* 66: 1–18.

Ruane, S., R. W. Bryson, Jr., R. A. Pyron, and F. T. Burbrink. 2014. Coalescent species delimitation in milksnakes (Genus *Lampropeltis*) and impacts on phylogenetic comparative analyses. *Systematic Biology* 63: 231–250.

Ruckdeschel, C., and C. R. Shoop. 2006. *Sea Turtles of the Atlantic and Gulf Coasts of the United States*. Athens: University of Georgia Press.

Rüppell, E. 1835. *Neue Wirbelthiere zu der Fauna von Abyssinien gehörig. Amphibien*. Frankfurt am Main, Germany: Siegmund Schmerber.

Russell, A. P., and A. M. Bauer. 2002. *Thecadactylus, T. rapicauda. Catalogue of American Amphibians and Reptiles* 753.1–753.16.

Ruthven, A. G. 1915. Description of a new subspecies of *Cnemidophorus lemniscatus* Laurenti. *Occasional Papers of the Museum of Zoology, University of Michigan* 16: 1–4, 1 plate.

Sabaj, M. H. 2016. Standard Symbolic Codes for Institutional Resource Collections in Herpetology and Ichthyology [Internet]. Version 6.5. Washington: American Society of Ichthyologists and Herpetologists; [cited 2016 Oct 29]. Available at: http://www.asih.org/.

Sackett, J. T. 1941. Preliminary report on results of the West Indies–Guatemala Expedition of 1940 for the Academy of Natural Sciences of Philadelphia. Part II.—A new teeid [sic] lizard of the genus *Cnemidophorus*. *Notulae Naturae* 77: 1–4.

Sagra, R. de la. 1840. *Histoire Physique, Politique et Naturelle de L'ile de Cuba*. Tome VIII. *Atlas. Reptiles*. Paris: Arthus Bertrand.

Sajdak, R. A., M. A. Nickerson, R. W. Henderson, and M. W. Moffett. 1980. Notes on the movements of *Basiliscus plumifrons* (Sauria: Iguanidae) in Costa Rica. *Contributions in Biology and Geology, Milwaukee Public Museum* 36: 1–8.

Sánchez, A., I. Oviedo, P. R. House, and D. Vreugdenhil. 2002. *Racionalización del Systema Nacional de Areas Protegidas de Honduras. Vol. V: Estado Legal de las Areas Protegidas de Honduras, Actualizacion. 2002*. Tegucigalpa, Honduras: World Institute for Conservation and Environment.

Sanders, K. L., M. S. Y. Lee, Mumpuni, T. Bertozzi, and A. R. Rasmussen. 2013. Multilocus phylogeny and recent rapid radiation of the viviparous sea snakes (Elapidae: Hydrophiinae). *Molecular Phylogenetics and Evolution* 66: 575–591.

Sasa, M., and J. Salvador Monrós. 2000. Dietary analysis of helmeted basilisks, *Corytophanes* (Reptilia: Corytophanidae). *The Southwestern Naturalist* 45: 358–361.

Sasa, M., and A. Solórzano. 1995. The reptiles and amphibians of Santa Rosa National Park, Costa Rica, with comments about the herpetofauna of xerophytic areas. *Herpetological Natural History* 3: 113–126.

Savage, J. M. 1963. Studies on the lizard family Xantusiidae IV. The genera. *Los Angeles County Museum, Contributions in Science* 71: 1–38.

Savage, J. M. 2002. *The Amphibians and Reptiles of Costa Rica. A Herpetofauna between Two Continents, between Two Seas*. Chicago, Illinois: The University of Chicago Press.

Savage, J. M. 2011. The correct species-group name for an *Oxyrhopus* (Squamata: Dipsadidae) variously called *Coluber petalarius, C. pethola, C. petola*, or *C. petolarius* by early authors. *Proceedings of the Biological Society of Washington* 124: 223–225.

Savage, J. M. 2017. Crocodilian confusion: the order-group names Crocodyli, Crocodilia, Crocodylia, and the authorship of the family-group name Crocodilidae or Crocodylidae. *Herpetological Review* 48: 110–114.

Savage, J. M., and K. R. Lips. 1994 (dated 1993). A review of the status and biogeography of the lizard genera *Celestus* and *Diploglossus* (Squamata: Anguidae), with description of two new species from Costa Rica. *Revista de Biología Tropical* 41: 817–842.

Savage, J. M., K. R. Lips, and R. Ibáñez D. 2008. A new species of *Celestus* from west-central Panama, with consideration of the status of the genera of the Anguidae: Diploglossinae (Squamata). *Revista de Biología Tropical* 56: 845–859.

Schlegel, H. 1827. Erpetologische Nachrichten. *Isis* 20: cols. 281–294.

Schmid, K. 1819. *Naturhistorische Beschreibung der Amphibien. Systematisch bearbeitet zum gemeinnützigen Gebrauche*. Munich: Kunst–Anstalt Feyertags–Schule.

Schmidt, A. A. 2011. Die "Mutter aller Schlangen" *Diploglossus monotropis*—eine prachtvolle skinkähnliche Schleiche aus Mittelamerika. *Elaphe* 2011: 36–40.

Schmidt, K. P. 1924. Notes on Central American crocodiles. *Zoological Series of Field Museum of Natural History* 12: 79–92, plates V–IX.

Schmidt, K. P. 1928a. Reptiles collected in Salvador for the California Institute of Technology. *Zoological Series of Field Museum of Natural History* 12: 193–201.

Schmidt, K. P. 1928b. Notes on South American caimans. *Zoological Series of Field Museum of Natural History* 12: 205–231, plates XVI–XXI.

Schmidt, K. P. 1928c. Amphibians and land reptiles of Porto [sic] Rico, with a list of those reported from the Virgin Islands. *New York Academy of Sciences, Scientific Survey of Puerto Rico and the Virgin Islands* 10: 1–160, plates 1–4.

Schmidt, K. P. 1933. New reptiles and amphibians from Honduras. *Zoological Series of Field Museum of Natural History* 20: 15–22.

Schmidt, K. P. 1936. New amphibians and reptiles from Honduras in the Museum of Comparative Zoology. *Proceedings of the Biological Society of Washington* 49: 43–50.

Schmidt, K. P. 1941. The amphibians and reptiles of British Honduras. *Zoological Series of Field Museum of Natural History* 22: 475–510.

Schmidt, K. P. 1942. A cloud forest camp in Honduras. *The Chicago Naturalist* 5: 23–30.

Schmidt, K. P. 1944. Crocodiles. *Fauna* 6: 67–73.

Schmidt, K. P. 1946. Turtles collected by the Smithsonian Biological Survey of the Panamá Canal Zone. *Smithsonian Miscellaneous Collections* 106: 1–9, plate 1.

Schmidt, K. P. 1952. Crocodile-hunting in Central America. *Chicago Natural History Museum, Popular Series* 15: 1–23.

Schmidt, K. P. 1953. *A Check List of North American Amphibians and Reptiles.* Sixth Edition. Chicago: American Society of Ichthyologists and Herpetologists.

Schmidt, K. P., and R. F. Inger. 1951. Amphibians and reptiles of the Hopkins-Branner Expedition to Brazil. *Fieldiana: Zoology* 31: 439–465.

Schmidt, W., and F.-W. Henkel. 1995. *Leguane. Biologie, Haltung und Zucht.* Stuttgart: Verlag Eugene Ulmer.

Schmitz, A., P. Mausfield, and D. Embert. 2004. Molecular studies on the genus *Eumeces* Wiegmann, 1834: phylogenetic relationships and taxonomic implications. *Hamadryad* 28: 73–89.

Schneider, J. G. 1801. *Historiae Amphibiorum naturalis et literariae Fasciculus Primus; continens Seaindus, Crocodilos, Scincos, Chamaesauras, Boas, Pseudoboas, Elapes, Angues, Amphisbaenas et Caecilias descriptos notisque suis distinctos.* Ienae [Jena], Germay: Friederici Frommanni.

Schoepff, J. D. 1792–1801. *Historia Testudinum Iconibus. Illustrata.* Fasciculus I. et II–VI. Praefatione. Erlangae [Erlangen], Germany: Joannis Jacobi Palm.

Schulte, U. 2007a. Beobachtungen zur Hybridisierung zwischen *Ctenosaura similis* (Gray, 1831) und *Ctenosaura bakeri* Stejneger, 1901 auf Utila, Honduras. *Elaphe* 15: 55–59.

Schulte, U. 2007b. Zur arborikolen Lebensweise des Utila-Leguans (*Ctenosaura bakeri* Stejneger, 1901) während der Trocken-und Regenzeit auf der honduranischen Karibikinsel Utila. *Iguana Rundschreiben* 20: 6–12.

Schulte, U., and G. Köhler. 2010. Microhabitat selection in the spiny-tailed iguana *Ctenosaura bakeri* on Utila Island, Honduras. *Salamandra* 46: 141–146.

Schwartz, A. 1966 (dated 1965). Geographic variation in *Sphaerodactylus notatus* Baird. *Revista de Biologia Tropical* 13: 161–185.

Schwartz, A. 1970. *Sphaerodactylus notatus. Catalogue of American Amphibians and Reptiles* 90.1–90.2.

Schwartz, A. 1973. *Sphaerodactylus. Catalogue of American Amphibians and Reptiles* 142.1–142.2.

Schwartz, A. 1975. New subspecies of *Sphaerodactylus copei* Steindachner (Sauria, Gekkonidae) from Hispaniola. *Herpetologica* 31: 1–18.

Schwartz, A., and R. I. Crombie. 1975. A new species of the genus *Aristelliger* (Sauria: Gekkonidae) from the Caicos Islands. *Proceedings of the Biological Society of Washington* 88: 305–313.

Schwartz, A., and O. H. Garrido. 1981. Las salamanquitas cubanas del género *Sphaerodactylus* (Sauria: Gekkonidae) 1. El grupo *copei. Poeyana* 230: 1–27.

Schwartz, A., and R. W. Henderson. 1988. West Indian amphibians and reptiles: a check-list. *Milwaukee Public Museum, Contributions in Biology and Geology* 74: 1–264.

Schwartz, A., and R. W. Henderson. 1991. *Amphibians and Reptiles of the West Indies: Descriptions, Distributions, and Natural History.* Gainesville: University of Florida Press.

Schwartz, A., and R. Thomas. 1975. A check-list of West Indian amphibians and reptiles. *Carnegie Museum of Natural History Special Publication* 1: 1–216.

Schwartz, A., R. Thomas, and L. D. Ober. 1978. First supplement to a check-list of West Indian amphibians and reptiles. *Carnegie Museum of Natural History Special Publication* 5: 1–35.

Schweigger, A. F. 1812. Prodromus monographiae Cheloniorum. *Königsberger Archiv für Naturwissenschaften und Mathematik* 1: 271–368, 406–462.

Schwenk, K., S. K. Sessions, and D. M. Peccinini Seale. 1982. Karyotypes of the basiliscine lizards *Corytophanes cristatus* and *Corytophanes hernandesii*, with comments on the relationship between chromosomal and morphological evolution in lizards. *Herpetologica* 38: 493–501.

Seba, A. 1734. *Locupletissimi Rerum Naturalium Thesauri Accurata Descriptio, et Iconibus Artificiosissimis Expressio, per Universam Physices Historiam. Opus, cui, in Hoc Rerum Genere, Nullum par Exstitit. Ex Toto Terrarum orbe collegit, Digessit, Descripsit, et Depingendum*

Curavit. Amsterdam: Janssonio-Waesbergios & J. Wetstenium & J. Smith.

Seidel, M. E. 1988. Revision of the West Indian emydid turtles (Testudines). *American Museum Novitates* 2918: 1–41.

Seidel, M. E. 2002. Taxonomic observations on extant species and subspecies of slider turtles, genus *Trachemys*. *Journal of Herpetology* 36: 285–292.

Seidel, M. E., and C. H. Ernst. 2006. *Trachemys scripta*. *Catalogue of American Amphibians and Reptiles* 831.1–831.94.

Seidel, M. E., and C. H. Ernst. 2012. *Trachemys*. *Catalogue of American Amphibians and Reptiles* 891.1–891.17.

Seidel, M. E., and C. H. Ernst. 2017. A systematic review of the turtle family Emydidae. *Vertebrate Zoology* 67: 1–122.

Seidel, M. E., and R. Franz. 1994. Amphibians and reptiles (exclusive of marine turtles) of the Cayman Islands, PP. 407–433. IN: M. A. Brunt and J. E. Davies, editors. *The Cayman Islands: Natural History and Biogeography*. Dordrecht, The Netherlands: Kluwer Academic Publishers.

Seminoff, J. A., and B. P. Wallace, editors. 2012. *Sea Turtles of the Eastern Pacific. Advances in Research and Conservation*. Tucson: The University of Arizona Press.

Seminoff, J. A., J. Alfaro-Shigueto, D. Amorocho, R. Arauz, A. B. Gallegos, D. Chacón Chaverri, A. R. Gaos, S. Kelez, J. C. Mangel, J. Urteaga, and B. P. Wallace. 2012. Biology and conservation of sea turtles in the eastern Pacific Ocean, PP. 11–38. IN: J. A. Seminoff and B. P. Wallace, editors. *Sea Turtles of the Eastern Pacific. Advances in Research and Conservation*. Tucson: The University of Arizona Press.

Sexton, O. J., and O. Turner. 1971. The reproductive cycle of a neotropical lizard. *Ecology* 52: 159–164.

Shaffer, H. B., D. E. Starkey, and M. K. Fujita. 2008. Molecular insights into the systematics of the snapping turtles (Chelydridae), PP. 44–49. IN: A. C. Steyermark, M. S. Finkler, and R. J. Brooks, editors. *Biology of the Snapping Turtle (Chelydra serpentina)*. Baltimore, Maryland: The Johns Hopkins University Press.

Shaw, G. 1802. *General Zoology, or Systematic Natural History*. Vol. III. Part I. *Amphibia*. London: Thomas Davison.

Shi, H-T., editor. 2013. *Identification Manuel for the Conservation of Turtles in China*. Beijing: Encyclopedia of China Publishing House.

Siebenrock, F. 1907. Die Schildkrötenfamilie Cinosternidae m. Monographisch bearbeitet. *Sitzungsberichte der Akademie der Wissenschaften mathematisch-naturwissenschaftliche Klasse* 116: 527–599, 2 foldout maps, 1 plate.

Sinclair, E. A., J. B. Pramuk, R. L. Bezy, K. A. Crandall, and J. W. Sites, Jr. 2009. DNA evidence for nonhybrid origins of parthenogenesis in natural populations of vertebrates. *Evolution* 64: 1346–1357.

Sites, J. W., Jr., and J. R. Dixon. 1982. Geographic variation in *Sceloporus variabilis*, and its relationship to *S. teapensis* (Sauria: Iguanidae). *Copeia* 1982: 14–27.

Sites, J. W., Jr., J. W. Archie, C. J. Cole, and O. Flores-Villela. 1992. A review of phylogenetic hypotheses for lizards of the genus *Sceloporus* (Phrynosomatidae): implications for ecological and evolutionary studies. *Bulletin of the American Museum of Natural History* 213: 1–110.

Sites, J. W., Jr., S. K. Davis, T. Guerra, J. B. Iverson, and H. L. Snell. 1996. Character congruence and phylogenetic signal in molecular and morphological data sets: a case study in the living iguanas (Squamata: Iguanidae). *Molecular Biology and Evolution* 13: 1087–1105.

Sites, J. W., Jr., I. F. Greenbaum, and J. W. Bickham. 1981. Biochemical systematics of neotropical turtles of the genus *Rhinoclemmys* (Emydidae: Batagurinae). *Herpetologica* 37: 256–264.

Sletto, B. 1999. Getting it on the map. *Wildlife Conservation* 102: 24–29.

Smith, C. A., and K. L. Krysko. 2007. Distributional comments on the teiid lizards (Squamata: Teiidae) of Florida with a key to species. *Caribbean Journal of Science* 43: 260–265.

Smith, E. N. 2001. Species Boundaries and Evolutionary Patterns of Speciation among the Malachite Lizards (formosus group) of the genus Sceloporus (Squamata: Phrynosomatidae) [dissertation]. Arlington: The University of Texas.

Smith, H. M. 1936. Descriptions of new species of lizards of the genus *Sceloporus* from Mexico. *Proceedings of the Biological Society of Washington* 49: 87–96.

Smith, H. M. 1937. A synopsis of the *variabilis* group of the lizard genus *Sceloporus*, with descriptions of new subspecies. *Occasional Papers of the Museum of Zoology, University of Michigan* 358: 1–14.

Smith, H. M. 1939. The Mexican and Central American lizards of the genus *Sceloporus*. *Zoological Series of Field Museum of Natural History* 26: 1–397, plates 1–31.

Smith, H. M. 1941a. A new genus of Mexican snakes related to *Rhadinaea*. *Copeia* 1941: 7–10.

Smith, H. M. 1941b. A new name for the Mexican snakes of the genus *Dendrophidion*. *Proceedings of the Biological Society of Washington* 54: 73–76.

Smith, H. M. 1942. Mexican herpetological miscellany. *Proceedings of the United States National Museum* 92: 349–395.

Smith, H. M. 1971. The status of Wilhelm Deppe's herpetological names. *Journal of Herpetology* 5: 74–76.

Smith, H. M. 1987. Current nomenclature for the names and material cited in Günther's Reptilia and Batrachia volume of the *Biologia Centrali-Americana*, PP. xxiii–li. IN: A. C. L. G. Günther. Biologia Centrali-Americana. Reptilia and Batrachia. *Society for the Study of Amphibians and Reptiles, Facsimile Reprints Herpetology* (Facsimile number not given).

Smith, H. M. 2005. *Plestiodon*: a replacement name for most members of the genus *Eumeces* in North America. *Journal of Kansas Herpetology* 14: 15–16.

Smith, H. M., and M. Álvarez del Toro. 1962. Notulae herpetologicae Chiapasiae III. *Herpetologica* 18: 101–107.

Smith, H. M., and R. B. Smith. 1976. *Synopsis of the Herpetofauna of Mexico. Vol. III. Source Analysis and Index for Mexican Reptiles.* North Bennington, Vermont: John Johnson.

Smith, H. M., and R. B. Smith. 1977. *Synopsis of the Herpetofauna of Mexico. Vol. V. Guide to Mexican Amphisbaenians and Crocodilians. Bibliographic Addendum II.* North Bennington, Vermont: John Johnson.

Smith, H. M., and R. B. Smith. 1980 (dated 1979). *Synopsis of the Herpetofauna of Mexico. Vol. VI. Guide to Mexican Turtles. Bibliographic Addendum III.* North Bennington, Vermont: John Johnson.

Smith, H. M., and E. H. Taylor. 1950a. Type localities of Mexican reptiles and amphibians. *The University of Kansas Science Bulletin* 33: 313–380.

Smith, H. M., and E. H. Taylor. 1950b. An annotated checklist and key to the reptiles of Mexico exclusive of the snakes. *United States National Museum Bulletin* 199: i–v, 1–253.

Smith, H. M., and E. H. Taylor. 1966. Annotated checklists and keys, PP. 1–29. IN: *Herpetology of Mexico. Annotated Checklists and Keys to the Amphibians and Reptiles. A Reprint of Bulletins 187, 194 and 199 of the U. S. National Museum with a List of Subsequent Taxonomic Innovations.* Ashton, Maryland: Eric Lundberg.

Smith, H. M., and P. V. Terentjev. 1963. *Sphaerodactylus argus continentalis* Werner, 1896 (Reptilia: Lacertilia): request for a ruling on interpretation. *The Bulletin of Zoological Nomenclature* 20: 367–369.

Smith, H. M., D. Chiszar, and R. Humphrey. 2001. The distribution of *Sceloporus acanthinus* (Reptilia: Sauria) and its relationships. *Bulletin of the Maryland Herpetological Society* 37: 3–9.

Smith, H. M., D. A. Langebartel, and K. L. Williams. 1964. Herpetological type-specimens in the University of Illinois Museum of Natural History. *Illinois Biological Monographs* 32: 1–80.

Smith, H. M., G. Pérez-Higareda, and D. Chiszar. 1993. A review of the members of the *Sceloporus variabilis* lizard complex. *Bulletin of the Maryland Herpetological Society* 29: 85–125.

Smith, M. A. 1933a. Amphibia and Reptilia. *Zoological Record* 69: 1–39.

Smith, M. A. 1933b. Amphibia and Reptilia. *Zoological Record* 70: 1–36.

Smith, M. A. 1935. *The Fauna of British India, including Ceylon and Burma. Reptilia and Amphibia.* Vol. II.—*Sauria.* London: Taylor & Francis Ltd.

Smith, M. A. 1937. Amphibia and Reptilia. *Zoological Record* 73: 1–47.

Smith, M. A. 1941. Amphibia and Reptilia. *Zoological Record* 77: 1–39.

Smith, P. W. 1950. *Thecadactylus rapicaudus* in Honduras. *Herpetologica* 6: 55.

Smith, P. W., and W. L. Burger. 1950. Herpetological results of the University of Illinois Field Expedition, Spring 1949. III. Sauria. *Transactions of the Kansas Academy of Science* 53: 165–175.

Smith, R. E. 1968. Studies on reproduction in Costa Rican *Ameiva festiva* and *Ameiva quadrilineata* (Sauria: Teiidae). *Copeia* 1968: 236–239.

Smithe, F. B. 1975–1981. *Naturalist's Color Guide. Part I. Color Guide.* New York: American Museum of Natural History.

Solís, J. M., R. M. Valle, L. A. Herrera, C. M. O'Reilly, and R. Downing. 2015. Range extensions and new departmental records for amphibians and reptiles in Honduras. *Mesoamerican Herpetology* 2: 557–561.

Solís, J. M., L. D. Wilson, and J. H. Townsend. 2014. An updated list of the amphibians and reptiles of Honduras, with comments on their nomenclature. *Mesoamerican Herpetology* 1: 123–144.

Sonnini, C. S., and P. A. Latreille. 1801. *Histoire Naturelle des Reptiles, avec figures dessinées d'après nature.* Tome IV. Seconde Partie. Serpens. Paris: Deterville.

Sowerby, J. de and E. Lear. 1872. *Tortoises, Terrapins, and Turtles. Drawn from Life* (text by J. E. Gray). London: Henry Sotheran, Joseph Baer & Co.

Sparman, A. 1784. *LACERTA Sputator* och *LACERTA bimaculata*, två nya Ödlor från America; befkrifne. *Königlchen Vetenskaps Academiens, Handlingar* 5: 164–167.

Spinks, P. Q., R. C. Thomson, M. Gidis, and H. B. Shaffer. 2014. Multilocus phylogeny of the New-World mud turtles (Kinosternidae) supports the traditional classification of the group. *Molecular Phylogenetics and Evolution* 76: 254–260.

Spix, J. B. de. 1824. *Animalia Nova sive species novae Testudinum et Ranarum, quas in itinere per Brasiliam annis MDCCCXVII–MDCCCXX jussu et auspiciis Maximiliani Josephi I. Bavariae Regis.* Monachii [Munich]: Typis Franc. Seraph. Hübschmanni.

Spix, J. B. de. 1825. *Animalia Nova sive species novae Lacertarum, quas in itinere per Brasiliam annis MDCCCXVII — MDCCCXX jussu et auspiciis Maximiliani Josephi I. Bavariae Regis.* Monachii [Munich]: Typis Franc. Seraph. Hübschmanni.

Spotila, J. R. 2004. *Sea Turtles. A Complete Guide to their Biology, Behavior, and Conservation.* Baltimore, Maryland: The Johns Hopkins University Press.

Spotila, J. R. 2011. *Saving Sea Turtles. Extraordinary Stories from the Battle against Extinction.* Baltimore, Maryland: The Johns Hopkins University Press.

Sprackland, R. G. 1999. Species, species everywhere: where do new species come from? *The Vivarium* 10: 7–8, 36–38.

Stafford, P. J. 1991. Amphibians and reptiles of the Joint Services Scientific Expedition to the Upper Raspaculo, Belize, 1991. *The British Herpetological Society Bulletin* 38: 10–17.

Stafford, P. J. 1994. Amphibians and reptiles of the Upper Raspaculo River Basin, Maya Mountains, Belize. *The British Herpetological Society Bulletin* 47: 23–29.

Stafford, P. J., and E. P. Mallory. 2002. Egg-laying habitats of the Middle American arboreal lizard *Laemanctus longipes*, with particular reference to nest site selection. *The Herpetological Bulletin* 79: 30–32.

Stafford, P. J., and J. R. Meyer. 1999 (dated 2000). *A Guide to the Reptiles of Belize.* San Diego, California: Academic Press.

Steindachner, F. 1867. *Reise der Österreichischen Fregatte Novara um die Erde in den Jahren 1857, 1858, 1859 unter den Befehlendes Commodore B. von Wüllerstorf-Urbair. Reptilien. Zoologischer Theil.* Band I. Vienna: K. Gerold'Sohn.

Stejneger, L. 1899. Description of a new species of spiny-tailed iguana from Guatemala. *Proceedings of the United States National Museum* 21: 381–383.

Stejneger, L. 1901. On a new species of spiny-tailed iguana from Utilla [sic] Island, Honduras. *Proceedings of the United States National Museum* 23: 467–468.

Stejneger, L. 1904. The herpetology of Puerto Rico. *Annual Report of the National Museum 1902*, Part 2, No. 2. 1902: 549–724, plate 1.

Stejneger, L. 1907. Herpetology of Japan and adjacent territory. *Smithsonian Institution, United States National Museum Bulletin* 58: i–xx, 1–577, plates I–XXXV.

Stejneger, L. 1917. Cuban amphibians and reptiles collected for the United States National Museum from 1899 to 1902. *Proceedings of the United States National Museum* 53: 259–291.

Stejneger, L. 1925. New species and subspecies of American turtles. *Journal of the Washington Academy of Sciences* 15: 462–463.

Stejneger, L. 1936. Types of the amphibian and reptilian genera proposed by Laurenti in 1768. *Copeia* 1936: 133–141.

Stejneger, L. 1941. Notes on Mexican turtles of the genus *Kinosternon*. *Proceedings of the United States National Museum* 90: 457–459.

Stephen, C., S. Pasachnik, A. Reuter, P. Mosig, L. Ruyle, and L. Fitzgerald. 2011. *Survey of Status, Trade, and Exploitation of Central American Iguanas.* Washington: Traffic.

Sterling, E. J., K. W. McFadden, K. E. Holmes, E. C. Vintinner, F. Arengo, and E. Naro-Maciel. 2013. Ecology and conservation of marine turtles in a central Pacific foraging ground. *Chelonian Conservation and Biology* 12: 2–16.

Steyermark, A. C., M. S. Finkler, and R. J. Brooks, editors. 2008. *Biology of the Snapping Turtle (Chelydra serpentina).* Baltimore, Maryland: The Johns Hopkins University Press.

Strahm, M. H., and A. Schwartz. 1977. Osteoderms in the anguid lizard subfamily Diploglossinae and their taxonomic importance. *Biotropica* 9: 58–72.

Strauch, A. 1887. Bemerkugen über die Geckoniden-Sammlung im Zoologischen Museum der Kaiserlichen Akademie der Wissenschaften zu St. Petersburg. *Mémoires de L'Académie Impériale des Sciences de St.-Pétersburgh* 35: 1–72, i–ii, 1 plate.

Streicher, J. W., G. R. Fajardo, and C. R. Vásquez-Almazán. 2011. Natural history notes. *Plestiodon sumichrasti* (Sumichrast's Skink). Predation. *Herpetological Review* 42: 431–432.

Stroud, J. T., and K. Krysko. 2014. Natural history notes. *Ctenosaura similis* (Gray's Spiny-Tailed Iguana). Non-native diet. *Herpetological Review* 44: 322.

Stuart, L. C. 1934. A contribution to a knowledge of the herpetological fauna of El Peten, Guatemala. *Occasional Papers of the Museum of Zoology, University of Michigan* 292: 1–18.

Stuart, L. C. 1939. A description of a new *Gymnophthalmus* from Guatemala, with notes on other members of the genus. *Occasional Papers of the Museum of Zoology, University of Michigan* 409: 1–10, plate I.

Stuart, L. C. 1940. Notes on the "lampropholis" group of Middle American *Lygosoma* (Scincidae) with descriptions of two new forms. *Occasional Papers of the Museum of Zoology, University of Michigan* 421: 1–16.

Stuart, L. C. 1948. The amphibians and reptiles of Alta Verapaz Guatemala. *Miscellaneous Publications Museum of Zoology, University of Michigan* 69: 1–109.

Stuart, L. C. 1951. The herpetofauna of the Guatemalan Plateau, with special reference to its distribution on the southwestern highlands. *Contributions from the Laboratory of Vertebrate Biology, University of Michigan* 49: 1–71, plates I–VII.

Stuart, L. C. 1954. A description of a subhumid corridor across northern Central America, with comments on its herpetofaunal indicators. *Contributions from the Laboratory of Vertebrate Biology, University of Michigan* 65: 1–26, plates I–VI.

Stuart, L. C. 1963. A checklist of the herpetofauna of Guatemala. *Miscellaneous Publications Museum of Zoology, University of Michigan* 122: 1–150.

Stuart, L. C. 1970. A brief review of the races of *Sceloporus serrifer* Cope with special reference to *Sceloporus serrifer prezygus* Smith. *Herpetologica* 26: 141–149.

Stuart, L. C. 1971. Comments on the malachite *Sceloporus* (Reptilia: Sauria: Iguanidae) of southern Mexico and Guatemala. *Herpetologica* 27: 235–259.

Sumichrast, F. 1873. Notes sur quelques reptiles Mexicains peu connus. *Archives des Sciences Physiques et Naturelles* 46: 251–262.

Sumichrast, F. 1880. Contribution a l'histoire naturelle du Mexique. I. Notes sur une collection des reptiles et de batraciens de la partie occidentale de l'Isthme de Tehuantepec. *Bulletin de la Société Zoologique de France* 5: 162–190.

Sumichrast, F. 1882. Enumeracion de las especies de reptiles observados en la parte meridional de la República Mexicana. *La Naturaleza* 6: 31–45.

Sunyer, J., and G. Köhler. 2007. New country and departmental records of herpetofauna in Nicaragua. *Salamandra* 43: 57–62.

Sunyer, J., R. García-Roa, and J. H. Townsend. 2013. First country record of *Norops wermuthi* Köhler & Obermeier, 1998, for Honduras. *Herpetozoa* 26: 103–106.

Sunyer, J., K. E. Nicholson, J. G. Phillips, J. A. Gubler, and L. A. Obando. 2013. Lizards (Reptilia: Squamata) of the Corn Islands, Caribbean Nicaragua. *Check List* 9: 1383–1390.

Sunyer, J., J. H. Townsend, L. D. Wilson, S. L. Travers, L. A. Obando, G. Páiz, D. M. Griffith, and G. Köhler. 2009. Three new country records of reptiles from Nicaragua. *Salamandra* 45: 186–190.

Sunyer, J., M. Vesely, and G. Köhler. 2008. Morphological variation in *Anolis wermuthi* (Köhler & Obermeier 1998), a species endemic to the highlands of north-central Nicaragua (Reptilia, Squamata, Polychrotidae). *Senckenbergiana biologica* 88: 335–343.

Tamsitt, J. R., and D. Valdivieso. 1963. The herpetofauna of the Caribbean islands San Andres and Providencia. *Revista de Biologia Tropical* 11: 131–139.

Taylor, E. H. 1922. *The Lizards of the Philippine Islands.* Manila: Bureau of Printing.

Taylor, E. H. 1936 (dated 1935). A taxonomic study of the cosmopolitan scincoid lizards of the genus *Eumeces* with an account of the distribution and relationships of its species. *The University of Kansas Science Bulletin* 23: 1–643.

Taylor, E. H. 1940a (dated 1939). Mexican snakes of the genus *Typhlops*. *The University of Kansas Science Bulletin* 26: 441–444.

Taylor, E. H. 1940b (dated 1939). Herpetological miscellany. No. I. *The University of Kansas Science Bulletin* 26: 489–571.

Taylor, E. H. 1942. Some geckoes of the genus *Phyllodactylus*. *The University of Kansas Science Bulletin* 28: 91–112.

Taylor, E. H. 1947. A review of the Mexican forms of the lizard genus *Sphaerodactylus*. *The University of Kansas Science Bulletin* 31: 299–309.

Taylor, E. H. 1953. A review of the lizards of Ceylon. *The University of Kansas Science Bulletin* 35: 1525–1585.

Taylor, E. H. 1955. Additions to the known herpetological fauna of Costa Rica with comments on other species. No. II. *The University of Kansas Science Bulletin* 37: 499–575.

Taylor, E. H. 1956a. *Sphaerodactylus lineolatus* (Reptilia: Lacertilia) in Mexico. *Herpetologica* 12: 283–284.

Taylor, E. H. 1956b. A review of the lizards of Costa Rica. *The University of Kansas Science Bulletin* 38: 3–322.

Taylor, E. H. 1969. Wiegmann and the herpetology of México, PP. iii–vi. IN: A. F. A. Wiegmann. Herpetologia Mexicana, seu Descriptio Amphibiorum Novae Hispaniae, quae Itineribus comitis de Sack, Ferdinandi Deppe et Chr. Guil. Schiede in Museum Zoologicum Berolinense Pervenerunt. Pars Prima, Saurorum Species Amplectens. Adiecto Systematis Saurorum Prodromo, Additisque multis in hunc Amphibiorum Ordinem Observationibus. *Facsimile Reprints in Herpetology, Society for the Study of Amphibians and Reptiles* 23: i–vi, i–vi, 1–54 pp., plates I–X.

Taylor, E. H., and H. M. Smith. 1943. A review of American Sibynophine snakes, with a proposal of a new genus. *The University of Kansas Science Bulletin* 29: 301–337, plates XXI–XXV.

Telford, S. R., Jr. 1971. Reproductive patterns and relative abundance of two microteiid lizard species in Panama. *Copeia* 1971: 670–675.

Telford, S. R., Jr. 1977. The distribution, incidence and general ecology of saurian malaria in Middle America. *International Journal for Parasitology* 7: 299–314.

Telford, S. R., Jr., and H. W. Campbell. 1970. Ecological observations on an all female population of the lizard *Lepidophyma flavimaculatum* (Xantusiidae) in Panamá. *Copeia* 1970: 379–381.

Terán-Juaréz, S. A., E. García-Padilla, F. E. Leyto-Delgado, and L. J. García-Morales. 2015. New records and distributional range extensions for amphibians and reptiles from Tamaulipas, Mexico. *Mesoamerican Herpetology* 2: 208–214.

Teska, W. R. 1965. Terrestrial movements of the mud turtle *Kinosternon scorpioides* in Costa Rica. *Copeia* 1976: 579–580.

Theobald, W. 1868. Catalogue of reptiles in the Museum of the Asiatic Society of Bengal. *Journal of the Asiatic Society*, Extra Number i–vi + 1–88 + i–iii + plates i–iv.

Thomas, R. 1965. The smaller teiid lizards (*Gymnophthalmus* and *Bachia*) of the southeastern Caribbean. *Proceedings of the Biological Society of Washington* 78: 141–154.

Thomas, R. 1975. The *argus* group of West Indian *Sphaerodactylus* (Sauria: Gekkonidae). *Herpetologica* 31: 177–195.

Thorbjarnarson, J. B. 1988. The status and ecology of the American crocodile in Haiti. *Bulletin of the Florida State Museum, Biological Sciences* 33: 1–86.

Thorbjarnarson, J. B. 1989. Ecology of the American crocodile, *Crocodylus acutus*, PP. 228–259. IN: *Crocodiles, Their Ecology, Management, and Conservation. A Special Publication of the Crocodile Specialist Group of the Species Survival Commission of the International Union for Conservation of Nature and Natural Resources.* Gland, Switzerland: IUCN Publication, New Series.

Thorbjarnarson, J. [B.], compiler. 1992. *Crocodiles. An Action Plan for their Conservation.* Gland, Switzerland: International Union for Conservation of Nature and Natural Resources.

Thorbjarnarson, J. B. 1993. Diet of the Spectacled Caiman (*Caiman crocodilus*) in the central Venezuelan Llanos. *Herpetologica* 49: 108–117.

Thorbjarnarson, J. B. 1994. Reproductive ecology of the Spectacled Caiman (*Caiman crocodilus*) in the Venezuelan Llanos. *Copeia* 1994: 907–919.

Thorbjarnarson, J. [B.], F. Mazzotti, E. Sanderson, F. Buitrajo, M. Lazcano, K. Minkowski, M. Muñiz, P. Ponce, L. Sigler, R. Soberon, A. M. Trelancia, and A. Velasco. 2006. Regional habitat conservation priorities for the American Crocodile. *Biological Conservation* 128: 25–36.

Tihen, J. A. 1949. The genera of gerrhonotine lizards. *American Midland Naturalist* 41: 580–601.

Toral, C. E. 2004. Natural history notes. *Ameiva festiva* (Central American Racerunner). Predation. *Herpetological Review* 35: 266.

Townsend, J. H. 2005. Geographic distribution. *Sphenomorphus incertus* (Stuart's Forest Skink). *Herpetological Review* 36: 337–338.

Townsend, J. H. 2006a. Inventory and Conservation Assessment of the Herpetofauna of the Sierra de Omoa, Honduras, with a Review of the Geophis (Squamata: Colubridae) of Eastern Nuclear Central America [thesis]. Gainesville: The University of Florida.

Townsend, J. H. 2006b. *Celestus montanus. Catalogue of American Amphibians and Reptiles* 834.1–834.3.

Townsend, J. H. 2009. Morphological variation in *Geophis nephodrymus* (Squamata: Colubridae), with comments on conservation of *Geophis* in Eastern Nuclear Central America. *Herpetologica* 65: 292–302.

Townsend, J. H., and L. D. Wilson. 2006. Denizens of the dwarf forest: the herpetofauna of the elfin forests of Cusuco National Park, Honduras. *Iguana* 13: 242–251.

Townsend, J. H., and L. D. Wilson. 2008. *Guide to the Amphibians & Reptiles of Cusuco National Park, Honduras = Guía de los Anfibios y Reptiles del Parque Nacional Cusuco, Honduras.* P. M. Kulstad, translator. Salt Lake City, Utah: Bibliomania!

Townsend, J. H., and L. D. Wilson. 2009. New species of cloud forest *Anolis* (Squamata: Polychrotidae) in the *crassulus* group from Parque Nacional Montaña de Yoro, Honduras. *Copeia* 2009: 62–70.

Townsend, J. H., and L. D. Wilson. 2010a. Conservation of the Honduran herpetofauna: issues and imperatives, PP. 460–487. IN: L. D. Wilson, J. H. Townsend, and J. D. Johnson, editors. *Conservation of Mesoamerican Amphibians and Reptiles.* Eagle Mountain, Utah: Eagle Mountain Publishing, LC.

Townsend, J. H., and L. D. Wilson. 2010b. Biogeography and conservation of the Honduran subhumid forest herpetofauna, PP. 686–705. IN: L. D. Wilson, J. H. Townsend, and J. D. Johnson, editors. *Conservation of Mesoamerican Amphibians and Reptiles.* Eagle Mountain, Utah: Eagle Mountain Publishing, LC.

Townsend, J. H., H. C. Aldrich, L. D. Wilson, and J. R. McCranie. 2005. First record of sporangia of a myxomycete (*Physarum pusillum*) on the body of a living animal, the lizard *Corytophanes cristatus*. *Mycologia* 97: 346–348.

Townsend, J. H., J. M. Butler, L. D. Wilson, and J. D. Austin. 2010. A distinctive new species of moss salamander (Caudata: Plethodontidae: *Nototriton*) from an imperiled Honduran endemic hotspot. *Zootaxa* 2434: 1–16.

Townsend, J. H., S. M. Hughes, J. J. Hines, D. J. Carter, and G. Sandoval. 2005. Notes on a juvenile *Celestus montanus* Schmidt, 1933, a rare lizard from Parque Nacional El Cusuco, Honduras. *Herpetozoa* 18: 67–68.

Townsend, J. H., J. R. McCranie, and L. D. Wilson. 2004a. *Corytophanes. Catalogue of American Amphibians and Reptiles* 788.1–788.4.

Townsend, J. H., J. R. McCranie, and L. D. Wilson. 2004b. *Corytophanes cristatus. Catalogue of American Amphibians and Reptiles* 789.1–789.6.

Townsend, J. H., J. R. McCranie, and L. D. Wilson. 2004c. *Corytophanes percarinatus. Catalogue of American Amphibians and Reptiles* 791.1–791.3.

Townsend, J. H., M. Medina-Flores, L. D. Wilson, R. C. Jadin, and J. D. Austin. 2013. A relic lineage and new species of green palm-pitviper (Squamata, Viperidae, *Bothriechis*) from the Chortís Highlands of Mesoamerica. *ZooKeys* 298: 77–105.

Townsend, J. H., L. D. Wilson, L. P. Ketzler, and I. R. Luque-Montes. 2008. The largest blindsnake in Mesoamerica: a new species of *Typhlops* (Squamata: Typhlopidae) from an isolated karstic mountain in Honduras. *Zootaxa* 1932: 18–26.

Townsend, J. H., L. D. Wilson, M. Medina Flores, E. Aguilar-Urbina, B. K. Atkinson, C. A. Cerrato-Mendoza, A. Contreras-Castro, L. N. Gray, L. A. Herrera-B. I. R. Luque-Montes, M. McKewy-Mejiá, A. Portillo-Avilez, A. L. Stubbs, and J. D. Austin. 2012. A Premontane hotspot for herpetological endemism on the windward side of Refugio de Vida Silvestre Texíguat, Honduras. *Salamandra* 48: 92–114.

Townsend, J. H., L. D. Wilson, M. Medino-Flores, and L. A. Herrera-B. 2013. A new species of centipede snake in the *Tantilla taeniata* group (Squamata: Colubridae) from premontane rainforest in Refugio de Vida Silvestre Texíguat Honduras. *Journal of Herpetology* 47: 191–200.

Townsend, J. H., L. D. Wilson, and J. Restrepo. 2007. Informe preliminar. Investigaciones sobre la herpetofauna en el Parque Nacional Montaña de Yoro y la Reserva Biológica Cerro Uyuca, Honduras. Gainesville: Unpublished Report, The University of Florida.

Townsend, J. H., L. D. Wilson, B. L. Talley, D. C. Fraser, T. L. Plenderleith, and S. M. Hughes. 2006. Additions to the herpetofauna of Parque Nacional El Cusuco, Honduras. *The Herpetological Bulletin* 96: 29–39.

Townsend, T. M., D. G. Mulcahy, B. P. Noonan, J. W. Sites, Jr., C. A. Kuczynski, J. J. Wiens, and T. W. Reeder. 2011. Phylogeny of iguanian lizards inferred from 29 nuclear loci, and a comparison of concatenated and species-tree approaches for an ancient, rapid radiation. *Molecular Phylogenetics and Evolution* 61: 363–380.

Traveset, A. 1990. *Ctenosaura similis* Gray (Iguanidae) as a seed disperser in a Central American deciduous forest. *American Midland Naturalist* 123: 402–404.

Troschel, F. H. 1860. Bericht über die Leistungen in der Herpetologie während des Jahres 1859. *Archiv für Naturgeschichte* 26: 265–278.

Trutnau, L. 1986. *Krokodile und Echsen in Farbe*. Rüschlikon-Zürich: Albert Müller Verlag.

Trutnau, L., and R. Sommerlad. 2006. *Crocodilians. Their Natural History & Captive Husbandry*. Frankfurt am Main, Germany: Edition Chimaira.

Tucker, J. K., J. T. Lamer, C. R. Dolan, and E. A. Dustman. 2006. Natural history notes. Chelonian species. Record carapace lengths for Illinois. *Herpetological Review* 37: 453–455.

[TTWG] Turtle Taxonomy Working Group (P. P. van Dijk, J. B. Iverson, H. B. Shaffer, R. Bour, and A. G. J. Rhodin). 2012. Turtles of the World, 2012 update: annotated checklist of taxonomy, synonymy, distribution, and conservation status, PP. 000.243–000.328. IN: A. G. J. Rhodin, P. C. H. Pritchard, P. P. van Dijk, R. A. Saumure, K. A. Buhlmann, J. B. Iverson, and R. A. Mittermeier, editors. *Conservation Biology of Freshwater Turtles and Tortoises: A Compilation Project of the IUCN/SSC Tortoise and Freshwater Turtle Specialist Group*. Lunenburg, Massachusetts: Chelonian Research Foundation. *Chelonian Research Monographs* 5.

TTWG (A. G. J. Rhodin, J. B. Iverson, R. Bour, U. Fritz, A. Georges, H. B. Shaffer, and P. P. van Dijk). 2017. Turtles of the World. Annotated Checklist and Atlas of Taxonomy, Synonymy, Distribution, and Conservation Status (8th ed.). *In* A. G. J. Rhodin, J. B. Iverson, P. P. van Dijk, R. A. Saumure, K. A. Buhlmann, P. C. H. Pritchard, and R. A. Mittermeier (eds.). Conservation Biology of Freshwater Turtles and Tortoises: a Compilation Project of the IUCN/SSC Tortoise and Freshwater Turtle Specialist Group. *Chelonian Research Monographs* 7: 1–292.

TTWG (P. P. van Dijk, J. B. Iverson, A. G. J. Rhodin, H. B. Shaffer, and R. Bour). 2014. Turtles of the World, 7th ed: annotated checklist of taxonomy, synonymy, distribution with maps, and conservation status, PP. 329–479. IN: A. G. J. Rhodin, P. C. H. Pritchard, P. P. van Dijk, R. A. Saumure, K. A. Buhlmann, J. B. Iverson, and R. A. Mittermeier, editors. *Conservation Biology of Freshwater Turtles and Tortoises: A Compilation Project of the IUCN/SSC Tortoise and Freshwater Turtle Specialist Group*. Lunenburg, Massachusetts: Chelonian Research Monographs 5.

Ugueto, G. N., and M. B. Harvey. 2012 (dated 2011). Revision of *Ameiva ameiva* Linnaeus (Squamata: Teiidae) in Venezuela: recognition of four species and status of introduced populations in southern Florida, USA. *Herpetological Monographs* 25: 113–170.

Ugueto, G. N., and G. A. Rivas. 2010. *Amphibians and Reptiles of Margarita, Coche, and Cubagua.* Frankfurt am Main, Germany: Edition Chimaira.

Ugueto, G. N., P. Velozo, L. E. Sanchez, L. A. Bermúdez Villapol, O. Lasso-Alcalá, T. R. Barros, and G. A. Rivas. 2013. Noteworthy new records of squamate reptiles (Reptilia: Squamata) from various Venezuelan Caribbean islands, including a new addition to the herpetofauna of Venezuela. *Check List* 9: 1075–1080.

Underwood, G. 1954. On the classification and evolution of geckos. *Proceedings of the Zoological Society of London* 124: 469–492.

Valdés Orellana, L., and J. R. McCranie. 2011a. Geographic distribution. *Kinosternon leucostomum* (White-lipped Mud Turtle; Pochitoque). *Herpetological Review* 42: 565.

Valdés Orellana, L., and J. R. McCranie. 2011b. Geographic distribution. *Gonatodes albogularis* (Yellow-headed Gecko; Geco Cabeza Amarilla). *Herpetological Review* 42: 568.

Valdés Orellana, L., J. R. McCranie, and A. Gutsche. 2011a. Geographic distribution. *Hemidactylus frenatus* (Common House Gecko). *Herpetological Review* 42: 240–241.

Valdés Orellana, L., J. R. McCranie, and A. Gutsche. 2011b. Geographic distribution. *Mesoscincus managuae* (NCN). *Herpetological Review* 42: 242.

Vandelli, D. 1761. *Epistola de Holothurio, et Testudine coriacea ad celeberrimum Carolum Linnaeum. Equitem Naturae Curiosorum Dioscoridem II.* Patavii [Padua]: Typographia Conzatti.

Van Denburgh, J. 1898 (dated 1897). Reptiles from Sonora, Sinaloa and Jalisco, Mexico, with a description of a new species of *Sceloporus. Proceedings of the Academy of Natural Sciences of Philadelphia* 49: 460–464.

Van Devender, R. W. 1982. Growth and ecology of spiny-tailed and green iguanas in Costa Rica, with comments on the evolution of herbivory and large body size, PP. 162–183. IN: G. M. Burghardt and A. S. Rand, editors. *Iguanas of the World. Their Behavior, Ecology, and Conservation.* Park Ridge, New Jersey: Noyes Publications.

Vanzolini, P. E. 1978. On South American *Hemidactylus* (Sauria, Gekkonidae). *Papéis Avulsos de Zoologia, São Paulo* 31: 307–343.

Vanzolini, P. E., and C. M. de Carvalho. 1991. Two sibling and sympatric species of *Gymnophthalmus* in Roraima, Brasil (Sauria, Teiidae). *Papéis Avulsos de Zoologia, São Paulo* 37: 173–226.

Vanzolini, P. E., and C. W. Myers. 2015. The herpetological collection of Maximilian, Prince of Wied (1782–1867), with special reference to Brazilian materials. *Bulletin of the American Museum of Natural History* 395: 1–155.

Vanzolini, P. E., and E. E. Williams. 1962. Jamaican and Hispaniolan *Gonatodes* and allied forms (Sauria, Gekkonidae). *Bulletin of the Museum of Comparative Zoology* 127: 479–498.

Vargas-Ramírez, M., J. L. Carr, and U. Fritz. 2013. Complex phylogeography in *Rhinoclemmys melanosterna*: conflicting mitochondrial and nuclear evidence suggests past hybridization (Testudines: Geoemydidae). *Zootaxa* 3670: 238–254.

Vargas-Ramírez, M., C. del Valle, C. P. Ceballos, and U. Fritz. 2017. *Trachemys medemi* n. sp. from northwestern Colombia turns the biogeography of South American slider turtles upside down. *Journal of Zoological Systematics and Evolutionary Research* 55: 326–339.

Vaughan, C., O. Ramirez, G. Herrera, E. Fallas, and R. W. Henderson. 2007. Home range and habitat use of *Basiliscus plumifrons* (Squamata: Corytophanidae) in an active Costa Rican cacao farm. *Applied Herpetology* 4: 217–226.

Venegas-Anaya, M., A. J. Crawford, A. H. Escobedo Galván, O. I. Sanjur, L. D. Densmore, III, and E. Bermingham. 2008. Mitochondrial DNA phylogeography of *Caiman crocodilus* in Mesoamerica and South America. *Journal of Experimental Zoology* 309A: 614–627.

Vesely, M., and G. Köhler. 2001. Zur Kenntnis von *Mesaspis moreletii* (Bocourt, 1871) in El Salvador. *Salamandra* 37: 185–192.

Vetter, H. 2004. *Turtles of the World*, Vol. 2. *North America. Schildkröten der Welt*, Band 2. *Nordamerika.* Frankfurt am Main, Germany: Edition Chimaira.

Vetter, H. 2005. *Turtles of the World*, Vol. 3. *Central and South America. Schildkröten der Welt*, Band 3. *Mittel- und Südamerika.* Frankfurt am Main, Germany: Edition Chimaira.

Vidal, N., and S. B. Hedges. 2005. The phylogeny of squamate reptiles (lizards, snakes, and amphisbaenians) inferred from nine nuclear protein-coding genes. *Comptes Rendus Biologies* 328: 1000–1008.

Vidal, N., and S. B. Hedges. 2009. The molecular evolutionary tree of lizards, snakes, and amphisbaenians. *Comptes Rendus Biologies* 332: 129–139.

Vidal, N., A.-S. Delmas, P. David, C. Cruaud, A. Couloux, and S. B. Hedges. 2007. The phylogeny and classification of caennophidian snakes inferred from seven nuclear protein-coding genes. *Comptes Rendus Biologies* 330: 182–187.

Vidal, N., M. Dewynter, and D. J. Gower. 2010. Dissecting the major American snake radiation: a molecular phylogeny of the Dipsadidae Bonaparte (Serpentes, Caenophidia). *Comptes Rendus Biologies* 333: 48–55.

Villa, J. 1994. Geographic distribution. *Cnemidophorus deppei deppei. Herpetological Review* 25: 33.

Villa, J., and N. J. Scott, Jr. 1967. The iguanid lizard *Enyaliosaurus* in Nicaragua. *Copeia* 1967: 474–476.

Villa, J., and L. D. Wilson. 1988. *Celestus bivittatus*. Catalogue of American Amphibians and Reptiles 423.1–423.2.

Villa, J., L. D. Wilson, and J. D. Johnson. 1988. *Middle American Herpetology. A Bibliographic Checklist*. Columbia: University of Missouri Press.

Vitt, L. J. 1996. Geckos of the Amazon River Basin. *Reptile & Amphibian Magazine* 43: 12–20.

Vitt, L. J., and J. P. Caldwell. 2014. *Herpetology*. 4th ed. *An Introductory Biology of Amphibians and Reptiles*. Amsterdam, The Netherlands: Academic Press.

Vitt, L. J., and P. A. Zani. 1996. Ecology of the lizard *Ameiva festiva* (Teiidae) in southeastern Nicaragua. *Journal of Herpetology* 30: 110–117.

Vitt, L. J., and P. A. Zani. 1997. Ecology of the nocturnal lizard *Thecadactylus rapicauda* (Sauria: Gekkonidae) in the Amazon region. *Herpetologica* 53: 165–179.

Vitt, L. J., P. A. Zani, J. P. Caldwell, and R. D. Durtsche. 1993. Ecology of the whiptail lizard *Cnemidophorus deppii* on a tropical beach. *Canadian Journal of Zoology* 71: 2391–2400.

Vogt, R. C. 1990. Reproductive parameters of *Trachemys scripta venusta* in southern Mexico, PP. 162–168. IN: J. W. Gibbons, editor. *Life History and Ecology of the Slider Turtle*. Washington: Smithsonian Institution Press.

Vogt, R. C. 1997. *Staurotypus triporcatus* (tres lomos, guao, galápago), PP. 494–495. IN: E. González Soriano, R. Dirzo, and R. C. Vogt, editors. *Historia Natural de Los Tuxtlas*. México City: Universidad Nacional Autónoma de México.

Vogt, R. C., and S. G. Guzman. 1988. Food partitioning in a neotropical freshwater turtle community. *Copeia* 1988: 37–47.

Vogt, R. C., S. G. Platt, and T. R. Rainwater. 2009. *Rhinoclemmys areolata* (Duméril and Bibron 1851)—Furrowed Wood Turtle, Black-Bellied Turtle, Mojena, PP. 022.1–022.7. IN: A. G. J. Rhodin, P. C. H. Pritchard, P. P. van Dijk, R. A. Saumure, K. A. Buhlmann, J. B. Iverson, and R. A. Mittermeier, editors. *Conservation Biology of Freshwater Turtles and Tortoises:A Compilation Project of the IUCN/SSC Tortoise and Freshwater Turtle Specialist Group*. Lunenburg, Massachusetts: Chelonian Research Foundation. *Chelonian Research Monographs* 5.

Vogt, R. C., J. R. Polisar, D. Moll, and G. Gonzalez-Porter. 2011. *Dermatemys mawii* Gray 1847—Central American River Turtle, Tortuga Blanca, Hickatee, PP. 058.1–058.12. IN: A. G. J. Rhodin, P. C. H. Pritchard, P. P. van Dijk, R. A. Saumure, K. A. Buhlmann, J. B. Iverson, and R. A. Mittermeier, editors. *Conservation Biology of*

Freshwater Turtles and Tortoises: A Compilation Project of the IUCN/SSC Tortoise and Freshwater Turtle Specialist Group. Lunenburg, Massachusetts: Chelonian Research Foundation. *Chelonian Research Monographs* 5.

Wagler, J. 1824. *Serpentum Brasiliensium Species Novae ou Histoire Naturelle des Espècies Nouvelles de Serpens, Recueillies et Observées Pendant le Voyage dans l'Intérieur du Brésil dans les Années 1817, 1818, 1819, 1820, Exécuté par Orde de sa Majesté le Roi de Baviére*. Monachii [Munich]: Typis Franc. Seraph. Hübschmanni.

Wagler, J. 1830a. *Natürliches System der Amphibien, mit vorangehender Classification der Säugthiere und Vögel. Ein Beitrag zur vergleichenden Zoologie*. Munich: J. G. Cotta Buchhandlung.

Wagler, J. 1830b. *Natürliches System der Amphibien. Tafeln*. Erstes Heft. Taf. (Plates) I–VII + 2 unnumbered Plates. Munich: J. G. Cotta Buchhandlung.

Wagler, J. 1833. *Descriptiones et Icones Amphibiorum*. Parte Tres. Monachii [Munich]: J. G. Cotttae.

Wallach, V. 2016. Morphological review and taxonomic status of the *Epictia phenops* species group of Mesoamerica, with description of six new species and discussion of South American *Epictia albifrons*, *E. goudotii*, and *E. tenella* (Serpentes: Leptotyphlopidae: Epictinae). *Mesoamerican Herpetology* 3: 216–374.

Wallach, V., K. L. Williams, and J. Boundy. 2014. *Snakes of the World. A Catalogue of Living and Extinct Species*. Boca Raton, Florida: CRC Press.

Wallbaum, J. J. 1782. *Chelonographia oder Beschreibung einiger Schildkröten nach natürlichen Urbildern*. Lübeck, Germany: Johann Friedrich Gleditsche.

Wallin, L. 1985. A survey of Linnaeus's material of *Chelone mydas*, *Caretta caretta* and *Eretmochelys imbricata* (Reptilia: Cheloniidae). *Zoological Journal of the Linnean Society* 85: 121–130.

Wartenberg, L. 2013. Bedrohung und Schutz der Schwarzleguane (*Ctenosaura*). *Iguana Rundschreiben* 26: 20–33.

Watling, J. I., J. H. Waddle, D. Kizirian, and M. A. Donnelly. 2005. Reproductive phenology of three lizard species in Costa Rica, with comments on seasonal reproduction of neotropical lizards. *Journal of Herpetology* 39: 341–348.

Webb, R. G. 1958. The status of the Mexican lizards of the genus *Mabuya*. *The University of Kansas Science Bulletin* 38: 1303–1313.

Wegener, J. E., G. E. A. Gartner, and J. B. Losos. 2014. Lizard scales in an adaptive radiation: variation in scale number follows climatic and structural habitat diversity in *Anolis* lizards. *Biological Journal of the Linnean Society* 113: 570–579.

Weigel, E. P. 1973. Great Swan Island—hurricane sentry in the Caribbean. *NOAA* 1973: 20–27.

Weiss, A. J., and S. B. Hedges. 2007. Molecular phylogeny and biogeography of the Antillean geckos *Phyllodactylus wirshingi, Tarentola americana,* and *Hemidactylus haitianus* (Reptilia, Squamata). *Molecular Phylogenetics and Evolution* 45: 409–416.

Welch, K. R. G. 1982. Herpetology of the Old World II. Preliminary comments on the classification of skinks (family Scincidae) with specific reference to those genera found in Africa, Europe and southwest Asia. *Herptile* 7: 25–27.

Wermuth, H. 1965. Liste de rezenten Amphibien und Reptilien. Gekkonidae, Pygopodidae, Xantusiidae. *Das Tierreich* 80: i–xxii, 1–246.

Wermuth, H. 1969. Liste der rezenten Amphibien und Reptilien. Anguidae, Anniellidae, Xenosauridae. *Das Tierreich* 90: i–xii, 1–41.

Wermuth, H., and R. Mertens. 1961. *Schildkröten, Krokodile, Brückenechsen.* Jena, Germany: Gustav Fischer Verlag.

Wermuth, H., and R. Mertens. 1977. Liste der rezenten Amphibien und Reptilien. Testudines, Crocodylia, Rhynchocephalia. *Das Tierreich* 100: i–xxvii, 1–174.

Werner, F. 1896. Beiträge zur Kenntniss der Reptilien und Batrachier von Centralamerika und Chile, sowie einiger seltenerer Schlangenarten. *Verhandlungen der zoologisch-botanischen Gesellschaft in Wien* 46: 344–365, plate VI.

Werner, F. 1899. Reptilia und Amphibia für 1896. (Inhaltsverzeichnis am Schlusse). *Archiv für Naturgeschichte* 2: 1–78.

Werner, F. 1904. Reptilia und Amphibia für 1901. (Inhaltsverzeichnis am Schlusse des Berichts.). *Archiv für Naturgeschichte* 7: 1–72.

Werning, H. 2002. *Stachelleguane.* Münster, Germany: Natur und Tier—Verlag.

Wettstein, O. 1934. Ergebnisse der österreichischen biologischen Costa Rica-Expedition 1930. Die Amphibien und Reptilien. *Akademie der Wissenschaften, Wien* 143: 1–39.

Whiting, A. S., J. W. Sites, Jr., K. C. M. Pellegrino, and M. T. Rodrigues. 2006. Comparing alignment methods for inferring the history of the new world lizard genus *Mabuya* (Squamata: Scincidae). *Molecular Phylogenetics and Evolution* 38: 719–730.

Wied, M. zu. 1824. Verzeichnis der Amphibien, welche im zweyten Bande der Naturgeschichte Brasiliens vom Prinz Max von Neuwied werden beschrieben werden. *Isis* 14(6): cols. 661–674 (according to Vanzolini and Myers, 2015: 20, the title of this publication was written by Oken and the manuscript is a synopsis of the Beiträge of Wied, 1825).

°Wied, M. zu. 1825. Beiträge zur Naturgeschichte von Brasililien, von Maximilian, Prizen zu Wied. Volumn 1. Weimar, Germany: Landes-Industrie-Comptoirs.

Wied, M. zu. 1839. *Reise in Dans Innere Nord—America in den Jahren 1832 bis 1834.* Erster Band. Coblenz, Germany: J. Hoelscher.

Wiegmann, A. F. A. 1828. Beyträge zur Amphibienkunde. *Isis* 21: cols. 364–383.

Wiegmann, A. F. A. 1831. Ueber den Cuapapalcatl oder *Chamaeleo mexicanus* des Hernandez. *Isis* 24: cols. 296–299.

Wiegmann, A. F. A. 1833. Ueber die mexicanischen Kroten. *Isis* 26: cols. 651–662.

Wiegmann, A. F. A. 1834a. Beiträge zur Zoologie, Gesammelt auf einer Reise um die Erde, von Dr. F. J. F. Meyen. Siebente Abhandlung. Amphibien. *Nova Acta Academiae Caesareae Leopoldino Carolinae, Halle* 17: 183–268 (268a–d), plates XIII–XXII.

Wiegmann, A. F. A. 1834b. *Herpetologia Mexicana, seu Descriptio Amphibiorum Novae Hispaniae, quae Itineribus comitis de Sack, Ferdinandi Deppe et Chr. Guil. Schiede in Museum Zoologicum Berolinense Pervenerunt. Pars Prima, Saurorum Species Amplectens. Adiecto Systematis Saurorum Prodromo, Additisque multis in hunc Amphibiorum Ordinem Observationibus.* Berolini [Berlin]: Sumptibus C. G. Lüderitz.

Wiens, J. J. 1993. Phylogenetic relationships of phrynosomatid lizards and monophyly of the *Sceloporus* group. *Copeia* 1993: 287–299.

Wiens, J. J., and T. W. Reeder. 1997. Phylogeny of the spiny lizards (*Sceloporus*) based on molecular and morphological evidence. *Herpetological Monographs* 11: 1–101.

Wiens, J. J., C. A. Kuczynski, S. Arif, and T. W. Reeder. 2010. Phylogenetic relationships of phrynosomatid lizards based on nuclear and mitochondrial data, and a revised phylogeny for *Sceloporus. Molecular Phylogenetics and Evolution* 54: 150–161.

Wiewandt, T. A. 1982. Evolution of nesting patterns in iguanine lizards, PP. 119–141. IN: G. M. Burghardt and A. S. Rand, editors. *Iguanas of the World. Their Behavior, Ecology, and Conservation.* Park Ridge, New Jersey: Noyes Publications.

Wilson, L. D., and G. A. Cruz Díaz. 1993. The herpetofauna of the Cayos Cochinos, Honduras. *Herpetological Natural History* 1: 13–23.

Wilson, L. D., and D. E. Hahn. 1973. The herpetofauna of the Islas de la Bahía, Honduras. *Bulletin of the Florida State Museum, Biological Sciences* 17: 93–150.

Wilson, L. D., and J. R. McCranie. 1982. A new cloud forest *Anolis* (Sauria: Iguanidae) of the *schiedei* group from Honduras. *Transactions of the Kansas Academy of Science* 85: 133–141.

Wilson, L. D., and J. R. McCranie. 1994a. Comments on the occurrence of a salamander and three lizard species in Honduras. *Amphibia-Reptilia* 15: 416–421.

Wilson, L. D., and J. R. McCranie. 1994b. Second update on the list of amphibians and reptiles known from Honduras. *Herpetological Review* 25: 146–150.

Wilson, L. D., and J. R. McCranie. 1998. The biogeography of the herpetofauna of the subhumid forests of Middle America (Isthmus of Tehuantepec to northwestern Costa Rica). *Royal Ontario Museum, Life Sciences Contributions* 163: 1–50.

Wilson, L. D., and J. R. McCranie. 2002. Update on the list of reptiles known from Honduras. *Herpetological Review* 33: 90–94.

Wilson, L. D., and J. R. McCranie. 2003. Herpetofaunal indicator species as measures of environmental stability in Honduras. *Caribbean Journal of Science* 39: 50–67.

Wilson, L. D., and J. R. McCranie. 2004a. The conservation status of the herpetofauna of Honduras. *Amphibian & Reptile Conservation* 3: 6–33.

Wilson, L. D., and J. R. McCranie. 2004b. The herpetofauna of the cloud forests of Honduras. *Amphibian & Reptile Conservation* 3: 34–48.

Wilson, L. D., and J. R. McCranie. 2004c. The herpetofauna of Parque Nacional El Cusuco, Honduras (Reptilia, Amphibia). *The Herpetological Bulletin* 87: 13–24.

Wilson, L. D., and J. R. Meyer. 1969. A review of the colubrid snake genus *Amastridium. Bulletin of the Southern California Academy of Sciences* 68: 145–159.

Wilson, L. D., and J. H. Townsend. 2006. The herpetofauna of the rainforests of Honduras. *Caribbean Journal of Science* 42: 88–113.

Wilson, L. D., and J. H. Townsend. 2007. Biogeography and conservation of the herpetofauna of the upland pine-oak forests of Honduras. *Biota Neotropica* 7: 137–148.

Wilson, L. D., I. R. Luque-Montes, A. B. Alegría, and J. H. Townsend. 2013. El componente endémico de la herpetofauna Hondureña en peligro crítico: priorización y estrategias de conservación. *Revista Latinoamericana de Conservación* 2–3: 47–67.

Wilson, L. D., J. R. McCranie, and M. R. Espinal. 2001. The ecogeography of the Honduran herpetofauna and the design of biotic reserves, PP. 109–158. IN: J. D. Johnson, R. G. Webb, and O. A. Flores-Villela, editors. Mesoamerican Herpetology: Systematics, Zoogeography, and Conservation. *Centennial Museum, Special Publication, University of Texas at El Paso* 1: i–iv, 1–200.

Wilson, L. D., J. R. McCranie, S. Gotte, and J. H. Townsend. 2003. Distributional comments on some members of the herpetofauna of the Mosquitia, Honduras. *The Herpetological Bulletin* 84: 15–19.

Wilson, L. D., J. R. McCranie, and L. Porras. 1979a. New departmental records for reptiles and amphibians from Honduras. *Herpetological Review* 10: 25.

Wilson, L. D., J. R. McCranie, and L. Porras. 1979b. *Rhadinaea montecristi* Mertens: an addition to the snake fauna of Honduras. *Herpetological Review* 10: 62.

Wilson, L. D., J. R. McCranie, and K. L. Williams. 1986. The identity of the crocodile of Lago de Yojoa, Honduras. *Journal of Herpetology* 20: 87–88.

Wilson, L. D., J. R. McCranie, and K. L. Williams. 1991. Additional departmental records for the herpetofauna of Honduras. *Herpetological Review* 22: 69–71.

Wilson, L. D., B. Myton, and G. Cruz [Díaz]. 1976. New distributional records for reptiles from Honduras. *Herpetological Review* 7: 179.

Wilson, L. D., L. Porras, and J. R. McCranie. 1986. Distributional and taxonomic comments on some members of the Honduran herpetofauana [sic]. *Milwaukee Public Museum, Contributions in Biology and Geology* 66: 1–18.

Wirth, M. 2012a. Basilisken. Spektakuläre Vertreter der Leguane. *Elaphe* 2012: 14–23.

Wirth, M. 2012b. Ökologie der Basilisken und Freilandbeobachtungen in Costa Rica. *Elaphe* 2012: 24–30, 32–35.

Witzell, W. N. 1983. Synopsis of biological data on the Hawksbill Turtle *Eretmochelys imbricata* (Linnaeus, 1766). Rome: Food and Agriculture Organization of the United Nations, *FAO Fisheries Synopsis* 137.

Wright, J. W. 1993. Evolution of the lizards of the genus *Cnemidophorus*, PP. 27–81. IN: J. W. Wright and L. J. Vitt, editors. *Biology of Whiptail Lizards (Genus* Cnemidophorus*)*. Norman: The Oklahoma Museum of Natural History and The University of Oklahoma.

Yarrow, H. C. 1875. Report upon the collections of batrachians and reptiles made in portions of Nevada, Utah, California, Colorado, New Mexico, and Arizona, during the years 1871, 1872, 1873, and 1874, PP. 509–584. IN: *Report on the Geography and Geology of the Exploration and Survey West of the 100th Meridian, Wheeler Survey*, Vol. 5. Washington: Government Printing Office.

Zaher, H., F. G. Grazziotin, J. E. Cadle, R. W. Murphy, J. Cesar de Moura-Leite, and S. L. Bonatto. 2009. Molecular phylogeny of advanced snakes (Serpentes, Caenophidia) with an emphasis on South American xenodontines: a revised classification and description of new taxa. *Papéis Avulsos de Zoologia, São Paulo* 49: 115–153.

Zaher, H., F. G. Grazziotin, R. Graboski, R. G. Fuentes, P. Sánchez-Martinez, G. G. Montinglli, Y-P. Zhang, and R. W. Murphy. 2012. Phylogenetic relationships of the genus *Sibynophis* (Serpentes: Colubroidea). *Papéis Avulsos de Zoologia, São Paulo* 52: 141–149.

Zanchi-Silva, D., and D. M. Borges-Nojosa. 2017. Natural History Notes. *Hemidactylus mabouia* (Tropical House Gecko). Predation. *Herpetological Review* 48: 438–439.

Zug, G. R. 1966. The penial morphology and the relationships of cryptodiran turtles. *Occasional Papers of the Museum of Zoology, University of Michigan* 647: 1–24.

Zug, G. R. 1991. The lizards of Fiji: natural history and systematics. *Bishop Museum Bulletins in Zoology* 2: i–xii, 1–136 pp.

Zug, G. R. 2013. *Reptiles and Amphibians of the Pacific Islands. A Comprehensive Guide.* Berkeley: The University of California Press.

Zug, G. R., C. H. Ernst, and R. V. Wilson. 1998. *Lepidochelys olivacea. Catalogue of American Amphibians and Reptiles* 653.1–653.13.

Zug, G. R., L. J. Vitt, and J. P. Caldwell. 2001. *Herpetology. An Introductory Biology of Amphibians and Reptiles.* 2nd ed. San Diego: Academic Press.

Index to Scientific Names

Index to Authors